An Introduction to DIFFERENTIAL EQUATIONS

Written according to the latest syllabi (CBCS) of Honours Degree Course [BA and BSc] in Mathematics of all Indian universities.

An Introduction to
DIFFERENTIAL EQUATIONS

Ram Krishna Ghosh

Kantish Chandra Maity

AND

Dipak Kumar Ghosh

Author of

A Textbook of Complex Analysis, Introduction to Partial Differential Equation, Laplace Transform, A Short Journey into Mathematical Logic, Sets and Boolean Algebra, and Introduction to Graph Theory

New Central Book Agency (P) Ltd

Co-Published by

Bijay Publications

E-mail : bijaypublications@gmail.com

Exclusively Distributed by

Bijay Pustak Bhandar (P) Ltd

90/6A M. G. Road, Bhawani Dutta Lane, Kolkata-700007

✆ : (033) 4605-7856, 8777898576, 8274916104

E-mail : bijaypustakbhandar@gmail.com

NCBA

REGD OFFICE
8/1 Chintamoni Das Lane, Kolkata 700 009, India
email: ncbakolkata@gmail.com

OVERSEAS
NCBA (UK) Ltd, 149 Park Avenue North, Northampton, NN3 2HY, UK
email: ncbauk@yahoo.com

CORPORATE OFFICE
212 Shahpur Jat, New Delhi 110 049
email: ncbaexports@ncbapvtltd.com

BIJAY PUBLICATIONS
90/6A M. G. Road, Bhawani Dutta Lane, Kolkata-7
E-mail : bijaypublications@gmail.com

BIJAY PUSTAK BHANDAR (P) LTD
90/6A M. G. Road, Bhawani Dutta Lane, Kolkata-7
E-mail : bijaypustakbhandar@gmail.com

AN INTRODUCTION TO DIFFERENTIAL EQUATIONS • Ramkrishna Ghosh, Kantish Chandra Maity

First Published: 1971
Revised and Enlarged Ninth Edition: 2005
Reprinted: 2006, 2007, 2009, 2011, 2012, 2013, 2014, 2015, 2016, 2017
Thoroughly Revised and Enlarged Tenth Edition: 2018
Reprinted with Addition: 2019
Reprinted: 2020, 2021
Thoroughly Revised, Enlarged and Updated Eleventh Edition: **2023**

PUBLISHER • TYPESETTER • PRINTER
New Central Book Agency (P) Ltd
8/1 Chintamoni Das Lane, Kolkata 700 009

CO-PUBLISHED BY
Bijay Publications
90/6A M G Road, Bhawani Dutta Lane, Kolkata 700007

TECHNICAL EDITOR
Dipan Roy, Pradip Kr Chowdhury

PROJECT SUPERVISOR
Abhijit Sen

ISBN: 978 81 7381 237 8

Price ₹ 685.00

Dedicated to
the true worshippers of Mathematics

Preface to the Eleventh Edition

I felt blessed, honoured and privileged when I was bestowed with the opportunity to revise *An Introduction to Differential Equations* by R.K. Ghosh and K.C. Maity, an undisputed classic, so to say, to all those associated with teaching and learning of Mathematics over the last 52 years.

In this revised edition, four chapters have been annexed to the category of Ordinary Differential Equations. They are (i) Vector Differential Equations, (ii) Autonomous System, (iii) Power Series Solutions of the Second-order Linear Differential Equations, and (iv) Mathematical Modelling. Two chapters have been added to the Partial Differential Equations category. They are (i) Classification of Partial Differential Equations of Second-order and Canonical forms, and (ii) Initial and Boundary Value Problems on Wave Equations, Heat Equations and Laplace Equations. Part III contains multiple choice questions with answers.

The specific strengths of the revised edition of the book are (i) the inclusion of an introduction to each chapter, (ii) the inclusion of numerous examples along with their graphical representations, (iii) models on mathematical biology, physics, chemistry, economics, etc., (iv) the inclusion of existence of solutions, Lipschitz condition, convex sets, initial-value problems, boundary-value problems, well-posed problems, existence and uniqueness theorem, Picard's theorem, Picard's approximation and a large number of problems in chapter 2, (v) the inclusion of general theory of linear differential equations, Wronskian and related theorems, linearly independent solutions, fundamental set of solutions, Abel's theorem, reduction of order, formation of differential equations from a given fundamental set of solutions, principle of superposition for homogeneous and non-homogeneous equations and numerous examples in chapter 5, (vi) the inclusion of method of solution of simultaneous linear equations with constant coefficients: operator method, normal form, homogeneous linear systems, theorems and examples in chapter 9, (vii) the inclusion of Liner equation, Semi-linear equation, Quasi-linear equation, Non-linear equation in chapter 1 of partial differential equation section. All other parts of the original edition have been kept intact.

For the focus on conceptual understanding, I have tried my best to implement this goal through the rule of three – "Topics should be presented geometrically, numerically and analytically". Hope the text, in the new avatar, will be acceptable to all for the CBCS syllabus. I think, the book in new form will help my dearest and nearest readers in grasping the several topics independently. I hope, readers can find everything they need in this book to practise the concepts of differential equations. Feedback in the form of constructive criticism and positive advices are always welcome.

Those who inspired me to accept this project are Shri Sudarshan Ghosh, Shrimati Meera Ghosh, Dr Shyamal Sen, Sanjukta Pramanik and my wife, Shrimati Bithika Ghosh. Special thanks to Mr Abhijit Sen for his wonderful works in LaTeX typesettings, beautiful new formatting and almost 100 % accurate graphics. Last but not the least, I am grateful to Mr Amitabha Sen, Director of New Central Book Agency (P) Ltd for putting so much faith in me and selecting me for this project.

Srijani, Krishnanagar, Nadia, WB
25 September, 2022
dkghosh3010@gmail.com

Dipak Kumar Ghosh

Preface to the First Edition

The present treatise may be considered as a supplement to our earlier works on Calculus: Maity & Ghosh—Differential Calculus: An Introduction to Analysis and Maity and Bagchi—Integral Calculus: An Introduction to Analysis. Like the previous works, the present book on Ordinary Differential Equations is intended to serve as a textbook on the subject for the B.A. and B.Sc. Honours courses.

Our main aim is to explain the devices employed in solving ordinary differential equations. However, the need of emphasis on gaining insight and understanding as opposed to pure manipulative skill has constantly been kept in mind. For those who wish to use the subject as a tool we have given a fairly long chapter on Applications dealing with Geometrical and Physical problems. The students in general courses in Arts and Science will also find it extremely useful.

The authors do not claim the book to be original but the expositions (particularly the topics like exact differential equations of first as well as higher orders, the use of transformations for solving equations of second order, etc.) will be, no doubt, found illuminating to the students.

We are indebted to the authorities on the subject, to many of our colleagues and to our former students for their kind assistance, direct and indirect, in the preparation of the book. An exception ... ...

In spite of our best efforts some printing errors may have crept in. The authors will be thankful if such mistakes are brought to their notice. Suggestions for the improvement of the book will be gratefully accepted by the authors.

Kolkata
the 15th November, 1971

Authors

Contents

I Ordinary Differential Equations 1

1 Differential Equations: Preliminary Notions 3

1.1 Basic definitions . . . 3
1.2 Formation of differential equations . . . 6
1.3 Solutions of a differential equation: Complete primitive, particular solution and singular solution . 9
1.4 Genesis of an ordinary differential equation . . . 10
1.5 Further illustrative examples . . . 12
1.6 Family of curves represented by ordinary differential equations . . . 18

2 Equations of First Order and First Degree, Existence of Solutions 21

2.1 Equations of first order and first degree . . . 21
2.2 Continuity . . . 22
2.3 Bounded function . . . 22
2.4 Lipschitz condition . . . 22
2.5 Convex set . . . 22
2.6 Initial-value problem . . . 23
2.7 Well-posed problem . . . 24
2.8 Boundary-value problem . . . 24
2.9 Picard's approximation . . . 29
2.10 Exact equations . . . 33
2.11 Integrating factors . . . 37
2.12 Equations solvable by separation of variables . . . 42
2.13 Homogeneous equations . . . 47
2.13.1 Equations which are both homogeneous and exact . . . 50
2.13.2 Equations reducible to the homogeneous form . . . 54
2.14 Linear equations . . . 57
2.15 Rules for determining integrating factors . . . 70

3 Equations of the First Order but not of the First Degree 86

3.1 Equations of the first order but not of the first degree . . . 86
3.2 Equations solvable for p . . . 86
3.3 Equations solvable for y . . . 88
3.4 Equations solvable for x . . . 90

3.5 Clairaut's form . . . 92
3.6 Miscellaneous problems . . . 94

4 **Singular Solutions: Extraneous Loci** **102**
4.1 Singular solutions: Extraneous loci . . . 102

5 **Linear Equations with Constant Coefficients** **110**
5.1 Equations of nth order . . . 110
5.2 Equations of first order: the simplest case . . . 110
5.3 Equations of second order . . . 111
5.4 Equations of orders higher than the second . . . 112
5.5 General theory of linear differential equations . . . 115
5.6 The Wronskian . . . 117
5.7 Linear operator with constant coefficients: Complementary function . . . 130
5.7.1 Particular integral . . . 133
5.7.2 Short methods of finding particular integrals in special cases: operator D . . . 136
5.7.3 Method of undetermined coefficients . . . 146
5.7.4 Method of variation of parameters . . . 158
5.8 The homogeneous linear equation . . . 170

6 **Exact Differential Equations: A few Special Forms** **178**
6.1 Exact equations . . . 178
6.2 Criterion of an exact differential equation . . . 178
6.3 Nonlinear exact equations: Equations that are not exact . . . 181
6.4 Miscellaneous types of linear equations . . . 185

7 **Equations of the Second Order** **193**
7.1 Change of dependent variable . . . 193
7.2 Change of dependent variable (continued) . . . 201
7.3 Change of independent variable . . . 205
7.4 Factorisation of the operator . . . 209
7.5 Methods of solving linear equations of second order . . . 210

8 **Simple Eigenvalue Problems** **212**
8.1 Eigenfunctions and eigenvalues . . . 212
8.2 Regular Sturm-Liouville problems . . . 215
8.3 Conversion to Sturm-Liouville form . . . 215
8.4 Reduction to Cauchy-Euler form . . . 217
8.5 Related problems . . . 218

9 **Simultaneous Linear Equations** **222**
9.1 Method of solution of simultaneous linear equations with constant coefficients . . . 222
9.2 Simultaneous equations of the form . . . 228
9.3 Total differential equations . . . 240

10 Geometrical and Physical Applications **258**
10.1 Geometrical data: A recapitulation 258
10.2 Illustrative examples 260
10.2.1 Applications of first order and first degree differential equations 260
10.2.2 Applications of first order and higher degree differential equations 262
10.3 Trajectories 266
10.3.1 Rectangular Cartesian coordinates 266
10.3.2 Polar coordinates 267
10.3.3 ω–trajectories: Rectangular Cartesian coordinates 268
10.3.4 Illustrative examples 269
10.3.5 Further geometrical applications 275

11 Vector Differential Equations **279**
11.1 Homogeneous and nonhomogeneous systems 279
11.2 Wronskian of the vector functions 280
11.3 Fundamental matrix 281
11.4 Conversion of linear differential equation to a normal linear system 282
11.5 First order linear homogeneous systems with constant coefficients (Euler's method) 290
11.6 First order Linear nonhomogeneous system with constant coeıficients 297

12 Autonomous System **301**
12.1 Autonomous system 301
12.2 Two-dimensional autonomous system 302
12.2.1 Critical point and ordinary point 302
12.2.2 Isolated critical point 302
12.2.3 Stability of a critical point 302
12.3 Linear two-dimensional autonomous system (linear plane autonomous system) 303
12.3.1 Real and distinct eigenvalues 303
12.3.2 Repeated real eigenvalues 306
12.3.3 Complex eigenvalues 308
12.4 Linearisation of a nonlinear plane autonomous system at a critical point 313

13 Power Series Solutions of Ordinary Differential Equations **316**
13.1 Power series 316
13.2 Analytic function 317
13.3 Ordinary point and Singular point 317
13.4 Power series solution near the ordinary points 318
13.5 Illustrative examples 318
13.6 Series solution about regular singular point 325
13.7 Series solution about a point at infinity 334
13.8 Bessel's equation 335
13.9 Legendre's equation 337

14 Mathematical Modelling 34
14.1 Mathematical modelling 34
14.2 Models of population growth 34
14.2.1 Population growth 34
14.2.2 Malthus population model (unlimited growth) 34
14.2.3 Logistic population model (limited growth) 34
14.2.4 A mathematical model of limited population growth with a threshold level 34
14.2.5 Lotka-Volterra model 34
14.2.6 Mutualism model 34
14.3 Dynamics in classical mechanics 35
14.3.1 Models on a falling object 35
14.3.2 Model on resulting motion of a connected system 35
14.3.3 Models on a particle moving along a straight line 35
14.3.4 Models on simple pendulum 35
14.3.5 Models on horizontal beams 35
14.4 Miscellaneous types of mathematical models 35
14.4.1 Models on radioactive decay 35
14.4.2 Models on heat conduction 35
14.4.3 Models on electric circuits 35

II Partial Differential Equations 36

1 Partial Differential Equations 36
1.1 Definitions 36
1.2 Construction of partial differential equations by the process of elimination of arbitrary constants . 36
1.3 Construction of partial differential equations by the process of elimination of arbitrary functions . 37
1.4 Further solved problems 37

2 Lagrange's Method of Solving PDE 37
2.1 The general solution of a linear equation 37
2.2 An equation that is equivalent to $Pp + Qq = R$ 37
2.3 Lagrange's method of solving $Pp + Qq = R$ 37
2.4 Integral surfaces through a given curve 39

3 Nonlinear Partial Differential Equations of Order One 39
3.1 Definitions of various classes of integrals 39
3.2 General method of solving the PDE of first order in two independent variables x and y: Charpit's method 39

4 Classification of Partial Differential Equations of Second Order and Canonical Forms 40
4.1 Second-order linear partial differential equations 40
4.2 Classification of linear partial differential equations of second order 40
4.3 Transformation of coordinates 40
4.4 Canonical forms 40
4.4.1 Canonical forms of hyperbolic equations 40
4.4.2 Canonical forms of the parabolic equations 41
4.4.3 Canonical forms of the elliptic equations 41
4.5 Further solved problems 41

5 **Initial and Boundary Value Problems** 421
5.1 The principle of superposition . . . 421
5.2 One-dimensional wave equation . . . 422
5.3 The Fourier's method of separation of variables . . . 423
5.4 The Cauchy problem . . . 426
5.4.1 Cauchy problem for one-dimensional homogeneous wave equation . . . 426
5.4.2 Cauchy problems for one-dimensional nonhomogeneous wave equation . . . 428
5.4.3 Problem of a semi-infinite string with a fixed end . . . 430
5.4.4 Problem of a string of finite length . . . 431
5.4.5 Uniqueness of the solution . . . 432
5.4.6 Solution of the two-dimensional wave equation by the method of separation of variables (Cartesian coordinates) . . . 433
5.5 One-dimensional heat equation (diffusion equation) . . . 435
5.5.1 The nature and characteristics . . . 435
5.5.2 Solution of one-dimensional heat equation by the method of separation of variables (Cartesian coordinates) . . . 436
5.5.3 Solution of heat equation for finite rod . . . 436
5.5.4 Solution of two-dimensional heat equation by the method of separation of variables (Cartesian coordinates) . . . 439
5.6 Laplace's equation (in Cartesian coordinates) . . . 440
5.6.1 Boundary value problems . . . 440
5.6.2 Solution of two-dimensional Laplace's equation by separation of variables (Cartesian coordinates) . . . 441

III MCQ and Answers on Ordinary Differential Equations 449

1 **MCQ on chapter 1 and chapter 2** 451

2 **MCQ on chapter 3 and chapter 4** 456

3 **MCQ on chapter 5 and chapter 6** 460

4 **MCQ on chapter 7 and chapter 8** 463

5 **MCQ on chapter 9 and chapter 10** 465

6 **MCQ on chapter 11 and chapter 12** 467

7 **MCQ on chapter 13** 471

8 **MCQ and Answers on PDE** 473

For Revision: O.D.E. Test Exercises 477

Index 485

Notations and Symbols

1. $\dfrac{\partial z}{\partial x} = p$.
2. $\dfrac{\partial z}{\partial y} = q$.
3. $\dfrac{\partial^2 z}{\partial x^2} = r$.
4. $\dfrac{\partial^2 z}{\partial x \partial y} = s$.
5. $\dfrac{\partial^2 z}{\partial y^2} = t$.
6. Auxiliary equation: A.E.
7. Belongs to: $\in$.
8. Boundary-value problem: BVP.
9. C-discriminant of $\phi(x, y, c)$: $\text{Disct}_\text{C}\phi(x, y, c)$.
10. Complementary function: CF.
11. Critical point: (x^*, y^*).
12. Differential operator: $D \equiv \frac{d}{dx}$.
13. Does not belong to: $\notin$.
14. Equation: eq.
15. First-order derivative of x with respect to t : $\dot{x}$, $\dfrac{dx}{dt}$.
16. First-order derivative of y with respect to x : y', $y_1, Dy, p, \dfrac{dy}{dx}$.
17. For all: $\forall$.
18. General solution: GS.
19. Initial-value problem: IVP.
20. Integrating factor: I.F.
21. Intersection: $\cap$.
22. Jacobian: J.
23. Laplace operator for three independent variables): $\nabla^2 \equiv \dfrac{\partial^2}{\partial x^2} + \dfrac{\partial^2}{\partial y^2} + \dfrac{\partial^2}{\partial z^2}$.
24. Legendre's function of the first kind: $P_n(x)$.
25. Legendre's function of the second kind: $Q_n(x)$.
26. Null set: $\varnothing$.
27. Partial differential coefficient of f w.r.t. x : f_x.
28. Partial differential coefficient of f w.r.t. y : f_y.
29. Partial differential equation: PDE.
30. Particular integral: PI.
31. Particular solution: PS.
32. p-discriminant of $\phi(x, y, p)$: $\text{Disct}_\text{p}\phi(x, y, p)$.
33. Second-order derivative of x with respect to t : $\ddot{x}$, $\dfrac{d^2x}{dt^2}$.
34. Second-order derivative of y with respect to x : $y'', y_2, D^2y, \dfrac{d^2y}{dx^2}$.
35. Set of all integers: $\mathbb{Z}$.
36. Set of all natural numbers: $\mathbb{N}$.
37. Set of all positive integers: $\mathbb{Z}^+$.
38. Set of all positive rational numbers: $\mathbb{Q}^+$.
39. Set of all rational numbers: $\mathbb{Q}$.
40. Set of all real numbers: $\mathbb{R}$.
41. Singular solution: SS.
42. Stable degenerated node: SDN.
43. Subset: $\subseteq$.
44. There exists: $\exists$.
45. Union: $\cup$.
46. Unstable degenerated node: UDN.
47. With respect to: w.r.t.
48. Wronskian of $u_1, u_2, \ldots, u_n$: $W(u_1, u_2, \ldots, u_n)$.

Part I

Ordinary Differential Equations

Chapter 1

Differential Equations: Preliminary Notions

Differential equations were introduced first by Newton and Leibniz on the way of invention of Calculus. It is a large and very important branch of mathematical science. A large number of the general laws of nature can be expressed in terms of the language of differential equations. There are three major aspects of this subject, namely, theory, method and application. In application, generally the functions represent physical quantities, the derivatives of the functions represent their rate of changes and the differential equations define relationships between the functions and their derivatives. Differential equations play very important role in several disciplines, namely, Physics, Chemistry, Economics, Biology, Engineering, etc. It plays also an important role in mathematical modelling. Study of differential equations consists of two parts, (i) *study of their solutions and* (ii) *study of the properties of their solutions. The main aim of the first chapter is to introduce the readers to the preliminary notions of differential equations.*

1.1 Basic definitions

Any relation between an independent variable x, a dependent variable y, and one or more of the derived functions

$$\frac{dy}{dx}, \frac{d^2y}{dx^2}, \ldots, \frac{d^ny}{dx^n}, \ldots$$

is called an *ordinary differential equation.* The variables x or y may, or may not, enter explicitly into the equation. Instead of derivatives, differentials may appear in the equation. The following are examples of ordinary differential equations:

$$\frac{dy}{dx} = m \tag{1.1.1}$$

$$\frac{d^2y}{dx^2} = -a^2y \tag{1.1.2}$$

$$\frac{d^3y}{dx^3} + 5\frac{d^2y}{dx^2} + 6\frac{dy}{dx} = 7x^2 + 8x + 9 \tag{1.1.3}$$

$$3x^3\frac{d^3y}{dx^3} + 4x^2\frac{d^2y}{dx^2} + x\frac{dy}{dx} - 10y = xe^x \tag{1.1.4}$$

$$x\,dy - y\,dx + 2xydy = 0. \tag{1.1.5}$$

The word *ordinary* is used to emphasize that only one independent variable is involved; if, on the other hand, more than one independent variable and partial derivatives appear, then the equation is called a *partial differential equation*, examples of which are the following:

$$\frac{\partial^2 z}{\partial x^2} + \frac{\partial^2 z}{\partial y^2} = 0 \tag{1.1.6}$$

$$\frac{\partial^2 \theta}{\partial r^2} = 4rt\frac{\partial \theta}{\partial t}. \tag{1.1.7}$$

In part I of the present treatise, with very few exceptions, only *ordinary differential equations* will be considered. The word ordinary, therefore, will be omitted in most cases.

The *order* of a differential equation is the *order* of the highest-ordered derivative involved in the equation. The *degree* of a differential equation is the power of the highest ordered derivative involved in the equation after it has been made rational and integral as far as derivatives are concerned. That means if an ordinary differential equation can be written as a polynomial in the unknown function and its derivatives, then its degree is the power to which the highest-ordered derivative is raised. Thus, eqs (1.1.2), (1.1.6) and (1.1.7) are of second order and first degree, eqs (1.1.1) and (1.1.5) are of first order and first degree, eqs (1.1.3) and (1.1.4) are of third order and first degree. We next consider the two equations:

$$y = x\frac{dy}{dx} + c\frac{dx}{dy} \tag{1.1.8}$$

and $$\left\{1 + \frac{d^2y}{dx^2}\right\}^{3/2} = a\frac{d^2y}{dx^2}. \tag{1.1.9}$$

The eq. (1.1.8) is of first order and *second degree;* the eq. (1.1.9) is of second order and *third degree.* For, (1.1.8) can be written as

$$x\left(\frac{dy}{dx}\right)^2 - y\frac{dy}{dx} + c = 0 \tag{1.1.10}$$

and eq. (1.1.9), when squared, becomes

$$\left(1 + \frac{d^2y}{dx^2}\right)^3 = a^2\left(\frac{d^2y}{dx^2}\right)^2. \tag{1.1.11}$$

The degree can be, then, easily read from (1.1.10) and (1.1.11). In general, an ordinary differential equation of order n can be exhibited as:

$$f\left(x, y, \frac{dy}{dx}, \frac{d^2y}{dx^2}, \ldots, \frac{d^ny}{dx^n}\right) = 0. \tag{1.1.12}$$

Definition 1.1.1. *If a differential equation is of first degree in the dependent variable y and its derivative (consequently, there cannot be any term involving the product of y and its derivatives), then it is called a* linear differential equation; *otherwise, it is nonlinear.*

The coefficients of a linear equation are either constants or functions of the independent variable x.

A general form of a linear ordinary differential equation of order n in the unknown function y (or the dependent variable y) and the independent variable x is

$$f_0(x)\frac{d^ny}{dx^n} + f_1(x)\frac{d^{n-1}y}{dx^{n-1}} + \cdots + f_{n-1}(x)\frac{dy}{dx} + f_n(x)y = f(x), \tag{1.1.13}$$

where the functions $f_i(x), i = 0, 1, 2, \ldots, n$ and $f(x)$ are presumed known and depend only on the variable x.

Differential equations that cannot be put into this form are said to be nonlinear.

It should be observed that a linear equation is always of first degree but not conversely: thus the equation

$$\frac{dy}{dx} = 1 + x^2 y^2 \tag{1.1.14}$$

is of first degree but *nonlinear* ($\because$ y^2 is present).

The following are examples of linear differential equations:

$$(1 - x^2)\frac{d^2 y}{dx^2} - 2x\frac{dy}{dx} + 5y = 0 \tag{1.1.15}$$

$$\frac{\partial^2 V}{\partial x^2} + \frac{\partial^2 V}{\partial y^2} + \frac{\partial^2 V}{\partial z^2} = 0 \tag{1.1.16}$$

$$\frac{d^3 y}{dx^3} + 3\frac{d^2 y}{dx^2} + 3\frac{dy}{dx} + y = \cos x. \tag{1.1.17}$$

In particular, a linear differential equation of first order can be written as

$$\frac{dy}{dx} + Py = Q, \tag{1.1.18}$$

where P and Q are either constants or functions of x.

Below are given two examples of nonlinear equations:

$$(x + y)^2 \frac{dy}{dx} = 1 \tag{1.1.19}$$

$$\left(\frac{dy}{dx}\right)^2 + \frac{dy}{dx} - 12 = 0. \tag{1.1.20}$$

Some problems

Example 1.1.1. *Determine the order, degree, linearity, the independent variable and the dependent variable of the ordinary differential equation*

$$\frac{d^3 y}{dx^3} - 15x\frac{dy}{dx} = e^x + 2.$$

Solution: This ordinary differential equation has been written as a polynomial in the unknown function (or, the dependent variable) y and its derivatives. Also its highest-ordered derivative is $\frac{d^3y}{dx^3}$, which is of order 3 and degree 1. Hence, the given equation is a third order first degree differential equation.

It is also *linear*, since in the notation of (1.1.13),

$$f_0(x) = 1,\ f_1(x) = 0 = f_3(x),$$
$$f_2(x) = -15x \text{ and } f(x) = e^x + 2;$$

and these depend only on x which is the independent variable.

Example 1.1.2. *Determine the order, degree, linearity, unknown function and the independent variable of the differential equation*

$$t\frac{d^2 y}{dt^2} + t^2\frac{dy}{dt} - (\cos t)\sqrt{y} = 2t^2 - 3t + 4.$$

Solution: The differential equation is of order 2, since the highest-ordered derivative is of second order. The unknown function is y and the independent variable is t. Because of the term $\sqrt{y}$, the equation cannot be written as a polynomial in y and its derivatives, hence no degree. It is also *nonlinear*, since the unknown function y is raised to a power other than the first.

Example 1.1.3. *Determine the order, degree, linearity, the unknown function and the dependent variable of*

$$\left(\frac{d^2x}{dt^2}\right)^2 + \frac{d^2x}{dt^2} + t\frac{dx}{dt} = 0.$$

Solution: As before, it is of order 2, degree 2. The unknown function is x and the independent variable is t.

It is *nonlinear*, since one of the derivatives $\frac{d^2x}{dt^2}$ has been raised to a power 2 other than the first.

Example 1.1.4. *Determine linearity of the following differential equations:*

(i) $\frac{d^2y}{dx^2} + x^2\frac{dy}{dx} + x\sin y = 0.$

(iii) $\frac{d^2y}{dx^2} + y\frac{dy}{dx} + y^2 = 0.$

(ii) $x^3\frac{d^2y}{dx^2} + \cos x\frac{dy}{dx} + (\sin x)y = 0.$

Solution: In all cases, refer to notations of (1.1.13):

(i) nonlinear, since the unknown function y is the argument of a transcendental function, i.e. the sine function.
Here the nonlinear term is $\sin y$.

(ii) Linear, since $f_0(x) = x^3, f_1(x) = \cos x, f_2(x) = \sin x, f(x) = 0$ and the equation is of the form (1.1.13).

(iii) nonlinear, since $\frac{dy}{dx}$ is multiplied by y and the last term is y^2 and not y.

1.2 Formation of differential equations

A differential equation is a concise form of expressing the properties of a family or a group of functions. Many general laws of nature–in Physics, Chemistry, Biology, Economics, Space Science and, in fact, in any branch of human knowledge–find their most general expression in the language of differential equations.

We illustrate through the following examples:

Geometric problems

Example 1.2.1. *Suppose, a curve is defined by the condition that the sum of x- and y-intercepts of its tangents is always equal to a. Express this fact by a differential equation.*

Solution: The equation of the tangent at any point (x, y) on the curve is

$$Y - y = \frac{dy}{dx}(X - x) \text{ or, } Y - \frac{dy}{dx}X = y - x\frac{dy}{dx} \text{ or, } \frac{X}{\dfrac{y - x\frac{dy}{dx}}{-\frac{dy}{dx}}} + \frac{Y}{y - x\frac{dy}{dx}} = 1.$$

The differential equation expressing the given condition is

$$\left(x - y\frac{dx}{dy}\right) + \left(y - x\frac{dy}{dx}\right) = a \text{ or, } x\left(\frac{dy}{dx}\right)^2 - (x + y - a)\frac{dy}{dx} + y = 0.$$

This is an ordinary differential equation of first order and second degree.

Example 1.2.2. *What is the differential equation of the curves determined by the condition: At each point (r, θ) the trigonometrical tangent of the angle between the radius vector and the tangent is equal to 1/3 of the trigonometrical tangent of the vectorial angle?*

Solution: Here $\tan\phi = \frac{1}{3}\tan\theta$

$$\text{or, } r\frac{d\theta}{dr} = \frac{1}{3}\tan\theta \text{ (reqd. diff. eqn.)}$$

Physical problems

Example 1.2.3. *A particle of mass m moves along a straight line (x-axis) subject to a force proportional to its displacement x from the fixed point O in its path and directed towards O and also to a resisting force proportional to its velocity. Express the total force in the form of a differential equation.*

Solution: The first force (being proportional to x) may be represented by $-kx$ and the second by $-r\frac{dx}{dt}$, where k and r are constants of proportionality. The first force is directed opposite to the direction in which x increases and the second force retards the motion. In each case, therefore, a minus sign has appeared.

Newton's law of motion gives

$$\text{Total force} = \text{mass} \times \text{acceleration} = m\frac{d^2x}{dt^2}.$$

$$\therefore\ m\frac{d^2x}{dt^2} = -kx - r\frac{dx}{dt}.$$

Using the notations $\dot{x} = \dfrac{dx}{dt}$ and $\ddot{x} = \dfrac{d^2x}{dt^2}$, the differential equation takes the form:

$$m\ddot{x} + r\dot{x} + kx = 0.$$

Example 1.2.4. *Five hundred grammes of sugar in water are being converted into dextrose at a rate which is proportional to the amount unconverted. Form a differential equation expressing the rate of conversion after t minutes.*

Solution: Let y denote the number of gram converted in t minutes. Then $(500 - y)$ is the number of gram unconverted and the rate of conversion is

$$\frac{dy}{dt} = k(500 - y), \text{ (}k\text{ is the constant of variation).}$$

Elimination of arbitrary constants

Example 1.2.5. *Eliminate the arbitrary constants A and B from the relation* $v = \frac{A}{r} + B$ *and obtain a differential equation of second order.*

Solution: Given: $v = \frac{A}{r} + B$.

Differentiating w.r.t. r, we get

$$\frac{dv}{dr} = -\frac{A}{r^2}.$$

Differentiating again we get

$$\frac{d^2v}{dr^2} = \frac{2A}{r^3}.$$

$$\therefore\ \frac{1}{2}r^3\frac{d^2v}{dr^2} = A = -r^2\frac{dv}{dr} \text{ or, } \frac{d^2v}{dr^2} + \frac{2}{r}\frac{dv}{dr} = 0 \text{ (reqd. differential equation)}$$

Example 1.2.6. *Construct a differential equation by the elimination of the arbitrary constants a and b from the equation* $ax^2 + by^2 = 1$.

Solution: Differentiating,

$$2ax + 2by\frac{dy}{dx} = 0, \text{ or, } \frac{y}{x}\frac{dy}{dx} = -\frac{2a}{2b}.$$

Again differentiating and simplifying,

$$xy\frac{d^2y}{dx^2} + x\left(\frac{dy}{dx}\right)^2 - y\frac{dy}{dx} = 0.$$

Example 1.2.7. *Obtain the differential equation corresponding to the primitive*

$$(x - \alpha)^2 + (y - \beta)^2 = r^2, \tag{1.2.1}$$

where r is a fixed constant and α, β are arbitrary constants. Give a geometrical interpretation of the result.

Solution: Differentiating (1.2.1) w.r.t. x, we obtain

$$(x - \alpha) + (y - \beta)\frac{dy}{dx} = 0. \tag{1.2.2}$$

Differentiating again,

$$1 + \left(\frac{dy}{dx}\right)^2 + (y - \beta)\frac{d^2y}{dx^2} = 0. \tag{1.2.3}$$

From (1.2.3), we get

$$y - \beta = -\frac{1 + \left(\frac{dy}{dx}\right)^2}{\frac{d^2y}{dx^2}}$$

and hence (1.2.2) gives

$$x - \alpha = \frac{\frac{dy}{dx}\left\{1 + \left(\frac{dy}{dx}\right)^2\right\}}{\frac{d^2y}{dx^2}}.$$

These values when substituted in (1.2.1) lead to the required equation

$$r^2\left(\frac{d^2y}{dx^2}\right)^2 = \left\{1 + \left(\frac{dy}{dx}\right)^2\right\}^3. \tag{1.2.4}$$

Geometrical interpretation. If we vary α and β in (1.2.1), we get a system of circles of given radius r, having their centres anywhere in the xy-plane. The differential eq. (1.2.4) expresses the fact that for every member of the system the radius of curvature has everywhere the constant value r.

Example 1.2.8. *Obtain the differential equation by eliminating the arbitrary constants a and b of the relation $x = a\cos nt + b\sin nt$. Give the physical interpretation.*

Solution: The primitive $x = a\cos nt + b\sin nt$ (a and b are arbitrary constants) represents a group of motions. Differentiating twice, one obtains easily $d^2x/dt^2 = -n^2x$. This differential equation asserts that in the group of motions represented by the primitive the acceleration is towards the origin and varies (in a given ratio n^2) as the distance from the origin.

Example 1.2.9. *The equation*

$$\frac{x^2}{a^2+\lambda}+\frac{y^2}{b^2+\lambda}=1, \tag{1.2.5}$$

(where a and b are fixed constants and λ is an arbitrary parameter which can assume all real values) represents a family of confocal conics. To obtain the differential equation of this family.

Solution: The required differential equation is obtained by eliminating λ between (1.2.5) and the derived equation

$$\frac{2x}{a^2+\lambda}+\frac{2yy'}{b^2+\lambda}=0\left(y'=\frac{dy}{dx}\right). \tag{1.2.6}$$

From (1.2.6) one gets

$$\frac{a^2+\lambda}{x}=\frac{b^2+\lambda}{-yy'}=k, \text{ (say)}.$$

Hence, from (1.2.5) it follows,

$$\frac{x^2}{kx}-\frac{y^2}{kyy'}=1, \text{ or, } k=(xy'-y)/y'.$$

This gives

$$a^2+\lambda=kx=\frac{xy'-y}{y'}\cdot x \text{ and } b^2+\lambda=-kyy'=\frac{xy'-y}{y'}\cdot(-yy').$$

Eliminating λ, we get

$$a^2-b^2=\frac{x(xy'-y)}{y'}+(xy'-y)y \text{ or, } (a^2-b^2)y'=(xy'-y)(x+yy'), \tag{1.2.7}$$

which is the required differential equation.

Some observations

A differential equation can be constructed by eliminating the parameters of a family of curves or of a group of motions. The equation so constructed represents a common property of the family. The order of the equation is equal to the number of parameters of the family.

We shall discuss the genesis of an ordinary differential equation in a more general way in art. 1.4.

1.3 Solutions of a differential equation: Complete primitive, particular solution and singular solution

A differential equation is considered as *solved* when one obtains a relation between the variables (dependent and independent) by means of which and the derivatives derived therefrom, the equation is identically satisfied. The relation obtained is called *a solution* (or *an integral*) of the differential equation.

Consider the differential equation

$$\frac{d^2y}{dx^2}=9y. \tag{1.3.1}$$

Suppose, by some means or other, we obtain from it the relation

$$y=e^{3x}. \tag{1.3.2}$$

Is (1.3.2) a solution of (1.3.1)? The answer is in the affirmative. For, putting $y = e^{3x}$ whereby $d^2y/dx^2 = 9e^{3x} = 9y$, we find that the eq. (1.3.1) is identically satisfied. But this may not be the only solution. In fact, we can verify that $y = e^{-3x}$ is also a solution. More generally, the following relation

$$y = Ae^{3x} + Be^{-3x} \tag{1.3.3}$$

also satisfies eq. (1.3.1) (A and B are independent, arbitrary constants). Observe that the solution (1.3.3) contains as many independent arbitrary constants as the *order of the differential equation, viz.* two.

The solution of a differential equation which contains the same number of independent arbitrary constants as the order of the equation, is called its *general solution* (or *complete integral* or *complete primitive*). Any solution obtained from the general solution by giving particular values to the arbitrary constants is called a *particular solution.*

Thus, (1.3.3) is the general solution of (1.3.1) and $y = e^{3x}$ (where $A = 1, B = 0$) and $y = e^{-3x}$ (where $A = 0, B = 1$) are *particular solutions.*

The general solution *does not always include* all possible solutions of a differential equation. In other words, solutions may exist which are not deducible from the general solution by giving particular values to the arbitrary constants. Such solutions are called *singular solutions.* We give an example:

The equation

$$y = px + \sqrt{a^2p^2 + b^2}\left(p = \frac{dy}{dx}\right), \tag{1.3.4}$$

where a and b are fixed constants, is satisfied by

$$y = cx + \sqrt{a^2c^2 + b^2}, \tag{1.3.5}$$

where c is an arbitrary constant. This is the general solution of the eq. (1.3.4). But the relation

$$\frac{x^2}{a^2} + \frac{y^2}{b^2} = 1 \tag{1.3.6}$$

also satisfies (1.3.4) and it is not deducible from (1.3.5). Hence, (1.3.6) is a *singular solution* of the eq. (1.3.4).

We shall discuss about singular solutions in a later chapter.

1.4 Genesis of an ordinary differential equation

Suppose, we are given the function

$$y = f(x; c_1, c_2, c_3, \dots). \tag{1.4.1}$$

This function depends on x and several *arbitrary* and *independent* constants $c_1, c_2, c_3, \dots$. Our problem is to determine a differential equation of which f is the *primitive.* The process involves successive differentiation and then elimination of the arbitrary constants.

We now take up the general process. Consider the equation

$$f(x, y;\ c_1, c_2, \dots, c_n) = 0, \tag{1.4.2}$$

in which x and y are variables and $c_1, c_2, \dots, c_n$ are *arbitrary* and *independent* (see the observations below) constants. This equation serves to determine y as a function of x (of course, under some specific restrictions) for a particular set of values for $c_1, c_2, \dots, c_n$. For a different set of values of c's a different function can be obtained. We can now form a differential equation which is satisfied by everyone of these functions. Differentiating n times in succession with respect to x, we obtain

$$\frac{\partial f}{\partial x} + \frac{\partial f}{\partial y} \cdot \frac{dy}{dx} = 0,$$

$$\frac{\partial^2 f}{\partial x^2} + 2\frac{\partial^2 f}{\partial x \partial y}\frac{dy}{dx} + \frac{\partial^2 f}{\partial y^2}\left(\frac{dy}{dx}\right)^2 + \frac{\partial f}{\partial y}\frac{d^2 y}{dx^2} = 0,$$

$$\vdots$$

$$\frac{\partial^n f}{\partial x^n} + \cdots + \frac{\partial f}{\partial y}\frac{d^n y}{dx^n} = 0.$$

[The existence of all the partial derivatives involved is assumed and also it is supposed $\partial f/\partial y \neq 0$].

From these n equations and the eq. (1.4.2) the n arbitrary constants $c_1, c_2, \ldots, c_n$ can be eliminated and the eliminant is the differential equation of which (1.4.2) is the primitive. Clearly, the differential equations will, in general, be of order n. We may write it as

$$F\left(x, y, \frac{dy}{dx}, \frac{d^2 y}{dx^2}, \ldots, \frac{d^n y}{dx^n}\right) = 0. \tag{1.4.3}$$

(1.4.2) is the primitive of (1.4.3) and (1.4.2) contains n arbitrary constants. Thus, (1.4.2) is the general solution of (1.4.3). The general solution is, then, essentially the same as the primitive of the differential equation.

Observations

The n arbitrary constants in (1.4.2) must be independent, otherwise the differential eq. (1.4.3) will not be of order n. If, suppose, two or more of the constants can be replaced by a single constant (e.g. $y = Ae^{x+B} = Ae^x \cdot e^B = Ce^x$) without essentially modifying the primitive, then the order of the resulting equation will be less than n.

Again, if the primitive is *reducible*, that is, if suppose (1.4.2) breaks up into two factors, each of which contains y, the order of the resulting equation may be less than n. For, if neither factor contains all the n constants, then each factor will give rise to a differential equation of order less than n. It may occur that these two differential equations are identical and, therefore, is satisfied by the primitive itself. We take an example:

Let the primitive be $y^2 - (a+b)xy + abx^2 = 0$. This breaks up into two factors which gives $y = ax$ and $y = bx$, from each of which we obtain $y/x = dy/dx$, which is an equation of order one (not two) though the primitive contains two arbitrary constants a and b.

The converse problem

Given a differential equation of order n, the process of integration leads to an equation of order $(n-1)$ and an arbitrary constant.

The latter equation is called the *first integral* of the given equation. For example:

Given

$$\frac{d^2 y}{dx^2} = \phi(y),$$

we multiply both sides by $2\frac{dy}{dx}$ and obtain

$$2\frac{dy}{dx} \cdot \frac{d^2 y}{dx^2} = 2\phi(y)\frac{dy}{dx}, \quad \text{or,} \quad \frac{d}{dx}\left\{\left(\frac{dy}{dx}\right)^2\right\} = 2\phi(y)\frac{dy}{dx}.$$

On integration, we obtain

$$\left(\frac{dy}{dx}\right)^2 = C + 2\int \phi(y)\,dy,$$

where C is arbitrary constant of integration. This gives the *first integral* of the second order equation

$$\frac{d^2y}{dx^2} = \phi(y).$$

Theoretically speaking, the process of constructing the primitive of a given differential equation is the construction of a first integral of the given equation; then a first integral of the latter (*second integral* of the given equation) and so on, until the differential coefficients cease to appear. The primitive will be obtained when this step-by-step process has been repeated the number of times equal to the order of the given equation.

Depending on the transformations used prior to integration different first integrals of the same equation can be obtained. For example, the equation

$$\frac{d^2y}{dx^2} + y = 0 \tag{1.4.4}$$

has the following first integrals:

$$\left(\frac{dy}{dx}\right)^2 + y^2 = A^2, \tag{1.4.5}$$

$$\frac{dy}{dx}\cos x + y\sin x = B, \tag{1.4.6}$$

$$-\frac{dy}{dx}\sin x + y\cos x = C, \tag{1.4.7}$$

$$\frac{dy}{dx} = y\cot(x+\alpha), \tag{1.4.8}$$

where A, B, C, α are arbitrary constants. These first integrals are not *all independent*. For the constants, we have the relations: $A\cos\alpha = B$, $A\sin\alpha = C$, $B^2 + C^2 = A^2$.

However, *if independent first integrals of a differential equation, equal in number to the order of the equation, have been obtained, all the differential coefficients can be eliminated so as to leave the primitive.* Thus, from (1.4.6) and (1.4.7), we get the primitive $y = B\sin x + C\cos x$ and from (1.4.5) and (1.4.8), we deduce $y = A\sin(x+\alpha)$. These two primitives are same on account of relations between the constants.

We now state the following *fundamental theorem* (without proof):

A differential equation of order n has n, and cannot have more than n independent first integrals. The equation cannot have, therefore, more than n arbitrary and independent constants in the general solution.

1.5 Further illustrative examples

Example 1.5.1. *Show that the substitution $x = e^z$ transforms the equation*

$$x^2\frac{d^2y}{dx^2} - x\frac{dy}{dx} + y = 2\ln x$$

into the equation:

$$\frac{d^2y}{dz^2} - 2\frac{dy}{dz} + y = 2z.$$

Solution: This is a problem of transformations of independent variable. They are often useful in solving a differential equation (Art. 7.4).

Here,

$$\frac{dy}{dx} = \frac{dy}{dz}\frac{dz}{dx} = \frac{1}{x}\frac{dy}{dz} \quad \left(\because z = \ln x, \frac{dz}{dx} = \frac{1}{x}\right)$$

$$\frac{d^2y}{dx^2} = \frac{d}{dx}\left(\frac{dy}{dx}\right) = \frac{d}{dx}\left(\frac{1}{x}\frac{dy}{dz}\right) = -\frac{1}{x^2}\frac{dy}{dz} + \frac{1}{x}\frac{d}{dx}\left(\frac{dy}{dz}\right) = -\frac{1}{x^2}\frac{dy}{dz} + \frac{1}{x}\frac{d}{dz}\left(\frac{dy}{dz}\right)\cdot\frac{dz}{dx}$$

$$= -\frac{1}{x^2}\frac{dy}{dz} + \frac{1}{x^2}\frac{d^2y}{dz^2} \quad \left(\text{again using } \frac{dz}{dx} = \frac{1}{x}\right)$$

or, $$x^2\frac{d^2y}{dx^2} = \frac{d^2y}{dz^2} - \frac{dy}{dz}.$$

Hence,

$$x^2\frac{d^2y}{dx^2} - x\frac{dy}{dx} + y = 2\ln x$$

becomes

$$\frac{d^2y}{dz^2} - 2\frac{dy}{dz} + y = 2z.$$

Example 1.5.2. *Transform the differential equation*

$$\frac{d^2y}{dx^2}\cos x + \frac{dy}{dx}\sin x - 2y\cos^3 x = 2\cos^5 x$$

into one having z as independent variable, where $z = \sin x$.

Solution:

$$\frac{dz}{dx} = \cos x, \quad \frac{dy}{dx} = \frac{dy}{dz}\frac{dz}{dx} = \cos x\frac{dy}{dz}.$$

$$\frac{d^2y}{dx^2} = \frac{d}{dx}\left(\cos x\frac{dy}{dz}\right) = -\sin x\frac{dy}{dz} + \cos x\frac{d}{dx}\left(\frac{dy}{dz}\right) = -\sin x\frac{dy}{dz} + \cos^2 x\frac{d^2y}{dz^2}.$$

∴ the given equation, then becomes

$$\cos x\left[-\sin x\frac{dy}{dz} + \cos^2 x\frac{d^2y}{dz^2}\right] + \sin x\left[\cos x\frac{dy}{dz}\right] - 2y\cos^3 x = 2\cos^5 x$$

or, $$\cos^3 x\frac{d^2y}{dz^2} - 2y\cos^3 x = 2\cos^5 x$$

or, $$\frac{d^2y}{dz^2} - 2y = 2\cos^2 x = 2(1 - z^2).$$

Example 1.5.3. *Show that if z satisfies* $\frac{d^2z}{dx^2} + P\frac{dz}{dx} = 0$, *by changing the independent variable from x to z, we can transform*

$$\frac{d^2y}{dx^2} + P\frac{dy}{dx} + Qy = R \text{ into } \frac{d^2y}{dz^2} + Sy = T,$$

where P, Q, R are functions of x and S, T are functions which are to be determined.

Solution: Let $z = f(x)$. Then,

$$\frac{dy}{dx} = \frac{dy}{dz}\frac{dz}{dx}, \quad \frac{d^2y}{dx^2} = \frac{d^2z}{dx^2}\frac{dy}{dz} + \frac{d^2y}{dz^2}\left(\frac{dz}{dx}\right)^2.$$

Then the given equation becomes

$$\frac{d^2y}{dz^2}\left(\frac{dz}{dx}\right)^2 + \left(P\frac{dz}{dx} + \frac{d^2z}{dx^2}\right)\frac{dy}{dz} + Qy = R$$

$$\text{or,}\quad \frac{d^2y}{dz^2} + Sy = T \left(\text{using } P\frac{dz}{dx} + \frac{d^2z}{dx^2} = 0\right),$$

where

$$S = \frac{Q}{\left(\frac{dz}{dx}\right)^2} \quad \text{and} \quad T = \frac{R}{\left(\frac{dz}{dx}\right)^2}.$$

Example 1.5.4. *Obtain the differential equation of all circles each of which touches the axis of x at the origin.*

Solution: The circles passes through the origin. Hence, its equations are of the form:

$$x^2 + y^2 + 2gx + 2fy = 0.$$

But the tangent at the origin is given to be x-axis, i.e. $y = 0$.
∴ the equation will be of the form:

$$x^2 + y^2 + 2fy = 0, \text{ (Fig. 1.1)}$$

where f is an arbitrary constant.

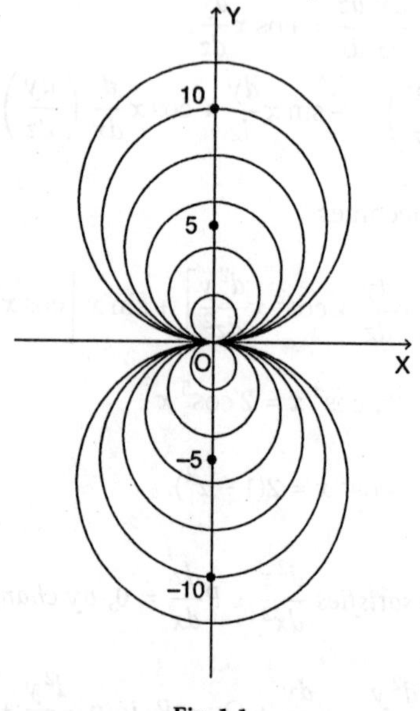

Fig. 1.1

(Lowest degree terms = 0 gives the equation of the tangent at the origin).

Differentiating with respect to x, we get

$$2x + 2y\frac{dy}{dx} + 2f\frac{dy}{dx} = 0.$$

$\therefore$ eliminating f, we obtain

$$(x^2 + y^2) + 2y\left\{-\frac{x + y\dfrac{dy}{dx}}{\dfrac{dy}{dx}}\right\} = 0 \quad \text{or,} \quad (x^2 + y^2)\frac{dy}{dx} - 2xy - 2y^2\frac{dy}{dx} = 0$$

$$\text{or,} \quad (x^2 - y^2)\frac{dy}{dx} = 2xy \text{ (reqd. differential equation).}$$

Examples I(A)

1. Form the differential equation of which each of the following functions is the primitive. State the order and degree of the equation; also say whether the equation is linear or not, [A, B, C, D are arbitrary constants]:
   (a) $y = Ax^2 + Bx$
   (b) $y = Ae^{kx} + Be^{-kx}$
   (c) $y = (A + Bx)e^{kx}$
   (d) $x = e^{-\frac{1}{2}kt}(A\cos nt + B\sin nt)$
   (e) $y = (A + Bx)\cos kx + (C + Dx)\sin kx$
   (f) $y = A\cos(\ln x) + B\sin(\ln x)$
   (g) $y = A\cosh kx + B\sinh kx + C\sin kx + D\cos kx$
   (h) $y = A\cos x + B\cosh x$
   (i) $y = Cx + \sqrt{a^2C^2 + b^2}$.
2. Eliminate α from the following equations. State the order and degree of the resulting differential equations; say whether linear or not:
   (a) $x^2 + y^2 - 2\alpha y = \alpha^2$
   (b) $\sqrt{(1 - x^2)} + \sqrt{(1 - y^2)} = \alpha(x - y)$
   (c) $y = \alpha x + \alpha - \alpha^3$
   (d) $x\cos\alpha + y\sin\alpha = a$.
3. Determine the order, degree, the independent variable, the dependent variable and linearity of the following differential equations:
   (a) $\frac{d^3y}{dx^3} - 15x\frac{dy}{dx} = e^x + 2$
   (b) $x^2\frac{d^2y}{dx^2} - x\frac{dy}{dx} = 1 - \sin x + \ln x$
   (c) $t\frac{d^2y}{dt^2} + t^2\frac{dy}{dt} - (\cos t)\sqrt{y} = 2t^2 - 3t + 4$
   (d) $3t^2\frac{d^3y}{dt^3} - (\sin t)\frac{d^2y}{dt^2} - (\cos t)y = 0$
   (e) $\frac{d^2y}{dx^2} + x^2\frac{dy}{dx} + x\sin y = 0$
   (f) $x^3\frac{d^2y}{dx^2} + (\cos x)\frac{dy}{dx} + (\sin x)y = 0$
   (g) $\frac{d^2y}{dx^2} + y\frac{dy}{dx} + y^2 = 0$
   (h) $\left(\frac{d^2x}{dt^2}\right)^2 + \frac{d^2x}{dt^2} + t\frac{dx}{dt} = 0.$
4. Eliminate the parameters a and b from the following primitives:
   (a) $xy = ae^x + be^{-x}$ (b) $y^2 = a(b^2 - x^2)$ (c) $v = a\ln r + b$.
5. Verify that the functions given on the right are general solutions (or primitives) of the corresponding differential equations on the left (A or B, whenever occurs are supposed to be parameters):

$$\left(\text{Notations: } y' = \frac{dy}{dx},\ y'' = \frac{d^2y}{dx^2}\right)$$

   (a) $xy'^2 - yy' + a = 0$; $y = mx + a/m$ (m, parameter)
   (b) $xy'^2 - 2yy' - x = 0$; $2cy + c^2 = x^2$ (c, parameter)
   (c) $x(y - 1)y'' = 2xy'^2 + (y - 1)y'$; $y = (x^2 + A^2)/(x^2 + B^2)$
   (d) $x^2y'' - xy' + 2y = x\ln x$; $y = Ax\cos(\ln x) + Bx\sin(\ln x) + x\ln x$

(e) $y'' = y\sec^2 x + y' \tan x$; $y = A\sec x + B\tan x$.

6. Prove that if $y = \frac{ax+b}{cx+d}$ (a, b, c, d are arbitrary constants), then $2\frac{dy}{dx}\frac{d^3y}{dx^3} = 3\left(\frac{d^2y}{dx^2}\right)^2$, and that if $a + d = 0$, then $(y - x)\frac{d^2y}{dx^2} = 2y\left(1 + \frac{dy}{dx}\right)$.

7. (a) Prove that the family of catenaries $y = c\cosh(x/c)$ satisfies the equation

$$y\sinh^{-1}(y) = x\left(1 + y'^2\right)^{1/2}, \left(y' = \frac{dy}{dx}\right).$$

(b) Show that the differential equation satisfied by the family of curves given by $c^2 + 2cy - x^2 + 1 = 0$, where c is the parameter of the family, is $(1 - x^2)p^2 + 2xyp + x^2 = 0$, where $p = \frac{dy}{dx}$.

8. (a) Find the differential equation associated with the primitive $y = a + be^{5x} + ce^{-7x}$; a, b, c being parameters.

(b) Construct a differential equation by the elimination of the arbitrary constants a and b from each of the relations:

(i) $ax^2 + by^2 = 1$ (ii) $x = a\cos nt + b\sin nt$.

9. (a) Show that the equation $ax + by + c = 0$ leads to $\frac{d^2y}{dx^2} = 0$ after the elimination of the three constants a, b and c. How do you explain this equation being of second order?

(b) Show that the primitive $y = ae^{bx+c}$ leads to $\frac{d^2}{dx^2}(\ln y) = 0$.

How do you explain this equation being of the second order, whereas there are apparently three constants a, b, c?

10. (a) Show that $x = e^z$ transforms $x^2\frac{d^2y}{dx^2} + x\frac{dy}{dx} + 4y = 0$ into $\frac{d^2y}{dz^2} + 4y = 0$.

(b) Show that the substitution $x = \sinh z$ transforms the equation $(1 + x^2)\frac{d^2y}{dx^2} + x\frac{dy}{dx} = 4y$ into $\frac{d^2y}{dz^2} = 4y$.

(c) Prove that $z = \sin x$ transforms $\frac{d^2y}{dx^2} + \frac{dy}{dx}\tan x + y\cos^2 x = 0$ into $\frac{d^2y}{dz^2} + y = 0$.

(Geometric and Physical Problems)

11. A curve is defined by the condition that the sum of the x and y intercepts of its tangents is always equal to a. Express this property by means of a differential equation.

12. (a) Find the differential equation of the family of parabolas $y = cx^2$; when would the equation be meaningful?

(b) Find the differential equation of the system of circles of *constant radius a* with their centres on the x-axis. Obtain the equation if a is also allowed to vary.

13. (a) Find the differential equation of all circles in the xy-plane.

(b) Obtain the differential equation of

(i) all parabolas each of which has a latus rectum $4a$, and whose axes are parallel to the x-axis;

(ii) all conics whose axes coincide with the axes of coordinates;

(iii) all circles each of which touches the axis of x at the origin;

(iv) all hyperbolas which pass through the origin and have their asymptotes parallel to the coordinate axes;

(v) all parabolas having their axes parallel to the y-axis.

14. Verify that all circles which touch the y-axis at the origin are included in $x^2 + y^2 = 2cx$. Deduce that the differential equation of the family is $y^2 - x^2 = 2xy\, dy/dx$.

15. The motion of a particle of mass m moving along a straight line is governed by the equation $m\ddot{x} + r\dot{x} + kx = 0$, where r and k are constants. Verify that $x = Ae^{-(r/2m)t}\cos\left(\frac{\omega}{2m}t + B\right)$, $(\omega^2 = 4mk - r^2)$ satisfies the differential equation, A and B being arbitrary constants.
16. Prove that the equations $2y - \alpha x = xy'$ and $y + \beta x^2 = xy'$ have a common complete primitive.
17. A point describing rectilinear simple harmonic motion of frequency $\omega/2\pi$ per second has displacement x at time t given by either $x = A\cos\omega t + B\sin\omega t$ or, $x = C\sin(\omega t + \alpha)$. Show that in both cases the corresponding differential equation is $\ddot{x} + \omega^2 x = 0$.
18. Prove that the adiabatic law $pv^\gamma = c$, where γ is a fixed and c is an arbitrary constant, leads to $v\frac{dp}{dv} + \gamma p = 0$.
19. For a certain substance the rate of change of vapour pressure (P) w.r.t. temperature (T) is proportional to the vapour pressure and inversely proportional to the square of the temperature. Express this condition in the form of a differential equation. Show that $P = Ae^{-k/T}$, where A is an arbitrary constant which satisfies the differential equation.

Answers

$\left[y' = \frac{dy}{dx},\ y'' = \frac{d^2y}{dx^2},\ y''' = \frac{d^3y}{dx^3},\ y^{iv} = \frac{d^4y}{dx^4},\ \dot{x} = \frac{dx}{dt},\ \ddot{x} = \frac{d^2x}{dt^2},\ \text{etc.}\right]$

1. (a) $y'' - (2/x)y' + (2/x^2)y = 0$
   (b) $y'' - k^2y = 0$
   (c) $y'' - 2ky' + k^2y = 0$
   (d) $\ddot{x} + k\dot{x} + \left(n^2 + \frac{1}{4}k^2\right)x = 0$
   [(a)–(d) are all of second order, first degree and linear]
   (e) $y^{iv} + 2k^2y'' + k^4y = 0$ (fourth order, first degree and linear)
   (f) $x^2y'' + xy' + y = 0$ (second order, first degree and linear)
   (g) $y^{iv} = k^4y$ (fourth order, first degree and linear)
   (h) $(\tanh x + \tan x)y'' - 2y' + (\tanh x - \tan x)y = 0$ (Second order, first degree, linear)
   (i) $y = xy' + \sqrt{a^2y'^2 + b^2}$ (first order, second degree and nonlinear).
2. (a) $(x^2 - 2y^2)y'^2 - 4xyy' - x^2 = 0$ (order one, degree two, nonlinear)
   (b) $\sqrt{1 - x^2}\,dy = \sqrt{1 - y^2}\,dx$ (order one, degree one, nonlinear)
   (c) $y = xy' + y' - y'^3$ (order one, degree three, nonlinear)
   (d) $(y - xy')^2 = a^2(1 + y'^2)$ (order one, degree two, nonlinear).
3. (a) 3, 1, x, y, linear (b) 2, 1, x, y, linear (c) 2, no, t, y, nonlinear
   (d) 3, 1, t, y, linear (e) 2, 1, x, y, nonlinear (f) 2, 1, x, y, linear
   (g) 2, 1, x, y, nonlinear (h) 2, 2, t, x, nonlinear.
4. (a) $xy'' + 2y' - xy = 0$ (b) $xyy'' + xy'^2 - yy' = 0$ (c) $\frac{d^2v}{dr^2} + \frac{1}{r}\frac{dv}{dr} = 0$.
8. (a) $35\frac{dy}{dx} - 2\frac{d^2y}{dx^2} = \frac{d^3y}{dx^3}$ (b) (i) $xyy'' + x(y')^2 - yy' = 0$ (ii) $\frac{d^2x}{dt^2} = -n^2x$.
11. $xy'^2 - (x + y - a)y' + y = 0$.
12. (a) $xy' = 2y$. Everywhere except when c is indeterminate and when $x = 0$, $y = 0$.
    (b) $y^2y'^2 + y^2 = a^2$; $yy'' + y'^2 + 1 = 0$.
13. (a) $(1 + y'^2)y''' - 3y'(y'')^2 = 0$;
    (b) (i) $y'^2 + 2ay'' = 0$ (ii) $xyy'' + xy'^2 = yy'$ (iii) $(x^2 - y^2)y' = 2xy$ (iv) $xyy'' - 2xy'^2 + 2yy' = 0$ (v) $y''' = 0$.
19. $\frac{dP}{dT} = k\frac{P}{T^2}$ (k is the constant of proportionality).

1.6 Family of curves represented by ordinary differential equations

The primitive of an ordinary *differential equation of first order* is a relation between the two variables x and y, and a parameter c. For each numerical value of c, we get a separate curve and hence the primitive gives a *one-parameter family of curves.* The differential equation is then said to represent a one-parameter family of curves. Each curve of the family is called an *integral curve.*

I. Let the equation be of first order and first degree:

$$\frac{dy}{dx} = f(x, y); \tag{1.6.1}$$

and let D be the domain in the (x, y)-plane throughout which $f(x, y)$ is *single-valued* and *continuous.* Let (x_1, y_1) be an interior point of D. At (x_1, y_1) there will be a corresponding value of dy/dx, say p_1, determined by the eq. (1.6.1). For each point (x_1, y_1) of D, we have three elements (x_1, y_1, p_1), p_1 being determined by the eq. (1.6.1).

Suppose, now we take an adjacent point (x_2, y_2) in the direction of p_1. At (x_2, y_2), we have again three elements (x_2, y_2, p_2), p_2 being determined by eq. (1.6.1). By continuing this process a broken line is obtained which may be considered as an approximation to the integral curve passing through (x_1, y_1). Since $f(x, y)$ is supposed to be *continuous* and *single-valued* at every point of D, there will pass one and only one integral curve through each point of D. Outside D there may be points at which $f(x, y)$ is not continuous or single-valued. Such points are known as *singular points.* The behaviour of integral curves at singular points may be exceptional. The equation of each curve is thus a particular solution of the differential equation; the whole system of curves is the general solution (*the locus of the differential equation*).

II. Let the first-order equation $f\left(x, y, \frac{dy}{dx}\right) = 0$ be of *second degree* in $\frac{dy}{dx}$. Then there will be, in general, two values of $\frac{dy}{dx}$ at each point (x_1, y_1) of its domain of definition D. Hence, two curves of the system, which is the locus of the general solution, pass through each point of D.

The general solution, $\phi(x, y, c) = 0$, say, must have then two different values of c for each point and hence c must appear in the solution in the second degree.

III. More generally, a differential equation of first order, say, $f(x, y, dy/dx) = 0$ which is of degree n in dy/dx and which has $\phi(x, y, c) = 0$ for its general solution, has a one-parameter family of integral curves ($\because$ there is only one arbitrary constant); but, in general, n of these curves pass through each point in the domain of definition of f. The constant c must appear in the nth degree in the general solution.

IV. Now consider an equation of second order $\frac{d^2y}{dx^2} = f\left(x, y, \frac{dy}{dx}\right)$, where $f(x, y, dy/dx)$ is continuous and single-valued for a certain range of values of its arguments. The value of dy/dx at a point (x_1, y_1) can be chosen arbitrarily within certain limits. Hence, through the point (x_1, y_1) there passes a one-fold infinity of integral curves. The general solution involves two arbitrary constants, and therefore the aggregate of integral curves forms a two-parameter family (We say that there are ∞^2 curves).

V. More generally, the integral curves of an ordinary equation of order n form an n-parameter family, and through each point (nonsingular) there passes in general an $(n-1)$-fold infinity of integral curves. In this case, the set of curves represented by the general solution are ∞^n in number.

Example 1.6.1. *The primitive of the equation $dy/dx = y/x$ is $y = cx$. Find the equation of the integral curve through $(2, 3)$ and also through $(0, 0)$.*

Solution:

(i) When $x = 2$, $y = 3$; $c = 3/2$ and the required equation is $2y = 3x$.

(ii) When $x = 0$, $y = 0$; c is not determined. All the integral curves pass through $(0, 0)$. Note that here $f(x, y) = y/x$ is not continuous at the origin. In other words, $(0, 0)$ is a singular point and hence such an exceptional behaviour at that point.

Note 1.6.1. *The geometrical meaning of $dy/dx = y/x$ is that the radius vector and the tangent have the same gradient, which can only happen for straight lines through the origin (i.e. $y = cx$ is the primitive). The general solution represents a family of lines passing through the origin.*

Example 1.6.2. *The primitive of the first-order and first-degree equation $dy/dx = -y/x$ is $y = c/x$. Find a particular solution that will satisfy the following initial condition $y = 1$, when $x = 1$. Interpret geometrically.*

Solution: Putting $x = 1, y = 1$ in $y = c/x$, we have $c = 1$ whereby the function $y = 1/x$, or, $xy = 1$ becomes the particular solution we are interested to find.

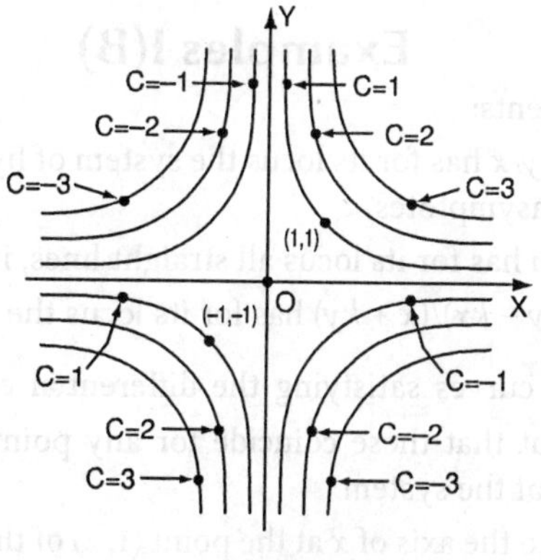

Fig. 1.2

Geometrically, the general solution or the complete primitive is a family of curves (incidentally, these are rectangular hyperbolas) in a coordinate plane (Fig. 1.2), which family depends on a single arbitrary constant *c*. These curves are called *integral curves* of the given differential equation. The particular solution (or, the particular integral) defined by the given initial condition is depicted by that particular hyperbola which passes through the point (1, 1), as shown in the figure.

Example 1.6.3. *From the geometrical meaning of the equation $dy/dx = -x/y$ obtain the locus of the differential equation, and see that it can be expressed as $x^2 + y^2 = c^2$. Find also the equation of the integral curve through* (3, 4).

Solution: Writing the equation in the form $\frac{y}{x} \cdot \frac{dy}{dx} = -1$, we see that the radius vector and the tangent have gradients whose product is −1, i.e. they are perpendicular. The integral curves are, therefore, circles of any radius with the origin as centre. That is, the locus of the differential equation is made up of all circles (infinite in number) that have the origin for centre (Fig. 1.3).

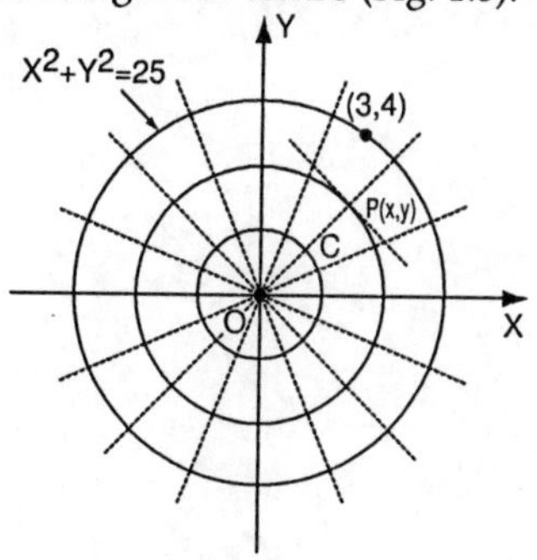

Fig. 1.3

For example, the radius-vector *OP* is perpendicular to the tangent to the circle through *P*, as shown in Fig. 1.3. Thus, the general solution (i.e. the locus of the differential equation) is a family of circles with centre at (0, 0) and radius equal to *c* (where *c* is an arbitrary parameter) which is given by $x^2 + y^2 = c^2$. And the integral curve through (3, 4) is given by $x^2 + y^2 = 25$, as shown in the figure. In this case, the singular point $(0, 0)$ may be regarded as a circle of zero radius, the limiting form of the integral curves near it, but no integral curve of finite size passes through it.

Example 1.6.4. *From the geometrical significance of the equation* $\frac{d^2y}{dx^2} = 0$ *suggest the locus of the differential equation.*

Solution: $d^2y/dx^2 = 0$ if and only if the curve described by the point (x, y) has the curvature zero; it can, therefore, be any straight line. The general solution represents straight lines, ∞^2 in number. The primitive $y = mx + c$ involves two arbitrary constants m and c.

Examples I(B)

1. Verify the following statements:
   (a) The equation $dy/dx = -y/x$ has for its locus the system of hyperbolas, infinite in number, that have x- and y-axes for asymptotes.
   (b) The equation $dy/dx = m$ has for its locus all straight lines, infinite in number, of gradient m.
   (c) The equation $dy/dx = (y - kx)/(x + ky)$ has for its locus the system of equiangular spirals.
2. Prove that in general two curves satisfying the differential equation $x\left(\frac{dy}{dx}\right)^2 - y\frac{dy}{dx} + 1 = 0$ pass through any point, but that these coincide for any point on a certain parabola, which is the envelope of the curves of the system.
3. Prove that the inclinations to the axis of x at the point $(1, 2)$ of the two curves passing through this point and satisfying $(dy/dx)^2 = y^2 - 2x + x^2$ are $60°$ and $-60°$. Prove that the radius of curvature of either of the curves at the point $(1, 2)$ is 4.
4. Prove that if the family of integral curves of the linear differential equation of the first order $\frac{dy}{dx} + f(x)y = g(x)$ is cut by the line $x = \alpha$, the tangents at the points of intersection are concurrent. In particular, let $f(x) = -1/x$ and $g(x) = -1/x^3$. Prove that for varying α the locus of the point of concurrence is a straight line.

Chapter 2

Equations of First Order and First Degree, Existence of Solutions

In the present chapter, our main aim is to introduce several equations of the first order and first degree and the methods of finding their solutions. It is quite natural to ask whether the solution of an equation exists or not. If it exists, the solution of an equation may not be unique. In this context, we may consider initial-value and boundary-value problems. Here we will discuss about several types of initial-value and boundary-value problems, Basic existence and uniqueness theorem and Picard's theorem that guarantees the existence of unique solution. By Picard's approximation, one can find the approximate solution of an initial-value problem.

The method of finding solutions of the equations of first order and first degree is very easy if the equation is exact; otherwise we have to find an integrating factor to solve it. For exact equation, method of separation of variables is remarkable. This method is also applicable for homogeneous equations of the first order and first degree. First order non-homogeneous linear differential equations can be solved by using undetermined coefficients with considerably less effort. Now the question arises: How is an integrating factor of non-exact equation found? It depends on the nature of the equations. One can find it by inspection. But it needs a considerable amount of knowledge and skill. In this chapter, we have introduced some methods of finding integrating factors of the non-exact equations of the first order and first degree. Other special types of equations are linear equations, which always have integrating factors of a certain special form that may easily be found. Bernoulli equations have been also introduced in this chapter. This type of equations can be reduced in linear form.

Before going to our main topic, we would like to develop some fundamental concepts on continuity of functions of two independent variables, bounded functions, Lipschitz condition, convex sets, initial-value and boundary value problems.

2.1 Equations of first order and first degree

An ordinary differential equation of first order and first degree can be written as

$$\frac{dy}{dx} = f(x, y) \tag{2.1.1}$$

or in the more symmetric form:

$$M\,dx + N\,dy = 0, \tag{2.1.2}$$

where $M = M(x, y)$ and $N = N(x, y)$ are functions of x and y.

2.2 Continuity

Let $f(x, y)$ be a real function defined on a domain $D \subseteq \mathbb{R}^2$ and let (x_0, y_0) be an interior point of D. Then the function $f(x, y)$ is said to be continuous at (x_0, y_0) if, for given any $\varepsilon > 0$, $\exists\, \delta > 0$, such that $|f(x, y) - f(x_0, y_0)| < \varepsilon\ \forall\, (x, y) \in D$ whenever $|x - x_0| < \delta$ and $|y - y_0| < \delta$.

Example 2.2.1. *Let* $f(x, y) = \begin{cases} x \sin \frac{1}{y} + y \sin \frac{1}{x}, & \text{if } xy \neq 0 \\ 0, & \text{if } xy = 0. \end{cases}$

Then show that $f(x, y)$ *is continuous at the point* $(0, 0)$.

Solution: For given any $\varepsilon > 0$, we can find $\delta \left(= \frac{\varepsilon}{2}\right) > 0$ such that

$$|f(x, y) - f(0, 0)| = \left|x \sin \tfrac{1}{y} + y \sin \tfrac{1}{x}\right| \leq \left|x \sin \tfrac{1}{y}\right| + \left|y \sin \tfrac{1}{x}\right| = |x| \left|\sin \tfrac{1}{y}\right| + |y| \left|\sin \tfrac{1}{x}\right| \leq |x| + |y| < \varepsilon,$$

whenever $|x - 0| < \delta$ and $|y - 0| < \delta$.

Hence, the function $f(x, y)$ is continuous at the point $(0, 0)$.

2.3 Bounded function

Let $f(x, y)$ be a real function defined on $D \subseteq \mathbb{R}^2$, D being either a domain or a closed domain of $f(x, y)$. Then $f(x, y)$ is said to be a bounded function on D if there exists $K > 0$ such that $|f(x, y)| \leq K\ \forall\, (x, y) \in D$.

2.4 Lipschitz condition

Let $f(x, y)$ be a real function defined on $D \subseteq \mathbb{R}^2$, D being either a domain or a closed domain of $f(x, y)$. Then $f(x, y)$ is said to satisfy a Lipschitz condition with respect to y in D if there exists a constant $L > 0$ such that $|f(x, y_1) - f(x, y_2)| \leq L|y_1 - y_2|$ for all $(x, y_1), (x, y_2) \in D$. Here L is known as the Lipschitz constant.

Note 2.4.1. *It is also known as Cauchy-Lipschitz condition.*

Example 2.4.1. *Show that the function* $f(x, y)$ *defined by* $f(x, y) = 2y^2$, $(x, y) \in D$, *where* $D = \{(x, y) \in \mathbb{R}^2 : |x| \leq 2, |y| \leq 1\}$ *satisfies a Lipschitz condition.*

Solution: Let (x, y_1), (x, y_2) be any two points in D. Now $|f(x, y_1) - f(x, y_2)| = |2y_1^2 - 2y_2^2| = 2|y_1 + y_2|\,|y_1 - y_2| \leq 2(|y_1| + |y_2|)|y_1 - y_2| \leq 4|y_1 - y_2|$. It confirms that $f(x, y)$ satisfies a Lipschitz condition.

Example 2.4.2. *Show that the function* $f(x, y)$ *defined by* $f(x, y) = \begin{cases} xy \sin \frac{1}{x} & \text{if } x \neq 0 \\ 0, & \text{if } x = 0 \end{cases}$

on $D = \{(x, y) \in \mathbb{R}^2 : |x| \leq a, |y| \leq b\}$, *satisfies a Lipschitz condition.*

Solution: Let (x, y_1), (x, y_2) be any two points in D. Then

$$|f(x, y_1) - f(x, y_2)| = \left|xy_1 \sin \tfrac{1}{x} - xy_2 \sin \tfrac{1}{x}\right| = |x| \left|\sin \tfrac{1}{x}\right| |y_1 - y_2| \leq a|y_1 - y_2|,$$

which shows that $f(x, y)$ satisfies a Lipschitz condition.

2.5 Convex set

A non-empty set $D \subset \mathbb{R}^2$ is said to be convex set if for any pair of points $(x_1, y_1), (x_2, y_2) \in D$, $(1 - \lambda)x_1 + \lambda x_2, (1 - \lambda)y_1 + \lambda y_2) \in D$ for all $\lambda \in [0, 1]$. In other words, a non-empty set $D \subset \mathbb{R}^2$ is said to be convex set if the straight line segment joining any two points in D lies entirely within D.

Theorem 2.5.1. *Let* $f(x, y)$ *be a real function defined on a convex set* $D \subseteq \mathbb{R}^2$ *and let* $f_y(x, y)$ *exist and it is bounded in* D. *Then* $f(x, y)$ *satisfies a Lipschitz condition with respect to* y *in* D, *where the Lipschitz constant is given by* $L = \max\limits_{(x,y) \in D} \{|f_y(x, y)|\}$.

Example 2.5.1. *Let us consider the function $f(x, y)$ defined by $f(x, y) = y^2$ on the convex set $D = \{(x, y) \in \mathbb{R}^2 : -a \leq x \leq a, -b \leq y \leq b\}$. Then $f_y(x, y) = 2y$ and $|f_y(x, y)| = 2|y| \leq 2b$ ($\because$ $|y| \leq b$). So $f_y(x, y)$ exists and is bounded in D. Hence, $f(x, y)$ satisfies a Lipschitz condition in D with Lipschitz constant $L = \max_{(x,y)\in D} |f_y(x, y)| = 2b$.*

Example 2.5.2. *Show that the function $f(x, y)$ defined by $f(x, y) = 2 + x\sin(xy)$, on $D = \{(x, y) \in \mathbb{R}^2 : 0 \leq x \leq 2, -\infty < y < \infty\}$, satisfies a Lipschitz condition. Hence, find the Lipschitz constant.*

Solution: Observe that D is a convex set in which $f(x, y)$ is continuous and $f_y(x,y)$ exists and it is bounded; because, $|f_y(x, y)| = |x^2 \cos(xy)| \leq |x|^2 |\cos(xy)| \leq |x|^2 \leq 2^2 = 4$. It confirms that the function satisfies a Lipschitz condition on D with Lipschitz constant, $L = \max_{(x,y)\in D} |f_y(x, y)| = \max_{(x,y)\in D} |x|^2 = 4$.

Alternative: Let y_1, y_2 be any two distinct real numbers so that $y_2 > y_1$. Then considering x as a constant and applying the mean value theorem to the function $f(x, y) = 2 + x\sin(xy)$ on the closed interval $[y_1, y_2]$,we can find $\xi \in (y_1, y_2)$ for which

$$\frac{f(x, y_2) - f(x, y_1)}{y_2 - y_1} = f_y(x, \xi) = x^2 \cos(x\xi).$$

$$\therefore\ |f(x, y_2) - f(x, y_1)| = |y_2 - y_1||x^2 \cos(x\xi)| = |y_2 - y_1||x^2||\cos(x\xi)| \leq 4|y_2 - y_1|.$$

Hence, the given function satisfies Lipschitz condition with respect to the variable y with Lipschitz constant, $L = \max_{(x,y)\in D} |f_y(x, y)| = \max_{(x,y)\in D} |x|^2 = 4$.

Example 2.5.3. *Show that the function $f(x, y)$ defined by $f(x, y) = x^2|y|$ on the convex set $D = \{(x, y) \in \mathbb{R}^2 : |x| \leq a, |y| \leq b\}$, satisfies a Lipschitz condition. Examine whether the partial derivative $f_y(x, y)$ exists in D.*

Solution: Let (x, y_1), (x, y_2) be any two points in D. Then $|f(x, y_1) - f(x, y_2)| = \big||x^2|y_1| - x^2|y_2|\big| \leq |x|^2|y_1 - y_2| \leq a^2|y_1 - y_2|$, which shows that $f(x, y)$ satisfies a Lipschitz condition. Now let us examine whether $f_y(x, y)$ exists or not in D.

$$\lim_{k\to 0} \frac{f(x, 0 + k) - f(x, 0)}{k} = \lim_{k\to 0} \frac{x^2|(0 + k)| - x^2 \cdot 0}{k} = \lim_{k\to 0} \frac{x^2|k|}{k} = \pm x^2.$$

It follows that $f_y(x, y)$ does not exist at any point $(x, 0)$ in D, except the origin.

Note 2.5.1. *The above example ensures that the sufficient condition of the theorem 2.1 is not necessary for the function $f(x, y)$ to satisfy a Lipschitz condition.*

Example 2.5.4. *Show that the function $f(x, y) = \frac{2y}{x} + x^2 e^{2x}$ satisfies a Lipschitz condition on the rectangle $R : [1, 2] \times [-2, 5]$.*

Solution: Let $(x, y_1), (x, y_2)$ be any two points in R. Then

$$|f(x, y_1) - f(x, y_2)| = \left|\left(\frac{2y_1}{x} + x^2 e^{2x}\right) - \left(\frac{2y_2}{x} + x^2 e^{2x}\right)\right| = \frac{2}{|x|}|y_1 - y_2| \leq 2|y_1 - y_2|,$$

because $x \in [1, 2]$. Hence, $f(x, y)$ satisfies a Lipschitz condition with respect to y on D.

2.6 Initial-value problem

An ordinary differential equation together with a specified value that is the initial condition of the unknown function at a given point of a domain of the solution is known as *initial-value problem.* Let $f(x, y)$ be a real function defined on a domain $D \subseteq \mathbb{R}^2$ and (x_0, y_0) be an interior point of D. Then a differential equation of the form $\frac{dy}{dx} = f(x, y)$ along with the initial condition $y(x_0) = y_0$, is called an initial-value problem. In an initial-value problem, all the conditions are specified at the same value of the independent variable and this value is at the lower boundary of the domain; this is why the problem is called the initial-value problem. A solution to an initial-value problem is a function $y = \phi(x)$ that is a solution of the differential equation $\frac{dy}{dx} = f(x, y)$ satisfying the initial condition $y(x_0) = y_0$. For example, let us consider the initial-value problem: $\frac{dy}{dx} = 1 + xy$, $y(0) = 1$. Here $y(0) = 1$ is the initial condition corresponding to the initial-value problem.

Note 2.6.1. *Initial-value problems are also known as Cauchy's problems.*

Example 2.6.1. *Solve the initial-value problem* $\frac{dy}{dx} = \frac{x}{y}$, $y(5) = 4$.

Solution: The general solution of the differential equation

$$\frac{dy}{dx} = \frac{x}{y} \text{ is } x^2 - y^2 = c^2 \tag{2.6.1}$$

which represents one parameter family of rectangular hyperbolas. Using the initial-condition, we obtain $c^2 = 9$. Putting the value of c^2 in (2.6.1), we have $x^2 - y^2 = 9$.

$\therefore$ $y = \pm\sqrt{x^2 - 9}$. Negative sign cannot be accepted, because $y = 4$ at $x = 5$. Thus, the solution of the initial-value problem is given by $f(x) = \sqrt{x^2 - 9}$, $x \in \mathbb{R} - [-3, 3]$ (Fig. 2.1).

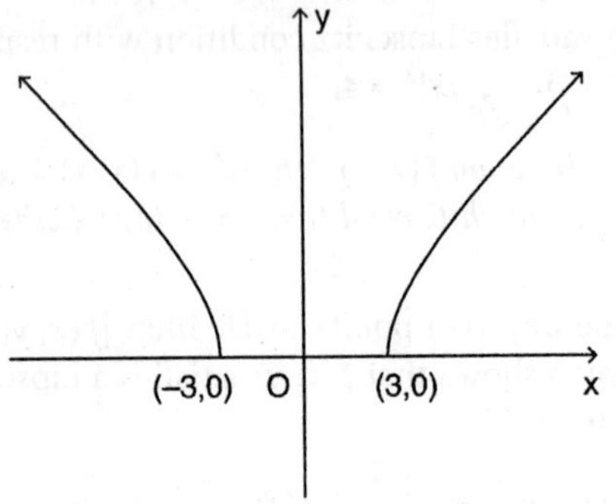

Fig. 2.1

2.7 Well-posed problem

An initial-value problem (IVP) is said to be well-posed if the following three conditions are satisfied: (i) A solution of IVP exists, (ii) The solution is unique, (iii) The solution's behaviour changes continuously with the initial condition.

The concept of well-posed IVP was introduced by Jacques Salomon Hadamard (8 December 1865 to 17 October 1963), a French mathematician. An IVP that is not well-posed is called ill-posed.

For example, let us consider the initial-value problem $\frac{dy}{dx} = 3y$, $y(0) = 2$. It has unique solution: $y = 2e^{3x}$. The solutions behaviour changes continuously with the initial condition. So it is a well-posed problem. Let us consider another initial-value problem $\frac{dy}{dx} = \frac{2y}{x}$, $y(0) = 0$. The general solution of the equations $\frac{dy}{dx} = \frac{2y}{x}$ is $y = cx^2$, c being an arbitrary constant. Observe that for all $c \in \mathbb{R}$, $y = cx^2$ is a solution of the IVP. Thus, it has infinitely many solutions. Hence, $\frac{dy}{dx} = \frac{2y}{x}$, $y(0) = 0$ is an ill-posed problem.

2.8 Boundary-value problem

An ordinary differential equation together with a set of additional constraints called the boundary conditions is known as boundary-value problem. A boundary-value problem contains the conditions specified at the boundaries of the independent variable in the differential equation; this is why the problem is called the boundary-value problem. For example, let us consider the following boundary-value problem:

$$\frac{d^2y}{dx^2} = 2y, \; y(0) = 0, \; y(\pi) = 0.$$

Here $y(0) = 0$ and $y(\pi) = 0$ are two boundary conditions corresponding to the boundary-value problem.

Theorem 2.8.1 (Basic existence and uniqueness). *Let $f(x, y)$ be continuous in some domain $D \subset \mathbb{R}^2$ and let $f_y(x, y)$ exist and be continuous in the domain D containing the point (x_0, y_0). Then there exists unique solution, namely $y = \phi(x)$, of the initial-value problem $\frac{dy}{dx} = f(x, y)$, $y(x_0) = y_0$ in the open interval $(x_0 - h, x_0 + h)$, $h(> 0)$ being sufficiently small.*

Example 2.8.1. *Show that the following initial-value problem has unique solution:*

$$\frac{dy}{dx} = x\sqrt{y-3}, \; y(-2) = 12.$$

Also find the solution.

Solution: Let $f(x, y) = x\sqrt{y-3}$. Then $f_y(x, y) = \frac{x}{2\sqrt{y-3}}$. Observe that $f(x, y)$ and $f_y(x, y)$ are both continuous around the point (−2, 12). Therefore, by the basic existence and uniqueness theorem, the initial-value problem has unique solution around the point (−2, 12).

$$\frac{dy}{dx} = x\sqrt{y-3}, \text{ or, } \frac{dy}{\sqrt{y-3}} = x\,dx.$$

$$\therefore \int \frac{dy}{\sqrt{y-3}} = \int x\,dx + c, \; c \text{ being arbitrary constant}$$

$$\text{or, } 2\sqrt{y-3} = \frac{x^2}{2} + c.$$

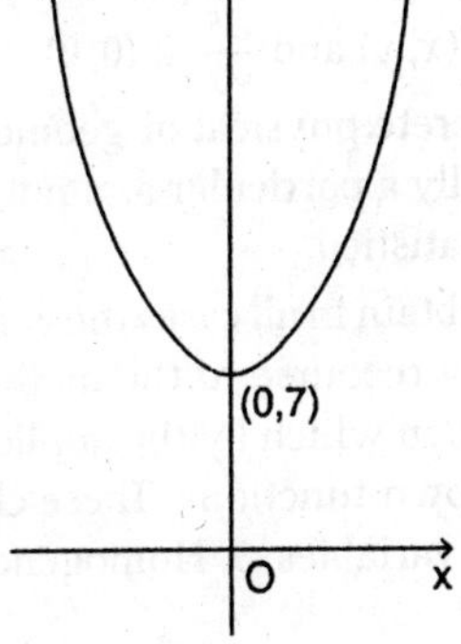

Fig. 2.2

Using given initial condition, we obtain $c = 4$.

$$\therefore \; 2\sqrt{y-3} = \frac{x^2}{2} + 4 \text{ or, } \sqrt{y-3} = \frac{x^2}{4} + 2 \text{ or, } y - 3 = \left(\frac{x^2}{4} + 2\right)^2 \text{ or, } y = 3 + \left(\frac{x^2}{4} + 2\right)^2,$$

which is the required solution of the initial-value problem (Fig. 2.2).

Theorem 2.8.2 (Picard). *Let $f(x, y)$ be a real function defined on a domain $D \subseteq \mathbb{R}^2$ and (x_0, y_0) be an interior point of D so that D contains the rectangle $R : |x - x_0| \leq a, |y - y_0| \leq b$. Let $M = \sup_{(x,y)\in R} |f(x, y)|$ and $h = \min(a, \frac{b}{M})$ and let $f(x, y)$ satisfy the following conditions:*

(i) $f(x, y)$ is continuous in D, (ii) $f(x, y)$ satisfies a Lipschitz condition with respect to y in D.

Then the initial-value problem $\frac{dy}{dx} = f(x, y), y(x_0) = y_0$ possesses unique solution on the closed interval $[x_0 - h, x_0 + h]$.

Note 2.8.1. *Picard's theorem is also known as Picard-Lindelöf theorem or, Cauchy-Lipschitz's theorem.*

Definitions 2.1.1

(a) Particular Solution: The solution of the eq. (2.1.1) satisfying the given initial condition (say, $y = y_0$, when $x = x_0$) is called a *Particular Solution* and the corresponding integral a *Particular Integral.*

(b) General Solution: The general solution of the first order differential eq. (2.1.1) is the solution $y = \phi(x, C)$, from which *given any initial condition $y = y_0$ when $x = x_0$*, a unique $C = C_0$ can be found such that $\phi(x_0, C_0) = y_0$. The equation $g(x, y, C) = 0$ connecting the independent variable and the general solution $\phi(x, C)$, is called the *general integral* or *complete primitive* of eq. (2.1.1).

The general solution (integral) is represented geometrically by a *one-parameter family of integral curves*, with not more than one curve of the family passing through each point of the region of the xy-plane where $f(x, y)$ and $\partial f/\partial y$ are continuous. We shall not, in the present text, include the proof of Existence Theorem but we shall mention some simple examples for which the theorem does not hold:

(i) *Consider the equation:* $\frac{dy}{dx} = \frac{y}{x}$.

We get the general solution (on separating the variables, $dy/y = dx/x$ and integrating) $y = Cx$ from which it is impossible to find a unique particular solution satisfying *the initial condition $y = 0$, when $x =$* 0. An infinite set of particular solution exists (Take C, any arbitrary constant). In fact, any straight line through the origin is an integral curve through the origin $(0, 0)$. The uniqueness property does not hold here. The explanation is that the right-hand side of $f(x, y) = y/x$ is not continuous and $\partial f/\partial y$ does not exist at $x = 0$.

(ii) *Consider the equation:* $\frac{dy}{dx} = -\frac{x}{y}$.

Integration gives the general solution $x^2 + y^2 = C$, from which it is clear that the differential equation has, in general, no solution satisfying the initial condition $y = 0$ when $x = 0$. The family of integral curves is the set of concentric circles with centre at the origin. The *absence* of a particular solution of $(0, 0)$ can be explained by the discontinuity of $f(x, y)$ and $\frac{\partial f}{\partial y}$ at $(0, 0)$.

We observe here that when a concrete physical or geometrical problems is being investigated with the aid of a differential equation, usually a particular and not the general solution is required, such that all the conditions of the problem are satisfied.

Unfortunately it is not possible to obtain in all cases the general solution in terms of known functions; in many cases one is obliged to have a recourse to the method of numerical approximation. We shall, however, discuss certain simple classes in which by the application of elementary processes the general solutions can be found in terms of known functions. These classes are classified as: 1. Exact equations, 2. Equations solvable by separation of variables, 3. Homogeneous equations, 4. Linear equations of first order.

Example 2.8.2. *Let us consider the initial-value problem $\frac{dy}{dx} = 1 + y^2, y(0) = 0$ on the domain $D \supseteq R : \{(x, y) \in \mathbb{R}^2 : |x| \leq 2, |y| \leq 1\}$. Observe that $f(x, y) = 1 + y^2$ is continuous in D (in fact, everywhere continuous) and $(0, 0)$ is an interior point of D. Again, for any two points (x, y_1), (x, y_2) in R, $|f(x, y_1) - (x, y_2)| = |y_1^2 - y_2^2| = |y_1 - y_2||y_1 + y_2| \leq (|y_1| + |y_2|)|y_1 - y_2| \leq 2|y_1 - y_2|$. Thus, $f(x, y)$ satisfies a Lipschitz condition in D. Here $M = \sup_{(x,y)\in R} |f(x, y)| = 2$ and $h = \min\{2, \frac{1}{M}\} = \frac{1}{2}$. Hence, the Picard's theorem guarantees the existence of unique solution in the closed interval $\left[-\frac{1}{2}, \frac{1}{2}\right]$.*

Solving the differential equation $\frac{dy}{dx} = 1 + y^2$, we obtain $\tan^{-1} y = x + c$, c being arbitrary constant. Using the initial condition $y(0) = 0$, we get $c = 0$.

$\therefore$ $\tan^{-1} y = x$, i.e. $y = \tan x$, which is the solution of the given initial-value problem (Fig. 2.3).

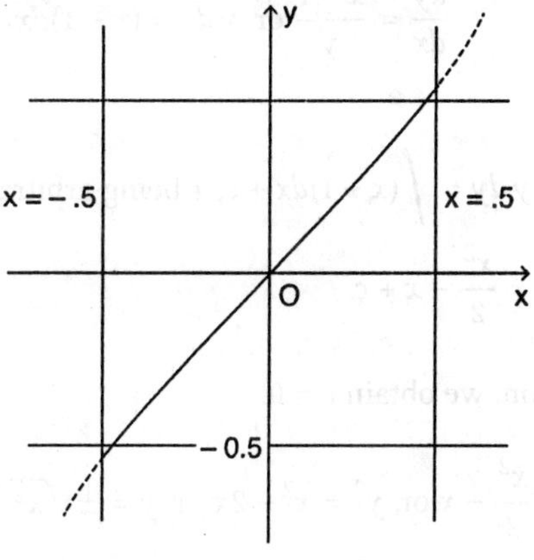

Fig. 2.3

Example 2.8.3. *Show that the initial-value problem, $\frac{dy}{dx} = -\frac{1}{y}$, $y(0) = 0$, has no solution in $(0, \infty)$.*

Solution: Let $f(x, y) = -\frac{1}{y}$. Observe that $f(x, y)$ is not continuous at the point $(0, 0)$. So Picard's theorem does not hold. Thus, there is no guarantee for the existence and uniqueness of the solution to the initial-value problem. The general solution of the differential equation $\frac{dy}{dx} = -\frac{1}{y}$ is given by $y^2 = -2x + c$, c being an arbitrary constant. Using the initial condition $y(0) = 0$, we obtain $c = 0$. Consequently, $y^2 = -2x$, which does not hold for $x \in (0, \infty)$. Again, $y = 0$ fails to satisfy the given equation. So there is no missing solution. Hence, the initial-value problem cannot have any solution.

Example 2.8.4. *Examine whether each of the following initial-value problems has unique solution or more than one solutions or, no solution. Also find the solution(s) if exists.*

(a) $\frac{dy}{dx} = \frac{x-1}{y}$, $y(0) = 1$, (b) $\frac{dy}{dx} = \frac{x-1}{y}$, $y(2) = 0$, (c) $\frac{dy}{dx} = \frac{y-1}{x}$, $y(0) = 1$.

Solution: (a) Let $f(x, y) = \frac{x-1}{y}$. Then $f_y(x, y) = -\frac{x-1}{y^2}$. Observe that $f(x, y)$ and $f_y(x, y)$ are both continuous around the point $(0, 1)$. Therefore, by the basic existence and uniqueness theorem, the initial-value problem has unique solution. To find the solution, we have

$$\frac{dy}{dx} = \frac{x-1}{y}, \text{ or, } y\,dy = (x-1)dx.$$

$$\therefore \int y\,dy = \int (x-1)dx + c, \; c \text{ being arbitrary constant}$$

$$\text{or, } \frac{y^2}{2} = \frac{x^2}{2} - x + c.$$

Using given initial condition, we obtain $c = \frac{1}{2}$.

$$\therefore \frac{y^2}{2} = \frac{x^2}{2} - x + \frac{1}{2} \text{ or, } y^2 = x^2 - 2x + 1 \text{ or, } y^2 = (x-1)^2 \text{ or, } y = \pm(x-1).$$

Observe that only $y = -(x-1)$ satisfies the initial condition $y(0) = 1$. Hence, required unique solution is given by $y = -(x-1)$.

(b) Let $f(x, y) = \frac{x-1}{y}$. Then $f_y(x, y) = -\frac{x-1}{y^2}$. Observe that $f(x, y)$ and $f_y(x, y)$ are not continuous at the point $(2, 0)$. In this stage we cannot conclude whether the solution(s) exists or not.

$$\frac{dy}{dx} = \frac{x-1}{y} \text{ or, } y\,dy = (x-1)dx.$$

$$\therefore \int y\,dy = \int (x-1)dx + c, \; c \text{ being arbitrary constant}$$

$$\text{or, } \frac{y^2}{2} = \frac{x^2}{2} - x + c.$$

Using given initial condition, we obtain $c = 0$.

$$\therefore \frac{y^2}{2} = \frac{x^2}{2} - x \text{ or, } y^2 = x^2 - 2x \text{ or, } y = \pm\sqrt{x^2 - 2x} \text{ (Fig. 2.4).}$$

Observe that $y = \pm\sqrt{x^2 - 2x}$ satisfy the initial condition $y(2) = 0$. The given initial-value problem has no missing solution, because $y(x) = 0$ fails to satisfy the given equation. Hence, the initial-value problem has more than one solutions; the solutions are given by $y = \pm\sqrt{x^2 - 2x}$.

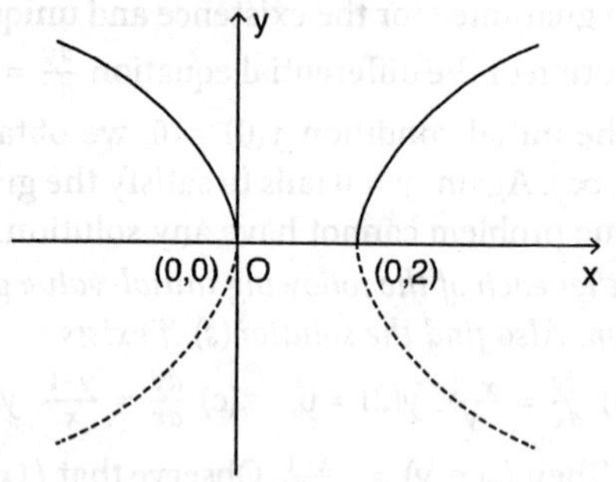

Fig. 2.4

(c) Let $f(x, y) = \frac{y-1}{x}$. Observe that $f(x, y)$ is not continuous at the point $(0, 1)$. In this stage we cannot conclude whether the solution(s) exists or not. Now

$$\frac{dy}{dx} = \frac{y-1}{x}.$$

$$\therefore \int \frac{1}{y-1}dy = \int \frac{1}{x}dx + \ln c, \; c \text{ being arbitrary constant}$$

$$\text{or, } \ln(y-1) = \ln x + \ln c \text{ or, } y - 1 = cx.$$

$\therefore \; y = cx + 1.$

We see that the initial condition $y(0) = 1$ satisfies the equation for all real values of c, that means for all real values of c, $y = cx + 1$ will be the solution of the initial-value problem. It confirms that the given problem has infinitely many solutions, which represents a family of straight lines (shown in the following Fig. 2.5) passing through the point $(0, 1)$ on the xy-plane. It is remarkable that Y-axis does not belong to this family (why?).

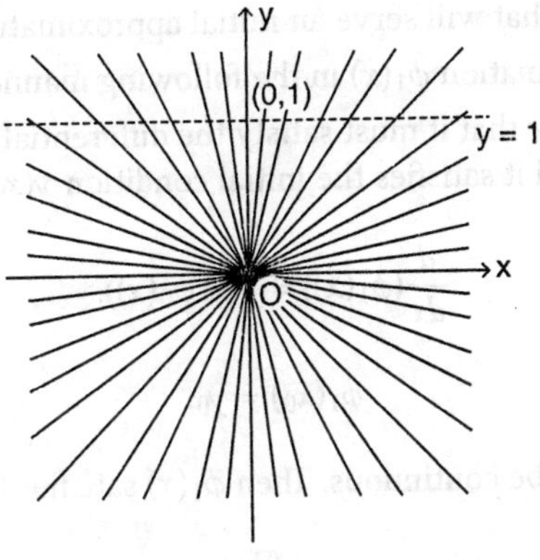

Fig. 2.5

Example 2.8.5. *Examine whether the following initial-value problems have unique solution or more than one solution or no solution. Also find the solution(s) if exists:*

$$\frac{dy}{dx} = e^x \sin y, \; y(0) = 0.$$

Solution: Let $f(x, y) = e^x \sin y$. Then $f_y(x, y) = e^x \cos y$. Observe that $f(x, y)$ and $f_y(x, y)$ are both continuous around the point $(0, 0)$. Thus, the initial-value problem has unique solution.

$$\frac{dy}{dx} = e^x \sin y \text{ or, } \operatorname{cosec}(y)dy = e^x\, dx.$$

$$\therefore \int \operatorname{cosec}(y)dy = \int e^x\, dx + k, \; k \text{ being arbitrary constant}$$

$$\text{or, } \ln(\operatorname{cosec}(y) - \cot(y)) = e^x + \ln c, \text{ where } k = \ln c$$

$$\text{or, } \operatorname{cosec}(y) - \cot(y) = ce^{e^x}.$$

Using initial condition $y(0) = 0$, we obtain $\operatorname{cosec}(0) - \cot(0) = c \cdot e$, which shows that c does not exist. We see that $y(x) = 0$ is the only solution of the given initial-value problem.

2.9 Picard's approximation

We have seen that an initial-value problem may or may not have a solution. If a solution exists, it may or may not be unique. Here we would like to introduce a method of finding the approximate solutions of initial-value problems. It is known as the method of successive approximation or Picard's method named after Charles Emile Picard (24 July 1856–11 December 1941), a French mathematician.

Let us consider the initial-value problem of the form

$$\frac{dy}{dx} = f(x, y) \tag{2.9.1}$$

with initial condition

$$y(x_0) = y_0 \tag{2.9.2}$$

Here, our main aim is to find a solution of this problem which is valid on the interval $(x_0 - \delta, x_0 + \delta)$ for some $\delta > 0$. First we choose a function $\phi_0(x)$ that is known as an initial approximation to the actual solution of the initial-value problem; but several different choices could be made to satisfy the initial condition $y(x_0) = y_0$, though it is not necessary to satisfy the condition $y(x_0) = y_0$ to guess the initial approximation. Generally we choose $\phi_0(x)$ as a constant function that gives the value y_0 for all x.

Step 1. Choose a function $\phi_0(x)$ that will serve an initial approximation.
Step 2. Determine a first approximation $\phi_1(x)$ in the following manner:

Determine $\phi_1(x)$ in such a way that it must satisfy the differential equation obtained from (2.9.2) by replacing y in $f(x, y)$ by $\phi_0(x)$ and it satisfies the initial condition $y(x_0) = y_0$, i.e.

$$\frac{d}{dx}(\phi_1(x)) = f(x, \phi_0(x)). \tag{2.9.3}$$

$$\phi_1(x_0) = y_0. \tag{2.9.4}$$

Let us assume that $f(x, \phi_0(x))$ be continuous. Then $\phi_1(x)$ satisfies (2.9.3) and (2.9.4), if and only if

$$\phi_1(x) = y_0 + \int_{x_0}^{x} f(t, \phi_0(t))dt. \tag{2.9.5}$$

Equation (2.9.5) is an integral equation that gives the 1st approximation $\phi_1(x)$.
Step 3. In a similar manner, determine 2nd approximation $\phi_2(x)$ such that

$$\frac{d}{dx}(\phi_2(x)) = f(x, \phi_1(x)). \tag{2.9.6}$$

$$\phi_2(x_0) = y_0. \tag{2.9.7}$$

Let us assume that $f(x, \phi_1(x))$ be continuous. Then $\phi_2(x)$ satisfies (2.9.6) and (2.9.7), if and only if

$$\phi_2(x) = y_0 + \int_{x_0}^{x} f(t, \phi_1(t))dt. \tag{2.9.8}$$

Equation (2.9.8) is another integral equation that serves the 2nd approximation $\phi_2(x)$. In order to determine a 3rd approximation $\phi_3(x)$, a 4th approximation $\phi_4(x)$ and so on, we have to proceed in like manner.
Step 4. The nth approximation, namely $\phi_n(x)$, is determined by

$$\phi_n(x) = y_0 + \int_{x_0}^{x} f(t, \phi_{n-1}(t))dt, \tag{2.9.9}$$

where $\phi_{n-1}(x)$ is the $(n-1)$th approximation.
Step 5. We obtain a sequence of functions$\{\phi_n(x)\}$. Under certain general conditions and for all $x \in (x_0 - \delta, x_0 + \delta)$ (for some positive δ that is sufficiently small), the sequence $\{\phi_n(x)\}$ of functions is convergent to a limit function $\phi(x)$, say, satisfying (2.9.1) and (2.9.2). Thus the function $\phi(x) = \lim_{n\to\infty} \phi_n(x)$ is the exact solution of the initial-value problem under suitable restriction.
Note 2.9.1 (Integral equation). *An equation containing an unknown function, which is to be determined and which appears under one or more integral signs, is called integral equation. Example of an integral equation:*

$$\phi(x) = \lambda \int_a^b K(x, t)\phi(t)dt,$$

ϕ being an unknown function.
Example 2.9.1. *Solve the initial-value problem $\frac{dy}{dx} = y$, $y(0) = 1$ using Picard's method.*

Solution: Here $f(x, y) = y$, $x_0 = 0$, $y_0 = 1$. Let us consider the initial approximation, $\phi_0(x) = 1$. Then the 1st approximation is given by

$$\phi_1(x) = y_0 + \int_{x_0}^{x} f(t, \phi_0(t))dt = 1 + \int_0^x 1 \cdot dt = 1 + x.$$

The 2nd approximation is given by

$$\phi_2(x) = y_0 + \int_{x_0}^{x} f(t, \phi_1(t))dt = 1 + \int_0^x (1 + t)dt = 1 + x + \frac{x^2}{2} = 1 + \frac{x}{1!} + \frac{x^2}{2!}.$$

The 3rd approximation is given by

$$\phi_3(x) = y_0 + \int_{x_0}^{x} f(t, \phi_2(t))dt = 1 + \int_0^x \left(1 + t + \frac{t^2}{2}\right) dt = 1 + x + \frac{x^2}{2} + \frac{x^3}{6} = 1 + \frac{x}{1!} + \frac{x^2}{2!} + \frac{x^3}{3!} \text{ and so on.}$$

Thus, we obtain a sequence $\{\phi_n(x)\}$ of functions $\phi_1(x), \phi_2(x), \phi_3(x), \phi_4(x), \ldots$, where

$$\phi_n(x) = 1 + \frac{x}{1!} + \frac{x^2}{2!} + \frac{x^3}{3!} + \cdots + \frac{x^n}{n!}.$$

The sequence $\{\phi_n(x)\}$ is convergent to the limit function

$$\phi(x) = \lim_{n\to\infty} \phi_n(x) = 1 + \frac{x}{1!} + \frac{x^2}{2!} + \frac{x^3}{3!} + \cdots + \frac{x^n}{n!} + \cdots = e^x,$$

which is the required solution of the given initial-value problem.

Example 2.9.2. *Using Picard's method solve the following initial-value problem up to 4th approximation:* $\frac{dy}{dx} = x + y$, $y(0) = 1$.

Solution: Here $f(x, y) = x + y$, $x_0 = 0$, $y_0 = 1$. Let us consider the initial approximation, $\phi_0(x) = 1$. Then the 1st approximation is given by

$$\phi_1(x) = y_0 + \int_{x_0}^{x} f(t, \phi_0(t))dt = 1 + \int_0^x f(t, 1)dt = 1 + \int_0^x (t + 1)dt = 1 + x + \frac{x^2}{2}.$$

The 2nd approximation is given by

$$\phi_2(x) = y_0 + \int_{x_0}^{x} f(t, \phi_1(t))dt = 1 + \int_0^x \left(t + 1 + t + \frac{t^2}{2}\right) dt = 1 + x + x^2 + \frac{x^3}{6}.$$

The 3rd approximation is given by

$$\phi_3(x) = y_0 + \int_{x_0}^{x} f(t, \phi_2(t))dt = 1 + \int_0^x \left(t + 1 + t + t^2 + \frac{t^3}{6}\right) dt = 1 + x + x^2 + \frac{x^3}{3} + \frac{x^4}{24}.$$

The 4th approximation is given by

$$\phi_4(x) = y_0 + \int_{x_0}^{x} f(t, \phi_3(t))dt = 1 + \int_0^x \left(t + 1 + t + t^2 + \frac{t^3}{3} + \frac{t^4}{24}\right) dt = 1 + x + x^2 + \frac{x^3}{3} + \frac{x^4}{12} + \frac{x^5}{120}.$$

Example 2.9.3. *Using Picard's method solve the following initial-value problem up to 3rd approximation:* $\frac{dy}{dx} = 1 + xy$, $y(0) = 1$.

Solution: Here $f(x, y) = 1 + xy$, $x_0 = 0$, $y_0 = 1$. Let us consider the initial approximation, $\phi_0(x) = 1$. Then the 1st approximation is given by

$$\phi_1(x) = y_0 + \int_{x_0}^{x} f(t, \phi_0(t))dt = 1 + \int_0^x f(t, 1)dt = 1 + \int_0^x (1 + t)dt = 1 + x + \frac{x^2}{2}.$$

The 2nd approximation is given by

$$\phi_2(x) = y_0 + \int_{x_0}^{x} f(t, \phi_1(t))dt = 1 + \int_0^x \left(1 + t\left(1 + t + \frac{t^2}{2}\right)\right) dt$$
$$= 1 + \int_0^x \left(1 + t + t^2 + \frac{t^3}{2}\right) dt = 1 + x + \frac{x^2}{2} + \frac{x^3}{3} + \frac{x^4}{8}.$$

The 3rd approximation is given by

$$\phi_3(x) = y_0 + \int_{x_0}^{x} f(t, \phi_2(t))dt = 1 + \int_0^x \left(1 + t\left(1 + t + \frac{t^2}{2} + \frac{t^3}{3} + \frac{t^4}{8}\right)\right) dt$$
$$= 1 + x + \frac{x^2}{2} + \frac{x^3}{3} + \frac{x^4}{8} + \frac{x^5}{15} + \frac{x^6}{48}.$$

Examples II(A)

Solve the following differential equations:

1. Show that the function $f(x, y)$ defined by $f(x, y) = x|y|$ on the convex set $D = \{(x, y) \in \mathbb{R}^2 : |x| \leq a, |y| \leq b\}$, satisfies a Lipschitz condition. Examine whether the partial derivative $f_y(x, y)$ exists or not in D.
2. Show that the function $f(x, y) = x^2 + y^2$ satisfies a Lipschitz condition in the rectangle $R : [-a, a] \times [-b, b]$.
3. Show that the function $f(x, y) = |y|$ satisfies a Lipschitz condition everywhere, but $f_y(x, 0)$ does not exist.
4. Examine whether the function $f(x, y) = x\sqrt[3]{y}$ satisfies a Lipschitz condition over the rectangle $R : [1, 2] \times [0, 1]$.
5. Show that the function $f(x, y) = x \sin y + y \cos x$ satisfies a Lipschitz condition on the rectangle $R : [-a, a] \times [-b, b]$.
6. Solve the initial-value problem $\frac{dy}{dx} = -\frac{x}{y}$, $y(3) = 4$.
7. Show that the initial-value problem $\frac{dy}{dx} = x \sin y$, $y(0) = 1$ has at most one solution.
8. Show that the initial-value problem $\frac{dy}{dx} = y^{\alpha}$, $y(0) = 0$, where $\alpha \in (0, 1)$ has at least two solutions.
9. Show that the initial-value problem $\frac{dy}{dx} = x^2 + y^2$, $y(1) = 3$ has unique solution.
10. Examine whether the following initial-value problem has unique solution or, more than one solution or no solution. Also find the solution(s) if exists $\frac{dy}{dx} = \frac{x-1}{y}$, $y(3) = -5$.
11. Examine whether each of the following initial-value problems has unique solution or more than one solution or no solution. Also find the solution(s) if exists.

(a) $\frac{dy}{dx} = e^x \cos y,\ y(0) = \frac{\pi}{2}$ (c) $\frac{dy}{dx} = -\frac{x}{y},\ y(0) = 1$ (e) $\frac{dy}{dx} = y^{1/3},\ y(0) = 1$

(b) $\frac{dy}{dx} = x\sqrt{y-3},\ y(4) = 3$ (d) $\frac{dy}{dx} = y^{1/3},\ y(0) = 0$ (f) $\frac{dy}{dx} = y^{2/3},\ y(0) = 0.$

12. Apply Picard's method to solve the following initial-value problems up to 3rd approximation:
 (a) $\frac{dy}{dx} = x + y^2,\ y(0) = 0$ (b) $\frac{dy}{dx} = 2xy,\ y(0) = 1$ (c) $\frac{dy}{dx} = 2 - \frac{y}{x},\ y(1) = 2.$

13. Using Picard's method solve the following initial-value problems up to 4th approximation:
 (a) $\frac{dy}{dx} = 1 + xy,\ y(0) = 2$ (c) $\frac{dy}{dx} = x - y,\ y(0) = 1$ (e) $\frac{dy}{dx} = \sin x + y^2,\ y(0) = 0$
 (b) $\frac{dy}{dx} = x^2 - y,\ y(0) = 0$ (d) $\frac{dy}{dx} = xy,\ y(0) = 1$ (f) $\frac{dy}{dx} = e^x + y^2,\ y(0) = 0.$

14. Using Picard's method solve the following initial-value problems:
 (a) $\frac{dy}{dx} = y - x,\ y(0) = 2$ (b) $\frac{dy}{dx} = x - y,\ y(0) = 1.$

Answers

4. Yes.
6. $y = \sqrt{25 - x^2},\ x \in (-5, 5).$
10. Unique solution exists: $y = -\sqrt{x^2 - 2x + 22}.$
11. (a) Unique solution; $y = \frac{\pi}{2}$
 (b) More than one solution; $y = 3,\ y = 3 + \left(\frac{x^2}{4} - 4\right)^2$
 (c) Unique solution: $x^2 + y^2 = 1.$
 (d) The solution is not unique; $y = 0,\ y = \left(\frac{2x}{3}\right)^{3/2}$
 (e) Unique solution; $y = \left(1 + \frac{2x}{3}\right)^{3/2}$
 (f) The solution is not unique; $y = 0,\ y = \left(\frac{x}{3}\right)^3.$
14. (a) $\phi(x) = (1 + x) + e^x$ (b) $\phi(x) = (x - 1) + 2e^{-x}.$

2.10 Exact equations

A differential eq. (2.1.1) of first order and of first degree can be expressed in the form: $M(x, y)\,dx + N(x, y)\,dy = 0$, or simply, $M\,dx + N\,dy = 0$, where M and N are functions of x and y.

If the differential $M\,dx + N\,dy$ without multiplication by any factor, is immediately expressible in the form du, where u is a function of x and y, it is said to be exact and the equation $M\,dx + N\,dy = 0$ is then called an exact differential equation.

If the equation $M\,dx + N\,dy = 0$ is exact, i.e. if a function $u(x, y)$ exists such that $du = M\,dx + N\,dy$, then the general solution of the equation is clearly $u = c$, where c is an arbitrary constant.

Example 2.10.1. *The equation $y\,dx + x\,dy = 0$ is an exact differential equation, since $y\,dx + x\,dy = d(xy)$. Its general solution is $xy = c$.*

Example 2.10.2. *$2x^2y\,dy + 2y^2x\,dx = 0$ is an exact differential equation, since $2x^2y\,dy + 2y^2x\,dx = d\left(x^2y^2\right)$. Its primitive is $x^2y^2 = c$.*

Theorem 2.10.1. *The necessary and sufficient condition that*

$$M\,dx + N\,dy = 0 \tag{2.10.1}$$

be exact is

$$\frac{\partial M}{\partial y} = \frac{\partial N}{\partial x}. \tag{2.10.2}$$

This theorem gives a test for exactness. Thus, if the condition (2.10.2) is satisfied, then the eq. (2.10.1) is exact. We often call this condition *the condition of integrability* of the eq. (2.10.1).

Proof. *To prove that the condition is necessary.*

If the eq. (2.10.1) is exact, then for some function $u(x, y)$ we have $M\,dx + N\,dy$ = a perfect differential of $u = du$.

The expression for du is given by $du = \frac{\partial u}{\partial x}dx + \frac{\partial u}{\partial y}dy$.

The two expressions must be identical and hence, dx, dy being independent increments $\frac{\partial u}{\partial x} = M$, $\frac{\partial u}{\partial y} = N$.

$$\therefore \frac{\partial M}{\partial y} = \frac{\partial}{\partial y}\left(\frac{\partial u}{\partial x}\right) = \frac{\partial^2 u}{\partial y \partial x} \text{ and } \frac{\partial N}{\partial x} = \frac{\partial}{\partial x}\left(\frac{\partial u}{\partial y}\right) = \frac{\partial^2 u}{\partial x \partial y}.$$

Then $\frac{\partial M}{\partial y} = \frac{\partial N}{\partial x}$, provided that the equivalent expressions $\frac{\partial^2 u}{\partial y \partial x}$ and $\frac{\partial^2 u}{\partial x \partial y}$ are continuous. The tacit assumption that M and N have continuous partial derivatives should not be overlooked.

The necessity of the condition is thus justified under these tacit assumptions.

To prove that the condition is sufficient.

Suppose that the condition (2.10.2), namely $\frac{\partial M}{\partial y} = \frac{\partial N}{\partial x}$ holds.

Put $F = \int M\,dx$, where the integration is performed on the supposition that y is constant.

Then $\frac{\partial F}{\partial x} = M$ and $\frac{\partial^2 F}{\partial x \partial y} = \frac{\partial^2 F}{\partial y \partial x} = \frac{\partial M}{\partial y} = \frac{\partial N}{\partial x}$.

So $\frac{\partial}{\partial x}\left(N - \frac{\partial F}{\partial y}\right) = 0$.

$\therefore$ $N - \frac{\partial F}{\partial y}$ is constant so far as x is concerned, that is, a function of $y = \phi(y)$, say.

Hence, $N = \phi(y) + \frac{\partial F}{\partial y}$.

We next take function $u(x, y)$ so that $u = F + \int \phi(y)\,dy$.

Then $\frac{\partial u}{\partial y} = N$.

Also

$$M = \frac{\partial F}{\partial x} \text{ (by our construction of } F)$$

$$= \frac{\partial u}{\partial x}\left(\because\ u = F + \int \phi(y)\,dy\right).$$

Thus, $M\,dx + N\,dy = \frac{\partial u}{\partial x}dx + \frac{\partial u}{\partial y}dy = du$, a perfect differential.

So the eq. (2.10.1) is exact and the condition (2.10.2) is, therefore, sufficient.

Example 2.10.3. *Is the equation* $(x^2 - y)\,dx + (y^2 - x)\,dy = 0$ *an exact equation?*

Solution: Here $M = x^2 - y, N = y^2 - x$. Then $\frac{\partial M}{\partial y} = -1 = \frac{\partial N}{\partial x}$, i.e. the condition of exactness is satisfied. Hence, the equation is exact.

When the condition of integrability is satisfied, the *general solution* (or the *primitive*) can be obtained by the following rules:

1. Integrate $M\,dx$ as if y is constant.
2. Integrate $N\,dy$ as if x is constant but, then *delete those terms already obtained in* 1.
3. The sum of these integrals will give a function $u(x, y)$ such that $du = M\,dx + N\,dy$.

[The Justification of these rules can be found in the proof of the sufficiency part of Theorem 2.10.1]

4. Once the function u is obtained, then $u = c$ is the primitive.

Example 2.10.4. *Solve:* $(x^2 - y)\ dx + (y^2 - x)\ dy = 0.$

Solution: The equation is exact (see Example 2.10.3). We now apply the rules to find u such that $du = (x^2 - y)\ dx + (y^2 - x)\ dy.$

1. $\int (x^2 - y)\ dx = \frac{1}{3}x^3 - xy$ (as if y is constant).

2. $\int (y^2 - x)\ dy = \frac{1}{3}y^3 - xy$ (as if x is constant).

3. $u = \left(\frac{1}{3}x^3 - xy\right) + \frac{1}{3}y^3$ [deleting the term $-xy$ from 2 which has already been obtained in 1].

4. The required primitive is $\frac{1}{3}\left(x^3 + y^3\right) - xy = c$ or, $x^3 + y^3 - 3xy = k$ (where $3c = k$), k being an arbitrary constant (Fig. 2.6).

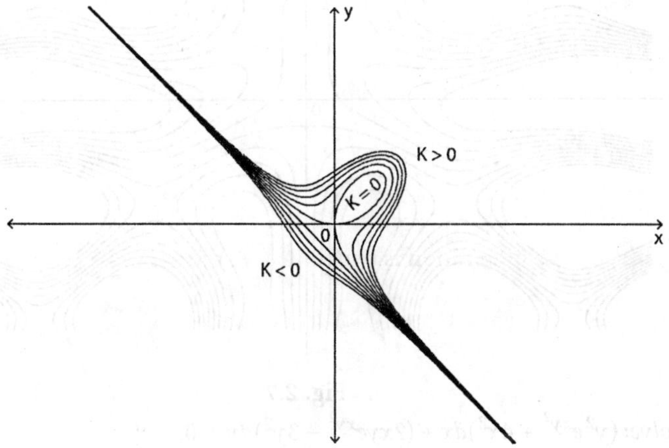

Fig. 2.6

Important differentials

The following differentials are easily seen to be exact:

(a) $dx \pm dy = d(x \pm y)$

(b) $x\,dx + y\,dy = d\left(\frac{1}{2}x^2 + \frac{1}{2}y^2\right)$

(c) $x^2\,dx + y^2\,dy = d\left(\frac{1}{3}x^3 + \frac{1}{3}y^3\right)$

(d) $X\,dx + Y\,dy = d\left\{\int X\,dx + \int Y\,dy\right\}$, where X is a function of x alone and Y is a function of y alone.

Sometimes instead of using the condition of integrability an equation may be seen to be exact after a regrouping of its terms.

Example 2.10.5. *Solve:* $(\cos y + y\cos x)\,dx + (\sin x - x\sin y)\,dy = 0.$

Solution: Method 1. Using the condition of integrability, we find $\frac{\partial M}{\partial y} = -\sin y + \cos x = \frac{\partial N}{\partial x}.$

Hence, the equation is exact.

$$\int M\,dx = \int (\cos y + y\cos x)\,dx = x\cos y + y\sin x \text{ (assuming } y \text{ to be constant)}$$

$$\int N\,dy = \int (\sin x - x\sin y)\,dy = y\sin x + x\cos y \text{ (assuming } x \text{ to be constant).}$$

Both the terms in $\int N\,dy$ are already obtained in $\int M\,dx.$

$\therefore$ $u = x\cos y + y\sin x$ and hence the primitive is $x\cos y + y\sin x = c$.

Method 2. $(\cos y + y\cos x)\,dx + (\sin x - x\sin y)\,dy = 0$.

Regrouping the terms we write $(\cos y\,dx - x\sin y\,dy) + (y\cos x\,dx + \sin x\,dy) = 0$ or, $d(x\cos y) + d(y\sin x) = 0$.

Integrating, $x\cos y + y\sin x = c$, is the required primitive (Fig. 2.7).

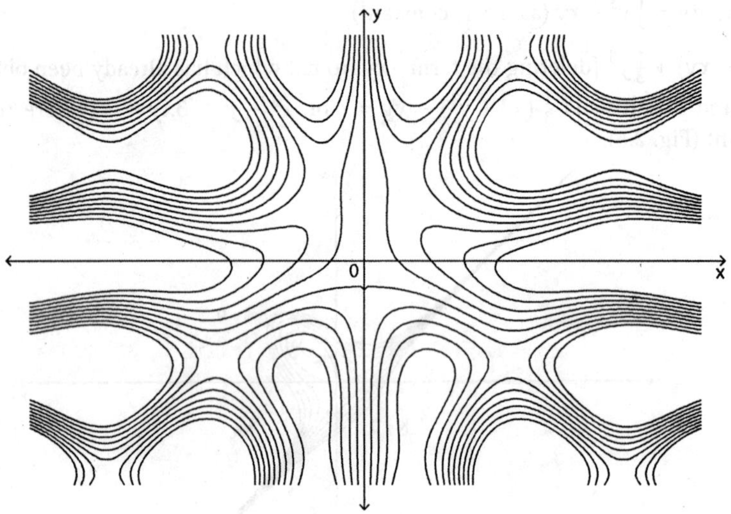

Fig. 2.7

Example 2.10.6. *Solve:* $(y^2 e^{xy^2} + 4x^3)dx + (2xye^{xy^2} - 3y^2)dy = 0$.

Solution: As in the previous example, the equation is exact, since

$$\frac{\partial M}{\partial y} = 2y\,e^{xy^2} + 2xy^3 e^{xy^2} = \frac{\partial N}{\partial x}.$$

$$\therefore \quad \int M dx = \int (y^2 e^{xy^2} + 4x^3)dx = e^{xy^2} + x^4, \quad y = \text{constant}.$$

$$\int N dy = \int (2xye^{xy^2} - 3y^2)dy = e^{xy^2} - y^3, \quad x = \text{constant}.$$

The required solution is $e^{xy^2} + x^4 - y^3 = c$, where c is an arbitrary constant.

Example 2.10.7. *Show that* $(4x + 3y + 1)dx + (3x + 2y + 1)dy = 0$ *is an exact equation. Hence, show that it represents a family of curves having* $x + y = 0$ *as one asymptote.*

Solution: Let $M = 4x + 3y + 1$ and $N = 3x + 2y + 1$.

Then $\frac{\partial M}{\partial y} = 3 = \frac{\partial N}{\partial x}$, which shows that the equation is exact.

Now $\int M\,dx = \int (4x + 3y + 1)dx = 2x^2 + 3xy + x$ (assuming y to be constant) and

$\int N\,dy = \int (3x + 2y + 1)dy = 3xy + y^2 + y$ (assuming x to be constant).

Therefore, the general solution of the given equation is $2x^2 + 3xy + y^2 + x + y + c = 0$, c being an arbitrary constant.

This solution can be expressed as $(x + y)(2x + y) + (x + y) + c = 0$ (Fig. 2.8).

Thus the asymptote parallel to the straight line $x + y = 0$ is given by

$$(x+y) + \lim_{\substack{|x|\to\infty \\ y/x\to -1}} \left(\frac{x+y}{2x+y}\right) + \lim_{\substack{|x|\to\infty \\ y/x\to -1}} \frac{c}{2x+y} = 0 \text{ or, } x+y=0.$$

Hence, given differential equation represents a family of curves having $x + y = 0$ as one asymptote.

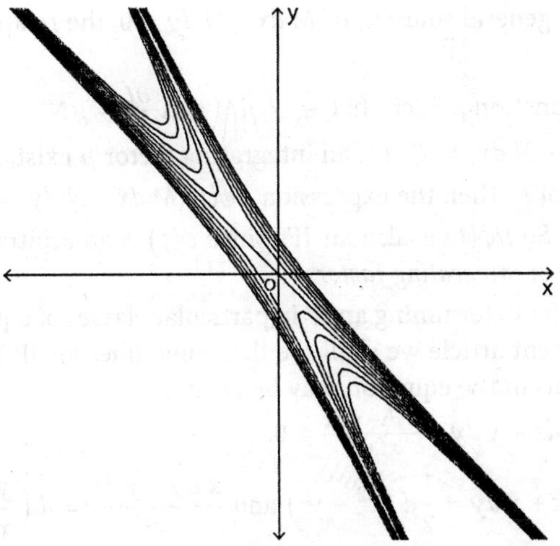

Fig. 2.8

Examples II(A)

Solve the following differential equations:

1. $(y + x)\,dx + x\,dy = 0$.
2. $(y + x^2)\,dx + x\,dy = 0$.
3. $y\,dx + (x + \cos y)\,dy = 0$.
4. $y\,dx + (x + e^y)\,dy = 0$.
5. $(2ax + by + g)\,dx + (2fy + bx + e)\,dy = 0$.
6. $(4x^3y^3 - 2xy)\,dx + (3x^4y^2 - x^2)\,dy = 0$.
7. $(a^2 - 2xy - y^2)\,dx - (x + y)^2\,dy = 0$.
8. $(x^2 - 4xy - 2y^2)\,dx + (y^2 - 4xy - 2x^2)\,dy = 0$.
9. $(1 + 6y^2 - 3x^2y)\frac{dy}{dx} = 3xy^2 - x^2$.
10. $y(x^2 + y^2 + a^2)\frac{dy}{dx} + x(x^2 + y^2 - a^2) = 0$.

Answers

1. $x^2 + 2xy = c$.
2. $x^3 + 3xy = c$.
3. $xy + \sin y = c$.
4. $xy + e^y = c$.
5. $ax^2 + bxy + fy^2 + gx + ey = k$ (k being an arbitrary constant).
6. $x^4y^3 - x^2y = c$.
7. $3a^2x - y^3 - 3xy^2 - 3x^2y = c$.
8. $x^3 - 6x^2y - 6xy^2 + y^3 = c$.
9. $y + 2y^3 + \frac{1}{3}x^3 - \frac{3}{2}x^2y^2 = c$.
10. $y^4 + 2a^2y^2 + x^4 - 2a^2x^2 + 2x^2y^2 = c$.

2.11 Integrating factors

The equation $\tan y\,dx + \tan x\,dy = 0$ is not exact as can be easily verified. But, if we multiply by $\cos x \cos y$, the equation becomes $\sin y \cos x\,dx + \sin x \cos y\,dy = 0$ or, $d(\sin x \sin y) = 0$.

Consequently the primitive is $\sin x \sin y = c$.

Definition 2.11.1. *A function $\mu(x, y)$ is said to be an Integrating Factor (IF) of the equation $M\,dx + N\,dy = 0$, if it is possible to obtain a function $u(x, y)$ such that $\mu(M\,dx + N\,dy) = du$. In other words, an IF is a multiplying factor by which the equation can be made exact.*

The question arises as to whether or not integrating factors exist. Again, if an IF does exist, is it unique? These questions are answered by the following theorem:

Theorem 2.11.1. *If the equation* $M\,dx + N\,dy = 0$ *has one and only one solution, then there exists an infinity of integrating factors.*

Proof. Let the general solution be $f(x, y) = c$, where c is the arbitrary constant. Taking differential, we obtain $df = \frac{\partial f}{\partial x}\,dx + \frac{\partial f}{\partial y}\,dy = 0$.

Since $f(x, y) = c$ is the general solution of $M\,dx + N\,dy = 0$, the relation $\frac{\partial f}{\partial x}/M = \frac{\partial f}{\partial y}/N$ must hold identically.

Hence, there exists a function μ such that $\frac{\partial f}{\partial x} = \mu M$ and $\frac{\partial f}{\partial y} = \mu N$.

Consequently, $\mu(M\,dx + N\,dy) = df$, i.e. an integrating factor μ exists.

Let $\phi(f)$ be a function of f. Then the expression $\mu\phi(f)(M\,dx + N\,dy) = \phi(f)df$.

The right side is exact. So $\mu\phi(f)$ is also an IF. Since $\phi(f)$ is an arbitrary function of f we conclude that *there exists an infinity of integrating factors.*

There are several rules for determining an IF in particular classes of equations. We shall discuss them in Art. 2.15. But in the present article we shall see that sometimes an IF can be found by inspection. A suitable regrouping of terms in the equation may be needed.

Example 2.11.1. *Solve:* $x\,dx + y\,dy + \frac{x\,dy - y\,dx}{x^2+y^2} = 0$.

Solution: We see that $x\,dx + y\,dy = \frac{1}{2}d\left(x^2 + y^2\right)$ and $\frac{x\,dy - y\,dx}{x^2} = d\left(\frac{y}{x}\right)$.

Therefore,

$$\frac{x\,dy - y\,dx}{x^2 + y^2} = \frac{\dfrac{x\,dy - y\,dx}{x^2}}{\dfrac{x^2 + y^2}{x^2}} = \frac{d(y/x)}{1 + (y/x)^2} = d\left\{\tan^{-1}(y/x)\right\}.$$

Hence, the given equation becomes $d\left\{\frac{1}{2}\left(x^2 + y^2\right)\right\} + d\left\{\tan^{-1}\frac{y}{x}\right\} = 0.$

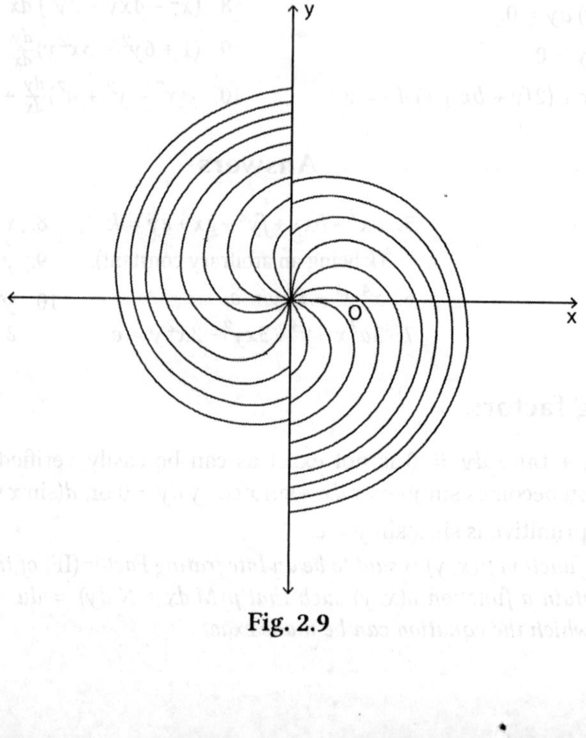

Fig. 2.9

∴ the primitive is

$$\tfrac{1}{2}\left(x^2 + y^2\right) + \tan^{-1}(y/x) = k \text{ (arbitrary constant)}$$

$$\text{or,}\quad x^2 + y^2 + 2\tan^{-1}(y/x) = c, \text{ where } c = 2k \text{ (Fig. 2.9).}$$

Example 2.11.2. *Solve:* $x\cos\left(\frac{y}{x}\right)(y\,dx + x\,dy) = y\sin\left(\frac{y}{x}\right)(x\,dy - y\,dx)$.

Solution: We note that $y\,dx + x\,dy = d(xy)$ and $\dfrac{x\,dy - y\,dx}{x^2} = d\left(\dfrac{y}{x}\right)$.

The equation can be written as

$$\frac{x\cos\left(\frac{y}{x}\right)d(xy)}{x^2} = y\sin\left(\frac{y}{x}\right)\frac{x\,dy - y\,dx}{x^2} \text{ or, } \frac{d(xy)}{xy} = \frac{\sin\left(\frac{y}{x}\right)}{\cos\left(\frac{y}{x}\right)}d\left(\frac{y}{x}\right).$$

Integrating, $\ln xy = -\ln\cos(y/x) + \ln c$.

$$\therefore\ \ln xy = \ln\frac{c}{\cos(y/x)},\ \text{or, } xy\cos(y/x) = c \text{ (Fig. 2.10).}$$

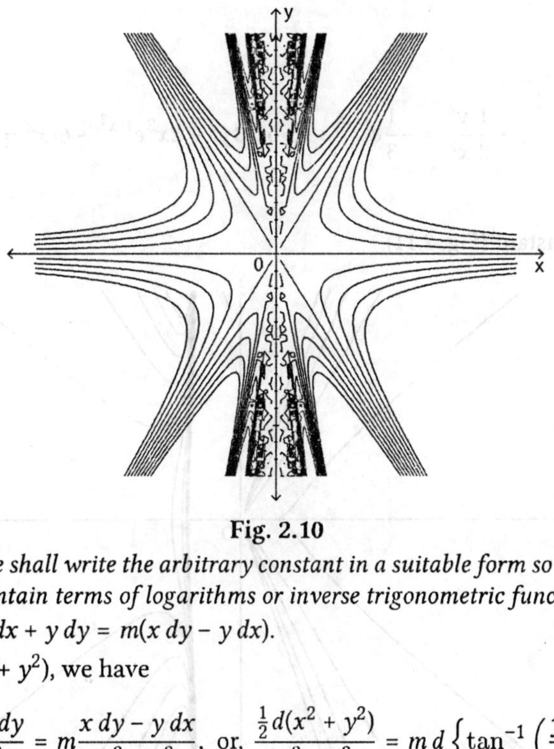

Fig. 2.10

Note 2.11.1. *Sometimes we shall write the arbitrary constant in a suitable form so that the final expression of the primitive may not contain terms of logarithms or inverse trigonometric functions.*

Example 2.11.3. *Solve:* $x\,dx + y\,dy = m(x\,dy - y\,dx)$.

Solution: Dividing by $(x^2 + y^2)$, we have

$$\frac{x\,dx + y\,dy}{x^2 + y^2} = m\frac{x\,dy - y\,dx}{x^2 + y^2},\ \text{or, } \frac{\frac{1}{2}d(x^2 + y^2)}{x^2 + y^2} = m\,d\left\{\tan^{-1}\left(\frac{y}{x}\right)\right\}.$$

Integrating, $\frac{1}{2}\ln(x^2 + y^2) = m\tan^{-1}\left(\frac{y}{x}\right) + c$.

Example 2.11.4. *Show that the equation* $(x^3 - 3x^2y + 2xy^2)dx - (x^3 - 2x^2y + y^3)dy = 0$ *is exact and find the solution, if* $y = 1$*, when* $x = 1$.

Solution: Here $M = x^3 - 3x^2y + 2xy^2$, $N = -x^3 + 2x^2y - y^3$.

$$\therefore\ \frac{\partial M}{\partial y} = -3x^2 + 4xy = \frac{\partial N}{\partial x}.$$

$$\int M\,dx = \int (x^3 - 3x^2y + 2xy^2)dx = \tfrac{1}{4}x^4 - x^3y + x^2y^2, \; y = \text{constant}.$$

$$\int N\,dy = \int (-x^3 + 2x^2y - y^3)dy = -x^3y + x^2y^2 - \tfrac{1}{4}y^4, \; x = \text{constant}.$$

$\therefore$ the general solution is $\frac{1}{4}x^4 - x^3y + x^2y^2 - \frac{1}{4}y^4 = c$, repeated terms rejected.

When $x = 1$, $y = 1$, $c = 0$, then required solution is $x^4 - 4x^3y + 4x^2y^2 - y^4 = 0$.

Example 2.11.5. *Solve:* $(xy^2 - e^{1/x^3})dx - x^2y\,dy = 0$.

Solution: $xy(y\,dx - x\,dy) = e^{1/x^3}\,dx$. Dividing by x^4,

$$\frac{xy}{x^2}\cdot\frac{y\,dx - x\,dy}{x^2} = \frac{e^{1/x^3}}{x^4}dx, \text{ or, } -\int \frac{y}{x}d\left(\frac{y}{x}\right) = \int \frac{e^{1/x^3}}{x^4}dx.$$

$$\therefore \; -\frac{1}{2}\frac{y^2}{x^2} = -\frac{1}{3}e^{1/x^3} + c, \text{ or, } 3y^2 - 2x^2e^{1/x^3} + 6cx^2 = 0,$$

c being an arbitrary constant (Fig. 2.11).

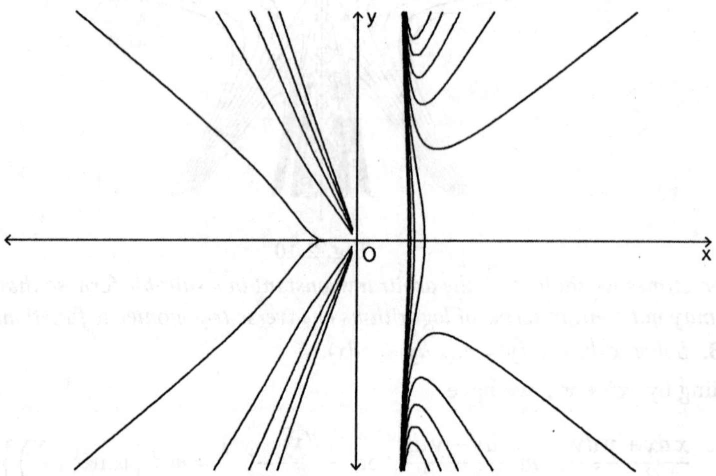

Fig. 2.11

Example 2.11.6. *Solve:* $x\frac{dy}{dx} - y = x\sqrt{x^2 + y^2}$.

Solution: $\dfrac{xdy - ydx}{x^2} = \sqrt{1 + \left(\dfrac{y}{x}\right)^2}\,dx$, or, $\dfrac{d\left(\frac{y}{x}\right)}{\sqrt{1 + \left(\frac{y}{x}\right)^2}} = dx$.

Integrating, $\sinh^{-1}\left(\frac{y}{x}\right) = x + c$, or, $y = x\sinh(x + c)$ (Fig. 2.12).

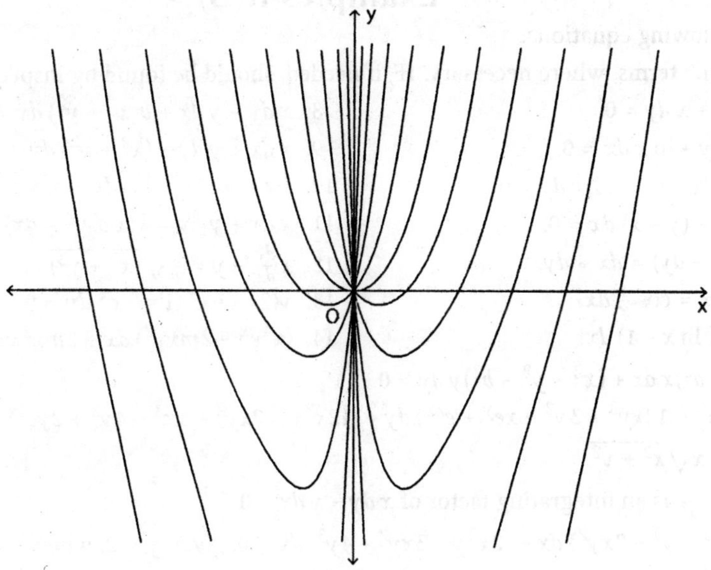

Fig. 2.12

Below is given a table which will be helpful to the students for determining an IF by inspection.

Table I

	Given Expression	Integrating Factor	Exact Differential
1.	$dx \pm dy$	—	$d(x \pm y)$
2.	$x\,dx \pm y\,dy$	—	$\frac{1}{2} d(x^2 \pm y^2)$
3.	$x\,dy + y\,dx$	—	$d(xy)$
4.	$x\,dy - y\,dx$	$\frac{1}{x^2}$	$\frac{x\,dy - y\,dx}{x^2} = d\left(\frac{y}{x}\right)$
5.	$x\,dy - y\,dx$	$\frac{1}{x^2 + y^2}$	$\frac{x\,dy - y\,dx}{x^2 + y^2} = d\left\{\tan^{-1}\frac{y}{x}\right\}$
6.	$x\,dy - y\,dx$	$\frac{1}{xy}$	$\frac{x\,dy - y\,dx}{xy} = \frac{dy}{y} - \frac{dx}{x} = d\left\{\ln\frac{y}{x}\right\}$
7.	$x\,dx + y\,dy$	$\frac{1}{x^2 + y^2}$	$\frac{x\,dx + y\,dy}{x^2 + y^2} = \frac{\frac{1}{2}d(x^2 + y^2)}{x^2 + y^2} = \frac{1}{2} d\{\ln(x^2 + y^2)\}$
8.	$y\,dx - x\,dy$	$\frac{1}{y^2}$	$\frac{y\,dx - x\,dy}{y^2} = d\left(\frac{x}{y}\right)$

Examples II(B)

Solve the following equations:

[Rearrange the terms, where necessary. IF, if needed, should be found by inspection]

1. $(y + x)\,dx + x\,dy = 0.$
2. $y\,dx - x\,dy + \ln x\,dx = 0.$
3. $(1 + xy)y\,dx + (1 - xy)x\,dy = 0.$
4. $(x + y)\,dy + (y - x)\,dx = 0.$
5. $(x + y)(dx - dy) = dx + dy.$
6. $x\,dy - y\,dx = \cos\frac{1}{x}\,dx.$
7. $x\,dy = y(y\ln x - 1)\,dx.$
8. $x\,dy - y\,dx + a(x^2 + y^2)\,dx = 0.$
9. $x\,dx + y\,dy + (x^2 + y^2)\,dy = 0.$
10. $\left\{x + y\cos\left[\frac{y}{x}\right]\right\}\,dx = x\cos\left[\frac{y}{x}\right]\,dy.$
11. $x\,dx + y\,dy = k(x\,dy - y\,dx).$
12. $x\frac{dy}{dx} - y = \frac{x}{a}\sqrt{(x^2 + y^2)}.$
13. $y(2xy + e^x)\,dx - e^x\,dy = 0.$
14. $(x^4e^x - 2mxy^2)\,dx + 2mx^2y\,dy = 0.$
15. $(x^2 + y^2 - a^2)x\,dx + (x^2 - y^2 - b^2)y\,dy = 0.$
16. $(2x^2y + 4x^3 - 12xy^2 + 3y^2 - xe^y + e^{2x})\,dy + (12x^2y + 2xy^2 + 4x^3 - 4y^3 + 2ye^{2x} - e^y)\,dx = 0.$
17. $x\frac{dy}{dx} - y = x\sqrt{x^2 + y^2}.$
18. Show that $\frac{1}{x^2}$ is an integrating factor of $x\,dy - y\,dx = 0.$
19. Solve: $(4x^3 - y^3 + 2xy^2)\,dx + (2x^2y - 3xy^2 + 4y^3)\,dy = 0$; given $y = 2$, when $x = 1$.

Answers

1. $x^2 + 2xy = c.$
2. $cx + y + \ln x + 1 = 0.$
3. $x = cye^{1/xy}.$
4. $x^2 - 2xy - y^2 = c.$
5. $x - y = \ln(x + y) + c.$
6. $\frac{y}{x} + \sin\frac{1}{x} = c.$
7. $1 + cxy = y(1 + \ln x).$
8. $ax + \tan^{-1}\frac{y}{x} = c.$
9. $\frac{1}{2}\ln\left(x^2 + y^2\right) + y = c.$
10. $\ln x + c = \sin(y/x).$
11. $\ln\left(x^2 + y^2\right) = 2k\tan^{-1}(y/x) + c.$
12. $y = x\sinh\frac{x+c}{a}.$
13. $x^2 + e^x/y = c.$
14. $x^2e^x + my^2 = cx^2.$
15. $x^4 - y^4 + 2x^2y^2 - 2a^2x^2 - 2b^2y^2 = c.$
16. $x^2y^2 + 4x^3y - 4xy^3 + y^3 - xe^y + e^{2x}y + x^4 = c.$
17. $y = x\sinh(x + c).$

19. $x^4 - xy^3 + x^2y^2 + y^4 - 13 = 0.$

2.12 Equations solvable by separation of variables

1. A special case of an exact equation occurs when, in the equation $M\,dx + N\,dy = 0$, M is a function of x alone [$= f_1(x)$, say] and N is a function of y alone [$= f_2(y)$, say]. The equation $f_1(x)\,dx + f_2(y)\,dy = 0$ is then said to have *separated variables.* Its general solution is

$$\int f_1(x)\,dx + \int f_2(y)\,dy = c,$$

c being an arbitrary constant.

Example 2.12.1. *Solve:* $\dfrac{dy}{\sqrt{1 - y^2}} + \dfrac{dx}{\sqrt{1 - x^2}} = 0,\ (|x| < 1, |y| < 1).$

Solution: The general solution is clearly

$$\sin^{-1} y + \sin^{-1} x = c \text{ or, } \sin^{-1}\left\{x\sqrt{1 - y^2} + y\sqrt{1 - x^2}\right\} = c \text{ or, } x\sqrt{1 - y^2} + y\sqrt{1 - x^2} = k,$$

where $k = \sin c$ is an arbitrary constant.

2. If it is possible to factorise M and N such that $M = f_1(x) \times \phi_1(y)$ and $N = f_2(y) \times \phi_2(x)$, then also the variables are *separable*. For, the equation $M\,dx + N\,dy = 0$ becomes

$$f_1(x)\phi_1(y)\,dx + f_2(y)\phi_2(x)\,dy = 0. \tag{2.12.1}$$

This can be written in the separated form:

$$\frac{f_1(x)}{\phi_2(x)}\,dx + \frac{f_2(y)}{\phi_1(y)}\,dy = 0, \text{ [Dividing by } \phi_1(y) \times \phi_2(x)], \tag{2.12.2}$$

which now can be easily integrated.

Example 2.12.2. *Solve:* $x\sqrt{1-y^2}\,dx + y\sqrt{1-x^2}\,dy = 0$, $(|x| < 1, |y| < 1)$.

Solution: Divide both sides by $\sqrt{1-x^2} \times \sqrt{1-y^2}$ ($\neq 0$, suppose).

Then we get $\frac{x\,dx}{\sqrt{1-x^2}} + \frac{y\,dy}{\sqrt{1-y^2}} = 0$, which on integration, gives $\sqrt{1-x^2} + \sqrt{1-y^2} = c$, c being an arbitrary constant.

Important observations

It must be observed that a number of solutions may be lost in the division of the eq. (2.12.1) by $\phi_1(y) \times \phi_2(x)$. If, for example, $x = \alpha$ is root of $\phi_2(x) = 0$ it would give a solution of eq. (2.12.1) but not *necessarily* of the eq. (2.12.2). We make the point clear by means of an example.

Example 2.12.3. $(x^2+1)(y^2-1)\,dx + xy\,dy = 0$.

Solution: Dividing by $x(y^2-1)$ the equation will have separated variables; thus

$$\frac{x^2+1}{x}\,dx + \frac{y}{y^2-1}\,dy = 0.$$

Integrating:

$$\tfrac{1}{2}x^2 + \ln x + \tfrac{1}{2}\ln(y^2-1) = k, \; k \text{ being an arbitrary constant}$$

$$\text{or,} \quad x^2 + \ln x^2 + \ln(y^2-1) = \ln c, \text{ where } k = \ln c.$$

$$\therefore \quad x^2 = \ln\frac{c}{x^2(y^2-1)} \quad \text{or, } y^2 = 1 + \left(ce^{-x^2}\right)/x^2,$$

which is the general solution (Fig. 2.13).

Now it can be easily verified that $x = 0, y = 1, y = -1$ [obtained by equating $x(y^2-1) = 0$] are real solutions of the given equation. Putting $c = 0$ in the general solution, we see that $y = 1, y = -1$ are particular solutions. But $x = 0$ is not included in the general solution. This is a *singular solution*. So, division by $x(y^2-1)$ deprived us from the solution $x = 0$.

At times a convenient transformation of the dependent variable will facilitate separation of variables. The following example will be instructive:

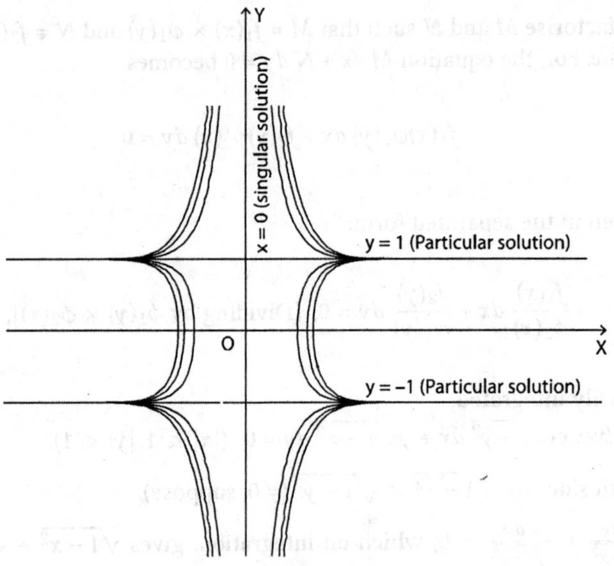

Fig. 2.13

Example 2.12.4. *Solve:* $\frac{dy}{dx} = \sin(x + y)$.

Solution: It is not easy to separate the variables, as the equation now stands. However, if we make the substitution $x + y = V$ so that $dy/dx = dV/dx - 1$, then the equation becomes separable. Thus,

$$\frac{dV}{dx} - 1 = \sin V, \text{ or, } \frac{dV}{dx} = 1 + \sin V, \text{ or, } \frac{dV}{1 + \sin V} = dx.$$

∴ integrating we get

$$x + c = \int \frac{dV}{1 + \sin V} = \int \frac{1 - \sin V}{\cos^2 V} dV = \tan V - \sec V.$$

∴ the general solution is $\tan(x + y) - \sec(x + y) = x + c$, c being an arbitrary constant (Fig. 2.14).

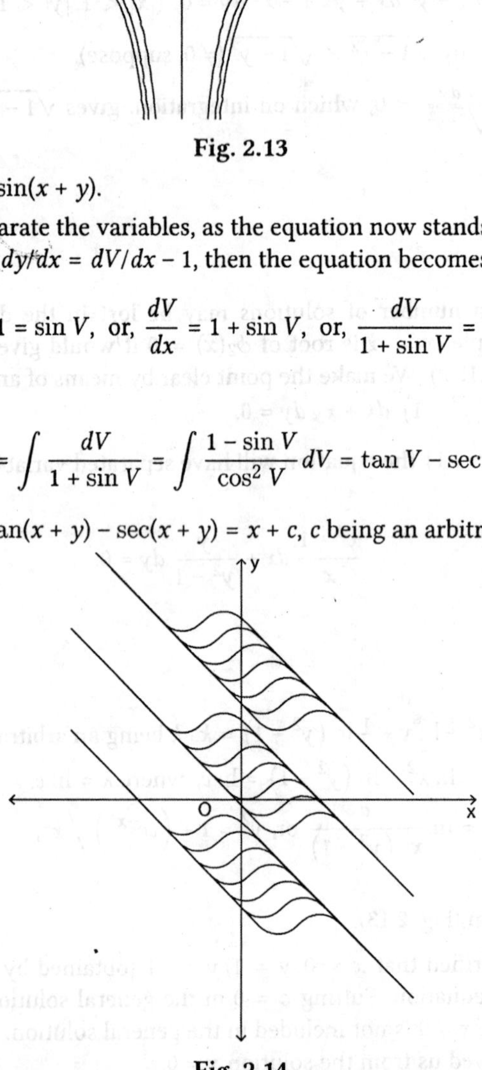

Fig. 2.14

Example 2.12.5. *Solve the equation* $(1 - x^2)\, dy = 2y\, dx$ *and show that if* $y = 1$, *when* $x = 2$, *then* $3xy = x + 3y + 1$.

Solution: Here,

$$\int \frac{dy}{y} = \int \frac{2}{1-x^2}\,dx, \text{ or, } \ln cy = 2\cdot\frac{1}{2}\cdot\ln\left|\frac{1+x}{1-x}\right|.$$

Thus,

$$cy = \frac{1+x}{1-x}, \text{ or, } c(1-x)y = 1+x \text{ (Fig. 2.15).}$$

But $y = 1$, when $x = 2$. Hence, $c(-1) = 3$, or, $c = -3$. Thus, the solution becomes $-3(1-x)y = 1+x$, or, $3xy = x + 3y + 1$.

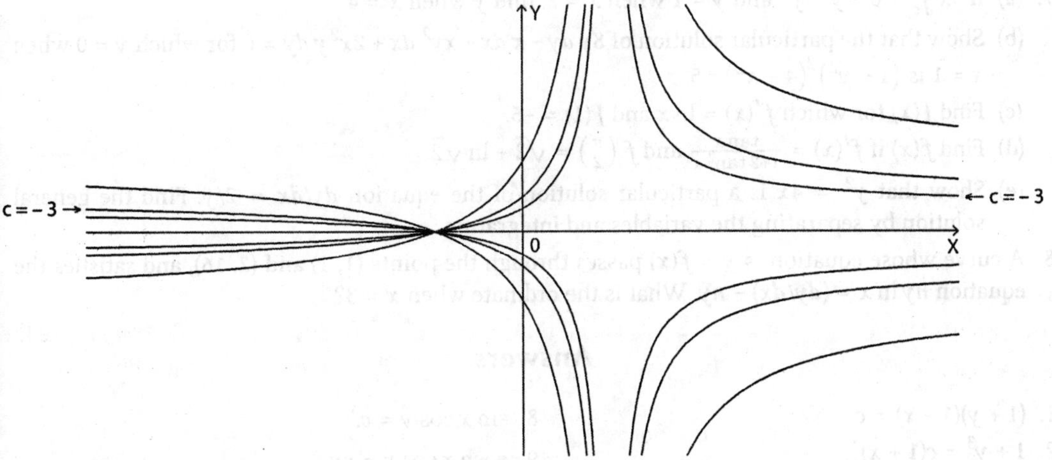

Fig. 2.15

Examples II(C)

Solve (Nos. 1–12):

1. $(1-x)\,dy - (1+y)\,dx = 0$.
2. $xy^2\,dy - y^3\,dx + y^2\,dy = dx$.
3. $\frac{dy}{dx} = \frac{xy^2-x}{yx^2+y}$.
4. $\left(y - x\frac{dy}{dx}\right) = a\left(y^2 + \frac{dy}{dx}\right)$.
5. $x\left(1+y^2\right)\,dx + y\left(1+x^2\right)\,dy = 0$.
6. $2y\,dx - x\,dy = xy^3\,dy$.
7. (a) $dr + r\tan\theta\,d\theta = 0$; (b) $y\,dy = e^{x+2y}\sin x\,dx$.
8. $\frac{dy}{dx} = \cot x\cot y$.
9. $\frac{1}{y}\frac{dy}{dx} = \tanh x - \tan x$.
10. $\sec^2 x\tan y\,dx + \sec^2 y\tan x\,dy = 0$.
11. $3e^x\tan y\,dx + (1-e^x)\sec^2 y\,dy = 0$.
12. (a) $\frac{dy}{dx} + \frac{y^2+y+1}{x^2+x+1} = 0$; (b) $(x+1)\frac{dy}{dx} + 1 = 2e^{-y}$.
13. (a) If $\frac{dy}{dx} = f(ax+by+c)$, show that the substitution $ax+by+c = v$ will change it to a separable equation.
    (b) Using (a) solve the following equations:
    (i) $(x-y)^2\frac{dy}{dx} = a^2$.
    (ii) $\frac{dy}{dx} = \sqrt{y-x}$.
    (iii) $\frac{dy}{dx} = (x+y)^2$.
    (iv) $\left(x-y^2\right)\,dx + 2xy\,dy = 0$ (Put $y^2 = v$).
    (v) $(x+y)^2\frac{dy}{dx} = a^2$.
    (vi) $(x-y)^2\frac{dy}{dx} = a^2$.
    (c) Solve: $\frac{dy}{dx} + 2xy = x^2 + y^2$.
14. Solve: (a) $\left(1+y^2\right)\,dx - \left\{y + \sqrt{(1+y)}\right\}(1+x)^{3/2}\,dy = 0$; (b) $(y-x)\left(1+x^2\right)^{1/2}\frac{dy}{dx} = n\left(1+y^2\right)^{3/2}$.
15. Solve: (a) $y - x\frac{dy}{dx} = b\left(1 + x^2\frac{dy}{dx}\right)$; (b) $x^3\frac{dy}{dx} + 3y^2 = xy^2$.

16. Solve with the conditions given:
   (a) $\tan x\,dy - \tan y\,dx = 0$; $y = \pi/2$ when $x = \pi/4$.
   (b) $\tan x\frac{dy}{dx} = 1 + y^2$; $y = 1$ when $x = \frac{\pi}{2}$.
   (c) $2y + \frac{1}{x}\frac{dy}{dx} = 3y^2$; $y = 1$ when $x = 0$.
   (d) $\frac{dy}{dx} = \frac{\sqrt{1+y^2}}{x}$; $y = 0$ when $x = 2$.
   (e) $\frac{ds}{dt} = \frac{1+s^2}{\sqrt{t}}$; $s = 0$ when $t = 0$.

17. (a) If $5x\frac{dy}{dx} = 6 - y - y^2$ and $y = 1$ when $x = 2$, find y when $x = 4$.
   (b) Show that the particular solution of $8y\,dy - x\,dx + xy^2\,dx + 2x^2y\,dy = 0$ for which $y = 0$ when $x = 1$ is $(1 - y^2)^2(4 + x^2) = 5$.
   (c) Find $f(x)$ for which $f'(x) = \ln x$ and $f(1) = -5$.
   (d) Find $f(x)$ if $f'(x) = \frac{\tan x}{1+2\tan^2 x}$ and $f\left(\frac{\pi}{2}\right) = \sqrt{2} + \ln\sqrt{2}$.
   (e) Show that $y^2 = 4x$ is a particular solution of the equation $dy/dx = 2/y$. Find the general solution by separating the variables and integrating.

18. A curve whose equation is $y = f(x)$ passes through the points (1, 1) and (2, 16), and satisfies the equation $ny\ln x = (dy/dx) - ny$. What is the ordinate when $x = 3$?

Answers

1. $(1 + y)(1 - x) = c$.
2. $1 + y^3 = c(1 + x)^3$.
3. $y^2 - 1 = c\left(x^2 + 1\right)$.
4. $y = c(a + x)(1 - ay)$.
5. $\left(1 + x^2\right)\left(1 + y^2\right) = c$.
6. $\ln\frac{x^2}{y} = \frac{y^3}{3} + c$.
7. (a) $r = c\cos\theta$;
   (b) $e^x(\sin x - \cos x) + e^{-2y}\left(y + \frac{1}{2}\right) = c$.
8. $\sin x\cos y = c$.
9. $\cosh x\cos x = cy$.
10. $\tan x\tan y = c$.
11. $\tan y = c(1 - e^x)^3$.
12. (a) $2xy + x + y + c(x + y + 1) = 1$;
   (b) $(x + 1)e^y = 2x + c$.
13. (b) (i) $\frac{\alpha^2}{2}\ln\frac{x-y+\alpha}{x-y-\alpha} + y = c$.
   (ii) $\sqrt{y-x} + \ln\left(\sqrt{y-x} - 1\right) = \frac{1}{2}x + c$.
   (iii) $\tan^{-1}(x + y) = x + c$.
   (iv) $xe^{y^2/x} = c$.
   (v) $(x + y)/a = \tan(y + c)/a$.
   (vi) $\frac{a}{2}\ln\frac{x-y+a}{x-y-a} + y = c$.
   (c) $y - x - 1 = Ae^{2x}(y - x + 1)$.
14. (a) $\int\frac{(1+y)^{\frac{1}{2}}}{1+y^2}dy + \frac{1}{2}\ln\left(1 + y^2\right) + 2\sqrt{(1 + x)} = c$. [To evaluate this integral put $1 + y = z^2$]
   (b) Substitute $y = \tan\phi$; $x = \tan\theta$, $\phi - \theta = \lambda$; the integral is $\phi - n\int\frac{d\lambda}{n - \sin\lambda} = c$.
15. (a) $c(y - b) = x/(1 + bx)$;
   (b) $2x^3 - 2xy + 3y + 2cx^2y = 0$.
16. (a) $\sin y = \sqrt{2}\sin x$.
   (b) $\tan^{-1} y = \ln\sin x + \frac{1}{4}\pi$.
   (c) $y\left(3 - e^{x^2}\right) = 2$.
   (d) $x^2 = 4xy + 4$.
   (e) $\tan^{-1} s = 2\sqrt{t}$.
17. (a) $\frac{13}{9}$.
   (c) $f(x) = x(\ln x - 1) - 4$.
   (d) $f(x) = \frac{1}{2}\ln\left(1 + \sin^2 x\right) + \sqrt{2}$.
   (e) $y^2 = 4x + c$.
18. 729.

2.13 Homogeneous equations

We say that a function $f(x, y)$ is *homogeneous* of *degree n* if it can be expressed in the form $x^n\phi(y/x)$. If M and N are homogeneous functions of x and y of the *same degree n*, then the equation $M\,dx + N\,dy = 0$ is said to be a *homogeneous equation of degree n*. In this case, we may write

$$M = x^n\phi(y/x),\ N = x^n\psi(y/x).$$

On the substitution of $y = vx$ so that v may be considered a new dependent variable, the equation $M\,dx + N\,dy = 0$ becomes

$$x^n\phi(v)\,dx + x^n\psi(v)\{x\,dv + v\,dx\} = 0$$

$$\text{or,}\quad x^n\{[\phi(v) + v\psi(v)]\,dx + x\psi(v)dv\} = 0 \text{ or, } \frac{dv}{F(v)} + \frac{dx}{x} = 0\ (\because\ x^n \neq 0),$$

where $F(v) = \frac{v\psi(v)+\phi(v)}{\psi(v)} = v + \frac{\phi(v)}{\psi(v)}$.

Thus, the variables are separated and the solution is

$$\int \frac{dv}{F(v)} = \ln\frac{c}{x}.$$

The primitive will be given by the substitution of y/x for v after the integration has been performed.

Otherwise

We see that the equation $M\,dx + N\,dy = 0$, when homogeneous, can be written as

$$\frac{dy}{dx} = \frac{-M}{N} = -\frac{\phi(y/x)}{\psi(y/x)} = f(y/x), \text{ say.}$$

The substitution $y = vx$ gives

$$\frac{dy}{dx} = x\frac{dv}{dx} + v, \text{ or, } x\frac{dv}{dx} + v = f(v).$$

Separating the variables we then obtain

$$\frac{dv}{f(v) - v} = \frac{dx}{x}, \text{ or, } \int \frac{dv}{f(v) - v} = \ln cx.$$

After performing the integration we replace v by y/x.

Example 2.13.1. *Solve:* $2xy\frac{dy}{dx} = y^2 - x^2$.

Solution: Here the equation is homogeneous, for

$$\frac{dy}{dx} = \frac{y^2 - x^2}{2xy} = \frac{(y/x)^2 - 1}{2(y/x)}.$$

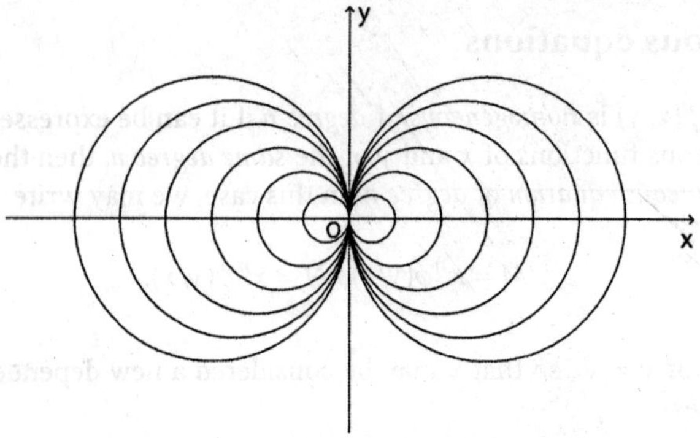

Fig. 2.16

The substitution $y = vx$ gives

$$v + x\frac{dv}{dx} = \frac{v^2 - 1}{2v} \text{ or, } x\frac{dv}{dx} = \frac{-1 - v^2}{2v}, \text{ or, } \frac{2v\,dv}{1 + v^2} + \frac{dx}{x} = 0.$$

The variables are thus separated. Integration gives $\ln\left(v^2 + 1\right) + \ln x = \ln c$ (say) or, $x\left(v^2 + 1\right) = c$.

The primitive is $x\left(\frac{y^2}{x^2} + 1\right) = c$, or, $cx = y^2 + x^2$ (Fig. 2.16).

Example 2.13.2. *Solve:* $x^2dy + y(x + y)dx = 0$.

Solution: $\frac{dy}{dx} = -\frac{y(x+y)}{x^2}$ a homogeneous equation.

Put $y = vx$, where v is a variable. Then,

$$v + x\frac{dv}{dx} = -\frac{vx(x + vx)}{x^2} = -v(1 + v)$$

$$\text{or, } \quad x\frac{dv}{dx} = -v(v + 2), \text{ or, } \frac{dx}{x} + \frac{dv}{v(v + 2)} = 0.$$

Integrating,

$$\ln x + \frac{1}{2}\ln\frac{v}{v + 2} = \ln c, \text{ or, } \ln x^2 + \ln\frac{y}{y + 2x} = \ln c, \text{ or, } \frac{x^2y}{2x + y} = c.$$

Therefore, solution is: $x^2y = c(2x + y)$, c being an arbitrary constant (Fig. 2.17).

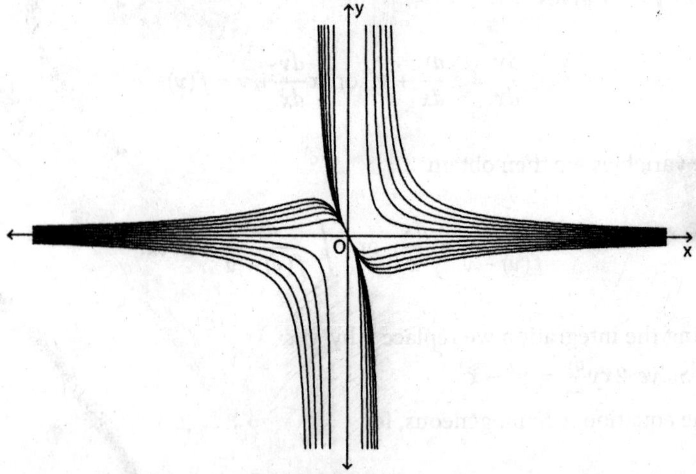

Fig. 2.17

Example 2.13.3. *Solve:* $(y^4 - 2x^3y)\ dx + (x^4 - 2xy^3)\ dy = 0.$

Solution: Here $M = y^4 - 2x^3y = x^4\{(y/x)^4 - 2(y/x)\}$

$$N = x^4 - 2xy^3 = x^4\{1 - 2(y/x)^3\}.$$

$\therefore\ M\,dx + N\,dy = 0$ becomes $\{(y/x)^4 - 2(y/x)\}\ dx + \{1 - 2(y/x)^3\}\ dy = 0\ (\because\ x^4 \neq 0).$

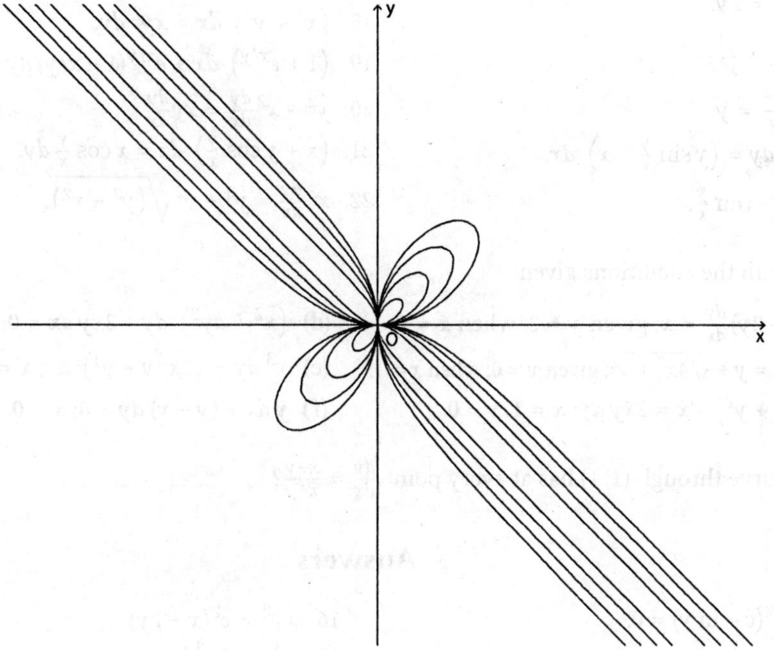

Fig. 2.18

The substitution $y = vx$ gives

$$(v^4 - 2v)\ dx + (1 - 2v^3)\ (x\,dv + v\,dx) = 0$$

$$\text{or,}\quad (v^4 - 2v + v - 2v^4)\ dx + x\,(1 - 2v^3)\ dv = 0$$

$$\text{or,}\quad \frac{dx}{x} = \frac{1 - 2v^3}{v + v^4}\,dv = \left(\frac{1}{v} - \frac{3v^2}{1 + v^3}\right) dv.$$

The variables are thus separated. Integration gives

$$\ln x = \ln v - \ln\,(1 + v^3) + \ln c$$

$$\text{or,}\quad x\,(1 + v^3) = cv$$

i.e. $x^3 + y^3 = cxy$ is the required solution (Fig. 2.18),

c being an arbitrary constant.

Examples II(D)

Solve:

1. $(x^2 - y^2)\ dy = xy\,dx.$
2. $2xy\,dx + (y^2 - x^2)\ dy = 0.$
3. $x\frac{dy}{dx} = y + \sqrt{(x^2 + y^2)}.$
4. $x\,dy - y\,dx = \sqrt{(x^2 + y^2)}\ dx.$
5. $(x + y)\frac{dy}{dx} = y - x.$

6. $(4y+3x)\frac{dy}{dx}+y-2x=0.$
7. $(x+y)^2=xy\frac{dy}{dx}.$
8. $\left(x^2+y^2\right)dx-2xy\,dy=0.$
9. $x+y\frac{dy}{dx}=2y.$
10. $2\frac{dy}{dx}=\frac{y}{x}+\frac{y^2}{x^2}.$
11. $x\frac{dy}{dx}+\frac{y^2}{x}=y.$
12. $x\sin\frac{y}{x}\,dy=\left(y\sin\frac{y}{x}-x\right)dx.$
13. $\frac{dy}{dx}=\frac{y}{x}+\tan\frac{y}{x}.$
14. $2x^2\frac{dy}{dx}=x^2+y^2.$
15. $x^2\,dy+\left(xy+y^2\right)dx=0.$
16. $y^2\,dx+\left(xy+x^2\right)dy=0.$
17. $\left(x^3-y^3\right)dx+xy^2\,dy=0.$
18. $\left(x^3+y^3\right)dx=xy^2\,dy.$
19. $\left(1+e^{x/y}\right)dx+e^{x/y}(1-x/y)\,dy=0.$
20. $y^2+x^2\frac{dy}{dx}=xy\frac{dy}{dx}.$
21. $\left(x+y\cos\frac{y}{x}\right)dx=x\cos\frac{y}{x}\,dy.$
22. $x^3\frac{dy}{dx}=y^3+y^2\sqrt{\left(y^2-x^2\right)}.$

23. Solve with the conditions given:

(a) $(x+2y)\frac{dy}{dx}=x$; given $y=2$, when $x=1$.
(b) $x\frac{dy}{dx}=y+\sqrt{4x^2+y^2}$; given $y=0$, when $x=1$.
(c) $\left(x^2+y^2\right)dx=2xy\,dy$; $x=1, y=0$.
(d) $\left(x^2+ay^2\right)dy+2xy\,dx=0$; $x=0, y=a$.
(e) $x^3\,dy=\left(2x^2y+y^3\right)dx$; $x=1, y=2$.
(f) $y\,dx+(y-x)\,dy=0$; $x=0, y=1$.

24. What curve through (1, 1) has at every point $\frac{dy}{dx}=\frac{x-y}{x+y}$?

Answers

1. $x^2+2y^2(c+\ln y)=0.$
2. $x^2+y^2=cy.$
3. & 4. $2cy=c^2x^2-1$ or $y+\sqrt{\left(x^2+y^2\right)}=cx^2.$
5. $\ln\left(y^2+x^2\right)+2\tan^{-1}(y/x)+c=0.$
6. $c\left(2y^2+2xy-x^2\right)^{2\sqrt{3}}=\frac{(\sqrt{3}+1)x+2y}{(1-\sqrt{3})x+2y}.$
7. $x^3(x+2y)=ce^{2y/x}.$
8. $x^2-y^2-cx.$
9. $c/(y-x)=e^{x/(x-y)}.$
10. $xy^2=c(x-y)^2.$
11. $cx=e^{x/y}.$
12. $\ln x=\cos(y/x)+c.$
13. $x=c\sin(y/x).$
14. $cx=e^{2x/(x-y)}.$
15. $x^2y=c(2x+y).$
16. $xy^2=c^2(x+2y).$
17. $y^3=-3x^3\ln cx.$
18. $y^3=x^3\ln cx^3.$
19. $x+ye^{x/y}=c.$
20. $cy=e^{y/x}.$
21. $\ln x=\sin(y/x)+c.$
22. $cxy=y+\sqrt{\left(y^2-x^2\right)}.$
23. (a) $(x-2y)^2(x+y)=27.$
(b) $2x^2=y+\sqrt{\left(4x^2+y^2\right)}.$
(c) $y^2=x^2-x.$
(d) $ay^3+3x^2y=a^4.$
(e) $5y^2=4x^2\left(x^2+y^2\right).$
(f) $x+y\ln y=0.$
24. $y^2+2xy=x^2+2.$

2.13.1 Equations which are both homogeneous and exact

Theorem 2.13.1. *Suppose that M and N are homogeneous functions of x and y of degree n(≠ −1). Further suppose that M dx + N dy = 0 is exact. Then the primitive can be obtained without the introduction of a quadrature. The primitive is simply Mx + Ny = c.*

Proof. Let $u = Mx + Ny$. Since M and N are homogeneous of degree n, we obtain, from Euler's Theorem on homogeneous functions:

$$x\frac{\partial M}{\partial x} + y\frac{\partial M}{\partial y} = nM \tag{2.13.1}$$

$$x\frac{\partial N}{\partial x} + y\frac{\partial N}{\partial y} = nN \tag{2.13.2}$$

Also it is given that $M\,dx + N\,dy = 0$ is exact. Hence, by Theorem 2.13.1,

$$\frac{\partial M}{\partial y} = \frac{\partial N}{\partial x}. \tag{2.13.3}$$

Now

$$\begin{aligned}\frac{\partial u}{\partial x} = \frac{\partial}{\partial x}(Mx + Ny) &= M + x\frac{\partial M}{\partial x} + y\frac{\partial N}{\partial x} = M + x\frac{\partial M}{\partial x} + y\frac{\partial M}{\partial y} \quad \text{[using (2.13.3)]}\\ &= M + nM \quad \text{[using (2.13.1)]}\\ &= (n+1)M.\end{aligned}$$

Similarly, using (2.13.3) and (2.13.2) we get $\frac{\partial u}{\partial y} = (n+1)N$.

Consequently,

$$du = \frac{\partial u}{\partial x}\,dx + \frac{\partial u}{\partial y}\,dy = (n+1)(M\,dx + N\,dy)$$

$$\text{or,}\quad M\,dx + N\,dy = \frac{1}{n+1}du = \frac{1}{n+1}d\{Mx + Ny\}, \text{ if } n \neq -1.$$

$\therefore$ the primitive is $Mx + Ny = c$.

Example 2.13.4. *Solve:* $(x^3 + 3y^2x)\,dx + (y^3 + 3x^2y)\,dy = 0.$

Solution: $M = x^3 + 3y^2x,\ N = y^3 + 3x^2y.$

$$\therefore \frac{\partial M}{\partial y} = 6xy = \frac{\partial N}{\partial x}.$$

Thus, the equation is exact; M and N are both homogeneous of degree 3. Therefore, the primitive is $Mx + Ny = c$, or, $x^4 + 6x^2y^2 + y^4 = c$, $c\ (> 0)$ being an arbitrary constant (Fig. 2.19).

Next we consider a homogeneous equation $M\,dx + N\,dy = 0$. Suppose, it is not exact. In the following theorem we shall see that the equation can be made exact by introducing the *Integrating Factor*

$$\frac{1}{Mx + Ny}, \text{ provided, of course, } Mx + Ny \neq 0.$$

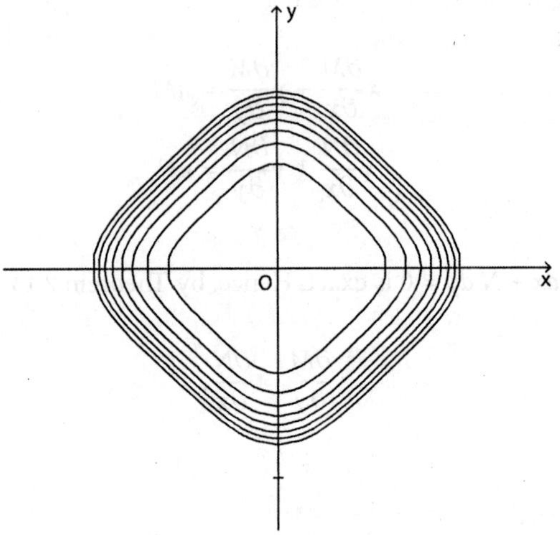

Fig. 2.19

Theorem 2.13.2. *If M and N are both homogeneous functions of x and y of same degree n, then*

$$\frac{M\,dx + N\,dy}{Mx + Ny} \quad (Mx + Ny \neq 0)$$

satisfies the condition of integrability.

Proof. The condition of integrability, viz.

$$\frac{\partial}{\partial y}\left(\frac{M}{Mx + Ny}\right) = \frac{\partial}{\partial x}\left(\frac{N}{Mx + Ny}\right)$$

is satisfied, if

$$\frac{(Mx + Ny)\frac{\partial M}{\partial y} - M\left(x\frac{\partial M}{\partial y} + N + y\frac{\partial N}{\partial y}\right)}{(Mx + Ny)^2} = \frac{(Mx + Ny)\frac{\partial N}{\partial x} - N\left(M + x\frac{\partial M}{\partial x} + y\frac{\partial N}{\partial x}\right)}{(Mx + Ny)^2},$$

which is true, if

$$Ny\frac{\partial M}{\partial y} - MN - My\frac{\partial N}{\partial y} = Mx\frac{\partial N}{\partial x} - NM - Nx\frac{\partial M}{\partial x} \text{ i.e. if } N\left(x\frac{\partial M}{\partial x} + y\frac{\partial M}{\partial y}\right) = M\left(x\frac{\partial N}{\partial x} + y\frac{\partial N}{\partial y}\right).$$

But the last relation is true, because each side will be equal to nMN, using Euler's Theorem on homogeneous functions. Hence, the theorem.

Note 2.13.1. $\frac{M\,dx+N\,dy}{Mx+Ny}$ *is homogeneous of degree −1 and also exact. We cannot apply theorem* 2.13.1. *That is, in general, integration of a homogeneous equation will involve a quadrature.*

We summarise the previous results in the form of a rule:

Rule I

(i) If $M\,dx + N\,dy = 0$ is both homogeneous and exact, then its primitive is $Mx + Ny = c$, provided the degree of homogeneity is not equal to −1.

(ii) If $M\,dx + N\,dy = 0$ is homogeneous (but not exact) and if $Mx + Ny \neq 0$, then $\frac{1}{Mx+Ny}$ is an integrating factor of the equation.

Other rules for determining Integrating Factors will be given in article 2.15.

Example 2.13.5. *Solve:* $x^2 y\, dx - (x^3 + y^3)\, dy = 0.$

Solution: This equation is homogeneous but not exact.

Here $M = x^2 y$, $N = -x^3 - y^3$. *Rule* I(ii)

$$\frac{\partial M}{\partial y} = x^2, \quad \frac{\partial N}{\partial x} = -3x^2. \text{ Not Exact.}$$

Also $Mx + Ny = -y^4 (\neq 0)$; hence IF $= -1/y^4$.

Multiplying by this IF, we get,

$$-\frac{x^2}{y^3}\,dx + \frac{x^3}{y^4}\,dy + \frac{1}{y}\,dy = 0 \text{ or, } -\frac{x^2}{y^2}\left\{\frac{y\,dx - x\,dy}{y^2}\right\} + \frac{1}{y}\,dy = 0 \text{ or, } -\frac{x^2}{y^2}\,d\left(\frac{x}{y}\right) + \frac{1}{y}\,dy = 0.$$

Integrating,

$$-\frac{x^3}{3y^3} + \ln y = c, \text{ or, } y = Ae^{x^3/3y^3} \quad (e^c = A).$$

Particular Case: Theorem 2.13.2 fails if $Mx + Ny = 0$ identically. But, then, the equation $M\,dx + N\,dy = 0$ becomes $y\,dx - x\,dy = 0$ for which $1/(xy)$ is an integrating factor. The solution will be $y = cx$.

Example 2.13.6. *Solve:* $(x^2 y - 2xy^2)dx - (x^3 - 3x^2 y)dy = 0.$

Solution: Here $M = x^2 y - 2xy^2$, $N = -x^3 + 3x^2 y$. *Rule* I(ii)

$$\frac{\partial M}{\partial y} = x^2 - 4xy, \quad \frac{\partial N}{\partial x} = -3x^2 + 6xy. \text{ Not Exact.}$$

Also, $Mx + Ny = x^2 y^2 (\neq 0)$; hence IF $= 1/x^2 y^2$.

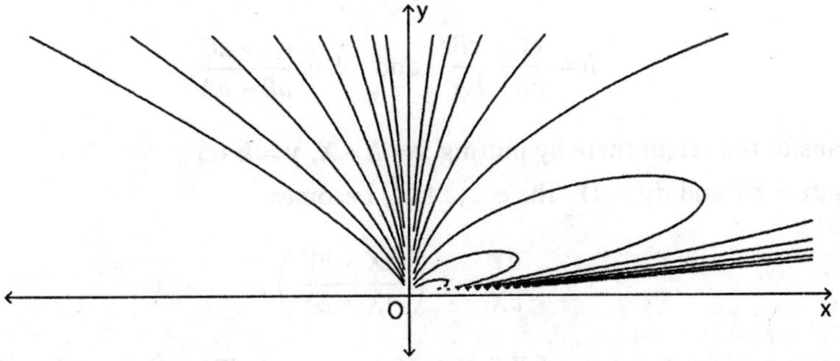

Fig. 2.20

Multiplying by the IF and simplifying,

$$\frac{3}{y}dy - \frac{2}{x}dx + \left(\frac{1}{y}dx - \frac{x}{y^2}dy\right) = 0, \text{ or, } \frac{3}{y}dy - \frac{2}{x}dx + d\left(\frac{x}{y}\right) = 0.$$

Integrating,

$$3\ln y - 2\ln x + \frac{x}{y} = c, \text{ or, } \ln\frac{y^3}{x^2} + \frac{x}{y} = c,$$

c being an arbitrary constant (Fig. 2.20).

Examples II(E)

Solve the following equations:

1. $y^2 + x^2\frac{dy}{dx} = xy\frac{dy}{dx}$.
2. $(x^4 + y^4)\ dx - xy^3\ dy = 0$.
3. $y^2\ dx + (x^2 - xy - y^2)\ dy = 0$.
4. $3x^2y\ dx + (x^3 + y^3)\ dy = 0$.

Answers

1. $cy = e^{y/x}$.
2. $y^4 = 4x^4 \ln x + cx^4$.
3. $(x - y)y^2 = c(x + y)$.
4. $4x^3y + y^4 = c$.

2.13.2 Equations reducible to the homogeneous form

Certain non-homogeneous equations can be made homogeneous by a linear transformation of the variables.

Theorem 2.13.3. *An equation of the form:*

$$\frac{dy}{dx} = F\left(\frac{ax + by + c}{Ax + By + C}\right), \tag{2.13.4}$$

in which a, b, c, A, B, C are fixed constants such that $aB - bA \neq 0$ can be made homogeneous and hence can be solved by separation.

Proof. We first note that under the given restriction (*viz.* $aB - bA \neq 0$), the equations

$$ax + by + c = 0$$
$$Ax + By + C = 0$$

will represent two intersecting straight lines. The point of intersection (h, k) is given by

$$h = \frac{bC - cB}{aB - bA} \quad \text{and} \quad k = \frac{cA - aC}{aB - bA}. \tag{2.13.5}$$

We now transfer the origin there by putting $x = h + X$, $y = k + Y$.

This leaves $dx = dX$ and $dy = dY$. The eq. (2.13.4) becomes

$$\frac{dY}{dX} = F\left(\frac{aX + bY}{AX + BY}\right),$$

so that F is a homogeneous function of X and Y of degree zero. The values of h and k are determinate since $aB - bA \neq 0$. Now proceeding as in art. 2.13, we put $Y = vX$ and obtain

$$v + X\frac{dv}{dX} = F\left(\frac{a + bv}{A + Bv}\right) = \phi(v), \text{ (say)}.$$

Separating the variables, we get

$$\frac{dv}{\phi(v) - v} = \frac{dX}{X}.$$

The primitive will then be

$$\int \frac{dv}{\phi(v) - v} = \ln C_1X,$$

where, after the integration is performed, the value of

$$v = \frac{Y}{X} = \frac{y-k}{x-h} \text{ [}h, k \text{ given by (2.13.5)]}$$

is to be substituted.

Particular Case: When $aB - bA = 0$, let X be a new dependent variable defined by

$$X = x + \frac{by}{a} = x + \frac{By}{A} \left(\text{see that } \frac{b}{a} = \frac{B}{A}\right),$$

so that

$$\frac{dX}{dx} = 1 + \frac{b}{a}\frac{dy}{dx}.$$

The eq. (2.13.4) then becomes

$$\frac{a}{b}\left(\frac{dX}{dx} - 1\right) = F\left(\frac{aX + c}{AX + C}\right) \text{ or, } \frac{dX}{dx} = 1 + \frac{b}{a}F\left(\frac{aX + c}{AX + C}\right).$$

The variables X and x are now separable.

In practice, the following procedure is adopted:

Since $a/A = b/B = K$, say, we observe that $ax + by = K(Ax + By)$. Now we put $ax + by = v$ so that $dv/dx = a + b\, dy/dx$. Hence, the eq. (2.13.4) becomes

$$\frac{1}{b}\left(\frac{dv}{dx} - a\right) = F\left(\frac{v + c}{Kv + C}\right) = \phi(v), \text{ say.}$$

The variables v and x can then be easily separated (See Ex. 2.13.8).

Example 2.13.7. *Solve:* $\frac{dy}{dx} = \frac{y-x+1}{y+x+5}$.

Solution: The two lines $y - x + 1 = 0$ and $y + x + 5 = 0$ intersect at the point $(-2, -3)$. The substitution $x = X - 2$, $y = Y - 3$ gives

$$\frac{dY}{dX} = \frac{Y - X}{Y + X} = \frac{Y/X - 1}{Y/X + 1}.$$

Put $Y = vX$. Then the equation becomes

$$v + X\frac{dv}{dX} = \frac{v-1}{v+1} \text{ or, } X\frac{dv}{dX} = \frac{v-1}{v+1} - v = -\frac{v^2+1}{v+1}.$$

Separating the variables, we get

$$\frac{(v+1)dv}{v^2+1} + \frac{dX}{X} = 0.$$

Integrating,

$$\frac{1}{2}\ln\left(v^2+1\right) + \tan^{-1} v + \ln X = \ln c$$

$$\text{or, } \ln\left[\left(\frac{y+3}{x+2}\right)^2 + 1\right] + 2\tan^{-1}\frac{y+3}{x+2} + 2\ln(x+2) = K \text{ (say)}$$

$$\text{or, } \ln\left[(y+3)^2 + (x+2)^2\right] + 2\tan^{-1}\frac{y+3}{x+2} = K \left(\text{Putting } v = \frac{Y}{X} = \frac{y+3}{x+2}\right).$$

Example 2.13.8. *Solve:* $\frac{dy}{dx} = \frac{3x-4y-2}{6x-8y-5}$.

Solution: Here the two lines $3x-4y-2=0$ and $6x-8y-5=0$ are parallel. We observe, here that the equation may be written as

$$\frac{dy}{dx} = \frac{(3x-4y)-2}{2(3x-4y)-5}.$$

Put $3x-4y=v$, so that

$$\frac{dv}{dx} = 3 - 4\frac{dy}{dx}.$$

Hence, we obtain

$$\frac{3}{4} - \frac{1}{4}\frac{dv}{dx} = \frac{v-2}{2v-5}.$$

This reduces to

$$\frac{2v-7}{2v-5} = \frac{dv}{dx}.$$

Separating the variables,

$$\frac{(2v-5)dv}{2v-7} = dx.$$

Integrating, $v + \ln(2v-7) = x + c$ or, $\ln(6x-8y-7) = c - 2x + 4y$, which is the required solution.

Examples II(F)

Solve:

1. $\frac{dy}{dx} = \frac{2x+9y-20}{6x+2y-10}$.
2. $\frac{dy}{dx} = \frac{4y+5x-3}{10y-x-21}$.
3. $(3y-7x+7)\,dx + (7y-3x+3)\,dy = 0$.
4. $(2x+3y-5)\,dy + (3x+2y-5)\,dx = 0$.
5. $(2x+4y+3)\,dy = (2y+x+1)\,dx$.
6. $(3y+2x+4)\,dx = (4x+6y+5)\,dy$.
7. (a) $\frac{dy}{dx} = \frac{x+y+1}{x+y-1}$ (b) $\frac{dy}{dx} = \frac{x-y}{x+1}$.
8. $(x+2y)(dx-dy) = dx+dy$.
9. (a) $\frac{dy}{dx} = \frac{3x-4y-2}{3x-4y-3}$ (b) $\frac{dy}{dx} = \frac{x+2y-3}{2x+y-3}$.
10. $(y-3x+3)\frac{dy}{dx} = 2y-x-4$.
11. $(x+y+1)\,dx = (2x+2y+1)\,dy$.
12. $\frac{dy}{dx} = \frac{x+y+1}{x+y-1}$ when $y=\frac{1}{3}$ at $x=\frac{2}{3}$.
13. Show that the equation $\frac{dy}{dx} = \frac{ax+by+c}{Ax+By+C}$ is rendered homogeneous by the substitutions $ax+by+c=X$, $Ax+By+C=Y$.

Answers

1. $(2x-y)^2 = c(x+2y-5)$.
2. $(x-y+3)^3(x+2y-3)^2 = c$.
3. $(y-x+1)^2(y+x-1)^5 = c$.
4. $4xy+3y^2-10y+3x^2-10x = c$.
5. $\ln(4x+8y+5) = 4x-8y+c$.
6. $\ln(14x+21y+22) = \frac{7}{3}(2y-x)+c$.
7. (a) $y - \ln(x+y) = x+c$.
   (b) $(x+1)(x-2y-1) = c$.
8. $3x-3y+c = 2\ln(3x+6y-1)$.
9. (a) $x-y+c = \ln(3x-4y+1)$.
   (b) $c^2x^2(x-y)^3 = (x-1)^2(x+y-2)$.
10. $c\left(y^2-5xy+x^2+11x+4y-17\right) = \left\{\frac{2y-(5+\sqrt{21})x+2(2+\sqrt{21})}{2y-(5-\sqrt{21})x+2(2-\sqrt{21})}\right\}^{\frac{1}{\sqrt{21}}}$.
11. $\ln(3x+3y+2) = 6y-3x-c$.
12. $x-y+\ln(x+y) = \frac{1}{3}$.

2.14 Linear equations

If, in a differential equation, the dependent variable and its derivatives occur in the first degree only and not as higher powers or products, then the equation is said to be *linear* (art. 1.1). So the coefficients of a linear equation are either functions of the independent variable or constants (including zero). The most general linear equation of first order can be written in the form,

$$\frac{dy}{dx} + Py = Q, \tag{2.14.1}$$

where P and Q are either functions of x or constants (including zero).

If P and Q are both constants, then the variables are separable; the same is true if either P or Q is zero.

I. Let $Q = 0$, $P \neq 0$. Then

$$\frac{dy}{dx} + Py = 0, \text{ or, } \frac{dy}{y} = -P\,dx.$$

$$\therefore\ y = ce^{-\int P\,dx}, \tag{2.14.2}$$

where c is the constant of integration.

[Note that if $P = 0$, $Q \neq 0$, then $y = c + \int Q\,dx$. But, if $P = Q = 0$, then $y = c$.]

II. When $Q \neq 0$, we assume the same form [i.e. of the form (2.14.2)] for y but we do not restrict c to be a constant. Thus, we assume that $y = ve^{-\int P\,dx}$ is a solution of eq. (2.14.1), where $Q \neq 0$.

Then

$$\frac{dy}{dx} = \frac{dv}{dx}e^{-\int P\,dx} - Pve^{-\int P\,dx}. \tag{2.14.3}$$

Substituting this value of dy/dx in (2.14.1), we find

$$\frac{dv}{dx}e^{-\int P\,dx} - Pve^{\int P\,dx} + Pve^{-\int P\,dx} = Q \text{ or, } \frac{dv}{dx} = Qe^{\int P\,dx}.$$

Integrating,

$$v = \int \left\{Qe^{\int P\,dx}\right\} dx + c,$$

where c is the arbitrary constant of integration.

This gives the solution of (2.14.1) in the form:

$$y = ve^{-\int P\,dx} = e^{-\int P\,dx}\left[c + \int \left\{Qe^{\int P\,dx}\right\} dx\right] = ce^{-\int P\,dx} + e^{-\int P\,dx}\int \left\{Qe^{\int P\,dx}\right\} dx.$$

Observations

1. The method adopted here is to find the general solution of an equation by regarding the constant of integration of a simpler equation as variable and so determining it that the more general equation is satisfied. This method is known as the *method of variation of parameters*.
2. It should be observed from (2.14.3) that

$$\frac{dy}{dx} + Py = \frac{dv}{dx}e^{-\int P\,dx}.$$

$$\therefore\ e^{\int P\,dx}\left\{\frac{dy}{dx} + Py\right\} = \frac{dv}{dx}.$$

In other words, $e^{\int P\,dx}$ is an integrating factor of the equation (2.14.1).

3. That $e^{\int P\,dx}$ is an IF of the eq. (2.14.1) can also be obtained in the following manner:

III. We write the eq. (2.14.1) in the differential form; thus $dy + [Py - Q]\,dx = 0$.

The term $Q\,dx$ is by itself exact and will remain so if we multiply it by a function of x, say $R(x)$ (or R).

We, therefore, try to make $dy + Py\,dx = 0$ exact multiplying by R. Then $RPy\,dx + R\,dy = 0$ is exact (by using the condition of integrability) if

$$\frac{\partial}{\partial y}(RPy) = \frac{\partial(R)}{\partial x}, \text{ or, } RP = \frac{dR}{dx}$$

(since R and P are functions of x alone).

$\therefore$ separating the variables,

$$\frac{dR}{R} = P\,dx, \text{ or, } R = e^{\int P\,dx}.$$

That is, $e^{\int P\,dx}$ is an Integrating Factor (IF).

IV. Method of solution of an equation of the form:

$$\frac{dy}{dx} + Py = Q.$$

1. $e^{\int P\,dx}$ is an integrating factor of the equation, as the above analysis shows.
2. Multiply both sides by $e^{\int P\,dx}$. The equation becomes

$$\frac{dy}{dx}e^{\int P\,dx} + Pye^{\int P\,dx} = Qe^{\int P\,dx} \text{ or, } \frac{d}{dx}\left\{y\cdot e^{\int P\,dx}\right\} = Qe^{\int P\,dx}.$$

Integrating,

$$ye^{\int P\,dx} = \int\left\{Qe^{\int P\,dx}\right\}dx + C.$$

$\therefore$ the general solution $y = e^{-\int P\,dx}\left\{C + \int\left\{Qe^{\int P\,dx}\right\}dx\right\}$.

Note 2.14.1. *We very often require to apply the identity* $e^{a\ln x} = e^{\ln x^a} = x^a$.

Method of variation of parameters

Example 2.14.1. *Find a solution of the differential equation* $\frac{dy}{dx} + 2xy = 0$ *in the form* $y = y_1$. *Hence, solve* $\frac{dy}{dx} + 2xy = 4x$ *by the substitution* $y = vy_1$.

Solution: Now $\frac{dy}{dx} + 2xy = 0$ gives $\frac{dy}{y} + 2x\,dx = 0$.

Integrating, $\ln\frac{y}{c_1} = -x^2$, or, $y = c_1e^{-x^2}$.

We take $y = y_1 = e^{-x^2}$, which is a solution of the reduced equation $\frac{dy}{dx} + 2xy = 0$ disregarding the constant of integration.

Next take $y = vy_1 = ve^{-x^2}$, whereby $\frac{dy}{dx} = \frac{dv}{dx}e^{-x^2} - 2xve^{-x^2}$.

Putting in $\frac{dy}{dx} + 2xy = 4x$, we have

$$\frac{dv}{dx}e^{-x^2} - 2xve^{-x^2} + 2xve^{-x^2} = 4x, \text{ or, } \frac{dv}{dx} = 4xe^{x^2}.$$

Integrating, $v = \int 4xe^{x^2}\,dx$ (put $x^2 = u$) $= 2e^{x^2} + c$.

$\therefore$ general solution is $y = vy_1 = e^{-x^2}\left(2e^{x^2} + c\right) = 2 + ce^{-x^2}$ (Fig. 2.21)

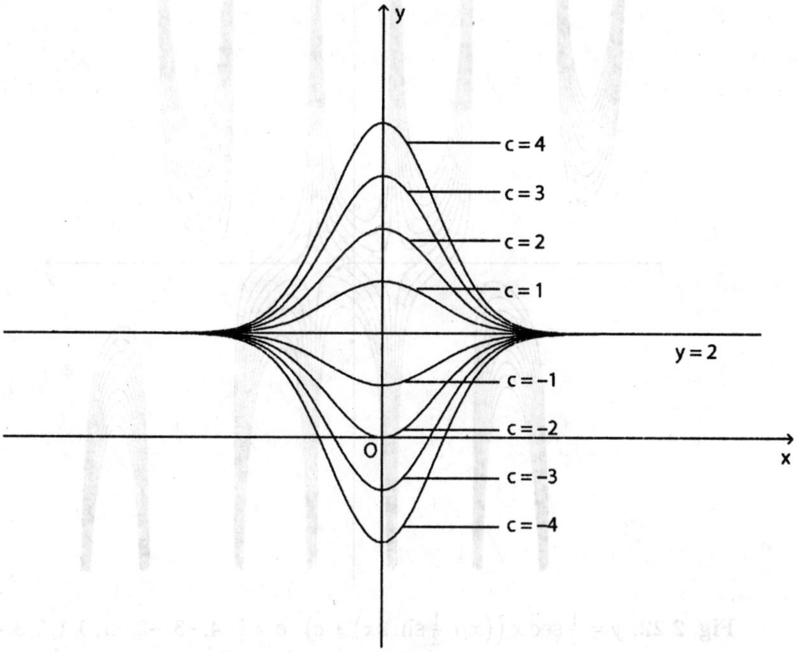

Fig. 2.21

Example 2.14.2. *Find a solution of the differential equation* $\frac{dy}{dx} - y\tan x = 0$ *in the form* $y = y_1(x)$, *say. Hence, solve* $\frac{dy}{dx} - y\tan x = \cos x$ *by the substitution* $y = y_1(x)v(x)$.

Solution: $\frac{dy}{dx} - y\tan x = 0$ gives $\frac{dy}{y} = \tan x\, dx$.

On integration, $\ln y = \ln c_1 \sec x$, or, $y = c_1 \sec x$.

Disregarding the constant of integration, we take $y = y_1 = \sec x$ as one of the solutions of the reduced equation.

Next take $y = v\sec x$, whereby $\frac{dy}{dx} = \frac{dv}{dx}\sec x + v\sec x\tan x$.

Putting in $\frac{dy}{dx} - y\tan x = \cos x$, we have

$$\frac{dv}{dx}\sec x + v\sec x\tan x - v\sec x\tan x = \cos x \text{ or, } \frac{dv}{dx} = \cos^2 x = \frac{1}{2}(1 + \cos 2x).$$

Integrating, we have $v = \frac{1}{2}\left(x + \frac{1}{2}\sin 2x\right) + c$, c being an arbitrary constant.

$\therefore$ general solution is $y = \frac{1}{2}\sec x\left\{\left(x + \frac{1}{2}\sin 2x\right) + c\right\}$.

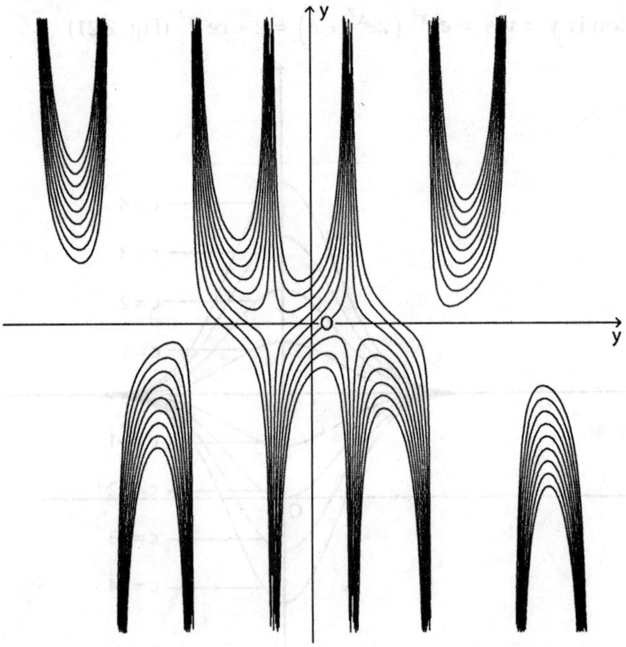

Fig. 2.22: $y = \frac{1}{2}\sec x\left(\left(x + \frac{1}{2}\sin 2x\right) + c\right)$, $c = \{-4, -3, -2, -1, 0, 1, 2, 3, 4\}$.

Example 2.14.3. *Solve by the method of variation of parameters* $\frac{dy}{dx} - 5y = \sin x$.

Solution: The reduced equation is $\frac{dy}{dx} - 5y = 0$, or, $\frac{dy}{y} = 5dx$.

Integrating, $\ln \frac{y}{c_1} = 5x$, or, $y = c_1 e^{5x}$.

Disregarding the constant of integration, we take $y = y_1 = e^{5x}$ as one of the solutions of the reduced equation.

Next take

$$y = vy_1 = ve^{5x}, \quad \frac{dy}{dx} = \frac{dv}{dx}e^{5x} + 5ve^{5x}$$

and the given equation changes to

$$e^{5x}\frac{dv}{dx} = \sin x \text{ or, } v = \int e^{-5x}\sin x\, dx = e^{-5x}\left(-\frac{5}{26}\sin x - \frac{1}{26}\cos x\right) + c,$$

c being an arbitrary constant.

$\therefore$ general solution is

$$y = ce^{5x} - \frac{5}{26}\sin x - \frac{1}{26}\cos x \text{ (Fig. 2.23).}$$

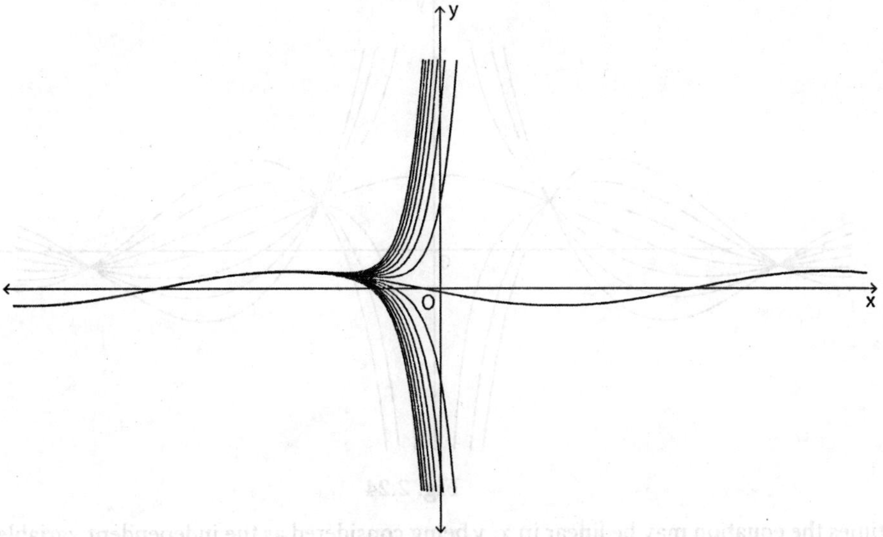

Fig. 2.23

Use of integrating factors (IV)

Example 2.14.4. *Solve:* $x\cos x\frac{dy}{dx} + y(x\sin x + \cos x) = 1$.

Solution: We write the equation in the standard form:

$$\frac{dy}{dx} + \frac{x\sin x + \cos x}{x\cos x}y = \frac{1}{x\cos x}. \qquad (2.14.4)$$

$$\text{IF} = e^{\int \frac{x\sin x+\cos x}{x\cos x}dx} = e^{\int \frac{\sin x}{\cos x}dx + \int \frac{dx}{x}} = e^{-\ln\cos x + \ln x} = e^{\ln\frac{x}{\cos x}} = \frac{x}{\cos x}.$$

Multiplying (2.14.4) by the IF $\frac{x}{\cos x}$, we obtain

$$\frac{d}{dx}\left\{y\frac{x}{\cos x}\right\} = \frac{1}{x\cos x}\cdot\frac{x}{\cos x} = \sec^2 x.$$

Integrating,

$$y\frac{x}{\cos x} = \int \sec^2 x\,dx + c, \text{ or, } y\frac{x}{\cos x} = \tan x + c \text{ or, } y = \frac{\cos x}{x}[\tan x + c] \text{ or, } xy = \sin x + c\cos x,$$

which is the required general solution (Fig. 2.24).

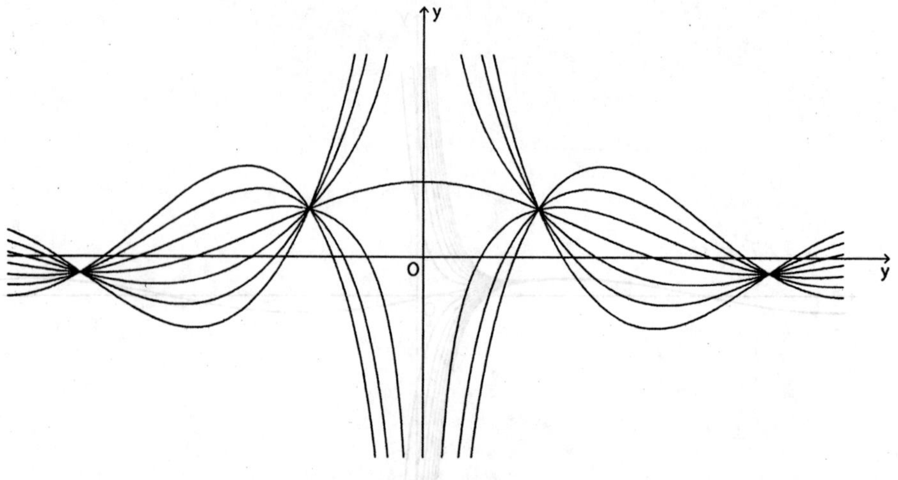

Fig. 2.24

V. Sometimes the equation may be linear in x; y being considered as the independent variable. The form of such an equation is $\frac{dx}{dy} + P_1 x = Q_1$, where P_1 and Q_1 are functions of y or constants.

The integrating factor is $e^{\int P_1\,dy}$. Multiplying by the IF,

$$\frac{d}{dy}\left\{x \cdot e^{\int P_1\,dy}\right\} = Q_1 e^{\int P_1\,dy}. \quad \therefore\ x = e^{-\int P_1\,dy}\left[C + \int \left\{Q_1 e^{\int P_1\,dy}\right\} dy\right].$$

Example 2.14.5. *Solve:* $y^2 + \left(x - \frac{1}{y}\right)\frac{dy}{dx} = 0.$

Solution: Note that the equation contains y^2 and so it cannot be linear in y. We try to see whether it is linear in x. With that end in view, we write the equation as below:

$$\frac{xy-1}{y}\frac{dy}{dx} = -y^2, \text{ or, } -\left(\frac{xy-1}{y^3}\right) = \frac{dx}{dy} \text{ or, } \frac{dx}{dy} + \frac{1}{y^2}\cdot x = \frac{1}{y^3}, \qquad (2.14.5)$$

which is a linear equation in x.

$$\text{IF} = e^{\int \frac{1}{y^2}dy} = e^{-\frac{1}{y}}.$$

Multiplying the eq. (2.14.5) by this IF, we find

$$\frac{d}{dy}\left\{xe^{-\frac{1}{y}}\right\} = e^{-\frac{1}{y}}\cdot\frac{1}{y^3}.$$

Integrating,

$$xe^{-\frac{1}{y}} = e^{-\frac{1}{y}}\left(1 + \frac{1}{y}\right) + c, \text{ or, } x = 1 + \frac{1}{y} + ce^{\frac{1}{y}},$$

which is the required general solution (Fig. 2.25).

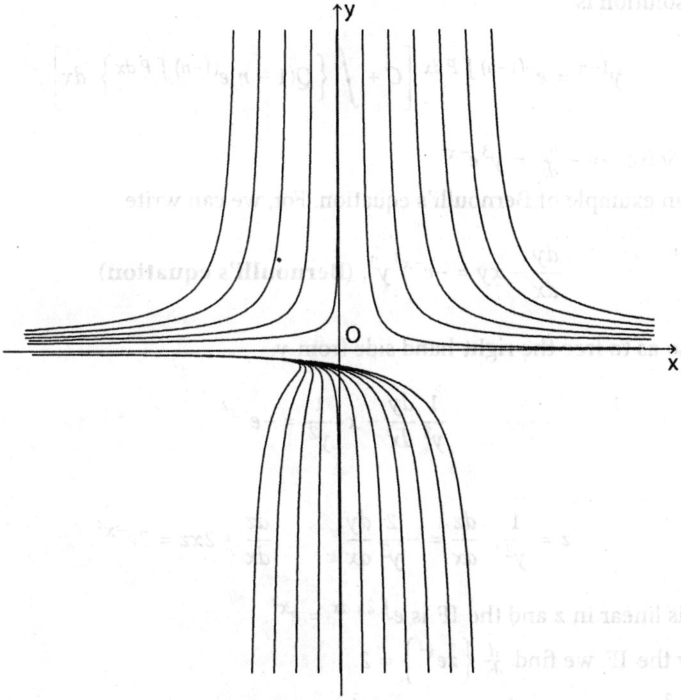

Fig. 2.25: $C = \{-5, -4, \ldots, 5\}$.

VI. Equations Reducible to Linear Form

Bernoulli's Equation.

The equation

$$\frac{dy}{dx} + Py = Qy^n,$$

in which P and Q are functions of x alone or constants, is known as *Bernoulli's equation*. It may be brought into the linear form by a change of dependent variable. Write the equation as

$$y^{-n}\frac{dy}{dx} + Py^{1-n} = Q \text{ (dividing by } y^n).$$

Let $z = y^{1-n}$. Then

$$\frac{dz}{dx} = (1-n)y^{-n}\frac{dy}{dx}.$$

So the equation becomes

$$\frac{1}{1-n}\frac{dz}{dx} + Pz = Q, \text{ or, } \frac{dz}{dx} + P(1-n)z = Q(1-n),$$

which is a linear equation in z. Its IF is $e^{(1-n)\int P\,dx}$.

Multiplying by the IF, we obtain

$$\frac{d}{dx}\left\{z \cdot e^{(1-n)\int P\,dx}\right\} = Q(1-n)e^{(1-n)\int P\,dx}.$$

Integrating,

$$z = e^{-(1-n)\int P\,dx}\left[C + \int \left\{Q(1-n)e^{(1-n)\int P\,dx}\right\} dx\right].$$

∴ the general solution is

$$y^{1-n} = e^{-(1-n)\int P\,dx}\left[C + \int \left\{Q(1-n)e^{(1-n)\int P\,dx}\right\} dx\right].$$

Example 2.14.6. *Solve:* $xy - \frac{dy}{dx} = y^3 e^{-x^2}$.

Solution: This is an example of Bernoulli's equation. For, we can write

$$\frac{dy}{dx} - xy = -e^{-x^2} y^3. \text{ (Bernoulli's equation)}$$

Divide by y^3, so as to free the right-hand side from y.

$$\frac{1}{y^3}\frac{dy}{dx} - x\cdot\frac{1}{y^2} = -e^{-x^2}.$$

Let

$$z = \frac{1}{y^2},\ \frac{dz}{dx} = -\frac{2}{y^3}\frac{dy}{dx}. \quad \therefore\ \frac{dz}{dx} + 2xz = 2e^{-x^2}.$$

This equation is linear in z and the IF is $e^{\int 2x\,dx} = e^{x^2}$.

Multiplying by the IF, we find $\frac{d}{dx}\left\{ze^{x^2}\right\} = 2$.

Integrating, $ze^{x^2} = 2x + c$, or, $\frac{1}{y^2} = (2x + c)e^{-x^2}$, which is the general solution (Fig. 2.26).

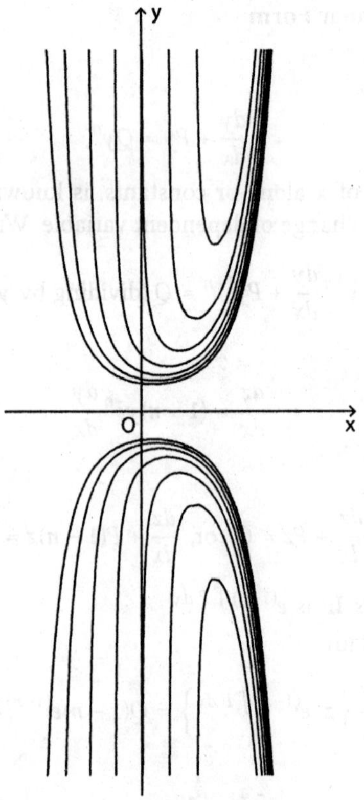

Fig. 2.26

Example 2.14.7. *Solve:* $\frac{dy}{dx} - \frac{\tan y}{1+x} = (1+x)e^x \sec y.$

Solution: We follow a similar method as that of Bernoulli, i.e. make the right-hand side free from functions of y. Thus, we divide both sides by $\sec y$ and obtain

$$\cos y \frac{dy}{dx} - \frac{1}{1+x} \sin y = (1+x)e^x.$$

Let $z = \sin y$, so that

$$\frac{dz}{dx} = \cos y \frac{dy}{dx}. \quad \therefore \quad \frac{dz}{dx} - \frac{1}{1+x} z = (1+x)e^x.$$

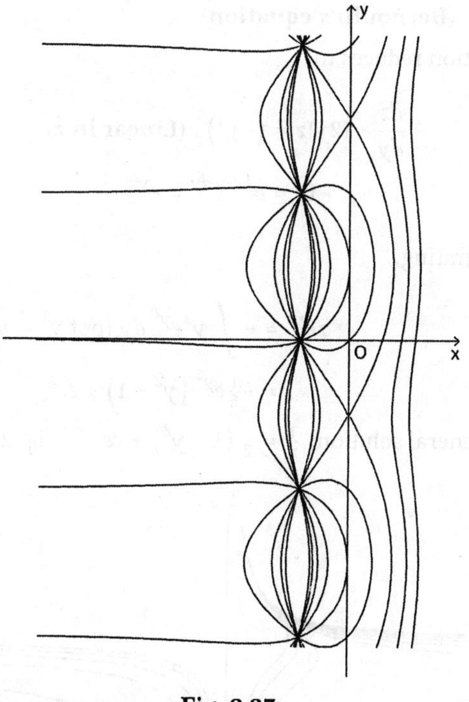

Fig. 2.27

This is linear in z. The integrating factor is

$$e^{-\int \frac{1}{1+x} dx} = e^{-\ln(1+x)} = \frac{1}{1+x}.$$

Multiplying by the IF, we obtain $\frac{d}{dx}\left\{z \cdot \frac{1}{1+x}\right\} = e^x$.

$\therefore$ on integration, the required solution $\sin y = z = (1+x)e^x + c(1+x)$ is obtained (Fig. 2.27).

Example 2.14.8. *Reduce the equation* $\sin y \frac{dy}{dx} = \cos x(2\cos y - \sin^2 x)$ *to a linear equation and hence solve it.*

Solution: To make it linear, put $\cos y = z$, then $-\sin y \frac{dy}{dx} = \frac{dz}{dx}$ and the given equation reduces to

$$\frac{dz}{dx} + (2\cos x)z = \sin^2 x \cos x \quad \textbf{[Linear equation]}$$

$$\text{IF} = e^{\int 2\cos x\, dx} = e^{2\sin x}.$$

Multiplying IF and integrating,

$$z \cdot e^{2\sin x} = \int e^{2\sin x} \cdot \sin^2 x \cos x \, dx \text{ (put } \sin x = u)$$

$$= \int u^2 e^{2u} du = e^{zu}\left(\frac{1}{2}u^2 - \frac{1}{2}u + \frac{1}{4}\right) + c,$$

c being an arbitrary constant.

General solution: $\cos y = \frac{1}{2}\sin^2 x - \frac{1}{2}\sin x + \frac{1}{4} + ce^{-2\sin x}$.

Example 2.14.9. *Solve:* $\left(x^2y^3 + 2xy\right) dy = dx$.

Solution: $\frac{dx}{dy} - (2y)x = y^3x^2$. **(Bernoulli's equation)**

Put $1/x = z$ and the equation reduces to

$$\frac{dz}{dy} + (2y)z = \left(-y^3\right). \text{ (Linear in } z)$$

$$\text{IF} = e^{\int 2y\, dy} = e^{y^2}.$$

Multiplying IF and integrating,

$$ze^{y^2} = -\int y^3 e^{y^2} dy \text{ (put } y^2 = u)$$

$$= -\tfrac{1}{2}e^{y^2}\left(y^2 - 1\right) + c.$$

General solution: $\frac{1}{x} = \frac{1}{2}\left(1 - y^2\right) + ce^{-y^2}$ (Fig. 2.28).

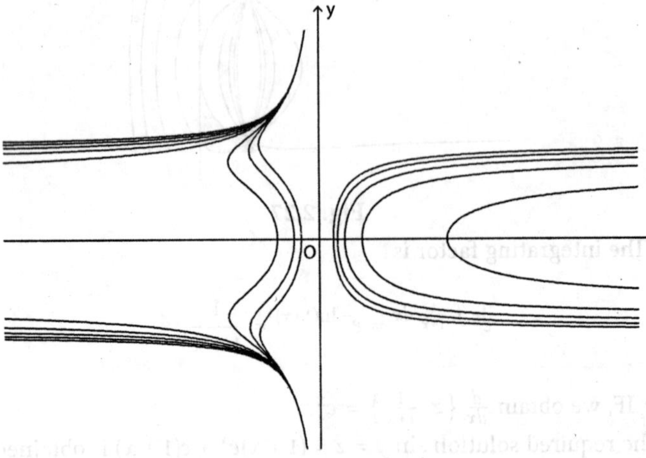

Fig. 2.28

Example 2.14.10. *Solve:* $y\left(2xy + e^x\right) dx - e^x dy = 0$.

Solution: $\frac{dy}{dx} - y = \frac{2x}{e^x}y^2$. **(Bernoulli's equation)**

Put $1/y = z$ and the equation reduces to

$$\frac{dz}{dx} + z = -\frac{2x}{e^x}. \text{ (Linear equation)}$$

$$\text{IF} = e^{\int dx} = e^x.$$

Multiplying IF and integrating, $ze^x = \int -2x\,dx = -x^2 + c$, c being an arbitrary constant.

$\therefore$ general solution is $e^x = y\left(c - x^2\right)$ (Fig. 2.29)

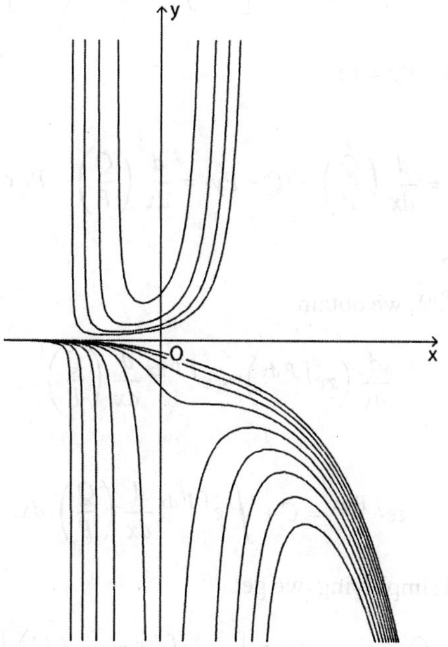

Fig. 2.29

Some interesting results

Example 2.14.11. *Show that the general solution of the equation $\frac{dy}{dx} + Py = Q$ can be written in the form $y = k(u - v) + v$, where k is a constant, and u and v are its two particular solutions.*

Solution: Since u and v are its solutions

$$\frac{du}{dx} + Pu = Q \text{ and } \frac{dv}{dx} + Pv = Q.$$

Subtracting

$$\frac{du}{dx} - \frac{dv}{dx} + P(u - v) = 0. \tag{2.14.6}$$

Also since

$$\frac{dy}{dx} + Py = Q \text{ and } \frac{dv}{dx} + Pv = Q,$$

on subtraction

$$\frac{dy}{dx} - \frac{dv}{dx} + P(y - v) = 0. \tag{2.14.7}$$

From (2.14.6) and (2.14.7),

$$\frac{d}{dx}(u - v) = -P(u - v) \text{ and } \frac{d}{dx}(y - v) = -P(y - v) \text{ or, } \frac{d(u - v)}{u - v} = -P\,dx = \frac{d(y - v)}{y - v}.$$

$\therefore \ln(y - v) = \ln(u - v) + \ln k$, or, $y - v = k(u - v)$, or, $y = k(u - v) + v$.

Example 2.14.12. *Show that the solution of* $\frac{dy}{dx} + Py = Q$ *can also be written in the form:*

$$y = \frac{Q}{P} - e^{-\int P\,dx}\left[C + \int e^{\int P\,dx}\, d\left(\frac{Q}{P}\right)\right].$$

Solution: Let $z = \frac{Q}{P} - y$, or, $Q - Py = Pz$.

Then

$$\frac{dz}{dx} = \frac{d}{dx}\left(\frac{Q}{P}\right) - \frac{dy}{dx} = \frac{d}{dx}\left(\frac{Q}{P}\right) - (Q - Py) = \frac{d}{dx}\left(\frac{Q}{P}\right) - Pz \text{ or, } \frac{dz}{dx} + Pz = \frac{d}{dx}\left(\frac{Q}{P}\right),$$

which is a linear equation in z.

Multiplying by the IF $e^{\int P\,dx}$, we obtain

$$\frac{d}{dx}\left(ze^{\int P\,dx}\right) = e^{\int P\,dx}\frac{d}{dx}\left(\frac{Q}{P}\right).$$

Hence, on integration,

$$ze^{\int P\,dx} = C + \int e^{\int P\,dx}\frac{d}{dx}\left(\frac{Q}{P}\right)\,dx.$$

Putting the value of z and simplifying, we get

$$\frac{Q}{P} - y = e^{-\int P\,dx}\left[C + \int e^{\int P\,dx}\, d\left(\frac{Q}{P}\right)\right].$$

Hence, the result as stated.

Example 2.14.13 (Use of Bernoulli's form). *Show that the equation* $(P + Qx)\frac{dy}{dx} = R + Qy$, *where* P, R *are homogeneous functions in* x *and* y *of degree* n *and* Q *is a homogeneous function in* x *and* y *of degree* m, *can be solved by the substitution* $y = vx$.

Solution: Let $P = x^n f_1(y/x) = x^n f_1(v)$, $R = x^n f_3(v)$ and $Q = x^m f_2(v)$.

From $y = vx$ we also obtain $\frac{dy}{dx} = v + x\frac{dv}{dx}$.

Then the given equation becomes

$$\{x^n f_1(v) + x^{m+1} f_2(v)\}\left\{v + x\frac{dv}{dx}\right\} = x^n f_3(v) + x^m f_2(v)\cdot v\cdot x$$

$$\text{or,}\quad x^n f_1(v)\left\{v + x\frac{dv}{dx}\right\} + x^{m+2} f_2(v)\frac{dv}{dx} = x^n f_3(v)$$

$$\text{or,}\quad \frac{dv}{dx}\{x^{n+1} f_1(v) + x^{m+2} f_2(v)\} = x^n f_3(v) - x^n f_1(v)\cdot v$$

$$\text{or,}\quad x^{n+1} f_1(v) + x^{m+2} f_2(v) = x^n \{f_3(v) - v f_1(v)\}\frac{dx}{dv}$$

$$\text{or,}\quad \frac{dx}{dv} - x\frac{f_1(v)}{f_3(v) - v f_1(v)} = \frac{f_2(v)}{f_3(v) - v f_1(v)}x^{m-n+2},$$

which is in Bernoulli's form and we can obtain solutions.

Example 2.14.14. *Show that, if* y_1 *and* y_2 *be solutions of the equation* $\frac{dy}{dx} + Py = Q$, *where* P *and* Q *are functions of* x *alone, and* $y_2 = y_1 z$, *then* $z = 1 + ae^{-\int \frac{Q}{y_1}dx}$ *(a is an arbitrary constant).*

Solution: Given: $\frac{dy_1}{dx} + Py_1 = Q$ and $\frac{dy_2}{dx} + Py_2 = Q$.

It follows that (eliminating P)

$$y_1\frac{dy_2}{dx} - y_2\frac{dy_1}{dx} = Q(y_1 - y_2).$$

Now from $z = \frac{y_2}{y_1}$, we get

$$\frac{dz}{dx} = \frac{y_1\frac{dy_2}{dx} - y_2\frac{dy_1}{dx}}{y_1^2}, \text{ i.e. } \frac{dz}{dx} = \frac{Q(y_1 - y_2)}{y_1^2} = \frac{Q(1-z)}{y_1},$$

whence

$$\frac{dz}{z-1} = -\frac{Q}{y_1}dx \text{ or, } \ln(z-1) = -\int\frac{Q}{y_1}dx + \text{constant or, } z = 1 + ae^{-\int\frac{Q}{y_1}dx},$$

where a is an arbitrary constant.

Examples II(G)

1. Solve the following equations:
   (a) Transform $\frac{dy}{dx} + \frac{y}{x} = \frac{y^2}{x^2}$ into a linear form and then solve it.
   (b) Solve: $\frac{dy}{dx} + x\sin 2y = x^3\cos^2 y$.
2. Solve by the method of variation of parameters (i.e. accepting a solution of reduced equation):
   (a) $\frac{dy}{dx} + 6y = 18e^{3x}$.
   (b) $\frac{dx}{dt} + \frac{1}{t}x = t^2$.
   (c) $\frac{dy}{dt} + e^t y = e^t$.
3. (a) $\frac{dy}{dx} + y\cot x = 2\cos x$. (b) $\frac{dy}{dx} + \frac{1-2x}{x^2}y = 1$.
4. (a) $\left(1-x^2\right)\frac{dy}{dx} - xy = 1$. (b) $\frac{dy}{dx} + \frac{y}{(1-x^2)^{3/2}} = \frac{x+\sqrt{(1-x^2)}}{(1-x^2)^2}$.
5. (a) $\frac{dy}{dx} + \frac{n}{x}y = \frac{a}{x^n}$. (b) $\frac{dy}{dx} + \frac{4x}{x^2+1}y = \frac{1}{(x^2+1)^3}$.
6. $x\left(1-x^2\right)\frac{dy}{dx} + \left(2x^2-1\right)y = ax^3$.
7. $\sqrt{\left(a^2+x^2\right)}\frac{dy}{dx} + y = \sqrt{\left(a^2+x^2\right)} - x$.
8. $\frac{dy}{dx} - \frac{2y}{x+1} = (x+1)^{\frac{5}{2}}$.
9. $a\frac{du}{d\theta} + a(\tan\theta)u = \tan\theta$.
10. $(x+a)\frac{dy}{dx} - 3y = (x+a)^5$.
11. $x\ln x\frac{dy}{dx} + y = 2\ln x$.
12. $(x+y+1)\,dy = dx$.
13. $\left(1+y^2\right)dx = \left(\tan^{-1}y - x\right)dy$.
14. $2(1-xy)\frac{dy}{dx} = y^2$.
15. $\left(x+2y^3\right)\frac{dy}{dx} = y$.
16. $\frac{dy}{dx} + \frac{1}{x}y = x^2y^6$.
17. $\frac{dy}{dx} + \frac{x}{1-x^2}y = x\sqrt{y}$.
18. (a) $3\frac{dy}{dx} + \frac{2}{x+1}y = \frac{x^3}{y^2}$.
    (b) $x\frac{dy}{dx} + y = xy^2$.
19. (a) $3x\left(1-x^2\right)y^2\frac{dy}{dx} + \left(2x^2-1\right)y^3 = ax^3$.
    (b) $2x^2\frac{dy}{dx} = xy + y^2$.
20. $x\frac{dy}{dx} + y = y^2\ln x$.
21. $\sec^2 y\frac{dy}{dx} + 2x\tan y = x^3$.
22. $x\frac{dy}{dx} + y = y^2x^3\cos x$.
23. $dx + x\,dy = e^{-y}\sec^2 y\,dy$.
24. $\left(x^2y^3 + xy\right)dy = dx$.
25. $\left(x^2y^3 + 2xy\right)dy = dx$.
26. $\frac{dy}{dx} + y\cos x = y^n\sin 2x$.
27. $y\,dx + \left(ax^2y^n - 2x\right)dy = 0$.
28. (a) $y\,dy + by^2\,dx = a\cos x\,dx$.

(b) $\cos x\frac{dy}{dx} - y\sin x + y^2 = 0$.

29. $xy - \frac{dy}{dx} = e^{-x^2}y^3$.

30. (a) $y(2xy + e^x)\,dx - e^x\,dy = 0$.

(b) $y + 2\frac{dy}{dx} = y^3(x-1)$.

31. (a) $x^2y - x^3\frac{dy}{dx} = y^4\cos x$.

(b) $\frac{dy}{dx} + \frac{y}{x}\ln y = \frac{y}{x^2}(\ln y)^2$.

32. If $\frac{dr}{d\theta} + 2r\tan\theta = \sin\theta$ and $r = 0$, when $\theta = \frac{1}{3}\pi$, show that the maximum value of r is 1/8.

33. Find the particular solutions for given values of x and y:

(a) $x\,dy = \left(2y + 2x^4 + x^2\right)dx;\ x = 1, y = 3$.

(b) $\left(1 + x^2\right)dy = (1 + xy)\,dx;\ x = 0, y = 2$.

(c) $\frac{dy}{dx} - \frac{2y}{x} = x^2e^x;\ x = 1, y = 0$.

(d) $\sin x\frac{dy}{dx} + 4y\cos x = \cot x;\ x = \frac{\pi}{2}, y = \frac{4}{3}$.

Answers

1. (a) $\frac{1}{xy} = \frac{1}{2x^2} + C$.
   (b) $\tan y = \frac{1}{2}(x^2 - 1) + e\cdot e^{-x^2}$.
2. (a) $y = ce^{-6x} + 2e^{3x}$.
   (b) $x = \frac{c}{t} + \frac{1}{4}t^3$.
   (c) $y = 1 + ce^{-e^t}$.
3. (a) $y = \sin x + c\operatorname{cosec} x$.
   (b) $y = x^2\left(1 + ce^{1/x}\right)$.
4. (a) $y = \left(\sin^{-1}x + c\right)/\sqrt{1 - x^2}$.
   (b) $y = ce^{-x/\sqrt{(1-x^2)}} + x/\sqrt{(1 - x^2)}$.
5. (a) $x^ny = ax + c$.
   (b) $y\left(x^2 + 1\right)^2 = \tan^{-1}x + c$.
6. $y = ax + cx\sqrt{(1 - x^2)}$.
7. $\left\{x + \sqrt{(a^2 + x^2)}\right\}y = a^2\ln\left\{x + \sqrt{(a^2 + x^2)}\right\} + c$.
8. $3y = 2(x + 1)^{\frac{7}{2}} + 3c(x + 1)^2$.
9. $au = 1 + c\cos\theta$.
10. $2y = (x + a)^5 + 2c(x + a)^3$.
11. $y\ln x = (\ln x)^2 + c$.
12. $x + y + 2 = ce^y$.
13. $x + 1 = \tan^{-1}y + ce^{-\tan^{-1}y}$.
14. $xy^2 = 2y + c$.
15. $x = y^3 + cy$.
16. $y^{-5} = cx^5 + \frac{1}{2}5x^3$.
17. $\sqrt{y} = c\left(1 - x^2\right)^{\frac{1}{4}} - \frac{1}{3}\left(1 - x^2\right)$.
18. (a) $60y^3(x + 1)^2 = 10x^6 + 24x^5 + 15x^4 + c$.
    (b) $cxy - xy\ln x = 1$.
19. (a) $y^3 = ax + cx\sqrt{(1 - x^2)}$.
    (b) $x - y = cy\sqrt{x}$.
20. $1 + cxy = y(1 + \ln x)$.
21. $2\tan y = x^2 - 1 + ce^{-x^2}$.
22. $xy(x\sin x + \cos x + c) + 1 = 0$.
23. $x = e^{-y}(c + \tan y)$.
24. $1/x = 2 - y^2 + ce^{-\frac{1}{2}y^2}$.
25. $\frac{1}{x} = \frac{1}{2}\left(1 - y^2\right) + ce^{-y^2}$.
26. $y^{-n+1} = ce^{(n-1)\sin x} + 2\sin x + 2/(n - 1)$.
27. $ay^{n+2}/(n + 2) = y^2/x + c$.
28. (a) $\left(4b^2 + 1\right)y^2 = 2a(\sin x + 2b\cos x) + ce^{-2bx}$.
    (b) $y^{-1} = \sin x + c\cos x$.
29. $\frac{1}{y^2} = (2x + c)e^{-x^2}$.
30. (a) $e^x = y\left(c - x^2\right)$.
    (b) $y^2\left(x + ce^x\right) = 1$.
31. (a) $x^3 = y^3(3\sin x + c)$.
    (b) $2x = \ln y\left(1 + 2cx^2\right)$.
33. (a) $y = x^4 + x^2\ln x + x^2$.
    (b) $y = x + 2\sqrt{x^2 + 1}$.
    (c) $y = x^2\left(e^x - e\right)$.
    (d) $y\sin^4 x = \frac{1}{3}\sin^3 x + 1$.

2.15 Rules for determining integrating factors

In article 2.13.1 we discussed the rules for finding integrating factors of a homogeneous equation (Theorems 2.13.1 and 2.13.2). We propose to give here a few more rules:

Rule II. If $\frac{1}{N}\left(\frac{\partial M}{\partial y} - \frac{\partial N}{\partial x}\right)$ is a function of x alone, $f(x)$ (say), then $e^{\int f(x)\,dx}$ is an integrating factor of $M\,dx + N\,dy = 0$.

Proof. Suppose, $\frac{1}{N}\left(\frac{\partial M}{\partial y} - \frac{\partial N}{\partial x}\right) = f(x)$.

Hence, $\frac{\partial M}{\partial y} = \frac{\partial N}{\partial x} + Nf(x)$.

Multiplying by the integrating factor, we have

$$\frac{\partial M}{\partial y} e^{\int f(x)\,dx} = \frac{\partial N}{\partial x} e^{\int f(x)\,dx} + Nf(x)e^{\int f(x)\,dx} \text{ or, } \frac{\partial}{\partial y}\left\{Me^{\int f(x)\,dx}\right\} = \frac{\partial}{\partial x}\left\{Ne^{\int f(x)\,dx}\right\}.$$

Taking $M' = Me^{\int f(x)\,dx}$, $N' = Ne^{\int f(x)\,dx}$, we see that $M'\,dx + N'\,dy = 0$ is exact, since $\frac{\partial M'}{\partial y} = \frac{\partial N'}{\partial x}$.

That is, $(M\,dx + N\,dy)e^{\int f(x)\,dx} = 0$ satisfies the condition for exactness.

Note 2.15.1. *It is easy to see (proceeding backwards) that if the condition for an exact equation is satisfied by* $(M\,dx + N\,dy)e^{\int f(x)\,dx} = 0$, *then* $\left(\frac{\partial M}{\partial y} - \frac{\partial N}{\partial x}\right)/N = f(x)$.

Example 2.15.1. *If in the differential equation* $M\,dx + N\,dy = 0$, $\frac{1}{N}\left(\frac{\partial M}{\partial y} - \frac{\partial N}{\partial x}\right) = f(x)$, *show that* $\mu = e^{\int f(x)\,dx}$ *is an IF of the equation. Hence, show that* $e^{\int P\,dx}$ *is an IF of* $\frac{dy}{dx} + P(x)y = Q(x)$.

Solution: The first part is *rule* II of art. 2.15.

From the second part, we have $dy + P(x)y\,dx = Q(x)\,dx$ or, $1.dy + \{P(x)y - Q(x)\}\,dx = 0$.

Here $M = P(x)\cdot y - Q(x)$ and $N = 1$.

Also

$$\frac{1}{N}\left(\frac{\partial M}{\partial y} - \frac{\partial N}{\partial x}\right) = 1\{P(x) - 0\} = P(x).$$

Hence, $e^{\int P\,dx}$ is an IF of the given equation.

Example 2.15.2. *Solve:* $(x^3 + xy^4)\,dx + 2y^3\,dy = 0$.

Solution: Writing $M = x^3 + xy^4$, $N = 2y^3$, we see that $\frac{\partial M}{\partial y} = 4xy^3$, $\frac{\partial N}{\partial x} = 0$.

Hence, the equation is not exact.

However,

$$\frac{1}{N}\left(\frac{\partial M}{\partial y} - \frac{\partial N}{\partial x}\right) = \frac{1}{2y^3}(4xy^3 - 0) = 2x = f(x),$$

and hence $e^{\int f(x)\,dx} = e^{\int 2x\,dx} = e^{x^2}$ is an IF.

Multiplying by the IF the equation becomes

$$e^{x^2}(x^3 + xy^4)\,dx + e^{x^2}2y^3\,dy = 0$$

$$\text{or, } 2x^3e^{x^2}\,dx + \left\{y^4e^{x^2}\cdot 2x\,dx + e^{x^2}\cdot 4y^3\,dy\right\} = 0 \text{ [Multiplying by 2]}$$

$$\text{or, } \int 2x^3e^{x^2}\,dx + \int d\left(y^4e^{x^2}\right) = 0$$

$$\text{or, } e^{x^2}(x^2 - 1) + e^{x^2}y^4 = c, \text{ c being an arbitrary constant}$$

$$\text{or, } e^{x^2}(x^2 - 1 + y^4) = c \text{ is the required solution.}$$

[To evaluate $\int 2x^3e^{x^2}\,dx$, put $x^2 = z$ and use integration by parts.]

Rule III. If $\frac{1}{M}\left(\frac{\partial M}{\partial y} - \frac{\partial N}{\partial x}\right)$ is a function of y alone, say $\phi(y)$, then $e^{-\int \phi(y)\,dy}$ is an integrating factor of $M\,dx + N\,dy = 0$.

Proof. Exactly similar to Rule II. In fact, $(M\,dx + N\,dy)e^{-\int \phi(y)\,dy} = 0$ is exact if and only if

$$\frac{\partial}{\partial y}\left\{Me^{-\int \phi(y)\,dy}\right\} = \frac{\partial}{\partial x}\left\{Ne^{-\int \phi(y)\,dy}\right\}$$

$$\text{i.e. iff} \quad \frac{\partial M}{\partial y}e^{-\int \phi(y)\,dy} - M\phi(y)e^{-\int \phi(y)\,dy} = \frac{\partial N}{\partial x}e^{-\int \phi(y)\,dy}$$

$$\text{i.e. iff} \quad \frac{\dfrac{\partial M}{\partial y} - \dfrac{\partial N}{\partial x}}{M} = \phi(y).$$

Example 2.15.3. *Solve:* $(2xy^4e^y + 2xy^3 + y)\;dx + (x^2y^4e^y - x^2y^2 - 3x)\;dy = 0.$

Solution: Here

$$\frac{\partial M}{\partial y} = 8xy^3e^y + 2xy^4e^y + 6xy^2 + 1, \quad \frac{\partial N}{\partial x} = 2xy^4e^y - 2xy^2 - 3.$$

∴ the equation is not exact.

However,

$$\frac{\partial M}{\partial y} - \frac{\partial N}{\partial x} = 8xy^3e^y + 8xy^2 + 4$$

$$\text{and} \quad \frac{1}{M}\left(\frac{\partial M}{\partial y} - \frac{\partial N}{\partial x}\right) = \frac{8xy^3e^y + 8xy^2 + 4}{y\,(2xy^3e^y + 2xy^2 + 1)} = \frac{4}{y} = \phi(y).$$

$$\therefore \quad e^{-\int \frac{4}{y}\,dy} = e^{-4\ln y} = e^{\ln y^{-4}} = \frac{1}{y^4} \text{ is an IF.}$$

Multiplying by the IF the equation becomes

$$\left(2xe^y + \frac{2x}{y} + \frac{1}{y^3}\right)dx + \left(x^2e^y - \frac{x^2}{y^2} - 3\frac{x}{y^4}\right)dy = 0.$$

We may now use the four steps indicated in article 2.10 to solve an exact equation. Thus,

1. $\displaystyle\int\left(2xe^y + 2\frac{x}{y} + \frac{1}{y^3}\right)dx = x^2e^y + \frac{x^2}{y} + \frac{x}{y^3}$ (y is supposed to be constant).
2. $\displaystyle\int\left(x^2e^y - \frac{x^2}{y^2} - \frac{3x}{y^4}\right)dy = x^2e^y + \frac{x^2}{y} + \frac{x}{y^3}$ (x is supposed to be constant).
3. $u = x^2e^y + \dfrac{x^2}{y} + \dfrac{x}{y^3}$ (All the terms of 2 are already obtained in 1).
4. The required solution is, therefore, $x^2e^y + \dfrac{x^2}{y} + \dfrac{x}{y^3} = c$, c being an arbitrary constant (Fig. 2.30).

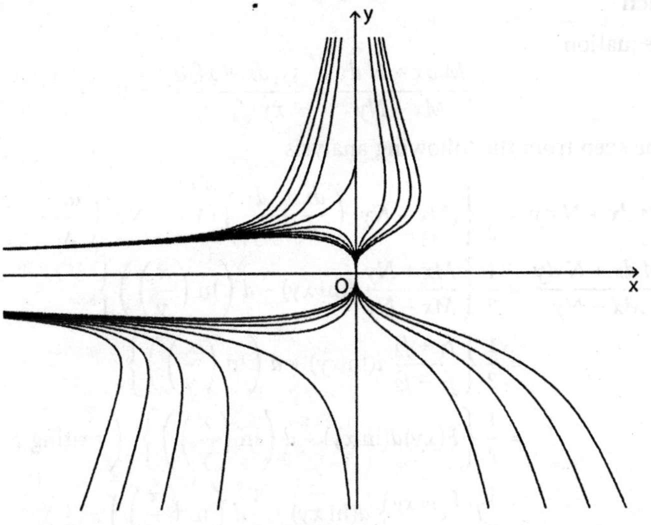

Fig. 2.30

Rule IV. If $Mx - Ny \neq 0$, then $\frac{1}{Mx-Ny}$ is an integrating factor of the equation $M\,dx + N\,dy = 0$, where M and N (functions of x and y) can be written in the form $M = yf_1(xy)$, $N = xf_2(xy)$.

Proof. Here $Mx - Ny = xy\{f_1(xy) - f_2(xy)\} \neq 0$.

$$\therefore \quad \frac{M\,dx + N\,dy}{Mx - Ny} = \frac{yf_1(xy)\,dx + xf_2(xy)\,dy}{xy\{f_1(xy) - f_2(xy)\}} = \frac{yf_1\,dx + xf_2\,dy}{xy(f_1 - f_2)}, \text{ say,}$$

is a perfect differential, if and only if

$$\frac{\partial}{\partial y}\left\{\frac{f_1}{x(f_1 - f_2)}\right\} = \frac{x(f_1 - f_2)\dfrac{\partial f_1}{\partial y} - f_1 x\left(\dfrac{\partial f_1}{\partial y} - \dfrac{\partial f_2}{\partial y}\right)}{x^2(f_1 - f_2)^2} = \frac{-f_2\dfrac{\partial f_1}{\partial y} + f_1\dfrac{\partial f_2}{\partial y}}{x(f_1 - f_2)^2}.$$

Similarly,

$$\frac{\partial}{\partial x}\left\{\frac{f_2}{y(f_1 - f_2)}\right\} = \frac{f_1\dfrac{\partial f_2}{\partial x} - f_2\dfrac{\partial f_1}{\partial x}}{y(f_1 - f_2)^2}$$

and
$$\frac{\partial}{\partial y}\left\{\frac{f_1}{x(f_1 - f_2)}\right\} - \frac{\partial}{\partial x}\left\{\frac{f_2}{y(f_1 - f_2)}\right\} = \frac{f_2\left(-y\dfrac{\partial f_1}{\partial y} + x\dfrac{\partial f_1}{\partial x}\right) + f_1\left(y\dfrac{\partial f_2}{\partial y} - x\dfrac{\partial f_2}{\partial x}\right)}{xy(f_1 - f_2)^2}.$$

This is identically zero, since
$$y\frac{\partial f(xy)}{\partial y} = x\frac{\partial f(xy)}{\partial x}.$$

Hence, the condition for exactness is satisfied.

Special case

If $Mx - Ny = 0$ identically, then $M/N = y/x$ and the equation reduces to $x\,dy + y\,dx = 0$ with solution $xy = c$.

Another approach

The fact that the equation

$$\frac{M\,dx + N\,dy}{Mx - Ny} = \frac{yf_1\,dx + xf_2\,dy}{xy\,(f_1 - f_2)} = 0$$

is exact can also be seen from the following analysis:

$$M\,dx + N\,dy = \frac{1}{2}\left\{(Mx + Ny)\left(\frac{dx}{x} + \frac{dy}{y}\right)(Mx - Ny)\left(\frac{dx}{x} - \frac{dy}{y}\right)\right\}$$

$$\therefore \quad \frac{M\,dx + N\,dy}{Mx - Ny} = \frac{1}{2}\left\{\frac{Mx + Ny}{Mx - Ny}d(\ln xy) + d\left(\ln\left(\frac{x}{y}\right)\right)\right\}$$

$$= \frac{1}{2}\left\{\frac{f_1 + f_2}{f_1 - f_2}d(\ln xy) + d\left(\ln\left(\frac{x}{y}\right)\right)\right\}$$

$$= \frac{1}{2}\left\{F(xy)d(\ln xy) + d\left(\ln\left(\frac{x}{y}\right)\right)\right\} \quad \left(\text{writing } F = \frac{f_1 + f_2}{f_1 - f_2}\right)$$

$$= \frac{1}{2}F\left\{e^{\ln xy}\right\} d(\ln xy) + \frac{1}{2}d\left(\ln\left(\frac{x}{y}\right)\right).$$

The students may try to verify in a similar way the truth of Theorem 2.13.2 (article 2.13.1).

Example 2.15.4. *Solve:* $(x^2y^2 + xy + 1)\,y\,dx - (x^2y^2 - xy + 1)\,x\,dy = 0.$

Solution: Here

$$\frac{M\,dx + N\,dy}{Mx - Ny} = \frac{1}{2}\left\{\frac{Mx + Ny}{Mx - Ny}d(\ln xy) + d\left(\ln\left(\frac{x}{y}\right)\right)\right\}$$

$$= \frac{1}{2}\left\{\frac{(x^2y^2 + xy + 1 - x^2y^2 + xy - 1)}{(x^2y^2 + xy + 1 + x^2y^2 - xy + 1)}d(\ln xy) + d\left(\ln\left(\frac{x}{y}\right)\right)\right\}$$

$$= \frac{1}{2}\left\{\frac{xy}{x^2y^2 + 1}\cdot\frac{1}{xy}d(xy) + d\left(\ln\left(\frac{x}{y}\right)\right)\right\}$$

$$= \frac{1}{2}\left\{\frac{d(xy)}{x^2y^2 + 1} + d\left(\ln\left(\frac{x}{y}\right)\right)\right\}.$$

The required solution is, therefore,

$$\tan^{-1} xy + \ln\left(\frac{x}{y}\right) = c,$$

c being an arbitrary constant.

Rule V.

1. The equation

$$py\,dx + qx\,dy = 0 \;(p, q \text{ are constants}) \tag{2.15.1}$$

has an integrating factor $x^{p-1}y^{q-1}$.

For, $d\,(x^p y^q) = px^{p-1}y^q dx + qx^p y^{q-1}\,dy = x^{p-1}y^{q-1}(py\,dx + qx\,dy)$.

Clearly, then $x^{p-1}y^{q-1}\phi\,(x^p y^q)$ [where ϕ is any arbitrary function of $x^p y^q$] is also an integrating factor of (2.15.1).

2. The equation

$$x^m y^n(ry\,dx + sx\,dy) = 0, \;\; (m, n, r, s \text{ are constants}) \tag{2.15.2}$$

has an integrating factor $x^{r-1-m}y^{s-1-n}$.

For, $d(x^r y^s) = x^{r-1-m} y^{s-1-n} [x^m y^n (ry\,dx + sx\,dy)]$.

More generally, $x^{r-1-m} y^{s-1-n} F(x^r y^s)$ [where F is any arbitrary function of $x^r y^s$] is an integrating factor of (2.15.2).

3. $x^h y^k$ **is an integrating factor of**

$$py\,dx + qx\,dy + x^m y^n (ry\,dx + sx\,dy) = 0, \tag{2.15.3}$$

if $\frac{h+1}{p} = \frac{k+1}{q}$ and $\frac{h+m+1}{r} = \frac{k+n+1}{s}$.

By 2.15.1, $x^{p-1} y^{q-1} \phi(x^p y^q)$ is an integrating factor of the expression $py\,dx + qx\,dy$, where ϕ is arbitrary.

By 2.15.2, $x^{r-m-1} y^{s-n-1} F(x^r y^s)$ is an integrating factor of the expression $x^m y^n (ry\,dx + sx\,dy)$, where F is arbitrary.

If, therefore, ϕ and F be so determined that $x^{p-1} y^{q-1} \phi(x^p y^q) = x^{r-m-1} y^{s-n-1} F(x^r y^s)$, then an integrating factor for (2.15.3) will be obtained.

Let $\phi(t) = t^\rho$, $F(t) = t^\mu$. Then $x^h y^k$ will be an IF if $x^{p-1} y^{q-1} (x^p y^q)^\rho = x^{r-m-1} y^{s-n-1} (x^r y^s)^\mu$; that is, if

$$\left.\begin{aligned} h &= p(\rho+1) - 1 = r(\mu+1) - m - 1 \\ \text{and} \quad k &= q(\rho+1) - 1 = s(\mu+1) - n - 1 \end{aligned}\right\}$$

that is, if

$$\left.\begin{aligned} \frac{h+1}{p} &= \frac{k+1}{q} = \rho + 1 \\ \text{and} \quad \frac{h+m+1}{r} &= \frac{k+n+1}{s} = \mu + 1 \end{aligned}\right\}. \tag{2.15.4}$$

Important note

The eq. (2.15.4) determines h and k, provided $sp - qr \neq 0$.

But suppose, $sp - qr = 0$, i.e. suppose $s/q = r/p = \alpha$ (say).

Then the eq. (2.15.3) will reduce to $(1 + \alpha x^m y^n)(py\,dx + qx\,dy) = 0$, an integrating factor of which is $\frac{1}{1+\alpha x^m y^n} \cdot x^{p-1} y^{q-1}$.

4. The equation

$$x^\alpha y^\beta (my\,dx + nx\,dy) + x^{\alpha_1} y^{\beta_1} (m_1 y\,dx + n_1 x\,dy) = 0 \tag{2.15.5}$$

[where $\alpha, \beta, \alpha_1, \beta_1, m, n, m_1, n_1$ are constants and $mn_1 - m_1 n \neq 0$] has an integrating factor of the form $x^h y^k$, where h and k are constants given by

$$\left.\begin{aligned} \frac{h_1 + 1 + \alpha}{m} &= \frac{k + 1 + \beta}{n} \\ \frac{h + 1 + \alpha_1}{m_1} &= \frac{k + 1 + \beta_1}{n_1} \end{aligned}\right\}. \tag{2.15.6}$$

In other words, for finding h, k, we are to solve for h, k from the following two equations

$$\left.\begin{aligned} nh - mk &= (m - n) + (m\beta - n\alpha) \\ \text{and} \quad n_1 h - m_1 k &= (m_1 - n_1) + (m_1 \beta_1 - n_1 \alpha_1) \end{aligned}\right\}. \tag{2.15.7}$$

Now h, k will have determined values, if

$$\begin{vmatrix} n & -m \\ n_1 & -m_1 \end{vmatrix} = mn_1 - m_1 n \neq 0.$$

Justification

If $x^h y^k$ be an IF of (2.15.5), then

$$\left(mx^{h+\alpha}y^{k+\beta+1} + m_1x^{h+\alpha_1}y^{k+\beta_1+1}\right)dx + \left(nx^{h+\alpha+1}y^{\beta+k} + n_1x^{h+\alpha_1+1}y^{\beta_1+k}\right)dy = 0$$

would be exact and hence

$$\frac{\partial}{\partial y}\left(mx^{h+\alpha}y^{k+\beta+1} + m_1x^{h+\alpha_1}y^{k+\beta_1+1}\right) = \frac{\partial}{\partial y}\left(nx^{h+\alpha+1}y^{\beta+k} + n_1x^{h+\alpha_1+1}y^{\beta_1+k}\right),$$

i.e. $mx^{h+\alpha}(k+\beta+1)y^{k+\beta} + m_1x^{h+\alpha_1}(k+\beta_1+1)y^{k+\beta_1} = n(h+\alpha+1)x^{h+\alpha}y^{k+\beta} + n_1(h+\alpha_1+1)x^{h+\alpha_1}y^{k+\beta_1}$ should hold $\forall\, x, y$.

Hence, comparing like coefficients, we get

$$\left.\begin{aligned} m(k+\beta+1) &= n(h+\alpha+1) \\ \text{or,}\quad \frac{h+\alpha+1}{m} &= \frac{k+\beta+1}{n} \\ \text{and}\quad m_1(k+\beta_1+1) &= n_1(h+\alpha_1+1) \\ \text{or,}\quad \frac{h+\alpha_1+1}{m_1} &= \frac{k+\beta_1+1}{n_1} \end{aligned}\right\}.$$

Example 2.15.5. *Solve:* $3y\,dx - 2x\,dy + x^2y^{-1}(10y\,dx - 6x\,dy) = 0.$

Solution: If x^hy^k is an IF, then $x^hy^k(3y\,dx - 2x\,dy) + x^hy^kx^2y^{-1}(10y\,dx - 6x\,dy) = 0$ must be exact and hence

$$\frac{\partial}{\partial y}\left\{3y^{k+1}x^h + 10x^{h+2}y^k\right\} = \frac{\partial}{\partial x}\left\{-2x^{h+1}y^k - 6x^{h+3}y^{k-1}\right\}$$

$$\text{or,}\quad 3(k+1)y^kx^h + 10kx^{h+2}y^{k-1} = -2(h+1)x^hy^k - 6(h+3)x^{h+2}y^{k-1}\ (\forall\, x, y).$$

$\therefore$ comparing coefficient of x^hy^k, we get $3(k+1) = -2(h+1)$, or, $2h+3k = -5$ and comparing coefficient of $x^{h+2}y^{k+1}$, we get $10k = -6(h+3)$ or, $6h + 10k = -18$.

Solving for h, k, we get $h = 2, k = -3$.

That is, x^2y^{-3} is an IF of the given equation. Multiplying by x^2y^{-3}, we get the first expression

$$\begin{aligned} &= x^2y^{-3}(3y\,dx - 2x\,dy) \\ &= 3x^2y^{-2}\,dx - 2x^3y^{-3}\,dy = d\left(x^3y^{-2}\right) \end{aligned}$$

and the second expression

$$\begin{aligned} &= x^2y^{-3}\left\{x^2y^{-1}(10y\,dx - 6x\,dy)\right\} \\ &= 10x^4y^{-3}\,dx - 6x^5y^{-4}\,dy \\ &= 2\left\{5x^4y^{-3}\,dx - 3y^{-4}x^5\,dy\right\} \\ &= 2d\left(x^5y^{-3}\right). \end{aligned}$$

$\therefore$ the required solution is $x^3y^{-2} + 2x^5y^{-3} = c$, c being an arbitrary constant.

Note 2.15.2. *If the formulae for h, k are obtained, then we can use them as in the following example 2.15.6.*

Example 2.15.6. *Solve:* $\left(y^3 - 2yx^2\right)dx + \left(2xy^2 - x^3\right)dy = 0.$

Solution: Rearranging the terms we write the equation in the form $y^2(y\,dx+2x\,dy)-x^2(2y\,dx+x\,dy)=0$.

This is of the form (2.15.5). Here $\alpha=0$, $\beta=2$, $m=1$, $n=2$; $\alpha_1=2$, $\beta_1=0$, $m_1=-2$, $n_1=-1$. Hence, the relations (2.15.6) give $\frac{h+1}{1}=\frac{k+3}{2}$ and $\frac{h+3}{-2}=\frac{k+1}{-1}$.

Solving for h and k, we get $h=k=1$.

$\therefore$ xy is an integrating factor of the given equation.

Multiplying by xy, we get the first expression $= xy^4dx + 2x^2y^3dy = \frac{1}{2}\left(2xy^4dx + 4x^2y^3dy\right) = \frac{1}{2}d\left(x^2y^4\right)$ and the second expression $= -2x^3y^2dx - x^4y\,dy = -\frac{1}{2}\left(4x^3y^2dx + 2x^4y\,dy\right) = -\frac{1}{2}d\left(x^4y^2\right)$.

$\therefore$ the required solution is $\frac{1}{2}x^2y^4 - \frac{1}{2}x^4y^2 = \frac{1}{2}c$ or, $x^2y^4 - x^4y^2 = c$, c being an arbitrary constant.

Examples II(H)

Solve the following eqs (1–20):

1. $y\left(xy + 2x^2y^2\right)\,dx + x\left(xy - x^2y^2\right)\,dy = 0$.
2. $y\left(x^2y^2 + 2\right)\,dx + x\left(2 - 2x^2y^2\right)\,dy = 0$.
3. $y(2xy + 1)\,dx + x\left(1 + 2xy - x^3y^3\right)\,dy = 0$.
4. $\left(x^3y^3 + x^2y^2 + xy + 1\right)y\,dx + x\left(x^3y^3 - x^2y^2 - xy + 1\right)\,dy = 0$.
5. $\left(x^2 + y^2 + 2x\right)\,dx + 2y\,dy = 0$.
6. $\left(x^2 + y^2\right)\,dx - 2xy\,dy = 0$.
7. $\left(3x^2y^4 + 2xy\right)\,dx + \left(2x^3y^3 - x^2\right)\,dy = 0$.
8. $\left(y^4 + 2y\right)\,dx + \left(xy^3 + 2y^4 - 4x\right)\,dy = 0$.
9. $\left(x^2 + y^2 + x\right)\,dx + xy\,dy = 0$.
10. $\left(2x^3y^2 + 4x^2y + 2xy^2 + xy^4 + 2y\right)\,dx + 2\left(y^3 + x^2y + x\right)\,dy = 0$.
11. $\left(x^2 + y^2 + 1\right)\,dx - 2xy\,dy = 0$.
12. $2xy\,dx + \left(y^2 - x^2\right)\,dy = 0$.
13. $\left(2x^2y - 3y^4\right)\,dx + \left(3x^3 + 2xy^3\right)\,dy = 0$.
14. $\left(y^2 + 2x^2y\right)\,dx + \left(2x^3 - xy\right)\,dy = 0$.
15. $\left(2x^2y^2 + y\right)\,dx - \left(x^3y - 3x\right)\,dy = 0$.
16. $x(4y\,dx + 2x\,dy) + y^3(3y\,dx + 5x\,dy) = 0$.
17. $(8y\,dx + 8x\,dy) + x^2y^3(4y\,dx + 5x\,dy) = 0$.
18. $x^3y^3(2y\,dx + x\,dy) - (5y\,dx + 7x\,dy) = 0$.

Answers

1. $2\ln x - \ln y = 1/xy + c$.
2. $x = cy^2e^{1/x^2y^2}$.
3. $y = ce^{-(3xy+1)/(3x^3y^3)}$.
4. $x^2y^2 - 2xy\ln cy = 1$.
5. $e^x\left(x^2 + y^2\right) = c$.
6. $x^2 - y^2 = cx$.
7. $x^3y^3 + x^2 = cy$.
8. $xy + y^2 + (2x)/y^2 = c$.
9. $3x^4 + 4x^3 + 6x^2y^2 = c$.
10. $\left(2x^2y^2 + 4xy + y^4\right)e^{x^2} = c$.
11. $x^2 - y^2 - 1 = cx$.
12. $x^2 + y^2 = cy$.
13. $5x^{-\frac{36}{13}}y^{\frac{24}{13}} - 12x^{-\frac{10}{13}}y^{-\frac{15}{13}} = c$.
14. $6\sqrt{(xy)} - x^{-3/2}y^{3/2} = c$.
15. $4x^2y = 5 + cx^{4/7}y^{12/7}$.
16. $x^4y^2 + x^3y^5 = c$.
17. $4x^2y^2 + x^4y^5 = c$.
18. $x^3y^3 + 2 = cx^{5/3}y^{7/3}$.

Solutions by special devices: Miscellaneous exercises:

Example 2.15.7. *Solve:* $dy/dx = (y + 3x)^2$.

Solution: Put $y + 3x = z$, whereby $dy/dx + 3 = dz/dx$ and hence

$$\frac{dz}{dx} = z^2 + 3, \quad \text{or,} \quad dx = \frac{dz}{z^2 + 3}.$$

Integrating,

$$x + c = \frac{1}{\sqrt{3}} \tan^{-1} \frac{z}{\sqrt{3}} = \frac{1}{\sqrt{3}} \tan^{-1} \frac{y + 3x}{\sqrt{3}}.$$

Example 2.15.8. *Solve:* $\left(xy^2 - e^{1/x^3}\right) dx - x^2 y\, dy = 0.$

Solution: $(xy^2 - e^{1/x^3})dx - x^2 y\, dy = 0.$

$$\therefore\ xy(x\,dy - y\,dx) + e^{1/x^3} dx = 0, \text{ or, } \frac{y}{x} d\left(\frac{y}{x}\right) + \frac{1}{x^4} e^{1/x^3}\, dx = 0. \qquad (2.15.8)$$

Since

$$\int \frac{1}{x^4} e^{1/x^3} dx = -\frac{1}{3} \int e^z dz \ \left(\text{putting } \frac{1}{x^3} = z\right) = -\frac{1}{3} e^z = -\frac{1}{3} e^{1/x^3},$$

we have after integration of (2.15.8)

$$\frac{1}{2}\left(\frac{y}{x}\right)^2 - \frac{1}{3} e^{1/x^3} = c, \ c \text{ being an arbitrary constant}$$

$$\text{or,}\quad 3y^2 - 2x^2 e^{1/x^3} = 6cx^2$$

as the required solution (Fig. 2.31).

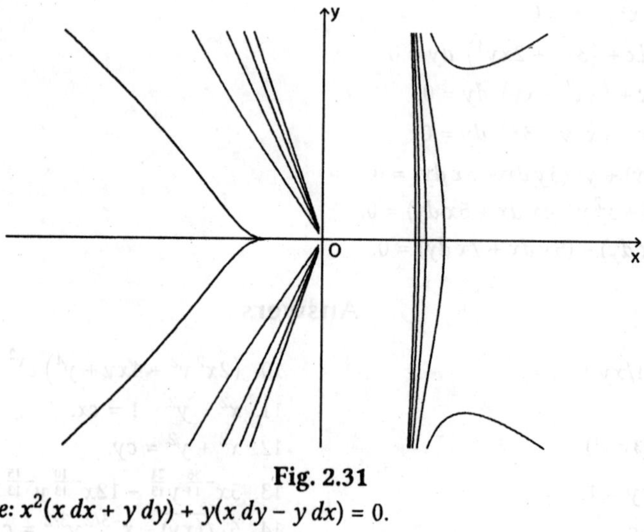

Fig. 2.31

Example 2.15.9. *Solve:* $x^2(x\,dx + y\,dy) + y(x\,dy - y\,dx) = 0$.

Solution: The equation is

$$x^2 \frac{1}{2} d\left(x^2 + y^2\right) + yx^2 d\left(\frac{y}{x}\right) = 0.$$

This suggests that we are to substitute

$$x^2 + y^2 = r^2, \ \frac{y}{x} = \tan\theta, \text{ or, } x = r\cos\theta,\ y = r\sin\theta.$$

Thus, $dx = \cos\theta dr - r\sin\theta d\theta$ and $dy = \sin\theta dr + r\cos\theta d\theta$.

The equation, then, reduces to

$$x^2 \cdot \tfrac{1}{2} d\left(r^2\right) + yx^2 d(\tan\theta) = 0$$

$$\text{or,} \quad r^2\cos^2\theta \cdot r\,dr + r^3\sin\theta\cos^2\theta\sec^2\theta\,d\theta = 0$$

$$\text{or,} \quad dr + \sec\theta\tan\theta\,d\theta = 0.$$

Integrating, $r + \sec\theta = c$, or, $r + r/x = c$, i.e. $r^2(x+1)^2 = c^2x^2$, or, $\left(x^2 + y^2\right)(x+1)^2 = c^2x^2$, which is the required general solution.

Example 2.15.10. *Solve:* $\left(4x^2y - 6\right) dx + x^3 dy = 0$.

Solution: This is same as $(x\,dy + y\,dx) + 3y\,dx = \frac{6}{x^2}\,dx$.

The first term suggests that we should put $xy = z$. Then

$$d(xy) + 3\frac{z}{x}\,dx = \frac{6}{x^2}\,dx, \quad \text{or,} \quad \frac{dz}{dx} + \frac{3}{x}z = \frac{6}{x^2}.$$

This is a linear equation and IF $= e^{\int (3/x)\,dx} = e^{3\ln x} = x^3$.

∴ general solution is

$$zx^3 = \int 6x\,dx, \quad \text{or,} \quad x^4y = 3x^2 + c,$$

c being an arbitrary constant (Fig. 2.32).

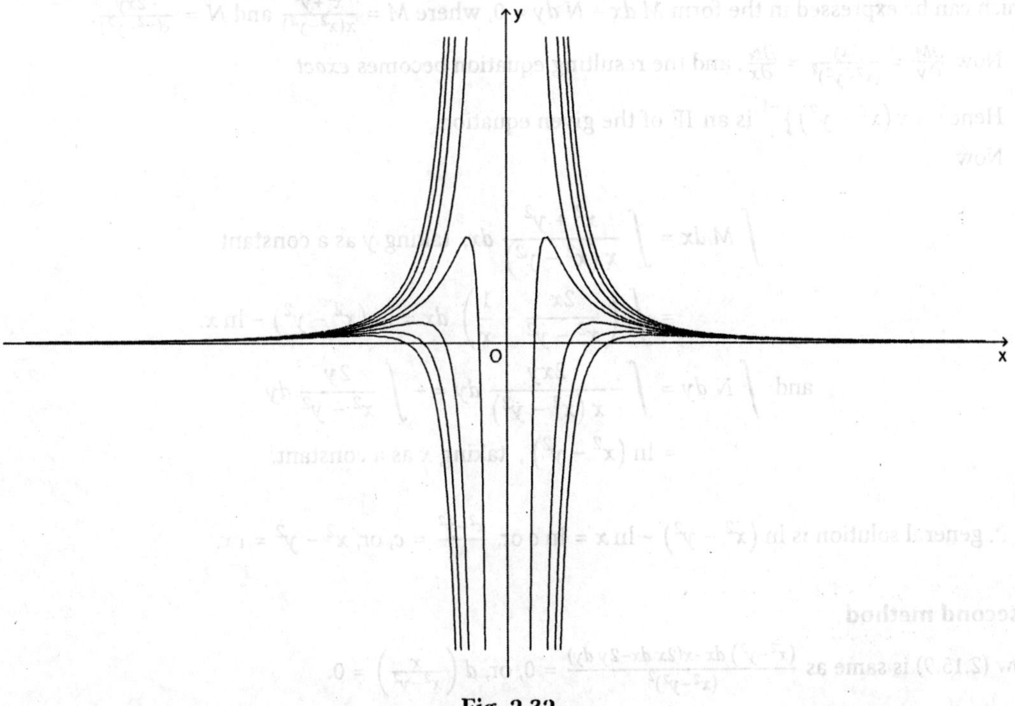

Fig. 2.32

Example 2.15.11. *Show that a constant a can be found so that* $(x+y)^a$ *is an integrating factor of* $\left(4x^2 + 2xy + 6y\right) dx + \left(2x^2 + 9y + 3x\right) dy = 0$ *and hence solve the equation.*

Solution: Multiplying by the IF $(x+y)^a$, we have $(x+y)^a(4x^2 + 2xy + 6y)\,dx + (x+y)^a\left(2x^2 + 9y + 3x\right) dy = 0$.

The equation is exact, if

$$\frac{\partial M}{\partial y} = \frac{\partial N}{\partial x}$$

or, $a(x+y)^{a-1}\left(4x^2+2xy+6y\right)+(x+y)^a(2x+6) = a(x+y)^{a-1}\left(2x^2+9y+3x\right)+(x+y)^a(4x+3).$

Simplifying, we have, $a = 1$.

Now the equation becomes an exact equation of the form:

$$(x+y)\left(4x^2+2xy+6y\right)\,dx+(x+y)\left(2x^2+9y+3x\right)\,dy = 0$$

$$\text{or,}\quad \left(4x^3+6x^2y+2xy^2+6xy+6y^2\right)\,dx+\left(2x^3+2x^2y+3x^2+12xy+9y^2\right)\,dy = 0$$

whose solution is: $x^4+2x^3y+x^2y^2+3x^2y+6xy^2+3y^3 = c$, c being an arbitrary constant.

Example 2.15.12. *Show that* $\left\{x\left(x^2-y^2\right)\right\}^{-1}$ *is an integrating factor of the differential equation* $\left(x^2+y^2\right)\,dx-2xy\,dy = 0$ *and hence solve the equation.*

Solution: Multiplying by the IF $\dfrac{1}{x\left(x^2-y^2\right)}(\neq 0)$, we have

$$\frac{\left(x^2+y^2\right)\,dx-2xy\,dy}{x\left(x^2-y^2\right)} = 0 \tag{2.15.9}$$

which can be expressed in the form $M\,dx+N\,dy = 0$, where $M = \frac{x^2+y^2}{x(x^2-y^2)}$ and $N = \frac{-2xy}{x(x^2-y^2)}$.

Now $\frac{\partial M}{\partial y} = \frac{4xy}{(x^2-y^2)^2} = \frac{\partial N}{\partial x}$, and the resulting equation becomes *exact*.

Hence, $\left\{x\left(x^2-y^2\right)\right\}^{-1}$ is an IF of the given equation.

Now

$$\int M\,dx = \int \frac{x^2+y^2}{x\left(x^2-y^2\right)}\,dx, \text{ taking } y \text{ as a constant}$$

$$= \int\left(\frac{2x}{x^2-y^2}-\frac{1}{x}\right)dx = \ln\left(x^2-y^2\right)-\ln x.$$

$$\text{and}\quad \int N\,dy = \int -\frac{2xy}{x\left(x^2-y^2\right)}\,dy = -\int\frac{2y}{x^2-y^2}\,dy$$

$$= \ln\left(x^2-y^2\right), \text{ taking } x \text{ as a constant.}$$

$\therefore$ general solution is $\ln\left(x^2-y^2\right)-\ln x = \ln c$ or, $\frac{x^2-y^2}{x} = c$, or, $x^2-y^2 = cx$.

A second method

Now (2.15.9) is same as $\frac{(x^2-y^2)\,dx-x(2x\,dx-2y\,dy)}{(x^2-y^2)^2} = 0$, or, $d\left(\frac{x}{x^2-y^2}\right) = 0$.

Integrating, we have $\frac{x}{x^2-y^2} = \frac{1}{c}$, or, $x^2-y^2 = cx$ as the required solution.

Example 2.15.13. *Find an integrating factor of the differential equation* $\left(y^2+2x^2y\right)\,dx+\left(2x^3-xy\right)\,dy = 0$ *and hence solve it.*

Solution:

First Method: One can use Rule V; namely, $x^h y^k$ is an IF of $x^\alpha y^\beta(my\,dx+nx\,dy)+x^{\alpha_1}y^{\beta_1}(m_1 y\,dx+n_1 x\,dy)$ if h, k satisfy

$$\left.\begin{aligned}\frac{h+1+\alpha}{m}&=\frac{k+1+\beta}{n}\\ \frac{h+1+\alpha_1}{m_1}&=\frac{k+1+\beta_1}{n_1}\end{aligned}\right\}(mn_1-m_1 n\neq 0).$$

The given equation can be written as $y(y\,dx-x\,dy)+x^2(2y\,dx+2x\,dy)=0$.
Here $\alpha=0, \beta=1, m=1, n=-1, \alpha_1=2, \beta_1=0, m_1=2, n_1=2$.

$$\left.\begin{aligned}\therefore\quad \frac{h+1}{1}&=\frac{k+2}{-1}, \quad\text{or,}\quad k+h+3=0\\ \text{or,}\quad \frac{h+3}{2}&=\frac{k+1}{2}, \quad\text{or,}\quad k-h-2=0\end{aligned}\right\}.$$

$$\therefore\ h=-\frac{5}{2},\ k=-\frac{1}{2}$$

$\therefore$ the equation when multiplied by the IF $x^{-5/2}\,y^{-1/2}$ becomes

$$x^{-5/2}y^{1/2}(y\,dx-x\,dy)+x^{-1/2}y^{-1/2}(2y\,dx+2x\,dy)=0$$

$$\text{or,}\quad \left\{x^{-5/2}y^{3/2}\,dx-x^{-3/2}y^{1/2}\,dy\right\}+\left\{2x^{-1/2}y^{1/2}\,dx+2x^{1/2}y^{-1/2}\,dy\right\}=0$$

$$\text{or,}\quad -\tfrac{2}{3}d\left(x^{-3/2}y^{3/2}\right)+4d\left(x^{1/2}y^{1/2}\right)=0.$$

On integration we get $-\frac{1}{3}x^{-3/2}y^{3/2}+2x^{1/2}y^{1/2}=$ constant, i.e. $6\sqrt{xy}-x^{-3/2}y^{3/2}=$ **C (reqd. solution).**

Second Method: (*Without using the formula for h and k*).

Multiplying the given equation by $x^h y^k$ we get $x^h y^k\left(y^2+2x^2y\right)dx+x^h y^k\left(2x^3-xy\right)dy=0$, i.e. of the form $M\,dx+N\,dy=0$.

If $x^h y^k$ is an IF, then $\frac{\partial M}{\partial y}=\frac{\partial N}{\partial x}$, i.e. $x^h(k+2)y^{k+1}+2x^{h+2}(k+1)y^k=2(h+3)x^{h+2}y^k-y^{k+1}(h+1)x^h$.

Comparing the coefficients $k+2=-(h+1)$ and $2(k+1)=2(h+3)$, i.e. $k+h+3=0$ and $k-h-2=0$.

This again gives $h=-\frac{5}{2}$, $k=-\frac{1}{2}$ and now proceed as above.

Example 2.15.14. *Prove that* $\frac{1}{(x+y+1)^4}$ *is an integrating factor of* $\left(2xy-y^2-y\right)dx+\left(2xy-x^2-x\right)dy=0$ *and hence solve it.*

Solution: Multiplying by $\frac{1}{(x+y+1)^4}$, we get

$$\frac{2xy-y^2-y}{(x+y+1)^4}\,dx+\frac{2xy-x^2-x}{(x+y+1)^4}\,dy=0$$

i.e. of the form: $M\,dx+N\,dy=0$.

If $(x+y+1)^{-4}$ be an IF, then the equation is exact and so $\partial M/\partial y=\partial N/\partial x$, where

$$\frac{\partial M}{\partial y}=\frac{\partial}{\partial y}\left\{\frac{2xy-y^2-y}{(x+y+1)^4}\right\}=\frac{(2x-2y-1)(x+y+1)^4-4(x+y+1)^3\left(2xy-y^2-y\right)}{(x+y+1)^8}$$

$$=\frac{\left\{(2x-2y-1)(x+y+1)-4\left(2xy-y^2-y\right)\right\}}{(x+y+1)^5}=\left(2x^2+2y^2-8xy+x+y-1\right)/(x+y+1)^5$$

$$\frac{\partial N}{\partial x}=\frac{\partial}{\partial x}\left\{\frac{2xy-x^2-x}{(x+y+1)^4}\right\}=\frac{2x^2+2y^2-8xy+x+y-1}{(x+y+1)^5}.$$

Thus, $1/(x+y+1)^4$ is an IF.

$$\therefore \int M\,dx = \int \frac{2xy - y^2 - y}{(x+y+1)^4}\,dx \text{ (as if } y \text{ is constant)}$$

$$= 2y\int \frac{x}{(x+y+1)^4}\,dx - y(y+1)\int \frac{dx}{(x+y+1)^4}$$

$$= 2y\int \frac{x+y+1-(y+1)}{(x+y+1)^4}\,dx - y(y+1)\int \frac{dx}{(x+y+1)^4}$$

$$= 2y\int \frac{dx}{(x+y+1)^3} - 3y(y+1)\int \frac{dx}{(x+y+1)^4}$$

$$= 2y\frac{(x+y+1)^{-2}}{-2} - 3y(y+1)\frac{(x+y+1)^{-3}}{-3}$$

$$= -\frac{y}{(x+y+1)^2} + \frac{y(y+1)}{(x+y+1)^3}$$

$$\text{and } \int N\,dy = \int \frac{2xy - x^2 - x}{(x+y+1)^4}\,dy \text{ (as if } x \text{ is constant)}$$

$$= \frac{-x}{(x+y+1)^2} + \frac{x(x+1)}{(x+y+1)^3}.$$

$\therefore$ the solution is $\frac{-(x+y)}{(x+y+1)^2} + \frac{x(x+1)+y(y+1)}{(x+y+1)^3} + k = 0$, or, $xy = k(x+y+1)^3$, where k is an arbitrary constant.

Example 2.15.15. *Show that $(1/3x^3y^3)$ is an integrating factor of $y(xy + 2x^2y^2)dx + x(xy - x^2y^2)dy = 0$.*

Solution: Multiply by IF as in the previous example. We have

$$M = \frac{1}{3x^2y} + \frac{2}{3x},\quad N = \frac{1}{3xy^2} - \frac{1}{3y};\quad \frac{\partial M}{\partial y} = -\frac{1}{3x^2y^2} = \frac{\partial N}{\partial x}.$$

Hence, the IF is verified.

Example 2.15.16. *Prove that e^{x^2} is an integrating factor of $(x^2 + xy^4)\,dx + 2y^3dy = 0$ and hence solve it.*

Solution: Multiplying, $e^{x^2}(x^2 + xy^4)\,dx + 2e^{x^2}y^3dy = 0$ is of the form $M\,dx + N\,dy = 0$, where $M = e^{x^2}(x^2 + xy^4)$ and $N = 2e^{x^2}y^3$ and $(\partial M/\partial y) = 4xy^3e^{x^2} = \partial N/\partial x$.

Thus, e^{x^2} is an IF of the given equation.

Now,

$$\int M\,dx = \int e^{x^2}(x^2 + xy^4)\,dx, \text{ taking } y \text{ as a constant.}$$

$$= \int x^2e^{x^2}\,dx + y^4\int xe^{x^2}\,dx \text{ (put } x^2 = u)$$

$$= \int x^2e^{x^2}\,dx + \tfrac{1}{2}y^4e^{x^2}$$

$$\text{and } \int N\,dy = \int 2e^{x^2}y^3\,dy, \text{ taking } x \text{ as a constant.}$$

$$= 2e^{x^2}\cdot\tfrac{1}{4}y^4 = \tfrac{1}{2}y^4e^{x^2}.$$

$\therefore$ general solution is $\int x^2e^{x^2}\,dx + \tfrac{1}{2}y^4e^{x^2} = C$, where C is an arbitrary constant.

Example 2.15.17. *Solve: $(2xy + e^x)ydx - e^xdy = 0$.*

Solution: $\frac{e^x y dx - e^x dy}{y^2} + 2x dx = 0$, or, $d\left(\frac{e^x}{y}\right) + 2x dx = 0$.

Integrating, we have $\frac{e^x}{y} + x^2 = c$, which is the general solution.

Examples II(I)

(Textual Questions)

1. (a) State the existence and uniqueness theorem for the ordinary differential equation $\frac{dy}{dx} = f(x, y)$.

   (b) Determine the *necessary* and *sufficient* condition for a differential equation of first order and first degree to be *exact*.

   (c) If $M\,dx + N\,dy = 0$ has one and only one solution, then prove that there exists an infinity of integrating factors.

2. (a) If in the differential equation $M\,dx + N\,dy = 0$, $\frac{1}{N}\left(\frac{\partial M}{\partial y} - \frac{\partial N}{\partial x}\right) = f(x)$, show that $\mu = e^{\int f(x)\,dx}$ is an IF of the equation. Hence, show that $e^{\int P\,dx}$ is an IF of $\frac{dy}{dx} + P(x)y = Q(x)$.

   (b) If $\frac{1}{M}\left(\frac{\partial M}{\partial y} - \frac{\partial N}{\partial x}\right)$ be a function of y alone, say $\phi(y)$, prove that $e^{-\int \phi(y)dy}$ is an IF of $M\,dx + N\,dy = 0$.

   (c) If $Mx - Ny \neq 0$, then show that $\frac{1}{Mx - Ny}$ is an IF of the equation $Mdx + Ndy = 0$, where M and N (functions of x and y) can be written in the form $M = yf_1(xy)$, $N = xf_2(xy)$.

(Miscellaneous Types)

Solve (3–9):

3. (a) $dy = \tan^2(x + y)\,dx$.

   (b) $2dx = (x + 2y - 3)\,dy$.

   (c) $\left(2x^2 + 3y^2 - 7\right)x\,dx - \left(3x^2 + 2y^2 - 8\right)y\,dy = 0$.

4. $\left(x^2 + y^2 - a^2\right)x\,dx + \left(x^2 - y^2 - b^2\right)y\,dy = 0$.

5. $\left(x + 2y^3\right)\,dy = y\,dx$.

6. $\left(x^2 + y^2 + 2x\right)dx + 2y\,dy = 0$.

7. $\sin y\frac{dy}{dx} = (1 - x\cos y)\cos y$.

8. $x\,dy - y\,dx + 3x^3 y\,dx - x^2\,dx = 0$.

9. $x\sin y\,dy + \left(x^3 - 2x^2\cos y + \cos y\right)dx = 0$.

10. Using the transformation $x^2\sqrt{y} = v$, solve: $(2 + 2x^2\sqrt{y})\,y\,dx + (x^2\sqrt{y} + 2)\,x\,dy = 0$.

11. Solve: $(x + 1)\,dy = (y - 1)\,dx$; y is bounded, when $x \to +\infty$.

12. (a) An equation of the form $f'(y)\frac{dy}{dx} + f(y)P(x) = Q(x)$ is a linear equation of the first order $\frac{dv}{dx} + vP(x) = Q(x)$ in the new variable $v = f(y)$. Apply it to solve the equation: $\frac{dy}{dx} + 1 = 4e^{-y}\sin x$.

    (b) Solve: $\sin x\frac{dy}{dx} + 2y\cos x = \cos x$; given $y = -\frac{1}{2}$, when $x = \left(\frac{1}{2}\right)\pi$.

    (c) Solve: $2y\frac{dy}{dx} + xy^2 = \frac{e^{-x^2/2}}{1+x^2}$; given $y = 1$, when $x = 0$.

    (d) Solve: $\left(5x^2 + xy - 1\right)dx + \left(\frac{1}{2}x^2 - y + 2y^2\right)dy = 0$; given $y = 1$, when $x = 0$.

13. (a) Reduce the differential equation $\frac{dy}{dx} + f(x)y = g(x)y^n$ into linear form and find the general solution.

    (b) Solve: $\left(x^2y^3 + 2xy\right)dy = dx$.

14. Solve:

    (a) $\left(x^2 - yx^2\right)dy + \left(y^2 + xy^2\right)dx = 0$.

    (b) $2y\,dx - x\,dy = xy^3\,dy$.

    (c) $\frac{dy}{dx} = x^3y^3 - xy$.

    (d) $\frac{dy}{dx} = (y + 3x)^2$.

    (e) $x^2(x\,dx + y\,dy) + y(x\,dy - y\,dx) = 0$.

    (f) $\left(4x^2y - 6\right)dx + x^3dy = 0$.

15. (a) Solve: $4x^2y\frac{dy}{dx} = 3x\left(3y^2 + 2\right) + 2\left(3y^2 + 2\right)^3$, using $3y^2 + 2 = v$.

(b) Using the substitution $y^3 = vx$, solve: $(xy^3 - y^3 - x^2e^x)\, dx + 3xy^2 dy = 0$.

(c) Solve: $(4x^2y - 6)dx + x^3 dy = 0$, by putting $xy = z$.

16. (a) Verify that $1/xy^2$ is an integrating factor of $(y^2 + xy)\, dx - x^2 dy = 0$.

(b) Find a so that $\left(\frac{1}{x} + x^a y^4\right) dx = y^3 x^{a+1} dy$ may be exact.

(c) Find an integrating factor of the equation: $(x^4y^2 - y)\, dx + (x^2y^4 - x)\, dy = 0$ and hence solve it.

(d) Show that a constant a can be found so that $(x + y)^a$ is an integrating factor of

$$(4x^2 + 2xy + 6y)\, dx + (2x^2 + 9y + 3x)\, dy = 0$$

and hence solve the equation.

17. (a) Solve: $\left(xy^2 - e^{1/x^3}\right) dx - x^2 y\, dy = 0$.

(b) What is meant by an integrating factor of a differential equation of first order and first degree? Show that $\{x(x^2 - y^2)\}^{-1}$ is an IF of $(x^2 + y^2)\, dx - 2xy\, dy = 0$ and hence solve the equation.

(c) Find an integrating factor of the differential equation: $(y^2 + 2x^2y)\, dx + (2x^3 - xy)\, dy = 0$ and hence solve it.

(d) Prove that $\frac{1}{(x+y+1)^4}$ is an integrating factor of $(2xy - y^2 - y)\, dx + (2xy - x^2 - x)\, dy = 0$ and hence solve it.

(e) Show that $\frac{1}{3x^3y^3}$ is an integrating factor of $y(xy + 2x^2y^2)dx + x(xy - x^2y^2)dy = 0$. Hence solve.

(f) Prove that e^{x^2} is an integrating factor of $(x^2 + xy^4)\, dx + 2y^3 dy = 0$ and hence solve it.

18. (a) If x^a be an integrating factor of $(x - y^2)\, dx + 2xy\, dy = 0$, find a and hence solve the equation. [See also sum No. 13(b)(iv) of Examples II(C)]

(b) If $x^\alpha y^\beta$ be an IF of $2y\, dx - 3xy^2\, dx - x\, dy = 0$ find α and β and use it to solve the equation.

(c) If $(x^2 + y^2)^\alpha$ be an IF of $(x + y)\, dx - (x - y)\, dy = 0$, find α and hence solve it.

Hints and Answers

3. (a) Put $x + y = z$; $2(x - y) = c + \sin 2(x + y)$.

(b) $x + 2y - 3 = v$; $(x + 2y - 3) - 4\ln(x + 2y + 1) = x + c$.

(c) $x^2 = u$, $y^2 = v$; $(x^2 + y^2 - 1)^3 = c(x^2 + y^2 - 3)$.

4. $x^4 - y^4 + 2x^2y^2 - 2a^2x^2 - 2b^2y^2 = c$.

5. $x = y^3 + cy$.

6. $(x^2 + y^2)\, e^x = c$ [put $y^2 = v$].

7. $v = \sec y$; $\sec y = x + 1 + ce^x$.

8. $y = vx$; $y = xe^{-x^3} \int e^{x^3} dx + cxe^{-x^3}$.

9. $xv = \cos y$; $2\cos y = x + cxe^{-x^2}$.

10. $xy\left(x^2\sqrt{y} + 3\right) = c$.

11. $y = 1$.

12. (a) $v = e^y$; $e^y = 2(\sin x - \cos x) + ce^{-x}$.

(b) $y = \frac{1}{2} - \operatorname{cosec}^2 x$.

(c) $y^2 e^{x^2/2} = 1 + \tan^{-1} x$.

(d) $10x^3 + 3x^2y - 6x - 3y^2 + 4y^3 - 1 = 0$.

13. (a) Theorem.

(b) $\frac{1}{x} = \frac{1}{2}(1 - y^2) + ce^{-y^2}$.

14. (a) $\ln \frac{x}{y} = c + \frac{x+y}{xy}$.

(b) $\ln \frac{x^2}{y} = \frac{y^3}{3} + c$.

(c) $\frac{1}{y^2} = x^2 + 1 + ce^{x^2}$.

(d) $x + c = \frac{1}{\sqrt{3}} \tan^{-1} \frac{y+3x}{\sqrt{3}}$.

(e) $(x^2 + y^2)(x + 1)^2 = c^2x^2$.

(f) $x^4y = 3x^2 + c$.

15. (a) $4x^9 = (c - 3x^8)(3y^2 + 2)^2$.

(b) $2y^3e^x = xe^{2x} + cx$.

(c) $x^4y = 3x^2 + c$.

16. (b) -5.

(c) IF $= \frac{1}{x^2y^2}$; $\frac{x^3}{3} + \frac{y^3}{3} + \frac{1}{xy} = c$.

(d) $a = 1$, $x^4 + 2x^3y + x^2y^2 + 3x^2y + 6xy^2 + 3y^3 = c$.

17. (a) $3y^2 - 2x^2 e^{1/x^3} = 6cx^2$.
(b) $x^2 - y^2 = cx$.
(c) IF $= x^{-5/2} y^{-1/2}; 6\sqrt{xy} - x^{3/2} y^{3/2} = c$.
(d) $xy = c(x + y + 1)^3$.
(e) $2 \ln x - \ln y = \frac{1}{xy} + c$.
(f) $\int x^2 e^{x^2} dx + \frac{1}{2} y^4 e^{x^2} = c$.

18. (a) $-2; y^2 + x \ln x = cx$.
(b) $\alpha = 1, \beta = -2; x^2 - x^3 y = cy$.
(c) $-1; \ln(x^2 + y^2) - 2\tan^{-1}(y/x) = c$.

Chapter 3

Equations of the First Order but not of the First Degree

In this chapter we would like to consider only the equations of the first order but not of the first degree. These types of equations can be classified primarily into two categories: (i) Equations that can be resolved into component equations of the first degree and (ii) Equations that cannot be resolved into component equations. The equations solvable for p belong to the first category. On the other hand, the equations solvable for y and the equations solvable for x belong to the second category. In this category, there is another important form of equation, namely, Clairaut's form. It is named after Alexis Claude Clairaut (13 May 1713–17 May 1765), a French mathematician. The equations of Clairaut form generally give two types of solutions, namely, general solution and singular solution. General solution represents a family of curves while singular solution represents the envelope of the family of curves represented by general solution. Here we are interested only for general solutions. Singular solutions will be introduced in next chapter.

3.1 Equations of the first order but not of the first degree

An equation of first order and nth degree may be written as

$$p^n + P_1 p^{n-1} + P_2 p^{n-2} + \cdots + P_{n-1} p + P_n = 0, \tag{3.1.1}$$

where $p = dy/dx$ and $P_1, P_2, \ldots, P_n$ are given functions of x and y. We shall, in this chapter, discuss three special types of such equations: 1. Those solvable for p. 2. Those solvable for y. 3. Those solvable for x.

3.2 Equations solvable for p

If we look upon eq. (3.1.1) as an algebraical equation in p, which has n roots $p_1, p_2, \ldots, p_n$, these being functions of x and y, we may write (3.1.1) in the form $(p - p_1)(p - p_2)\cdots(p - p_n) = 0$.

This can be true only if one or more of the factors on the left-hand side vanish; and hence any relation between x and y which makes a factor vanish will be a solution of (3.1.1), while no relation which does not make some factor vanish can be a solution.

Let $\phi_r(x, y, c_r) = 0$ be the primitives of $p - p_r = 0$ $(r = 1, 2, 3, \ldots, n)$.

All possible solutions of the eq. (3.1.1) will then be included in the relation

$$\phi_1(x, y, c_1)\,\phi_2(x, y, c_2)\cdots\phi_n(x, y, c_n) = 0. \tag{3.2.1}$$

At this point we meet with a difficulty. The general solution apparently contains n arbitrary constants, whereas we expect only one, as the equation is only of the first order. See that the generality will still be maintained if all the constants $c_1, c_2, \ldots, c_n$ be made the same, say c; c, of course, may have *any numerical value*; for, a particular solution of eq. (3.1.1) is obtained by solving one or other of the equations $\phi_r(x, y, c) = 0$, c may have any arbitrary value. Hence, the complete primitive of (3.1.1) is the product $\phi_1(x, y, c)\phi_2(x, y, c)\phi_3(x, y, c)\cdots\phi_n(x, y, c) = 0$, where c is any arbitrary constant.

Example 3.2.1. *Solve the equation* $p^2 + px + py + xy = 0$.

Solution: The above equation may be written as $(p+x)(p+y) = 0$, from which $p+x = 0$ gives $2y = -x^2 + c_1$ and $p + y = 0$ gives $x = -\ln y + c_2$.

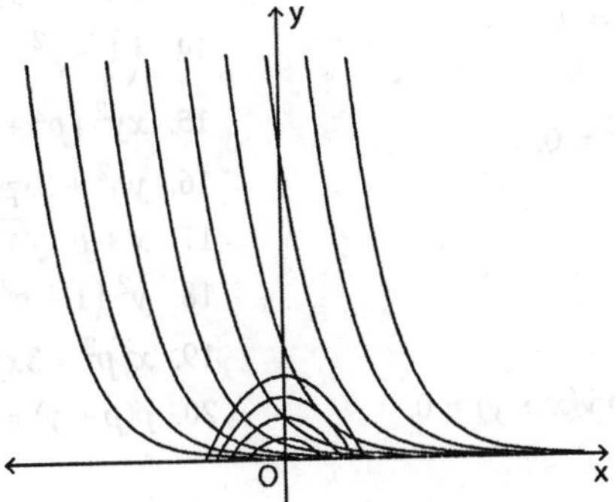

Fig. 3.1

The complete primitive can be written in the form $\left(2y + x^2 - c\right)(x + \ln y - c) = 0$, where c is an arbitrary constant (Fig. 3.1).

Example 3.2.2. *Solve the equation* $x^2p^2 - 2xyp + y^2 = x^2y^2 + x^4$.

Solution: The above equation can be written as

$$(xp - y)^2 = x^2\left(x^2 + y^2\right), \text{ or, } p = \frac{y \pm x\sqrt{x^2 + y^2}}{x}.$$

Putting $y = vx$, we see that $\frac{dv}{\sqrt{1+v^2}} = \pm dx$ and hence $v = \sinh(x + c_1)$ and $v = \sinh(c_2 - x)$.

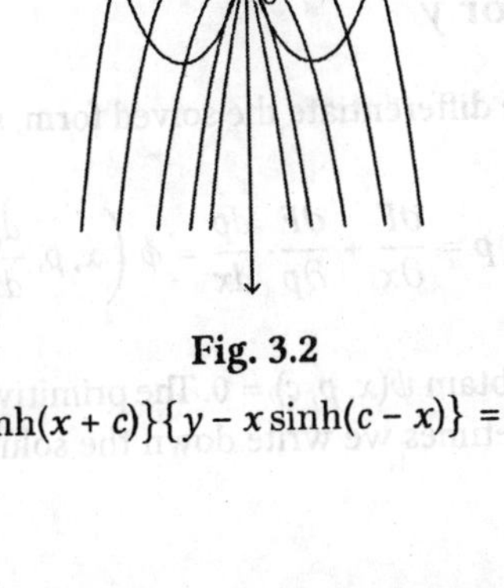

Fig. 3.2

$\therefore$ the general solution is $\{y - x\sinh(x + c)\}\{y - x\sinh(c - x)\} = 0$, where c is an arbitrary constant (Fig. 3.2).

Examples III(A)

Solve the following differential equations:

1. $p^2 + p - 2 = 0$.
2. $p^2 - 9p + 18 = 0$.
3. $p^2 = ax^3$.
4. $p^2 - 2p\cosh x + 1 = 0$.
5. $x^2p^2 - 2xyp + 2y^2 - x^2 = 0$.
6. $p^2 + 2py\cot x = y^2$.
7. $3p^2y^2 - 2xyp + 4y^2 - x^2 = 0$.
8. $p - 1/p = x/y - y/x$.
9. $y = b + \left(1 + p^2\right)^{-\frac{1}{2}}$.
10. $p^2 + 2xp - 3x^2 = 0$.
11. $p^3 - p\left(x^2 + xy + y^2\right) + xy(x + y) = 0$.
12. $p^3 - p^2\left(x^2 + xy + y^2\right) + p\left(x^3y + x^2y^2 + xy^3\right) - x^3y^3 = 0$.
13. (a) $\left(p^2 - \frac{1}{a^2 - x^2}\right)\left(p - \sqrt{\frac{y}{x}}\right) = 0$.
    (b) $x + yp^2 = p(1 + xy)$.
14. $\left(1 - y^2 + \frac{y^4}{x^2}\right)p^2 - 2\frac{y}{x}p + \frac{y^2}{x^2} = 0$.
15. $xy^2\left(p^2 + 2\right) = 2py^3 + x^3$.
16. $yp^2 + 2xp - y = 0$.
17. $x + p/\sqrt{1 + p^2} = a$.
18. $y^2\left(1 - p^2\right) = b$.
19. $x^2p^2 + 3xyp + 2y^2 = 0$.
20. $p(p + y) = x(x + y)$.

Answers

1. $(y - x + c)(y + 2x + c) = 0$.
2. $(y - 6x + c)(y - 3x + c) = 0$.
3. $25(y + c)^2 = 4ax^5$.
4. $(y - e^x + c)(y + e^{-x} + c) = 0$.
5. $y = x\sin\{\ln cx\}$.
6. $y(1 \pm \cos x) = c$.
   [**Hints:** $(p + y\cot x)^2 = y^2\operatorname{cosec}^2 x$]
7. $x^2 + y^2 + 3c^2 - 4cx = 0$.
   [**Hints:** Multiply by 3; obtain $(3py - x)^2 = 4\left(x^2 - 3y^2\right)$; put $x^2 - 3y^2 = v^2$]
8. $\left(x^2 - y^2 - c\right)(xy - c) = 0$.
9. $(x - c)^2 + (y - b)^2 = 1$.
10. $\left(2y - x^2 - c\right)\left(2y + 3x^2 - c\right) = 0$.
11. $\left(2y - x^2 - c\right)\left(y - ce^x\right)\left(y + x - 1 - ce^{-x}\right) = 0$.
12. $\left(x^3 - 3y + c\right)\left(e^{x^2/2} + cy\right)(xy + cy + 1) = 0$.
13. (a) $\left(y - \sin^{-1}x/a - c\right)\left(y - \cos^{-1}x/a - c\right)\left(\sqrt{y} - \sqrt{x} - c\right) = 0$
    (b) $\left(2y - x^2 - c\right)\left(2x - y^2 - c\right) = 0$.
14. $y/\left\{x + \sqrt{x^2 - y^2}\right\} = ce^{\pm y}$.
    [**Hints:** Write $p^2 - 2\frac{y}{x}p + \frac{y^2}{x^2} = p^2y^2\left(1 - \frac{y^2}{x^2}\right)$]
15. $\left(x^2 - y^2 + c\right)\left(x^2 - y^2 + cx^4\right) = 0$.
16. $\left\{y^2 - c\left(x + \sqrt{x^2 + y^2}\right)\right\}\left\{x + \sqrt{x^2 + y^2} - c\right\} = 0$.
17. $(y + c)^2 + (x - a)^2 = 1$.
18. $y^2 - b = (x + c)^2$.
19. $(xy - c)\left(x^2y - c\right) = 0$.
20. $\left(y - x^2/2 + c\right)\left(y + x - 1 - ce^{-x}\right) = 0$.

3.3 Equations solvable for y

If the equation is solvable for y we differentiate the solved form, say $y = F(x, p)$, with respect to x. Thus, we obtain

$$\frac{dy}{dx} = p = \frac{\partial F}{\partial x} + \frac{\partial F}{\partial p}\cdot\frac{dp}{dx} = \phi\left(x, p, \frac{dp}{dx}\right) \text{ (say).}$$

We solve $p = \phi(x, p, dp/dx)$ to obtain $\psi(x, p, c) = 0$. The primitive is obtained by eliminating p between $y = F(x, p)$ and $\psi(x, p, c) = 0$. Sometimes we write down the solution by expressing x and y separately as functions of p, a parameter.

Example 3.3.1. $y = px + p^2x$.

Solution: Differentiating with respect to x, we get

$$p = p + x\frac{dp}{dx} + p^2 + 2xp\frac{dp}{dx} \text{ or, } -p^2 = x(1+2p)\frac{dp}{dx}, \text{ or, } \frac{dx}{x} + \frac{1+2p}{p^2}dp = 0.$$

Integrating, we obtain

$$\ln x + \ln c = \frac{1}{p} - 2\ln p, \text{ or, } cxp^2 = e^{1/p}.$$

$$\therefore \quad x = \frac{e^{1/p}}{cp^2} \text{ and } y = x\left(p + p^2\right) = \frac{e^{1/p}}{cp^2}\left(p + p^2\right)$$

constitute the parametric solution of the given equation (p is the parameter).

Example 3.3.2. $y + px = p^2x^4$.

Solution: Differentiating w.r.t. x, we get

$$p + p + x\frac{dp}{dx} = 2px^4\frac{dp}{dx} + p^2 \cdot 4x^3 \text{ or, } 2p\left(1 - 2x^3p\right) + x\frac{dp}{dx}\left(1 - 2x^3p\right) = 0 \text{ or, } \left(1 - 2x^3p\right)\left(2p + x\frac{dp}{dx}\right) = 0.$$

$$\therefore\ 2p + x\frac{dp}{dx} = 0, \text{ if } 1 - 2x^3p \neq 0.$$

This gives, on integration, $p = c/x^2$.

Eliminating p between $p = c/x^2$ and the given equation, we obtain the complete primitive as $y = -c/x + c^2$, or, $xy + c = c^2x$ (Fig. 3.3)

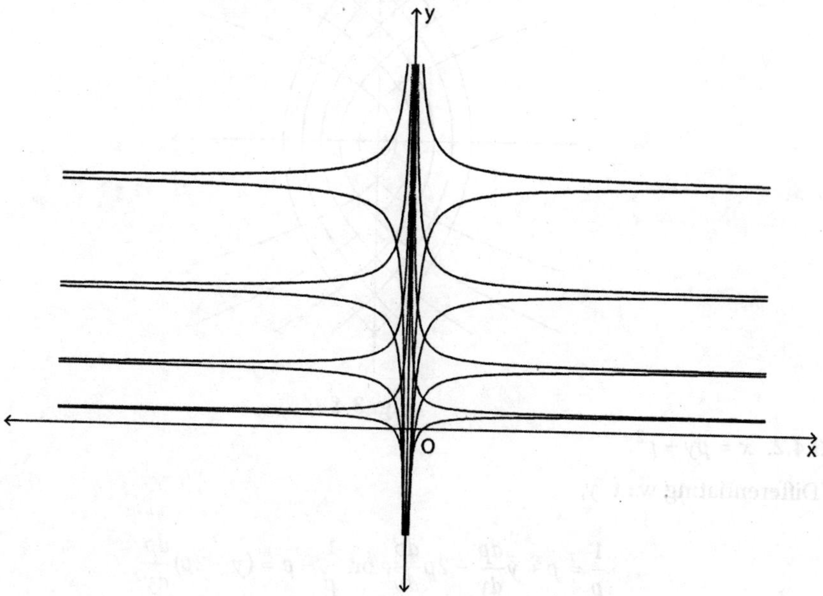

Fig. 3.3

An Important Observation. In the previous example we left out the possibility that $1 - 2x^3p = 0$. In case, this happens, i.e. in case $p = \frac{1}{2x^3}$, we get from the given equation $y = -px + p^2x^4 = -\frac{1}{4x^2}$, or, $4x^2y + 1 = 0$.

This relation between x and y satisfies the given differential equation; hence it is a solution. But this solution is not a particular solution deducible from the general solution $xy + c = c^2x$ by giving a particular value to c. Such a solution is known as *singular solution* of the given differential equation.

3.4 Equations solvable for x

If the solved form is $x = F(y, p)$, we differentiate with respect to y (and write $dx/dy = 1/p$, wherever necessary). We deduce a relation like $\frac{1}{p} = \phi\left(y, p, \frac{dp}{dy}\right)$.

Solve this equation and obtain a relation of the form $\psi(y, p, c) = 0$.

Eliminate p between this last relation and the given equation. The eliminant is the complete primitive. As before, the complete primitive may be given in the parametric form.

Example 3.4.1. $p^2y + 2px = y$.

Solution: Solving for x, $x = \frac{y}{2p}(1 - p^2)$.

Differentiating w.r.t. y,

$$\frac{1}{p} = \frac{1-p^2}{2p} + y\left\{-\frac{1}{2}\frac{1+p^2}{p^2}\cdot\frac{dp}{dy}\right\} \text{ or, } \frac{1+p^2}{2p} = -\frac{1}{2}\frac{1+p^2}{p^2}y\frac{dp}{dy},$$

$$\text{or, } 1 = -\frac{y}{p}\frac{dp}{dy} \; (\because 1 + p^2 \neq 0) \text{ or, } \frac{dy}{y} + \frac{dp}{p} = 0. \therefore py = c.$$

Eliminating p between this relation $p = c/y$ and the given equation $x = y\left(1 - p^2\right)/2p$, we obtain $y^2 = c^2 + 2cx$ (Fig. 3.4)

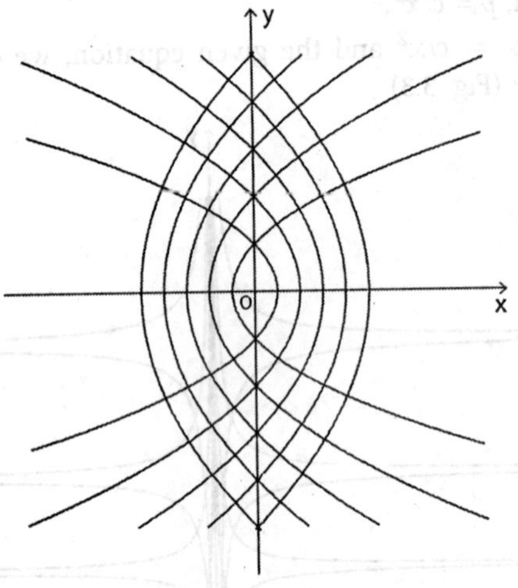

Fig. 3.4

Example 3.4.2. $x = py - p^2$.

Solution: Differentiating w.r.t. y,

$$\frac{1}{p} = p + y\frac{dp}{dy} - 2p\frac{dp}{dy}, \text{ or, } \frac{1}{p} - p = (y - 2p)\frac{dp}{dy}$$

$$\text{or, } \frac{dy}{dp}\left(\frac{1}{p} - p\right) = y - 2p \text{ or, } \frac{dy}{dp} + \frac{y}{p - 1/p} = \frac{2p}{p - 1/p},$$

which is a linear equation in y with p as its independent variable. The integrating factor is

$$e^{\int \frac{p}{p^2-1}dp} = e^{\frac{1}{2}\ln(p^2-1)} = \sqrt{p^2-1}.$$

Multiplying by the integrating factor and then integrating, we find

$$y\sqrt{p^2-1} = \int \frac{2p^2}{\sqrt{p^2-1}}dp + c = 2\int \frac{p^2-1+1}{\sqrt{p^2-1}}dp + c = 2\int \sqrt{p^2-1}\,dp + 2\int \frac{dp}{\sqrt{p^2-1}} + c$$

$$= p\sqrt{p^2-1} + \cosh^{-1} p + c.$$

$$\therefore\ y = p + \left(c + \cosh^{-1} p\right)\left(p^2-1\right)^{-\frac{1}{2}}.$$

This relation and the given equation $x = py - p^2 = p\left(c + \cosh^{-1} p\right)\left(p^2-1\right)^{-\frac{1}{2}}$ constitute the parametric solution of the given equation.

Note 3.4.1. *This problem may be considered as an equation solvable for y, and hence can also be solved by the method of art.* 3.3.

Examples III(B)

Solve:

1. $y = p + x/p$.
2. $y = x + a\tan^{-1} p$.
3. $xp^2 - 2yp + ax = 0$.
4. $y = x\left(p + p^3\right)$.
5. $y = (1+p)x + ap^2$.
6. $p^3x - p^2y - 1 = 0$.
7. $y = 2px + p^2$.
8. $y = 2px + 4xp^2$.
9. $3p^5 - py + 1 = 0$.
10. $y = 2px + \tan^{-1}\left(xp^2\right)$.
11. $y = 2px + f\left(xp^2\right)$.
12. $p^3 + mp^2 = a(y + mx)$.
13. $e^{p-y} = p^2 - 1$.
14. $y = 2px + p^n$.
15. $y = xp^2 + p$.
16. $y = x + p^3$.
17. $p^3 - 4xyp + 8y^2 = 0$.
18. $x = y + p^2$.
19. $x = y + a\ln p$.
20. $p = \tan\left(x - \frac{p}{1+p^2}\right)$.
21. $y = 2px + y^2p^3$. [Solve for x]
22. $8p^3x = y\left(12p^2 - 9\right)$.
23. $y = p\tan p + \ln\cos p$.
24. $y = \ln\left(p^3 + p\right)$.
25. $x\left(1+p^2\right) = 1$.
26. $(x-a)p^2 + (x-y)p - y = 0$.

Answers

1. $\left.\begin{array}{l} x = p\left(c + \cosh^{-1} p\right)\left(p^2-1\right)^{-\frac{1}{2}} \\ y = p + \left(c + \cosh^{-1} p\right)\left(p^2-1\right)^{-\frac{1}{2}} \end{array}\right\}.$
2. $x + c = \frac{a}{2}\left\{\ln\left(1+p^2\right)^{-\frac{1}{2}}(p-1) - \tan^{-1} p\right\}$ and the given equation.
3. $2y = cx^2 + a/c$.
4. $cxp^3 = e^{1/2p^2}$ and the given equation.
5. $x = 2a(1-p) + ce^{-p}$ and the given relation.
6. $\left(c^3x - c^2y - 1\right)\left(y^3 - cx^2\right) = 0$.
7. $\left(3xy + 2x^3 + c\right)^2 = 4\left(x^2 + y\right)^3$.
8. $(y - 4c)^2 = 4cx$.
9. $\left.\begin{array}{l} x = 4p^3 + \frac{1}{2}p^{-2} + c \\ y = 3p^4 + p^{-1} \end{array}\right\}.$
10. $y = 2\sqrt{cx} + \tan^{-1} c$.
11. $y = 2c\sqrt{x} + f\left(c^2\right)$.
12. $ax + c = \frac{3}{2}p^2 - mp + m^2\ln(p+m)$ and the given relation.
13. $p(p+1) = c(p-1)e^x$ and the given equation.
14. $x = \frac{c}{p^2} - \frac{n}{n+1}p^{n-1}$ and the given equation.
15. $x = (\ln p - p + c)(p-1)^{-2}$ and the given relation.
16. $x = \frac{3}{2}p^2 + 3p + 3\ln(p-1)$ and the given relation.

17. $y = c(x-c)^2$.

18. $\left.\begin{array}{l} y = c - [p^2 + 2p + 2\ln(p-1)] \\ x = c - [2p + 2\ln(p-1)] \end{array}\right\}$.

19. $\left.\begin{array}{l} y = c - a\ln(p-1) \\ x = c + a\ln p/(p-1) \end{array}\right\}$.

20. $\left.\begin{array}{l} y = c - 1/\left(1+p^2\right) \\ x = p/\left(1+p^2\right) + \tan^{-1} p \end{array}\right\}$.

21. $y^2 = 2cx + c^3$.

22. $(8cx+9)^3 = 1728c^2y^2$.

23. $x = \tan p + c$ and the given relation.

24. $x + (1/p) + c = 2\tan^{-1} p$ and the original.

25. $y - c = \sqrt{x - x^2} - \tan^{-1}\sqrt{\{(1-x)/x\}}$.

26. GS: $y = cx - \frac{ac^2}{1+c}$;
SS: $(x+y)^2 = 4ay$.

3.5 Clairaut's form

A differential equation of the form,

$$y = px + f(p),\ p = dy/dx \tag{3.5.1}$$

is called *Clairaut's equation.* Its complete primitive is $y = cx + f(c)$ (replacing p by c in the given equation).

Justification

Differentiate the given equation w.r.t. x; then

$$p = p + \{x + f'(p)\}\frac{dp}{dx}.$$

This gives either

$$\frac{dp}{dx} = 0 \tag{3.5.2}$$

$$\text{or,}\quad x + f'(p) = 0. \tag{3.5.3}$$

The relation (3.5.2) gives

$$p = \text{constant} = c\,(\text{say}). \tag{3.5.4}$$

From (3.5.1) and (3.5.4) we eliminate p and obtain the complete primitive in the form $y = cx + f(c)$. **Geometrically** this represents a family of straight lines.

Now from (3.5.1) and (3.5.3), we obtain

$$\left.\begin{array}{l} x = -f'(p) \\ y = -pf'(p) + f(p) \end{array}\right\} \tag{3.5.5}$$

We see at once that it contains no arbitrary constant and therefore is not a *general solution.* But it is a solution of (3.5.1) as can be easily verified. In fact, (3.5.5) gives a *singular solution* of (3.5.1).

An observation

The general solution $y = cx + f(c)$ represents a family of straight lines. If this family has an envelope it is found by differentiating the equation w.r.t. c (in fact, this is same as to give c a pair of equal value for the same values of x and y). Then we have $x + f'(c) = 0$. The result of elimination of c between these two equations is the same as that of eliminating p in (3.5.5). Therefore, the curve represented by (3.5.5) is the envelope of the family of lines $y = cx + f(c)$, these lines should have an envelope. Then the singular solution of the Clairaut's equation becomes the envelope of the family of lines given by the general solution. From our knowledge of the concept of envelope we can state that *the curve given by the singular solution touches every member of the family of lines given by the general solution and also at each point of the curve it is touched by some member of that family of lines.*

An alternative method for finding an envelope

If $f(x, y, p) = A(x, y)p^2 + B(x, y)p + c(x, y)$ be the equation in Clairaut's form, then the envelope is given by the *p-discriminant relation* $B^2 - 4AC = 0$ which is the *condition* that $f(x, y, p) = 0$ has two equal roots. The same result comes from $f(x, y, c) = 0$, that is from the *c-discriminant relation*.

Example 3.5.1. $y = px + \frac{a}{p}$.

Solution: Differentiating w.r.t. x, we get

$$p = p + x\frac{dp}{dx} - \frac{a}{p^2}\frac{dp}{dx}.$$

$\therefore$ either $\frac{dp}{dx} = 0$, or, $x - \frac{a}{p^2} = 0$.

The first one gives $p = c$ and the general solution is $y = cx + a/c$ (Fig. 3.5).

The second one gives $x = \frac{a}{p^2}$, $y = \frac{2a}{p}$ and eliminating p, we obtain $y^2 = 4ax$. This is a solution of the given equation but not deducible from the general solution by giving a particular value to c. So it is the singular solution.

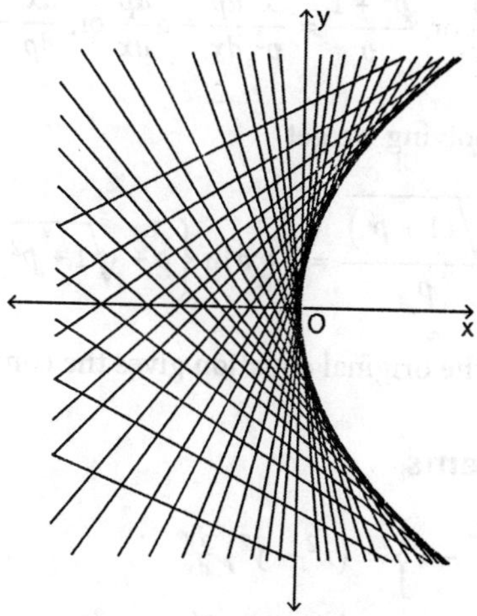

Fig. 3.5

Note 3.5.1. *See that every member of* $y = cx + a/c$ *touches the parabola* $y^2 = 4ax$ *and at any point of the parabola it is touched by some member of* $y = cx + a/c$. *The singular solution is thus the envelope of the family of lines given by the general solution.*

Sometimes a change of variables transforms a given equation to Clairaut's equation and hence the general solution can be easily written down

Example 3.5.2. *Solve:* $x^2(y - px) = p^2 y$.

Solution: We put $x^2 = u$, $y^2 = v$, so that $2x\,dx = du$, $2y\,dy = dv$.

$$\therefore\ p = \frac{dy}{dx} = \frac{\frac{1}{2y}dv}{\frac{1}{2x}du} = \frac{x\,dv}{y\,du}.$$

Substituting this value of p in the given differential equation, we get

$$x^2\left(y - \frac{x^2}{y}\cdot\frac{dv}{du}\right) = \frac{x^2}{y^2}\left(\frac{dv}{du}\right)^2 y \text{ or, } y^2 - x^2\frac{dv}{du} = \left(\frac{dv}{du}\right)^2 \text{ or, } v = u\frac{dv}{du} + \left(\frac{dv}{du}\right)^2$$

which is in Clairaut's form: The solutions are given below:

Complete primitive: $v = cu + c^2$, or, $y^2 = cx^2 + c^2$;

Singular solution: $u^2 + 4v = 0$, or, $x^4 + 4y^2 = 0$.

Extended form of Clairaut's equation

To *Solve*: $y = xf(p) + \phi(p)$.

Differentiate w.r.t. x and obtain

$$p = f(p) + [xf'(p) + \phi'(p)]\frac{dp}{dx}, \text{ or, } \frac{dx}{dp} + x\frac{f'(p)}{f(p) - p} = \frac{\phi'(p)}{p - f(p)},$$

which is linear in x and comes under art. 2.14. Suppose that the integral is $F(x, p, c) = 0$.

The result of eliminating p between this equation and the original equation is the complete primitive.

Example 3.5.3. *Solve*: $x + yp = ap^2$, *or*, $y = -\frac{x}{p} + ap$.

Solution: Differentiating w.r.t. x, we get

$$p = -\frac{1}{p} + \frac{x}{p^2}\frac{dp}{dx} + a\frac{dp}{dx} \text{ or, } \frac{p^2+1}{p} = \frac{x}{p^2}\frac{dp}{dx} + a\frac{dp}{dx} \text{ or, } \frac{dx}{dp} - \frac{x}{p\,(p^2+1)} = \frac{ap}{1+p^2}.$$

This is a linear equation in x. Solving we get

$$\frac{x\sqrt{(1+p^2)}}{p} = c + a\ln\left\{p + \sqrt{1+p^2}\right\}.$$

This equation combined with the original equation gives the complete primitive.

3.6 Miscellaneous problems

Example 3.6.1. *Solve*: $xy\left\{\left(\frac{dy}{dx}\right)^2 - 1\right\} = \left(x^2 - y^2\right)\frac{dy}{dx}$.

Solution: Taking $(dy/dx) = p$, we have $xy\left(p^2 - 1\right) = \left(x^2 - y^2\right)p$, or, $(x - py)(y + px) = 0$.

$\therefore$ $x - py = 0$ gives $x\,dx - y\,dy = 0$, integrating, $x^2 - y^2 = 2c$

and $y + px = 0$ gives $x\,dy + y\,dx = 0$, integrating $xy = c$.

Hence, the solution is $\left(x^2 - y^2 - 2c\right)(xy - c) = 0$, c being an arbitrary constant.

Example 3.6.2. *Solve*: $y\left\{x(2x+1)\frac{dy}{dx} - y\left(\frac{dy}{dx}\right)^2\right\} = 2x^3$.

Solution: Writing $\frac{dy}{dx} = p$ and simplifying $(py - x)\left(py - 2x^2\right) = 0$ and by the previous method, the solution becomes $\left(y^2 - x^2 - 2c\right)\left(3y^2 - 4x^3 - 6c\right) = 0$.

Example 3.6.3. *Obtain the complete primitive and singular solution of* $y = px + \sqrt{1+p^2}$.

Solution: This is in Clairaut's form: Differentiating w.r.t. x,

$$p = p + x\frac{dp}{dx} + \frac{p}{\sqrt{1+p^2}}\frac{dp}{dx} \text{ or, } \frac{dp}{dx}\left(x + \frac{p}{\sqrt{1+p^2}}\right) = 0.$$

$(dp/dx) = 0$ gives $p = c$. Complete primitive is $y = cx + \sqrt{1+c^2}$ (Fig. 3.6).

However, $x + \frac{p}{\sqrt{1+p^2}} = 0$ gives $y = \frac{1}{\sqrt{1+p^2}}$.

Hence, the singular solution is

$$x^2 + y^2 = \frac{p^2}{1+p^2} + \frac{1}{1+p^2}, \text{ or, } x^2 + y^2 = 1.$$

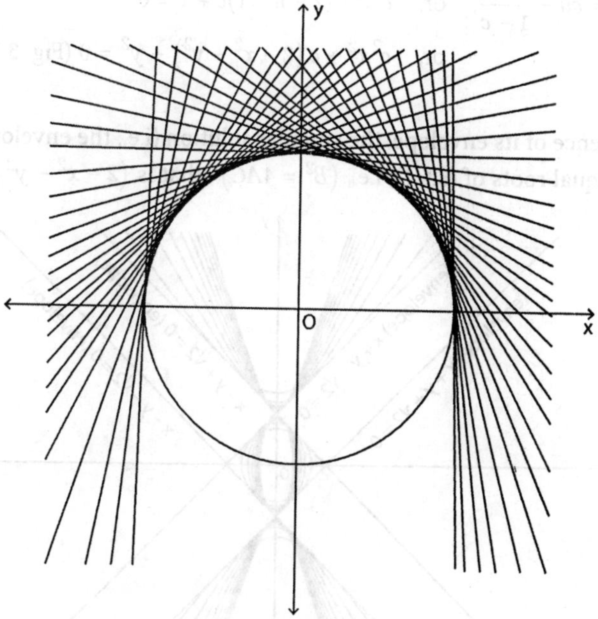

Fig. 3.6

Example 3.6.4. *Solve and find the singular solution of the differential equation*

$$\sin\left(x\frac{dy}{dx}\right)\cos y = \cos\left(x\frac{dy}{dx}\right)\sin y + \frac{dy}{dx}.$$

Solution: Putting $dy/dx = p$, the given equation reduces to

$$\sin px \cos y - \cos px \sin y = p$$

or, $\sin(px - y) = p$, or, $y = px - \sin^{-1} p$. **[Clairaut's Eqn.]**

Differentiating with respect to x

$$p = p + x\frac{dp}{dx} - \frac{1}{\sqrt{1-p^2}}\frac{dp}{dx} \text{ or, } \frac{dp}{dx}\left(x - \frac{1}{\sqrt{1-p^2}}\right) = 0$$

$\frac{dp}{dx} = 0$ gives $p = c$. General solution: $y = cx - \sin^{-1} c$.

However, $x - \frac{1}{\sqrt{1-p^2}} = 0$ gives $p^2 = \frac{x^2-1}{x^2}$.

From $y = px - \sin^{-1} p$, we have $y = \sqrt{x^2-1} - \sin^{-1}\frac{\sqrt{x^2-1}}{x}$ as the singular solution.

Example 3.6.5. *Solve and find the singular solution of the differential equation* $(px - y)(x - py) = 2p$, *where* $p = dy/dx$.

Solution: Put $x^2 = u$, $y^2 = v$; then taking $\frac{dv}{du} = q$,

$$2x\,dx = du,\ 2y\,dy = dv \text{ and } p = \frac{dy}{dx} = \frac{x}{y}\frac{dv}{du} = \frac{x}{y}q.$$

The equation then transforms into $\left(\frac{x}{y}qx - y\right)(x - xq) = 2\frac{x}{y}q$ or, simplifying $v = uq - \frac{2q}{1-q}$, which is in *Clairaut's form.*

The general solution is, therefore,

$$v = cu - \frac{2c}{1-c}, \quad \text{or,} \quad c^2u + (2 - u - v)c + v = 0$$

$$\text{or,} \quad c^2x^2 + c\left(2 - x^2 - y^2\right) + y^2 = 0 \text{ (Fig. 3.7).} \tag{3.6.1}$$

Accepting the existence of its envelope, the singular solution (i.e., the envelope solution) is given by the condition for two equal roots of (3.6.1), i.e., $\left(B^2 = 4AC\right)$, that is $\left(2 - x^2 - y^2\right)^2 = 4x^2y^2$ (Fig. 3.7).

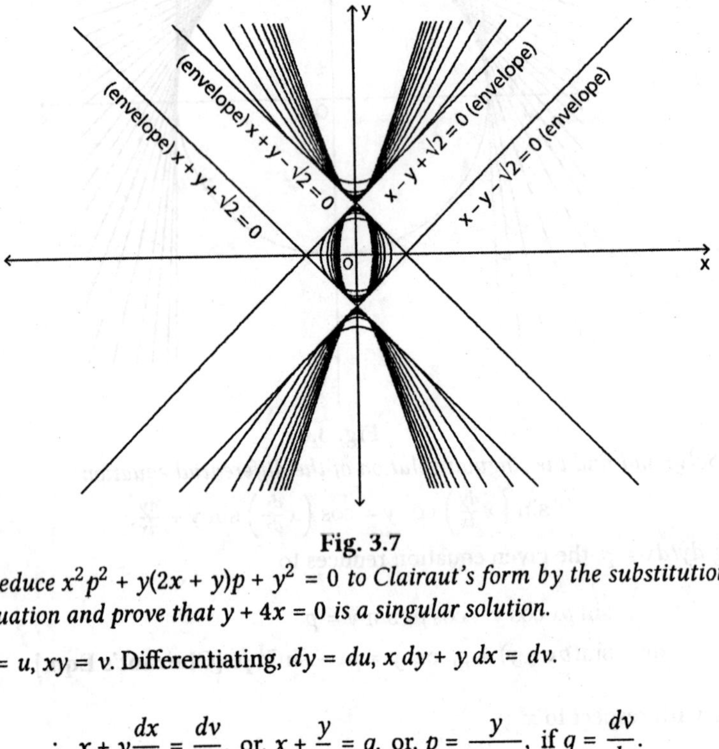

Fig. 3.7

Example 3.6.6. *Reduce $x^2p^2 + y(2x + y)p + y^2 = 0$ to Clairaut's form by the substitution $y = u$, $xy = v$. Hence, solve the equation and prove that $y + 4x = 0$ is a singular solution.*

Solution: Here $y = u$, $xy = v$. Differentiating, $dy = du$, $x\,dy + y\,dx = dv$.

$$\therefore\ x + y\frac{dx}{dy} = \frac{dv}{du}, \text{ or, } x + \frac{y}{p} = q, \text{ or, } p = \frac{y}{q - x}, \text{ if } q = \frac{dv}{du}.$$

Putting in the given relation $v = uq + q^2$ which is in *Clairaut's form.*

Differentiating w.r.t. u,

$$q = q + u\frac{dq}{du} + 2q\frac{dq}{du}, \quad \text{or,} \quad \frac{dq}{du}(u + 2q) = 0$$

which gives $\frac{dq}{du} = 0$, or, $u + 2q = 0$.

When $\frac{dq}{du} = 0$, $q = c$. Thus, the general solution is $v = cu + c^2$, or, $xy = cy + c^2$ (Fig. 3.8).

When $u + 2q = 0$, $q = -\frac{1}{2}u$. Thus, the singular solution is $v = -\frac{1}{2}u^2 + \frac{1}{4}u^2$, or, $v = -\frac{1}{4}u^2$, $u^2 + 4v = 0$, $y^2 + 4xy = 0$, or, $y + 4x = 0$ (Fig. 3.8).

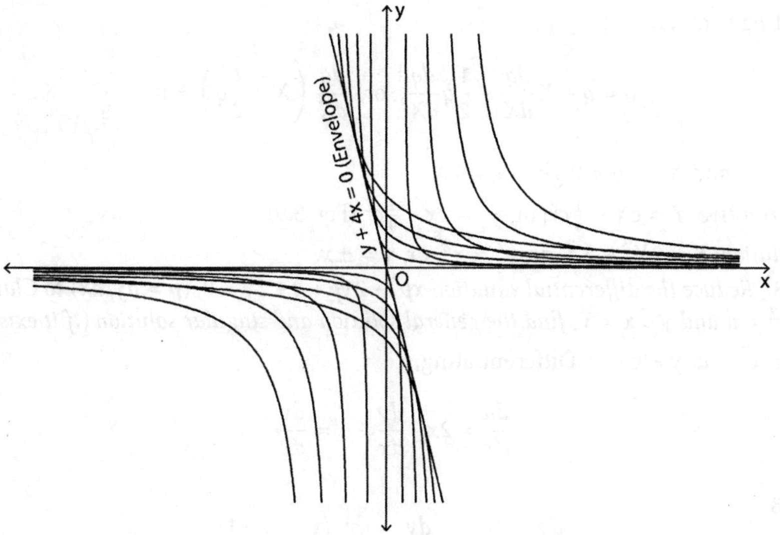

Fig. 3.8

Example 3.6.7. *Reduce the differential equation $y = 2px - p^2y$ to Clairaut's form by the substitution $y^2 = Y$, $x = X$ and then obtain its complete primitive and singular solution, if any.*

Solution: We have

$$p = \frac{dy}{dx} \text{ and let } q = \frac{dY}{dX}.$$

Differentiating $y^2 = Y$ and $x = X$, we have

$$\therefore 2py = \frac{dY}{dx},\ dx = dX \text{ whereby } p = \frac{q}{2y}.$$

The given equation reduces to $Y = qX - \frac{1}{4}q^2$, which is in Clairaut's form.

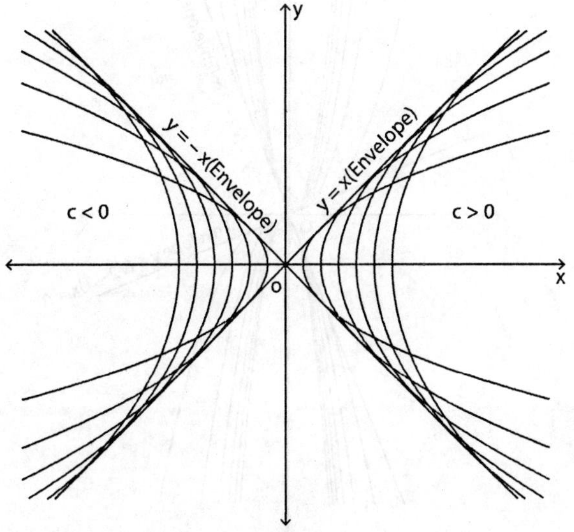

Fig. 3.9

Differentiating w.r.t. X,

$$q = q + X\frac{dq}{dX} - \frac{1}{2}q\frac{dq}{dX}, \text{ or, } \frac{dq}{dX}\left(X - \frac{1}{2}q\right) = 0.$$

$\frac{dq}{dX} = 0$ gives $q = c$, and $X - \frac{1}{2}q = 0$ gives $q = 2X$.

Complete primitive: $Y = cX - \frac{1}{4}c^2$, or, $y^2 = cx - \frac{1}{4}c^2$ (Fig. 3.9).

Singular solution: $Y = 2X^2 - X^2$, or, $y^2 = x^2$, or, $y = \pm x$.

Example 3.6.8. *Reduce the differential equation $xp^2 - 2yp + x + 2y = 0$, ($p = dy/dx$) to Clairaut's form by transforming $x^2 = u$ and $y - x = v$, find the general solution and singular solution (if it exists).*

Solution: Here $x^2 = u$, $y - x = v$. Differentiating,

$$\frac{du}{dx} = 2x, \quad \frac{dy}{dx} - 1 = \frac{dv}{dx}.$$

Now taking

$$\frac{dv}{du} = q, \; q = \frac{dv}{du} = \frac{dv/dx}{du/dx} = \frac{p-1}{2x},$$

where $p = \frac{dy}{dx}$.

Putting $p = 1 + 2xq$ in the given equation,

$$x(1 + 2qx)^2 - 2y(1 + 2qx) + x + 2y = 0$$

$$\text{or, } 4q^2x^2 - 4q(y - x) + 2 = 0, \text{ since } x \neq 0$$

$$\text{or, } 4q^2u - 4qv + 2 = 0$$

$$\text{or, } v = qu + \frac{1}{2q}. \textbf{ (Clairaut's Equation).}$$

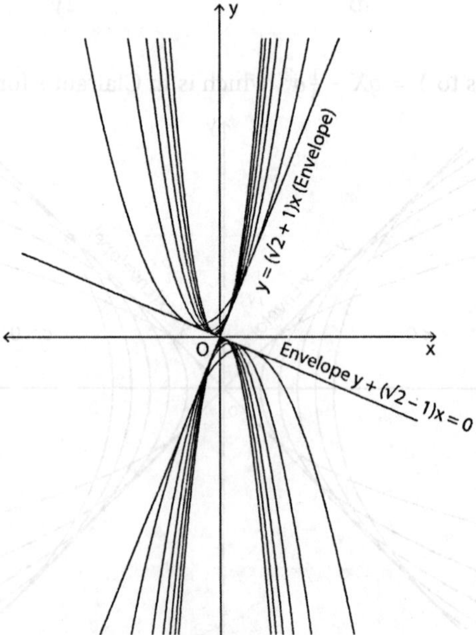

Fig. 3.10

Differentiating w.r.t. u,

$$q = q + u\frac{dq}{du} - \frac{1}{2q^2}\frac{dq}{du}, \quad \text{or,} \quad \frac{dq}{du}\left(u - \frac{1}{2q^2}\right) = 0.$$

When $\frac{dq}{du} = 0$, $q = c$ and the general solution becomes $v = cu + 1/2c$, or, $2c^2x^2 - 2c(y - x) + 1 = 0$ (Fig. 3.10).

Again, when $u = \frac{1}{2q^2}$, taking $v = uq + \frac{1}{2q}$ and simplifying, singular solution is $v^2 = 2u$, or, $(y - x)^2 = 2x^2$.

Example 3.6.9. *Find the general and singular solutions of* $16x^2 + 2p^2y - p^3x = 0$.

Solution: Let us consider the equation as solvable for y.

Simplifying,

$$2y = \frac{p^3x - 16x^2}{p^2} = px - 16\frac{x^2}{p^2}.$$

Differentiating with respect to x, we get

$$2p = p + x\frac{dp}{dx} - 16\frac{p^2 \cdot 2x - x^2 \cdot 2p\, dp/dx}{p^4}.$$

Simplifying,

$$(p^3 + 32x)\left(p - x\frac{dp}{dx}\right) = 0.$$

Either $p^3 + 32x = 0$, or, $p - x\frac{dp}{dx} = 0$.

When $p - x\frac{dp}{dx} = 0$, then $\frac{dp}{p} - \frac{dx}{x} = 0$. Integrating, $p = cx$.

General solution is

$$16x^2 + 2c^2x^2y - c^3x^4 = 0. \tag{3.6.2}$$

If $p^3 + 32x = 0$, then simplifying with $16x^2 + 2p^2y - p^3x = 0$ we have $27x^6 + 2x^2y^3 = 0$ as the singular solution, since it is not contained in the general solution.

Otherwise. Reduction to CLAIRAUT'S FORM.

Let $x^2 = u$, $y = v$, then

$$q = \frac{dv}{du} = \frac{dy}{2xdx} = \frac{p}{2x} = \frac{p}{2\sqrt{u}}.$$

Putting in the given equation $v = uq - \frac{2}{q^2}$, which is in *Clairaut's form.*

Differentiating w.r. to u,

$$q = q + u\frac{dq}{du} + \frac{4}{q^3}\frac{dq}{du}, \quad \text{or,} \quad \frac{dq}{du}\left(u + \frac{4}{q^3}\right) = 0.$$

When $\frac{dq}{du} = 0$, $q = c$. Thus, the general solution is

$$v = cu - \frac{2}{c^2}, \quad \text{or,} \quad 2 + c^2y - c^3x^2 = 0. \tag{3.6.3}$$

Again, when $u + \frac{4}{q^3} = 0$, $q^3 = -\frac{4}{u}$. Thus, the singular solution is $vq^2 = uq^3 - 2 = -6$.

Cubing, $v^3 \cdot 16 + 216u^2 = 0$, or, $2y^3 + 27x^4 = 0$.

Note 3.6.1. *See that* 3.6.2 *and* 3.6.3 *are the same, only that* 3.6.2 *will change to* 3.6.3 *by replacing the constant c by* 2*c. Similar for singular solution.*

Examples III(C)

1. Obtain the complete primitives and singular solutions of the following Clairaut's equations:

(a) $y = px + \dfrac{m}{p}$.

(b) $y = px + p - p^2$.

(c) $y = p(x - a + ap)$.

(d) $y = p(x - b) + a/p$.

(e) $(y - 1)p - xp^2 + 2 = 0$.

(f) $(x - a)p^2 + (x - y)p - y = 0$.

(g) $y = px + \sin^{-1} p$.

(h) $(y - px)(p - 1) = p$.

(i) $y = px + \sqrt{a^2p^2 + b^2}$.

(j) $y = px + \sqrt{1 + p^2}$.

(k) $p = \ln(px - y)$.

(l) $\sin px \cos y = \cos px \sin y + p$.

(m) $p^2 (x^2 - a^2) - 2pxy + y^2 - b^2 = 0$.

2. By the substitution $x^2 = u$, $y^2 = v$ (or otherwise) reduce the equation $x^2 + y^2 - (p + p^{-1})\,xy = c^2$ to Clairaut's form and find the general integral and singular solution.

3. The following equations are deducible to Clairaut's form under suitable substitutions. Obtain the complete primitives:

(a) $x^2(y - px) = p^2y$. [Put $x^2 = u$, $y^2 = v$]

(b) $(px - y)(py + x) = h^2p$.

(c) $(x^2 + y^2 - 1)\, p = xy\,(1 + p^2)$.

(d) $axyp^2 + (x^2 - ay^2 - b)\, p - xy = 0$.

(e) $xy(y - px) = x + py$.

(f) $y^2 = pxy + f(py/x)$.

(g) $y = 2px + y^2p^3$. [Put $y^2 = v$]

(h) $e^{3x}(p - 1) + p^3e^{2y} = 0$. [$e^x = u$, $e^y = v$]

(i) $(x^2 + y^2)(1 + p)^2 - 2(x + y)(1 + p)(x + yp) + (x + yp)^2 = 0$. [$x^2 + y^2 = v$, $x + y = u$]

(j) $(px^2 + y^2)(px + y) = (p + 1)^2$. [$u = xy$, $v = x + y$]

(k) $(xp - y)(xp - 2y) + x^3 = 0$. [$y = vx$]

(l) $x^3p^2 + x^2yp + a^3 = 0$. [$y = u$, $x = 1/v$]

(m) $yp^2 - 2xp + y = 0$. [$x = u$, $y^2 = v$]

(n) $(x^2 - a^2)\, p^2 - 2xyp - x^2 = 0$. [$x^2 = u$, $y = v$]

(o) $(px + y)^2 = py^2$. [$u = y$, $v = xy$]

(p) $y^2(y - xp) = x^4p^2$. [$x = 1/u$, $y = 1/v$].

Answers

1.

Complete Primitive	**Singular Solution**
(a) $y = cx + m/c$;	$y^2 = 4mx$.
(b) $y = cx + c - c^2$;	$4y = (x + 1)^2$.
(c) $y = cx + ac^2 - ac$;	$(x + a)^2 + 4ay = 0$.
(d) $y = cx + a/c - bc$;	$y^2 = 4a(x - b)$.
(e) $y = cx + 1 - 2/c$;	$8x + (1 - y)^2 = 0$.
(f) $y = cx - ac^2/(c + 1)$;	$(x + y)^2 = 4ay$.
(g) $y = cx + \sin^{-1} c$;	$y = \sqrt{x^2 - 1} + \sin^{-1}\left(\frac{\sqrt{x^2-1}}{x}\right)$.
(h) $y = cx + c/(c - 1)$;	$(x + y - 1)^2 = 4xy$.
(i) $y = cx + \sqrt{a^2c^2 + b^2}$;	$x^2/a^2 + y^2/b^2 = 1$.
(j) $y = cx + \sqrt{1 + c^2}$;	$x^2 + y^2 = 1$.
(k) $y = cx - e^c$;	$y = x(\ln x - 1)$.
(l) $y = cx - \sin^{-1} c$;	$\pm y = \sqrt{(x^2 - 1)} - \sin^{-1}\sqrt{(1 - 1/x^2)}$.
(m) $y = cx \pm \sqrt{a^2c^2 + b^2}$;	$x^2/a^2 + y^2/b^2 = 1$.

2. $y^2 = Ax^2 + \{c^2A/(A - 1)\}$, where A is an arbitrary constant; $(y - x - c)(y + x + c)(y - x + c)(y + x - c) = 0$.

3. (a) $y^2 = cx^2 + c^2$. S.S. $x^4 + 4y^2 = 0$
(b) $y^2 = cx^2 - \frac{c}{c+1}h^2$.
S.S. $(x^2 - y^2 - h^2)^2 + 4x^2y^2 = 0$
(c) $y^2 = cx^2 + \frac{c}{c-1}$.
(d) $y^2 = cx^2 - \frac{bc}{1+ac}$.
(e) $y^2 = cx^2 + 1 + c$.
(f) $y^2 = cx^2 + f(c)$.
(g) $y^2 = cx + c^3/8$. S.S. $32x^3 + 27y^4 = 0$
(h) $e^y = ce^x + c^3$.
(i) $x^2 + y^2 = c(x + y) - c^2/4$.
(j) $x + y = cxy + c^2$. S.S. $x^2y^2 + 4(x + y) = 0$
(k) $cy = c^2x^2 + x$.
(l) $cxy - a^3c^2x = 1$. S.S. $x(xy^2 - 4a^3) = 0$
(m) $4y^2 = 4cx - c^2$. S.S. $y = \pm x$
(n) $2cy = x^2 - a^2 - c^2$.
(o) $xy = cy - c^2$.
(p) $x = yc + xyc^2$. S.S. $y(y + 4x^2) = 0$

Examples III(D)

(Miscellaneous)

Solve:

1. $x + py^2 - p(1 + xy) = 0$.
2. $p^3 + 2xp^2 - y^2p^2 - 2xy^2p = 0$.
3. $4y^2p^2 + 2pxy(3x + 1) + 3x^3 = 0$.
4. $xp^2 + (y - x)p - y = 0$.
5. $p^2y + (x - y)p - x = 0$.
6. $y^2 + xyp - x^2p^2 = 0$.
7. $p(p + x) = y(x + y)$.
8. $xyp^2 - (x^2 - y^2)p - xy = 0$.
9. $y = 2p + 3p^2$.
10. $y = Ap^2 + Bp^3$.
11. $xp = 1 + p^2$.
12. $(xp - y)^2 = p^2 - 2\frac{y}{x}p + 1$.
13. $y = p^2x + p^3$.
14. $(py + nx)^2 = (y^2 + nx^2)(1 + p^2)$.
15. $(xp - y)^2 = a(1 + p^2)(x^2 + y^2)^{3/2}$.
16. $ayp^2 + (2x - b)p - y = 0$.
17. $y = px + \sqrt{1 + p^2}\phi(x^2 + y^2)$.
18. $p^3 + mp^2 = a(y + mx)$.
19. $y = 3px + 6p^2y^2$.
20. $\cos^2 yp^2 + \sin x \cos x \cos yp - \sin y \cos^2 x = 0$.
21. $p^2x(x - 2) + p(2y - 2xy - x + 2) + y^2 + y = 0$.

Answers

1. $(2y - x^2 - c)(2x - y^2 - c) = 0$.
2. $(y - c)(xy + cy + 1)(y + x^2 - c) = 0$.
3. $(x^2 + 2y^2 - c)(x^3 + y^2 - c) = 0$.
4. $(y - x + c)(xy + c) = 0$.
5. $(y - x + c)(x^2 + y^2 - c^2) = 0$.
6. $\left(y^2 - cx^{1+\sqrt{5}}\right)\left(y^2 - cx^{1-\sqrt{5}}\right) = 0$.
7. $(y - ce^x)(y + x - 1 - ce^{-x}) = 0$.
8. $(xy - c)(y^2 - x^2 - 2c) = 0$.
9. $x = \ln p^2 + 6p + c,\ y = 2p + 3p^2$.
10. $x = 2Ap + \frac{3}{2}Bp^2 + C,\ y = Ap^2 + Bp^3$.
11. $y = \frac{p^2}{2} - \ln p + c,\ x = p + 1/p$.
12. $\sin^{-1} y/x + \sin^{-1} 1/x = c$.
13. $x = \frac{2c+3p^2-2p^3}{2(p-1)^2},\ y = \frac{2cp^2+2p^3-p^4}{2(p-1)^2}$.
14. $y + \sqrt{y^2 + nx^2} = cx^{1\pm\sqrt{(n-1)/n}}$.
15. $\tan^{-1} y/x + c = \text{vers}^{-1} 2a\sqrt{x^2 + y^2}$ ($\text{vers}\,\theta = 1 - \cos\theta$), or, $1 - 2a\sqrt{x^2 + y^2} = \cos\left(\tan^{-1}\frac{y}{x} + c\right)$.
16. $ac^2 + c(2x - b) - y^2 = 0$.
17. $v + c = \int \frac{\phi(r^2)\,dr}{r\sqrt{r^2 - \phi^2(r^2)}}$. $\left.\begin{array}{l}\theta = \tan^{-1} y/x\\ r^2 = x^2 + y^2\end{array}\right\}$.
18. $ax + c = \frac{3}{2}p^2 - mp + m^2\ln(p + m)$ and the given equation.
19. $y^3 = 3cx + 6c^2$.
20. $\sin y = c\sin x + c^2$.
21. $(y - cx + 1)(y - cx + 2c) = 0$.

Chapter 4

Singular Solutions: Extraneous Loci

In the last chapter we have observed that a solution of differential equation of first order and higher degree can sometimes be found such that it is not deducible from the complete primitive by attributing a special value to the arbitrary constant and, moreover, it does not contain any constant in its expression; such a solution was called a Singular Solution. *We have seen that a singular solution exists for a* Clairaut's equation *and its locus is the* envelope *of the family of straight lines given by the general solution. In fact, one definition of singular solution is that it is the envelope of the system of curves given by the complete primitive. The analytical theory of singular solution is beyond the scope of this treatise. Our main task will be the determination of singular solutions without going into deep analytical arguments.*

4.1 Singular solutions: Extraneous loci

1. In the following discussions we shall consider only differential equations of first order. Unless otherwise mentioned we shall denote a differential equation of first order and nth degree by $f(x, y, p) = 0$ and its complete primitive by $\phi(x, y, c) = 0$; $p = dy/dx$ and c is an arbitrary constant.

We have observed in art. 1.6, **III** that $\phi(x, y, c) = 0$ represents a one-parameter family of curves and in general, n of these curves pass through a particular point in the domain of definition of f; the constant c appears in the nth degree in $\phi(x, y, c)$.

2. The condition that c has two or more equal values is that $\partial\phi/\partial c = 0$. If we eliminate c between this and $\phi(x, y, c) = 0$, we call it a **c-discriminant relation** and denote it by $\text{Disct}_c\phi(x, y, c) = 0$.

Similarly, we define the **p-discriminant relation** $\text{Disct}_p f(x, y, p) = 0$ obtained by eliminating p between $f(x, y, p) = 0$ and $\partial f/\partial p = 0$. This relation expresses the condition that p has multiple roots in $f(x, y, p) = 0$.

3. We recall a few definitions.

A point of a curve in a plane is called a *double point* if two branches of the curve pass through it. In general, we expect two tangents (one for each branch) at a double point. The double point is a **Node** or a **Cusp** according as the two tangents are *real and distinct* or *real and coincident.*

A point through which r branches of a curve pass is called a *multiple point* of rth order (in particular, a *triple point* is a multiple point of 3rd order). A multiple point is also called a *singular point.* A multiple point is a *Node* or a *Cusp* according as the tangents are distinct or coincident.

4. The c-discriminant $\text{Disct}_c\phi(x, y, c) = 0$ gives the locus of all points in the plane at which the parameter c has two or more equal values. In it there will, therefore, be included the following:

(i) **The Nodal Locus:** The locus of all the *nodal points* (double, triple, etc.) of the system of curves $\phi(x, y, c) = 0$; for at such a point there are as many equal values of c as there are branches of the curve through the point;

(ii) **The Cuspidal Locus:** The locus of all the *cusps* of the system of curves $\phi(x, y, c) = 0$, for similar reasons;

(iii) **The Envelope Locus:** The envelope of the system of curves $\phi(x, y, c) = 0$; for any point on the envelope may be considered as belonging to two separate but consecutive curves of the system, the constants of these consecutive curves being ultimately equal.

We shall add here (analytical theory omitted) that the *c*-discriminant contains the envelope locus as a factor once, the nodal locus twice and the cusp locus thrice; thus: $\mathbf{Disct_c\phi(x, y, c) = EN^2C^3}$.

5. We shall see the truth of following three statements:

(i) The *Envelope* of the system of curves $\phi(x, y, c) = 0$ when exists, satisfies the differential equation $f(x, y, p) = 0$; such a solution of the differential equation is known as its *singular solution;*

(ii) The *Nodal locus* will not, in general, be a solution of the differential equation;

(iii) The *Cuspidal locus* will not also, in general, be a solution of the differential equation.

[(i) The envelope satisfies the differential equation; for at any point on the envelope (which is a point on two consecutive curves) the direction of the tangent is the same as that of the tangent to either of these curves at that point; and since the differential equation is satisfied by the quantities which are connected with the element of the system of curves, it must be satisfied by these (unaltered) quantities, which are connected with the element of the envelope.

(ii) But the nodal locus is, in general, not a solution of the differential equation; for if it were, then for the values of x and y at any node, the equation would be satisfied by the corresponding value of p at this point on the nodal locus. Remembering that the nodal locus is formed by a series of points on our system of curves, we know that the values of p at any such point which satisfy the differential equation are those given by that curve of the system which passes through the point. But as the tangent to the nodal locus at such a point will not, in general, be a tangent to any of the branches of the curve of the system at the point, it follows that the value of p for the nodal locus differs from those values of p for the curve of the system, which satisfy the equation when substituted in it with the co-ordinates of the point. And in exceptional cases the values of p for the nodal locus could coincide with any of the remaining values of p, which do not belong to the curve on which the node lies, but are furnished by other curves of the system through that point. Hence, the value of p for the nodal locus at the point will usually be such that will not satisfy the differential equation; and the nodal locus will therefore, in general, not be a solution of the differential equation:

Exactly similar considerations, applied to the cuspidal locus, lead to the conclusion (iii).]

6. The envelope of the system of curves $\phi(x, y, c) = 0$ [General solution of $f(x, y, p) = 0$] can be obtained directly from the differential equation without first obtaining the primitive.

At any point of the envelope, at least two of the branches of the different curves coincide in direction and hence for such a point we shall have equal values of p belonging to *different but consecutive curves.*

$\therefore$ the p-discriminant relation $\text{Disct}_p f(x, y, p) = 0$ includes as a factor the equation of the envelope (provided envelope exists).

Besides the envelope locus the p-discriminant includes as a factor

(i) the equation of the **cuspidal locus** once;

(The equation of the cuspidal locus does not satisfy the differential equation unless it is also a particular solution or a singular solution);

(ii) the equation of the **tac-locus** twice;

(The equation of the tac-locus does not satisfy the differential equation unless it is also a singular solution or a particular solution).

[*Tac-locus* is the locus of all points at which two curves of the system, which are different but not consecutive, touch; See Fig. 4.3]

We remember **Disct$_p f(x, y, p) = ET^2C$.**

7. Thus, of all the loci, the only one which satisfies the differential equation is the envelope locus and this solution of the differential equation is called the '*Singular Solution*'. This locus occurs as a factor of both p- and c-discriminants.

Example 4.1.1. *Examine for singular solution* $8ap^3 = 27y$.

Solution: Its general solution is $ay^2 = (x - c)^3$ (Fig. 4.1).

The p-discriminant relation [from the given equation] is $y = 0$ and the c-discriminant relation [from general solution] is $y^4 = 0$.

Here $y = 0$ occurs in both the discriminants and also satisfies the differential equation. Hence, $y = 0$ is a singular solution.

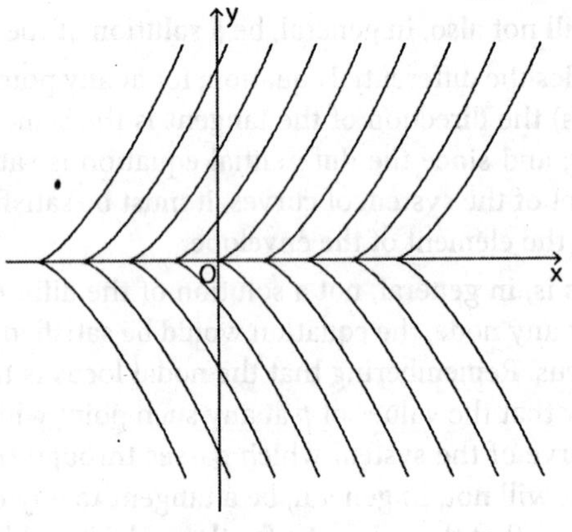

Fig. 4.1

Example 4.1.2. *Find the singular solution, if any of* $y = 2px + p^2$.

Solution: The p-discriminant relation is $4x^2 + 4y = 0$, or, $x^2 + y = 0$ and see that it does not satisfy the differential equation, hence there cannot be any singular solution of the equation.

Example 4.1.3. *Find the general and singular solution (if at all exists) of* $xp^2 - 2py + 4x = 0$.

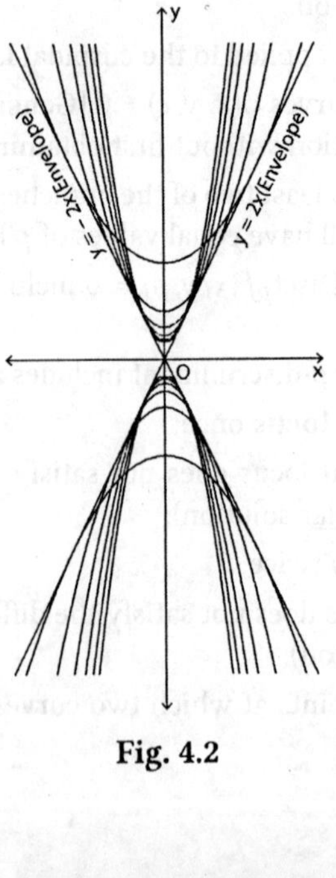

Fig. 4.2

Solution: $y = \frac{p^2x+4x}{2p}$. Differentiating w.r.t. x and simplifying $dp/p = dx/x$. Integrating $p = cx$ and hence the general solution becomes $c^2x^2 - 2cy + 4 = 0$ (Fig. 4.2).

For *singular solution,* $y = \pm 2x$ occurs in both p-discriminant and c-discriminant relations and also satisfies the differential equation. Hence, $y = \pm 2x$ are the singular solutions.

Example 4.1.4. *Solve:* $4xp^2 - (3x - 1)^2 = 0$ *and examine for singular solutions and extraneous loci.*

Solution: Solving, $p = \pm\left(\frac{3}{2}\sqrt{x} - \frac{1}{2\sqrt{x}}\right)$.

Integrating, the complete primitive is $(y + c)^2 = x(x - 1)^2$.

Now $x = 0$ occurs in both the p-discriminant relation $x(3x - 1)^2 = 0$ and the c-discriminant relation $x(x - 1)^2 = 0$. Also $x = 0$ satisfies the differential equation.

$\therefore$ $x = 0$ is a *singular solution* (Envelope Locus).

Again, $(3x - 1) = 0$ which occurs twice in the p-discriminant, and does not satisfy the differential equation (verify) and does not occur in the c-discriminant.

$\therefore$ $3x - 1 = 0$ is a *Tac-Locus.*

Also $x - 1 = 0$ is a *Nodal Locus* since it occurs twice in the c-discriminant, does not satisfy the differential equation and does not occur in the p-discriminant relation.

Note 4.1.1. *The primitive represents a family of cubics obtained by moving* $y^2 = x(x-1)^2$ *along the y-axis. These curves are tangents to the y-axis and have a double point at* $x = 1$. *Through each point on* $x = 1/3$ *there pass two curves of the family having a common tangent there. See Figure 4.3.*

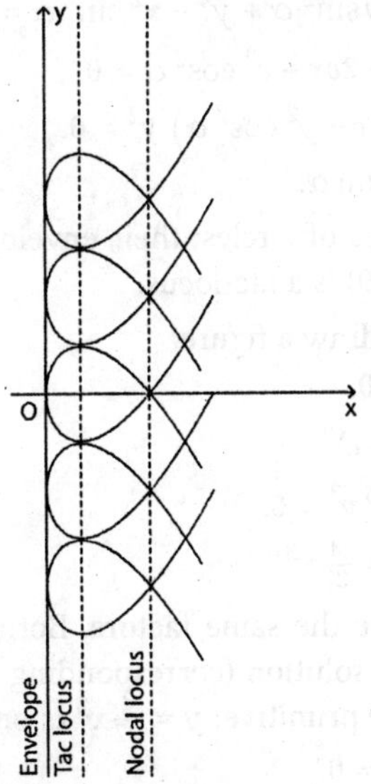

Fig. 4.3 Family of cubics $(y + c)^2 = x(x - 1)^2$.

Example 4.1.5. *Solve:* $9yp^2 + 4 = 0$ *and examine for singular solutions and extraneous loci.*

Solution: We have $y = -\frac{4}{9p^2}$.

Differentiating w.r.t. x,

$$p = \frac{8}{9p^3}\frac{dp}{dx}, \quad \text{or,} \quad dx = \frac{8}{9}\frac{1}{p^4}dp, \quad \therefore x + c = -\frac{8}{27p^3}.$$

Eliminating p between this last relation and the given equation, we obtain the primitive in the form:

$$y^3 + (x + c)^2 = 0.$$

$$\left.\begin{array}{ll} \text{The } p\text{-discriminant is} & y = 0; \\ \text{The } c\text{-discriminant is} & y^3 = 0; \\ y = 0 \text{ does not satisfy the equation.} & \end{array}\right\}$$

Since $y = 0$ occurs once in the p-discriminant, thrice in the c-discriminant and does not satisfy the differential equation, it is a *Cuspidal Locus.*

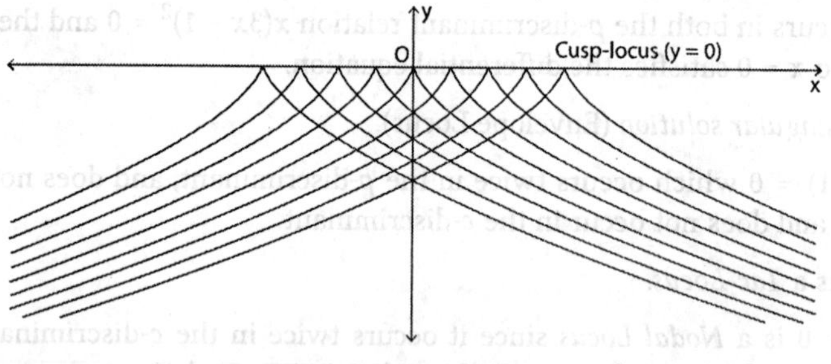

Fig. 4.4 Family of semi-cubical parabolas $y^3 + (x + c)^2 = 0$.

The primitive represents a family of semi-cubical parabolas obtained by moving $y^3 + x^2 = 0$ along the x-axis. Each curve has a cusp at the point where it cuts the x-axis and x-axis is the locus of these cusps. See Fig. 4.4.

Example 4.1.6. $p^2y^2\cos^2\alpha - 2pxy\sin^2\alpha + y^2 - x^2\sin^2\alpha = 0.$

Solution: Primitive: $x^2 + y^2 - 2cx + c^2\cos^2\alpha = 0.$

p-discriminant: $(x^2\sin^2\alpha - y^2\cos^2\alpha)\,y^2 = 0.$

c-discriminant: $y = \pm x\tan\alpha.$

The primitive represents a series of circles; their envelope is the two straight lines $y = \pm x\tan\alpha$ (singular solution). The x-axis ($y = 0$) is a tac-locus.

Verify all these statements and draw a figure.

Example 4.1.7. $p^3 - 4xyp + 8y^2 = 0.$

Solution: Primitive: $y = c(x - c)^2.$

p-discriminant: $y^4 - \frac{4}{27}x^3y^3 = 0.$

c-discriminant: $y = 0,\ y = \frac{4}{27}x^3,$

so that both the discriminants have the same factors. Both $y = 0$, $y = \frac{4}{27}x^3$ satisfy the differential equation. But $y = 0$ is a particular solution (corresponding to $c = 0$) and is therefore both a singular solution and a particular case of the primitive; $y = \frac{4}{27}x^3$ is only a singular solution.

Example 4.1.8. $p^2y + p(x - y) - x = 0.$

Solution: p-discriminant: $(x - y)^2 + 4xy = 0$, or, $(x + y)^2 = 0$, i.e. $y = -x$.

We can see that it is not a solution of the differential equation; hence it cannot be a singular solution.

Now the equation can be written as $(p - 1)(py + x) = 0$, the solutions of which are $y - x = c$, and $y^2 + x^2 = c$.

This is an example to show that if a first-order equation be reducible to linear and rational factors, it has no singular solution.

Example 4.1.9. *Solve and find the singular solution of* $x^3p^2 + x^2py + a^3 = 0.$

Solution: $y = -\frac{x^3p^2+a^3}{x^2p} = -px - \frac{a^3}{px^2}$.

$$\therefore \frac{dy}{dx} = p = -p - x\frac{dp}{dx} + \frac{a^3}{p^2x^2}\frac{dp}{dx} + \frac{2a^3}{px^3}.$$

Simplifying, $(p^2x^3 - a^3)\left(\frac{2}{px^3} + \frac{1}{p^2x^2}\frac{dp}{dx}\right) = 0$.

From $\frac{2}{px^3} + \frac{1}{p^2x^2}\frac{dp}{dx} = 0$, we have $\frac{2}{x}dx + \frac{1}{p}dp = 0$.

Integrating, $2\ln x + \ln p = \ln c$, or, $p = c/x^2$.

$\therefore$ from given relation, the general solution becomes

$$x^3 \cdot \frac{c^2}{x^4} + x^2y \cdot \frac{c}{x^2} + a^3 = 0, \quad \text{or,} \quad a^3x + cxy + c^2 = 0.$$

Next from $p^2x^3 - a^3 = 0$, we have from the given relation, the singular solution $a^3 + x^2py + a^3 = 0$, or, $x^4p^2y^2 = 4a^6$, or, $a^3xy^2 = 4a^6$, or, $xy^2 = 4a^3$.

Example 4.1.10. *Obtain the differential equation satisfied by the family of curves $c^2 + 2cy - x^2 + 1 = 0$, where c is the parameter. Find also the singular solution of this equation.*

Solution: Differentiating the given equation w.r.t. x,

$$2c\frac{dy}{dx} - 2x = 0, \quad \text{or,} \quad c = \frac{x}{p}.$$

Therefore, the differential equation is

$$\frac{x^2}{p^2} + 2\frac{x}{p}y - x^2 + 1 = 0, \quad \text{or,} \quad x^2 + 2xyp + (1 - x^2)p^2 = 0.$$

Here c-discriminant relation is $4y^2 = 4(1 - x^2)$ and p-discriminant relation is $4x^2y^2 = 4x^2(1 - x^2)$.

Hence, the singular solution is the common $y^2 = 1 - x^2$, or, $x^2 + y^2 = 1$.

Example 4.1.11. *Reduce $x^2p^2 + py(2x + y) + y^2 = 0$ to Clairaut's form by the substitution $y = u$, $xy = v$. Hence, solve the equation and prove that $y + 4x = 0$ is a singular solution.*

Solution: $y = u$, $xy = v$. Differentiating $dy = du$, $x\,dy + y\,dx = dv$ and $x + y/p = q$. Put $p = \frac{y}{q-x}$ in the given relation; then $qy - xy + q^2 = 0$, or, $uq - v + q^2 = 0$.

Thus, $v = uq + q^2$ is in Clairaut's form.

$\therefore$ general solution is $v = cu + c^2$, or, $xy = cy + c^2$.

The singular solution (c-discriminant) is given by $u^2 + 4v = 0$, i.e. $y^2 + 4xy = 0$, or, $y + 4x = 0$ and $y = 0$.

Here $y = 0$ is also a particular solution corresponding to $c = 0$.

Examples IV

Solve the following differential equations and examine for singular solutions (No. 1–6):

1. $y = 2xp - yp^2$.
2. $p^3 + px - y = 0$.
3. $6p^2y^2 + 3px - y = 0$.
4. $xp^2 - 2yp + x + 2y = 0$.
5. $y = 2px + p^2$.
6. $8a(1 + p)^3 = 27(x + y)(1 - p)^3$.
7. Examine for singular solutions, if any, of the following: (a) $p^2 + 2px = 3x^2$ (b) $x^2p^2 - 3xyp + 2y^2 + x^2 = 0$.

8. Show that the complete primitive of $8p^3x = y\left(12p^2 - 9\right)$ is $(x + c)^3 = 3y^2c$, the p-discriminant is $y^2\left(9x^2 - 4y^2\right) = 0$, and the c-discriminant is $y^4\left(9x^2 - 4y^2\right) = 0$. Interpret these discriminants.
9. Reduce $x^2p^2 + py(2x + y) + y^2 = 0$ to Clairaut's form by the substitution $y = u$, $xy = v$. Hence, solve the equation and prove that $y + 4x = 0$ is a singular solution.
10. Show that the system of curves $y^2 - 2cx^2y + c^2\left(x^4 - x^3\right) = 0$, all have a cusp at the origin, touching the axis of x. Obtain the corresponding differential equation in the form: $4p^2x^2(x - 1) - 4pxy(4x - 3) + (16x - 9)y^2 = 0$.

    Verify that both p- and c-discriminants take the form $x^3y^2 = 0$, but that $x = 0$ is not a solution while $y = 0$ satisfies the equation.
11. Show that the complete primitive of $r^4 + r^2\left(\frac{dr}{d\theta}\right)^2 = a^4$ represents the family of equal lemniscates of Bernoulli $r^2 = a^2\cos(2\theta - 2\alpha)$ inscribed in the circle $r = a$, which is a singular solution, with the point $r = 0$ as a nodal locus.
12. Solve the following equations and obtain the singular solutions where these exist; specify the nature of the extraneous loci:

    (a) $xp^2 - 2yp + 4x = 0$.
    (b) $\left(x^2 - a^2\right)p^2 - 2xyp - x^2 = 0$.
    (c) $p^4 = 4y(xp - 2y)^2$.
    (d) $x^3p^2 + x^2yp + a^3 = 0$.
13. (a) Solve $x^3p^2 + x^2py + 1 = 0$ and discuss about its singular solution.

    (b) Reduce the differential equation $y = 2px - yp^2$, ($p = dy/dx$) to Clairaut's form by the substitution $y^2 = Y$, $x = X$ and then obtain its complete primitive and singular solution, if any.

    (c) Find the general and singular solutions of $16x^2 + 2p^2y - p^3x = 0$, ($p = dy/dx$).
14. Solve and examine for singular solutions:

    (a) $\left(8p^3 - 27\right)x = 12p^2y$.
    (b) $p^2\left(x^2 - a^2\right) - 2pxy + y^2 - b^2 = 0$.
    (c) $(px - y)(x - py) = 2p$.
15. Solve the following equations and examine for singular solutions and extraneous loci:

    (a) $\left(x^2 - 4\right)p^2 - 2xyp - x^2 = 0$.
    (b) $x^3p^2 + x^2yp + 1 = 0$.
    (c) $p^3x - 2p^2y - 16x^2 = 0$.
    (d) $y(3 - 4y)^2p^2 = 4(1 - y)$.
    (e) $(3y - 1)^2p^2 = 4y$.
    (f) $y = -px + x^4p^2$.
    (g) $2y = p^2 + 4xp$.
16. Use $x = r\cos\theta$, $y = r\sin\theta$ to solve $\left(p^2 + 1\right)(x - y)^2 = (x + yp)^2$ and then prove that $xy = 0$ constitutes singular solutions and that $y = x$ is the tac-locus.
17. Show that the primitive of the equation $3y = 2px - 2p^2/x$ represents a family of semi-cubical parabolas with their cusps on the axis of y. Verify that $x = 0$ is the cusp locus and $6y = x^3$ represents the envelope. Draw the graph.
18. Investigate for singular solutions and other loci $4x(x - 1)(x - 2)p^2 - \left(3x^2 - 6x + 2\right)^2 = 0$. Discuss the graph.

Answers

[C.P. = Complete Primitive, S.S. = Singular Solution, T.L. = Tac Locus, N.L. = Nodal Locus, C.L. = Cusp Locus]

	C.P.	S.S.
1.	$y^2 = 2cx - c^2$.	$y = \pm x$.

2. $y = cx + c^3$. $\quad 4x^3 + 27y^2 = 0$.
3. $y^3 = 3cx + 6x^2$. $\quad 3x^2 + 8y^3 = 0$.
4. $2c^2x^2 - 2c(y - x) + 1 = 0$. $\quad x^2 + 2xy - y^2 = 0$.
5. $(3xy + 2x^3 + c)^2 = 4(x^2 + y)^3$. No. W.O. Ex. 4.2.2.
6. $(x - y + c)^3 = a(x + y)^2$. $\quad x + y = 0$.
7. (a) No. S.S. (b) No.
9. C.P.: $xy = cy + c^2$.
12. (a) C.P.: $c^2x^2 - 2cy + 4 = 0$; S.S.: $y = \pm 2x$.
   (b) C.P.: $c^2 + 2cy + a^2 = x^2$; S.S.: $x^2 + y^2 = a^2$; T.L.: $x = 0$.
   (c) C.P.: $y = c^2(x - c)^2$; S.S.: $x^4 - 16y = 0$. S.S. also particular, $y = 0$.
   (d) C.P.: $a^3x + cxy + c^2 = 0$; S.S.: $(xy^2 - 4a^2) = 0$; T.L.: $x = 0$.
13. (a) C.P.: $x + cxy + c^2 = 0$; S.S.: $xy^2 = 4$.
   (b) C.P.: $y^2 = cx - \frac{1}{4}c^2$; S.S.: $y = \pm x$.
   (c) C.P.: $2 + c^2y - c^3x^2 = 0$; S.S.: $2y^3 + 27x^4 = 0$.
14. (a) $x\left\{x^3 - c(y + c)^2\right\} = 0$; S.S.: $4y^3 + 27x^3 = 0$; $x = 0$ is a particular solution and also a cusp locus.
   (b) $y = cx + \sqrt{a^2c^2 + b^2}$; S.S.: $b^2x^2 + a^2y^2 = a^2b^2$.
   (c) $x^2 + y^2 - c\left(x^2 - y^2\right) - 1 + c^2 = 0$; S.S.: $x^4 - 2x^2y^2 + y^4 - 4x^2 - 4y^2 + 4 = 0$.
15. (a) $c^2\left(x^2 - 4\right) - 2cy - 1 = 0$; S.S.: $x^2 + y^2 = 4$. T.L.: $x = 0$.
   (b) $c^2 + cxy + x = 0$; S.S.: $xy^2 = 4$; $x = 0$ is a particular solution.
   (c) $c^3x^2 - c^2y - 2 = 0$; S.S.: $2y^3 + 27x^4 = 0$. T.L.: $x = 0$. $x = 0$ is also a particular solution (How?).
   (d) $(x - c)^2 = y^3(1 - y)$; S.S.: $y = 1$; C.L.: $y = 0$; T.L.: $y = 3/4$.
   (e) $(x + c)^2 = y(y - 1)^2$; S.S.: $y = 0$; T.L.: $y = 1/3$; N.L.: $y = 1$.
   (f) $xy = c + c^2x$; S.S.: $4x^2y + 1 = 0$; T.L.: $x = 0$.
   (g) $\left(4x^3 + 3xy + c\right)^2 = 2\left(2x^2 + y\right)^3$; no S.S.; C.L.: $2x^2 + y = 0$.
18. $(y + c)^2 = x(x - 1)(x - 2)$ is the C.P.; S.S. $x = 0$, $x = 1$, $x = 2$; $x = 1 \pm \frac{1}{\sqrt{3}}$ are T.L. The curve when $c = 0$ consists of an oval cutting the axis of x at the origin and at $x = 1$, and a curve resembling a parabola in shape, having its vertex at the point for which $x = 2$.

Chapter 5

Linear Equations with Constant Coefficients

In this chapter, we would like to consider only linear equations of an order higher than the first. In this class of equations, the dependent variable and its derivatives appear only in the first degree and they are not multiplied together. These type of equations originates in a variety of applications to engineering and science. Here our main aim is to study the method of solving the linear equations with constant coefficients. In order to study the explicit methods of solution, we need some basic theorems. One of the most important parts of this chapter is Wronskian by which we can examine whether a set of solutions of an equation is a fundamental set or not. In order to find the general solution of a linear equation of higher order, first we need a fundamental set of solutions. Then using the principle of superposition, the general solution is obtained. General solution can be found also determining complementary function and particular integral. In addition, we have two other methods, namely, method of undetermined coefficients and method of variation of parameters for solving such equations. A special type of linear equation with variable coefficients can be reduced in a linear equation with constant coefficients by a suitable substitution. This type of equations is known as Cauchy-Euler equations named after two mathematicians Cauchy and Euler.

5.1 Equations of nth order

Linear equations occur so frequently in applications (particularly in Mathematical Physics) as to call for a somewhat detailed treatment which we propose to do.

The general form of a linear equation of nth order is

$$\frac{d^n y}{dx^n} + P_1 \frac{d^{n-1} y}{dx^{n-1}} + P_2 \frac{d^{n-2} y}{dx^{n-2}} + \cdots + P_{n-1} \frac{dy}{dx} + P_n y = Q, \tag{5.1.1}$$

where $P_1, P_2, \ldots, P_n$ and Q are constants or functions of x (but do not contain y). In the present chapter, however, we shall assume that the coefficients $P_1, P_2, \ldots, P_n$ are constants. Further in articles 5.2, 5.3 and 5.4 we shall consider simpler equations where Q is supposed to be 0.

5.2 Equations of first order: the simplest case

In (5.1.1) let us take $n = 1$ and $Q = 0$. We then obtain

$$\frac{dy}{dx} + P_1 y = 0. \tag{5.2.1}$$

Separating the variables we easily get $y = Ae^{-P_1 x}$, where A is an arbitrary constant.

5.3 Equations of second order

If we take $n = 2$ and $Q = 0$, eq. (5.1.1) becomes

$$\frac{d^2y}{dx^2} + P_1\frac{dy}{dx} + P_2 y = 0. \tag{5.3.1}$$

The solution of eq. (5.2.1) suggests that $y = Ae^{mx}$, where m is a constant and may satisfy (5.3.1). With this value of y, eq. (5.3.1) reduces to $Ae^{mx}\left(m^2 + P_1 m + P_2\right) = 0$.

Thus, if m is a root of

$$m^2 + P_1 m + P_2 = 0, \tag{5.3.2}$$

$y = Ae^{mx}$ is a solution of eq. (5.3.1), whatever may be the value of A.

Let the eq. (5.3.2) have two *distinct real roots* α and β. Then we have two solutions of eq. (5.3.1), namely $y = Ae^{\alpha x}$ and $y = Be^{\beta x}$, where A and B are arbitrary constants.

Now if we substitute $y = Ae^{\alpha x} + Be^{\beta x}$ in the left side of eq. (5.3.1), we shall obtain

$$Ae^{\alpha x}\left(\alpha^2 + P_1\alpha + P_2\right) + Be^{\beta x}\left(\beta^2 + P_1\beta + P_2\right),$$

which is obviously 0 as α and β are the roots of eq. (5.3.2).

Thus, $y = Ae^{\alpha x} + Be^{\beta x}$ is a solution of equation (5.3.1) and since it contains two independent arbitrary constants A and B, we may regard it as the *general solution* or *complete primitive* of eq. (5.3.1).

We shall call $y = Ae^{mx}$ as a *trial solution* and the equation (5.3.2) as the *auxiliary equation.*

Example 5.3.1. *Solve:* $\frac{d^2y}{dx^2} + 3\frac{dy}{dx} - 54y = 0.$

Solution: We take $y = Ae^{mx}$ as a trial solution. This leads to $Ae^{mx}\left(m^2 + 3m - 54\right) = 0$, which is satisfied by $m = 6, -9$.

The general solution is, therefore, $y = Ae^{6x} + Be^{-9x}$, where A and B are arbitrary constants.

Two special cases

I. *When the auxiliary equation has imaginary roots*

When the auxiliary eq. (5.3.2) has complex roots of the form $\alpha + i\beta$ and $\alpha - i\beta(\alpha, \beta$ real), then the general solution is

$$\begin{aligned} y &= Ae^{(\alpha+i\beta)x} + Be^{(\alpha-i\beta)x} = e^{\alpha x}\left[Ae^{i\beta x} + Be^{-i\beta x}\right] \\ &= e^{\alpha x}\{A(\cos\beta x + i\sin\beta x) + B(\cos\beta x - i\sin\beta x)\} \\ &= e^{\alpha x}[E\cos\beta x + F\sin\beta x], \end{aligned}$$

writing E for $A + B$ and F for $i(A - B)$.

We note that E and F are arbitrary constants, just as A and B are. E and F need not be imaginary; take, for example $A = 1 + 3i$, $B = 1 - 3i$.

Example 5.3.2. *Solve:* $\frac{d^2y}{dx^2} - 6\frac{dy}{dx} + 13y = 0.$

Solution: If the trial solution is taken as $y = Ae^{mx}$, the auxiliary equation becomes $m^2 - 6m + 13 = 0$ whose roots are $3 \pm 2i$.

The general solution may be written in the form:

$$y = Ae^{(3+2i)x} + Be^{(3-2i)x} \tag{5.3.3}$$

or, in the form:

$$y = e^{3x}(E\cos 2x + F\sin 2x) \tag{5.3.4}$$

or, in the form:

$$y = Ce^{3x}\cos(2x - \varepsilon), \tag{5.3.5}$$

where $E = C\cos\varepsilon$ and $F = C\sin\varepsilon$.

Note 5.3.1. *The students are advised to give the result either in the form* (5.3.4) *or in the form* (5.3.5).

II. *Case of Equal Roots*

When the auxiliary equation has equal roots α and α, the solution becomes

$$y = Ae^{\alpha x} + Be^{\alpha x} = (A + B)e^{\alpha x}.$$

Now $A+B$ is really only a single arbitrary constant. Hence, it cannot be regarded as the most general solution. We shall prove later (Art. 5.7) that the general solution in such a case will be $y = (A + Bx)e^{\alpha x}$, A and B being arbitrary constants.

Example 5.3.3. *Solve:* $\frac{d^2y}{dx^2} + 4\frac{dy}{dx} + 4y = 0$.

Solution: The trial solution is $y = Ae^{mx}$.

The auxiliary equation is $m^2 + 4m + 4 = 0$, which gives $m = -2, -2$. Hence, the general solution is $y = (A + Bx)e^{-2x}$.

Thus, introducing the operators D for $\frac{d}{dx}$, D^2 for $\frac{d^2}{dx^2}$, we may come to the following table:

Table II

Differential equation $\left(D^2 + P_1 D + P_2\right) y = 0$

Auxiliary equation: $m^2 + P_1 m + P_2 = 0$

Nature of the roots of the auxiliary equation	Condition on the coefficients of the auxiliary equation	General solution of the differential equation
Real and unequal $m_1 \neq m_2$	$P_1^2 - 4P_2 > 0$	$y = c_1 e^{m_1 x} + c_2 e^{m_2 x}$
Real and equal $m_1 = m_2 = \alpha$	$P_1^2 - 4P_2 = 0$	$y = e^{\alpha x}\{c_1 + c_2 x\}$
Conjugate complex $\alpha \pm i\beta$	$P_1^2 - 4P_2 < 0$	$y = e^{\alpha x}\{A\cos\beta x + B\sin\beta x\}$ or, $= Re^{\alpha x}\cos(\beta x + \phi)$ or, $= R'e^{\alpha x}\sin(\beta x + \theta)$

5.4 Equations of orders higher than the second

The methods of article 5.3 apply to any linear equation with constant coefficients, the right-hand member being 0. The students should be acquainted with problems which can be solved with the help of the standard rules given below. The justification of the rules and the general theory will be given in the next few articles.

Let the equation be

$$\frac{d^n y}{dx^n} + P_1\frac{d^{n-1}y}{dx^{n-1}} + \cdots + P_{n-1}\frac{dy}{dx} + P_n y = 0 \tag{5.4.1}$$

and suppose that the trial solution be $y = Ae^{mx}$.

The auxiliary equation (AE) is

$$m^n + P_1 m^{n-1} + \cdots + P_{n-1}m + P_n = 0. \tag{5.4.2}$$

Rule I. Let $m_1, m_2, m_3, \ldots, m_n$ be n *distinct* real roots of (5.4.2). The general solution is then

$$y = c_1 e^{m_1 x} + c_2 e^{m_2 x} + \cdots + c_n e^{m_n x},$$

where $c_1, c_2, \ldots, c_n$ are arbitrary constants.

Rule II. If two roots m_1 and m_2 of the auxiliary equation are equal, each equal to m, the corresponding part of the general solution is $(c_1 + c_2 x)\, e^{mx}$, and if three roots m_1, m_2, m_3 are each equal to m, the corresponding part of the solution is $\left(c_1 + c_2 x + c_3 x^2\right) e^{mx}$; and so on.

Rule III. If the AE has a pair of imaginary roots, say $m_1 = \alpha + i\beta$, $m_2 = \alpha - i\beta$, the corresponding part of the general solution is $e^{\alpha x}(c_1 \cos\beta x + c_2 \sin\beta x)$ or, $c_3 e^{\alpha x}\cos(\beta x + c_4)$ or, $c_5 e^{\alpha x}\sin(\beta x + c_6)$.

Any one of these forms may be used.

Rule IV. If a pair of imaginary roots $\alpha \pm i\beta$ occur twice, the corresponding part of the general solution is $e^{\alpha x}\left[(c_1 + c_2 x)\cos\beta x + (c_3 + c_4 x)\sin\beta x\right]$.

[For, $(A + Bx)e^{(\alpha+i\beta)x} + (C + Dx)e^{(\alpha-i\beta)x}$ can be put in the said form.]

Rule V. If the two real roots be m and $-m$, the corresponding part of the solution is $c_1 e^{mx} + c_2 e^{-mx}$ or, $c_1(\cosh mx + \sinh mx) + c_2(\cosh mx - \sinh mx)$ or, $c_3 \cosh mx + c_4 \sinh mx$.

Example 5.4.1. *Solve:* $\frac{d^3y}{dx^3} - 3\frac{d^2y}{dx^2} + 4y = 0.$

Solution: The auxiliary equation is

$m^3 - 3m^2 + 4 = 0$ or, $(m+1)\left(m^2 - 4m + 4\right) = 0$ or, $(m+1)(m-2)(m-2) = 0$, whence $m = -1, 2, 2$.

$\therefore$ the general solution is $y = (c_1 + c_2 x)\, e^{2x} + c_3 e^{-x}$, where c_1, c_2 and c_3 are arbitrary constants.

Example 5.4.2. *Solve:* $\frac{d^4y}{dx^4} + a^4 y = 0.$

Solution: The auxiliary equation is $m^4 + a^4 = 0$. This gives $m = \pm\frac{1}{\sqrt{2}}(1+i)a, \pm\frac{1}{\sqrt{2}}(1-i)a$.

Hence, the general solution is

$$\begin{aligned} y &= Ae^{\frac{1}{\sqrt{2}}(1+i)ax} + Be^{-\frac{1}{\sqrt{2}}(1+i)ax} + Ce^{\frac{1}{\sqrt{2}}(1-i)ax} + De^{-\frac{1}{\sqrt{2}}(1-i)ax} \\ &= Ae^{\frac{1}{\sqrt{2}}ax}\left(\cos\frac{ax}{\sqrt{2}} + i\sin\frac{ax}{\sqrt{2}}\right) + Be^{-\frac{1}{\sqrt{2}}ax}\left(\cos\frac{ax}{\sqrt{2}} - i\sin\frac{ax}{\sqrt{2}}\right) \\ &= \left(C_1\cosh\frac{ax}{\sqrt{2}} + C_2\sinh\frac{ax}{\sqrt{2}}\right)\cos\frac{ax}{\sqrt{2}} + \left(C_3\cosh\frac{ax}{\sqrt{2}} + C_4\sinh\frac{ax}{\sqrt{2}}\right)\sin\frac{ax}{\sqrt{2}}. \end{aligned}$$

Example 5.4.3. *Solve:* $\frac{d^4y}{dx^4} + 8\frac{d^2y}{dx^2} + 16y = 0.$

Solution: The auxiliary equation is $m^4 + 8m^2 + 16 = 0$, or, $\left(m^2 + 4\right)^2 = 0$, which gives two pairs of imaginary roots $\pm 2i$.

$\therefore$ the general solution is

$$y = e^{0x}\left[(c_1 + c_2 x)\cos 2x + (c_3 + c_4 x)\sin 2x\right] = (c_1 + c_2 x)\cos 2x + (c_3 + c_4 x)\sin 2x.$$

Example 5.4.4. *Solve:* $\left(D^4 - n^4\right)y = 0$ *completely. Now if* $Dy = y = 0$, *when* $x = 0$ *and* $x = l$, *prove that* $y = A(\cos nx - \cosh nx) + B(\sin nx - \sinh nx)$ *and* $\cos nl \cosh nl = 1$.

Solution: The auxiliary equation is $m^4 - n^4 = 0$, hence $m = n, -n, \pm in$.

General solution is $y = c_1 e^{nx} + c_2 e^{-nx} + A\cos nx + B\sin nx$.

Second part

$Dy = n\left\{c_1 e^{nx} - c_2 e^{-nx} - A\sin nx + B\cos nx\right\}$.

Putting $Dy = y = 0$, when $x = 0$ and $x = l$, $c_1 + c_2 + A = 0$ and $c_1 - c_2 + B = 0$.

Also

$$\left.\begin{aligned} c_1 e^{nl} + c_2 e^{-nl} + A\cos nl + B\sin nl = 0 \\ c_1 e^{nl} - c_2 e^{-nl} - A\sin nl + B\cos nl = 0 \end{aligned}\right\}. \tag{5.4.3}$$

Solving $c_1 = -\frac{1}{2}(A+B),\ c_2 = \frac{1}{2}(B-A)$.

$$\begin{aligned} \therefore\ y &= -\tfrac{1}{2}(A+B)e^{nx} + \tfrac{1}{2}(B-A)e^{-nx} + A\cos nx + B\sin nx \\ &= -\tfrac{A}{2}\left(e^{nx}+e^{-nx}\right) - \tfrac{B}{2}\left(e^{nx}-e^{-nx}\right) + A\cos nx + B\sin nx \\ &= A(\cos nx - \cosh nx) + B(\sin nx - \sinh nx), \end{aligned}$$

since $e^x + e^{-x} = 2\cosh x$ and $e^x - e^{-x} = 2\sinh x$.

Also putting c_1 and c_2 in (5.4.3) and simplifying,

$$A(\cos nl - \cosh nl) + B(\sin nl - \sinh nl) = 0$$

$$A(-\sin nl - \sinh nl) + B(\cos nl - \cosh nl) = 0.$$

Eliminating A, B, we have $\cos nl \cosh nl = 1$.

Examples V(A)

Solve:

1. $2\frac{d^2y}{dx^2} - 3\frac{dy}{dx} + y = 0.$
2. $\frac{d^3y}{dx^3} - 6\frac{d^2y}{dx^2} + 11\frac{dy}{dx} - 6y = 0.$
3. $\frac{d^4y}{dx^4} = m^4 y.$
4. $\frac{d^3y}{dx^3} + 2\frac{d^2y}{dx^2} - 5\frac{dy}{dx} - 6y = 0.$
5. $\frac{d^4y}{dx^4} + 2\frac{d^3y}{dx^3} + 3\frac{d^2y}{dx^2} + 2\frac{dy}{dx} + y = 0.$
6. $\frac{d^4y}{dx^4} - y = 0.$
7. $\frac{d^4y}{dx^4} - 4\frac{d^3y}{dx^3} + 8\frac{d^2y}{dx^2} - 8\frac{dy}{dx} + 4y = 0.$
8. $\frac{d^5y}{dx^5} - 13\frac{d^3y}{dx^3} + 26\frac{d^2y}{dx^2} + 82\frac{dy}{dx} + 104y = 0.$
9. (a) $\frac{d^2y}{dx^2} + y = 0$, given $y = 2$ for $x = 0$ and $y = -2$ for $x = \frac{1}{2}\pi$.

   (b) $\frac{d^2y}{dx^2} - 5\frac{dy}{dx} + 6y = 0$, where $y(0) = 3, y(\ln 2) = 8$.
10. $\frac{d^2x}{dt^2} - 3\frac{dx}{dt} + 2x = 0$, given that when $t = 0, x = 0$ and $\frac{dx}{dt} = 0$.
11. $l\frac{d^2\theta}{dt^2} + g\theta = 0$, given that $\theta = \alpha$ and $\frac{d\theta}{dt} = 0$, when $t = 0$; g, l being positive.
12. If $\frac{d^2V}{dx^2} + p^2V = 0, p > 0$ and $V = V_0$ at $x = 0$ and $V = 0$ at $x = l$, prove that $V = V_0 \sin p(l-x)/\sin pl$.
13. If $\frac{d^2x}{dt^2} + 4x = 0$ and $x = 3$, $\frac{dx}{dt} = 8$ at $t = 0$; find x in terms of t. What is the maximum value of x?
14. $L\frac{d^2Q}{dt^2} + R\frac{dQ}{dt} + \frac{Q}{C} = 0$, given that $Q = Q_0$ and $\frac{dQ}{dt} = 0$, when $t = 0$ and that $CR^2 < 4L$.

    [The charge Q of a condenser having capacity C, discharged in a circuit of resistance R and self-inductance L satisfies this equation.]
15. $m\frac{d^2s}{dt^2} + k\frac{ds}{dt} + cs = 0$, given that $k^2 < 4mc$.

    Also find the solution when k is so small that k^2/mc is negligible.

    [The equation is the equation of motion of a particle of mass m, attracted towards a fixed point in its line of motion by a force of c times its distance from that point, and damped by a frictional resistance of k times its velocity.]
16. Solve: $\left(D^4 - n^4\right) y = 0$ completely.

    Now if $Dy = y = 0$, when $x = 0$ and $x = l$, prove that $y = c_1(\cos nx - \cosh nx) + c_2(\sin nx - \sinh nx)$ and $\cos nl \cosh nl = 1$.

Answers

1. $y = c_1 e^x + c_2 e^{\frac{1}{2}x}$.
2. $y = c_1 e^x + c_2 e^{2x} + c_3 e^{3x}$.
3. $y = c_1 e^{mx} + c_2 e^{-mx} + c_3 \cos mx + c_4 \sin mx$.
4. $y = c_1 e^{-x} + c_2 e^{2x} + c_3 e^{-3x}$.
5. $y = e^{-\frac{1}{2}x}\left[(c_1 + c_2 x)\cos\frac{\sqrt{3}}{2}x + (c_3 + c_4 x)\sin\frac{\sqrt{3}}{2}x\right]$.
6. $y = c_1 \cos x + c_2 \sin x + c_3 \cosh x + c_4 \sinh x$.
7. $y = e^x (c_1 + c_2 x)\sin x + e^x (c_3 + c_4 x)\cos x$.
8. $y = c_1 e^{-x}\cos(x + c_2) + c_3 e^{3x}\cos(2x + c_4) + c_5 e^{-4x}$.
9. (a) $y = 2(\cos x - \sin x)$
   (b) $y = 4e^{2x} - e^{3x}$.
10. $x = 0$.
11. $\theta = \alpha \cos t\sqrt{(g/l)}$.
13. $x = 3\cos 2t + 4\sin 2t$; 5.
14. $Q = Q_0 e^{-Rt/2L}\left(\cos nt + \frac{R}{2Ln}\sin nt\right)$, where $n = \sqrt{\left(\frac{1}{LC} - \frac{R^2}{4L^2}\right)}$.
15. $s = c_1 e^{-\frac{k}{2m}t}\cos\left(\frac{a}{2m}t + c_2\right)$, where $a = \sqrt{4mc - k^2}$.
    $s = c_1 e^{-\frac{k}{2m}t}\cos\left(\sqrt{\frac{c}{m}}t + c_2\right)$.

5.5 General theory of linear differential equations

The most general linear differential equation of nth order is

$$\frac{d^n y}{dx^n} + P_1\frac{d^{n-1}y}{dx^{n-1}} + \cdots + P_{n-1}\frac{dy}{dx} + P_n y = Q, \tag{5.5.1}$$

where the coefficients $P_1, P_2, \ldots, P_n$ and the right-hand member Q are continuous functions of x in $a \leq x \leq b$. The fundamental existence theorem proves that there exists a *unique* continuous solution $y(x)$ which assumes a given value y_0 at a point x_0 within $[a, b]$ and whose first $(n-1)$ derivatives are continuous and assume respectively the values $y_0', y_0'', \ldots, y_0^{(n-1)}$ at $x = x_0$.

Introducing the operators D for $\frac{d}{dx}$, D^2 for $\frac{d^2}{dx^2}$, etc., we write the eq. (5.5.1) in the form:

$$\{D^n + P_1 D^{n-1} + \cdots + P_{n-1}D + P_n\}\, y = Q.$$

Then the expression $D^n + P_1 D^{n-1} + \cdots + P_{n-1}D + P_n$ will be called a *linear differential operator of order n* and will be denoted by $F(D)$. The eq. (5.5.1) can then be written as

$$F(D)y = Q. \tag{5.5.2}$$

The equation

$$F(D)y = 0, \tag{5.5.3}$$

where the right-hand member Q is supposed to be 0 is called the *Reduced Equation* (RE) of eq. (5.5.1).

Definition 5.5.1. *Let $f_1, f_2, f_3, \ldots, f_m$ be m functions and let $c_1, c_2, c_3, \ldots, c_m$ be m constants. Then the expression $c_1 f_1 + c_2 f_2 + c_3 f_3 + \cdots + c_m f_m$ is called a linear combination of $f_1, f_2, f_3, \ldots, f_m$.*

Definition 5.5.2. *Let $f_1, f_2, f_3, \ldots, f_m$ be m functions defined on the closed interval $[a, b]$. The functions $f_1, f_2, f_3, \ldots, f_m$ are called linearly dependent on $[a, b]$, if there exist constants $c_1, c_2, c_3, \ldots, c_m$, not all zero, such that $c_1 f_1 + c_2 f_2 + c_3 f_3 + \cdots + c_m f_m = 0$ for all $x \in [a, b]$.*

Example 5.5.1. *Let $f_1 = 3x, f_2 = 2x, x \in [0, 1]$. Then we can find two non-zero constants $c_1 = 2, c_2 = -3$, such that $c_1 f_1 + c_2 f_2 = 0$ for all $x \in [0, 1]$. It confirms that f_1, f_2 are linearly dependent.*

Definition 5.5.3. *Let $f_1, f_2, f_3, \ldots, f_m$ be m functions defined on the closed interval $[a, b]$. The functions $f_1, f_2, f_3, \ldots, f_m$ are called linearly independent on $[a, b]$ if there exist constants $c_1, c_2, c_3, \ldots, c_m$ such that $c_1 f_1 + c_2 f_2 + c_3 f_3 + \cdots + c_m f_m = 0$ for all $x \in [a, b]$ implies that $c_1 = c_2 = c_3 = \cdots = c_m = 0$.*

Example 5.5.2. *Let $f_1 = 3x, f_2 = 2x^2, x \in [0, 1]$. Let c_1, c_2 be two constants such that $c_1 f_1 + c_2 f_2 = 0$ for all $x \in [0, 1]$. Then $c_1.3x + c_2.2x^2 = 0$, i.e. $3c_1 x + 2c_2 x^2 = 0$.*

Differentiating with respect to x, we have $3c_1 + 4c_2x = 0$, which also holds for all $x \in [0,1]$. It follows that $3c_1x + 4c_2x^2 = 0$.

Thus, for all $x \in [0,1]$, $3c_1x + 2c_2x^2 = 0$ and $3c_1x + 4c_2x^2 = 0$.

On subtraction we have $2c_2x^2 = 0$ for all $x \in [0,1]$. It confirms that $c_2 = 0$. Using this result in $3c_1x + 2c_2x^2 = 0$, we obtain $3c_1x = 0$ for all $x \in [0,1]$. It gives $c_1 = 0$. Hence, the functions f_1, f_2 are linearly independent on the closed interval $[0,1]$.

Theorem 5.5.1. *The nth-order homogeneous linear differential equation*

$$\frac{d^ny}{dx^n} + P_1\frac{d^{n-1}y}{dx^{n-1}} + \cdots + P_{n-1}\frac{dy}{dx} + P_ny = 0$$

always possesses n linearly independent solutions. If $f_1, f_2, f_3, \ldots, f_n$ are n linearly independent solutions of the equation, every solution of the equation can be expressed as a linear combinations of $f_1, f_2, f_3, \ldots, f_n$.

Properties of F(D)

I. *If $y = y_1$ is a solution of the reduced equation* (5.5.3), *then $y = Cy_1$, is also a solution, where C is any arbitrary constant.*

Proof. Clearly, $D^r(Cy_1) = CD^ry_1$.

Then $F(D)Cy_1 = D^n(Cy_1) + P_1D^{n-1}(Cy_1) + \cdots + P_{n-1}D(Cy_1) + P_nCy_1$

$= C\{D^ny_1 + P_1D^{n-1}y_1 + \cdots + P_{n-1}Dy_1 + P_ny_1\}$

$= CF(D)y_1$.

$\therefore$ when $y = y_1$ is a solution of $F(D)y = 0$, the right-hand member vanishes and consequently $F(D)Cy_1 = 0$, i.e. Cy_1 is also a solution of $F(D)y = 0$.

II. *If $y = y_1, y_2, \ldots, y_m$ are m solutions of the reduced equation $F(D)y = 0$, then $y = C_1y_1 + C_2y_2 + \cdots + C_my_m$ is a solution of the same equation ($C_1, C_2, \ldots, C_m$ are arbitrary constants).*

Proof. The result follows from the fact $D^r\{C_1y_1 + C_2y_2 + \cdots + C_my_m\} = C_1D^ry_1 + C_2D^ry_2 + \cdots + C_mD^ry_m$.

Now argue similarly as in I.

Important Note. If n linearly independent solutions $y_1, y_2, \ldots, y_n$ of the Reduced Equation $F(D)y = 0$ are known, then the solution $y = C_1y_1 + C_2y_2 + \cdots + C_ny_n$ containing n arbitrary constants, is the *complete primitive* of the equation $F(D)y = 0$.

III. *Let $y = y_0(x)$, be* **any solution** *of eq.* (5.5.2), *namely, $F(D)y = Q$. Then if $u(x)$ be the complete primitive of the reduced equation $F(D)y = 0$, $y = y_0(x) + u(x)$ will be the most general solution of $F(D)y = Q$.*

Proof. Since the operator D^r is distributive, the linear operator $F(D)$ is distributive, that is to say, $F(D)\{y_0(x) + u(x)\} = F(D)\{y_0(x)\} + F(D)\{u(x)\} = Q + 0 = Q$.

Moreover, the solution $y = y_0(x) + u(x)$ involves n arbitrary constants, it is, therefore, the most general solution of $F(D)y = Q$.

Important Note. The general solution of $F(D)y = Q$ consists of two parts, viz.,

(i) The *Complementary Function* (CF), which is the complete primitive of the Reduced Equation and is of the form $y = C_1y_1 + C_2y_2 + \cdots + C_ny_n$ containing n arbitrary constants;

(ii) The *Particular Integral* (PI), which is **any solution** of $F(D)y = Q$ contains no arbitrary constant.

Example 5.5.3. *Solve:* $\frac{d^2y}{dx^2} + 4y = 4x$.

Solution: The equation $\frac{d^2y}{dx^2} + 4y = 4x$ has the reduced equation $\frac{d^2y}{dx^2} + 4y = 0$, the complete primitive of which is $y = A\cos 2x + B\sin 2x$. Clearly, $y = x$ is *some* solution of the given equation. The general solution is, therefore, $y = \text{CF} + \text{PI} = (A\cos 2x + B\sin 2x) + x$ (Fig. 5.1).

Any particular solution is obtained by assigning to A and B definite numerical values.

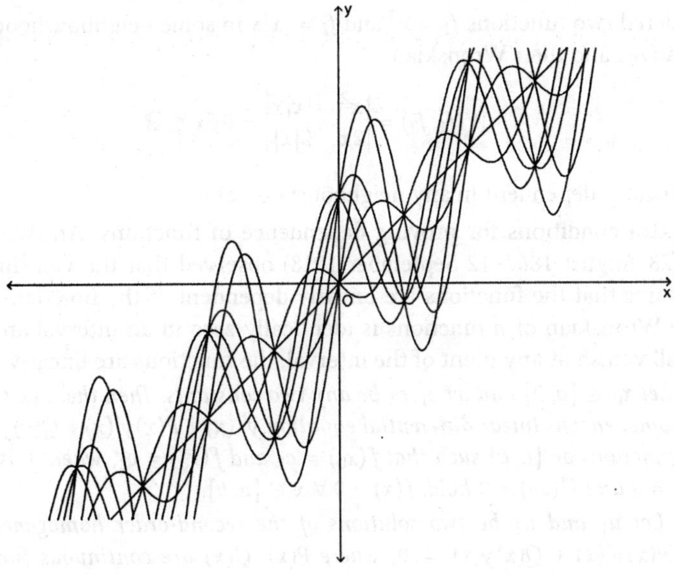

Fig. 5.1

5.6 The Wronskian

Wronskian was first introduced by a Polish mathematician Józef Maria Hoëné-Wroński (23 August 1776–9 August 1853). It is useful in the study of differential equations to show linear independence in a set of solutions.

Let $f_1, f_2, f_3, \ldots, f_n$ be n real functions defined on the closed interval $[a, b]$ and let each of the functions have $(n-1)$th derivative on $[a, b]$. Then the determinant

$$W(f_1, f_2, \ldots, f_n) = \begin{vmatrix} f_1 & f_2 & \cdots & f_n \\ f_1' & f_2' & \cdots & f_n' \\ \vdots & \vdots & \cdots & \vdots \\ f_1^{(n-1)} & f_2^{(n-1)} & \cdots & f_n^{(n-1)} \end{vmatrix},$$

where primes denote derivatives of the functions, is called the *Wronskian of the functions* $f_1, f_2, f_3, \ldots, f_n$. Here $W(f_1, f_2, \ldots, f_n)$ is a real-valued function defined on $[a, b]$. The value of Wronskian at any point x in $[a, b]$ is often denoted by $W(f_1, f_2, \ldots, f_n)(x)$ or, by $W(f_1(x), f_2(x), \ldots, f_n(x))$.

Theorem 5.6.1. *The n solutions $y_1, y_2, \ldots, y_n$ of the nth order homogeneous linear differential equation $\frac{d^n y}{dx^n} + P_1 \frac{d^{n-1}y}{dx^{n-1}} + \cdots + P_{n-1}\frac{dy}{dx} + P_n y = 0$ are linearly independent on the closed interval $[a, b]$, if and only if the Wronskian of $y_1, y_2, \ldots, y_n$ is non-zero for some x on $[a, b]$.*

Important notes

Nonvanishing Wronskian of the n solutions $y_1, y_2, \ldots, y_n$ on $[a, b]$ ensures that the solutions are linearly independent. What will happen if $W(y_1, y_2, \ldots, y_n) = 0$ holds? There is a common misconception about it. The misconception is that $W(y_1, y_2, \ldots, y_n) = 0$ for all $x \in [a, b]$ implies linear dependence of $y_1, y_2, \ldots, y_n$. If $y_1, y_2, \ldots, y_n$ are polynomial functions defined on $[a, b]$, $W(y_1, y_2, \ldots, y_n) = 0$ becomes the necessary and sufficient condition for the dependence of n solutions $y_1, y_2, \ldots, y_n$. In general, $W(y_1, y_2, \ldots, y_n) = 0$ for all $x \in [a, b]$ is the necessary condition for linear dependence of the solutions $y_1, y_2, \ldots, y_n$ but it is not sufficient. An Italian mathematician, Giuseppe Peano (27 August 1858–20

April 1932) considered two functions $f_1 = x^2$ and $f_2 = |x|x$ in some neighbourhood of 0. Here f_1, f_2 have continuous derivatives and their Wronskian

$$W(f_1, f_2) = \begin{vmatrix} x^2 & |x|x \\ 2x & 2|x| \end{vmatrix} = 0 \forall x \in \mathbb{R},$$

yet they are not linearly dependent in any neighbourhood of 0.

So we need extra conditions for proving dependence of functions. An American mathematician Maxime Bôcher (28 August 1867–12 September 1918) observed that the vanishing of the Wronskian in an interval ensures that the functions are linearly dependent, if the functions are analytic. He also proved that if the Wronskian of n functions is identically zero in an interval and Wronskian of $n-1$ functions do not all vanish at any point of the interval, the functions are linearly dependent.

Theorem 5.6.2. *Let $x_0 \in [a, b]$ and let c_1, c_2 be any two constants. Then there exists unique solution f of the second-order homogeneous linear differential equation $y''(x) + P(x)y'(x) + Q(x)y(x) = 0$, $P(x)$ and $Q(x)$ being continuous functions on $[a, b]$ such that $f(x_0) = c_1$ and $f(x_0) = c_2$, where f is defined over $[a, b]$. In particular, if $f(x_0) = 0$ and $f'(x_0) = 0$ hold, $f(x) = 0 \; \forall \; x \in [a, b]$.*

Theorem 5.6.3. *Let u_1 and u_2 be two solutions of the second-order homogeneous linear differential equation $y''(x) + P(x)y'(x) + Q(x)y(x) = 0$, where $P(x)$, $Q(x)$ are continuous functions on $[a, b]$. Then u_1 and u_2 are linearly dependent on the closed interval $[a, b]$, if and only if the value of the Wronskian $W(u_1, u_2)(x) = 0 \; \forall \; x \in [a, b]$.*

Proof. Let $W(u_1, u_2)(x) = 0 \; \forall \; x \in [a, b]$, i.e. $\begin{vmatrix} u_1(x) & u_2(x) \\ u_1'(x) & u_2'(x) \end{vmatrix} = 0 \; \forall \; x \in [a, b]$.

Let $x_0 \in [a, b]$. Then $\begin{vmatrix} u_1(x_0) & u_2(x_0) \\ u_1'(x_0) & u_2'(x_0) \end{vmatrix} = 0$. So we can find two constants c_1, c_2, not both zero, such that

$$c_1 u_1(x_0) + c_2 u_2(x_0) = 0$$
$$c_1 u_1'(x_0) + c_2 u_2'(x_0) = 0.$$

Now let us consider a function $f : [a, b] \to \mathbb{R}$ defined by $f(x) = c_1 u_1(x) + c_2 u_2(x)$. Then $f(x)$ is also a solution of the equation $y''(x) + P(x)y'(x) + Q(x)y(x) = 0$. Again, $f(x_0) = 0$ and $f'(x_0) = 0$.

Therefore, $f(x) = c_1 u_1(x) + c_2 u_2(x) = 0 \; \forall \; x \in [a, b]$, where c_1, c_2 are two constants not both zero. It confirms that u_1 and u_2 are linearly dependent on the closed interval $[a, b]$.

Conversely, let u_1 and u_2 be two linearly dependent solutions of the equation on $[a, b]$. Then by definition, there exist constants c_1, c_2, not all zero, such that $c_1 u_1(x) + c_2 u_2(x) = 0 \; \forall \; x \in [a, b]$.

Differentiating we have $c_1 u_1'(x) + c_2 u_2'(x) = 0 \; \forall \; x \in [a, b]$.

Let $x_0 \in [a, b]$ be an arbitrary point. Then $c_1 u_1(x_0) + c_2 u_2(x_0) = 0$ and $c_1 u_1'(x_0) + c_2 u_2'(x_0) = 0$. Since c_1, c_2 are not both zero, we must have $\begin{vmatrix} u_1(x_0) & u_2(x_0) \\ u_1'(x_0) & u_2'(x_0) \end{vmatrix} = 0$.

Since $x_0 \in [a, b]$ be an arbitrary point, $\begin{vmatrix} u_1(x) & u_2(x) \\ u_1'(x) & u_2'(x) \end{vmatrix} = 0 \; \forall \; x \in [a, b]$, i.e. $W(u_1, u_2)(x) = 0 \; \forall \; x \in [a, b]$. Hence, the proof is complete.

Linearly independent solutions

We mentioned in the last article that if $y_1, y_2, \ldots, y_n$ are n linearly independent solutions of $F(D)y = 0$, then $y = c_1 y_1 + c_2 y_2 + \cdots + c_n y_n$, where $c_1, c_2, \ldots, c_n$ are n arbitrary constants, is its complete primitive.

By linearly independent solutions we mean that it is impossible to choose constants $c_1, c_2, \ldots, c_n$, not all zero, so that the expression $c_1 y_1 + c_2 y_2 + \cdots + c_n y_n$ is identically zero. The conditions that the

n functions $y_1, y_2, \ldots, y_n$, which are supposed to be derivable $(n-1)$ times in some interval $[a, b]$, be linearly independent will now be obtained.

Suppose that these n functions are not linearly independent. Then by definition, n constants $c_1, c_2, \ldots, c_n$, not all zero, may be determined so that $c_1 y_1 + c_2 y_2 + \cdots + c_n y_n = 0$ identically. Since this relation is identically satisfied in $[a, b]$, it may be differentiated any number of times up to $(n-1)$ in that interval. Thus,

$$c_1 y_1' + c_2 y_2' + \cdots + c_n y_n' = 0$$

$$\vdots$$

$$c_1 y_1^{(n-1)} + c_2 y_2^{(n-1)} + \cdots + c_n y_n^{(n-1)} = 0.$$

We have thus n equations to determine the coefficients $c_1, c_2, \ldots, c_n$. If these equations are consistent,

$$W(y_1, y_2, \ldots, y_n) = \begin{bmatrix} y_1 & y_2 & \cdots & y_n \\ y_1' & y_2' & \cdots & y_n' \\ \vdots & \vdots & \cdots & \vdots \\ y_1^{(n-1)} & y_2^{(n-1)} & \cdots & y_n^{(n-1)} \end{bmatrix} = 0.$$

The identical vanishing of Wronskian of the functions $y_1, y_2, \ldots, y_n$ is a necessary condition for the linear dependence of $y_1, y_2, \ldots, y_n$; therefore, its nonvanishing is sufficient for the linear independence of $y_1, y_2, \ldots, y_n$.

Example 5.6.1. *Show that the equation $x^3 \frac{d^3y}{dx^3} - 6x\frac{dy}{dx} + 12y = 0$ has three linearly independent solutions of the form $y = x^r$.*

Solution: Since $\frac{dy}{dx} = rx^{r-1}$ and $\frac{d^3y}{dx^3} = r(r-1)(r-2)x^{r-3}$, from the given equation, we have

$$x^r\{r(r-1)(r-2) - 6r + 12\} = 0,$$

which is satisfied only when $r^3 - 3r^2 - 4r + 12 = 0$, i.e. when $r = 2, 3, -2$.

The corresponding solutions are $y = x^2$, $y = x^3$, $y = x^{-2}$. They are linearly independent, since the Wronskian

$$W(x^2, x^3, x^{-2}) = \begin{vmatrix} x^2 & x^3 & x^{-2} \\ 2x & 3x^2 & -2x^{-3} \\ 2 & 6x & 6x^{-4} \end{vmatrix} = 20 \neq 0.$$

Now we can write that the complete primitive of the given equation as $y = c_1 x^2 + c_2 x^3 + c_3 x^{-2}$, where c_1, c_2, c_3 are three arbitrary constants.

Fundamental set of solutions

Let $S = \{u_1(x), u_2(x), u_3(x), \ldots, u_n(x)\}$ be a set of linearly independent solutions of the nth order homogeneous linear differential equation

$$a_0(x)\frac{d^n y}{dx^n} + a_1(x)\frac{d^{n-1}y}{dx^{n-1}} + \cdots + a_{n-1}(x)\frac{dy}{dx} + a_n(x)y = 0.$$

Then the set S is called a fundamental set of n solutions of the equation. Thus, the condition that the n solutions of the equation given above, should be a fundamental set (or, a fundamental system) of the solutions is that the Wronskian, $W(u_1(x), u_2(x), \ldots, u_n(x)) \neq 0$.

Example 5.6.2. *Find a fundamental set of solutions of the equation $(D^3 - 2D^2 - D + 2)y = 0$.*

Solution: Let $y = e^{mx}$ be a trial solution of the given equation. Then the auxiliary equation is

$$m^3 - 2m^2 - m + 2 = 0$$
$$\text{or, } m^2(m-2) - 1\cdot(m-2) = 0$$
$$\text{or, } (m+1)(m-1)(m-2) = 0.$$

$\therefore\ m = -1, 1, 2.$

Therefore, e^{-x}, e^{x}, e^{2x} are three solutions of the equation. Now we would like to examine whether these solutions are linearly independent or not. The Wronskian of these solution is given by

$$W(e^{-x}, e^{x}, e^{2x}) = \begin{vmatrix} e^{-x} & e^{x} & e^{2x} \\ -e^{-x} & e^{x} & 2e^{2x} \\ e^{-x} & e^{x} & 4e^{2x} \end{vmatrix} = e^{2x}\begin{vmatrix} 1 & 1 & 1 \\ -1 & 1 & 2 \\ 1 & 1 & 4 \end{vmatrix} = -6e^{2x} \neq 0$$

for all $x \in \mathbb{R}$.

It confirms that the solutions are linearly independent. Hence, a fundamental set of solutions corresponding to the given equation is $\{e^{-x}, e^{x}, e^{2x}\}$.

Theorem 5.6.4. *Let $\{u_1(x), u_2(x)\}$ be a fundamental set of solutions of the differential equation $y''(x) + P(x)y'(x) + Q(x)y(x) = 0$, where $P(x)$, $Q(x)$ are either functions of x or constants. Then the general solution to the differential equation is $y = c_1u_1(x) + c_2u_2(x)$, where c_1, c_2 are two arbitrary constants.*

Theorem 5.6.5 (Abel). *Let $\{u_1(x), u_2(x)\}$ be a fundamental set of solutions of the differential equation $y''(x)+P(x)y'(x)+Q(x)y(x) = 0$, where $P(x)$, $Q(x)$ are either functions of x or constants. Then $W(u_1, u_2)(x) = ce^{-\int P(x)dx}$, where c is an arbitrary constant and $W(u_1, u_2)(x)$ is the Wronskian of u_1, u_2.*

Proof. Since $\{u_1(x), u_2(x)\}$is a fundamental set of solutions of the equation, the Wronskian $W(u_1(x), u_2(x)) \neq 0$ and

$$u_1''(x) + P(x)u_1'(x) + Q(x)u_1(x) = 0, \text{ i.e. } u_1''(x) = -(P(x)u_1'(x) + Q(x)u_1(x)) \tag{5.6.1}$$

$$\text{and } u_2''(x) + P(x)u_2'(x) + Q(x)u_2(x) = 0, \text{ i.e. } u_2''(x) = -(P(x)u_2'(x) + Q(x)u_2(x)). \tag{5.6.2}$$

Now

$$W(u_1(x), u_2(x)) = \begin{vmatrix} u_1(x) & u_2(x) \\ u_1'(x) & u_2'(x) \end{vmatrix} = u_1(x)u_2'(x) - u_2(x)u_1'(x).$$

$$\therefore\ \frac{d}{dx}(W(u_1(x), u_2(x))) = u_1'(x)u_2'(x) + u_1(x)u_2''(x) - u_2'(x)u_1'(x) - u_2(x)u_1''(x) = u_1(x)u_2''(x) - u_2(x)u_1''(x)$$
$$= -u_1(x)((P(x)u_2'(x) + Q(x)u_2(x))) + u_2(x)((P(x)u_1'(x) + Q(x)u_1(x)))$$
[by (5.6.1) and (5.6.2)]
$$= -P(x)(u_1(x)u_2'(x) - u_2(x)u_1'(x))$$
$$= -P(x)W(u_1(x), u_2(x)).$$

$$\therefore\ \frac{1}{W(u_1(x), u_2(x))}\frac{d}{dx}(W(u_1(x), u_2(x))) = -P(x).$$

Integrating, we get

$$\ln\{W(u_1, u_2)(x)\} = -\int P(x)dx + \ln c.$$

$$\therefore\ W(u_1, u_2)(x) = ce^{-\int P(x)dx},$$

c being an arbitrary constant.

Hence, the theorem is proved.

Note 5.6.1. *Let W_0 be the value of W at $x = x_0$. Then relation $W(u_1, u_2)(x) = ce^{-\int P(x)dx}$ becomes*

$$W = W_0 \exp\left\{-\int_{x_0}^{x} P(t)dt\right\};$$

that is known as Abel's identity (or Abel's formula) named after a Norwegian mathematician Niels Henrik Abel (5 August 1802–6 April 1829).

Example 5.6.3. *Let W be the Wronskian of two independent solutions of the differential equation $(1 + x^2)y'' - (2 + x)y' + x^3y = 0$. Find $\left[\frac{1}{W}\frac{dW}{dx}\right]_{x=0}$.*

Solution: Since W is the Wronskian of two independent solutions of the given differential equation, $W \neq 0$. Again, the normalized form of the equation is

$$y'' - \frac{2+x}{1+x^2}y' + \frac{x^3}{1+x^2}y = 0. \quad \therefore \left[\frac{1}{W}\frac{dW}{dx}\right]_{x=0} = \left[\frac{2+x}{1+x^2}\right]_{x=0} = 2.$$

Theorem 5.6.6. *The value of the Wronskian of two solutions u_1, u_2 of the differential equation $y''(x) + P(x)y'(x) + Q(x)y(x) = 0$ is either zero for all $x \in [a, b]$ or zero for no $x \in [a, b]$.*

Proof. Case 1: Let u_1, u_2 be two linearly dependent solutions of the equation. Then by theorem (5.6.3), $W(u_1, u_2)(x) = 0 \ \forall \ x \in [a, b]$.

Case 2: Let u_1, u_2 be two linearly independent solutions of the equation. By Abel's theorem we have $W(u_1, u_2)(x) = ce^{-\int P(x)dx}$. Let W_0 be the value of W at $x = x_0 \in [a, b]$. Then $c = W(x_0) = W_0$. Thus, we obtain the Abel's identity as

$$W(x) = W_0 \exp\left\{-\int_{x_0}^{x} P(t)dt\right\} \ \forall \ x \in [a, b].$$

Observe that

$$W(x) = W_0 \exp\left\{-\int_{x_0}^{x} P(t)dt\right\} = 0 \ \forall \ x \in [a, b],$$

only when $W_0 = 0$. In this case, the solutions u_1, u_2 are linearly dependent on $[a, b]$, which contradicts our hypothesis. Thus, $W_0 \neq 0$. Since x_0 is an arbitrary point in $[a, b]$, $W(u_1, u_2)(x) = 0$ for no $x \in [a, b]$.

Hence, the theorem is proved.

Reduction of Order: We can use Abel's theorem to solve a second order linear homogeneous differential equation with only one known fundamental solution. Abel's theorem is able to reduce the problem of finding a solution of a second order equation to that of solving a related first order equation. This is why the method is known as the reduction of order.

Let u_1 be a known nontrivial solution and u_2 be an unknown solution of the equation

$$y''(x) + P(x)y'(x) + Q(x)y(x) = 0.$$

Then by Abel's theorem, $W(u_1, u_2)(x) = ce^{-\int P(x)dx}$.

Again, by the definition of Wronskian, $W(u_1, u_2)(x) = \begin{vmatrix} u_1 & u_2 \\ u_1' & u_2' \end{vmatrix} = u_1u_2' - u_2u_1'$.

$\therefore \ u_1u_2' - u_2u_1' = ce^{-\int P(x)dx}$, which is a first order linear differential equation. Solving it we will get the second fundamental solution.

Example 5.6.4. *Given that $u(x) = e^x$ is a solution of the equation $y'' - 2y' + y = 0$. Find a linearly independent solution by reducing the order. Hence, find its general solution.*

Solution: Let $v(x)$ be an unknown solution of the equation. Then by Abel's formula

$$W(u,v)(x) = ce^{2\int dx} = c \cdot e^{2x}.$$

By the definition of Wronskian, $W(u,v)(x) = e^x(v' - v)$.

$$\therefore\ e^x(v' - v) = c \cdot e^{2x} \text{ or, } \frac{dv}{dx} - v = c \cdot e^x.$$

$\text{IF} = e^{-\int dx} = e^{-x}$.

Multiplying the above equation by IF, we obtain $\frac{d}{dx}(ye^{-x}) = c$.

$$\therefore\ ye^{-x} = c\int dx + c',$$

c' being an integrating constant or, $y = cxe^x + c'e^x$.

Choosing $c = 1$, $c' = 0$, we obtain $y = xe^x$, which is another solution of the equation.

Now $W(u,v)(x) = \begin{vmatrix} e^x & xe^x \\ e^x & xe^x + e^x \end{vmatrix} = xe^{2x} + e^{2x} - xe^{2x} = e^{2x} \neq 0$.

Therefore, $\{e^x, xe^x\}$ is a set of two linearly independent solutions of the equation.

Hence, the general solution of the equation is $y = c_1e^x + c_2xe^x$, where c_1, c_2 are two arbitrary constants.

Theorem 5.6.7. *Let $u(x)$ be a nontrivial solution of the second-order homogeneous linear differential equation $y''(x) + P(x)y'(x) + Q(x)y(x) = 0$. Then the transformation $y = u(x)v(x)$ reduces the equation to a first-order homogeneous linear differential equation in the dependent variable $w = \frac{dv}{dx}$.*

Proof. Here $u(x)$ is a known solution (nontrivial) of the equation

$$y''(x) + P(x)y'(x) + Q(x)y(x) = 0 \tag{5.6.3}$$

and $v(x)$ is an unknown function of x that will be determined. Then

$$u''(x) + P(x)u'(x) + Q(x)u(x) = 0. \tag{5.6.4}$$

Let us consider the transformation $y = u(x)v(x)$. Differentiating, we obtain

$$\frac{dy}{dx} = u\frac{dv}{dx} + v\frac{du}{dx} = u\frac{dv}{dx} + vu' \text{ and } \frac{d^2y}{dx^2} = u\frac{d^2v}{dx^2} + 2\frac{du}{dx}\cdot\frac{dv}{dx} + v\frac{d^2u}{dx^2} = u\frac{d^2v}{dx^2} + 2u'\cdot\frac{dv}{dx} + vu''.$$

Then eq. (5.6.3) becomes

$$u(x)\frac{d^2v}{dx^2} + 2u'(x)\cdot\frac{dv}{dx} + vu''(x) + P(x)(u(x)\frac{dv}{dx} + vu'(x)) + Q(x)u(x)v = 0$$

$$\text{or, } u(x)\frac{d^2v}{dx^2} + (2u'(x) + P(x)u(x))\frac{dv}{dx} + (u''(x) + P(x)u'(x) + Q(x)u(x))v = 0.$$

$$\therefore\ u(x)\frac{d^2v}{dx^2} + (2u'(x) + P(x)u(x))\frac{dv}{dx} = 0 \text{ by (5.6.4)}$$

$$\text{or, } u(x)\frac{dw}{dx} + (2u'(x) + P(x)u(x))w = 0, \text{ since } w = \frac{dv}{dx}.$$

It is a first-order homogeneous linear differential equation in the dependent variable w.

Hence, the proof is complete.

Note 5.6.2. *Using the above theorem, we can find the general solution of the differential eq.* (5.6.3). *It is proposed by a French mathematician Jean-Baptiste le Rond d'Alembert* (16 November 1717–29 October 1783). *The method is given below:*

Let us start from the reduced equation

$$u(x)\frac{dw}{dx} + (2u'(x) + P(x)u(x))w = 0. \tag{5.6.5}$$

Then $\dfrac{dw}{w} = -\left(\dfrac{2u'(x)}{u(x)} + P(x)\right) dx.$

$$\therefore \int \frac{dw}{w} = -\int \left(\frac{2u'(x)}{u(x)} + P(x)\right) dx + \ln c \text{ or, } \ln w = -\ln[u(x)]^2 - \int p(x)dx + \ln c$$

$$\text{or, } w = \frac{c\exp\{-\int p(x)dx\}}{[u(x)]^2}$$

which is the general solution of the eq. (5.6.5). Let us choose $c = 1$. Then

$$w = \frac{\exp\{-\int p(x)dx\}}{[u(x)]^2}$$

is a particular solution of (5.6.5).

Now $\dfrac{dv}{dx} = w \Rightarrow \dfrac{dv}{dx} = \dfrac{c\exp\{-\int p(x)dx\}}{[u(x)]^2}.$

$$\therefore \int dv = \int \frac{c\exp\{-\int p(x)dx\}}{[u(x)]^2} dx + c',$$

c' being another integrating constant.

Choosing $c' = 1$, we get a particular solution

$$v(x) = \int \frac{c\exp\{-\int p(x)dx\}}{[u(x)]^2} dx$$

of the eq. (5.6.3).

Let

$$u(x) \int \frac{c\exp\{-\int p(x)dx\}}{[u(x)]^2} dx = u_1(x).$$

Then

$$W(u, u_1)(x) = \begin{vmatrix} u(x) & u_1(x) \\ u'(x) & u_1'(x) \end{vmatrix} = \begin{vmatrix} u(x) & u(x)v \\ u'(x) & u'(x)v + u(x)v' \end{vmatrix}$$

$$= u(x)(u'(x)v + u(x)v') - u'(x)u(x)v = [u(x)]^2 v'.$$

$$\therefore W(u, u_1)(x) = \exp\left\{-\int p(x)dx\right\} \neq 0.$$

It confirms that $u(x)$, $u_1(x)$ are linearly independent solutions of the eq. (5.6.3).

Hence, the general solution of (5.6.3) is $c_1 u(x) + c_2 u_1(x)$, where c_1, c_2 are two arbitrary constants.

Theorem 5.6.8. *Let $u(x)$ be a nontrivial solution of the nth-order homogeneous linear differential equation*

$$\frac{d^n y}{dx^n} + P_1(x)\frac{d^{n-1}y}{dx^{n-1}} + P_2(x)\frac{d^{n-2}y}{dx^{n-2}} + \cdots + P_{n-1}(x)\frac{dy}{dx} + P_n(x)y = 0.$$

Then the transformation $y = u(x)v(x)$ reduces the equation to an $(n-1)$th order homogeneous linear differential equation in the dependent variable $w = \frac{dv}{dx}$.

Example 5.6.5. *Given that $y = x$ is a nontrivial solution of the equation $x^2\frac{d^2y}{dx^2} - 4x\frac{dy}{dx} + 4y = 0,\ x \neq 0.$ Find a linearly independent solution by reducing the order. Hence, find its general solution.*

Solution: We see that $y = x$ satisfies the equation

$$x^2\frac{d^2y}{dx^2} - 4x\frac{dy}{dx} + 4y = 0. \tag{5.6.6}$$

Let $y = xv$ be another solution of (5.6.6). Then $\frac{dy}{dx} = x\frac{dv}{dx} + v$ and $\frac{d^2y}{dx^2} = x\frac{d^2v}{dx^2} + 2\frac{dv}{dx}$. Using these results in (5.6.6), we get

$$x^2\left(x\frac{d^2v}{dx^2} + 2\frac{dv}{dx}\right) - 4x\left(x\frac{dv}{dx} + v\right) + 4xv = 0 \text{ or, } x^3\frac{d^2v}{dx^2} - 2x^2\frac{dv}{dx} = 0 \text{ or, } x\frac{d^2v}{dx^2} - 2\frac{dv}{dx} = 0.$$

Let $\frac{dv}{dx} = w$. Then the above equation reduces to the form $x\frac{dw}{dx} - 2w = 0.$

$$\therefore \frac{dw}{w} = \frac{2dx}{x} \text{ or, } \int\frac{dw}{w} = 2\int\frac{dx}{x} + \ln c, \text{ } c \text{ being an arbitrary constant}$$

or, $\ln w = \ln x^2 + \ln c.$

$\therefore\ w = cx^2.$

Choosing $c = 1$, we have $\frac{dv}{dx} = x^2.$

$$\therefore \int dv = \int x^2 dx + c', \text{ } c' \text{ being an arbitrary constant}$$

$$\text{or, } v = \frac{x^3}{3} + c'.$$

Choosing $c' = 1$, we get another particular solution $y = xv$, i.e. $y = \frac{x^4}{3}$ of the eq. (5.6.6).

Now

$$W\left(x, \frac{x^4}{3}\right) = \begin{vmatrix} x & x^4/3 \\ 1 & 4x^3/3 \end{vmatrix} = x^4 \neq 0\ \forall\, x \in \mathbb{R} - \{0\}.$$

Thus, $\left\{x, \frac{x^4}{3}\right\}$ is a set of linearly independent solutions of the eq. (5.6.6) on the domain $\mathbb{R} - \{0\}$.

Hence, the general solution of the eq. (5.6.6) is $y = c_1x + c_2x^4$, where c_1, c_2 are two arbitrary constants.

Example 5.6.6. *Given that $y = x$ is a nontrivial solution of the equation*

$$(x+1)\frac{d^2y}{dx^2} + x\frac{dy}{dx} - y = 0,\ x \neq -1.$$

Find a linearly independent solution by reducing the order. Hence, find its general solution.

Solution: We see that $y = x$ satisfies the equation

$$(x+1)\frac{d^2y}{dx^2} + x\frac{dy}{dx} - y = 0. \tag{5.6.7}$$

Let $y = xv$ be another solution of (5.6.7). Then

$$\frac{dy}{dx} = x\frac{dv}{dx} + v \text{ and } \frac{d^2y}{dx^2} = x\frac{d^2v}{dx^2} + 2\frac{dv}{dx}.$$

Using these results in (5.6.7), we get

$$(x+1)\left(x\frac{d^2v}{dx^2} + 2\frac{dv}{dx}\right) + x\left(x\frac{dv}{dx} + v\right) - xv = 0 \text{ or, } x(x+1)\frac{d^2v}{dx^2} + (x^2+2x+2)\frac{dv}{dx} = 0.$$

Let $\frac{dv}{dx} = w$. Then the above equation reduces to the form:

$$x(x+1)\frac{dw}{dx} + (x^2+2x+2)w = 0.$$

$$\therefore \frac{dw}{w} = -\frac{x^2+2x+2}{x(x+1)}dx$$

or, $\displaystyle\int \frac{dw}{w} = -\int \frac{x^2+2x+2}{x(x+1)}dx + c$, c being an arbitrary constant

or, $\displaystyle\int \frac{dw}{w} = -\int \frac{x(x+1)+(x+1)+1}{x(x+1)}dx + \ln c$

or, $\displaystyle\int \frac{dw}{w} = -\int dx - 2\int \frac{dx}{x} + \int \frac{dx}{x+1} + \ln c$

or, $\ln w = -x - \ln x^2 + \ln(x+1) + \ln c$

or, $\displaystyle\ln\left(\frac{wx^2}{c(x+1)}\right) = -x.$

$\displaystyle\therefore w = c\left(\frac{x+1}{x^2}\right)e^{-x}.$

Choosing $c = 1$, we have $\displaystyle\frac{dv}{dx} = \left(\frac{x+1}{x^2}\right)e^{-x}.$

$\displaystyle\therefore \int dv = \int\left(\frac{x+1}{x^2}\right)e^{-x}dx + c'$, c' being an arbitrary constant

or, $\displaystyle v = -\frac{e^{-x}}{x} + c'.$

Choosing $c' = 1$, we get another particular solution $y = xv$, i.e. $y = -e^{-x}$ of the eq. (5.6.7).

Now

$$W(x, -e^{-x}) = \begin{vmatrix} x & -e^{-x} \\ 1 & e^{-x} \end{vmatrix} = (x+1)e^{-x} \neq 0 \;\forall\, x \in \mathbb{R} - \{-1\}.$$

Thus, $\{x, -e^{-x}\}$ is a set of linearly independent solutions of the eq. (5.6.7) on the domain $\mathbb{R} - \{-1\}$. Hence, the general solution of the eq. (5.6.7) is $y = c_1x + c_2e^{-x}$, where c_1, c_2 are two arbitrary constants.

Formation of a differential equation from a given fundamental set of solutions

Let $S = \{u_1(x), u_2(x), u_3(x), \ldots, u_n(x)\}$ be a fundamental set of solutions of an unknown nth order linear homogeneous differential equation and let $u_k(x)$, $k = 1, 2, 3, \ldots, n$, are functions of x having nth order derivatives in some closed interval $[\alpha, \beta]$. Then the equation having S as the fundamental set of solutions is given by

$$\begin{vmatrix} u_1(x) & u_2(x) & \ldots & u_n(x) & y(x) \\ u_1'(x) & u_2'(x) & \ldots & u_n'(x) & y'(x) \\ \ldots & \ldots & \ldots & \ldots & \ldots \\ u_1^{(n)}(x) & u_2^{(n)}(x) & \ldots & u_n^{(n)}(x) & y^{(n)}(x) \end{vmatrix} = 0,$$

where $y(x)$is an unknown function of x.

Proof. Since $S = \{u_1(x), u_2(x), u_3(x), \ldots, u_n(x)\}$ be a fundamental set of solutions of the required nth order linear homogeneous differential equation, the general solution of the equation is given by

$$y(x) = c_1 u_1(x) + c_2 u_2(x) + c_3 u_3(x) + \cdots + c_n u_n(x), \tag{5.6.8}$$

where $c_1, c_2, c_3, \ldots, c_n$ are arbitrary constants. Now differentiating both sides of (5.6.8) n times successively, we obtain

$$y'(x) = c_1 u_1'(x) + c_2 u_2'(x) + c_3 u_3'(x) + \cdots + c_n u_n'(x) \tag{5.6.9}$$

$$y''(x) = c_1 u_1''(x) + c_2 u_2''(x) + c_3 u_3''(x) + \cdots + c_n u_n''(x) \tag{5.6.10}$$

$$\vdots$$

$$y^{(n)}(x) = c_1 u_1^{(n)}(x) + c_2 u_2^{(n)}(x) + c_3 u_3^{(n)}(x) + \cdots + c_n u_n^{(n)}(x). \tag{5.6.11}$$

Eliminating $c_1, c_2, c_3, \ldots, c_n$ from (5.6.8) and above n equations, we obtain

$$\begin{vmatrix} u_1(x) & u_2(x) & \ldots & u_n(x) & -y(x) \\ u_1'(x) & u_2'(x) & \ldots & u_n'(x) & -y'(x) \\ \ldots & \ldots & \ldots & \ldots & \ldots \\ u_1^{(n)}(x) & u_2^{(n)}(x) & \ldots & u_n^{(n)}(x) & -y^{(n)}(x) \end{vmatrix} = 0 \text{ or, } \begin{vmatrix} u_1(x) & u_2(x) & \ldots & u_n(x) & y(x) \\ u_1'(x) & u_2'(x) & \ldots & u_n'(x) & y'(x) \\ \ldots & \ldots & \ldots & \ldots & \ldots \\ u_1^{(n)}(x) & u_2^{(n)}(x) & \ldots & u_n^{(n)}(x) & y^{(n)}(x) \end{vmatrix} = 0,$$

which represents the nth order linear homogeneous differential equation having

$$S = \{u_1(x), u_2(x), u_3(x), \ldots, u_n(x)\}$$

as a fundamental set of solutions.

Example 5.6.7. *Find the homogeneous linear differential equation having a fundamental set of solutions* $S = \{e^{2x}, e^{-x}\}$.

Solution: The homogeneous linear differential equation having $S = \{e^{2x}, e^{-x}\}$ as a fundamental set of solutions is given by

$$\begin{vmatrix} e^{2x} & e^{-x} & y \\ 2e^{2x} & -e^{-x} & y' \\ 4e^{2x} & e^{-x} & y'' \end{vmatrix} = 0 \text{ or, } \begin{vmatrix} 1 & 1 & y \\ 2 & -1 & y' \\ 4 & 1 & y'' \end{vmatrix} = 0 \text{ or, } y'' - y' - 2y = 0.$$

Theorem 5.6.9 (**Principle of superposition for homogeneous equations**).

Statement: Let $u_k(x)$, $k = 1, 2, \ldots, m$ be m solutions of the nth order homogeneous linear differential equation

$$a_0(x)\frac{d^n y}{dx^n} + a_1(x)\frac{d^{n-1}y}{dx^{n-1}} + a_2(x)\frac{d^{n-2}y}{dx^{n-2}} + \cdots + a_{n-1}(x)\frac{dy}{dx} + a_n(x)y = 0, \ a_0(x) \neq 0.$$

Then

$$y(x) = \sum_{k=1}^{m} c_k u_k(x)$$

is also a solution of the equation, where $c_1, c_2, \ldots, c_m$ are constants and $a_0(x), a_1(x), a_2(x), \ldots, a_n(x)$ are single-valued continuous functions of x defined on some closed interval$[a, b]$.

Proof. Since $u_k(x)$, $k = 1, 2, \ldots, m$ are m solutions of the nth order homogeneous linear equation

$$a_0(x)\frac{d^n y}{dx^n} + a_1(x)\frac{d^{n-1} y}{dx^{n-1}} + a_2(x)\frac{d^{n-2} y}{dx^{n-2}} + \cdots + a_{n-1}(x)\frac{dy}{dx} + a_n(x)y = 0, \ a_0(x) \neq 0,$$

we have

$$a_0(x)\frac{d^n u_k(x)}{dx^n} + a_1(x)\frac{d^{n-1} u_k(x)}{dx^{n-1}} + a_2(x)\frac{d^{n-2} u_k(x)}{dx^{n-2}} + \cdots + a_{n-1}(x)\frac{du_k(x)}{dx} + a_n(x)u_k(x) = 0,$$
$$a_0(x) \neq 0, \ k = 1, 2, 3, \ldots, m. \qquad (5.6.12)$$

Now

$$a_0(x)\frac{d^n}{dx^n}\left(\sum_{k=1}^{m} c_k u_k(x)\right) + a_1(x)\frac{d^{n-1}}{dx^{n-1}}\left(\sum_{k=1}^{m} c_k u_k(x)\right) + a_2(x)\frac{d^{n-2}}{dx^{n-2}}\left(\sum_{k=1}^{m} c_k u_k(x)\right) + \ldots$$
$$+ a_{n-1}(x)\frac{d}{dx}\left(\sum_{k=1}^{m} c_k u_k(x)\right) + a_n(x)\left(\sum_{k=1}^{m} c_k u_k(x)\right)$$
$$= c_1\left\{a_0(x)\frac{d^n u_1(x)}{dx^n} + a_1(x)\frac{d^{n-1} u_1(x)}{dx^{n-1}} + a_2(x)\frac{d^{n-2} u_1(x)}{dx^{n-2}} + \cdots + a_{n-1}(x)\frac{du_1(x)}{dx} + a_n(x)u_1(x)\right\}$$
$$+ c_2\left\{a_0(x)\frac{d^n u_2(x)}{dx^n} + a_1(x)\frac{d^{n-1} u_2(x)}{dx^{n-1}} + a_2(x)\frac{d^{n-2} u_2(x)}{dx^{n-2}} + \cdots + a_{n-1}(x)\frac{du_2(x)}{dx} + a_n(x)u_2(x)\right\}$$
$$+ c_3\left\{a_0(x)\frac{d^n u_3(x)}{dx^n} + a_1(x)\frac{d^{n-1} u_3(x)}{dx^{n-1}} + a_2(x)\frac{d^{n-2} u_3(x)}{dx^{n-2}} + \cdots + a_{n-1}(x)\frac{du_3(x)}{dx} + a_n(x)u_3(x)\right\} + \ldots$$
$$+ c_m\left\{a_0(x)\frac{d^n u_m(x)}{dx^n} + a_1(x)\frac{d^{n-1} u_m(x)}{dx^{n-1}} + a_2(x)\frac{d^{n-2} u_m(x)}{dx^{n-2}} + \cdots + a_{n-1}(x)\frac{du_m(x)}{dx} + a_n(x)u_m(x)\right\}$$
$$= c_1 \cdot 0 + c_2 \cdot 0 + c_3 \cdot 0 + \cdots + c_m \cdot 0 = 0 \text{ [by (5.6.12)]}.$$

It confirms that given equation is satisfied by

$$y(x) = \sum_{k=1}^{m} c_k u_k(x).$$

Hence, the proof is complete.

Theorem 5.6.10 (Principle of superposition for non-homogeneous equations).

Statement: Let $u_k(x)$ be a solution of the nth order linear differential equation

$$a_0(x)\frac{d^n y}{dx^n} + a_1(x)\frac{d^{n-1} y}{dx^{n-1}} + a_2(x)\frac{d^{n-2} y}{dx^{n-2}} + \cdots + a_{n-1}(x)\frac{dy}{dx} + a_n(x)y = f_k(x), \ a_0(x) \neq 0, \ k = 1, 2, 3, \ldots, m.$$

Then

$$y(x) = \sum_{k=1}^{m} c_k u_k(x)$$

is also a solution of the equation

$$a_0(x)\frac{d^n y}{dx^n} + a_1(x)\frac{d^{n-1} y}{dx^{n-1}} + a_2(x)\frac{d^{n-2} y}{dx^{n-2}} + \cdots + a_{n-1}(x)\frac{dy}{dx} + a_n(x)y = \sum_{k=1}^{m} c_k f_k(x),$$

where $c_1, c_2, c_3, \ldots, c_m$ are constants and $a_1(x), a_2(x), a_3(x), \ldots, a_n(x)$ and $f_k(x)$ are single-valued continuous functions of x defined on some closed interval $[\alpha, \beta]$.

Proof. Since $u_k(x)$ is a solution of the nth order linear differential equation

$$a_0(x)\frac{d^n y}{dx^n} + a_1(x)\frac{d^{n-1} y}{dx^{n-1}} + a_2(x)\frac{d^{n-2} y}{dx^{n-2}} + \cdots + a_{n-1}(x)\frac{dy}{dx} + a_n(x)y = f_k(x), \quad a_0(x) \neq 0,$$

$$k = 1, 2, 3, \ldots, m$$

$$a_0(x)\frac{d^n u_k(x)}{dx^n} + a_1(x)\frac{d^{n-1} u_k(x)}{dx^{n-1}} + a_2(x)\frac{d^{n-2} u_k(x)}{dx^{n-2}} + \cdots + a_{n-1}(x)\frac{du_k(x)}{dx} + a_n(x)u_k(x) = f_k(x),$$

$$k = 1, 2, 3, \ldots, m. \qquad (5.6.13)$$

Now

$$\begin{aligned}
&a_0(x)\frac{d^n}{dx^n}(c_1u_1(x) + c_2u_2(x) + c_3u_3(x) + \cdots + c_mu_m(x))\\
&+ a_1(x)\frac{d^{n-1}}{dx^{n-1}}(c_1u_1(x) + c_2u_2(x) + c_3u_3(x) + \cdots + c_mu_m(x))\\
&+ a_2(x)\frac{d^{n-2}}{dx^{n-2}}(c_1u_1(x) + c_2u_2(x) + c_3u_3(x) + \cdots + c_mu_m(x)) + \ldots\\
&+ a_{n-1}(x)\frac{d}{dx}(c_1u_1(x) + c_2u_2(x) + c_3u_3(x) + \cdots + c_mu_m(x))\\
&+ a_n(x)(c_1u_1(x) + c_2u_2(x) + c_3u_3(x) + \cdots + c_mu_m(x))\\
&= c_1\left\{a_0(x)\frac{d^n u_1(x)}{dx^n} + a_1(x)\frac{d^{n-1} u_1(x)}{dx^{n-1}} + a_2(x)\frac{d^{n-2} u_1(x)}{dx^{n-2}} + \cdots + a_{n-1}(x)\frac{du_1(x)}{dx} + a_n(x)u_1(x)\right\}\\
&+ c_2\left\{a_0(x)\frac{d^n u_2(x)}{dx^n} + a_1(x)\frac{d^{n-1} u_2(x)}{dx^{n-1}} + a_2(x)\frac{d^{n-2} u_2(x)}{dx^{n-2}} + \cdots + a_{n-1}(x)\frac{du_2(x)}{dx} + a_n(x)u_2(x)\right\}\\
&+ c_3\left\{a_0(x)\frac{d^n u_3(x)}{dx^n} + a_1(x)\frac{d^{n-1} u_3(x)}{dx^{n-1}} + a_2(x)\frac{d^{n-2} u_3(x)}{dx^{n-2}} + \cdots + a_{n-1}(x)\frac{du_3(x)}{dx} + a_n(x)u_3(x)\right\} + \ldots\\
&+ c_m\left\{a_0(x)\frac{d^n u_m(x)}{dx^n} + a_1(x)\frac{d^{n-1} u_m(x)}{dx^{n-1}} + a_2(x)\frac{d^{n-2} u_m(x)}{dx^{n-2}} + \cdots + a_{n-1}(x)\frac{du_m(x)}{dx} + a_n(x)u_m(x)\right\}\\
&= c_1f_1(x) + c_2f_2(x) + c_3f_3(x) + \cdots + c_mf_m(x) \text{ [by (5.6.13)]}\\
&= \sum_{k=1}^{m} c_k f_k(x).
\end{aligned}$$

Hence, the proof is complete.

Note 5.6.3. *For $f_k(x) = 0$, $k = 1, 2, 3, \ldots, m$, the superposition principle reduces to the following form:*

Let $u_1(x), u_2(x), u_3(x), \ldots, u_m(x)$ be m solutions of the nth order linear homogeneous differential equation

$$a_0(x)\frac{d^n y}{dx^n} + a_1(x)\frac{d^{n-1} y}{dx^{n-1}} + a_2(x)\frac{d^{n-2} y}{dx^{n-2}} + \cdots + a_{n-1}(x)\frac{dy}{dx} + a_n(x)y = 0.$$

Then

$$y(x) = \sum_{k=1}^{m} c_k u_k(x)$$

is also a solution of the equation, where $c_1, c_2, c_3, \ldots, c_m$ are constants and $a_0(x), a_1(x), a_2(x), a_3(x), \ldots, a_n(x)$ are single-valued continuous functions of x defined on some closed interval $[\alpha, \beta]$.

Example 5.6.8. *Solve the following equation using the principle of superposition:*

$$\frac{d^2y}{dx^2} - 6\frac{dy}{dx} + 9y = 4e^x - 16e^{3x}.$$

Solution: Let $y = e^{mx}$ be a trial solution of the equation

$$\frac{d^2y}{dx^2} - 6\frac{dy}{dx} + 9y = 0.$$

Then the auxiliary equation is $m^2 - 6m + 9 = 0$. Thus, $m = 3, 3$. Consequently the complementary function is $y_c = (c_1 + c_2x)e^{3x}$, where c_1, c_2 are two arbitrary constants.

In order to determine the particular solution of the given equation we would like to find the particular solutions of the following two equations using the method of undetermined coefficients:

$$\frac{d^2y}{dx^2} - 6\frac{dy}{dx} + 9y = 4e^x \tag{5.6.14}$$

$$\frac{d^2y}{dx^2} - 6\frac{dy}{dx} + 9y = -16e^{3x}. \tag{5.6.15}$$

Clearly eq. (5.6.14) has a particular solution of the form $y_p = Ae^x$, A being a constant. Substituting $y_p = Ae^x$ in (5.6.14), we obtain $A = 1$. Thus, $y_p = e^x$.

Again, eq. (5.6.15) has a particular solution of the form $y_p = Bx^2e^{3x}$, B being a constant. Substituting $y_p = Bx^2e^{3x}$ in the second eq. (5.6.15), we obtain $B = -8$. It follows that $y_p = -8x^2e^{3x}$. Therefore, by the principle of superposition, a particular solution of the given equation is $y_p = e^x - 8x^2e^{3x}$. Hence, required general solution of the given equation is $y = (c_1 + c_2x)e^{3x} + e^x - 8x^2e^{3x}$.

Examples V(B)

1. Show that the following sets of functions are linearly independent:
   (a) $x^2, \frac{1}{x^2}$, where $x \in \mathbb{R}^+$.
   (b) $1, x, x^2$, where $x \in \mathbb{R}$.
   (c) $e^{ax}, e^{bx}, e^{cx} (a \neq b \neq c)$, where $x \in \mathbb{R}$.
2. Show that the equation $\frac{d^2y}{dx^2} + \frac{dy}{dx} - 6y = 0$ has two linearly independent solutions of the form $y = e^{mx}$. Hence, write down the complete primitive.
3. Show that the equation $(D^4 - D^3 - 3D^2 + 5D - 2)y = 0$ has only two linearly independent solutions of the form $y = e^{mx}$. But verify that $y = e^x, y = xe^x, y = x^2e^x, y = e^{-2x}$ are four linearly independent solutions of the given equation. Hence, write down the complete primitive.
4. Show that $x \sin x$ and $x \cos x$ are solutions of the equation $(D^4 + 2D^2 + 1)y = 0$. What is the general solution?
5. What do you understand by linearly independent functions? Show that e^{ax}, xe^{ax} and x^2e^{ax} are solutions of $\frac{d^3y}{dx^3} - 3a\frac{d^2y}{dx^2} + 3a^2\frac{dy}{dx} - a^3y = 0$. Hence, find the general solution of the equation.
6. Find a fundamental set of solutions of each of the following equations:

(a) $(D^3 + 2D^2 - D - 2)y = 0$
(b) $(D^2 + 2D + 2)y = 0$
(c) $(D^2 + D - 6)y = 0$
(d) $(D^3 - 6D^2 + 5D + 12)y = 0$.

7. Show that the set S forms a fundamental set of solutions of the following equations in $(0, \infty)$. Hence, find the general solution of each of the equations.
(a) $(D^2 - 2D + 1)y = 0;\ S = \{e^x, xe^x\}$
(b) $(x^2D^2 + xD - 4)y = 0;\ S = \{x^2, x^{-2}\}$
(c) $(x^2D^2 - 2xD + 2)y = 0;\ S = \{x, x^2\}$
(d) $(x^3D^3 - 4x^2D^2 + 8xD - 8)y = 0;\ S = \{x, x^2, x^4\}$
(e) $(1 + x)\frac{d^2y}{dx^2} + x\frac{dy}{dx} - y = 0;\ S = \{x, e^{-x}\}$
(f) $(x^2D^2 + xD - 1)y = 0;\ S = \{x, \frac{1}{x}\}$
(g) $(x^2D^2 - x(x+2)D + x + 2)y = 0;\ S = \{x, xe^x\}$
(h) $(x^3D^3 + 3x^2D^2)y = 0;\ S = \{1, x, x^{-1}\}$.

8. Find the homogeneous linear differential equation having a fundamental set of solutions
(a) $\{e^x, e^{-x}\}$ (b) $\{e^x, x\}$ (c) $\{1, x, x^2\}$ (d) $\{x, \cos x, \sin x\}$.

9. (a) Given that $y = x$ is a solution of the equation $(x^2 + 1)\frac{d^2y}{dx^2} - 2x\frac{dy}{dx} + 2y = 0$. Find a linearly independent solution by reducing the order. Hence, find its general solution.
(b) Given that $y = e^{2x}$ is a solution of the equation $(2x+1)\frac{d^2y}{dx^2} - 4(x+1)\frac{dy}{dx} + 4y = 0$. Find a linearly independent solution by reducing the order. Hence, find its general solution.
(c) Show that $y = \frac{1}{x}$ is a solution of the equation $2x^2\frac{d^2y}{dx^2} + 3x\frac{dy}{dx} - y = 0,\ x > 0$. Find a second fundamental solution by reducing the order. Hence, find its general solution.
(d) Show that $y = e^x$ is a solution of the equation $(x - 1)\frac{d^2y}{dx^2} - x\frac{dy}{dx} + y = 0,\ x > 1$. Find a second fundamental solution by reducing the order. Hence, find its general solution.

10. Show that the equation $x^3\frac{d^3y}{dx^3} - 6x\frac{dy}{dx} + 12y = 0$ has three linearly independent solutions of the form $y = x^r$. Hence, write down the complete primitive.

Answers

5. $y = (c_1 + c_2x + c_3x^2)e^{ax}$.
6. (a) $\{e^{-x}, e^x, e^{-2x}\}$ (b) $\{e^{-x}\cos x, e^{-x}\sin x\}$ (c) $\{e^{2x}, e^{3x}\}$ (d) $\{e^{-x}, e^{3x}, e^{4x}\}$.
7. (a) $S = \{e^x, xe^x\}$ (b) $S = \{x^2, x^{-2}\}$ (c) $S = \{x, x^2\}$ (d) $S = \{x, x^2, x^4\}$ (e) $S = \{x, e^{-x}\}$ (f) $S = \{x, \frac{1}{x}\}$ (g) $S = \{x, xe^x\}$ (h) $S = \{1, x, x^{-1}\}$.
8. (a) $y'' - y = 0$ (b) $(1 - x)y'' + xy' - y = 0$ (c) $y''' = 0$ (d) $xy''' - y'' + xy' - y = 0$.
9. (a) $x - \frac{1}{x}, c_1x + c_2(x^2 - 1)$ (b) $x + 1, c_1e^{2x} + c_2(x + 1)$ (c) $2\sqrt{x}, y = c_1x^{-1} + c_2\sqrt{x}$ (d) $x, y = c_1x + c_2e^x$.
10. $y = c_1x^2 + c_2x^3 + c_3x^{-2}$.

5.7 Linear operator with constant coefficients: Complementary function

Before proceeding further we list below some important properties of the operator D and powers of D. The symbol D is subject to the fundamental laws of algebra.

Thus, if u and v are functions of x, we have

1. $D(u + v) = Du + Dv$ (Distributive Law).
2. $(D^r + D^s)u = (D^s + D^r)u$.
3. $D^r(D^nu) = D^n(D^ru) = D^{n+r}u$ (Index Law).
4. $D(cu) = cDu$ (c is a constant).

This last law does not hold if c is a variable. We have assumed that r, n and s are positive integers.

The operator D thus satisfies the fundamental laws of algebra except in that it is not commutative with variables. Thus, if α, β be any two constants, then we have

$$\begin{aligned}(D-\alpha)(D-\beta)u &= (D-\alpha)\{(D-\beta)u\} = (D-\alpha)\{Du-\beta u\}\\ &= D\{Du-\beta u\} - \alpha\{Du-\beta u\}\\ &= D^2u-(\alpha+\beta)Du+\alpha\beta u.\end{aligned}$$

It will be necessary to deal with negative indices of D; thus, if we have $Du = v$, and if, as in algebra, we write $u = D^{-1}v$, we have $v = Du = D\left(D^{-1}v\right)$, so that $D.D^{-1} = 1$.

Thus, D^{-1} represents such an operation on any function that if the operation D be subsequently performed, the function is left unaltered.

It follows that these symbols with negative indices also follow the laws of algebra; and an operation with a negative index is equivalent to an operation of integration; e.g. $D^{-1}(3x) = \frac{3}{2}x^2$;

$$D^{-2}(\sin x) = D^{-1}\left\{D^{-1}(\sin x)\right\} = D^{-1}(-\cos x) = -\sin x.$$

In what follows these inverse operations will be used to find *an integral* but *not the complete integral*; and the arbitrary constant which arises in integration will, therefore, be omitted.

Complementary function

The general linear equation with constant coefficients $P_1, P_2, \ldots, P_n$ is

$$\frac{d^ny}{dx^n} + P_1\frac{d^{n-1}y}{dx^{n-1}} + P_2\frac{d^{n-2}y}{dx^{n-2}} + \cdots + P_{n-1}\frac{dy}{dx} + P_ny = f(x) \qquad (5.7.1)$$

$$\text{or,} \quad \left(D^n + P_1D^{n-1} + P_2D^{n-2} + \cdots + P_{n-1}D + P_n\right)y = f(x).$$

Writing $F(D)$ for the linear operator with constant coefficients, i.e. $F(D) \equiv D^n + P_1D^{n-1} + \cdots + P_{n-1}D + P_n$, the eq. (5.7.1) may be written as

$$F(D)y = f(x). \qquad (5.7.2)$$

To find the Complementary Function of (5.7.2) we require to obtain the general solution of its Reduced Equation (RE), making the right-hand member zero.

We, therefore, proceed to find the general solution of

$$F(D)y = 0. \qquad (5.7.3)$$

We consider an auxiliary equation (AE)

$$F(m) = m^n + P_1m^{n-1} + \ldots P_{n-1}m + p_n = 0. \qquad (5.7.4)$$

Let it be assumed that the coefficients $P_1, P_2, \ldots, P_n$ are real numbers, so that the roots of the AE $F(m) = 0$ are either real or conjugate imaginaries.

Case 5.7.1. *Suppose,* $F(m) = 0$ *has n distinct real roots* $m_1, m_2, \ldots, m_n$.

Since the operator D obeys the fundamental laws of algebra the linear operator $F(D)$ can be factorised as $(D-m_1)(D-m_2)\ldots(D-m_n)$.

Since $m_1, m_2, \ldots, m_n$ are constants the factors $(D-m_1), (D-m_2), \ldots, (D-m_n)$ are permutable.

Hence, the eq. (5.7.4) is satisfied by the solutions of each of the n equations of first order:

$$(D-m_1)y = 0, \; (D-m_2)y = 0, \ldots, (D-m_n)y = 0.$$

Let y_r be the general solution of $(D - m_r)\, y = 0$.
Then $y_r = C_r e^{m_r x} (r = 1, 2, 3, \ldots, n)$.
$\therefore$ the general solution of eq. (5.7.3) is

$$y = C_1 e^{m_1 x} + C_2 e^{m_2 x} + \cdots + C_n e^{m_n x}, \tag{5.7.5}$$

where $C_1, C_2, \ldots, C_n$ are n arbitrary constants.

We note here that the n functions $e^{m_1 x}, e^{m_2 x}, \ldots, e^{m_n x}$ are linearly independent. For the Wronskian

$$W = e^{sx} \begin{vmatrix} 1 & 1 & 1 & \ldots & 1 \\ m_1 & m_2 & m_3 & \ldots & m_n \\ m_1^2 & m_2^2 & m_3^2 & \ldots & m_n^2 \\ \ldots & \ldots & \ldots & \ldots & \ldots \\ m_1^{n-1} & m_2^{n-1} & m_3^{n-1} & \ldots & m_n^{n-1} \end{vmatrix}$$

$$= e^{sx} \prod_{i>j} (m_i - m_j) \quad (\text{where } s = m_1 + m_2 + m_3 + \cdots + m_n).$$

Now W cannot vanish, since $m_i \neq m_j$.

Case 5.7.2. *Suppose that the roots of $F(m) = 0$ include one or more pairs of conjugate complex numbers. Let m_r and m_s be conjugate complex numbers, say, $m_r = \alpha + i\beta$, $m_s = \alpha - i\beta$.*

Then the terms $C_r e^{m_r x} + C_s e^{m_s x}$ in the general solution (5.7.5) may be written as

$$\begin{aligned} C_r e^{(\alpha+i\beta)x} + C_s e^{(\alpha-i\beta)x} &= e^{\alpha x} \left\{ C_r e^{i\beta x} + C_s e^{-i\beta x} \right\} \\ &= e^{\alpha x} \{ C_r(\cos\beta x + i\sin\beta x) + C_s(\cos\beta x - i\sin\beta x) \} \\ &= e^{\alpha x} \left\{ C_r' \cos\beta x + C_s' \sin\beta x \right\}, \end{aligned}$$

where $C_r' = C_r + C_s$, $C_s' = i(C_r - C_s)$.

The number of arbitrary constants, therefore, remains as before.

The constants can be adjusted so as to write the expression either in the form $Re^{\alpha x} \cos(\beta x \pm \varepsilon)$ or in the form $Re^{\alpha x} \sin(\beta x \pm \varepsilon)$.

Case 5.7.3 (Repeated Root). *Suppose that $F(m) = 0$ has a root m, repeated p times. Then $F(D)$ has the factor $(D - m)^p$. Then the general solution of $(D - m)^p y = 0$ will be included in the general solution of $F(D)y = 0$.*

One solution corresponding to this factor is known, namely, $y = Ce^{mx}$, where C is a constant; to determine the general solution, the method of variation of parameters is applied (See art. 2.14). We write $y = e^{mx} v$, where v is a function of x to be determined. Then

$$\begin{aligned} (D - m)^p e^{mx} v &= (D - m)^{p-1} \{(D - m)e^{mx} v\} = (D - m)^{p-1} \left\{ D\left(e^{mx} v\right) - me^{mx} v \right\} \\ &= (D - m)^{p-1} \left\{ e^{mx} Dv \right\} = (D - m)^{p-2} \left\{ e^{mx} D^2 v \right\} \\ &= \ldots \\ &= e^{mx} D^p v. \end{aligned}$$

Consequently, $y = e^{mx} v$ is a solution of $(D - m)^p y = 0$, provided that v is a solution of $D^p v = 0$, which means that v is an arbitrary polynomial in x of degree $p - 1$.

Thus, the solution of $(D - m)^p y = 0$ is $y = \left(c_1 + c_2 x + \cdots + c_p x^{p-1}\right) e^{mx}$, and contains the required number (viz., p) of arbitrary constants.

Case 5.7.4. *If $F(m) = 0$ contains two conjugate imaginary roots (say, $\alpha \pm i\beta$), each repeated p times, then the corresponding part of the solution will be of the form:*

$$y = \left(c_1 + c_2x + \cdots + c_px^{p-1}\right) e^{\alpha x} \cos \beta x + \left(c_1' + c_2'x + \cdots + c_p'x^{p-1}\right) e^{\alpha x} \sin \beta x,$$

with appropriate number, 2p, of arbitrary constants.

e.g. the general solution of $\left(D^4 + 2D^2 + 1\right) y = 0$ is $y = (c_1 + c_2x) \cos x + (c_3 + c_4x) \sin x$, $(m = i, i, -i, -i)$.

The rules given in article 5.4 are now established.

5.7.1 Particular integral

The complete primitive of $F(D)y = f(x)$ [eq. (5.7.2) of art. 5.7] is the sum of the Complementary Function (CF) [i.e. the complete primitive of $F(D)y = 0$] obtained in the previous article and any particular solution of $F(D)y = f(x)$, called the *Particular Integral* (PI) to be denoted by y_p.

At times a particular integral may be found by inspection. But very often it is not an easy affair. So we give below two general procedures for obtaining a particular integral.

Our determination of a particular integral will depend upon the properties of operators D^{-1}, $(D-m)^{-1}$, etc. As mentioned earlier, D^{-1} is the operation of simple indefinite integration; similarly, D^{-p} is the operation of p-ple integration. In fact, $\{F(D)\}^{-1}$ is defined by $\{F(D)\}^{-1} F(D)y = y$.

First Method. To obtain a PI of $F(D)y = f(x)$.

Apply the operator $\{F(D)\}^{-1}$, or, $\frac{1}{F(D)}$.

We have then $\frac{1}{F(D)} \cdot F(D)y = y = \frac{1}{F(D)} f(x)$.

Suppose, in the first place that $F(D)$ has no repeated factors. Let $F(D) = (D - m_1)(D - m_2) \ldots (D - m_n)$.

So $\frac{1}{F(D)}$ can be expressed as the sum of n partial fractions.

$$\frac{1}{F(D)} = \frac{A_1}{D - m_1} + \frac{A_2}{D - m_2} + \cdots + \frac{A_n}{D - m_n}.$$

Then

$$y = \frac{1}{F(D)} f(x) = \sum_{r=1}^{n} \frac{A_r}{D - m_r} f(x) \text{ (using distributive law)}$$

$$\text{or, } y = \sum_{r=1}^{n} A_r e^{m_r x} \int f(x) e^{-m_r x} dx \quad \text{(See art. 2.6)}$$

We note here that in obtaining the PI the arbitrary constant which arises in integration need not be written down; for the appropriate number of arbitrary constants are included in the CF.

Example 5.7.1. *Solve the following differential equation* $\frac{d^2y}{dx^2} + a^2y = \sec ax$ *with the symbolic operator D.*

Solution:

First Part. To find the Complementary Function.

The linear operator is $F(D) = D^2 + a^2 = (D + ia)(D - ia)$. The two roots of the auxiliary equation $(m + ia)(m - ia) = 0$ are $-ia, +ia$. The general solution of the reduced equation $\frac{d^2y}{dx^2} + a^2y = 0$ is $y = c_1e^{-iax} + c_2e^{iax} = A \cos ax + B \sin ax$.

This gives the required CF $= A \cos ax + B \sin ax$.

Second Part. To find the Particular Integral (PI)

We write

$$\frac{1}{F(D)} = \frac{1}{(D+ia)(D-ia)} = \frac{1}{2ia}\left[\frac{1}{D-ia} - \frac{1}{D+ia}\right].$$

Also

$$\frac{1}{D-ia}\sec ax = e^{iax}\int e^{-iax}\sec ax\,dx = e^{iax}\int \frac{(\cos ax - i\sin ax)}{\cos ax}dx = e^{iax}\left[x + \frac{i}{a}\ln\cos ax\right]$$

$$= (\cos ax + i\sin ax)\left(x + \frac{i}{a}\ln\cos ax\right)$$

$$= \left(x\cos ax - \frac{1}{a}\sin ax\ln\cos ax\right) + i\left(x\sin ax + \frac{1}{a}\cos ax\ln\cos ax\right).$$

Similarly,

$$\frac{1}{D+ia}\sec ax = \left(x\cos ax - \frac{1}{a}\sin ax\ln\cos ax\right) - i\left(x\sin ax + \frac{1}{a}\cos ax\ln\cos ax\right).$$

$$\therefore\ \text{PI} = \frac{1}{F(D)}\sec ax = \frac{1}{2ia}\frac{1}{D-ia}\sec ax - \frac{1}{2ia}\frac{1}{D+ia}\sec ax = \frac{1}{2ia}\left[2i\left(x\sin ax + \frac{1}{a}\cos ax\ln\cos ax\right)\right]$$

$$= \frac{x\sin ax}{a} + \frac{\cos ax\ln\cos ax}{a^2}.$$

Conclusion: The complete primitive of the given equation is

$$y = \text{CF} + \text{PI} = A\cos ax + B\sin ax + \frac{x\sin ax}{a} + \frac{\cos ax\ln\cos ax}{a^2},$$

where A and B are arbitrary constants.

Consider now the case in which $F(D)$ contains the factor $(D-m)^p$. The part of the expression of $1/F(D)$ in partial fractions which corresponds to this repeated factor is $\sum_{r=1}^{p}\frac{B_r}{(D-m)^r}$ and the corresponding contribution to the particular integral is

$$\sum_{r=1}^{p}\frac{B_r}{(D-m)^r}f(x) = \sum_{r=1}^{p}\left\{B_r\frac{1}{(D-m)^{r-1}}\frac{1}{D-m}f(x)\right\} = \sum_{r=1}^{p}\left\{B_r\frac{1}{(D-m)^{r-1}}\cdot e^{mx}\int e^{-mx}f(x)\,dx\right\}$$

$$= \sum_{r=1}^{p}\left\{B_r\frac{1}{(D-m)^{r-2}}e^{mx}\int\left\{\int e^{-mx}f(x)\,dx\right\}dx\right\} = \sum_{r=1}^{p}B_r e^{mx}\int\ldots\int e^{-mx}f(x)\,dx^r.$$

Second Method. This method consists of solving a succession of linear differential equations of order one, as follows:

$$y = \frac{1}{F(D)}f(x) = \frac{1}{(D-m_1)(D-m_2)\ldots(D-m_n)}f(x)$$

Let $u = \frac{1}{D-m_n}f(x)$.

Then $\frac{du}{dx} - m_n u = f(x)$.

$$\therefore\ u = e^{m_n x}\int e^{-m_n x}f(x)\,dx.$$

Now put $V = \frac{1}{D-m_{n-1}}u$.

Then

$$V = e^{m_{n-1}x}\int e^{-m_{n-1}x}u\,dx = e^{m_{n-1}x}\int\left\{e^{(m_n-m_{n-1})x}\int e^{-m_nx}f(x)\,dx\right\}dx.$$

We proceed in this way and finally obtain

$$y = e^{m_1x}\int e^{(m_2-m_1)x}\int e^{(m_3-m_2)x}\cdots\int e^{(m_n-m_{n-1})x}\int e^{-m_nx}f(x)\times(dx)^n.$$

Example 5.7.2. *Solve:* $\frac{d^3y}{dx^3}+3\frac{d^2y}{dx^2}-4y = xe^{-2x}$.

Solution: The equation is $(D-1)(D+2)^2y = xe^{-2x}$. The Complementary function is $C_1e^x+(C_2+C_3x)\,e^{-2x}$, where C_1, C_2, C_3 are arbitrary constants.

A particular integral is $y_p = \frac{1}{D-1}\cdot\frac{1}{D+2}\cdot\frac{1}{D+2}xe^{-2x}$.

Let $u = \frac{1}{D+2}xe^{-2x}$.

Then $\frac{du}{dx}+2u = xe^{-2x}$, or, $u = e^{-2x}\int xe^{-2x}\cdot e^{2x}dx = \frac{1}{2}x^2e^{-2x}$.

Let $V = \frac{1}{D+2}u$.

Then $\frac{dV}{dx}+2V = \frac{1}{2}x^2e^{-2x}$.

$$\therefore\ V = e^{-2x}\int\frac{1}{2}x^2e^{-2x}\cdot e^{2x}dx = \frac{1}{6}x^3e^{-2x}.$$

Now $y = \frac{1}{D-1}V$.

Then

$$\frac{dy}{dx}-y = \frac{1}{6}x^3e^{-2x},\ \text{or, } y_p = e^x\int\frac{1}{6}x^3e^{-2x}\cdot e^{-x}dx = \frac{1}{6}e^x\int x^3e^{-3x}dx = -\frac{1}{18}e^{-2x}\left(x^3+x^2+\frac{2}{3}x+\frac{2}{9}\right).$$

$\therefore$ the primitive is $y = C_1e^x+C_2e^{-2x}+C_3xe^{-2x}-\frac{1}{18}\left(x^3+x^2\right)e^{-2x}$, the remaining terms of the PI being absorbed by the CF.

Examples V(C)

Evaluate, omitting the arbitrary constants:

1. $\frac{1}{D+1}\left(x^2+1\right)$.
2. $\frac{1}{D+2}e^{-2x}\sin 3x$.
3. $\frac{1}{D^2-1}4xe^x$.
4. $\frac{1}{D^2-1}xe^{2x}$.
5. (a) $\frac{1}{D^2-5D+6}e^{4x}$. (b) $\frac{1}{D^2+3D+2}e^{e^x}$.
6. $\frac{1}{D^2-1}\sin^2 x$.
7. $\frac{1}{D^2-1}\left(1+e^{-x}\right)^{-2}$.
8. $\frac{1}{(D-3)(D-2)}\ln x$.
9. $\frac{1}{D^2+1}\operatorname{cosec} x$.
10. $\frac{1}{D^2+1}\sec x$.
11. $\frac{1}{(D-2)(D-1)}\left(\sin e^{-x}\right)$.
12. $\frac{1}{D^3-5D^2+8D-4}e^{2x}$.
13. $\frac{1}{D}(x\sin x)$.
14. $\frac{1}{D^2}x^2$.
15. $\frac{1}{D^2}\cos x$.

Solve the following differential equations with the symbolic operator D:

16. $\frac{d^2y}{dx^2}+4y = \sec 2x$.
17. $\frac{d^2y}{dx^2}+a^2y = \operatorname{cosec} ax$.
18. $\frac{d^2y}{dx^2}+4y = \tan 2x$.
19. $\frac{d^2y}{dx^2}+y = \operatorname{cosec} x$.
20. $\frac{d^2y}{dx^2}+a^2y = \tan ax$.

Answers

1. $x^2 - 2x + 3$.
2. $-\frac{1}{3}e^{-2x}\cos 3x$.
3. $e^x\left(x^2 - x + \frac{1}{2}\right)$.
4. $\frac{1}{9}e^{2x}(3x-4)$.
5. (a) $\frac{1}{2}e^{4x}$ (b) $e^{e^x} - 2x$.
6. $\frac{1}{10}\cos 2x - \frac{1}{2}$.
7. $-1 + e^{-x}\ln(1+e^x)$.
8. $e^{3x}\int e^{-3x}\ln x\,dx - e^{2x}\int e^{-2x}\ln x\,dx$.
9. $\sin x\ln\sin x - x\cos x$.
10. $x\sin x + \cos x\ln\cos x$.
11. $-e^{2x}\sin e^{-x}$.
12. $\frac{1}{2}e^{2x}\left(x^2 - 2x + 2\right)$.
13. $\sin x - x\cos x$.
14. $x^4/12$.
15. $-\cos x$.
16. $y = A\cos 2x + B\sin 2x + \frac{x}{2}\sin ax + \frac{1}{4}\cos ax\ \ln\cos ax$.
17. $y = A\cos ax + B\sin ax - \frac{x}{a}\cos ax + \frac{1}{a^2}\sin ax\ \ln\sin ax$.
18. $y = A\cos 2x + B\sin 2x - \frac{1}{4}\cos 2x\ln(\sec 2x + \tan 2x)$.
19. $y = A\cos x + B\sin x + \sin x\ln\sin x - x\cos x$.
20. $y = A\cos ax + B\sin ax - \frac{1}{a^2}\cos ax\ln|\sec ax + \tan ax|$.

5.7.2 Short methods of finding particular integrals in special cases: operator D

We know that a particular integral of a linear differential equation $F(D)y = f(x)$ with constant coefficients is given by $y = \frac{1}{F(D)}f(x)$, evaluation of which can be made by the methods described in art. 5.7.1. For certain special forms of $f(x)$, much shorter methods can be used:

I. Suppose, $f(x) = e^{mx}$. Then

$$y = \frac{1}{F(D)}e^{mx} = \frac{1}{F(m)}e^{mx}, \text{ if } F(m) \neq 0.$$

The exceptional case $F(m) = 0$ will be treated under **II** below.

Justification. Clearly, $D^r e^{mx} = m^r e^{mx}$ $(r = 1, 2, 3, \ldots)$.

$$\begin{aligned} F(D)e^{mx} &= [D^n + P_1D^{n-1} + P_2D^{n-2} + \cdots + P_{n-1}D + P_n]e^{mx} \\ &= [m^n + P_1m^{n-1} + P_2m^{n-2} + \cdots + P_{n-1}m + P_n]e^{mx} \\ &= F(m)e^{mx}. \end{aligned}$$

$$\therefore \frac{1}{F(D)}\{F(D)e^{mx}\} = \frac{1}{F(D)}F(m)e^{mx} = F(m)\frac{1}{F(D)}e^{mx} \quad [\because F(m) \text{ is constant}]$$

$$\text{or, } e^{mx} = F(m)\frac{1}{F(D)}e^{mx}.$$

$$\therefore \frac{1}{F(D)}e^{mx} = \frac{e^{mx}}{F(m)} \quad (\text{if } F(m) \neq 0).$$

Example 5.7.3. *Solve:* $(D^2 - 1)\,y = e^{2x}$.

Solution: The complementary function is $y = C_1e^x + C_2e^{-x}$, since the linear operator is

$$F(D) \equiv (D-1)(D+1).$$

For the particular integral, $y_p = \frac{1}{D^2-1}e^{2x} = e^{2x}/3$ [replace D by 2, since $F(2) \neq 0$].

$\therefore$ the complete solution is $y = C_1e^x + C_2e^{-x} + e^{2x}/3$ (Fig. 5.2).

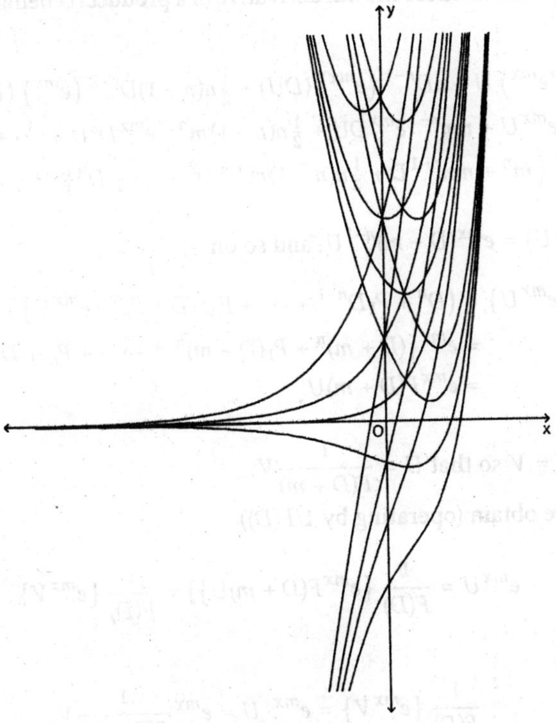

Fig. 5.2

Example 5.7.4. *Solve:* $(D^3 - 2D^2 - 5D + 6)\, y = (e^{2x} + 3)^2 + e^{3x}\cosh x$.

Solution: The complementary function is $c_1e^x + c_2e^{3x} + c_3e^{-2x}$; for the linear operator

$$F(D) \equiv (D-1)(D-3)(D+2).$$

Note that

$$(e^{2x}+3)^2 + e^{3x}\cosh x = e^{4x} + 6e^{2x} + 9e^{0.x} + e^{3x}.\frac{e^x + e^{-x}}{2} = e^{4x} + 6e^{2x} + 9e^{0.x} + \frac{1}{2}e^{4x} + \frac{1}{2}e^{2x}$$

$$= \frac{3}{2}e^{4x} + \frac{13}{2}e^{2x} + 9e^{0.x}.$$

A particular integral is

$$\frac{1}{(D-1)(D-3)(D+2)}\{(e^{2x}+3)^2 + e^{3x}\cosh x\}$$

$$= \frac{3}{2}\frac{1}{(D-1)(D-3)(D+2)}e^{4x} + \frac{13}{2}\frac{1}{(D-1)(D-3)(D+2)}e^{2x} + 9\frac{1}{(D-1)(D-3)(D+2)}e^{0.x}$$

$$= \frac{3}{2}\frac{1}{3.1.6}e^{4x} + \frac{13}{2}\frac{1}{1(-1)\cdot 4}e^{2x} + \frac{9}{(-1)(-3)2}e^{0.x}$$

$$= \frac{1}{12}e^{4x} - \frac{13}{8}e^{2x} + \frac{3}{2}.$$

The complete primitive is

$$y = c_1e^x + c_2e^{3x} + c_3e^{-2x} + \frac{1}{12}e^{4x} - \frac{13}{8}e^{2x} + \frac{3}{2}.$$

II. Suppose, $f(x) = e^{mx}V$, where V is any function of x. Then a particular integral of $F(D)y = f(x)$ is

$$y = \frac{1}{F(D)}\left(e^{mx}V\right) = e^{mx}\frac{1}{F(D+m)}V.$$

Justification. By Leibniz's theorem for the *n*th derivative of a product (*U* being any function of *x* derivable up to *n* times),

$$D^n\left(e^{mx}U\right) = \left(D^n e^{mx}\right)U + nD^{n-1}\left(e^{mx}\right)(DU) + \tfrac{1}{2}n(n-1)D^{n-2}\left(e^{mx}\right)\left(D^2U\right) + \cdots + e^{mx}D^nU$$
$$= m^n e^{mx}U + nm^{n-1}e^{mx}DU + \tfrac{1}{2}n(n-1)m^{n-2}e^{mx}D^2U + \cdots + e^{mx}D^nU$$
$$= e^{mx}\left\{m^n + nm^{n-1}D + \tfrac{1}{2}n(n-1)m^{n-2}D^2 + \cdots + D^n\right\}U = e^{mx}(D+m)^nU.$$

Similarly, $D^{n-1}\left(e^{mx}U\right) = e^{mx}(D+m)^{n-1}U$, and so on.

$$\therefore\ F(D)\left\{e^{mx}U\right\} = \left(D^n + P_1D^{n-1} + \cdots + P_{n-1}D + P_n\right)\left\{e^{mx}U\right\}$$
$$= e^{mx}\{(D+m)^n + P_1(D+m)^{n-1} + \cdots + P_{n-1}(D+m) + P_n\}U$$
$$= e^{mx}F(D+m)U. \tag{5.7.6}$$

Next, let $F(D+m)U = V$ so that $U = \dfrac{1}{F(D+m)}V.$

Then from (5.7.6), we obtain (operating by $1/F(D)$)

$$e^{mx}U = \frac{1}{F(D)}\{e^{mx}F(D+m)U\}] = \frac{1}{F(D)}\{e^{mx}V\}.$$

That is,

$$\frac{1}{F(D)}\{e^{mx}V\} = e^{mx}.\,U = e^{mx}\frac{1}{F(D+m)}V.$$

Particular Case. If $F(D) = (D-m)^p\phi(D)$ [i.e. if $(D-m)$ is a factor of $F(D)$ repeated p times and there is no factor of the form $D-m$ in $\phi(D)$] and if $V = 1$, then it follows that

$$\frac{1}{F(D)}\{e^{mx}\} = \frac{1}{(D-m)^p}\frac{1}{\phi(D)}\{e^{mx}\} = \frac{1}{(D-m)^p}\frac{e^{mx}}{\phi(m)}\quad(\because\ \phi(m)\neq 0)$$
$$= \frac{1}{\phi(m)}\frac{1}{(D-m)^p}\{e^{mx}.1\} = \frac{1}{\phi(m)}e^{mx}\frac{1}{D^p}\{1\} = \frac{e^{mx}}{\phi(m)}\cdot\frac{x^p}{p!}$$

and thus the exceptional case left over in **I** is accounted for.

Example 5.7.5. *Solve:* $\frac{d^3y}{dx^3} - 5\frac{d^2y}{dx^2} + 8\frac{dy}{dx} - 4y = e^{2x} + e^x + 3e^{-x}.$

Solution: The linear operator $F(D) \equiv D^3 - 5D^2 + 8D - 4 = (D-1)(D-2)^2.$

$\therefore$ the CF is $c_1e^x + (c_2 + c_3x)\,e^{2x}.$

A particular integral is

$$= \frac{1}{(D-1)(D-2)^2}\left\{e^{2x} + e^x + 3e^{-x}\right\} = \frac{1}{(D-1)(D-2)^2}e^{2x} + \frac{1}{(D-1)(D-2)^2}e^x + \frac{3}{(D-1)(D-2)^2}e^{-x}$$
$$= \frac{e^{2x}}{2-1}\frac{1}{D^2}\cdot 1 + \frac{e^x}{(1-2)^2}\cdot\frac{1}{D}\cdot 1 + \frac{3e^{-x}}{(-1-1)(-1-2)^2} = e^{2x}.\frac{1}{2}x^2 + e^x.\,x - \frac{1}{6}e^{-x}.$$

$$\therefore\ y = c_1e^x + (c_2 + c_3x)\,e^{2x} + \frac{1}{2}x^2e^{2x} + xe^x - \frac{1}{6}e^{-x}.$$

III. Suppose, $f(x) = \sin ax$ or $\cos ax$. Further, suppose that $F(D)$ is an even polynomial in D. We may write $F(D) = \phi(D^2)$, so that $\phi(D^2)$ is a polynomial in D^2. Then a particular integral of $F(D)y = \sin ax$ or $\cos ax$ is

$$y = \frac{1}{F(D)} \sin ax = \frac{1}{\phi(-a^2)} \sin ax \quad (\text{if } \phi(-a^2) \neq 0)$$

$$\text{or, } y = \frac{1}{F(D)} \cos ax = \frac{1}{\phi(-a^2)} \cos ax \quad (\text{if } \phi(-a^2) \neq 0).$$

Justification. Since $D^2 \sin ax = -a^2 \sin ax$, $D^4 \sin ax = (-a^2)^2 \sin ax$, $D^6 \sin ax = (-a^2)^3 \sin ax$ and so on.

Since $\phi(D^2)$ is a polynomial in D^2,
$\phi(D^2) \sin ax = \phi(-a^2) \sin ax$ or, $F(D) \sin ax = \phi(-a^2) \sin ax$.

$$\therefore \sin ax = \frac{1}{F(D)} \phi(-a^2) \sin ax = \phi(-a^2) \frac{1}{F(D)} \sin ax \text{ or, } \frac{1}{F(D)} \sin ax = \frac{1}{\phi(-a^2)} \sin ax \quad (\text{if } \phi(-a^2) \neq 0).$$

Similarly,

$$\frac{1}{F(D)} \cos ax = \frac{1}{\phi(-a^2)} \cos ax \quad (\text{if } \phi(-a^2) \neq 0).$$

Important observations

1. In the most general case, the polynomial $F(D)$ is not even; but suppose if it has an even polynomial factor $G(D)$, then let $F(D) = G(D)H(D)$.

Then

$$\frac{1}{F(D)} \sin ax = \frac{1}{G(D)H(D)} \sin ax = \frac{H(-D)}{G(D)H(D)H(-D)} \sin ax.$$

Now $G(D)H(D)H(-D)$ is an even polynomial in D and may be written $\phi(D^2)$; thus

$$\frac{1}{F(D)} \sin ax = \frac{H(-D)}{\phi(D^2)} \sin ax = \frac{1}{\phi(-a^2)} H(-D) \sin ax \; \{\phi(-a^2) \neq 0\}.$$

Similarly, $\frac{1}{F(D)} \cos ax$ may be evaluated provided that $\phi(-a^2) \neq 0$.

2. Combining the Rules II and III, particular integrals of the form $\frac{1}{F(D)} e^{mx} \sin ax$, or, $\frac{1}{F(D)} e^{mx} \cos ax$ may be evaluated.

Example 5.7.6. *Solve:* $(D^2 - 4) y = \sin 2x$.

Solution: The linear operator $F(D) \equiv D^2 - 4 = (D-2)(D+2)$.

Therefore, the auxiliary equation is $(m-2)(m+2) = 0$, or, $m = 2, -2$.

The complementary function is $C_1 e^{2x} + C_2 e^{-2x}$.

A particular integral is $y_p = \frac{1}{D^2-4} \sin 2x = -\frac{1}{8} \sin 2x$ [replace D^2 by -2^2].

$\therefore$ the complete solution is $y = C_1 e^{2x} + C_2 e^{-2x} - \frac{1}{8} \sin 2x$.

Example 5.7.7. *Solve:* $(3D^2 + 2D - 8)\, y = 5 \cos x$.

Solution: The linear operator $F(D) \equiv 3D^2 + 2D - 8 = (3D-4)(D+2)$.

The Complementary Function is $C_1 e^{\frac{4}{3}x} + C_2 e^{-2x}$.

A Particular Integral is

$$\frac{1}{(3D-4)(D+2)} 5 \cos x = \frac{(-3D-4)(-D+2)}{(9D^2-16)(D^2-4)} 5 \cos x = 5 \cdot \frac{3D^2 - 2D - 8}{\{9(-1) - 16\}\{-1-4\}} \cos x$$

[Replacing D^2 by -1^2 in the denominator]

$$= \frac{1}{25}\{3D^2 - 2D - 8\} \cos x = \frac{1}{25}\{2 \sin x - 11 \cos x\}.$$

9|B

The complete primitive is $y = C_1 e^{\frac{4}{3}x} + C_2 e^{-2x} + \frac{1}{25}\{2\sin x - 11\cos x\}$.

IV. Suppose, $f(x) = x^m$ (m, a is positive integer).

Then $\frac{1}{F(D)}x^m = (a_0 + a_1 D + \cdots + a_m D^m)\, x^m$, where $a_0 + a_1 D + \cdots + a_m D^m$ is the formal expansion of $\frac{1}{F(D)}$ in a series of ascending powers of D, and suppressing all the terms beyond D^m, since $D^r x^m = 0$ whenever $r > m$.

Justification. We easily see that

$$(D-a)x^m = mx^{m-1} - ax^m$$

$$(D-a)\frac{m}{a}x^{m-1} = \frac{m(m-1)}{a}x^{m-2} - mx^{m-1}$$

$$(D-a)\frac{m(m-1)}{a^2}x^{m-2} = \frac{m(m-1)(m-2)}{a^2}x^{m-3} - \frac{m(m-1)}{a}x^{m-2}$$

$$\vdots$$

$$(D-a)\frac{m(m-1)\ldots 3.2}{a^{m-1}}x = \frac{m!}{a^{m-1}} - \frac{m(m-1)\ldots 3.2}{a^{m-2}}x$$

$$(D-a)\frac{m!}{a^m} = -\frac{m!}{a^{m-1}}.$$

Hence, by addition,

$$(D-a)\left\{x^m + \frac{m}{a}x^{m-1} + \cdots + \frac{m!}{a^m}\right\} = -ax^m.$$

$$\therefore \quad \frac{1}{D-a}x^m = -\frac{1}{a}\left\{x^m + \frac{m}{a}x^{m-1} + \cdots + \frac{m!}{a^m}\right\} = -\sum_{r=0}^{m}\left\{\frac{m!}{r!\,a^{m-r+1}}x^r\right\}.$$

The result would have been same if we would formally expand the operator $(D-a)^{-1}$ in ascending powers of D and perform the differentiations.

It follows that if $f(x)$ is a polynomial in x of degree n,

$$\frac{1}{F(D)}f(x) = \left(a_0 + a_1 D + \cdots + a_n D^n\right) f(x),$$

where $a_0 + a_1 D + \cdots + a_n D^n$ is the formal expansion of $F^{-1}D$ to $(n+1)$ terms.

Example 5.7.8. *Solve:* $\left(D^4 - 2D^3 + D^2\right) y = x^3$.

Solution: The linear operator $F(D) = D^4 - 2D^3 + D^2 = D^2\left(D^2 - 2D + 1\right) = D^2(D-1)^2$.

$\therefore$ the CF is $(c_1 + c_2 x) + (c_3 + c_4 x)\, e^x$ (the roots of the AE are 0, 0, 1, 1).

Next we have, for a particular integral,

$$y_p = \frac{1}{D^2(1-D)^2}x^3 = \frac{1}{D^2}\left(1 + 2D + 3D^2 + 4D^3\right)x^3.$$

[Expanding $(1-D)^{-2}$ formally and suppressing all the terms beyond D^3 since $D^r x^3 = 0$, if $r > 3$]

$$= \frac{1}{D^2}\left(x^3 + 6x^2 + 18x + 24\right) = \frac{1}{D}\left\{\frac{x^4}{4} + 2x^3 + 9x^2 + 24x\right\} \text{ (operating } D^{-1}\text{)}$$

$$= \frac{x^5}{20} + \frac{x^4}{2} + 3x^3 + 12x^2 \text{ (operating } D^{-1} \text{ again)}$$

$\therefore$ the complete primitive is

$$y = (c_1 + c_2x) + (c_3 + c_4x)\, e^x + \frac{x^5}{20} + \frac{x^4}{2} + 3x^3 + 12x^2.$$

Practical Hints: Before expanding $\frac{1}{F(D)}$ in a formal manner write it in the form:

$$\frac{1}{D^p}\{1 + \phi(D)\}^{-n} = \frac{1}{D^p}\left\{1 - n\phi(D) + \frac{n(n+1)}{2}[\phi(D)]^2 + \dots\right\},$$

where p is the lowest power of D in $F(D)$, e.g.

$$\frac{1}{D^3 + 3D^2 + 2D} = \frac{1}{2D}\left(1 + \frac{D^2 + 3D}{2}\right)^{-1} = \frac{1}{2D}\left\{1 - \left(\frac{D^2 + 3D}{2}\right) + \left(\frac{D^2 + 3D}{2}\right)^2 + \dots\right\}.$$

The following illustration should be carefully observed.

Example 5.7.9. *Find a particular integral of* $(D^2 + 4)\, y = \sin 2x$.

Solution: We cannot evaluate $\frac{1}{D^2+4}\sin 2x$ by writing -2^2 for D^2 (Rule III), for this gives zero in the denominator. But we write $\sin 2x = \frac{1}{2i}\left(e^{2ix} - e^{-2ix}\right)$.

Then

$$y_p = \frac{1}{D^2 + 4}\sin 2x = \frac{1}{2i}\frac{1}{D^2 + 4}e^{2ix} - \frac{1}{2i}\frac{1}{D^2 + 4}e^{-2ix}.$$

Now

$$\frac{1}{D^2 + 4}e^{2ix} = e^{2ix}\cdot\frac{1}{(D + 2i)^2 + 4}\cdot 1 \text{ (Rule II)}$$

$$= e^{2ix}\frac{1}{D^2 + 4Di}x^0 = e^{2ix}\frac{1}{4iD}\left(1 + \frac{D}{4i}\right)^{-1}x^0 = \frac{e^{2ix}}{4i}\frac{1}{D}\left(1 - \frac{D}{4i}\right)x^0 \text{ [expanding formally]}$$

$$= \frac{e^{2ix}}{4i}\cdot\frac{1}{D}\cdot 1 = \frac{x}{4i}e^{2ix}.$$

Similarly, $\frac{1}{D^2 + 4}e^{-2ix} = -\frac{x}{4i}e^{-2ix}$.

$$\therefore\ y_p = \frac{1}{D^2 + 4}\sin 2x = \frac{1}{2i}\frac{x}{4i}e^{2ix} + \frac{1}{2i}\frac{x}{4i}e^{-2ix} = -\frac{x}{4}\left(\frac{e^{2ix} + e^{-2ix}}{2}\right) = -\frac{x}{4}\cos 2x.$$

V. Suppose, $f(x) = xV$ or x^nV, where V is any function of x and n is a positive integer. Then a particular integral of $F(D)y = f(x)$ is

(i) $y_p = \frac{1}{F(D)}f(x) = \frac{1}{F(D)}xV = \left\{x - \frac{1}{F(D)}\cdot F'(D)\right\}\frac{1}{F(D)}V$

(ii) $y_p = \frac{1}{F(D)}f(x) = \frac{1}{F(D)}x^nV = \left\{x - \frac{1}{F(D)}\cdot F'(D)\right\}^n\frac{1}{F(D)}V.$

Note 5.7.1. *The operators must be used in the proper order. The process becomes laborious if n is a large positive integer.*

Justification. We begin with $\{F(D)\}^{-1}xU$, where U is any function of x to be defined later.

Now if $y = xU$, then $Dy = xDU + U$

$$D^2y = xD^2U + 2DU$$

$$\vdots$$

$$D^ry = xD^rU + rD^{r-1}U = xD^rU + \frac{d}{dD}\{D^r\}\,U.$$

Hence,

$$F(D)xU = \left(D^n + P_1D^{n-1} + \cdots + P_{n-1}D + P_n\right) xU$$

$$= xD^nU + P_1xD^{n-1}U + \cdots + P_{n-1}xDU + P_nxU + \frac{d}{dD}\left(D^n\right)U + P_1\frac{d}{dD}\left(D^{n-1}\right)U + \ldots$$

$$+ P_{n-1}\frac{d}{dD}(D)U$$

$$= xF(D)U + F'(D)U. \tag{5.7.7}$$

Let U be such a function that $F(D)U = V$ so that $U = \frac{1}{F(D)}V$ (Note that V is given).
Hence, (5.7.7) gives

$$F(D)\left\{x\frac{1}{F(D)}V\right\} = xF(D)\left\{\frac{1}{F(D)}V\right\} + F'(D)\left\{\frac{1}{F(D)}V\right\} = xV + F'(D)\left\{\frac{1}{F(D)}V\right\}$$

$$\text{or, } xV = F(D)\left\{x\frac{1}{F(D)}V\right\} - F'(D)\left\{\frac{1}{F(D)}V\right\}.$$

$$\therefore\ \frac{1}{F(D)}xV = \left\{x\frac{1}{F(D)}V\right\} - \frac{1}{F(D)}\cdot F'(D)\cdot\left\{\frac{1}{F(D)}V\right\} = \left\{x - \frac{1}{F(D)}F'(D)\right\}\frac{1}{F(D)}V.$$

This justifies the result (i). The result (ii) can be obtained by the successive applications of (i).

Example 5.7.10. *Find a particular integral of* $\left(D^2 + 4\right)y = x\sin^2 x$.

Solution:

$$y_p = \frac{1}{D^2+4}x\sin^2 x = \frac{1}{2}\frac{1}{D^2+4}x(1-\cos 2x) = \frac{1}{2}\frac{1}{D^2+4}x - \frac{1}{2}\frac{1}{D^2+4}x\cos 2x.$$

Now

$$\frac{1}{D^2+4}x = \frac{1}{4}\left(1+\frac{D^2}{4}\right)^{-1}x = \frac{1}{4}\left(1-\frac{D^2}{4}+\ldots\right)x = \frac{1}{4}x;$$

and let

$$V = \frac{1}{D^2+4}x\cos 2x = \left\{x - \frac{1}{D^2+4}2D\right\}\frac{1}{D^2+4}\cos 2x = \left\{x - \frac{1}{D^2+4}2D\right\}\frac{x\sin 2x}{4}$$

$$\left[\because \frac{1}{D^2+4}\cos 2x = \frac{x\sin 2x}{4};\ \text{compare Example 5.7.9 above}\right]$$

$$= \frac{1}{4}x^2\sin 2x - \frac{1}{2}\frac{1}{D^2+4}D(x\sin 2x) = \frac{1}{4}x^2\sin 2x - \frac{1}{2}\frac{1}{D^2+4}(2x\cos 2x + \sin 2x)$$

$$= \frac{1}{4}x^2\sin 2x - \frac{1}{D^2+4}x\cos 2x - \frac{1}{2}\frac{1}{D^2+4}\sin 2x = \frac{1}{4}x^2\sin 2x - V - \frac{1}{2}\left(-\frac{x\cos 2x}{4}\right)$$

[See Example 5.7.9]

This gives

$$2V = \frac{x^2\sin 2x}{4} + \frac{x\cos 2x}{8},\ \text{or, } V = \frac{x^2\sin 2x}{8} + \frac{x\cos 2x}{16}.$$

∴ the required PI is

$$\frac{x}{8} - \frac{1}{2}\left(\frac{x^2\sin 2x}{8} + \frac{x\cos 2x}{16}\right) = \frac{x}{8} - \frac{x^2\sin 2x}{16} - \frac{x\cos 2x}{32}.$$

Example 5.7.11. *Find a particular integral of* $(D^2-1)\,y = x^2\cos x$.

Solution:

$$y_p = \frac{1}{D^2-1}x^2\cos x = \left\{x - \frac{1}{D^2-1}\cdot 2D\right\}^2 \frac{1}{D^2-1}\cos = \left\{x-\frac{1}{D^2-1}\cdot 2D\right\}\left\{x-\frac{1}{D^2-1}\cdot 2D\right\}\left(-\frac{1}{2}\cos x\right)$$

$$= \left\{x-\frac{1}{D^2-1}\cdot 2D\right\}\left\{-\frac{x\cos x}{2}+\frac{1}{D^2-1}D(\cos x)\right\} = \left\{x-\frac{1}{D^2-1}\cdot 2D\right\}\left\{-\frac{x\cos x}{2}-\frac{1}{D^2-1}\sin x\right\}$$

$$= \left\{x-\frac{1}{D^2-1}\cdot 2D\right\}\left\{-\frac{x\cos x}{2}+\frac{1}{2}\sin x\right\}$$

$$= -\frac{x^2\cos x}{2}+\frac{1}{2}x\sin x+\frac{1}{D^2-1}D(x\cos x)-\frac{1}{D^2-1}D(\sin x)$$

$$= -\frac{x^2\cos x}{2}+\frac{x\sin x}{2}+\frac{1}{D^2-1}\{\cos x - x\sin x\}-\frac{1}{D^2-1}\cos x$$

$$= -\frac{x^2\cos x}{2}+\frac{x\sin x}{2}-\frac{1}{D^2-1}x\sin x.$$

But

$$\frac{1}{D^2-1}x\sin x = \left\{x-\frac{1}{D^2-1}\cdot 2D\right\}\frac{1}{D^2-1}\sin x = \left\{x-\frac{1}{D^2-1}\cdot 2D\right\}\left(-\frac{1}{2}\sin x\right)$$

$$= -\frac{x\sin x}{2}+\frac{1}{D^2-1}\cos x = -\frac{x\sin x}{2}-\frac{1}{2}\cos x.$$

$$\therefore\ y_p = -\frac{x^2\cos x}{2}+\frac{x\sin x}{2}+\frac{x\sin x}{2}+\frac{\cos x}{2} = -\frac{1}{2}x^2\cos x + x\sin x + \frac{1}{2}\cos x.$$

Example 5.7.12. *Find a particular integral of* $(D^3+1)\,y = e^{x/2}\sin\frac{x\sqrt{3}}{2}$.

Solution:

$$\text{PI} = \frac{1}{D^3+1}e^{x/2}\sin\frac{x\sqrt{3}}{2} = \text{Imaginary Part of } \frac{1}{D^3+1}e^{x/2}e^{\frac{ix\sqrt{3}}{2}}$$

$$= \text{Imaginary Part of } e^{x/2}.\,e^{\frac{ix\sqrt{3}}{2}}\frac{1}{\left(D+\frac{1}{2}+\frac{i\sqrt{3}}{2}\right)^3+1}\cdot 1$$

$$= \text{Imaginary Part of } e^{x/2}.\,e^{\frac{ix\sqrt{3}}{2}}\frac{1}{(D+w)^3+1}\cdot 1$$

$$\left[\text{where } w = \frac{1}{2}+\frac{i\sqrt{3}}{2} \text{ and } w^2 = -\frac{1}{2}+\frac{i\sqrt{3}}{2} \text{ and } w^3 = -1\right]$$

$$= \text{Imaginary Part of } e^{x/2}.\,e^{\frac{ix\sqrt{3}}{2}}\frac{1}{D^3+3D^2w+3Dw^2+w^3+1}\cdot 1$$

$$= \text{Imaginary Part of } e^{x/2}.\,e^{\frac{ix\sqrt{3}}{2}}\frac{1}{D^3+3D^2w+3Dw^2}\cdot 1$$

$$= \text{Imaginary Part of } e^{x/2}.\,e^{\frac{ix\sqrt{3}}{2}}\frac{1}{3w^2D}\cdot 1$$

$$= \text{Imaginary Part of } e^{x/2}\frac{\left(\cos\frac{x\sqrt{3}}{2}+i\sin\frac{x\sqrt{3}}{2}\right)w}{3w^3}\cdot x$$

$$= \text{Imaginary Part of } -\frac{xe^{x/2}}{3}\left(\cos\frac{x\sqrt{3}}{2} + i\sin\frac{x\sqrt{3}}{2}\right)\left(\frac{1}{2} + \frac{i\sqrt{3}}{2}\right)$$

$$= -\frac{xe^{x/2}}{3}\left\{\frac{\sqrt{3}}{2}\cos\frac{x\sqrt{3}}{2} + \frac{1}{2}\sin\frac{x\sqrt{3}}{2}\right\} = -\frac{xe^{x/2}}{6}\left\{\sqrt{3}\cos\frac{x\sqrt{3}}{2} + \sin\frac{x\sqrt{3}}{2}\right\}.$$

Examples V(D)

Solve (Use the Symbolic Operator D):

1. (a) $\frac{d^3y}{dx^3} + 2\frac{d^2y}{dx^2} + \frac{dy}{dx} = e^{2x}$ (b) $\frac{d^2y}{dx^2} + 9y = 5x^2$ (c) $\frac{d^2y}{dx^2} + y = x\cos x$.
2. (a) $\left(D^4 - 4D^3 + 8D^2 - 8D + 4\right) y = e^x$ (b) $\left(D^2 - D - 2\right) y = 4x$.
3. (a) $\left(D^3 + D^2 - D - 1\right) y = \cosh x$ (b) $D^2\left(D^2 + D + 1\right) y = x^2$.
4. $\left(D^4 + D^3 - 3D^2 - 5D - 2\right) y = 3xe^{-x}$.
5. $\left(D^4 - 8D\right) y = x^2 + e^{3x}$.
6. (a) $\left(D^3 - 3D^2 + 3D - 1\right) y = xe^x + e^x$ (b) $\left(D^4 - 4D^3 + 6D^2 - 4D + 1\right) y = e^{2x}\cos x$.
7. (a) $\left(D^2 - 3D + 2\right) y = e^x$, if $y = 3$ and $Dy = 3$, when $x = 0$.

   (b) $\frac{d^2y}{dx^2} - y = 1$, given that $y = 0$ when $x = 0$ and $y \to a$ finite limit when $x \to -\infty$.
8. $\frac{d^4y}{dx^4} + y = \cosh 4x \sinh 3x$.
9. $\frac{d^2y}{dx^2} - 2\frac{dy}{dx} + 2y = 5\cos x$, given that $y = \frac{dy}{dx} = 0$, if $x = 0$.
10. $\frac{d^2y}{dx^2} + a^2y = \cos ax$; $\frac{d^2y}{dx^2} + a^2y = \sin ax$.
11. $\left(D^3 - 2D^2 + 9D - 18\right) y = 6\cos 3x + 48\sin 3x$.
12. $\frac{d^2y}{dx^2} + 2\frac{dy}{dx} + 10y + 37\sin 3x = 0$, given that $y = 3$, $Dy = 0$, when $x = 0$.
13. $(D-1)^2\left(D^2 + 1\right) y = e^x + \sin^2 \frac{1}{2}x$.
14. (a) $\frac{d^2y}{dx^2} + y = 3\cos^2 x + 2\sin^3 x$ (b) $(D-1)^2\left(D^2 + 1\right)^2 y = e^x + x$.
15. $\frac{d^4y}{dx^4} + \left(m^2 + n^2\right)\frac{d^2y}{dx^2} + m^2n^2y = \cos\frac{1}{2}(m+n)x\cos\frac{1}{2}(m-n)x \left(m^2 > n^2\right)$.
16. Find the value of u which satisfies the equation $\frac{d^2u}{d\theta^2} + u = 2k\cos\theta$ and also the following conditions:

    (a) u has the same value when $\theta = \pm\frac{1}{2}\pi$ (b) $\int_0^{\pi/2} u\, d\theta = 0$.
17. (a) $\left(D^3 + 4D^2 + 4D\right) y = 8e^{-2x}$ (b) $\frac{d^3y}{dx^3} - \frac{d^2y}{dx^2} + 3\frac{dy}{dx} + 5y = e^x\cos 3x$.
18. $\left(D^2 + 5D + 6\right) y = e^{-2x}\sin 2x$.
19. (a) $\left(D^4 - 8D\right) y = x^2 + e^{2x}$ (b) $\left(D^5 - D\right) y = e^x + \sin x - x$.
20. $\frac{d^3y}{dx^3} - \frac{dy}{dx} = 2x + 1 - 4\cos x + 2e^x$.
21. $\frac{d^4y}{dx^4} + \frac{d^2y}{dx^2} + 16y = 16x^2 + 256$.
22. $\frac{d^5y}{dx^5} - \frac{dy}{dx} = 12e^x + 8\sin x - 2x$.
23. $\left(D^2 + 2\right) y = x^2e^{3x} + e^x\cos 2x$.
24. (a) $\frac{d^2y}{dx^2} - y = x\sinh x$ (b) $\left(D^2 - 2D + 2\right) y = e^x\sin 2x$.
25. (a) $\left(D^2 - 2D + 4\right) y = e^x\cos^2 x$ (b) $\left(D^3 + D^2 + D + 1\right) y = \sin 2x$.
26. (a) $\frac{d^2y}{dx^2} - 7\frac{dy}{dx} - 18y = x^2e^{-2x}$ (b) $\frac{d^2y}{dx^2} - 5\frac{dy}{dx} + 6y = x^2e^{3x}$.

27. (a) $\left(D^2+1\right)y = xe^{2x}$ (b) $\left(D^4+2D^3+3D^2+2D+1\right)y = xe^x$.
28. (a) $\frac{d^2y}{dx^2}+4y = x\sin x$ (b) $\left(D^3-1\right)y = x\sin x$.
29. (a) $\left(D^4+2D^2+1\right)y = x^2\cos x$ (b) $\left(D^3+1\right)y = x^2\sin x$.
30. (a) $\frac{d^2y}{dx^2}-7\frac{dy}{dx}+6y=(x-2)e^x$ (b) $\frac{d^2y}{dx^2}-y=x^2\cos x$ (c) $\frac{d^2y}{dx^2}-y=x^2\sin x$.
31. (a) $\left(D^3+1\right)y = e^{2x}\sin x + e^{x/2}\sin\frac{x\sqrt{3}}{2}$ (b) $\frac{d^2y}{dx^2}-y = x\sin x+\left(1+x^2\right)e^x$.
32. $\left(D^4+D^3+D^2\right)y = x^2(a+bx)$.
33. $\left(D^2-6D+25\right)y = 2e^{3x}\cos 4x + 8e^{3x}(1-2x)\sin 4x$.
34. $(D-a)^a y = a^x$, where a is a positive integer.
35. (a) Using $z=\sin x$, find y, where $\frac{d^2y}{dx^2}+\frac{dy}{dx}\tan x + y\cos^2 x = 0$ and for $x=0$, $y=1$, $dy/dx=0$.

    (b) Show that the transformation $x=\sinh z$ transforms $\left(1+x^2\right)\frac{d^2y}{dx^2}+x\frac{dy}{dx}=4y$ to $\frac{d^2y}{dz^2}=4y$, and hence solve it.

    (c) For $m\frac{d^2s}{dt^2}+\frac{mg}{e}(s-l)=0$, ($l, g, e$ constants), find s and ds/dt when $s=s_0$, $ds/dt=0$, when $t=0$.

    (d) Solve $\frac{d^2y}{dt^2}+2n\cos\alpha\frac{dy}{dt}+n^2y=a\cos nt$, given that $y=0$ and $\frac{dy}{dt}=0$ when $t=0$.

    (e) If $\frac{d^2x}{dt^2}+2h\frac{dx}{dt}+\left(h^2+p^2\right)x = ke^{-ht}\cos pt$, then prove that $x=c_1e^{-ht}\cos(pt+c_2)+\frac{k}{2p}te^{-ht}\sin pt$. Show that the particular integral represents an oscillation of variable amplitude which is maximum when $ht=1$. What is the maximum value and show that it is very large when $h\to 0$. What is the value of the amplitude when $t\to\infty$?
36. Solve $\left(D^4-n^4\right)y=0$ completely. Now if $Dy=y=0$ when $x=0$ and $x=l$, then prove that $y=c_1(\cos nx-\cosh nx)+c_2(\sin nx-\sinh nx)$ and $\cos nl\cosh nl = 1$.

Answers

1. (a) $y=c_1+(c_2+c_3x)e^{-x}+\frac{1}{18}e^{2x}$ (b) $y=c_1\cos 3x+c_2\sin 3x+\frac{5}{9}x^2-\frac{10}{81}$

   (c) $y=c_1\cos x+c_1\sin x+\frac{1}{4}x(x\sin x+\cos x)$.
2. (a) $y=e^x[(c_1+c_2x)\cos x+(c_3+c_4x)\sin x]+e^x$ (b) $y=c_1e^{-x}+c_2e^{2x}+1-2x$.
3. (a) $y=c_1e^x+(c_2+c_3x)e^{-x}+\left\{\frac{1}{8}xe^x-\frac{1}{8}x^2e^{-x}\right\}$

   (b) $y=c_1+c_2x+e^{-\frac{1}{2}x}\left\{c_3\cos\left(\frac{\sqrt{3}}{2}\right)x+c_4\sin\left(\frac{\sqrt{3}}{2}\right)x\right\}+\frac{1}{12}x^4-\frac{1}{8}x^3$.
4. $y=c_1e^{2x}+e^{-x}\left(c_2+c_3x+c_4x^2\right)-e^{-x}\left(x^4/24+x^3/18\right)$.
5. $y=c_1+c_2e^{2x}+e^{-x}\left(c_3\cos\sqrt{3}x+c_4\sin\sqrt{3}x\right)-\frac{x^3}{24}+\frac{e^{3x}}{57}$.
6. (a) $y=(c_1+c_2x+c_3x^2)e^x+\frac{x^3}{24}(x+4)e^x$

   (b) $y=e^x\left(c_1+c_2x+c_3x^2+c_4x^3\right)-\frac{1}{4}e^{2x}\cos x$.
7. (a) $y=(2-x)e^x+e^{2x}$ (b) $y=e^x-1$.
8. $y=c_1e^{x/\sqrt{2}}\cos\left(x/\sqrt{2}+c_2\right)+c_3e^{-x/\sqrt{2}}\cos\left(\frac{x}{\sqrt{2}}+c_4\right)+\frac{1}{4804}\sinh 7x-\frac{1}{4}\sinh x$.
9. $y=e^x(-\cos x+3\sin x)+\cos x-2\sin x$.
10. $y=c_1\cos ax+c_2\sin ax+\frac{x}{2a}\sin ax$; same CF $+\left(\frac{-x\cos ax}{2a}\right)$.
11. $y=c_1e^{2x}+c_2\cos 3x+c_3\sin 3x+x\cos 3x-2x\sin 3x$.
12. $y=-3e^{-x}\cos 3x+6\cos 3x-\sin 3x$.
13. $y=(c_1+c_2x)e^x+c_3\cos x+c_4\sin x+\frac{1}{4}x^2e^x+\frac{1}{2}-\frac{1}{8}x\cos x$.
14. (a) $y=c_1\cos x+c_2\sin x+\frac{3}{2}-\frac{1}{2}\cos 2x-\frac{3}{4}x\cos x+\frac{1}{16}\sin 3x$

(b) $y = e^x(c_1 + c_2x) + (c_3 + c_4x)\cos x + (c_5 + c_6x)\sin x + \frac{1}{8}x^2e^x + x + 2.$

15. $y = c_1\cos mx + c_2\sin mx + c_3\cos nx + c_4\sin nx + \frac{x}{4(m^2-n^2)}\left(\frac{1}{n}\sin nx - \frac{1}{m}\sin mx\right).$

16. $u = k(\theta\sin\theta - \cos\theta).$

17. (a) $y = c_1 + e^{-2x}\{c_2 + c_3x\} - 2x^2e^{-2x}$ (b) $y = c_1e^{-x} + e^x(c_2\cos 2x + c_3\sin 2x) - \frac{1}{65}e^x(3\sin 3x + 2\cos 3x).$

18. $y = c_1e^{-2x} + c_2e^{-3x} - \frac{1}{10}e^{-2x}(\cos 2x + 2\sin 2x).$

19. (a) $y = c_1 + c_2e^{2x} + e^{-x}\left(c_3\cos\sqrt{3}x + c_4\sin\sqrt{3}x\right) - \frac{x^3}{24} + \frac{xe^{2x}}{24}$

(b) $y = c_1 + c_2e^x + c_3e^{-x} + c_4\cos x + c_5\sin x + e^x\cdot\frac{1}{4}x + \frac{1}{8}x^2\cos x - \frac{1}{2}x^2.$

20. $y = c_1 + c_2e^x + c_3e^{-x} - x^2 - x + 2\sin x + xe^x.$

21. $y = e^{\frac{\sqrt{7}x}{2}}\left(e_1\cos\frac{3}{2}x + c_2\sin\frac{3}{2}x\right) + e^{-\frac{\sqrt{7}x}{2}}\left(c_3\cos\frac{3}{2}x + c_4\sin\frac{3}{2}x\right) + x^2 + \frac{127}{8}.$

22. $y = c_1 + c_2e^x + c_3e^{-x} + c_4\cos x + c_5\sin x + 3xe^x + 2x\sin x + x^2.$

23. $y = c_1\cos\sqrt{2}x + c_2\sin\sqrt{2}x + \frac{e^{3x}}{11}\left[x^2 - \frac{12x}{11} + \frac{50}{121}\right] - \frac{1}{17}e^x(-4\sin 2x + \cos 2x).$

24. (a) $y = c_1e^x + c_2e^{-x} + \frac{x^2}{4}\cosh x - \frac{x}{4}\sinh x$ (b) $y = e^x(c_1\cos x + c_2\sin x) - \frac{1}{3}e^x\sin 2x.$

25. (a) $y = c_1e^x\cos\left(\sqrt{3}x + c_2\right) + \frac{1}{6}e^x - \frac{1}{2}e^x\cos 2x$ (b) $y = c_1e^{-x} + c_2\cos x + c_3\sin x + \frac{2\cos 2x - \sin 2x}{15}.$

26. (a) $y = c_1e^{2x} + c_2e^{-2x} - \frac{1}{11}e^{-2x}\left(\frac{1}{3}x^3 + \frac{1}{11}x^2 + \frac{2}{121}x\right)$ (b) $y = c_1e^{2x} + c_2e^{3x} + e^{3x}\left(\frac{1}{3}x^3 - x^2 + 2x\right).$

27. (a) $y = c_1\cos x + c_2\sin x + \frac{e^{2x}}{25}(5x-4)$ (b) $y = e^{-\frac{1}{2}x}\left\{(c_1 + c_2x)\cos\frac{\sqrt{3}}{2}x + (c_3 + c_4x)\sin\frac{\sqrt{3}}{2}x\right\} + \frac{e^x}{9}(x-2).$

28. (a) $y = c_1\cos(2x + c_2) + \frac{1}{3}x\sin x - \frac{2}{9}\cos x$

(b) $y = c_1e^x + e^{-x/2}\left\{c_2\cos\left(\sqrt{3}/2\right)x + c_3\sin\left(\sqrt{3}/2\right)x\right\} + \frac{1}{2}\{x\cos x - x\sin x - 3\cos x)\}.$

29. (a) $y = (c_1 + c_2x)\sin x + (c_3 + c_4x)\cos x + \frac{x^3\sin x}{12} + \frac{9x^2 - x^4}{48}\cos x.$

30. (a) $y = c_1e^x + c_2e^{6x} - \frac{1}{5}e^x\left(\frac{x^2}{2} - \frac{2}{5}x\right)$ (b) $y = c_1e^x + c_2e^{-x} - \frac{1}{2}x^2\cos x + x\sin x + \frac{1}{2}\cos x$ (c) Similar to (b).

31. (a) $y = c_1e^{-x} + e^{x/2}\left(c_2\cos\frac{x\sqrt{3}}{2} + c_3\sin\frac{x\sqrt{3}}{2}\right) + \frac{e^{2x}}{130}(3\sin x - 11\cos x) - \frac{1}{6}xe^{x/2}\left(\sin\frac{x\sqrt{3}}{2} + \sqrt{3}\cos\frac{x\sqrt{3}}{2}\right)$

(b) $y = c_1e^x + c_2e^{-x} - \frac{1}{2}(x\sin x + \cos x) + \frac{1}{12}xe^x\left(2x^2 - 3x + 9\right).$

32. $y = c_1 + c_2x + e^{-x/2}\left(c_3\sin\frac{x\sqrt{3}}{2} + c_4\cos\frac{x\sqrt{3}}{2}\right) + 3bx^2 - \frac{1}{3}ax^3.$

33. $y = \left\{c_1\sin 4x + \left(c_2 - x + x^2\right)\cos 4x\right\}e^{3x}.$

34. $y = \left(c_0 + c_1x + c_2x^2 + \cdots + c_{a-1}x^{a-1}\right)e^{ax} + \frac{a^x}{(\ln a - a)^a}.$

35. (a) $y = \cos(\sin x)$

(b) $y = c_1\exp(2\sinh^{-1}x) + c_2\exp(-2\sinh^{-1}x)$

(c) $S = (S_0 - l)\cos\sqrt{\frac{g}{e}}t + l,\ \frac{ds}{dt} - \sqrt{\frac{g}{e}}(l - s_0)\sin\sqrt{\frac{g}{e}}t.$

(d) $y = e^{(-n\cos\alpha)t}\left\{-\frac{a}{n^2\sin 2\alpha}\sin(n\sin\alpha)t\right\} + \frac{a\sin nt}{2n^2\cos\alpha}.$

5.7.3 Method of undetermined coefficients

We consider the problem of solving an equation of the form

$$\frac{d^2y}{dx^2} + P(x)\frac{dy}{dx} + Q(x)y = R(x) \tag{5.7.8}$$

for those cases in which the complete primitive of the reduced equation (making RHS $R(x) = 0$)

$$\frac{d^2y}{dx^2} + P(x)\frac{dy}{dx} + Q(x)y = 0 \tag{5.7.9}$$

is known. In other words, the complementary function (CF) of (5.7.8) is already known.

We know, if $y_p(x)$ is any particular solution of (5.7.8), then the complete primitive of (1) is

$$y = \text{CF} + y_p(x).$$

But how do we find y_p? In art. 5.7.2 we have used the symbolic operator D to find y_p in some special type of functions $R(x)$.

The method of undetermined coefficients is a procedure to find y_p when in (5.7.8) $P(x)$ and $Q(x)$ are just constants, say p and q and $R(x)$ is an exponential, a sine or cosine, a polynomial or some combinations of such functions. As an example, we consider the equation

$$\frac{d^2y}{dx^2} + p\frac{dy}{dx} + qy = e^{ax}. \tag{5.7.10}$$

When we differentiate e^{ax}, we get the same function with some numerical coefficient. It is, therefore, justified to guess that

$$y_p(x) = Ae^{ax} \tag{5.7.11}$$

might be a particular solution of (5.7.10). Here A is the *undetermined coefficient* to be so chosen that (5.7.11) satisfies (5.7.10). We then get $A\left(a^2 + pa + q\right) e^{ax} = e^{ax}$.

Hence, $A = \dfrac{1}{a^2 + pa + q}$, except when Denominator = 0.

In this exceptional case a is a root of the Auxiliary Equation (AE) $m^2 + pm + q = 0$ and then (5.7.11) reduces the left side of (5.7.10) to zero and cannot possibly satisfy (5.7.10) as it stands, with the right-hand side different from zero. We take in this case

$$y_p = Axe^{ax} \tag{5.7.12}$$

and substitute the trial solution. Substituting (5.7.12) in (5.7.10), we get

$$A\left(a^2 + pa + q\right) xe^{ax} + A(2a + p)e^{ax} = e^{ax}.$$

The expression $a^2 + pa + q = 0$ since we have assumed that a is a root of the AE; so

$$A = \frac{1}{2a + p}, \text{ except when } p = -2a,$$

i.e. except when a is a double root of the AE. In this last exceptional case we shall take

$$y_p = Ax^2 e^{ax}. \tag{5.7.13}$$

Whence, from (5.7.10) we get $A\left(a^2 + pa + q\right) x^2 e^{ax} + 2A(2a + p)xe^{ax} + 2Ae^{ax} = e^{ax}$.

Since a is a double root of $m^2 + pm + q = 0$, we have $a^2 + pa + q = 0$ as well as $2a + p = 0$ and then we obtain $A = 1/2$.

Observation

This method can be extended to cover the cases of higher order linear differential equations, as we shall use them in a number of subsequent examples.

To summarise

If a is not a root of the AE, then (5.7.10) has a particular solution Ae^{ax}.

If a is a simple root of the AE, then (5.7.10) has a particular solution Axe^{ax}.

If a is a double root of the AE, then (5.7.10) has a particular solution Ax^2e^{ax}.

In each case we can determine the coefficient A.

We next consider another example

$$\frac{d^2y}{dx^2} + p\frac{dy}{dx} + qy = \sin bx. \tag{5.7.14}$$

Since the derivatives of $\sin bx$ are constant multiples of $\sin bx$ and $\cos bx$, we take a trial solution

$$y_p = A\sin bx + B\cos bx. \tag{5.7.15}$$

The undetermined coefficients A and B can now be obtained by substituting (5.7.15) in (5.7.14) and equating the resulting coefficients of $\sin bx$ and $\cos bx$ on the left and right. This method breaks down if (5.7.15) satisfies

$$\frac{d^2y}{dx^2} + p\frac{dy}{dx} + qy = 0.$$

When this happens, we take $y_p = x(A\sin bx + B\cos bx)$ as our trial solution and proceed to find A and B.

We consider a third example when the right side of (5.7.14) is replaced by a polynomial, e.g.

$$\frac{d^2y}{dx^2} + p\frac{dy}{dx} + qy = a_0 + a_1x + a_2x^2 + \cdots + a_nx^n. \tag{5.7.16}$$

Since the derivative of a polynomial is again a polynomial, we take

$$y_p = A_0 + A_1x + A_2x^2 + \cdots + A_nx^n \tag{5.7.17}$$

and find the undetermined coefficients by substituting (5.7.17) in (5.7.16) and equating like powers of x.

If the constant q happens to be zero, then this procedure gives x^{n-1} as the highest power of x on the left of (5.7.16), so in this case we take our trial solution in the form:

$$y = x\left(A_0 + A_1x + A_2x^2 + \cdots + A_nx^n\right) = A_0x + A_1x^2 + A_2x^3 + \cdots + A_nx^{n+1}.$$

If p, q be both zero, then (5.7.16) can be solved by direct integration.

Working rule

When the equation is given in the form:

$$\frac{d^2y}{dx^2} + P(x)\frac{dy}{dx} + Q(x)y = R(x),$$

the *complete primitive* will be taken as y = CF + y_p, where y_p is *any particular solution* of the given equation.

To find y_p, we proceed as follows:

I. When the RHS $R(x)$ **is an exponential**: $R(x) = e^{ax}$.

(A) When a is *not a root* of the auxiliary equation, that is, e^{ax} is not in the CF, take $y_p = Ae^{ax}$.

(B) When a is a *simple root* of the auxiliary equation, that is, e^{ax} occurs in the CF, assume $y_p = Axe^{ax}$.

(C) When a is a *double root* of the auxiliary equation, that is, e^{ax} and xe^{ax} occur in the CF, take $y_p = Ax^2 e^{ax}$.

II. When the RHS **$R(x)$ is a polynomial or a constant,** i.e. $R(x) = a_0 + a_1 x + a_2 x^2 + \cdots + a_n x^n$.

(A) Take $y_p = A_0 + A_1 x + A_2 x^2 + \cdots + A_n x^n$.

In particular, if $R(x) = a_0$, a **constant** only, assume $y_p = A_0$.

(B) When $Q(x) = 0$, take $y_p = x\left(A_0 + A_1 x + A_2 x^2 + \cdots + A_n x^n\right)$.

Extension: If also $P(x) = 0$, multiply by x^2. But it is easy to apply direct integration.

III. When the RHS $R(x)$ **contains sines and cosines,** i.e. $R(x) = \sin ax$, or, $\cos ax$.

(A) When $\sin ax$ or $\cos ax$ is *not in the CF* take $y_p = A\sin ax + B\cos ax$.

(B) When $\sin ax$ or $\cos ax$ is *in the CF*, take $y_p = x(A\sin ax + B\cos ax)$.

IV. When the RHS $R(x)$ contains **products** of an exponential, a polynomial, sines and cosines, that is $R(x) = e^{ax}\sin bx$, or, $e^{ax}(a_0 + a_1 x + \cdots + a_n x^n)$, or, $\sin bx\,(a_0 + a_1 x + \cdots + a_n x^n)$ etc. modify y_p accordingly with the help of I, II and III.

Similar is the case for higher order linear differential equations. These can also be applied in the case of a first order first degree linear equations.

Let us illustrate these methods one by one.

A. Equations with exponential right side

Example 5.7.13 (Application of I(A)). *Solve by the method of undetermined coefficients the following differential equation:*

$$\frac{d^2y}{dx^2} + 2\frac{dy}{dx} + 2y = 8e^{2x} \text{ or, } \left(D^2 + 2D + 2\right) y = 8e^{2x} \text{ if } D \equiv \frac{d}{dx}.$$

Solution:

To find CF

Let $y = e^{mx} \neq 0$ be a trial solution, then the auxiliary equation is $m^2 + 2m + 2 = 0$, whereby $m = -1 \pm i$. $\therefore$ CF is $e^{-x}\{C_1\cos x + C_2\sin x\}$.

To find PI

The RHS contains e^{2x} of which 2 is *not a root* of the auxiliary equation, hence we take a particular integral y_p in the form $y_p = Ae^{2x}$ [See rule I(A)].

But y_p is a root of the given equation, hence $D^2 y_p + 2Dy_p + 2y_p = 8e^{ex}$ and also since $y_p = Ae^{2x}$, $Dy_p = 2Ae^{2x}$, $D^2 y_p = 4Ae^{2x}$, the equation changes to $4Ae^{2x} + 2.2Ae^{2x} + 2Ae^{2x} = 8e^{2x}$ or, $10Ae^{2x} = 8e^{2x}$, whereby $10A = 8$, or, $A = 4/5$.

Thus, $y_p = \frac{4}{5}e^{2x}$ and the complete primitive becomes $y = e^{-x}\{C_1\cos x + C_2\sin x\} + \frac{4}{5}e^{2x}$.

Example 5.7.14 (Application of I(B)). *Solve by the method of undetermined coefficients:* $\frac{d^2y}{dx^2} - \frac{dy}{dx} - 6y = 20e^{-2x}$.

Solution:

To find CF

As before auxiliary equation is $m^2 - m - 6 = 0$, hence $m = 3, -2$. $\therefore$ CF is $C_1e^{3x} + C_2e^{-2x}$.

To find PI

See that (-2) in the RHS e^{-2x} is a *simple root* of the auxiliary equation, hence we are to take $y_p = Axe^{-2x}$ [See rule I(B)].

Proceeding as before, since $D^2 y_p - Dy_p - 6y_p = 20e^{-2x}$ and $Dy_p = (A - 2Ax)e^{-2x}$, $D^2 y_p = (-4A + 4Ax)e^{-2x}$, we have $(-4A + 4Ax)e^{-2x} - (A - 2Ax)e^{-2x} - 6Axe^{-2x} = 20e^{-2x}$ or, $-5Ae^{-2x} = 20e^{-2x}$, whereby $-5A = 20$, or, $A = -4$.

Thus, $y_p = -4xe^{-2x}$.

$\therefore$ complete solution is $y = C_1e^{3x} + C_2e^{-2x} - 4xe^{-2x}$ (Fig. 5.3).

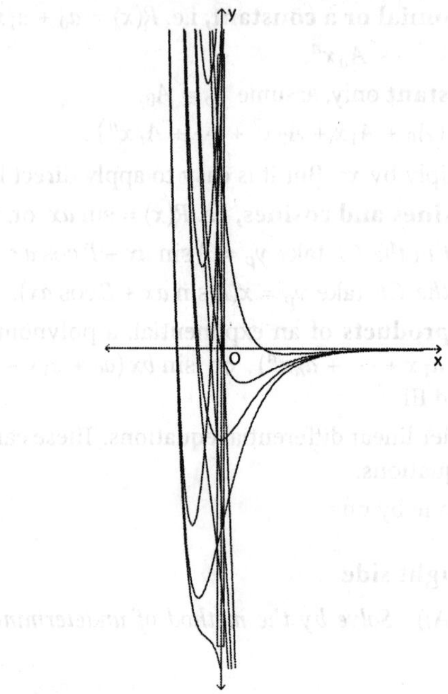

Fig. 5.3

Example 5.7.15 (Application of I(C)). *Solve by the method of undetermined coefficients:*

$$\frac{d^2y}{dx^2} + 6\frac{dy}{dx} + 9y = 24e^{-3x}.$$

Solution:

To find CF

Auxiliary equation: $m^2 + 6m + 9 = 0 = (m + 3)^2$, whereby $m = -3, -3$. $\therefore$ CF is $(C_1 + C_2x)\,e^{-3x}$.

To find PI

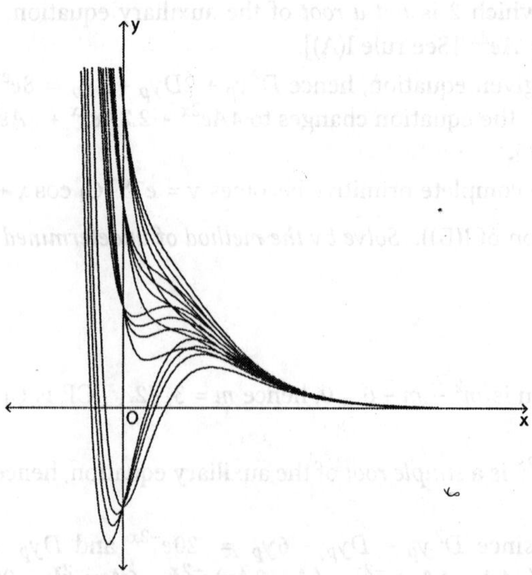

Fig. 5.4

Observe that (−3) in the RHS e^{-3x} is a *double root* of the auxiliary equation, hence we are to take $y_p = Ax^2e^{-3x}$ [See rule I(C)].

Now $D^2y_p + 6Dy_p + 9y_p = 24e^{-3x}$ and $Dy_p = Ae^{-3x}(2x - 3x^2)$, $D^2y_p = Ae^{-3x}(2 - 12x + 9x^2)$.

Putting $Ae^{-3x}(2 - 12x + 9x^2 + 12x - 18x^2 + 9x^2) = 24e^{-3x}$, or, $2Ae^{-3x} = 24e^{-3x}$, whereby $A = 12$.

$\therefore\ y_p = 12x^2e^{-3x}$.

Thus, CP is $y = (C_1 + C_2x)e^{-3x} + 12x^2e^{-3x}$ (Fig. 5.4).

B. Equations with constant or polynomial right side

Example 5.7.16 (Application of II(A)). *Solve by the method of undetermined coefficients:* $\frac{d^2y}{dx^2} - \frac{dy}{dx} - 2y = 8$.

Solution:

To find CF

The auxiliary equation is $m^2 - m - 2 = 0 = (m - 2)(m + 1)$.

Therefore, $m = 2, -1$, and CF is $C_1e^{2x} + C_2e^{-x}$.

To find PI

Since the right side is a constant, we assume a particular solution of the form: $y_p = A$. [See rule II(A)].

Now y_p cannot be obtained from CF by any choice of C_1 and C_2; hence there is no need to modify it. Also $D^2y_p - Dy_p - 2y_p = 8$.

Since $y_p = A$, $Dy_p = 0 = D^2y_p$, we have $-2A = 8$, or, $A = -4$, or, $y_p = -4$.

$\therefore$ complete solution is $y = C_1e^{2x} + C_2e^{-x} - 4$.

Example 5.7.17 (Application of II(A)). *Solve by the method of undetermined coefficients:*

$$\frac{d^2y}{dx^2} - 4\frac{dy}{dx} + y = x^2 - 2x + 2.$$

Solution:

To find CF

It is $(D^2 - 4D + 1)\,y = x^2 - 2x + 2$, if $D \equiv \frac{d}{dx}$.

Let $y = e^{mx} \neq 0$ be a trial solution, then auxiliary equation is $m^2 - 4m + 1 = 0$, whereby $m = 2 \pm \sqrt{3}$.

$\therefore$ CF is $C_1e^{(2+\sqrt{3})x} + C_2e^{(2-\sqrt{3})x}$.

To find PI

Since the RHS is a polynomial in x of degree 2, we assume a particular solution y_p of the form:

$$y_p = A + Bx + Cx^2. \qquad \text{[See rule II(A)]}$$

Now CF and y_p have no term in common, hence there is no need to modify y_p. Putting in the given equation $D^2y_p - 4Dy_p + y_p = x^2 - 2x + 2$, since $Dy_p = B + 2Cx$, $D^2y_p = 2C$, we have $2C - 4(B + 2Cx) + (A + Bx + Cx^2) = x^2 - 2x + 2$.

Equating coefficients of like powers of x: Coefficient of x^2 : $C = 1$

x : $B - 8C = -2$

constant : $A - 4B + 2C = 2$.

Solving $C = 1, B = 6, A = 24$, then $y_p = 24 + 6x + x^2$.

General solution is $y = C_1e^{(2+\sqrt{3})x} + C_2e^{(2-\sqrt{3})x} + x^2 + 6x + 24$.

Example 5.7.18 (Application of II(B): when $Q(x) = 0$). *Solve by the method of undetermined coefficients:*

$$\frac{d^2y}{dx^2} - 4\frac{dy}{dx} = x^2.$$

Solution:

To find CF

Auxiliary equation: $m^2 - 4m = 0 = m(m - 4)$, whereby $m = 0, 4$.

$\therefore$ CF is $C_1 e^{0x} + C_2 e^{4x} = C_1 + C_2 e^{4x}$.

To find PI

There is no term containing y, hence we are to take

$$y_p = x\left(A + Bx + Cx^2\right). \qquad \text{[See rule II(B)]}$$

Now $Dy_p = A + 2Bx + 3Cx^2$, $D^2 y_p = 2B + 6Cx$, and $D^2 y_p - 4Dy_p = x^2$ changes to

$$(2B + 6Cx) - 4\left(A + 2Bx + 3Cx^2\right) = x^2$$

$$\text{or,} \quad (2B - 4A) + 2(3C - 4B)x - 12Cx^2 = x^2.$$

Equating coefficients: $2B - 4A = 0$, $3C - 4B = 0$, $-12C = 1$.

Hence,

$$A = -\frac{1}{32},\ B = -\frac{1}{16},\ C = -\frac{1}{12} \text{ and } y_p = x\left(-\frac{1}{32} - \frac{1}{16}x - \frac{1}{12}x^2\right).$$

$\therefore$ general solution is

$$y = C_1 + C_2 e^{4x} - x\left(\frac{1}{32} + \frac{1}{16}x + \frac{1}{12}x^2\right).$$

C. Equations with right side containing sines and cosines

Example 5.7.19 (Application of III(A)). *Solve by the method of undetermined coefficients:*

$$\frac{d^2y}{dx^2} + 4y = 3\sin x.$$

Solution:

To find CF

Auxiliary equation: $m^2 + 4 = 0$ implies $m = \pm 2i$. $\therefore$ CF is, therefore, $C_1 \cos 2x + C_2 \sin 2x$.

To find PI

Here $\sin x$ is not in the complementary function, hence assume $y_p = A\sin x + B\cos x$, [See rule III(A)].

Then $Dy_p = A\cos x - B\sin x$, $D^2 y_p = -(A\sin x + B\cos x)$, and from $D^2 y_p + 4y_p = 3\sin x$, we have $3A\sin x + 3B\cos x = 3\sin x$, whereby $3A = 3$, $3B = 0$, or, $A = 1$, $B = 0$; $y_p = \sin x$.

$\therefore$ general solution is $y = C_1 \cos 2x + C_2 \sin 2x + \sin x$ (Fig. 5.5).

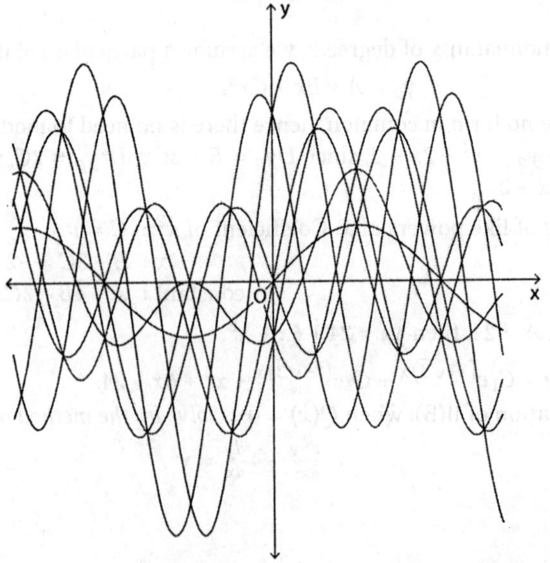

Fig. 5.5

Example 5.7.20 (Application of III(B)). *Solve by the method of undetermined coefficients the differential equation* $\frac{d^2y}{dx^2} + 4y = \sin 2x$.

Solution:

To find CF

As before, CF is $C_1 \cos 2x + C_2 \sin 2x$.

To find PI

Since the RHS, $\sin 2x$ appears in the CF, we take

$$y_p = x(A \sin 2x + B \cos 2x). \quad \text{[See rule III(B)]}$$

Then $Dy_p = A \sin 2x + B \cos 2x + 2x(A \cos 2x - B \sin 2x)$

$D^2 y_p = 4A \cos 2x - 4B \sin 2x - 4x(A \sin 2x + B \cos 2x)$.

And $D^2 y_p + 4y_p = \sin 2x$ changes to $4A \cos 2x - 4B \sin 2x = \sin 2x$.

Equating coefficients: $4A = 0$, $-4B = 1$, or, $A = 0$, $B = -\frac{1}{4}$.

Then $y_p = -\frac{1}{4} x \cos 2x$.

$\therefore$ general solution is $y = C_1 \cos 2x + C_2 \sin 2x - \frac{1}{4} x \cos 2x$ (Fig. 5.6).

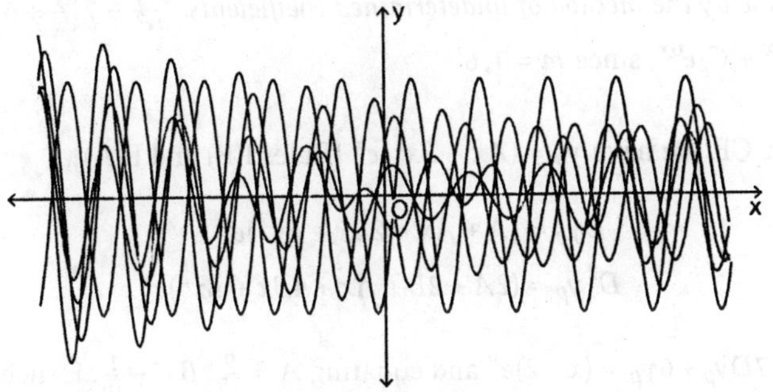

Fig. 5.6

D. Equations with right side containing products of an exponential, a polynomial, sines and cosines or any other. Higher orders are also included.

Here we are to modify y_p accordingly with the help of I, II, and III. The following examples will make things clear:

Example 5.7.21. *Solve by the method of undetermined coefficients:* $\frac{d^2y}{dx^2} - \frac{dy}{dx} - 2y = x^2 e^x$.

Solution:

To find CF

Auxiliary equation: $m^2 - m - 2 = 0 = (m+1)(m-2)$; $m = -1, 2$. $\therefore$ CF is $C_1 e^{-x} + C_2 e^{2x}$.

To find PI

Since the RHS, $x^2 e^x$ does not appear in the CF, we take the particular integral y_p in the usual form: $y_p = (A + Bx + Cx^2)\, e^x$.

$$Dy_p = (A + Bx + Cx^2)\, e^x + (B + 2Cx)e^x,$$

$$D^2 y_p = (A + Bx + Cx^2)\, e^x + 2(B + 2Cx)e^x + 2Ce^x,$$

and putting in the given equation $e^x \left\{(2C + B - 2A) + 2(C - B)x - 2Cx^2\right\} = x^2 \cdot e^x$.

Equating $A = -\frac{3}{4}$, $B = -\frac{1}{2} = C$. $\therefore$ $y_p = \left(-\frac{3}{4} - \frac{1}{2}x - \frac{1}{2}x^2\right) e^x$.

General solution is $y = C_1 e^{-x} + C_2 e^{2x} + e^x \left(-\frac{3}{4} - \frac{1}{2}x - \frac{1}{2}x^2\right)$.

Example 5.7.22. *Solve the following differential equation by the method of undetermined coefficients:*

$$\frac{d^2y}{dx^2} - 2\frac{dy}{dx} + y = xe^x.$$

Solution:

To find CF

Auxiliary equation: $m^2 - 2m + 1 = 0 = (m-1)^2$, $m = 1, 1$. ∴ CF is $e^x (C_1 + C_2x)$.

To find PI

See that e^x and xe^x in the RHS are in CF. Hence, we modify it as $y_p = (A + Bx)x^2e^x$ [See rules I(C) and II(A)].

$$Dy_p = 2Axe^x + (A + 3B)x^2e^x + Bx^3e^x$$
$$D^2y_p = 2Ae^x + (4A + 6B)xe^x + (A + 6B)x^2e^x + Bx^3e^x.$$

Putting in $D^2y_p - 2Dy_p + y_p = xe^x$, we have $2Ae^x + 6Bxe^x = xe^x$, whereby $A = 0, B = \frac{1}{6}$. Thus, $y_p = \frac{1}{6}x^3e^x$.

∴ general solution is $y = e^x(C_1 + C_2x) + \frac{1}{6}x^3e^x$.

Example 5.7.23. *Solve by the method of undetermined coefficients:* $\frac{d^2y}{dx^2} - 7\frac{dy}{dx} + 6y = (x-2)e^x$.

Solution: CF is $C_1e^x + C_2e^{6x}$, since $m = 1, 6$.

To find PI

Since e^x occurs in CF, we have $y_p = (Ax + Bx^2)e^x$ [Rules I(B) and II(A)].

$$Dy_p = (A + Ax + 2Bx + Bx^2)e^x$$
$$D^2y_p = (2A + 2B + Ax + 4Bx + Bx^2)e^x.$$

Putting in $D^2y_p - 7Dy_p + 6y_p = (x-2)e^x$ and equating $A = \frac{9}{25}, B = -\frac{1}{10}$. Hence, $y_p = \left(\frac{9}{25}x - \frac{1}{10}x^2\right)e^x$.

General solution is $y = c_1e^x + c_2e^{6x} + \left(\frac{9}{25}x - \frac{1}{10}x^2\right)e^x$.

Example 5.7.24. *Solve by the method of undetermined coefficients:* $\frac{d^3y}{dx^3} + 3\frac{d^2y}{dx^2} + 2\frac{dy}{dx} = x^2 + 4x + 8$.

Solution:

To find CF

Auxiliary equation: $m^3 + 3m^2 + 2m = 0 = m(m+1)(m+2)$; $m = 0, -1, -2$. CF is $C_1 + C_2e^{-x} + C_3e^{-2x}$.

To find PI

The equation contains no term containing y. Hence, $y_p = x\left(A + Bx + Cx^2\right)$ [Rule II(B)]

$$Dy_p = A + 2Bx + 3Cx^2$$
$$D^2y_p = 2B + 6Cx$$
$$D^3y_p = 6C.$$

Putting in $D^3y_p + 3D^2y_p + 2Dy_p = x^2 + 4x + 8$,

$$6C + 3(2B + 6Cx) + 2\left(A + 2Bx + 3Cx^2\right) = x^2 + 4x + 8.$$

Equating coefficients: $6C = 1$, $4B + 18C = 4$, $2A + 6B + 6C = 8$.

Solving, $C = \frac{1}{6}, B = \frac{1}{4}, A = \frac{11}{4}$. ∴ $y_p = x\left(\frac{11}{4} + \frac{1}{4}x + \frac{1}{6}x^2\right)$.

General solution is $y = C_1 + C_2e^{-x} + C_3e^{-2x} + x\left(\frac{11}{4} + \frac{1}{4}x + \frac{1}{6}x^2\right)$.

Example 5.7.25. *Solve by the method of undetermined coefficients:* $\frac{d^2y}{dx^2} = x^2$.

Solution:

To find CF

Auxiliary equation: $m^2 = 0$ implies $m = 0, 0$. $\therefore$ CF $e^{0x}(C_1 + C_2x) = C_1 + C_2x$.

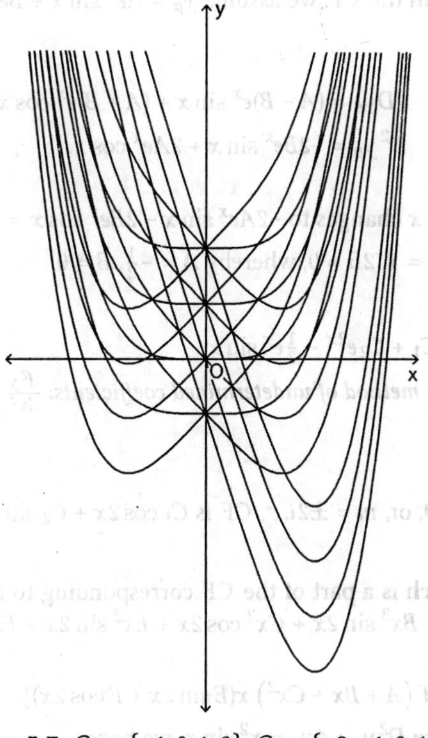

Fig. 5.7: $C_1 = \{-1, 0, 1, 2\}$; $C_2 = \{-3, -1, 0, 1\}$.

To find PI

Here both $P(x) = 0$, $Q(x) = 0$. Hence, we take $y_p = x^2(A + Bx + Cx^2)$ [See Rule II(B) Extension].

$$Dy_p = 2Ax + 3Bx^2 + 4Cx^3$$
$$D^2y_p = 2A + 6Bx + 12Cx^2.$$

$\therefore$ $2A + 6Bx + 12Cx^2 = x^2$, whereby $A = 0 = B, C = \frac{1}{12}$.

$\therefore$ $y_p = \frac{1}{12}x^4$.

$\therefore$ general solution is $y = C_1 + C_2x + \frac{1}{12}x^4$ (Fig. 5.7).

Observation. This could have been easily solved by direct integration:

$$\frac{d^2y}{dx^2} = x^2 \text{ implies } \frac{dy}{dx} = \frac{x^3}{3} + C_2$$
$$\text{implies } y = \frac{x^4}{12} + C_2x + C_1.$$

Example 5.7.26. *Solve by the method of undetermined coefficients:* $\frac{d^2y}{dx^2} - 2\frac{dy}{dx} = e^x \sin x$.

Solution:

To find CF

Auxiliary equation: $m^2 - 2m = 0 = m(m-2)$ implies $m = 0, 2$. ∴ CF is $C_1 + C_2e^{2x}$.

To find PI

Since the RHS $e^x \sin x$ is not in the CF, we assume $y_p = Ae^x \sin x + Be^x \cos x$. [Rules I(A) and III(A)].

Then

$$Dy_p = (A-B)e^x \sin x + (A+B)e^x \cos x$$
$$D^2 y_p = -2Be^x \sin x + 2Ae^x \cos x.$$

Now $D^2y_p - 2Dy_p = e^x \sin x$ changes to $-2Ae^x \sin x - 2Be^x \cos x = e^x \sin x$.

Equating coefficients: $-2A = 1$, $2B = 0$, whereby $A = -\frac{1}{2}$, $B = 0$.

∴ $y_p = -\frac{1}{2}e^x \sin x$.

∴ general solution is $y = C_1 + C_2e^{2x} - \frac{1}{2}e^x \sin x$.

Example 5.7.27. *Solve by the method of undetermined coefficients:* $\frac{d^2y}{dx^2} + 4y = x^2 \sin 2x$.

Solution:

To find CF

Auxiliary equation: $m^2 + 4 = 0$, or, $m = \pm 2i$. ∴ CF is $C_1 \cos 2x + C_2 \sin 2x$.

To find PI

The RHS contains $\sin 2x$ which is a part of the CF corresponding to a root of multiplicity one; hence we assume, $y_p = Ax^3 \cos 2x + Bx^3 \sin 2x + Cx^2 \cos 2x + Ex^2 \sin 2x + Fx \cos 2x + Gx \sin 2x$. [Rules II(A) and III(B)].

[It is the simplified form of $(A + Bx + Cx^2)\, x(E \sin 2x + F \cos 2x)$].

Obtain Dy_p, D^2y_p and since $D^2y_p + 4y_p = x^2 \sin x$, we have

$$12Bx^2 \cos 2x - 12Ax^2 \sin 2x + (6A + 8E)x \cos 2x + (6B - 8C)x \sin 2x$$
$$+ (2C + 4G) \cos 2x + (2E - 4F) \sin 2x = x^2 \sin 2x.$$

Equating coefficients: $A = -\frac{1}{12}$, $B = 0$, $C = 0$, $E = \frac{1}{16}$, $F = \frac{1}{32}$, $G = 0$ and hence, $y_p = -\frac{1}{12}x^3 \cos 2x + \frac{1}{16}x^2 \sin 2x + \frac{1}{32}x \cos 2x$.

∴ general solution is $y = C_1 \cos 2x + C_2 \sin 2x - \frac{1}{12}x^3 \cos 2x + \frac{1}{16}x^2 \sin 2x + \frac{1}{32}x \cos 2x$.

Example 5.7.28. *Solve by the method of undetermined coefficients:* $\frac{d^2y}{dx^2} - 2\frac{dy}{dx} + 3y = x^3 + \sin x$.

Solution:

To find CF

Auxiliary equation: $m^2 - 2m + 3 = 0$ gives $m = 1 \pm i\sqrt{2}$. ∴ CF is $e^x\left(C_1 \cos \sqrt{2}x + C_2 \sin \sqrt{2}x\right)$.

To find PI

x^3 and $\sin x$ are not in CF, hence we take $y_p = Ax^3 + Bx^2 + Cx + E + F \sin x + G \cos x$ [Rules II(A) and III(A)].

$$Dy_p = 3Ax^2 + 2Bx + C + F \cos x - G \sin x$$
$$D^2y_p = 6Ax + 2B - F \sin x - G \cos x$$

and $D^2y_p - 2Dy_p + 3y_p = x^3 + \sin x$ becomes

$$3Ax^3 + 3(B-2A)x^2 + (3C - 4B + 6A)x + (3E - 2C + 2B) + 2(F+G)\sin x + 2(G-F)\cos x = x^3 + \sin x.$$

Equating coefficients $A = \frac{1}{3},\ B = \frac{2}{3},\ C = \frac{2}{9},\ E = -\frac{8}{27},\ F = G = \frac{1}{4}$.

$$y_p = \tfrac{1}{3}x^3 + \tfrac{2}{3}x^2 + \tfrac{2}{9}x - \tfrac{8}{27} + \tfrac{1}{4}(\sin x + \cos x).$$

Complete solution is $y = e^x\left(C_1 \cos\sqrt{2}x + C_2 \sin\sqrt{2}x\right) + y_p$ (given).

Example 5.7.29. *Solve by the method of undetermined coefficients:* $\frac{d^3y}{dx^3} + 2\frac{d^2y}{dx^2} - \frac{dy}{dx} - 2y = e^x + x^2$.

Solution: CF is $C_1e^x + C_2e^{-x} + C_3e^{-2x}$, since $m = 1, -1, -2$.

To find PI

Since e^x occurs in CF, we have $y_p = Ax^2 + Bx + C + Exe^x$ [Rules I(B) and II(A)].

$$\therefore\ Dy_p = 2Ax + B + Ee^x + Exe^x$$

$$D^2y_p = 2A + 2Ee^x + Exe^x$$

$$D^3y_p = 3Ee^x + Exe^x.$$

Since $D^3y_p + 2D^2y_p - Dy_p - 2y_p = e^x + x^2$, we have $(4A - B - 2C) - 2(A+B)x - 2Ax^2 + 6Ee^x = e^x + x^2$.
Equating coefficients $A = -\frac{1}{2},\ B = \frac{1}{2},\ C = -\frac{5}{4},\ E = \frac{1}{6}$.

$$y_p = -\tfrac{1}{2}x^2 + \tfrac{1}{2}x - \tfrac{5}{4} + \tfrac{1}{6}xe^x.$$

$\therefore$ complete solution is $y = C_1e^x + C_2e^{-x} + C_3e^{-2x} + y_p$ (given).

Example 5.7.30. *Solve by the method of undetermined coefficients:*

$$(D^2 + D - 6)y = 10e^{2x} - 18e^{3x} - 6x - 11.$$

Solution: CF is $C_1e^{2x} + C_2e^{-3x}$, since $m = 2, -3$.

To find PI

Since e^{2x} occurs in CF, take $y_p = Axe^{2x} + Be^{3x} + Cx + D$ [Rules I(B) and II(A)].

Proceeding as before, $A = 2, B = -3, C = 1, D = 2$.

General solution is $y = C_1e^{2x} + C_2e^{-3x} + 2xe^{2x} - 3e^{3x} + x + 2$.

Example 5.7.31. *Solve by the method of undetermined coefficients:* $\frac{d^2y}{dx^2} - 4\frac{dy}{dx} + 4y = x^3e^{2x} + xe^{2x}$.

Solution: CF is $e^{2x}(c_1 + c_2x)$.

To find PI

See that the RHS contains e^{2x} which is a part of CF and 2 is a root of multiplicity 2 of the auxiliary equation. Assume, therefore, $y_p = x^2e^{2x}\left(Ax^3 + Bx^2 + Cx + E\right)$. [Rules I(C) and II(A)].

Proceeding as before, we see that $A = \frac{1}{20},\ B = 0,\ C = \frac{1}{6},\ E = 0$.

$\therefore\ y_p = \frac{1}{20}x^5e^{2x} + \frac{1}{6}x^3e^{2x}$.

General solution is $y = e^{2x}(c_1 + c_2x) + \frac{1}{20}x^5e^{2x} + \frac{1}{6}x^3e^{2x}$.

Example 5.7.32. *Solve by the method of undetermined coefficients* $(D^2 - 3D)y = x + e^x \sin x,\ D \equiv \frac{d}{dx}$.

Solution: CF is $c_1 + c_2e^{3x}$, since $m = 0, 3$.

To find PI

Take $y_p = x(A + Bx) + Ce^x \sin x + De^x \cos x$ [Rules II(B), I(A) and III(A)].

$$Dy_p = A + 2Bx + (C - D)e^x \sin x + (C + D)e^x \cos x$$
$$D^2 y_p = 2B - 2De^x \sin x + 2Ce^x \cos x.$$

Putting in $D^2 y_p - 3Dy_p = x + e^x \sin x$, $(2B-3A) - 6Bx + (D-3C)e^x \sin x - (C+3D)e^x \sin x = x + e^x \sin x$.

Equating coefficients $2B - 3A = 0, -6B = 1, D - 3C = 1, C + 3D = 0$ and hence $A = -\frac{1}{9}, B = -\frac{1}{6}, C = -\frac{3}{10}, D = \frac{1}{10}$.

$\therefore \quad y_p = -\frac{1}{9}x - \frac{1}{6}x^2 - \frac{3}{10}e^x \sin x + \frac{1}{10}e^x \cos x.$

The complete solution is $y = c_1 + c_2 e^{3x} - \frac{1}{3}\left(\frac{1}{2}x^2 + \frac{1}{3}x\right) + \frac{e^x}{10}(\cos x - 3 \sin x)$.

Examples V(E)

Solve, using the method of undetermined coefficients:

1. $\left(D^2 + 3D - 10\right) y = 6e^{4x}$.
2. $\frac{d^2y}{dx^2} + 10\frac{dy}{dx} + 25y = 14e^{-5x}$.
3. $\left(D^2 - 2D + 5\right) y = 25x^2 + 12$.
4. (a) $\frac{d^2y}{dx^2} - 3\frac{dy}{dx} + 2y = \sin 2x$
   (b) $\left(D^2 - 3D + 2\right) y = 14 \sin 2x - 18 \cos 2x$.
5. $\left(D^2 + 1\right) y = 2 \cos x$.
6. $\frac{d^2y}{dx^2} + k^2 y = \sin bx$ (k, b are positive constants).
7. $\left(D^2 + 2\right) y = e^x + 2$.
8. $\left(D^2 - 1\right) y = e^x \sin 2x$.
9. $\left(D^2 + 2D + 2\right) y = x^2 + \sin x$.
10. $\left(D^2 - 9\right) y = x + e^{2x} - \sin 2x$.
11. $\left(D^2 + 1\right) y = -2 \sin x + 4x \cos x$.
12. $\frac{d^2r}{dt^2} - \omega^2 r = -g \sin \omega t$ (g, ω positive constants)

Answers

1. $y = c_1 e^{2x} + c_2 e^{-5x} + \frac{1}{3}e^{4x}$.
2. $y = c_1 e^{-5x} + c_2 xe^{-5x} + 7x^2 e^{-5x}$.
3. $y = e^x (c_1 \cos 2x + c_2 \sin 2x) + 2 + 4x + 5x^2$.
4. (a) $y = c_1 e^x + c_2 e^{2x} - \frac{1}{20} \sin 2x + \frac{3}{20} \cos 2x$
   (b) $y = c_1 e^x + c_2 e^{2x} + 2 \sin 2x + 3 \cos 2x$.
5. $y = c_1 \sin x + c_2 \cos x + x \sin x$.
6. $y = c_1 \cos kx + c_2 \sin kx + \frac{\sin bx}{k^2 - b^2}$ unless $b = k$, in which case $y = c_1 \sin kx + c_2 \cos kx - \frac{x \cos kx}{2k}$.
7. $y = c_1 \cos \sqrt{2}x + c_2 \sin \sqrt{2}x + \frac{1}{3}e^x + 1$.
8. $y = c_1 e^x + c_2 e^{-x} - \frac{1}{8}e^x(\sin 2x + \cos 2x)$.
9. $y = e^{-x}(c_1 \cos x + c_2 \sin x) + \frac{1}{2}(x - 1)^2 + \frac{1}{5}(\sin x - 2 \cos x)$.
10. $y = c_1 e^{3x} + c_2 e^{-3x} - \frac{1}{9}x - \frac{1}{5}e^{2x} + \frac{1}{13} \sin 2x$.
11. $y = c_1 \cos x + c_2 \sin x + 2x \cos x + x^2 \sin x$.
12. $r = Ae^{\omega t} + Be^{-\omega t} + \frac{g}{2\omega^2} \sin \omega t$.

5.7.4 Method of variation of parameters

We have already introduced the method of variation of parameters in solving linear equations of first order and first degree in art. 2.14. Some examples have also been worked out there and some more have been set as exercises. Let us extend this case to a second order linear differential equation which in turn will be applicable to linear equations of higher order.

Second order linear differential equation

A more powerful method of finding a particular solution of the equation

$$\frac{d^2y}{dx^2} + P(x)\frac{dy}{dx} + Q(x)y = R(x), \tag{5.7.18}$$

when the general solution of the reduced equation

$$\frac{d^2y}{dx^2} + P(x)\frac{dy}{dx} + Q(x)y = 0. \tag{5.7.19}$$

[i.e. *complementary function of* (5.7.18)] is known, is provided by the Method of Variation of Parameters. Here $P(x)$, $Q(x)$ need not be constants—what is essential here is to know the complementary function of (5.7.18) in advance. Then a particular solution y_p of (5.7.18) can be obtained by the method of variation of parameters.

Assume then that the general solution of (5.7.19) is $y = c_1y_1(x) + c_2y_2(x)$, where c_1 and c_2 are constants, y_1 and y_2 are two linearly independent solution of (5.7.19).

We shall replace the constants c_1 and c_2 by unknown functions $v_1(x)$ and $v_2(x)$. Our problem is to determine v_1 and v_2 so that

$$y_p = v_1y_1 + v_2y_2 \tag{5.7.20}$$

satisfies (5.7.18). This will be a particular solution of (5.7.18). To determine v_1 and v_2 we require two equations:

We differentiate (5.7.20) and obtain

$$Dy_p = \left(v_1y_1' + v_2y_2'\right) + \left(v_1'y_1 + v_2'y_2\right). \tag{5.7.21}$$

We shall choose v_1 and v_2 in such a manner that

$$v_1'y_1 + v_2'y_2 = 0. \tag{5.7.22}$$

This will reduce (5.7.21) to

$$Dy_p = v_1y_1' + v_2y_2'. \tag{5.7.23}$$

$$\therefore\quad D^2y_p = \left(v_1y_1'' + v_2y_2''\right) + \left(v_1'y_1' + v_2'y_2'\right). \tag{5.7.24}$$

On substituting (5.7.20), (5.7.23), (5.7.24) into (5.7.18) and re- arranging, we get

$$v_1\left(y_1'' + Py_1' + Qy_1\right) + v_2\left(y_2'' + Py_2' + Qy_2\right) + v_1'y_1' + v_2'y_2' = R(x). \tag{5.7.25}$$

Since y_1 and y_2 are solutions of (5.7.19),

$$y_1'' + Py_1' + Qy_1 = 0$$
$$y_2'' + Py_2' + Qy_2 = 0$$

and (5.7.25) becomes

$$v_1'y_1' + v_2'y_2' = R(x). \tag{5.7.26}$$

We have thus two equations (5.7.22) and (5.7.26) from which we can solve for v_1' and v_2'. In fact,

$$v_1' = \frac{-y_2R(x)}{\begin{vmatrix} y_1 & y_2 \\ y_1' & y_2' \end{vmatrix}} \text{ and } v_2' = \frac{y_1R(x)}{\begin{vmatrix} y_1 & y_2 \\ y_1' & y_2' \end{vmatrix}}$$

[Denominators are the Wronskians $W(y_1, y_2)$ and $W(y_1, y_2) \neq 0$ because we have taken y_1 and y_2 linearly independent functions.]

To obtain v_1 and v_2 we are then to integrate and

$$v_1 = \int \frac{-y_2 R(x)}{W(y_1, y_2)} dx, \quad v_2 = \int \frac{y_1 R(x)}{W(y_1, y_2)} dx.$$

We then obtain a particular solution of (5.7.18), namely $y_p = v_1(x)y_1(x) + v_2(x)y_2(x)$.

Extension: *n*th order

To discuss the solution of an nth order linear differential equation of the form:

$$\frac{d^n y}{dx^n} + P_1(x)\frac{d^{n-1}y}{dx^{n-1}} + \cdots + P_{n-1}(x)\frac{dy}{dx} + P_n(x)y = R(x) \tag{5.7.27}$$

by the method of variation of parameters.

Let $y_1(x), y_2(x), \ldots, y_n(x)$ be a set of n *linearly independent* solutions to the associated homogeneous differential equation—that is the equation (5.7.27) itself with RHS $R(x) = 0$. The CF is $c_1y_1 + c_2y_2 + \cdots + c_ny_n$, where $c_1, c_2, \ldots, c_n$ are arbitrary constants. Here $P_1(x), P_2(x), \ldots, P_n(x)$ need not be constants only.

A particular solution y_p to (5.7.27) is $y_p = v_1y_1 + v_2y_2 + \cdots + v_ny_n$, where $v_1, v_2, \ldots, v_n$ are unknown function to be determined. As before these can be found by solving

$$\begin{aligned}
v_1'y_1 + v_2'y_2 + \cdots + v_n'y_n &= 0\\
v_1'y_1' + v_2'y_2' + \cdots + v_n'y_n' &= 0\\
&\vdots\\
v_1'y_1^{n-2} + v_2'y_2^{n-2} + \cdots + v_n'y_n^{n-2} &= 0\\
v_1'y_1^{n-1} + v_2'y_2^{n-1} + \cdots + v_n'y_n^{n-1} &= R(x)
\end{aligned}$$

since from these equations, we obtain $v_1', v_2', \ldots, v_n'$ and by integration we have the values $v_1, v_2, \ldots, v_n$.

Thus, the general solution is given by $y = c_1y_1 + c_2y_2 + \cdots + c_ny_n + y_p$.

Working rule

To solve

$$\frac{d^2y}{dx^2} + P(x)\frac{dy}{dx} + Q(x)y = R(x), \tag{5.7.28}$$

where $P(x)$ and $Q(x)$ are functions of x or constants.

To find CF

(A) When two linearly independent solutions $y_1(x)$ and $y_2(x)$ of

$$\frac{d^2y}{dx^2} + P(x)\frac{dy}{dx} + Q(x)y = 0$$

be given, the complementary function will be $c_1y_1 + c_2y_2$, where c_1 and c_2 are arbitrary constants.

(B) If the linearly independent solutions of the reduced homogeneous differential equation be not given, find by our previous method the CF in the form $c_1y_1 + c_2y_2$.

Note 5.7.2. *If y_1 and y_2 be not given as linearly independent, you are to show that they are so by proving that $W(y_1, y_2) \neq 0$.*

To find PI

Let the particular solution y_p of (5.7.28) be $y_p = v_1 y_1 + v_2 y_2$, where $v_1(x)$ and $v_2(x)$ are unknown functions.

Now v_1', v_2' are found by solving

$$\left.\begin{aligned} v_1' y_1 + v_2' y_2 &= 0 \\ \text{and } v_1' y_1' + v_2' y_2' &= R(x) \end{aligned}\right\}. \tag{5.7.29}$$

Integrating v_1' and v_2', we obtain v_1 and v_2.

General solution is $y = \text{CF} + y_p$.

Similar for any higher order.

Example 5.7.33. *Solve by the method of variation of parameters:* $\frac{d^2y}{dx^2} + y = \operatorname{cosec} x$.

Solution:

To find CF

Let $y = e^{mx} (\neq 0)$ be a trial solution. Then the auxiliary equation is $m^2 + 1 = 0$, whereby $m = \pm i$.

$\therefore$ CF is $c_1 \sin x + c_2 \cos x$.

To find PI

First show that $\sin x$ and $\cos x$ are *independent* solutions. Let $y_1 = \sin x$ and $y_2 = \cos x$.

The Wronskian of y_1 and y_2 is $\begin{vmatrix} \sin x & \cos x \\ \cos x & -\sin x \end{vmatrix} = -1 \neq 0 \; \forall \, x \in \mathbb{R}$.

$\therefore$ y_1 and y_2 are linearly independent.

Let the particular solution be

$$y_p = v_1(x) \sin x + v_2(x) \cos x \tag{5.7.30}$$

Then $Dy_p = \{v_1(x) \cos x - v_2(x) \sin x\} + \{v_1'(x) \sin x + v_2'(x) \cos x\}$.

Choose $v_1(x)$ and $v_2(x)$ such that

$$v_1'(x) \sin x + v_2'(x) \cos x = 0 \tag{5.7.31}$$

so that

$$\begin{aligned} Dy_p &= v_1(x) \cos x - v_2(x) \sin x \\ \text{and } \; D^2 y_p &= \{-v_1(x) \sin x - v_2(x) \cos x\} + v_1'(x) \cos x - v_2'(x) \sin x. \end{aligned}$$

Now

$$D^2 y_p + y_p = v_1'(x) \cos x - v_2'(x) \sin x = \operatorname{cosec} x. \tag{5.7.32}$$

From (5.7.31) and (5.7.32) we solve for $v_1'(x)$ and $v_2'(x)$: $v_1'(x) = \cot x$ and $v_2'(x) = -1$.

Hence,

$$v_1(x) = \int \frac{\cos x}{\sin x} dx = \ln \sin x \text{ and } v_2(x) = -x.$$

Accordingly, $y_p = \sin x \ln(\sin x) - x \cos x$.

The complete primitive is $y = c_1 \sin x + c_2 \cos x + \sin x \ln(\sin x) - x \cos x$.

Example 5.7.34. *Solve by the method of variation of parameters:* $\frac{d^2y}{dx^2} + a^2 y = \sec ax$, $a \neq 0$.

Solution:

To find CF

Auxiliary equation: $m^2 + a^2 = 0$, which gives $m = \pm ai$.

The CF is $c_1 \cos ax + c_2 \sin ax$.

To find PI

First show that $\cos ax$ and $\sin ax$ are *independent* solutions.

Let $y_1 = \cos ax$, $y_2 = \sin ax$.

The Wronskian of y_1 and y_2 is

$$W(y_1, y_2) = \begin{vmatrix} y_1 & y_2 \\ y_1' & y_2' \end{vmatrix} = \begin{vmatrix} \cos ax & \sin ax \\ -a\sin ax & a\cos ax \end{vmatrix} = a \neq 0 \; \forall x \in \mathbb{R}.$$

Therefore, y_1 and y_2 are independent solutions.

Let the particular solution be

$$y_p = v_1(x)\cos ax + v_2(x)\sin ax. \quad (5.7.33)$$

$\therefore\; Dy_p = \{-av_1(x)\sin ax + a\,v_2(x)\cos ax\} + \{v_1'(x)\cos ax + v_2'(x)\sin ax\}$.

Choose $v_1(x)$ and $v_2(x)$ such that

$$v_1'(x)\cos ax + v_2'(x)\sin ax = 0 \quad (5.7.34)$$

so that

$$Dy_p = -av_1(x)\sin ax + av_2(x)\cos ax$$
$$\text{and } D^2y_p = \{-a^2v_1(x)\cos ax - a^2v_2(x)\sin ax\} - av_1'(x)\sin ax + av_2'(x)\cos ax.$$

Now the given equation is

$$D^2y_p + a^2y_p = \sec ax$$
$$\text{or, } -a^2v_1(x)\cos ax - a^2v_2(x)\sin ax - av_1'(x)\sin ax + av_2'(x)\cos ax$$
$$+a^2v_1(x)\cos ax + a^2v_2(x)\sin ax = \sec ax$$
$$\text{or, } -av_1'(x)\sin ax + av_2'(x)\cos ax = \sec ax. \quad (5.7.35)$$

From (5.7.33) and (5.7.35),

$$v_1'(x)\cos ax + v_2'(x)\sin ax = 0$$
$$-av_1'(x)\sin ax + av_2'(x)\cos ax = \sec ax.$$

Solving, $v_1'(x) = -\frac{1}{a}\tan ax$, $v_2'(x) = \frac{1}{a}$.

$$\therefore\; v_1(x) = -\frac{1}{a}\int \tan ax\,dx = -\frac{1}{a^2}\ln(\sec ax) \text{ and } v_2(x) = \frac{1}{a}\int dx = \frac{x}{a}.$$

Accordingly, $y_p = -\frac{1}{a^2}\cos ax\ln(\sec ax) + \frac{1}{a}x\sin ax$.

The complete solution is, therefore, $y = c_1\cos ax + c_2\sin ax - \frac{1}{a^2}\cos ax\ln(\sec ax) + \frac{1}{a}x\sin ax$.

Example 5.7.35. *Solve* $(D^2 - 2D)\,y = e^x\cos x$ *by the method of variation of parameters.*

Solution: CF is $c_1 + c_2e^{2x}$ (c_1, c_2 arbitrary constants).

We take $y_1(x) = 1$ and $y_2(x) = e^{2x}$ and take $y_p = v_1(x) + v_2(x)e^{2x}$ (Replacing c_1 and c_2 by $v_1(x)$ and $v_2(x)$]. $y_1(x)$ and $y_2(x)$ are linearly independent, since

$$W(y_1, y_2) = \begin{vmatrix} y_1 & y_2 \\ y_1' & y_2' \end{vmatrix} = \begin{vmatrix} 1 & e^{2x} \\ 0 & 2e^{2x} \end{vmatrix} = 2e^{2x} \neq 0 \ \forall \ x \in \mathbb{R}.$$

Now $Dy_p = v_1'(x) + v_2'(x)e^{2x} + 2v_2(x)e^{2x}$.

Choose $v_1(x)$ and $v_2(x)$ such that $v_1'(x) + v_2'(x)e^{2x} = 0$.

Then

$$\begin{aligned} Dy_p &= 2v_2(x)e^{2x} \\ \text{and } D^2y_p &= 2v_2'(x)e^{2x} + 4v_2(x)e^{2x} \\ \left(D^2 - 2D\right) y_p &= 2v_2'(x)e^{2x} + 4v_2(x)e^{2x} - 4v_2(x)e^{2x} = 2v_2'(x)e^{2x} = e^x \cos x. \\ \therefore \ v_2'(x) &= \tfrac{1}{2}e^{-x} \cos x. \\ \therefore \ v_2(x) &= \tfrac{1}{2}\int e^{-x} \cos x \, dx = \tfrac{1}{4}e^{-x}(-\cos x + \sin x). \end{aligned}$$

Now

$$v_1'(x) = -v_2'(x)e^{2x} = -\tfrac{1}{2}e^x \cos x. \quad \therefore \ v_1(x) = -\tfrac{1}{4}e^x(\cos x + \sin x).$$

$\therefore$ a particular solution of the given equation is

$$y_p = v_1(x) + v_2(x)e^{2x} = -\tfrac{1}{4}e^x(\cos x + \sin x) + \tfrac{1}{4}(\sin x - \cos x)e^x = -\tfrac{1}{2}e^x \cos x$$

and the primitive is $y = c_1 + c_2e^{2x} - \frac{1}{2}e^x \cos x$.

Example 5.7.36. *Solve by the method of variation of parameters:* $\frac{d^2y}{dx^2} + 2\frac{dy}{dx} + y = \frac{e^{-x}}{x^2}$.

Solution:

To find CF

Auxiliary equation: $m^2 + 2m + 1 = 0 = (m + 1)^2$, whereby $m = -1, -1$.

CF is $e^{-x}(c_1 + c_2x) = c_1e^{-x} + c_2xe^{-x}$.

To find PI

First to show that e^{-x} and xe^{-x} are *independent* solutions.

Let $y_1 = e^{-x}$, $y_2 = xe^{-x}$.

The Wronskian of y_1 and y_2 is

$$W(y_1, y_2) = \begin{vmatrix} y_1 & y_2 \\ y_1' & y_2' \end{vmatrix} = \begin{vmatrix} e^{-x} & xe^{-x} \\ -e^{-x} & e^{-x}(1 - x) \end{vmatrix} = e^{-2x} \neq 0 \ \forall \ x \in \mathbb{R}.$$

Therefore, y_1 and y_2 are independent solutions.

Let the particular solution be

$$y_p = v_1(x)e^{-x} + v_2(x)xe^{-x}. \tag{5.7.36}$$

$\therefore \ Dy_p = \left\{-v_1(x)e^{-x} + (1 - x)v_2(x)e^{-x}\right\} + \left\{v_1'(x)e^{-x} + v_2'(x)xe^{-x}\right\}.$

Choose $v_1(x)$ and $v_2(x)$ such that

$$v_1'(x)e^{-x} + xv_2'(x)e^{-x} = 0 \tag{5.7.37}$$

so that

$$Dy_p = -v_1(x)e^{-x} + (1-x)v_2(x)e^{-x}$$
$$\text{and } D^2y_p = \{v_1(x)e^{-x} + (x-2)v_2(x)e^{-x}\} - v_1'(x)e^{-x} + (1-x)v_2'(x)e^{-x}.$$

Now the given equation is

$$D^2y_p + 2Dy_p + y_p = \frac{e^{-x}}{x^2}$$

or, $v_1(x)e^{-x} + (x-2)v_2(x)e^{-x} - v_1'(x)e^{-x} + (1-x)v_2'(x)e^{-x} - 2v_1(x)e^{-x} + 2(1-x)v_2(x)e^{-x} + v_1(x)e^{-x}$
$+v_2(x)xe^{-x} = \dfrac{e^{-x}}{x^2}$

or, $-v_1'(x)e^{-x} + (1-x)v_2'(x)e^{-x} = \dfrac{e^{-x}}{x^2}$. (5.7.38)

From (5.7.37) and (5.7.38)

$$v_1'(x)e^{-x} + xv_2'(x)e^{-x} = 0, \quad -v_1'(x)e^{-x} + (1-x)v_2'(x)e^{-x} = \frac{e^{-x}}{x^2}.$$

Solving, $v_1'(x) = -\frac{1}{x}$, $v_2'(x) = \frac{1}{x^2}$, whereby $v_1(x) = -\ln x$, $v_2(x) = -\frac{1}{x}$.
Accordingly, $y_p = -e^{-x}\ln x - e^{-x} = -e^{-x}(1+\ln x)$.
Complete solution is $y = e^{-x}(c_1 + c_2x) - e^{-x}(1+\ln x)$.

Example 5.7.37. *Solve by the method of variation of parameters:* $\frac{d^2y}{dx^2} + a^2y = \tan ax, a \neq 0.$

Solution: Proceed exactly as in Ex. 5.7.34.

To find CF

CF is $c_1\cos ax + c_2\sin ax$.

To find PI

Call $y_1 = \cos ax$, $y_2 = \sin ax$ and it has been observed that y_1, y_2 are independent solutions. Also the particular solution will be of the form $y_p = v_1(x)\cos ax + v_2(x)\sin ax$.

Next proceed as before where $v_1'(x), v_2'(x)$ can be found from

$$v_1'\cos ax + v_2'\sin ax = 0$$
$$\text{and } -av_1'\sin ax + av_2'\cos ax = \tan ax.$$

Solving,

$$v_1'(x) = -\frac{1}{a}\frac{\sin^2 ax}{\cos ax}$$
$$v_2'(x) = \frac{1}{a}\sin ax.$$
$$\therefore\ v_1(x) = -\frac{1}{a}\int(\sec ax - \cos ax)dx = -\frac{1}{a^2}\ln|\sec ax + \tan ax| + \frac{1}{a^2}\sin ax$$
$$\text{and } v_2(x) = \frac{1}{a}\int \sin ax dx = -\frac{1}{a^2}\cos ax.$$
$$\therefore\ y_p = -\frac{1}{a^2}\cos ax\ln|\sec ax + \tan ax|.$$

Complete solution is $y = c_1\cos ax + c_2\sin ax + y_p$.

Example 5.7.38. *Solve by the method of variation of parameters:* $\frac{d^3y}{dx^3} + \frac{dy}{dx} = \operatorname{cosec} x$.

Solution:

To find CF

Auxiliary equation: $m^3 + m = 0 = m(m+i)(m-i)$; then $m = 0, \pm i$. $\therefore$ CF is $c_1 + c_2 \cos x + c_3 \sin x$.

To find PI

Let the particular solution y_p be $y_p = v_1(x) + v_2(x)\cos x + v_3(x)\sin x$ [replace c_1, c_2, c_3 by v_1, v_2, v_3 respectively], where $v_1(x)$, $v_2(x)$, $v_3(x)$ are unknown functions.

Now v_1', v_2', v_3' are found by solving

$$v_1'(1) + v_2'(\cos x) + v_3'(\sin x) = 0,$$
$$v_1'(0) + v_2'(-\sin x) + v_3'(\cos x) = 0,$$
$$v_1'(0) + v_2'(-\cos x) + v_3'(-\sin x) = \operatorname{cosec} x,$$

which follows from Extension: nth order.

Solving, $v_1' = \operatorname{cosec} x$, $v_2' = -\cot x$, $v_3' = -1$.

Integrating, $v_1 = \ln(\operatorname{cosec} x - \cot x)$, $v_2 = -\ln \sin x$, $v_3 = -x$.

$\therefore$ $y_p = \ln(\operatorname{cosec} x - \cot x) - \cos x \ln \sin x - x \sin x$.

$\therefore$ complete solution is $y = c_1 + c_2 \cos x + c_3 \sin x + \ln(\operatorname{cosec} x - \cot x) - \cos x \ln \sin x - x \sin x$.

Example 5.7.39. *Solve by the method of variation of parameters:* $\frac{d^2y}{dx^2} - 3\frac{dy}{dx} + 2y = \frac{e^x}{1+e^x}$.

Solution:

To find CF

Auxiliary equation: $m^2 - 3m + 2 = 0 = (m-1)(m-2)$; then $m = 1, 2$. CF is $c_1 e^x + c_2 e^{2x}$.

To find PI

First we would like to show that e^x and e^{2x} are independent solutions.

Let $y_1 = e^x$, $y_2 = e^{2x}$.

The Wronskian of y_1 and y_2 is

$$W(y_1, y_2) = \begin{vmatrix} y_1 & y_2 \\ y_1' & y_2' \end{vmatrix} = \begin{vmatrix} e^x & e^{2x} \\ e^x & 2e^{2x} \end{vmatrix} = e^{3x} \neq 0 \ \forall\, x \in \mathbb{R}.$$

Therefore, y_1 and y_2 are independent solutions.

Let the particular solution be

$$y_p = v_1(x)e^x + v_2(x)e^{2x}. \tag{5.7.39}$$

$\therefore$ $Dy_p = \{v_1(x)e^x + 2v_2(x)e^{2x}\} + \{v_1'(x)e^x + v_2'(x)e^{2x}\}$.

Choose $v_1(x)$ and $v_2(x)$ such that

$$v_1'(x)e^x + v_2'(x)e^{2x} = 0 \tag{5.7.40}$$

so that

$$Dy_p = v_1(x)e^x + 2v_2(x)e^{2x}$$

and $$D^2y_p = \{v_1(x)e^x + 4v_2(x)e^{2x}\} + \{v_1'(x)e^x + 2v_2'(x)e^{2x}\}.$$

Now the given equation is

$$D^2y_p - 3Dy_p + 2y_p = \frac{e^x}{1+e^x} \text{ or, } v_1'(x)e^x + 2v_2'(x)e^{2x} = \frac{e^x}{1+e^x}. \tag{5.7.41}$$

From (5.7.40) and (5.7.41),

$$v_1'(x)e^x + v_2'(x)e^{2x} = 0$$

$$v_1'(x)e^x + 2v_2'(x)e^{2x} = \frac{e^x}{1+e^x}.$$

Solving, $v_1'(x) = -\frac{1}{1+e^x}$, $v_2'(x) = \frac{1}{e^x(1+e^x)}$
whereby $v_1(x) = \ln(1+e^x) - x$, $v_2(x) = \ln(1+e^x) - x - \frac{1}{e^x}$.
Accordingly, $y_p = (e^x + e^{2x})\ln(1+e^x) - (x+1)e^x - xe^{2x}$.
Complete solution is $y = C_1e^x + C_2e^{2x} + (e^x + e^{2x})\ln(1+e^x) - (x+1)e^x - xe^{2x}$.

When linearly independent solutions are given

Example 5.7.40. *Find the general solution of* $(1+x)\frac{d^2y}{dx^2} + x\frac{dy}{dx} - y = (1+x)^2$, $x \neq -1$ *by the method of variation of parameters; it is given that* $y = x$ *and* $y = e^{-x}$ *are two linearly independent solutions of the corresponding homogeneous equation.*

Solution:

To find CF

The equation is

$$\frac{d^2y}{dx^2} + \frac{x}{1+x}\frac{dy}{dx} - \frac{1}{1+x}y = 1 + x.$$

It is a second order differential equation and x and e^{-x} are the two linearly independent solutions of its reduced equation. Hence, the complementary function is $C_1x + C_2e^{-x}$.

To find PI

Replace C_1 by $v_1(x)$, C_2 by $v_2(x)$ and take y_p as a particular solution whereby $y_p = v_1(x)\cdot x + v_2(x)\cdot e^{-x}$.
Now $v_1'(x)$ and $v_2'(x)$ are found by solving:

$$v_1'(x)\cdot x + v_2'(x)\cdot e^{-x} = 0.$$

$$v_1'(x)\cdot 1 + v_2'(x)\cdot(-e^{-x}) = 1 + x.$$

Solving, $v_1'(x) = 1$, $v_2'(x) = -xe^x$, whereby $v_1(x) = x$, $v_2(x) = e^x(1-x)$.
Therefore, $y_p = x^2 + 1 - x$, and the complete solution is $y = C_1x + C_2e^{-x} + 1 - x + x^2$.

Example 5.7.41. *Solve by the method of variation of parameters* $\frac{d^2y}{dx^2} + \frac{1}{x}\frac{dy}{dx} - \frac{1}{x^2}y = \ln x (x > 0)$; *it being given that* $y = x$ *and* $y = 1/x$ *are two linearly independent solutions of the associated homogeneous differential equations.*

Solution:

To find CF

It is a second order differential equation and x and $\frac{1}{x}$ are the two linearly independent solutions of its reduced equation. Hence, the complementary function is $C_1x + C_2/x$.

To find PI

Replace C_1 by v_1, C_2 by v_2 and take y_p as a particular solution, whereby $y_p = v_1(x)\cdot x + v_2(x)\cdot(1/x)$.
Now v_1', v_2' are found by solving

$$v_1'\cdot x + v_2'\cdot(1/x) = 0$$

$$v_1'\cdot 1 + v_2'\cdot\left(-1/x^2\right) = \ln x.$$

Solving, $v_1' = \frac{1}{2}\ln x$ and $v_2' = -\frac{1}{2}x^2 \ln x$, whereby

$$v_1 = \frac{1}{2}\int \ln x\, dx = \frac{1}{2}x(\ln x - 1)$$

$$v_2 = -\frac{1}{2}\int x^2 \ln x\, dx = -\frac{1}{2}\left(\frac{x^3\ln x}{3} - \frac{x^3}{9}\right).$$

$$\therefore\ y_p = \frac{1}{2}x^2(\ln x - 1) - \frac{1}{2}\left(\frac{x^2\ln x}{3} - \frac{x^2}{9}\right) = x^2\left(\frac{1}{3}\ln x - \frac{4}{9}\right).$$

Complete solution is $y = c_1 x + \left(\frac{c_2}{x}\right) + x^2\left(\frac{1}{3}\ln x - \frac{4}{9}\right)$.

Example 5.7.42. *Find the general solution of* $(2x+1)(x+1)\frac{d^2y}{dx^2} + 2x\frac{dy}{dx} - 2y = (2x+1)^2$, $x \neq -1, -\frac{1}{2}$ *by the method of variation of parameters. It is given that* $y = x$ *and* $y = 1/(x+1)$ *are two linearly independent solutions of the corresponding homogeneous equation.*

Solution:

To find CF

The equation is

$$\frac{d^2y}{dx^2} + \frac{2x}{(2x+1)(x+1)}\frac{dy}{dx} - \frac{2}{(2x+1)(x+1)}y = \frac{2x+1}{x+1}.$$

It is given that $y = x$ and $y = 1/(x+1)$ are two linearly independent solutions of its corresponding homogeneous equation of this second order differential equation. Hence, CF is $c_1 x + c_2/(x+1)$.

To find PI

Let $y_p = v_1(x)\cdot x + v_2(x)\cdot \frac{1}{x+1}$.

Next v_1', v_2' are found by solving:

$$v_1'\cdot x + v_2'\cdot\frac{1}{x+1} = 0,\quad v_1'\cdot 1 + v_2'\cdot\left\{-\frac{1}{(x+1)^2}\right\} = \frac{2x+1}{x+1}.$$

Solving, $v_1' = 1$, $v_2' = -(x^2 + x)$. Integrating,

$$v_1 = x,\ v_2 = -\frac{1}{6}(2x^3 + 3x^2);\ y_p = \frac{x^2}{6}\cdot\frac{4x+3}{x+1}.$$

Complete solution is

$$y = c_1 x + \frac{c_2}{x+1} + \frac{x^2}{6}\cdot\frac{4x+3}{x+1}.$$

Example 5.7.43. *Solve* $x^2\frac{d^2y}{dx^2} - x(x+2)\frac{dy}{dx} + (x+2)y = x^3$, $x \neq 0$ *given that* $y = x$, $y = xe^x$ *are two linearly independent solutions of the corresponding homogeneous equation.*

Solution: $\frac{d^2y}{dx^2} - \frac{x+2}{x}\frac{dy}{dx} + \frac{x+2}{x^2}y = x.$

CF is $c_1 x + c_2 x e^x$.

To find PI

Let $y_p = v_1(x)\cdot x + v_2(x)\cdot xe^x$, where $v_1'(x)\cdot x + v_2'(x)\cdot xe^x = 0$, $v_1'(x) + v_2'(x)\{(x+1)e^x\} = x$.

Solving $v_1' = -1$, $v_2' = e^{-x}$ whereby $v_1(x) = -x$, $v_2(x) = -e^{-x}$.

$\therefore\ y_p = -x^2 - x$ and complete solution is $y = x(c_1 + c_2 e^x) - x(x+1)$.

Example 5.7.44. *Solve by the method of variation of parameters* $x^3\frac{d^3y}{dx^3}+3x^2\frac{d^2y}{dx^2}=1, x>0$, *it being given that* $y=x^{-1}$, $y=1$, $y=x$ *are three linearly independent solutions of its reduced equation.*

Solution:

To find CF

The equation is

$$\frac{d^3y}{dx^3}+\frac{3}{x}\frac{d^2y}{dx^2}=\frac{1}{x^3}. \quad \text{CF is } \frac{c_1}{x}+c_2+c_3x.$$

To find PI

Let $y_p=v_1\cdot\frac{1}{x}+v_2\cdot 1+v_3\cdot x$.

Then v_1', v_2', v_3' are found from

$$v_1'\cdot\frac{1}{x}+v_2'\cdot 1+v_3'\cdot x=0$$

$$v_1'\cdot\left(-\frac{1}{x^2}\right)+v_2'\cdot 0+v_3'\cdot 1=0$$

$$v_1'\cdot\frac{2}{x^3}+v_2'\cdot 0+v_3'\cdot 0=\frac{1}{x^3}.$$

Solving, $v_1'=\frac{1}{2}, v_2'=-\frac{1}{x}, v_3'=\frac{1}{2x^2}$.

Integrating, $v_1=\frac{1}{2}x, v_2=-\ln x, v_3=-\frac{1}{2x}$.

$\therefore \quad y_p=\frac{1}{2}-\ln x-\frac{1}{2}=-\ln x$.

And the complete primitive becomes $y=\frac{c_1}{x}+c_2+c_3x-\ln x$ (Fig. 5.8).

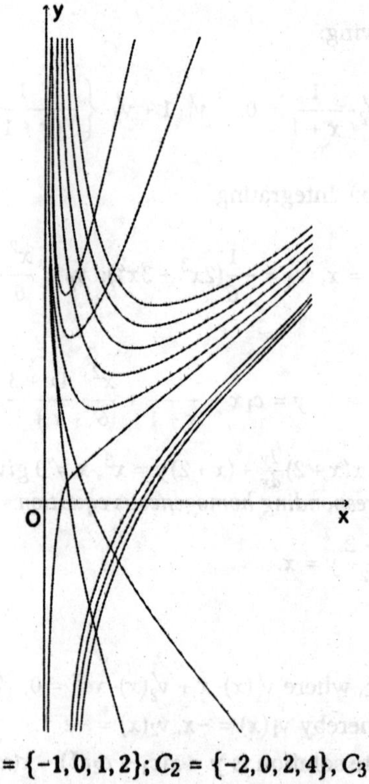

Fig. 5.8: $C_1=\{-1,0,1,2\}; C_2=\{-2,0,2,4\}, C_3=\{-2,2,3\}$.

Examples V(F)

Solve by the method of variation of parameters:

1. $\frac{d^2y}{dx^2} - 6\frac{dy}{dx} + 9y = \frac{e^{3x}}{x^2}$.

2. (a) $\frac{d^2y}{dx^2} + 4y = \sin 2x$

   (b) $\frac{d^2y}{dx^2} - 4\frac{dy}{dx} + 4y = e^{2x}$

   (c) $\frac{d^2y}{dx^2} + y = \cos x$.

3. (a) $\frac{d^2y}{dx^2} + 4y = \operatorname{cosec} 2x$

   (b) $\frac{d^2y}{dx^2} + 16y = 80$

4. (a) $\frac{d^2y}{dx^2} - \frac{dy}{dx} - 2y = 4x^2$

   (b) $\frac{d^2y}{dx^2} - 2\frac{dy}{dx} = e^x \sin x$.

Apply the method of variation of parameters to solve the following differential equations:

5. (a) $(D^2 + 4)\, y = \tan 2x$

   (b) $(D^2 + 2D + 1)\, y = e^{-x} \ln x$.

6. (a) $(D^2 - 2D - 3)\, y = 64xe^{-x}$

   (b) $(D^2 + 2D + 5)\, y = e^{-x} \sec 2x$.

7. (a) $(D^2 + 4)\, y = 4\sec^2 2x$

   (b) $(D^2 - 4D + 3)\, y = (1 + e^{-x})^{-1}$.

8. (a) $(D^3 - 3D^2 + 3D - 1)\, y = \frac{e^x}{x}$

   (b) $(D^3 + 6D^2 + 12D + 8)\, y = 12e^{-2x}$.

9. (a) $(D^4 - 9D^2)\, y = 54x^2$

   (b) $D^4 y = 5x$

   (c) $x^2\frac{d^2y}{dx^2} + x\frac{dy}{dx} - y = x^2 e^x$.

10. Solve by the method of variation of parameters:

    (a) $(x^2 + x)\frac{d^2y}{dx^2} + (2 - x^2)\frac{dy}{dx} - (2 + x)y = x(x + 1)^2, x \neq 0, -1,$
    if the complementary function is known to be $c_1 e^x + c_2 x^{-1}$.

    (b) $x^3\frac{d^3y}{dx^3} + x^2\frac{d^2y}{dx^2} - 2x\frac{dy}{dx} + 2y = x \ln x, (x > 0)$
    if the complementary function be given as $c_1 x^{-1} + c_2 x + c_3 x^2$.

    (c) $(1 - x^2)\frac{d^2y}{dx^2} - 4x\frac{dy}{dx} - (1 + x^2)y = x, x \neq \pm 1$, given that $\frac{\cos x}{1-x^2}$ and $\frac{\sin x}{1-x^2}$ are two independent solutions of its reduced equation.

Answers

1. $y = (c_1 + c_2 x)e^{3x} - 3^{3x} \ln x - e^{3x}$.
2. (a) $y = c_1 \cos 2x + c_2 \sin 2x - \frac{1}{4}x \cos 2x$

   (b) $y = e^{2x}\left(c_1 x + c_2 + \frac{1}{2}x^2\right)$

   (c) $y_p = \frac{1}{2}x \sin x + \frac{1}{4}\cos x$.
3. (a) $y = c_1 \cos 2x + c_2 \sin 2x - \frac{1}{2}x \cos 2x + \frac{1}{4}\sin 2x \ln|\sin 2x|$

   (b) $y = c_1 \cos 4x + c_2 \sin 4x + 5$
4. (a) $y = c_1 e^{-x} + c_2 e^{2x} - 2x^2 + 2x - 3$

   (b) $y = c_1 + c_2 e^{2x} - \frac{1}{2}e^x \sin x$.
5. (a) $y = c_1 \cos 2x + c_2 \sin 2x - \frac{1}{4}\cos 2x \ln|\sec 2x + \tan 2x|$

   (b) $y = (c_1 + c_2 x)\, e^{-x} + \frac{1}{2}x^2 e^{-x} \ln x - \frac{3}{4}x^2 e^{-x}$.
6. (a) $y = c_1 e^{-x} + c_2 e^{3x} - e^{-x}(8x^2 + 4x + 1)$

   (b) $y = e^{-x}\{c_1 \cos 2x + c_2 \sin 2x\} + \frac{1}{2}xe^{-x} \sin 2x + \frac{1}{4}e^{-x} \cos 2x \ln|\cos 2x|$.
7. (a) $y = c_1 \cos 2x + c_2 \sin 2x - 1 + \sin 2x. \ln|\sec 2x + \tan 2x|$

   (b) $y = c_1 e^x + c_2 e^{3x} + \frac{1}{2}(e^x - e^{3x}) \ln|1 + e^{-x}|$.
8. (a) $y = e^x(c_1 + c_2 x + c_3 x^2) + \frac{1}{2}x^2 e^x \ln|x|$

   (b) $y = e^{-2x}(c_1 + c_2 x + c_3 x^2 + 2x^3)$.
9. (a) $y = c_1 + c_2 x + c_3 e^{3x} + c_4 e^{-3x} - \frac{1}{2}x^4 - \frac{2}{3}x^2 - \frac{4}{27}$

   (b) $y = c_1 + c_2 x + c_3 x^2 + c_4 x^4 + \frac{1}{24}x^5$

(c) $y = c_1 x + \frac{c_2}{x} + \frac{x^2}{9}(3\ln x - 4)$.

10. (a) $y = c_1 e^x + c_2 x^{-1} - \frac{1}{3}x^2 - x - 1$.

(b) $y = c_1 x^{-1} + c_2 x + c_3 x^2 - \frac{1}{4}x\{(\ln x)^2 + \ln x\} - \frac{3}{8}x$

(c) $y = \frac{1}{1-x^2}(c_1 \cos x + c_2 \sin x + x)$.

5.8 The homogeneous linear equation

An equation of the type

$$x^n \frac{d^n y}{dx^n} + p_1 x^{n-1}\frac{d^{n-1}y}{dx^{n-1}} + p_2 x^{n-2}\frac{d^{n-2}y}{dx^{n-2}} + \cdots + p_{n-1}x\frac{dy}{dx} + p_n y = f(x) \quad (5.8.1)$$

in which $p_1, p_2, \ldots, p_{n-1}, p_n$ are constants, is known as *Euler's homogeneous linear equation* or a *Cauchy-Euler equation.* It may be transformed into a linear equation with constant coefficients by means of the substitution $x = e^t$, or, $t = \ln x$, so that $\frac{dx}{dt} = x$, or, $\frac{dt}{dx} = \frac{1}{x}$.

We give an illustration:

Example 5.8.1. *Solve:* $x^2\frac{d^2y}{dx^2} + 5x\frac{dy}{dx} + 13y = \ln x + x^2$.

Solution: We put $x = e^t$, i.e. $t = \ln x$ so that $\frac{dt}{dx} = \frac{1}{x}$.

Now

$$\frac{dy}{dx} = \frac{dy}{dt}\cdot\frac{dt}{dx} = \frac{1}{x}\cdot\frac{dy}{dt}, \text{ i.e. } x\frac{dy}{dx} = \frac{dy}{dt} \text{ or, } xDt = \theta y, \text{ where } D \equiv \frac{d}{dx} \text{ and } \theta \equiv \frac{d}{dt}.$$

Similarly,

$$\frac{d^2y}{dx^2} = \frac{d}{dx}\left(\frac{1}{x}\cdot\frac{dy}{dt}\right) = -\frac{1}{x^2}\cdot\frac{dy}{dt} + \frac{1}{x}\cdot\frac{d^2y}{dt^2}\cdot\frac{dt}{dx} = \frac{1}{x^2}\left(\frac{d^2y}{dt^2} - \frac{dy}{dt}\right).$$

$$\therefore\ x^2\frac{d^2y}{dx^2} = \frac{d^2y}{dt^2} - \frac{dy}{dt} = (\theta^2 - \theta)y \text{ i.e. } x^2D^2y = (\theta - 1)\theta y.$$

Therefore, the given equation reduces to the form:

$$\{\theta(\theta - 1) + 5\theta + 13\}y = t + e^{2t} \text{ or, } (\theta^2 + 4\theta + 13)y = t + e^{2t}.$$

Let $y = e^{mt}$ be a trial solution of the homogeneous equation $(\theta^2 + 4\theta + 13)y = 0$.
Then the auxiliary equation $m^2 + 4m + 13 = 0$ gives $m = -2 \pm 3i$.
$\therefore$ the CF is $e^{-2t}(A\cos 3t + B\sin 3t)$, where A and B are two arbitrary constants.
Again,

$$\begin{aligned}\text{PI} &= \frac{1}{\theta^2 + 4\theta + 13}(t + e^{2t}) = \frac{1}{13}\left\{1 + \frac{1}{13}(4\theta + \theta^2)\right\}^{-1} t + \frac{1}{\theta^2 + 4\theta + 13}e^{2t} \\ &= \frac{1}{13}\left(1 - \frac{4}{13}\theta - \ldots\right)t + \frac{1}{\theta^2 + 4\theta + 13}e^{2t} = \frac{1}{13}\left(t - \frac{4}{13}\right) + \frac{1}{25}e^{2t} \\ &= \frac{1}{169}(13t - 4) + \frac{1}{25}e^{2t}.\end{aligned}$$

Hence, the general solution is

$$y = e^{-2t}(A\cos 3t + B\sin 3t) + \frac{1}{169}(13t - 4) + \frac{1}{25}e^{2t}$$

$$\text{i.e. } y = \frac{1}{x^2}(A\cos(3\ln x) + B\sin(3\ln x)) + \frac{1}{169}(13\ln x - 4) + \frac{1}{25}x^2.$$

Example 5.8.2. *Solve:* $x^3\frac{d^3y}{dx^3} + 3x^2\frac{d^2y}{dx^2} + x\frac{dy}{dx} + 8y = 65\cos(\ln x)$.

Solution: We put $x = e^t$ so that $\frac{dt}{dx} = \frac{1}{x}$.

Now

$$\frac{dy}{dx} = \frac{dy}{dt}\cdot\frac{dt}{dx} = \frac{1}{x}\cdot\frac{dy}{dt}.$$

That is, $x\cdot\frac{dy}{dx} = \frac{dy}{dt}$, or, $xDt = \theta y$ where $D \equiv \frac{d}{dx}$ and $\theta \equiv \frac{d}{dt}$.

Similarly,

$$\frac{d^2y}{dx^2} = \frac{d}{dx}\left(\frac{1}{x}\cdot\frac{dy}{dt}\right) = -\frac{1}{x^2}\cdot\frac{dy}{dt} + \frac{1}{x}\cdot\frac{d^2y}{dt^2}\cdot\frac{dt}{dx} = \frac{1}{x^2}\left(\frac{d^2y}{dt^2} - \frac{dy}{dt}\right).$$

$$\therefore\; x^2\frac{d^2y}{dx^2} = \frac{d^2y}{dt^2} - \frac{dy}{dt} = (\theta^2 - \theta)\,y.$$

That is, $x^2D^2y = (\theta - 1)\theta y$.

Again,

$$\frac{d^3y}{dx^3} = \frac{d}{dx}\left\{\frac{1}{x^2}\cdot\frac{d^2y}{dt^2} - \frac{1}{x^2}\cdot\frac{dy}{dt}\right\} = -\frac{2}{x^3}\frac{d^2y}{dt^2} + \frac{2}{x^3}\cdot\frac{dy}{dt} + \frac{1}{x^2}\frac{d^3y}{dt^3}\cdot\frac{dt}{dx} - \frac{1}{x^2}\cdot\frac{d^2y}{dt^2}\cdot\frac{dt}{dx}$$

$$= \frac{1}{x^3}\left(\frac{d^3y}{dt^3} - 3\frac{d^2y}{dt^2} + 2\frac{dy}{dt}\right)$$

$$\therefore\; x^3\frac{d^3y}{dx^3} = \frac{d^3y}{dt^3} - 3\frac{d^2y}{dt^2} + 2\frac{dy}{dt} = (\theta - 2)(\theta - 1)\theta y.$$

This is, $x^3D^3y = (\theta - 2)(\theta - 1)\theta y$.

Making these replacements, we obtain from the given equation

$$\{(\theta - 2)(\theta - 1)\theta + 3(\theta - 1)\theta + \theta + 8\}\; y = 65\cos t.$$

$$\text{or,}\quad (\theta^3 + 8)\; y = 65\cos t,$$

which is a linear equation in y with constant coefficients.

The auxiliary equation $m^3 + 8 = 0$ gives $m = -2, 1 \pm i\sqrt{3}$.

$\therefore$ the CF is $Ae^{-2t} + e^t\left(B\cos\sqrt{3}t + C\sin\sqrt{3}t\right)$, where A, B, C are arbitrary constants.

Also the PI

$$= \frac{1}{\theta^3 + 8}65\cos t = \frac{1}{8 - \theta}65\cos t$$

$$= \frac{8 + \theta}{64 - \theta^2}65\cos t = (8 + \theta)\cos t$$

$$= 8\cos t - \sin t.$$

$\therefore$ the general solution is

$$y = Ae^{-2t} + e^t\left(B\cos\sqrt{3}t + C\sin\sqrt{3}t\right) + 8\cos t - \sin t$$

$$= Ax^{-2} + x\left\{B\cos\left(\sqrt{3}\ln x\right) + C\sin\left(\sqrt{3}\ln x\right)\right\} + 8\cos(\ln x) - \sin(\ln x).$$

The previous method can be used for solving any homogeneous equation of nth order. Suppose, we proceed to solve the eq. (5.8.1).

Then as shown before, we have (putting $x = e^t$),

$$\left.\begin{aligned} xDy &= \theta y \\ x^2D^2y &= (\theta-1)\theta y \\ x^3D^3y &= (\theta-2)(\theta-1)\theta y \\ &\vdots \\ x^nD^ny &= (\theta-n+1)(\theta-n+2)\ldots(\theta-1)\theta y \end{aligned}\right\}. \quad (5.8.2)$$

[This last result can be established by induction: for suppose,

$$x^nD^ny = xD(xD-1)(xD-2)\ldots(xD-n+1)y = \theta(\theta-1)(\theta-2)\ldots(\theta-n+1)y$$

and put $(\theta - n)v = y$ so that

$$D^ny = \frac{d^n}{dx^n}\left\{x\frac{dv}{dx} - nv\right\} \quad \left(\because \theta \equiv x\frac{d}{dx}\right)$$

$$= x\frac{d^{n+1}v}{dx^{n+1}} + n\frac{d^nv}{dx^n} - n\frac{d^nv}{dx^n} = xD^{n+1}v.$$

$$\therefore x^n\left\{xD^{n+1}v\right\} = \theta(\theta-1)(\theta-2)\ldots(\theta-n+1)(\theta-n)v$$

$$\text{or, } x^{n+1}D^{n+1}v = \theta(\theta-1)(\theta-2)\ldots(\theta-n+1)(\theta-n)v.$$

Now y is any general function; hence v is also a general function. The result, if true for n, is thus true for $n+1$, it is obviously true for $n = 1, 2$ and so it is true for all positive integers n.]

Using the results of (5.8.2) the equation is brought into the form:

$$F(\theta)y \equiv \left\{\theta^n + p_1'\theta^{n-1} + \cdots + p_{n-1}'\theta + p_n'\right\}y = f(e^t),$$

where p_1', p_2', etc., are constants. This is a linear equation with constant coefficients (independent variable t) and may be solved by the foregoing methods of art. 5.7.

More generally, the equation (known as *Legendre's equation*)

$$(a+bx)^nD^ny + p_1(a+bx)^{n-1}D^{n-1}y + \cdots + p_ny = f(x)$$

can be reduced to the homogeneous linear form by putting $z = a + bx$, giving

$$Dy = \frac{dy}{dx} = \frac{dy}{dz}\cdot\frac{dz}{dx} = b\frac{dy}{dz}, \quad D^ny = b^n\frac{d^ny}{dz^n}.$$

The equation can be directly transformed into a linear equation with constant coefficients by the substitution $a + bx = e^t$.

Example 5.8.3. *Solve:* $(5+2x)^2\frac{d^2y}{dx^2} - 6(5+2x)\frac{dy}{dx} + 8y = 8(5+2x)^2$.

Solution: Put $5 + 2x = e^t$ and so $2\frac{dx}{dt} = e^t = 5 + 2x$.

$$\frac{dy}{dx} = \frac{dy}{dt}\cdot\frac{dt}{dx} = \frac{2}{5+2x}\cdot\frac{dy}{dt} \text{ i.e. } (5+2x)\frac{dy}{dx} = 2\frac{dy}{dt} = 2\theta y,$$

where $\theta \equiv \frac{d}{dt}$.

$$\frac{d^2y}{dx^2} = \frac{d}{dx}\left(\frac{2}{5+2x}\cdot\frac{dy}{dt}\right) = \frac{-4}{(5+2x)^2}\cdot\frac{dy}{dt} + \frac{4}{(5+2x)^2}\cdot\frac{d^2y}{dt^2} \text{ or, } (5+2x)^2\frac{d^2y}{dx^2} = 4\left\{\frac{d^2y}{dt^2} - \frac{dy}{dt}\right\} = 4(\theta-1)\theta y$$

$\therefore$ the given equation becomes $\{4(\theta-1)\theta - 6.2\theta + 8\}y = 8e^{2t}$ or, $(4\theta^2 - 16\theta + 8)\,y = 8e^{2t}$ or, $(\theta^2 - 4\theta + 2)\,y = 2e^{2t}$.

$\therefore$ the general solution is

$$y = Ae^{(2+\sqrt{2})t} + Be^{(2-\sqrt{2})t} + \frac{1}{\theta^2 - 4\theta + 2}2e^{2t} = e^{2t}\left(Ae^{\sqrt{2}t} + Be^{-\sqrt{2}t}\right) - e^{2t}$$

$$= (5+2x)^2\left\{A(5+2x)^{\sqrt{2}} + B(5+2x)^{-\sqrt{2}}\right\} - (5+2x)^2.$$

Example 5.8.4. *Solve:*

$$x^4\frac{d^4y}{dx^4} + 6x^3\frac{d^3y}{dx^3} + 4x^2\frac{d^2y}{dx^2} - 2x\frac{dy}{dx} - 4y = x^2 + 2\cos(\ln x).$$

Solution: We put $x = e^t$, or, $t = \ln x$.

If the operator $x\frac{d}{dx}$ be denoted by θ, then we can deduce

$$x\frac{dy}{dx} = \frac{dy}{dt} = \theta y;$$

$$x^2\frac{d^2y}{dx^2} = \frac{d^2y}{dt^2} - \frac{dy}{dt} = \theta(\theta-1)y;$$

$$x^3\frac{d^3y}{dx^3} = \frac{d^3y}{dt^3} - 3\frac{d^2y}{dt^2} - 2\frac{dy}{dt} = \theta(\theta-1)(\theta-2)y;$$

$$\text{and } x^4\frac{d^4y}{dx^4} = \theta(\theta-1)(\theta-2)(\theta-3)y.$$

The given equation reduces to an equation with constant coefficients with independent variable t in place of x; thus

$$\{\theta(\theta-1)(\theta-2)(\theta-3) + 6\theta(\theta-1)(\theta-2) + 4\theta(\theta-1) - 2\theta - 4\}y = e^{2t} + 2\cos t,$$

$$\text{or, } (\theta^4 - 3\theta^2 - 4)\,y = e^{2t} + 2\cos t,$$

$$\text{or, } (\theta^2 - 4)(\theta^2 + 1)\,y = e^{2t} + 2\cos t.$$

$\therefore$ the CF is $Ae^{2t} + Be^{-2t} + C\cos t + D\sin t$, where $t = \ln x$.

$$= Ax^2 + \frac{B}{x^2} + C\cos(\ln x) + D\sin(\ln x),$$

A, B, C, D being arbitrary constants and PI is

$$\frac{1}{(\theta^2-4)(\theta^2+1)}e^{2t} + 2\frac{1}{(\theta^2-4)(\theta^2+1)}\cos t = \frac{1}{\theta^2-4}\cdot\frac{1}{5}e^{2t} + 2\cdot\frac{1}{\theta^2+1}\cdot\frac{1}{-5}\cos t$$

$$= \frac{1}{5}e^{2t}\frac{1}{(\theta+2)^2-4}\cdot 1 - \frac{2}{5}Re.\left\{\frac{1}{\theta^2+1}(\cos t + i\sin t)\right\}$$

$$= \frac{1}{5}e^{2t}\frac{1}{\theta^2+4\theta}\cdot 1 - \frac{2}{5}Re.\left\{\frac{1}{\theta^2+1}e^{it}\right\}$$

$$= \frac{1}{5}e^{2t}\frac{1}{4\theta}\left(1+\frac{\theta}{4}\right)^{-1}\cdot 1 - \frac{2}{5}Re\cdot\left\{e^{it}\frac{1}{(\theta+i)^2+1}\cdot 1\right\}$$

$$= \frac{1}{20}e^{2t}\frac{1}{\theta}\cdot 1 - \frac{2}{5}Re.\left\{e^{it}\frac{1}{2i\theta}\left(1+\frac{\theta}{2i}\right)^{-1}\cdot 1\right\} = \frac{1}{20}e^{2t}t - \frac{2}{5}Re\cdot\left\{\frac{\cos t + i\sin t}{2i}t\right\}$$

$$= \frac{1}{20}x^2\ln x - \frac{2}{5}\cdot\frac{1}{2}\sin t\cdot t = \frac{1}{20}x^2\ln x - \frac{1}{5}\ln x\sin(\ln x).$$

∴ the complete primitive is

$$y = Ax^2 + \frac{B}{x^2} + C\cos(\ln x) + D\sin(\ln x) + \frac{1}{20}x^2\ln x - \frac{1}{5}\ln x\sin(\ln x).$$

Example 5.8.5. *Solve:* $x^2\frac{d^2y}{dx^2} + 3x\frac{dy}{dx} + y = \frac{1}{(1-x)^2}$.

Solution: Let $x = e^t$ or, $t = \ln x$. Let $\theta \equiv \frac{d}{dt}$. Then the equation becomes

$$\{\theta(\theta-1) + 3\theta + 1\}y = \frac{1}{(1-e^t)^2} \text{ or, } (\theta+1)^2 y = \frac{1}{(1-e^t)^2}.$$

$$\therefore \ \text{CF} = (A + Bt)e^{-t} = \frac{1}{x}\{A + B\ln x\},$$

A and B being two arbitrary constants.

Let $u = \frac{1}{(\theta+1)}\frac{1}{(1-e^t)^2}$, or, $\frac{du}{d\theta} + u = \frac{1}{(1-e^t)^2}$ (linear in u).

$$\therefore \ \frac{d}{d\theta}\left(ue^t\right) = \frac{e^t}{(1-e^t)^2}, \text{ i.e. } u = e^{-t}\int\frac{e^t dt}{(1-e^t)^2} = e^{-t}\frac{1}{1-e^t}.$$

Again,

$$\frac{1}{\theta+1}\left\{\frac{1}{\theta+1}\frac{1}{(1-e^t)^2}\right\} = \frac{1}{\theta+1}\frac{e^{-t}}{1-e^t} = v\,(\text{say}), \text{ i.e. } \frac{dv}{d\theta} + v = \frac{e^{-t}}{1-e^t} \text{ (linear in } v).$$

$$\therefore \ \frac{d}{d\theta}\left(ve^t\right) = \frac{1}{1-e^t}$$

$$\text{or, } v = e^{-t}\left\{\int\frac{1}{1-e^t}dt\right\} = -e^{-t}\int\frac{dt}{e^t-1} = -e^{-t}\int\frac{1}{z}\frac{1}{z+1}dz, \text{ if we put } e^t - 1 = z$$

$$= -e^{-t}\left\{\ln\frac{z}{z+1}\right\} = \frac{1}{e^t}\ln\frac{z+1}{z} = \frac{1}{e^t}\ln\frac{e^t}{e^t-1} = \frac{1}{x}\ln\frac{x}{x-1}.$$

$$\therefore \ \text{PI} = \frac{1}{x}\ln\frac{x}{x-1}.$$

∴ the complete primitive is

$$y = \frac{1}{x}(A + B\ln x) + \frac{1}{x}\ln\frac{x}{x-1}.$$

Example 5.8.6. *Solve:* $x^3\frac{d^3y}{dx^3} + 2x^2\frac{d^2y}{dx^2} + 2y = 10\left(x + \frac{1}{x}\right)$.

Solution: Put $x = e^t$, or, $t = \ln x$ and let $x\frac{d}{dx} = \frac{d}{dt} = \theta$, then

$$\{\theta(\theta-1)(\theta-2) + 2\theta(\theta-1) + 2\}\,y = 10\left(e^t + e^{-t}\right)$$

$$\text{or, } \left(\theta^3 - \theta^2 + 2\right)y = (\theta+1)\left(\theta^2 - 2\theta + 2\right)y = 10\left(e^t + e^{-t}\right).$$

To find CF

$m = -1, 1 \pm i.$

$$\text{CF is } c_1e^{-t} + e^t\left(c_2\cos t + c_3\sin t\right) = \frac{c_1}{x} + x\left(c_2\cos(\ln x) + c_3\sin(\ln x)\right).$$

To find PI

$$y_p = \frac{10}{\theta^3-\theta^2+2}e^t + \frac{10}{\theta^3-\theta^2+2}e^{-t}\cdot 1 = 5e^t + 10e^{-t}\cdot\frac{1}{\theta^3-4\theta^2+5\theta}\cdot 1$$
$$= 5e^t + 10e^{-t}\cdot\frac{1}{5\theta}\left(1+\frac{\theta^2-4\theta}{5}\right)^{-1}\cdot 1 = 5e^t + 2e^{-t}\cdot\frac{1}{\theta}\cdot 1 = 5e^t + 2e^{-t}\cdot t = 5x + \frac{2}{x}\ln x.$$

Complete solution is

$$y = \frac{c_1}{x} + x\,(c_2\cos(\ln x) + c_3\sin(\ln x)) + 5x + \frac{2}{x}\ln x.$$

Example 5.8.7. *Solve:* $(x+a)^2\frac{d^2y}{dx^2} - 4(x+a)\frac{dy}{dx} + 6y = x.$

Solution: Put $x + a = u$, then

$$\frac{dy}{dx} = \frac{dy}{du},\ \frac{d^2y}{dx^2} = \frac{d^2y}{du^2}$$

and hence the equation changes to

$$u^2\frac{d^2y}{du^2} - 4u\frac{dy}{du} + 6y = u - a.$$

Now put $u = e^t$, or, $t = \ln u$ and let $u\frac{d}{du} = \frac{d}{dt} = \theta$.
Then $\{\theta(\theta-1) - 4\theta + 6\}y = e^t - a$, or, $(\theta^2 - 5\theta + 6)\ y = e^t - a$.
CF: $Ae^{3t} + Be^{2t} = Au^3 + Bu^2 = A(x+a)^3 + B(x+a)^2$.

$$\text{PI: } y_p = \frac{1}{\theta^2-5\theta+6}\cdot e^t - \frac{1}{\theta^2-5\theta+6}\cdot 1 = \frac{1}{2}e^t - \frac{1}{6}\left(1+\frac{\theta^2-5\theta}{6}\right)^{-1} a$$
$$= \tfrac{1}{2}e^t - \tfrac{1}{6}a = \tfrac{1}{2}u - \tfrac{1}{6}a = \tfrac{1}{2}(x+a) - \tfrac{1}{6}a$$

GS: $y = A(x+a)^3 + B(x+a)^2 + \frac{1}{2}(x+a) - \frac{1}{6}a.$

Example 5.8.8. *Solve:* $(2+x)^2\frac{d^2y}{dx^2} + (2+x)\frac{dy}{dx} + 4y = 2\sin\{2\ln(2+x)\}.$

Solution: Put $2 + x = u$, then put $u = e^t$, and $D = \frac{d}{dt}$; then $\{D(D-1) + D + 4\}y = (D^2+4)y = 2\sin 2t$.

From W.O. Ex. 5.7.9, $y = A\cos 2t + B\sin 2t - \frac{1}{2}t\cos 2t$, where $t = \ln(2+x)$.

Alternative method of solution of a homogeneous linear equation

We first establish the following results stated in the form of a theorem:

Theorem 5.8.1. *Suppose, the operators D and θ signify $\frac{d}{dx}$ and $x\frac{d}{dx}$ respectively. Let the linear operator $F(\theta) = \theta^n + p_1\theta^{n-1} + \cdots + p_{n-1}\theta + p_n$, where n is a positive integer and the p's are constants.*

Then

1. $x^nD^n = \theta(\theta-1)\ldots(\theta-n+1)$.
2. $F(\theta)x^m = F(m)x^m$.
3. $\frac{1}{F(\theta)}x^m = \frac{1}{F(m)}x^m$, provided $F(m) \neq 0$.
4. $\frac{1}{F(\theta)}(x^mV) = x^m\frac{1}{F(\theta+m)}V$, V being a function of x.

Proof.

1. This has been established by induction (art. 5.8).

2. We see that

$$\theta(x^m) = x\frac{d}{dx}(x^m) = mx^m$$
$$\theta^2(x^m) = \theta(mx^m) = m^2x^m$$
$$\vdots$$
$$\theta^n(x^m) = m^n x^m.$$

Hence, $F(\theta)x^m = \left(m^n + p_1 m^{n-1} + \cdots + p_n\right)x^m = F(m)x^m$.

3. From (2), operating on both sides by the inverse operator $1/F(\theta)$ the result follows.

4. Again, we observe that, if U be a function of x, then

$$\theta\left(x^m U\right) = x\frac{d}{dx}\left(x^m U\right) = mx^m U + x^m \cdot x\frac{dU}{dx} \text{ or, } \theta\left(x^m U\right) = x^m(\theta + m)U.$$

More generally, $\theta^n(x^m U) = x^m(\theta + m)^n U$.

Hence,, $F(\theta)(x^m U) = x^m F(\theta + m)U$.

Put $F(\theta + m)U = V$ so that $U = \frac{1}{F(\theta+m)}V$.

We then have

$$F(\theta)\left\{x^m \cdot \frac{1}{F(\theta + m)}V\right\} = x^m \cdot V.$$

Operating both sides by $\dfrac{1}{F(\theta)}$ the result follows.

Example 5.8.9. *Using the results of this theorem prove that the general solution of* $x^2\frac{d^2y}{dx^2} - 4x\frac{dy}{dx} + 6y = x^5$ *is* $y = \frac{1}{6}x^5 + Ax^2 + Bx^3$.

Solution: The given equation is $\{\theta(\theta - 1) - 4\theta + 6\}\, y = x^5$.

Since $F(\theta)x^m = x^m F(m)$, $y = x^m$ is a solution of $F(\theta)y = 0$, if m is a root of $F(m) = 0$; more generally, $y = Ax^m$ is a solution.

In the given equation: $F(\theta) = \theta^2 - 5\theta + 6$.

$\therefore$ $F(m) = 0$, or, $m^2 - 5m + 6 = 0$ gives $m = 3$, or, $m = 2$.

$\therefore$ CF of the given equation is $y = Ax^2 + Bx^3$.

Again, PI $= \frac{1}{\theta^2 - 5\theta + 6}x^5 = \frac{1}{6}x^5$, using (3) of the theorem.

$\therefore$ the general solution is $y = Ax^2 + Bx^3 + \dfrac{1}{6}x^5$.

Note 5.8.1. *The students should try to solve a few problems of Examples V(E) using the results of the previous theorem.*

Examples V(G)

Solve:

1. $x^2\frac{d^2y}{dx^2} - x\frac{dy}{dx} + y = 0.$
2. $x^3\frac{d^3y}{dx^3} - 3x^2\frac{d^2y}{dx^2} + 7x\frac{dy}{dx} - 8y = 0.$
3. $x^3\frac{d^3y}{dx^3} + 6x^2\frac{d^2y}{dx^2} + 4x\frac{dy}{dx} - 4y = 0.$
4. (a) $x^2\frac{d^2y}{dx^2} + y = 3x^2$

   (b) $x^2\frac{d^2y}{dx^2} - 3x\frac{dy}{dx} + 4y = 2x^2.$
5. (a) $x^2\frac{d^2y}{dx^2} + 5x\frac{dy}{dx} + 4y = x^4$

   (b) $x^2\frac{d^2y}{dx^2} + 5x\frac{dy}{dx} + 3y = e^{-x}.$
6. (a) $x^2\frac{d^2y}{dx^2} - 4x\frac{dy}{dx} + 6y = 6x^5$

   (b) $x^2\frac{d^2y}{dx^2} + 2x\frac{dy}{dx} - y = 3x^3\cos(\ln x)$

   (c) $x^3\frac{d^3y}{dx^3} - x^2\frac{d^2y}{dx^2} + 2x\frac{dy}{dx} - 2y = x^3.$

7. $x^2\frac{d^2y}{dx^2} + 2x\frac{dy}{dx} - 20y = (x+1)^2$.

8. $\frac{d^3y}{dx^3} - \frac{4}{x}\frac{d^2y}{dx^2} + \frac{5}{x^2}\frac{dy}{dx} - \frac{2y}{x^3} = 1$.

9. $\frac{d^2V}{dr^2} + \frac{1}{r}\frac{dV}{dr} = 4\pi\rho\ (\rho = \text{constant})$.

10. $x^2\frac{d^2y}{dx^2} - x\frac{dy}{dx} + y = \ln x$.

11. $x^4\frac{d^4y}{dx^4} + 2x^3\frac{d^3y}{dx^3} + x^2\frac{d^2y}{dx^2} - x\frac{dy}{dx} + y = \ln x$.

12. $x^2\frac{d^2y}{dx^2} + x\frac{dy}{dx} - y = x^m$.

13. $r^4\frac{d^4V}{dr^4} + 6r^3\frac{d^3V}{dr^3} + 9r^2\frac{d^2V}{dr^2} + 3r\frac{dV}{dr} + V = (1+\ln r)^2$

14. $x^2\frac{d^2y}{dx^2} + 4x\frac{dy}{dx} + 2y = e^x$.

15. $\left[x^2D^2 - (2m-1)xD + \left(m^2+n^2\right)\right] y = n^2x^m \ln x$.

16. $z^2\frac{d^2y}{dz^2} - 3z\frac{dy}{dz} + y = \frac{\ln z\cdot\sin(\ln z)+1}{z}$.

17. $(1+2x)^2\frac{d^2y}{dx^2} - 6(1+2x)\frac{dy}{dx} + 16y = 8(1+2x)^2$.

18. $(1+x)^2\frac{d^2y}{dx^2} + (1+x)\frac{dy}{dx} + y = 4\cos\ln(1+x)$.

19. $(x+1)^2\frac{d^2y}{dx^2} + (x+1)\frac{dy}{dx} = (2x+3)(2x+4)$.

20. $\frac{d^2y}{dx^2} + \frac{1}{x}\frac{dy}{dx} = \frac{12\ln x}{x^2}$.

21. $\frac{d^3y}{dx^3} = \frac{6y}{x^3}$.

22. $\frac{d^2y}{dx^2} + \frac{2}{x}\frac{dy}{dx} = 10$.

Answers

1. $y = (c_1 + c_2\ln x)\,x$.
2. $y = x^2\left\{c_1 + c_2\ln x + c_3(\ln x)^2\right\}$.
3. $y = x^{-2}(c_1 + c_2\ln x) + c_3x$.
4. (a) $y = \sqrt{x}\left(A\cos\frac{1}{2}\sqrt{3}\ln x + B\sin\frac{1}{2}\sqrt{3}\ln x\right) + x^2$ (b) $y = (c_1 + c_2\ln x)x^2 + (\ln x)^2x^2$.
5. (a) $y = x^{-2}(c_1 + c_2\ln x) + x^4/36$ (b) $y = c_1x^{-1}c_2x^{-3} + x^{-3}e^{-x}(x+1)$.
6. (a) $y = c_1x^2 + c_2x^3 + x^5$

   (b) $y = x^{-1/2}(c_1x^{\sqrt{5}/2} + c_2x^{-\sqrt{5}/2}) + \frac{3}{149}x^3[7\sin(\ln x) + 10\cos(\ln x)]$.

   (c) $y = x(c_1 + c_2\ln x) + c_3x^2 + \frac{1}{4}x^3$.
7. $y = c_1x^{-5} + c_2x^4 - \left(x^2/14 + x/9 + 1/20\right)$.
8. $y = c_1x^2 + x^{5/2}\left(c_2x^{\frac{1}{2}\sqrt{21}} + c_3x^{-\frac{1}{2}\sqrt{21}}\right) - x^3/5$.
9. $V = (c_1 + c_2\ln r) + \pi\rho r^2$.
10. $y = x(c_1 + c_2\ln x) + \ln x + 2$.
11. $y = 4 + \ln x + c_1x + c_2x\ln x + c_3x(\ln x)^2 + c_4x(\ln x)^3$.
12. $y = c_1x + c_2x^{-1} + x^m/\left(m^2-1\right)$.
13. PI $= (\ln r)^2 + 2\ln r - 3$.

    CF $= (c_1 + c_2\ln r)\sin(\ln r) + (c_3 + c_4\ln r)\cos(\ln r)$.
14. $y = c_1/x + c_2/x^2 + e^x/x^2$.
15. $y = x^m\left(c_1\sin\ln x^n + c_2\cos\ln x^n + \ln x\right)$.
16. $y = z^2\left(c_1z^{\sqrt{3}} + c_2z^{-\sqrt{3}}\right) + \frac{1}{6z} + \frac{\ln z}{61z}(5\sin\ln z + 6\cos\ln z) + \frac{2}{3721}\cdot\frac{1}{z}(27\sin\ln z + 191\cos\ln z)$.
17. $y = t^2\left[(\ln t)^2 + c_1\ln t + c_2\right];\ t = 1 + 2x$.
18. $y = c_1\cos\{\ln(1+x) - c_2\} + 2\ln(1+x)\sin\ln(1+x)$.
19. $y = c_1 + c_2\ln(x+1) + \{\ln(x+1)\}^2 + x^2 + 8x$.
20. $y = c_1 + c_2\ln x + 2(\ln x)^3$.
21. $y = c_1x^3 + c_2\cos\left\{\sqrt{2}\ln x - c_3\right\}$.
22. $3xy = 5x^3 + c_1x + c_2$.

Chapter 6

Exact Differential Equations: A few Special Forms

In the last chapter we discussed the method of solution of linear differential equations with constant coefficients. We shall now consider linear equations with variable coefficients, a special case (viz. homogeneous equations) of which was discussed in Art. 5.8. There is no general method of solving such equations. But there are a few special types which admit of easy integration. In this chapter linear differential equations that are exact will first be discussed, and then equations of a few other special forms will be considered.

6.1 Exact equations

Definition 6.1.1. *A differential equation*

$$f\left(\frac{d^n y}{dx^n}, \frac{d^{n-1} y}{dx^{n-1}}, \ldots, \frac{dy}{dx}, y, x\right) = Q(x)$$

is said to be exact if it can be obtained by merely differentiating *once an equation*

$$g\left(\frac{d^{n-1} y}{dx^{n-1}}, \frac{d^{n-2} y}{dx^{n-2}}, \ldots, \frac{dy}{dx}, y, x\right) = Q_1(x) + C$$

of one lower order.

For example,

$$x\frac{d^2y}{dx^2} + 2x\frac{dy}{dx} + 2y = 0$$

is an exact equation of second order for it can be obtained by *merely differentiating* once the equation

$$x\frac{dy}{dx} + (2x - 1)y = c.$$

6.2 Criterion of an exact differential equation

Theorem 6.2.1. *The linear equation*

$$P_0\frac{d^n y}{dx^n} + P_1\frac{d^{n-1} y}{dx^{n-1}} + P_2\frac{d^{n-2} y}{dx^{n-2}} + \cdots + P_n y = R, \tag{6.2.1}$$

where $P_0, P_1, \ldots, P_n, R$ are all functions of x, will be exact if and only if the successive derivatives of $P_0, P_1, P_2, \ldots, P_n$ satisfy the condition

$$P_n - P'_{n-1} + P''_{n-2} - \cdots + (-1)^n P_0^{(n)} = 0. \tag{6.2.2}$$

[$P_n^{(n)}$ *denotes the nth derivative of* P_n]

Proof. *The condition is necessary.*

Assume that the eq. (6.3.1) is exact. Then, by definition, there exists an equation of $(n-1)$th order which, when differentiated once, will yield eq. (6.3.1). Let that equation be

$$Q_0 y^{n-1} + Q_1 y^{n-2} + Q_2 y^{n-3} + \cdots + Q_{n-1} y = \int R\,dx + C.$$

Differentiating this equation, we actually obtain

$$\{Q_0 y^n + Q_0' y^{n-1}\} + \{Q_1 y^{n-1} + Q_1' y^{n-2}\} + \{Q_2 y^{n-2} + Q_2' y^{n-3}\} + \cdots + \{Q_{n-1} y_1 + Q_{n-1}' y\} = R. \quad (6.2.3)$$

Comparing (6.2.1) and (6.2.2), we obtain

$$P_0 = Q_0, P_1 = Q_0' + Q_1, P_2 = Q_1' + Q_2, \ldots, P_n = Q_{n-1}'.$$

Hence, the left side of (6.2.2)

$$P_n - P_{n-1}' + P_{n-2}'' - \cdots + (-1)^n P_0^{(n)}$$

becomes $Q_{n-1}' - (Q_{n-2}'' + Q_{n-1}') + (Q_{n-3}'' + Q_{n-2}') + \cdots + (-1)^{n-1}\left[Q_0^{(n)} + Q_1^{(n-1)}\right] + (-1)^n Q_0^{(n)} = 0$, identically, from which follows the necessity of the condition.

The condition is sufficient.

In what follows the successive derivatives of y will be indicated by $y_1, y_2, y_3, \ldots, y_n$ and those of the P's by $P', P'', \ldots, P^{(n)}$. The students may carefully study the following scheme:

$$P_0 y_n + P_1 y_{n-1} + P_2 y_{n-2} + \cdots + P_{n-1} y_1 + P_n y$$

$$\frac{d}{dx}(P_0 y_{n-1}) = P_0 y_n + P_0' y_{n-1}$$

On subtraction, $\quad (P_1 - P_0')\, y_{n-1} + P_2 y_{n-2} + \cdots + P_{n-1} y_1 + P_n y$

$$\frac{d}{dx}\{(P_1 - P_0')\, y_{n-2}\} = (P_1 - P_0')\, y_{n-1} + (P_1' - P_0'')\, y_{n-2}$$

On subtraction, $\quad (P_2 - P_1' + P_0'')\, y_{n-2} + \cdots + P_{n-1} y_1 + P_n y$

$$\cdots \quad \cdots \quad \cdots \quad \cdots$$

$$\cdots \quad \cdots \quad \cdots \quad \cdots$$

$$\left\{P_{n-1} - P_{n-2}' + P_{n-3}'' - \cdots + (-1)^{(n-1)} P_0^{n-1}\right\} y_1 + P_n y$$

$$\frac{d}{dx}\left\{\left(P_{n-1} - P_{n-2}' + \cdots + (-1)^{n-1} P_0^{(n-1)}\right) y\right\}$$

$$= \left\{P_{n-1} - P_{n-2}' + P_{n-3}'' - \cdots + (-1)^{n-1} P_0^{(n-1)}\right\} y_1$$

$$+ \left\{P_{n-1}' - P_{n-2}'' + \cdots + (-1)^{n-1} P_0^{(n)}\right\} y$$

On subtraction, $\left\{P_n - P_{n-1}' + P_{n-2}'' - \cdots + (-1)^n P_0^{(n)}\right\} y$

Hence, if the condition (6.2.2) holds, then the last remainder vanishes. Thus,

$$\frac{d}{dx}\left[P_0 y_{n-1} + (P_1 - P_0')\, y_{n-2} + \cdots + \left\{P_{n-1} - P_{n-2}' + \cdots + (-1)^{n-1} P_0^{(n-1)}\right\} y\right]$$

$$= P_0 y_n + P_1 y_{n-1} + \cdots + P_{n-1} y_1 + P_n y$$

$$= \text{left side of eq. (6.2.1).}$$

Thus, there exists a differential equation of $(n-1)$th order, *viz.*

$$P_0 y_{n-1} + \left(P_1 - P_0'\right) y_{n-2} + \cdots + \left\{P_{n-1} - P_{n-2}' + \cdots + (-1)^{n-1} P_0^{(n-1)}\right\} y = \int R\,dx + C, \qquad (6.2.4)$$

which, on differentiation, yields eq. (6.2.1).

Hence, the eq. (6.2.1) is exact and the condition is thus sufficient.

Note 6.2.1. *Equation (6.2.4) is now called a First Integral of equation (6.2.1). If (6.2.4) is exact, we may obtain its first integral in a similar manner. This will be called Second Integral of equation (6.2.1). The process can be repeated. It can be proved that any equation of nth order has exactly n independent first integrals. In general, if independent first integrals equal in number to the order of the equation have been obtained, all the differential coefficients can be eliminated from them so as to leave the primitive.*

Example 6.2.1. *Verify that the equation* $\left(2x^2+3x\right) y_2 + (6x+3)y_1 + 2y = (x+1)\,e^x$ *is exact and then solve it.*

Solution: Here $P_0 = 2x^2 + 3x$, $P_1 = 6x + 3$, $P_2 = 2$ and

$$P_2 - P_1' + P_0'' = 2 - \frac{d}{dx}(6x+3) + \frac{d^2}{dx^2}\left(2x^2+3x\right) = 2 - 6 + 4 = 0.$$

The condition for exactness is satisfied.

$$\left(2x^2+3x\right) y_2 + (6x+3)y_1 + 2y$$

$$\frac{d}{dx}\left\{\left(2x^2+3x\right) y_1\right\} = \left(2x^2+3x\right) y_2 + (4x+3)y_1$$

On subtraction, $\quad 2xy_1 + 2y$

$$\frac{d}{dx}\{2xy\} = 2xy_1 + 2y$$

$\therefore$ a first integral of the given equation is

$$\left(2x^2+3x\right) y_1 + 2xy = \int (x+1)\,e^x\,dx + c = xe^x + c \text{ or, } \frac{dy}{dx} + \frac{2}{2x+3}y = \frac{e^x}{2x+3} + \frac{c}{2x^2+3x}(x \neq 0).$$

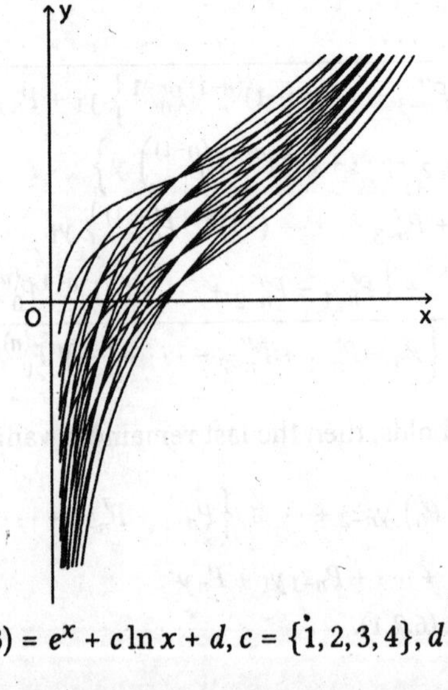

Fig. 6.1: $y(2x+3) = e^x + c\ln x + d,\ c = \{1,2,3,4\},\ d = \{-2,-1,0,1,2\}.$

This is a linear equation of first order; the integrating factor is $(2x + 3)$.

Multiplying by the IF, we get $\frac{d}{dx}\{y(2x+3)\} = e^x + \frac{c}{x}$.

Integrating, we obtain $y(2x + 3) = e^x + c\ln x + d$ (Fig. 6.1)

Example 6.2.2. $xy''' + (x^2 + x + 3)y'' + (4x + 2)y' + 2y = 0.$

Solution:

$$xy''' + (x^2 + x + 3)y'' + (4x + 2)y' + 2y$$

$$\frac{d}{dx}(xy'') = xy''' + y''$$

On subtraction, $\quad (x^2 + x + 2)y'' + (4x + 2)y' + 2y$

$$\frac{d}{dx}\{(x^2 + x + 2)y'\} = (x^2 + x + 2)y'' + (2x + 1)y'$$

On subtraction, $\quad (2x + 1)y' + 2y$

$$\frac{d}{dx}\{(2x + 1)y\} = (2x + 1)y' + 2y$$

$\therefore$ a first integral is: $xy'' + (x^2 + x + 2)y' + (2x + 1)y = c_1$.

Similarly, a second integral is $xy' + (x^2 + x + 1)y = c_1x + c_2$, which is a linear equation for which $xe^{\frac{1}{2}x(x+2)}$ is an integrating factor.

$\therefore$ the complete primitive is

$$yxe^{\frac{1}{2}x(x+2)} = c_1\int xe^{\frac{1}{2}x(x+2)}\,dx + c_2\int e^{\frac{1}{2}x(x+2)}\,dx + c_3.$$

6.3 Nonlinear exact equations: Equations that are not exact

If the eq. (6.2.1) of Art. 6.2 is not linear, we have no simple test for exactness. But we may see that the equation is exact by producing an equation of one lower order from which it may be obtained by a mere differentiation.

Again, if (6.2.1) is not exact we may sometimes find an Integrating Factor. But no general rule can be stated for determining such integrating factors.

Example 6.3.1. *Nonlinear but exact:* $2yy''' + 6y'y'' = -x^{-2}$.

Solution: $\quad 2yy''' + 6y'y''$

$$\frac{d}{dx}(2yy'') = 2yy''' + 2y'y''$$

On subtraction, $\quad 4y'y''$

$$\frac{d}{dx}(2y'^2) = 4y'y''$$

$\therefore$ the given equation is exact (though not linear); for it can be obtained by differentiating,

$$2yy'' + 2y'^2 = -\int x^{-2}\,dx + c_1 = x^{-1} + c_1$$

$$2yy'' + 2y'^2$$

$$\frac{d}{dx}\{2yy'\} = 2yy'' + 2y'^2$$

$\therefore$ a second integration yields $2yy' = \ln x + c_1x + c_2$.

Since $\frac{d}{dx}\{y^2\} = 2yy'$, we have the complete primitive

$$y^2 = -x + x\ln x + c_1\tfrac{1}{2}x^2 + c_2x + c_3 = x\ln x + Ax^2 + Bx + C.$$

Example 6.3.2. *Linear but not exact:* $x^3y''' + 5x^2y'' + (2x - x^3)\,y' - (2 + x^2)\,y = 40x^3 - 4x^5$.

Solution: We can verify that this linear equation is not exact;

$$\text{for } P_3 - P_2' + P_1'' - P_0''' = (-2 - x^2) - (2 - 3x^2) + (10) - (6) \neq 0.$$

However, an IF x^m can be deduced, which will make it an exact differential equation. Multiplication by x^m gives $x^{m+3}y''' + 5x^{m+2}y'' + (2x^{m+1} - x^{m+3})\,y' - (2x^m + x^{m+2})\,y = (40x^3 - 4x^5)x^m$.

If this new equation is exact, then the condition $P_3 - P_2' + P_1'' - P_0''' = 0$ must hold good.

i.e. $-(2x^m + x^{m+2}) - 2(m+1)x^m + (m+3)x^{m+2} + 5(m+2)(m+1)x^m - (m+3)(m+2)(m+1)x^m = 0$,

or, $(m+2)x^{m+2} + (m+2)\,(m - m^2)\,x^m = 0$,

for all values of x.

Then $m = -2$ and x^{-2} is an integrating factor.

Using it, we have

$$xy''' + 5y'' + \left(\frac{2}{x} - x\right)y' - \left(\frac{2}{x^2} + 1\right)y = 40x - 4x^3$$

$$\frac{d}{dx}(xy'') = xy''' + y''$$

$$4y'' + \left(\frac{2}{x} - x\right)y' - \left(\frac{2}{x^2} + 1\right)y$$

$$\frac{d}{dx}\left\{4y' + \left(\frac{2}{x} - x\right)y\right\} = 4y'' + \left(\frac{2}{x} - x\right)y' - \left(\frac{2}{x^2} + 1\right)y$$

$\therefore$ a first integral is $xy'' + 4y' + \left(\frac{2}{x} - x\right)y = 20x^2 - x^4 + c_1$.

This second order equation can be transformed into one with constant coefficients if we make the transformation $y = v/x^2$ and the complete solution will be $v = x^2y = Ae^x + Be^{-x} + Cx + x^5$. [This artifice will be explained in art. 7.2]

Example 6.3.3. *Nonlinear and Nonexact equation: Integrating factor by inspection* $x\cos y\,y''' - 3x\sin y\,y'y'' - \cos y\,y'' - x\cos y\,(y')^3 + \sin y\,(y')^2 + x\cos y\,y' - \sin y = 0$.

Solution: Since

$$\frac{d}{dx}\left(\frac{\sin y}{x}\right) = \frac{x\cos y\,y' - \sin y}{x^2},$$

the last two terms of the given equation suggest $\frac{1}{x^2}$ as a possible integrating factor.

Using it and integrating

$$\frac{\cos y\,y'' - \sin y\,(y')^2}{x} + \frac{\sin y}{x} = c_1 \text{ or, } \cos y\,y'' - \sin y\,(y')^2 + \sin y = c_1x.$$

The transformation $\sin y = z$ reduces it to $z'' + z = c_1x$.

$\therefore\ z = c_1x + c_2\cos x + c_3\sin x$ or, $\sin y = c_1x + c_2\cos x + c_3\sin x$.

Example 6.3.4. $x^2yy'' + (xy' - y)^2 - 3y^2 = 0$.

Solution:
$$x^2yy'' + \left(x^2y'^2 - 2xyy' + y^2\right) - 3y^2 = x^2yy' + x^2y'^2 - 2xyy' - 2y^2$$
$$\frac{d}{dx}\left(x^2yy'\right) = x^2yy'' + x^2y'^2 + 2xyy'$$
$$\overline{\qquad -4xyy' - 2y^2}$$
$$\frac{d}{dx}\left(-2xy^2\right) = \quad -4xyy' - 2y^2$$

$\therefore$ a first integral is $x^2yy' - 2xy^2 = c_1$.

If x^m is an IF, then multiplying by it we have $x^{m+2}yy' - 2x^{m+1}y^2 = c_1x^m$, or, $2x^{m+2}yy' - 4x^{m+1}y^2 = 2c_1x^m$.

But $\frac{d}{dx}\left(x^{m+2}y^2\right) = (m+2)x^{m+1}y^2 + x^{m+2}2yy'$.

$\therefore$ for exactness we may take $m + 2 = -4$, or, $m = -6$.

$\therefore$ the equation becomes $2x^{-4}yy' - 4x^{-5}y^2 = 2c_1x^{-6}$ or, $\frac{d}{dx}\left(x^{-4}y^2\right) = 2c_1x^{-6}$ or, $xy^2 = A + Bx^5$ (Fig. 6.2).

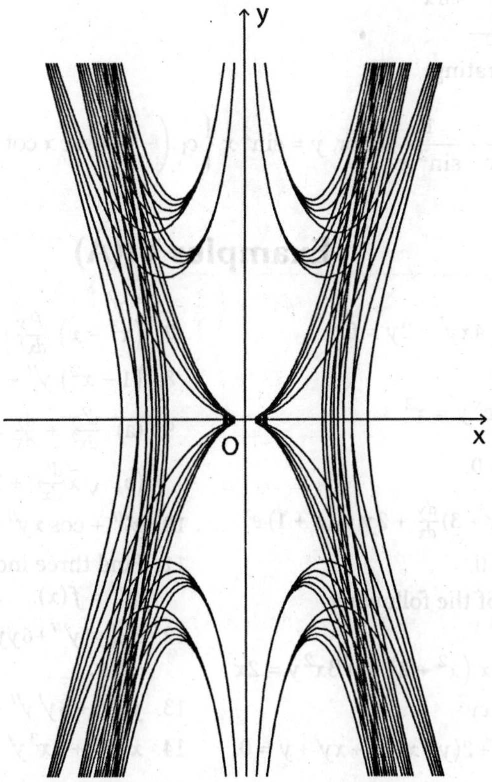

Fig. 6.2: $xy^2 = A + Bx^5, A = \{-2, -1, 0, 1\}, B = \{1, 2, 3, 4\}$.

Example 6.3.5. *Solve:* $\left(1 + x + x^2\right)\frac{d^3y}{dx^3} + (3 + 6x)\frac{d^2y}{dx^2} + 6\frac{dy}{dx} = 0$.

Solution: Here $P_0 = 1 + x + x^2$, $P_1 = 3 + 6x$, $P_2 = 6$, $P_3 = 0$.

Now $P_3 - P_2' + P_1'' - P_0''' = 0 - 0 + 0 + 0 = 0$. Exact.

The first integral is

$$P_0\frac{d^2y}{dx^2} + \left(P_1 - P_0'\right)\frac{dy}{dx} + \left(P_2 - P_1' + P_0''\right)y = c_1 \text{ or, } \underbrace{\left(1 + x + x^2\right)}_{P_0}\frac{d^2y}{dx^2} + \underbrace{(2 + 4x)}_{P_1}\frac{dy}{dx} + \underbrace{2y}_{P_2} = c_1.$$

This is again *exact*, since $P_2 - P_1' + P_0'' = 0$.
The second integral is, therefore,

$$\left(1+x+x^2\right)\frac{dy}{dx} + (1+2x)y = c_1x + c_2 \text{ or, } \frac{d}{dx}\left\{\left(1+x+x^2\right)y\right\} = c_1x + c_2.$$

Integrating, $\left(1+x+x^2\right)y = \frac{1}{2}c_1x^2 + c_2x + c_3$.

Example 6.3.6. *Show that* $\sin x\frac{d^2y}{dx^2} - \cos x\frac{dy}{dx} + 2y\sin x = 0$ *is exact and solve it completely.*

Solution: Here $P_0 = \sin x$, $P_1 = -\cos x$, $P_2 = 2\sin x$.
Since $P_2 - P_1' + P_0'' = 2\sin x - \sin x - \sin x = 0$, the equation is exact and the first integral is

$$P_0\frac{dy}{dx} + (P_1 - P_0')y = c_1, \text{ or, } \sin x\frac{dy}{dx} - 2y\cos x = c_1.$$

That is $\frac{dy}{dx} - (2\cot x)y = \frac{c_1}{\sin x}$.
IF $= e^{\int -2\cot x\, dx} = \frac{1}{\sin^2 x}$.
Multiplying and integrating,

$$y\cdot\frac{1}{\sin^2 x} = \int \frac{c_1}{\sin x}\cdot\frac{1}{\sin^2 x}\,dx \text{ or, } y = \sin^2 x\left\{c_1\left(-\frac{1}{2}\operatorname{cosec} x\cot x + \frac{1}{2}\ln\tan\frac{x}{2}\right) + c_2\right\}.$$

Examples VI(A)

Solve:

1. $xy''' + \left(x^2-3\right)y'' + 4xy' + 2y = 0$.
2. $xy'' + 2xy' + 2y = 0$.
3. (a) $\frac{d^2y}{dx^2} + 2e^x\frac{dy}{dx} + 2e^xy = x^2$
   (b) $y\frac{d^2y}{dx^2} + \left(\frac{dy}{dx}\right)^2 = 0$.
4. $\left(2x^2+3x\right)\frac{d^2y}{dx^2} + (6x+3)\frac{dy}{dx} + 2y = (x+1)e^x$.
5. $xyy'' + xy'^2 + yy' = 0$.
6. Find first integrals of the following equations:
   (a) $x^3y''' + 4x^2y'' + x\left(x^2+2\right)y' + 3x^2y = 2x$
   (b) $y'y'' - x^2yy' = xy^2$
   (c) $\left(y^2+2x^2y'\right)y'' + 2(y+x)y'^2 + xy' + y = 0$.
7. $\left(x^3-x\right)\frac{d^3y}{dx^3} + \left(8x^2-3\right)\frac{d^2y}{dx^2} + 14x\frac{dy}{dx} + 4y = \frac{2}{x^3}$.
8. $\left(1-x^2\right)y'' - xy' = 2$
9. (a) $\frac{d^2y}{dx^2} + \frac{dy}{dx} = e^x$
   (b) $\sqrt{x}\frac{d^2y}{dx^2} + 2x\frac{dy}{dx} + 3y = x$.
10. $y''' + \cos xy'' - 2\sin xy' - y\cos x = \sin 2x$
11. Find three independent first integrals of $y''' = f(x)$.
12. $3x[y^2y''' + 6yy'y'' + 2(y')^3] - 3y[yy'' + 2(y')^2] = -2/x$.
13. $yy''' + 3y'y'' - 2yy'' - 2(y')^2 + yy' = e^{2x}$.
14. $x^5y'' + 3x^3y' + (3-6x)x^2y = x^4 + 2x - 5$.

Answers

1. $\frac{e^{\frac{1}{2}x^2}y}{x^5} = c_1\int\frac{e^{\frac{1}{2}x^2}}{x^5}dx + c_2\int\frac{e^{\frac{1}{2}x^2}}{x^6}dx + c_3$.
2. $y = xe^{-2x}\left[\int c_1x^{-2}e^{2x}dx + c_2\right]$.
3. (a) $e^{2e^x}y = \frac{1}{2}\int e^{2e^x}\cdot x^3\,dx + c_1\int e^{2e^x}dx + c_2$
   (b) $y = \pm\sqrt{(c_1x+c_2)}$.
4. $(2x+3)y = e^x + c\ln x + d$.
5. $y = \pm\sqrt{(c_1\ln x + c_2)}$.
6. (a) $x^3y'' + x^2y' + x^3y = x^2 + c$
   (b) $y' = \sqrt{x^2y^2 + c}$
   (c) $y^2y' + x^2y'^2 + xy = c^2$.
7. $xy\sqrt{x^2-1} = \sec^{-1}x + c_1\sqrt{x^2-1} + c_2\ln\left(x+\sqrt{x^2-1}\right) + c_3$.

8. $y = c_1 \sin^{-1} x + (\sin^{-1} x)^2 + c_2$

9. (a) $y = c_1 e^{-x} + c_2 + \frac{1}{2} e^x$

(b) $e^{\frac{4}{3}x^{(3/2)}} \cdot \frac{y}{x} = \frac{1}{5} e^{(4/3)x^{(3/2)}} + c_1 \int e^{(4/3)x^{(3/2)}} \frac{dx}{x^2} + c_2$.

10. $y = e^{-\sin x} \int e^{\sin x} (c_1 x + c_2)\, dx + c_3 e^{-\sin x} - \frac{\sin x - 1}{2}$.

11. $y' = \int f(x)\, dx + c_1;\ xy'' - y' = \int x f(x)\, dx + c_2;$
$x^2 y'' - 2xy' + 2y = \int x^2 f(x)\, dx + c_3$.

12. $y^3 = c_1 x^3 + c_2 x + c_3 + x \ln x$.

13. $y^2 = c_1 + c_2 e^x + c_3 x e^x + e^{2x}$.

14. **First integral:**
$x^3 y' + 3x(1-x)y = x^3/3 + 2 \ln x + 5/x^2$,
which is a linear equation of first order.

6.4 Miscellaneous types of linear equations

(A) Equations of the form $\frac{d^n y}{dx^n} = f(x)$.

The equation is an exact differential equation. Integrating once, we get

$$\frac{d^{n-1} y}{dx^{n-1}} = \int f(x)\, dx + A_1;$$

a second integration gives

$$\frac{d^{n-2} y}{dx^{n-2}} = \int \left\{ \int f(x)\, dx \right\} dx + A_1 x + A_2.$$

Proceeding in this way the complete integral can be obtained by successive integrations.

Example 6.4.1. $\frac{d^2 y}{dx^2} = x^2 \sin x$.

Solution: Integrating once,

$$\frac{dy}{dx} = \int x^2 \sin x\, dx + A_1 = -x^2 \cos x + 2x \sin x + 2 \cos x + A_1.$$

Integrating again, $y = -x^2 \sin x - 4x \cos x + 6 \sin x + A_1 x + A_2$.

(B) Equations of the form $\frac{d^n y}{dx^n} = f(y)$.

(i) If $n = 1$, then the equation becomes

$$\frac{dy}{dx} = f(y). \quad \therefore \frac{dy}{f(y)} = dx, \text{ or, } \int \frac{dy}{f(y)} = x + A_1.$$

(ii) If $n = 2$, the equation becomes $\frac{d^2 y}{dx^2} = f(y)$.

Multiply both sides by $2\frac{dy}{dx}$. We then obtain

$$2\frac{dy}{dx}\frac{d^2 y}{dx^2} = 2f(y)\frac{dy}{dx}, \text{ or, } \frac{d}{dx}\left\{ \left(\frac{dy}{dx} \right)^2 \right\} = 2f(y)\frac{dy}{dx}.$$

Integrating,

$$\left(\frac{dy}{dx} \right)^2 = 2 \int f(y)\, dy + A_1 = \psi(y) + A_1 \text{ (say)},$$

whence

$$\pm\int \frac{dy}{\sqrt{\{\psi(y)+A_1\}}} = x + A.$$

This type of equation frequently occurs in Dynamics.

Example 6.4.2. $\frac{d^2y}{dx^2} = -\mu y$ (*μ is a constant*).

Solution: Multiplication of both sides by $2\frac{dy}{dx}$ gives

$$2\frac{dy}{dx}\frac{d^2y}{dx^2} = -2\mu y\frac{dy}{dx}.$$

Integrating,

$$\left(\frac{dy}{dx}\right)^2 = -2\mu\frac{y^2}{2} + \text{constant} = -\mu y^2 + \mu c^2.$$

(We write, for convenience, the constant $= \mu c^2$)

$$\therefore \quad \frac{dy}{dx} = \pm\sqrt{\mu}\sqrt{c^2 - y^2}, \text{ or, } \int \frac{dy}{\sqrt{c^2-y^2}} = \pm\int \sqrt{\mu}\, dx.$$

$$\therefore \quad \sin^{-1}\frac{y}{c} = \pm\sqrt{\mu}x + \alpha \ (\alpha = \text{arbitrary constant}).$$

$$\therefore \quad y = c\sin(\pm\sqrt{\mu}x + \alpha).$$

(C) Equations in which y appears in only two derivatives whose orders differ by either 1 or 2.

Case 6.4.1. *Equations of the type*

$$\frac{d^ny}{dx^n} = f\left(\frac{d^{n-1}y}{dx^{n-1}}\right), \text{ or, } \phi\left(\frac{d^ny}{dx^n}, \frac{d^{n-1}y}{dx^{n-1}}, x\right) = 0.$$

Put $q = \frac{d^{n-1}y}{dx^{n-1}}$. Then the equation takes the form:

$$\frac{dq}{dx} = f(q), \text{ or, } \int \frac{dq}{f(q)} = \int dx. \quad \therefore \psi(q) = x + A_1 \left[\text{assuming} \int \frac{dq}{f(q)} = \psi(q)\right].$$

Suppose that this equation can be solved for q and let

$$q = \phi(x + A_1); \ \therefore \ \frac{d^{n-1}y}{dx^{n-1}} = \phi(x + A_1).$$

This reduces to the case **(A)** given above.

Example 6.4.3. $-a\frac{d^2y}{dx^2} = \left\{1 + \left(\frac{dy}{dx}\right)^2\right\}^{3/2}$.

Solution: Put $\frac{dy}{dx} = q$. The equation becomes

$$-a\frac{dq}{dx} = (1+q^2)^{3/2} \text{ or, } \int \frac{dq}{(1+q^2)^{3/2}} = -\frac{1}{a}\int dx.$$

$$\therefore \quad \frac{q}{(1+q^2)^{1/2}} = -\frac{x}{a} + \text{constant} = -\frac{x}{a} + \frac{c}{a} \text{ (Take the constant} = c/a) \text{ or, } \frac{q^2}{1+q^2} = \frac{(c-x)^2}{a^2}.$$

$$\therefore\ q^2 = \frac{(c-x)^2}{a^2-(c-x)^2}. \quad \therefore\ q = \pm\frac{c-x}{\sqrt{a^2-(c-x)^2}}.$$

Integrating again,

$$y = \pm\int \frac{c-x}{\sqrt{a^2-(c-x)^2}}\,dx = \pm\sqrt{a^2-(c-x)^2} + \text{constant.}$$

We may write the solution in the form $(y-c_1)^2 = a^2-(x-c)^2$.

Case 6.4.2. *Equations of the type*

$$\frac{d^ny}{dx^n} = f\left(\frac{d^{n-2}y}{dx^{n-2}}\right), \text{ or, } \phi\left(\frac{d^ny}{dx^n}, \frac{d^{n-2}y}{dx^{n-2}}, x\right) = 0.$$

Put $\dfrac{d^{n-2}y}{dx^{n-2}} = q$. Then the equation becomes

$$\frac{d^2q}{dx^2} = f(q), \text{ or, } 2\frac{dq}{dx}\frac{d^2q}{dx^2} = 2f(q)\frac{dq}{dx}. \quad \therefore\ \frac{d}{dx}\left\{\left(\frac{dq}{dx}\right)^2\right\} = 2f(q)\frac{dq}{dx}.$$

Integrating,

$$\left(\frac{dq}{dx}\right)^2 = 2\int f(q)\,dq + c_1, \text{ or, } x + c_2 = \pm\int \frac{dq}{\left\{2\int f(q)\,dq + c_1\right\}^{1/2}}.$$

If after integrations have been carried out the equation can be solved for q in terms of x, say $q = \theta(x)$, where $\theta(x)$ will involve the constants c_1 and c_2, then $(n-2)$ direct integrations will yield the primitive.

Otherwise, using

$$\frac{dq}{dx} = \pm\left\{c_1 + 2\int f(q)\,dq\right\}^{1/2},$$

we get

$$\frac{d^{n-3}y}{dx^{n-3}} = \int q\,dx = \int \pm\frac{q\,dq}{\left\{c_1 + 2\int f(q)\,dq\right\}^{1/2}}$$

and we proceed in this manner till we get y as a function of q. The primitive will then be the eliminant with respect to q of the equation connecting y and q and the equation connecting x and q.

Example 6.4.4. $\dfrac{d^4y}{dx^4} - a^2\dfrac{d^2y}{dx^2} = 0.$

Solution: Put $\dfrac{d^2y}{dx^2} = q$. Then the equation takes the form:

$$\frac{d^2q}{dx^2} - a^2q = 0, \text{ or, } (D^2 - a^2)\,q = 0 \quad \left[\text{where } D^2 = \frac{d^2}{dx^2}\right].$$

$$\therefore\ q = c_1e^{ax} + c_2e^{-ax}, \text{ i.e. } \frac{d^2y}{dx^2} = c_1e^{ax} + c_2e^{-ax}.$$

Integrating once,

$$\frac{dy}{dx} = \frac{c_1}{a}e^{ax} - \frac{c_2}{a}e^{-ax} + c_3;$$

integrating again,

$$y = \frac{c_1}{a^2} e^{ax} + \frac{c_2}{a^2} e^{-ax} + c_3 x + c_4 = A + Bx + Ce^{ax} + De^{-ax},$$

where A, B, C, D are arbitrary constants.

Example 6.4.5. $x^2 \frac{d^4 y}{dx^4} + a^2 \frac{d^2 y}{dx^2} = 0.$

Solution: Put $q = \frac{d^2 y}{dx^2}$. The equation becomes a homogeneous equation of the form:

$$x^2 \frac{d^2 q}{dx^2} + a^2 q = 0.$$

Changing the independent variable by means of the substitution $e^z = x$, or, $z = \ln x$, we get

$$\{\theta(\theta - 1) + a^2\} q = 0, \quad \left[\text{where } \theta = \frac{d}{dz} = x\frac{d}{dx}\right].$$

The auxiliary equation $m^2 - m + a^2 = 0$ gives two values of m:

$$m = \frac{1 \pm \sqrt{1 - 4a^2}}{2}.$$

Case 6.4.3. *Suppose,* $4a^2 < 1$. *Then* $1 - 4a^2$ *is positive* $= k^2$ *(say) and*

$$m = \frac{1+k}{2}, \frac{1-k}{2}.$$

Then

$$q = c_1 e^{\left(\frac{1+k}{2}\right) z} + c_2 e^{\left(\frac{1-k}{2}\right) z}, \text{ or, } \frac{d^2 y}{dx^2} = c_1 x^{\frac{1+k}{2}} + c_2 x^{\frac{1-k}{2}}.$$

Integrating twice,

$$y = A + Bx + x^{\frac{5}{2}} \left(Cx^{\frac{k}{2}} + Dx^{-\frac{k}{2}} \right).$$

Case 6.4.4. *Suppose,* $4a^2 > 1$. *Then* $1 - 4a^2$ *is negative* $= -k^2$ *(say).*

And

$$m = \frac{1+ik}{2}, \frac{1-ik}{2}. \quad q = x^{\frac{1}{2}} \left[c_1 \cos \left(\frac{k}{2} \ln x + c_2 \right) \right],$$

whence we may obtain

$$y = A + Bx + Cx^{\frac{5}{2}} \cos \left(\frac{k}{2} \ln \frac{x}{D} \right).$$

Case 6.4.5. *Suppose,* $4a^2 = 1$. *Then* $m = \frac{1}{2}, \frac{1}{2}$. $q = x^{\frac{1}{2}} (c_1 + c_2 \ln x)$.

Integrating, twice the complete primitive can be obtained.

(D) Dependent variable absent.

(i) If the equation is free of y, i.e. of the form

$$f\left(\frac{d^n y}{dx^n}, \frac{d^{n-1} y}{dx^{n-1}}, \ldots, \frac{dy}{dx}, x \right) = 0, \qquad \textbf{(6.4.1)}$$

the substitution

$$\frac{dy}{dx} = p, \frac{d^2 y}{dx^2} = \frac{dp}{dx}, \ldots$$

will reduce the order by one. It may be possible to solve for p. Suppose that the solution is

$$p = \frac{dy}{dx} = F(x);$$

then $y = \int F(x)\, dx + c$ is the primitive.

(ii) If the given equation is of the form:

$$f\left(\frac{d^n y}{dx^n}, \frac{d^{n-1} y}{dx^{n-1}}, \ldots, \frac{d^k y}{dx^k}, x\right) = 0, \tag{6.4.2}$$

the substitution $\frac{d^k y}{dx^k} = q$ will reduce the order of the equation by k.

Now we solve for q and therefrom y by successive integrations.

Example 6.4.6. $\frac{d^2 y}{dx^2} = \sqrt{1 + \left(\frac{dy}{dx}\right)^2}$.

Solution: Put $\frac{dy}{dx} = p$. Then we get

$$\frac{dp}{dx} = \sqrt{1 + p^2}, \text{ or, } \frac{dp}{\sqrt{1 + p^2}} = dx$$

$$\text{or, } \sinh^{-1} p = x + a \text{ (} a \text{ is an arbitrary constant)}$$

$$\text{or, } \frac{dy}{dx} = p = \sinh(x + a).$$

$\therefore\ y = \cosh(x + a) + b$ (b is another arbitrary constant).

Example 6.4.7. $\frac{d^4 y}{dx^4} \cdot \frac{d^3 y}{dx^3} = 1$.

Solution: Put $p = \frac{d^3 y}{dx^3}$. The equation becomes $\frac{dp}{dx} \cdot p = 1$, or, $p\, dp = dx$.

$\therefore\ p^2 = 2x + c_1$.

$$\therefore\ \frac{d^3 y}{dx^3} = p = \pm\sqrt{2x + c_1}.$$

$$\therefore\ \frac{d^2 y}{dx^2} = \pm\frac{1}{3}(2x + c_1)^{\frac{3}{2}} + c_2 \text{ and } \frac{dy}{dx} = \pm\frac{1}{15}(2x + c_1)^{\frac{5}{2}} + c_2 x + c_3.$$

Hence,

$$y = \pm\frac{1}{105}(2x + c_1)^{\frac{7}{2}} + c_2\frac{x^2}{2} + c_3 x + c_4.$$

We may write $105y = \pm(2x + c_1)^{\frac{7}{2}} + Ax^2 + Bx + C$, where A, B, C and c_1 are arbitrary constants.

Example 6.4.8. $\left(\frac{d^3 y}{dx^3}\right)^2 + x\frac{d^3 y}{dx^3} - \frac{d^2 y}{dx^2} = 0$.

Solution: Put $\frac{d^2 y}{dx^2} = p$. Then

$$\left(\frac{dp}{dx}\right)^2 + x\frac{dp}{dx} - p = 0, \text{ or, } p = xq + q^2, \left[\text{where } q = \frac{dp}{dx}\right].$$

This is a Clairaut's equation. The general solution is

$$\frac{d^2 y}{dx^2} = p = c_1 x + c_1^2. \quad \therefore\ \frac{dy}{dx} = \frac{1}{2}c_1 x^2 + c_1^2 x + c_2, \text{ or, } y = \frac{1}{6}c_1 x^3 + \frac{c_1^2}{2}x^2 + c_2 x + c_3.$$

(E) Independent variable absent.

If the equation is free from x, i.e. of the form

$$f\left(\frac{d^n y}{dx^n}, \frac{d^{n-1} y}{dx^{n-1}}, \ldots, \frac{dy}{dx}, y\right) = 0;$$

the substitution $p = \frac{dy}{dx}$ gives

$$\frac{d^2y}{dx^2} = \frac{d}{dy}\left(\frac{dy}{dx}\right) \cdot \frac{dy}{dx} = \frac{dp}{dy} \cdot p,$$

$$\frac{d^3y}{dx^3} = \frac{d}{dy}\left(p\frac{dp}{dy}\right) \cdot \frac{dy}{dx} = \frac{d}{dy}\left(p\frac{dp}{dy}\right) \cdot p = \left(\frac{dp}{dy}\right)^2 p + p^2 \frac{d^2p}{dy^2} \text{ and so on.}$$

Thus, the equation will be reduced to one order lower than the original equation. Suppose, now it is possible to solve for p, say, $p = f(y)$, then

$$\int \frac{dy}{f(y)} = x + \text{constant is the primitive.}$$

Example 6.4.9. $\frac{d^2y}{dx^2} = \left(\frac{dy}{dx}\right)^3 + \frac{dy}{dx}$.

Solution: Put $\frac{dy}{dx} = p$. Then we have $p\frac{dp}{dy} = p^3 + p$.

Leaving the case $p = 0$ which is trivial, we have for $p \neq 0$,

$$\frac{dp}{dy} = p^2 + 1, \text{ or, } \frac{dp}{p^2 + 1} = dy.$$

$$\therefore \quad \tan^{-1} p = y + c_1, \text{ or, } p = \tan(y + c_1), \text{ or, } \frac{dy}{\tan(y + c_1)} = dx.$$

$\therefore$ $\ln \sin(y + c_1) = x + c_2$ is the primitive.

Example 6.4.10. $y(1 - \ln y)\frac{d^2y}{dx^2} + (1 + \ln y)\left(\frac{dy}{dx}\right)^2 = 0$.

Solution: Put $\frac{dy}{dx} = p$. Then the equation becomes

$$y(1 - \ln y)p\frac{dp}{dy} + (1 + \ln y)p^2 = 0 \text{ or, } \frac{dp}{p} = -\frac{1 + \ln y}{y(1 - \ln y)}dy \text{ (if } p \neq 0).$$

Integrating,

$$\ln p = \int \frac{1 - \ln y - 2}{y(1 - \ln y)}dy = \int \frac{dy}{y} - 2\int \frac{dy}{y(1 - \ln y)} = \ln y + 2\ln\{1 - \ln y\} + \text{constant.}$$

$\therefore$ $p = \frac{dy}{dx} = Cy(1 - \ln y)^2$, C is an arbitrary constant.

$$\therefore \int \frac{dy}{y(1 - \ln y)^2} = C\int dx + B. \quad \therefore \; Cx + B = -\frac{1}{1 - \ln y}$$

which is the required solution.

Examples VI(B)

Solve:

1. $\frac{d^2y}{dx^2} = \frac{b}{x}$.
2. $\frac{d^3y}{dx^3} = \sin^2 x$.
3. $x^2\frac{d^4y}{dx^4} + 1 = 0$.
4. $\frac{d^3s}{dt^3} = te^t$.
5. $\frac{d^3y}{dx^3} = x^4$; more generally, $\frac{d^ry}{dx^r} = x^k$ (k = a positive integer)
6. $\frac{d^2y}{dx^2} + m^2y = 0$.
7. $\frac{d^2y}{dx^2} = m^2y$.
8. $\frac{d^2y}{dx^2} + \frac{m^2}{y^2} = 0$.
9. $\frac{d^2y}{dx^2} = \frac{1}{\sqrt{a}}y^{-\frac{1}{2}}$.
10. $x\frac{d^2y}{dx^2} + \frac{dy}{dx} = 0$.
11. $m^2\frac{d^2y}{dx^2}\cdot\frac{dy}{dx} = x$.
12. $\frac{d^3y}{dx^3}\cdot\frac{d^2y}{dx^2} = 2$.
13. $\frac{d^5y}{dx^5} - m^2\frac{d^3y}{dx^3} = e^{ax}$.
14. $\frac{d^4r}{d\theta^4} + a^2\frac{d^2r}{d\theta^2} = 0$.
15. $\frac{d^2y}{dx^2} - \frac{1}{2}\left(\frac{dy}{dx}\right)^2 + 2 = 0$.
16. $x\frac{d^3y}{dx^3} - 2\frac{d^2y}{dx^2} = 0$.
17. $\left(1 + x^2\right)\frac{d^2y}{dx^2} + 1 + \left(\frac{dy}{dx}\right)^2 = 0$.
18. $2x\frac{d^3y}{dx^3}\cdot\frac{d^2y}{dx^2} = \left(\frac{d^2y}{dx^2}\right)^2 - a^2$.
19. $\frac{d^2y}{dx^2} + \frac{dy}{dx} = e^x$.
20. $\left(1 - x^2\right)\frac{d^2y}{dx^2} - x\frac{dy}{dx} = 2$.
21. $x\frac{d^3y}{dx^3} - x\frac{d^2y}{dx^2} - \frac{dy}{dx} = 0$.
22. $\frac{d^2y}{dx^2} + \frac{dy}{dx} + \left(\frac{dy}{dx}\right)^3 = 0$.
23. $\frac{d^2y}{dx^2} = a^2 + K^2\left(\frac{dy}{dx}\right)^2$.
24. $y\frac{d^2y}{dx^2} = 2\left(\frac{dy}{dx}\right)^2 - 2\frac{dy}{dx}$.
25. $y\frac{d^2y}{dx^2} - \left(\frac{dy}{dx}\right)^2 = y^2\ln y$.
26. $y\frac{d^2y}{dx^2} + \left(\frac{dy}{dx}\right)^2 = y^2$.
27. $\frac{d^2y}{dx^2} = e^{2y}$, given that $y = \frac{dy}{dx} = 0$, when $x = 0$.
28. $y^3\frac{d^2y}{dx^2} = k$.
29. $\frac{d^2y}{dx^2} - \frac{a^2}{x(a^2-x^2)}\frac{dy}{dx} = \frac{x^2}{a(a^2-x^2)}$.
30. $\left(\frac{dy}{dx}\right)^2 - y\frac{d^2y}{dx^2} = m\left\{\left(\frac{dy}{dx}\right)^2 + a^2\left(\frac{d^2y}{dx^2}\right)^2\right\}^{\frac{1}{2}}$.

Answers

1. $y = bx\ln x + c_1x + c_2$.
2. $y = c_1x^2 + c_2x + c_3 + \frac{x^3}{12} + \frac{\sin 2x}{16}$.
3. $y = c_1 + c_2x + c_3x^2 + c_4x^3 + \frac{1}{2}x^2\ln x$.
4. $s = te^t - 3e^t + c_1t^2 + c_2t + c_3$.
5. $y = c_1 + c_2x + c_3x^2 + \frac{4!}{7!}x^7$; $y = c_1 + c_2x + \cdots + c_rx^{r-1} + \frac{k!}{(k+r)!}x^{k+r}$.
6. $y = c_1\sin(mx + c_2)$.
7. $y = c_1e^{mx} + c_2e^{-mx}$.
8. $\sqrt{c_1y^2 + y} - \frac{1}{\sqrt{c_1}}\ln\left(\sqrt{c_1y} + \sqrt{1 + c_1y}\right) = mc_1\sqrt{2}x = c_2$.
9. $3x = 2a^{\frac{1}{4}}\left(y^{\frac{1}{2}} - 2c_1\right)\left(y^{\frac{1}{2}} + c_1\right)^{\frac{1}{2}} + c_2$.
10. $y = c_1\ln x + c_2$.
11. $my = \frac{1}{2}\left[x\sqrt{x^2 + c_1^2} + c_1^2\ln\left(x + \sqrt{x^2 + c_1^2}\right) + c_2\right]$.
12. $15y = 8\left(x + c_1\right)^{(5/2)} + c_2x + c_3$.
13. $y = c_1e^{mx} + c_2e^{-mx} + c_3 + c_4x + c_3x^2 + \frac{e^{ax}}{a^3(a^2-m^2)}$.
14. $r = c_1\sin a\theta + c_2\cos a\theta + c_3\theta + c_4$.
15. $y = 2x - 2\ln\left(1 - c_1e^{2x}\right) + c_2$.
16. $y = c_1x^4 + c_2x + c_3$.
17. $y + c_2 = -c_1x + \left(1 + c_1^2\right)\ln\left(x + c_1\right)$.
18. $15c_1^2y = 4\left(c_1x + a^2\right)^{\frac{5}{2}} + c_2x + c_3$.
19. $y = c_1e^{-x} + c_2 + \frac{1}{2}e^x$.
20. $y = c_1\sin^{-1}x + \left(\sin^{-1}x\right)^2 + c_2$.
21. $y = c_2e^x(x-1) + c_1x + c_1\int xe^x\int x^{-1}e^{-x}(dx)^2 + c_3$.
22. $y = c_2 - \sin^{-1}\left(c_1e^{-x}\right)$.
23. $y - c_2 = \frac{1}{K^2}\ln\sec aK\left(x - c_1\right)$.

24. $c_1 y = \tan(c_1 x + c_2)$.
25. $\ln y = c_1 e^x + c_2 e^{-x}$.
26. $y^2 = c_1 \sinh\left(\sqrt{2}x + c_2\right)$.
27. $e^{2y} = \sec^2 x$.
28. $(c_1 x + c_2)^2 + k = c_1 y^2$.
29. $2ay + x^2 = c_1\sqrt{a^2 - x^2} + c_2$.
30. $c_1 y = c_2 e^{c_1 x} - m\sqrt{1 + a^2 c_1^2}$.

Chapter 7

Equations of the Second Order

A linear equation of the second order has the form:

$$\frac{d^2y}{dx^2} + P\frac{dy}{dx} + Qy = R, \tag{7.0.1}$$

where P, Q, R are functions of x or constants.

If the coefficients P and Q are constants or, if the eq. (7.0.1) be homogeneous, the equation can be solved by the methods of Chapter 5. But, if P and Q are arbitrary functions of x, then no general method of solution can be suggested. In this chapter we shall give certain procedures which will yield solutions of certain equations of the form (7.0.1).

7.1 Change of dependent variable

If any solution of the eq. (7.0.1) with the right-hand side equated to zero be known, then the complete primitive of the eq. (7.0.1) can be found.

That means, if a solution of the *reduced equation* of (7.0.1) be known, then (7.0.1) can be solved.

Suppose that $y = u$ (where u is some function of x) satisfies the eq. (7.0.1) with its right-hand member zero. In other words, suppose

$y = u$ be a solution of the *reduced equation*

$$\frac{d^2y}{dx^2} + P\frac{dy}{dx} + Qy = 0$$

of (7.0.1). Then

$$\frac{d^2u}{dx^2} + P\frac{du}{dx} + Qu = 0. \tag{7.1.1}$$

We shall now find the complete primitive of (7.0.1).

For this, we put $y = uv$, where v is also a function of x. Then

$$\frac{dy}{dx} = u\frac{dv}{dx} + v\frac{du}{dx} \text{ and } \frac{d^2y}{dx^2} = u\frac{d^2v}{dx^2} + 2\frac{du}{dx}\cdot\frac{dv}{dx} + v\frac{d^2u}{dx^2}.$$

We, then, have from (7.0.1)

$$u\frac{d^2v}{dx^2} + \left(2\frac{du}{dx} + Pu\right)\frac{dv}{dx} + \left(\frac{d^2u}{dx^2} + P\frac{du}{dx} + Qu\right)v = R.$$

Using (7.0.1), this equation reduces to

$$\frac{d^2v}{dx^2} + P_1\frac{dv}{dx} = R_1, \tag{7.1.2}$$

where

$$P_1 = \frac{2}{u}\cdot\frac{du}{dx} + P \text{ and } R_1 = \frac{R}{u}.$$

Putting $\frac{dv}{dx} = p$, we have from (7.1.2)

$$\frac{dp}{dx} + P_1 p = R_1, \tag{7.1.3}$$

which is a linear equation in p of first order; its integrating factor is

$$e^{\int P_1 dx} = e^{\int\left(\frac{2}{u}\cdot\frac{du}{dx}+P\right)dx} = u^2 e^{\int P\,dx}.$$

∴ the eq. (7.1.3) becomes (on multiplying by the IF)

$$\frac{d}{dx}\left\{p\cdot u^2 e^{\int P\,dx}\right\} = \frac{R}{u}\cdot u^2 e^{\int P\,dx}$$

so that

$$u^2 p\, e^{\int P\,dx} = A + \int uR\, e^{\int P\,dx}\,dx.$$

Now, since $v = dp/dx$, we have

$$v = B + A\int \frac{1}{u^2}e^{-\int P\,dx}\,dx + \int\left\{\frac{1}{u^2}e^{-\int P\,dx}\int uR\,e^{\int P\,dx}\,dx\right\}dx.$$

Hence, the complete primitive of (7.0.1) is

$$y = uv = Bu + Au\int \frac{1}{u^2}e^{-\int P\,dx}\,dx + u\int\left\{\frac{1}{u^2}e^{-\int P\,dx}\int uR\,e^{\int P\,dx}\,dx\right\}dx$$

A and B being two arbitrary constants.

An important note

The problem of deducing the complete primitive is thus resolved into that of finding *some solution* of a simpler equation [cf. (7.1.1)]. In general, for arbitrary functions of P and Q it is not easy to know such a solution. But in the following special cases it is possible to determine such a solution by mere inspection

For the equation $(D^2 + PD + Q)\,y = 0$, where $D \equiv d/dx$,

(i) $y = x$ is a particular solution, if $P + xQ = 0$.
(ii) $y = e^x$ is a particular solution, if $1 + P + Q = 0$.
(iii) $y = e^{-x}$ is a particular solution, if $1 - P + Q = 0$.
(iv) $y = e^{mx}$ is a particular solution, if $m^2 + mP + Q = 0$.

The truth of these statements can be easily verified.

Example 7.1.1. *Verify that x is a solution of the reduced equation of* $x^2\frac{d^2y}{dx^2} - x\frac{dy}{dx} + y = x^2$. *Solve the equation after reducing it to a first order linear equation.*

Solution: The reduced equation of the given equation is

$$x^2\frac{d^2y}{dx^2} - x\frac{dy}{dx} + y = 0.$$

To verify that $y = x$ is a solution of this reduced equation.

Now

$$\frac{dy}{dx} = 1, \quad \frac{d^2y}{dx^2} = 0,$$

whereby

$$x^2\frac{d^2y}{dx^2} - x\frac{dy}{dx} + y = 0.$$

Next write $y = vx$ (v = variable) in the original equation, then

$$\frac{dy}{dx} = v + x\frac{dv}{dx}, \quad \frac{d^2y}{dx^2} = x\frac{d^2v}{dx^2} + 2\frac{dv}{dx}$$

and the equation transforms into

$$x^3\frac{d^2v}{dx^2} + x^2\frac{dv}{dx} = x^2.$$

Put $\frac{dv}{dx} = p$ and let $x \neq 0$, then

$$x\frac{dp}{dx} + p = 1, \text{ or, } \frac{dp}{dx} + \frac{1}{x}p = \frac{1}{x},$$

which is a first-order linear equation.

$$\text{IF} = e^{\int \frac{1}{x}dx} = e^{\ln x} = x.$$

Multiplying and integrating,

$$px = \int \frac{1}{x}\cdot x\,dx = x + c_1 \text{ or, } p = \frac{dv}{dx} = 1 + \frac{c_1}{x}.$$

Integrating, $v = x + c_1 \ln x + c_2$ and the complete solution becomes $y = vx = x^2 + c_1 x \ln x + c_2 x$.

Example 7.1.2. *Knowing that $y = x$ is a solution of the reduced equation of $x^2\frac{d^2y}{dx^2} - x(x+2)\frac{dy}{dx} + (x+2)y = x^3$, solve it after reducing it to a first order linear equation.*

Solution: Let $y = vx$ (v = a function of x), then the given equation becomes

$$x^2\frac{d^2v}{dx^2} - x^3\frac{dv}{dx} = x^3.$$

Put $\frac{dv}{dx} = p$ and let $x \neq 0$, then $\frac{dp}{dx} - p = 1$ (Linear).

Since

$$\text{IF} = e^{\int -dx} = e^{-x}, \quad pe^{-x} = \int e^{-x}dx = -e^{-x} + c_1.$$

Now $p = dv/dx = c_1e^x - 1$, integrating, $v = c_1e^x - x + c_2$.

The complete primitive is $y = c_1xe^x - x^2 + c_2x$ (Fig. 7.1).

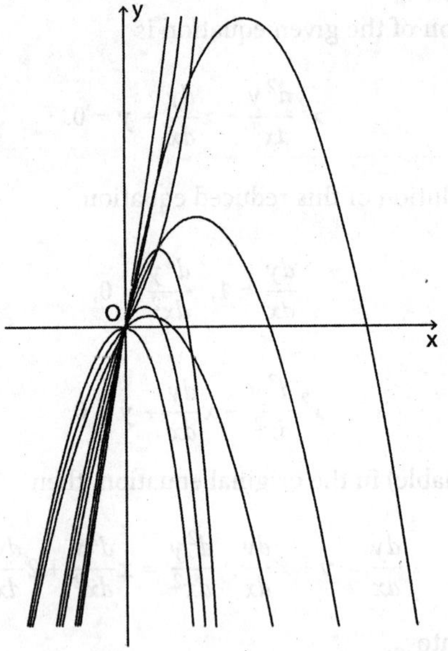

Fig. 7.1: $y = C_1xe^x - x^2 + C_2x$, $C_1 = \{-1, 0, 1\}$, $C_2 = \{1, 3, 5\}$.

Example 7.1.3. *Solve* $(1-x^2)\frac{d^2y}{dx^2} + x\frac{dy}{dx} - y = x(1-x^2)$, *given that* $y = x$ *is a solution of its reduced equation.*

Solution: As before, let $y = vx$, then the equation becomes

$$x(1-x^2)\frac{d^2v}{dx^2} + (2-x^2)\frac{dv}{dx} = x(1-x^2).$$

Put $\frac{dv}{dx} = p$, then

$$\frac{dp}{dx} + \frac{2-x^2}{x(1-x^2)}p = 1. \quad \text{(Linear)}$$

$$\text{IF} = e^{\int \frac{2-x^2}{x(1-x^2)}dx} = e^{\int \left(\frac{2}{x} + \frac{x}{1-x^2}\right)dx} = \frac{x^2}{\sqrt{1-x^2}}.$$

Hence,

$$p = \frac{dv}{dx} = \frac{1}{2}\frac{\sqrt{1-x^2}}{x^2}\sin^{-1}x + \frac{x^2-1}{2x} + c_1\frac{\sqrt{1-x^2}}{x^2}.$$

Integrating and putting $v = y/x$, we have the required solution as

$$y = -\tfrac{1}{2}\sqrt{1-x^2}\sin^{-1}x - \tfrac{1}{4}x(\sin^{-1}x)^2 + \tfrac{x^3}{4} - c_1\left\{\sqrt{1-x^2} + x\sin^{-1}x\right\} + c_2x.$$

Example 7.1.4. *Solve by the method of variation of parameters:* $x^2\frac{d^2y}{dx^2} + x\frac{dy}{dx} - y = x^2e^{2x}, x \neq 0.$

Solution: Apply Cauchy-Euler homogeneous linear equation of art. 5.8 to its reduced equation

$$x^2\frac{d^2y}{dx^2} + x\frac{dy}{dx} - y = 0.$$

Put $x = e^x$, or, $t = \ln x$ and $x\, d/dx = \theta$, it then reduces to $\{\theta(\theta-1) + \theta - 1\}y = 0$, or, $(\theta^2 - 1)y = 0$.
Auxiliary equation: $m^2 - 1 = 0$, $m = \pm 1$.

CF is $c_1 e^t + c_2 e^{-t} = c_1 x + c_2/x$.

The given equation reduces to

$$\frac{d^2y}{dx^2} + \frac{1}{x}\frac{dy}{dx} - \frac{1}{x^2}y = e^{2x}.$$

It being given that $y = x$ and $y = 1/x$ are two linearly independent solutions of the corresponding homogeneous equation.

See that x and $1/x$ are independent solutions, since

$$W(x, 1/x) = \begin{vmatrix} x & 1/x \\ 1 & -1/x^2 \end{vmatrix} = -\frac{2}{x} \neq 0 \; \forall \, x \in \mathbb{R} - \{0\}.$$

We have already CF is $c_1 x + c_2/x$.

PI: Replace c_1 by v_1, c_2 by v_2, then $y_p = v_1(x) \cdot x + v_2(x) \cdot (1/x)$.

Now v_1', v_2' are found by solving $v_1' \cdot x + v_2' \cdot (1/x) = 0$ and $v_1' \cdot 1 + v_2' \cdot (-1/x^2) = e^{2x}$.

Solving, $v_1' = \frac{1}{2}e^{2x}$, $v_2' = -\frac{1}{2}x^2 e^{2x}$; whereby

$$v_1 = \frac{e^{2x}}{4},\; v_2 = -\frac{1}{2}\left\{\frac{x^2 e^{2x}}{2} - \frac{xe^{2x}}{2} + \frac{e^{2x}}{4}\right\}.$$

Then

$$y_p = \frac{xe^{2x}}{4} - \frac{1}{2x}\left(\frac{x^2 e^{2x}}{2} - \frac{xe^{2x}}{2} + \frac{e^{2x}}{4}\right) = \frac{e^{2x}}{4} - \frac{e^{2x}}{8x}.$$

Complete solution is

$$y = c_1 x + \frac{c_2}{x} + \frac{e^{2x}}{4} - \frac{e^{2x}}{8x}.$$

Example 7.1.5. *Solve $x\frac{d^2y}{dx^2} - (x+2)\frac{dy}{dx} + 2y = x^3 e^x$ after the determination of a solution of its reduced equation.*

Solution: The equation can be written in the form:

$$\frac{d^2y}{dx^2} - \frac{x+2}{x}\frac{dy}{dx} + \frac{2}{x}y = x^2 e^x,$$

we see that

$$1 + P + Q = 1 - \frac{x+2}{x} + \frac{2}{x} = 0,$$

whereby $y = e^x$ is a solution of the reduced equations. Assume $y = e^x v$ as in Ex. 7.1.4 and hence

$$x\frac{d^2v}{dx^2} + (x-2)\frac{dv}{dx} = x^3, \quad \text{or,} \quad \frac{dp}{dx} + \left(1 - \frac{2}{x}\right)p = x^2,$$

where $p = dv/dx$ and the equation reducing to first order linear.

Multiplying by the IF, $e^{\int 1 - \frac{2}{x}\,dx} = \frac{e^x}{x^2}$ and integrating,

$$p \cdot \frac{e^x}{x^2} = \int x^2 \frac{e^x}{x^2}\,dx = e^x + c_1, \quad \text{or,} \quad dv = \left(x^2 + c_1 x^2 e^{-x}\right) dx.$$

Integrating, $v = \frac{1}{3}x^3 - c_1\left(x^2 + 2x + 2\right)e^{-x} + c_2$.

$\therefore$ general solution is $y = \frac{1}{3}x^3 e^x - c_1\left(x^2 + 2x + 2\right) + c_2 e^x$.

Example 7.1.6. *Solve* $x\frac{d^2y}{dx^2} - (2x-1)\frac{dy}{dx} + (x-1)y = 0.$

Solution: As in Ex. 7.1.5, $1 + P + Q = 0$ and hence e^x is its particular solution.

Let $y = e^x v$. Then the equation changes to $x\frac{d^2v}{dx^2} + \frac{dv}{dx} = 0$.

Then $\frac{dp}{dx} + \frac{1}{x}p = 0$, or, $\frac{dp}{p} + \frac{dx}{x} = 0$, if $p = \frac{dv}{dx}$.

Integrating,

$$px = c_1, \text{ or, } \frac{dv}{dx} = \frac{c_1}{x}, \text{ or, } v = c_1 \ln x + c_2.$$

$\therefore$ general solution is $y = e^x(c_1 \ln x + c_2)$.

Example 7.1.7. $x^2\frac{d^2y}{dx^2} - 2x(1+x)\frac{dy}{dx} + 2(1+x)y = x^3$.

Solution: As in Ex. 7.1.5, $P + xQ = 0$. Hence, $y = x$ is a particular solution. Putting $y = xv$ in the original equation,

$$\frac{d^2v}{dx^2} - 2\frac{dv}{dx} = 1, \quad \text{or,} \quad \frac{dp}{dx} - 2p = 1, \text{ if } p = \frac{dv}{dx}.$$

Multiplying by IF $= e^{\int -2dx} = e^{-2x}$ and integrating,

$$pe^{-2x} = \int e^{-2x}\,dx = -\frac{1}{2}e^{-2x} + A, \quad \text{or,} \quad p = \frac{dv}{dx} = Ae^{2x} - \frac{1}{2}.$$

Integrating,

$$v = \frac{1}{2}Ae^{2x} - \frac{1}{2}x + \frac{1}{2}B; \quad \text{or,} \quad 2y = x\left(Ae^{2x} + B - x\right),$$

which is the required general solution.

Example 7.1.8. $\frac{d^2y}{dx^2} - x^2\frac{dy}{dx} + xy = x^{m+1}$.

Solution: A particular solution of the given equation is evidently $y = x$ [here $P + xQ = 0$].

Writing $y = xv$ in the original equation, we get

$$x\frac{d^2v}{dx^2} + 2\frac{dv}{dx} - x^2\left(x\frac{dv}{dx} + v\right) + x^2v = x^{m+1} \text{ or, } \frac{d^2v}{dx^2} + \left(\frac{2}{x} - x^2\right)\frac{dv}{dx} = x^m.$$

$$\therefore \frac{d}{dx}\left\{px^2 e^{-\frac{1}{3}x^3}\right\} = x^{m+2}e^{-\frac{1}{3}x^3}, \text{ where } \frac{dv}{dx} = p \quad \left[\because \text{IF} = e^{\int\left(\frac{2}{x}-x^2\right)dx} = x^2 e^{-\frac{1}{3}x^3}\right].$$

$$\therefore p = \frac{dv}{dx} = \frac{A}{x^2 e^{-\frac{1}{3}x^3}} + \frac{1}{x^2}e^{\frac{1}{3}x^3}\int x^{m+2}e^{-\frac{1}{3}x^3}\,dx.$$

Hence,

$$v = B + \int \frac{A}{x^2}e^{\frac{1}{3}x^3}\,dx + \int\left\{\frac{1}{x^2}e^{\frac{1}{3}x^3}\int x^{m+2}e^{-\frac{1}{3}x^3}\,dx\right\}dx.$$

The complete primitive is

$$y = xv = Bx + Ax\int \frac{1}{x^2}e^{\frac{1}{3}x^3}\,dx + x\int\left\{\frac{1}{x^2}e^{\frac{1}{3}x^3}\int x^{m+2}e^{-\frac{1}{3}x^3}\,dx\right\}dx.$$

If $m = 0$ or any positive multiple of 3, the result can be further simplified.

Relation between the two known integrals

$$\frac{d^2y}{dx^2} + P\frac{dy}{dx} + Qy = 0. \tag{7.1.4}$$

Suppose, u and v are two known integrals of the aforesaid equation, then

$$\frac{d^2u}{dx^2} + P\frac{du}{dx} + Qu = 0, \quad \frac{d^2v}{dx^2} + P\frac{dv}{dx} + Qv = 0.$$

Multiplying the first by v and the second by u and then subtracting, we get

$$\left(u\frac{d^2v}{dx^2} - v\frac{d^2u}{dx^2}\right) + P\left\{u\frac{dv}{dx} - v\frac{du}{dx}\right\} = 0. \tag{7.1.5}$$

Observe that if

$$p = u\frac{dv}{dx} - v\frac{du}{dx},$$

then

$$\frac{dp}{dx} = u\frac{d^2v}{dx^2} - v\frac{d^2u}{dx^2},$$

so that from (7.1.5), we obtain

$$\frac{dp}{dx} + Pp = 0, \text{ or, } \frac{dp}{p} = -P\,dx \text{ or, } p = A\,e^{-\int P\,dx}.$$

Thus, the preceding analysis shows that any two particular solutions u and v of the eq. (7.1.4) are connected by the relation

$$u\frac{dv}{dx} - v\frac{du}{dx} = \alpha\,e^{-\int P\,dx}, \text{ where } \alpha(\neq 0) \text{ is a constant.}$$

Examples VII(A)

1. Verify that x is a solution of the reduced equation of $x^2\frac{d^2y}{dx^2} - x\frac{dy}{dx} + y = x^2$. Solve the equation after reducing it to a first order linear equation.
2. Verify that e^x is a solution of the reduced equation of $(1-x)y'' + xy' - y = (1-x)^2$. Solve the equation after reducing it to a first order linear equation.
3. Solve $x\frac{d^2y}{dx^2} - (x+2)\frac{dy}{dx} + 2y = x^3 e^x$ after the determination of a solution of its reduced equation.
4. (a) If y_1, y_2 are two linearly independent solutions of $(D^2+PD+Q)y = 0$, show that $y_1\frac{dy_2}{dx} - y_2\frac{dy_1}{dx} = ce^{-\int P dx}$, where $P(x)$ and $Q(x)$ are continuous and $c \neq 0$ is a constant.

   (b) Solve $\frac{d^2y}{dx^2} - 6\frac{dy}{dx} + 9y = x^3 e^{3x}$, assuming that e^{3x} is a solution of its reduced equation.
5. Solve by the method of variation of parameters: $x^2\frac{d^2y}{dx^2} + x\frac{dy}{dx} - y = x^2 e^{2x}$.
6. (a) Knowing that $y = x$ is a solution of the equation $x^2\frac{d^2y}{dx^2} - x(x+2)\frac{dy}{dx} + (x+2)y = 0$ reduce the equation $x^2\frac{d^2y}{dx^2} - x(x+2)\frac{dy}{dx} + (x+2)y = x^3$ to a first order first degree differential equation and find its complete primitive.

   (b) Solve $(1-x^2)\frac{d^2y}{dx^2} + x\frac{dy}{dx} - y = x(1-x^2)$, given that $y = x$ is a solution of its reduced equation.

(c) Knowing that $x^2\frac{d^2y}{dx^2} - 2x(x+2)\frac{dy}{dx} + 6(x+1)y = 0$ has a solution of the form x^n, reduce $x^2\frac{d^2y}{dx^2} - 2x(x+2)\frac{dy}{dx} + 6(x+1)y = x^4e^x$ to a first order differential equation.

Solve the following differential equations:

7. (a) $x\frac{d^2y}{dx^2} + (1-x)\frac{dy}{dx} - y = e^x$.

   $[1 + P + Q = 0;\ y = e^x$ is a particular solution of $xy_2 + (1-x)y_1 - y = 0]$

   (b) $(x+2)\frac{d^2y}{dx^2} - (2x+5)\frac{dy}{dx} + 2y = (x+1)\,e^x$.

   $[m^2 + mP + Q = 0$, if $m = 2;\ y = e^{2x}$ is a particular solution$]$

8. (a) $\frac{d^2y}{dx^2} - x^2\frac{dy}{dx} + xy = x$ (b) $\left(1-x^2\right)\frac{d^2y}{dx^2} + x\frac{dy}{dx} - y = x\left(1-x^2\right)^{3/2}$.

9. $x^2\frac{d^2y}{dx^2} + x\frac{dy}{dx} - y = 0$; given that $x + 1/x$ is some solution of the equation.

10. $x\frac{d^2y}{dx^2}(x\cos x - 2\sin x) + \left(x^2+2\right)\frac{dy}{dx}\sin x - 2y(x\sin x + \cos x) = 0$, given that $y = x^2$ is a solution.

11. $x^2\frac{d^2y}{dx^2} + x\frac{dy}{dx} - 9y = 0$; given that $y = x^3$ is a solution.

12. $x\frac{d^2y}{dx^2} - 2(x+1)\frac{dy}{dx} + (x+2)y = (x-2)\,e^{2x}$.

13. $(x-3)\frac{d^2y}{dx^2} - (4x-9)\frac{dy}{dx} + 3(x-2)y = 0$; e^x being a solution.

14. $(x\sin x + \cos x)\frac{d^2y}{dx^2} - x\cos x\frac{dy}{dx} + y\cos x = 0$; x being a solution.

15. $\left(1-x^2\right)\frac{d^2y}{dx^2} - x\frac{dy}{dx} - a^2y = 0$; $y = c\,e^{a\sin^{-1}x}$ is an integral.

16. $\frac{d^2y}{dx^2} - 2\tan x\frac{dy}{dx} + 3y = 2\sec x$; $\sec x$ being a solution.

17. $x^2\frac{d^3y}{dx^3} - x^2\frac{d^2y}{dx^2} - 2x\frac{dy}{dx} + 2y = 0$; $y = x$ is an integral.

Answers

1. $y = x^2 + c_1x\ln x + c_2x$.
2. $y = \left(x^2 + x + 1\right) + c_1x + c_2\,e^x$.
3. $y = \frac{1}{3}x^3\,e^x - c_1\left(x^2 + 2x + 2\right) + c_2\,e^x$.
4. (a) Theorem 7.2.1; (b) $y = \left(c_1x + c_2 + \frac{1}{20}x^5\right)e^{3x}$.
5. $y = c_1x + c_2/x + e^{2x}/4 - e^{2x}/8x$.
6. (a) $y = c_1xe^x - x^2 + c_2x$.

   (b) $y = -\frac{1}{2}\sqrt{1-x^2}\sin^{-1}x - \frac{1}{4}x(\sin^{-1}x)^2 + \frac{x^3}{4} - c_1\left\{\sqrt{1-x^2} + x\sin^{-1}x\right\} + c_2$.

   (c) $n = 3$, $x\frac{dp}{dx} + 2(1-x)p = e^x$, where $p = \frac{dv}{dx}$.
7. (a) $y = e^x\ln x + A\,e^x\int x^{-1}e^{-x}\,dx + B\,e^x$.

   (b) $y = -e^x - \frac{1}{4}A(2x+5) + B\,e^{2x}$.
8. (a) $y = Ax + Bx\int x^{-2}\,e^{\frac{1}{3}x^3}\,dx + 1$.

   (b) $y = Ax + B\left(x\sin^{-1}x + \sqrt{1-x^2}\right) - \frac{1}{9}x\left(1-x^2\right)^{-3/2}$.
9. $y = A/x + B(x + 1/x)$.
10. $y = Ax^2 + B\sin x$.
11. $y = Ax^3 + Bx^{-3}$.
12. $y = e^{2x} + \left(Ax^3 + B\right)e^x$.
13. $y = A\,e^x + B\,e^{3x}\left(4x^3 - 42x^2 + 150x - 183\right)$.

14. $y = Ax + B\cos x$.

15. $y = Ae^{a\sin^{-1}x} + Be^{-a\sin^{-1}x}$.

16. $y = \left(c_1\cos 2x + c_2\sin 2x + \frac{1}{2}\right)\sec x$.

17. $y = Ax^2 + Bx + C\left(x^2\int x^{-3}e^{-x}\,dx - x\int x^{-2}e^{-x}\,dx\right)$.

7.2 Change of dependent variable (continued)

In the previous article we have observed that under the transformation $y = uv$, $u = u(x)$, $v = v(x)$, the equation

$$\frac{d^2y}{dx^2} + P\frac{dy}{dx} + Qy = R \tag{7.2.1}$$

becomes

$$\frac{d^2v}{dx^2} + \left(\frac{2}{u}\frac{du}{dx} + P\right)\frac{dv}{dx} + \frac{1}{u}\left(\frac{d^2u}{dx^2} + P\frac{du}{dx} + Qu\right)v = \frac{R}{u}. \tag{7.2.2}$$

In art. 7.1 we assumed that $y = u$ is a known particular solution of (7.2.1) with $R = 0$. But suppose such a solution is not known. Then we choose the function u such that the coefficient of $\frac{dv}{dx}$ in (7.2.2), i.e. $\frac{2}{u}\frac{du}{dx} + P = 0$ so that $u = e^{-\frac{1}{2}\int P\,dx}$ [There is no necessity of adding a constant of integration as it will afterwards disappear].

With this value of u, the coefficient of v in (7.2.2) becomes

$$\frac{1}{u}\left\{-\frac{1}{2}u\frac{dP}{dx} - \frac{1}{2}P\frac{du}{dx} + P\frac{du}{dx} + Qu\right\} = \frac{1}{u}\left\{-\frac{1}{2}u\frac{dP}{dx} + \frac{1}{2}P\left(-\frac{Pu}{2}\right) + Qu\right\} = Q - \frac{1}{4}P^2 - \frac{1}{2}\frac{dP}{dx} = I \text{ (say)}.$$

Then the transformed eq. (7.2.2) becomes

$$\frac{d^2v}{dx^2} + Iv = Re^{\frac{1}{2}\int P\,dx}. \tag{7.2.3}$$

In case, $I = A$ (a constant) or A/x^2, then the eq. (7.2.3) can be solved by the methods of Chapter 5. Solving for v we can get the complete primitive of (7.2.1) $y = ve^{-\frac{1}{2}\int P\,dx}$.

Example 7.2.1. $\frac{d^2y}{dx^2} - \frac{2}{x}\frac{dy}{dx} + \left(1 + \frac{2}{x^2}\right)y = xe^x$.

Solution: $I = Q - \frac{1}{4}P^2 - \frac{1}{2}\frac{dP}{dx}$; here $P = -\frac{2}{x}$, $Q = 1 + \frac{2}{x^2}$.

$$\therefore\ I = 1 + \frac{2}{x^2} - \frac{1}{4}\left(\frac{4}{x^2}\right) - \frac{1}{2}\left(\frac{2}{x^2}\right) = 1.$$

$\therefore$ under the transformation $y = ve^{-\frac{1}{2}\int P\,dx} = vx$, the given equation reduces to

$$\frac{d^2v}{dx^2} + Iv = xe^x e^{\frac{1}{2}\int P\,dx} = xe^x\cdot x^{-1} = e^x \text{ or, } \frac{d^2v}{dx^2} + v = e^x\ (\because I = 1).$$

This is a linear equation in v with constant coefficients.

It is easy to obtain $v = A\cos x + B\sin x + \frac{1}{D^2+1}e^x = A\cos x + B\sin x + \frac{1}{2}e^x$.

$\therefore$ the complete primitive of the given equation is $y = ve^{-\frac{1}{2}\int P\,dx} = vx = Ax\cos x + Bx\sin x + \frac{1}{2}xe^x$ (Fig. 7.2).

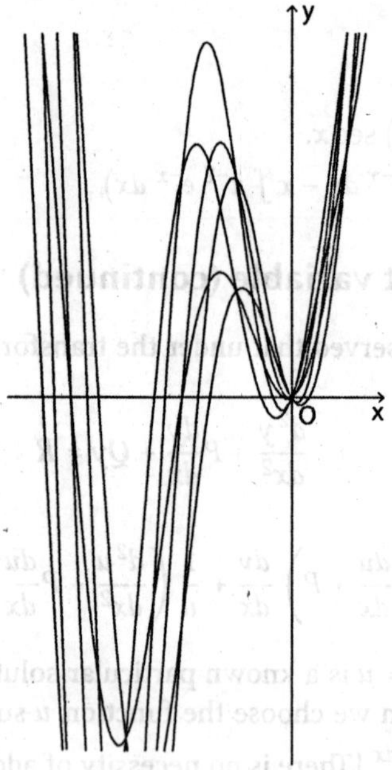

Fig. 7.2: $y = Ax\cos x + Bx\sin x + \frac{1}{2}xe^x, A = \{-2, 0, 2\}, B = \{2, 4\}$.

Example 7.2.2. $\frac{d^2y}{dx^2} - \frac{1}{x^{1/2}}\frac{dy}{dx} + \frac{y}{4x^2}\left(-8 + x^{1/2} + x\right) = 0.$

Solution: Here $P = -\frac{1}{x^{1/2}}$ and $Q = -\frac{2}{x^2} + \frac{1}{4x^{3/2}} + \frac{1}{4x}$, and therefore, $u = e^{-\frac{1}{2}\int P\,dx} = e^{x^{1/2}}$.

Also

$$I = \left(-\frac{2}{x^2} + \frac{1}{4x^{3/2}} + \frac{1}{4x}\right) - \frac{1}{4x^{3/2}} - \frac{1}{4x} = -\frac{2}{x^2}.$$

$\therefore$ the reduced equation becomes $\frac{d^2v}{dx^2} - \frac{2}{x^2}v = 0$.

The solution of this homogeneous equation is $v = Ax^2 + B/x$, and therefore, the complete primitive of the given equation is $y = \left(Ax^2 + Bx^{-1}\right)e^{x^{1/2}}$.

Examples VII(B)

Solve:

1. $\frac{d^2y}{dx^2} + x^{-\frac{1}{3}}\frac{dy}{dx} + \left(\frac{1}{4}x^{-2/3} - \frac{1}{6}x^{-4/3} - 6x^{-2}\right)y = 0.$
2. $\frac{d^2y}{dx^2} - 4x\frac{dy}{dx} + \left(4x^2 - 1\right)y = -3e^{x^2}\sin 2x.$
3. $\frac{d^2y}{dx^2} - 4x\frac{dy}{dx} + 4x^2y = xe^{x^2}.$
4. $\frac{d^2y}{dx^2} - \frac{3}{x}\frac{dy}{dx} + \frac{3}{x^2}y = 2x - 1.$
5. $\frac{d^2y}{dx^2} - 2x\frac{dy}{dx} + \left(x^2 + 2\right)y = e^{\frac{1}{2}(x^2+2x)}.$
6. $\frac{d^2y}{dx^2} - 2\tan x\frac{dy}{dx} + 5y = 0.$
7. $x^2\frac{d^2y}{dx^2} - 2\left(x^2 + x\right)\frac{dy}{dx} + \left(x^2 + 2x + 2\right)y = 0.$
8. $\frac{d^2y}{dx^2} + \frac{2}{x}\frac{dy}{dx} - k^2y = 0.$
9. $\frac{d^2y}{dx^2} + \frac{2}{x}\frac{dy}{dx} + k^2y = 0.$
10. $\frac{d^2y}{dx^2} - \frac{2}{x}\cdot\frac{dy}{dx} + \left(a^2 + \frac{2}{x^2}\right)y = 0.$
11. (a) $\frac{d^2y}{dx^2} - 2bx\frac{dy}{dx} + b^2x^2y = 0.$

    (b) $\frac{d^2y}{dx^2} + 4x\frac{dy}{dx} + 4x^2y = 0.$

Answers

1. $y = e^{-\frac{3}{4}x^{2/3}}\left(Ax^3 + Bx^{-2}\right)$.
2. $y = (A\cos x + B\sin x + \sin 2x)\,e^{x^2}$.
3. $y = e^{x^2}\left(A\cos\sqrt{2}x + B\sin\sqrt{2}x\right) + \frac{1}{2}x\,e^{x^2}$.
4. $y = Ax + Bx^3 + x^3\ln x + x^2$.
5. $y = e^{\frac{1}{2}x^2}(A\cos\sqrt{3}x + B\sin\sqrt{3}x) + \frac{1}{4}\,e^{\frac{1}{2}(x^2+2x)}$.
6. $y = \sec x(A\sin\sqrt{6}x + B\cos\sqrt{6}x)$.
7. $y = e^x\left(Ax^2 + Bx\right)$.
8. $y = x^{-1}\left(A\,e^{kx} + B\,e^{-kx}\right)$.
9. $xy = A\sin(kx + B)$.
10. $y = Ax\sin(Kx + B)$.
11. (a) $y = e^{\frac{1}{2}x^2 b}\{A\cos\sqrt{b}x + B\sin\sqrt{b}x\}$.
    (b) $y = e^{-x^2}\left(C_1\,e^{x\sqrt{2}} + C_2\,e^{-x\sqrt{2}}\right)$.

Normal form of a linear equation of second order

We have observed that the substitution

$$y = v\,e^{-\frac{1}{2}\int P\,dx} \tag{7.2.4}$$

transforms

$$\frac{d^2y}{dx^2} + P\frac{dy}{dx} + Qy = R, \tag{7.2.5}$$

into the form (called the *normal form*):

$$\frac{d^2v}{dx^2} + Iv = S, \tag{7.2.6}$$

where

$$I = Q - \frac{1}{2}\frac{dP}{dx} - \frac{1}{4}P^2 \quad \text{and} \quad S = R\,e^{\frac{1}{2}\int P\,dx}.$$

The advantage of using the normal form (7.2.6) instead of (7.2.5) as a typical linear differential equation of second order is that, for all substitutions such as $zf(x)$ for y in the eq. (7.2.5), I is a function of P and Q of such a form that, when the new equation

$$\frac{d^2z}{dx^2} + P_1\frac{dz}{dx} + Q_1 z = R_1 \tag{7.2.7}$$

has its second term removed by the substitution

$$z = w\,e^{-\frac{1}{2}\int P_1\,dx}, \tag{7.2.8}$$

it takes the form:

$$\frac{d^2w}{dx^2} + Iw = S_1. \tag{7.2.9}$$

Thus, I is exactly the same function of P_1 and Q_1 as it is of P and Q; and therefore, we call I an *invariant* of the coefficients of the differential eq. (7.2.5).

Any two linear equations, such as the equations in y and z—equations (7.2.5) and (7.2.7)—can be transformed into one another if the normal form of each be the same.

If the two equations are so transformable and the equation of substitution between the dependent variable be desired, this can be easily obtained by using the normal form as an intermediate transformed equation.

Thus, in (7.2.5), the equation in y becomes transformed to that in v by writing $y\,e^{\frac{1}{2}\int P\,dx} = v$, and the equation in v passes into that in z by writing $v = z\,e^{\frac{1}{2}\int P_1 dx}$ and hence the relation which transforms directly the y-equation into the z-equation is $y\,e^{\frac{1}{2}\int P\,dx} = z\,e^{\frac{1}{2}\int P_1 dx}$, i.e. the condition of equivalence is that the *invariant* I should be the same.

Example 7.2.3. *Show that the equations*

$$x^2\frac{d^2y}{dx^2}+2\left(x^3-x\right)\frac{dy}{dx}+\left(1-2x^2\right)y=0$$

$$\text{and}\quad x^2\frac{d^2z}{dx^2}+2\left(x^3+x\right)\frac{dz}{dx}-\left(1-2x^2\right)z=0$$

have the same invariant and find the relation that transforms one into the other. Verify by actually carrying out this transformation.

Solution: For the equation in y,

$$I=Q-\frac{1}{2}\frac{dP}{dx}-\frac{1}{4}P^2=\frac{1-2x^2}{x^2}-\frac{1}{2}\frac{d}{dx}\left\{2\left(x-\frac{1}{x}\right)\right\}-\frac{1}{4}\left\{2\left(x-\frac{1}{x}\right)\right\}^2=-\left(x^2+\frac{1}{x^2}+1\right).$$

For the equation in z,

$$I_1=Q_1-\frac{1}{2}\frac{dP_1}{dx}-\frac{1}{4}P_1^2=-\frac{\left(1-2x^2\right)}{x^2}-\frac{1}{2}\frac{d}{dx}\left\{2\left(x+\frac{1}{x}\right)\right\}-\frac{1}{4}\left\{2\left(x+\frac{1}{x}\right)\right\}^2=-\left(x^2+\frac{1}{x^2}+1\right).$$

Thus, they have the same invariant.

The equation in y transforms into the normal form $\frac{d^2v}{dx^2}+Iv=0$, where

$$v=y\,e^{\frac{1}{2}\int P\,dx}=y\,e^{\frac{1}{2}\int 2\left(x-\frac{1}{x}\right)dx}=yx^{-1}\,e^{\frac{1}{2}x^2}.$$

The equation in z transforms into same normal form ($\because I=I_1$) where

$$v=z\,e^{\frac{1}{2}\int P_1\,dx}=z\,e^{\frac{1}{2}\int 2\left(x+\frac{1}{x}\right)dx}=zx\,e^{\frac{1}{2}x^2}$$

$\therefore$ the relation is $yx^{-1}\,e^{\frac{1}{2}x^2}=zx\,e^{\frac{1}{2}x^2}$, or, $y=x^2z$.

Verification. Under the transformation $y=x^2z$, we have

$$\frac{dy}{dx}=2xz+x^2\frac{dz}{dx}\quad\text{and}\quad\frac{d^2y}{dx^2}=x^2\frac{d^2z}{dx^2}+4x\frac{dz}{dx}+2z.$$

$\therefore$ the equation in y becomes

$$x^2\left[x^2\frac{d^2z}{dx^2}+4x\frac{dz}{dx}+2z\right]+2\left(x^3-x\right)\left(2xz+x^2\frac{dz}{dx}\right)+\left(1-2x^2\right)x^2z=0.$$

On simplification, this gives

$$x^2\frac{d^2z}{dx^2}+2\left(x^3+x\right)\frac{dz}{dx}-\left(1-2x^2\right)z=0,$$

the given equation in z. Hence, the result is verified.

Example 7.2.4. *Prove that the equations*

$$\left(1-x^2\right)\frac{d^2y}{dx^2}+(1-3x)\frac{dy}{dx}+ky=0$$

$$\text{and}\quad\left(1-x^2\right)\frac{d^2z}{dx^2}-(1+x)\frac{dz}{dx}+(k+1)z=0$$

can be transformed into one another; and find the relation between y, z and x.

Solution: The invariant I in both the equations will be

$$\frac{k}{1-x^2} - \frac{(3x-5)(1+x)}{4\left(1-x^2\right)^2};$$

and the relation will be $z = y(1+x)$.

Example 7.2.5. *Find the value of Q which is such that the equation*

$$\frac{d^2y}{dx^2} + a\frac{dy}{dx} + Qy = 0$$

may be transformed, by the substitution $y = zf(x)$ into

$$\frac{d^2z}{dx^2} + \frac{1}{x}\frac{dz}{dx} + \left(1 - \frac{n^2}{x^2}\right)z = 0.$$

Solution: Obtain the value of $f(x)$.

The value of Q is $1 + \frac{1}{4}a^2 - n^2 - \frac{1}{4}$.

7.3 Change of independent variable

A linear differential equation of second order

$$\frac{d^2y}{dx^2} + P\frac{dy}{dx} + Qy = R, \tag{7.3.1}$$

(P, Q, R being functions of x) can sometimes be transformed into another equation which is easily integrable by changing the independent variable x. Let the new independent variable be z. Suppose the transformation connecting x and z be

$$z = f(x) \tag{7.3.2}$$

Then

$$\frac{dy}{dx} = \frac{dy}{dz}\frac{dz}{dx}, \frac{d^2y}{dx^2} = \frac{d^2y}{dz^2}\left(\frac{dz}{dx}\right)^2 + \frac{dy}{dz}\frac{d^2z}{dx^2}.$$

With these changes the eq. (7.3.1) becomes

$$\frac{d^2y}{dz^2}\left(\frac{dz}{dx}\right)^2 + \left(P\frac{dz}{dx} + \frac{d^2z}{dx^2}\right)\frac{dy}{dx} + Qy = R \text{ or, } \frac{d^2y}{dz^2} + P_1\frac{dy}{dz} + Q_1y = R_1, \tag{7.3.3}$$

where

$$P_1 = \frac{\frac{d^2z}{dx^2} + P\frac{dz}{dx}}{\left(\frac{dz}{dx}\right)^2}, \quad Q_1 = \frac{Q}{\left(\frac{dz}{dx}\right)^2}, \quad R_1 = \frac{R}{\left(\frac{dz}{dx}\right)^2}.$$

It is easy to see that P_1, Q_1, R_1 can be expressed in terms of z by using the relation (7.3.2).

As yet, z is quite arbitrary; it may be chosen to satisfy any assignable condition.

Case 7.3.1. *We may choose z so that in the transformed* eq. (7.3.3) *the coefficient of $\frac{dy}{dz}$ vanishes; then $\frac{d^2z}{dx^2} + P\frac{dz}{dx} = 0$ and hence,*

$$z = \int \left(e^{-\int P\,dx}\right) dx. \tag{7.3.4}$$

The eliminant of this relation between z and x and the transformed equation may furnish a differential equation which proves integrable.

Under this transformation if it so happens that

$$Q_1 = \text{a constant, or, } Q_1 = \frac{\text{a constant}}{z^2},$$

then the transformed equation will become an equation with constant coefficients or a homogeneous equation; that is, in either case, the equation will reduce to what may be called a *known form.*

Example 7.3.1. *Solve:* $(1-x^2)\frac{d^2y}{dx^2} - x\frac{dy}{dx} - c^2y = 0.$

Solution: The equation can be written as

$$\frac{d^2y}{dx^2} + P\frac{dy}{dx} + Qy = 0,$$

where $P = -x/\left(1-x^2\right)$ and $Q = -c^2/\left(1-x^2\right)$.

If we choose z so that

$$z = \int\left(e^{-\int P\,dx}\right)dx = \int\frac{1}{\sqrt{1-x^2}}\,dx = \sin^{-1}x,$$

then the transformed equation becomes $\frac{d^2y}{dz^2} + Q_1y = 0$, where $Q_1 = \frac{Q}{(dz/dx)^2} = -c^2$.

The equation is easily integrable and we have $y = c_1e^{cz} + c_2e^{-cz}$, where $z = \sin^{-1}x$.

Example 7.3.2. *Solve:* $x\frac{d^2y}{dx^2} - 3\frac{dy}{dx} + \frac{3y}{x} = 0.$

Solution:

$$P = -3/x,\ z = \int e^{-\int P\,dx}\,dx = \int x^3\,dx = \frac{1}{4}x^4$$

$$\text{and}\quad \frac{dz}{dx} = x^3,\ Q_1 = \frac{Q}{(dz/dx)^2} = \frac{3/x^2}{x^6} = \frac{3}{16z^2}.$$

∴ the transformed equation is

$$\frac{d^2y}{dz^2} + \frac{3}{16z^2}y = 0,\ \text{or,}\ z^2\frac{d^2y}{dz^2} + \frac{3}{16}y = 0.$$

This is a homogeneous equation. Using the method of art. 5.8, we obtain

$$y = Az^{1/4} + Bz^{3/4} = c_1x + c_2x^3.$$

Case 7.3.2. *We may choose z so that*

$$\left(\frac{dz}{dx}\right)^2 = \frac{\pm Q}{a^2}\ (\text{or, } Q_1 = a^2),$$

where a^2 is a constant (*we may take $a^2 = 1$*).

The sign of the right-hand side should be so taken that makes $\left(\frac{dz}{dx}\right)^2$ positive.

If now

$$P_1 = \frac{\frac{d^2z}{dx^2} + P\frac{dz}{dx}}{\left(\frac{dz}{dx}\right)^2} = \text{a constant} = K\ (\text{say}),$$

then the transformed equation becomes

$$\frac{d^2y}{dz^2} + k\frac{dy}{dz} + a^2y = R_1,$$

which is a linear equation with constant coefficients and can be integrated by the methods of Ch. 5.

Note that if, in this case, P_1 becomes zero, then the discussions will be similar to Case 7.3.1.

Example 7.3.3. *Solve:* $\frac{d^2y}{dx^2} - (1+4e^x)\frac{dy}{dx} + 3e^{2x}y = e^{2(x+e^x)}$.

Solution: Choose z so that

$$\left(\frac{dz}{dx}\right)^2 = \frac{3e^{2x}}{3} = e^{2x} \text{ (we take } a^2 = 3)$$

$$\text{and } \frac{dz}{dx} = +\sqrt{e^{2x}} = e^x,$$

then $z = e^x$ and

$$P_1 = \frac{P\frac{dz}{dx} + \frac{d^2z}{dx^2}}{\left(\frac{dz}{dx}\right)^2} = \frac{-(1+4e^x)\,e^x + e^x}{e^{2x}} = -4 \text{ (a const.)}$$

The transformed equation becomes

$$\frac{d^2y}{dz^2} - 4\frac{dy}{dz} + 3y = e^{2z} \text{ (Note: } Q_1 = a^2 = 3)$$

and hence

$$y = c_1 e^z + c_2 e^{2z} + \frac{1}{D^2 - 4D + 3}e^{2z}, \left(D \equiv \frac{d}{dz}\right)$$

where $z = e^x$.

Note 7.3.1. *The choice of $a^2 = 3$ is one of convenience only. One may consistently take $a^2 = 1$.*

Example 7.3.4. *Solve:* $\frac{d^2y}{dx^2}\cos x + \frac{dy}{dx}\sin x - 2y\cos^3 x = 2\cos^5 x$.

Solution: The given equation can be written in the form:

$$\frac{d^2y}{dx^2} + P\frac{dy}{dx} + Qy = R,$$

where $P = \tan x$, $Q = -2\cos^2 x$, $R = 2\cos^4 x$.

Using Case 7.3.1, $z = \int e^{-\int P\,dx}\,dx = \int e^{\ln\cos x}\,dx = \sin x$.

Using Case 7.3.2, $\left(\frac{dz}{dx}\right)^2 = \frac{2\cos^2 x}{2}$ (taking $a^2 = 2$) so that

$$\frac{dz}{dx} = \cos x, \text{ or, } z = \sin x.$$

Observe that in either case, the relation between z and x is given by $z = \sin x$. The transformed equation is

$$\frac{d^2y}{dz^2} + \frac{(-2y\cos^2 x)}{\cos^2 x} = \frac{2\cos^4 x}{\cos^2 x} \text{ or, } \frac{d^2y}{dz^2} - 2y = 2\left(1 - z^2\right) \ (\because z = \sin x).$$

This gives

$$y = c_1 \cosh\left(\sqrt{2}z + c_2\right) + \frac{1}{D^2 - 2}(2 - 2z^2) = c_1 \cosh\left(\sqrt{2}\sin x + c_2\right) + \sin^2 x,$$

where c_1 and c_2 are arbitrary constants.

Examples VII(C)

1. Solve the following equations:
   (a) $x\frac{d^2y}{dx^2} - \frac{dy}{dx} - 4x^3y = 8x^3 \sin x^2$. $[z = x^2]$
   (b) $(1 + x^2)^2 \frac{d^2y}{dx^2} + 2x(1 + x^2)\frac{dy}{dx} + 4y = 0$. $[x = \tan z]$
2. Show that if t satisfies the equation $\frac{d^2t}{dx^2} + P\frac{dt}{dx} = 0$, then by changing the independent variable from x to t, the equation $\frac{d^2y}{dx^2} + P\frac{dy}{dx} + Qy = R$ (P, Q, R are all functions of x) transforms into $\frac{d^2y}{dt^2} + Sy = T$. You are to write down the values of S and T in terms of P, Q, R.
   Hence solve: $\frac{d^2y}{dx^2} + \left(1 - \frac{1}{x}\right)\frac{dy}{dx} + 4x^2 e^{-2x}y = 4(x^2 + x^3)e^{-3x}$.
3. Solve:
   (a) $\frac{d^2y}{dx^2} + \frac{2}{x}\frac{dy}{dx} + \frac{a^2}{x^4}y = 0$.
   (b) $\frac{d^2y}{dx^2} - \cot x\frac{dy}{dx} - (\sin^2 x)\,y = \cos x - \cos^3 x$.
   (c) $\frac{d^2y}{dx^2} + \frac{2}{x}\frac{dy}{dx} + \frac{1}{x^4}y = \frac{2x^2+1}{x^6}$.
   (d) $\frac{d^2y}{dx^2} + \left(4x - \frac{1}{x}\right)\frac{dy}{dx} + 4x^2y = 3x\,e^{-x^2}$.
   (e) $\frac{d^2y}{dx^2} + \cot x\frac{dy}{dx} + 4y\,\mathrm{cosec}^2\, x = 0$.
   (f) $x\frac{d^2y}{dx^2} - \frac{dy}{dx} + 4x^2y = x^5$.
   (g) $x^6\frac{d^2y}{dx^2} + 3x^5\frac{dy}{dx} + a^2y = x^{-2}$.
4. Solve:
   (a) $(a^2 - x^2)\frac{d^2y}{dx^2} - \frac{a^2}{x}\frac{dy}{dx} + \frac{x^2}{a}y = 0$. $\left[\frac{dz}{dx} = -\sqrt{\frac{x^2}{a(a^2-x^2)}}\right]$
   (b) $(x^3 - x)\frac{d^2y}{dx^2} + \frac{dy}{dx} + n^2x^3y = 0$. $[z = \sqrt{x^2 - 1}]$
   (c) $x^4\frac{d^2y}{dx^2} + 2x^3\frac{dy}{dx} + n^2y = 0$.
5. Solve:
   (a) $x^2y\frac{d^2y}{dx^2} + \left(x\frac{dy}{dx} - y\right)^2 = 0$.
   (b) $y\frac{d^2y}{dx^2} + \left(\frac{dy}{dx}\right)^2 + 1 = 0$.

Answers

1. (a) $y = c_1 e^{x^2} + c_2 e^{-x^2} - \sin x^2$.
   (b) $y(1 + x^2) = c_1(1 - x^2) + c_2x$.
2. $y = c_1 \cos\{2(1 + x)e^{-x}\} + c_2 \sin\{2(1 + x)e^{-x}\} + (1 + x)e^{-x}$.
3. (a) $y = c_1 \cos(a/x) + c_2 \sin(a/x)$.
   (b) $y = c_1 e^{-\cos x} + c_2 e^{\cos x} - \cos x$.
   (c) $y = c_1 \cos z + c_2 \sin z + z^2$, where $z = -1/x$.
   (d) $y = c_1 e^{-x^2} + c_2x^2 e^{-x^2} + x^3 e^{-x^2}$.
   (e) $y = c_1 \sin(2 \ln \tan x/2 + c_2)$.
   (f) $y = c_1 \sin(x^2 + c_2) + x^2/4$.
   (g) $y = c_1 \cos(a/2x^2) + c_2 \sin(a/2x^2) + 1/a^2x^2$.
4. (a) $y = c_1 \sin\left(\sqrt{\frac{a^2-x^2}{a}} + c_2\right)$.
   (b) $y = c_1 \sin\left(n\sqrt{x^2 - 1} + c_2\right)$.
   (c) $y = c_1 \cos\left(\frac{n+c_2x}{x}\right)$.
5. (a) $y^2 = Ax + Bx^2$. (b) $y^2 + (x - c_1)^2 = c_2^3$.
   [Hints: In both cases, put $y^2 = v$. (a) reduces to $\frac{d^2v}{dx^2} - \frac{2}{x}\frac{dv}{dx} + \frac{2}{x^2}v = 0$; now the transformation is $z = \ln x$. But (b) reduces to $\frac{d^2v}{dx^2} + 2 = 0$; successive integrations.]

7.4 Factorisation of the operator

Consider the equation

$$\left(PD^2 + QD + R\right) y = S, \quad \left(D \equiv \frac{d}{dx}\right) \tag{7.4.1}$$

where P, Q, R and S are functions of x.

It may be possible, in some particular cases, to factorise the operator $PD^2 + QD + R$ in the form:

$$(pD + q) \times (rD + s),$$

in which p, q, r, s are functions of x or constants.

Then the eq. (7.4.1) can be written as

$$\{(pD + q)(rD + s)\} y = S. \tag{7.4.2}$$

It is important to stress upon the fact that *the factors must be written in the right order*, as they are not always commutative (See Ex. 7.4.1 below).

In order to solve (7.4.2), we get

$$(rD + s)y = v \tag{7.4.3}$$

$$\text{so that } (pD + q)v = S. \tag{7.4.4}$$

The eq. (7.4.4) is a linear equation of first order and hence can be easily solved for v. Then from (7.4.3), another linear equation of first order, y can be obtained.

Example 7.4.1. $(x + 2)y_2 - (2x + 5)y_1 + 2y = (x + 1)\, e^x$.

Solution: The left side is

$$\left\{(x+2)\frac{d^2}{dx^2} - (2x+5)\frac{d}{dx} + 2\right\} y = \left\{(x+2)\frac{d}{dx} - 1\right\}\left(\frac{d}{dx} - 2\right) y.$$

[For

$$\left\{\frac{d}{dx} - 2\right\} y = \frac{dy}{dx} - 2y$$

and hence

$$\left\{(x+2)\frac{d}{dx} - 1\right\}\left\{\frac{dy}{dx} - 2y\right\} = (x+2)\frac{d^2y}{dx^2} - 2(x+2)\frac{dy}{dx} - \frac{dy}{dx} + 2y = (x+2)y_2 - (2x+5)y_1 + 2y.]$$

Set

$$\left\{\frac{d}{dx} - 2\right\} y = v.$$

Then the given equation becomes

$$\left\{(x+2)\frac{d}{dx} - 1\right\} v = (x+1)\, e^x,$$

a linear equation of first order, whose IF is $1/(x + 2)$.

Hence,

$$\frac{d}{dx}\left\{\frac{1}{x+2} v\right\} = \frac{x+1}{(x+2)^2} e^x.$$

$$\therefore\ v\cdot\frac{1}{x+2}=\int e^x\left[\frac{1}{x+2}-\frac{1}{(x+2)^2}\right]dx+c_1=\frac{1}{x+2}e^x+c_1.$$

$\therefore\ v=c_1(x+2)+e^x.$

Consequently,

$$\left(\frac{d}{dx}-2\right)y=v=c_1(x+2)+e^x,$$

another linear equation of first order. Solving, we get $y=A(2x+5)+Be^{2x}-e^x$ [writing $A=-\frac{1}{4}c_1$.]

Thus, we obtain the complete primitive of the given equation.

Note 7.4.1. *The factors must be written in the right order. See that*

$$\left\{\frac{d}{dx}-2\right\}\left\{(x+2)\frac{d}{dx}-1\right\}y=\left\{\frac{d}{dx}-2\right\}\left\{(x+2)\frac{dy}{dx}-y\right\}=(x+2)\frac{d^2y}{dx^2}-2(x+2)\frac{dy}{dx}+2y$$

$$=(x+2)y_2-(2x+4)y_1+2y\neq \textit{left member of the given equation.}$$

In other words, the factors are not commutative.

Observations: In order that PD^2+QD+R may be expressible as the product $(pD+q)\times(rD+s)$ we must have three equations

$$pr=P,\ rq+p\left(\frac{dr}{dx}+s\right)=Q \text{ and } qs+p\left(\frac{dS}{dx}\right)=R,$$

to determine four quantities p, q, r, s; but we may consider p and r as known factors of P and use the remaining two equations to determine q and s. But clearly the equations cannot be solved in general and hence the method of factorisation of operators can be used only in some special cases.

7.5 Methods of solving linear equations of second order

A typical linear equation of second order is

$$\frac{d^2y}{dx^2}+P\frac{dy}{dx}+Qy=R, \tag{7.5.1}$$

in which P, Q, R are either functions of x or constants.

1. If the coefficients P, Q are constants, use arts. 5.7, 5.7.1 and 5.7.2.
2. For homogeneous equations, use art. 5.8.
3. For exact equations, use art. 6.3.
4. If the equation does not contain y explicitly, use art. 6.5 (D).
5. If the equation does not contain x explicitly, use art. 6.5. (E).
6. For the special type $\frac{d^2y}{dx^2}=f(y)$ occurring in Dynamics, use art 6.5. [B(ii)].
7. Find, by inspection or otherwise, a particular integral $u=u(x)$ of the eq. (7.0.1) in which $R=0$. Then use art. 7.1.
8. If such a particular integral cannot be found, then compute

$$Q-\frac{1}{4}P^2-\frac{1}{2}\frac{dP}{dx}.$$

If this is a constant or a constant/x^2, the transformation $y=v\,e^{-\frac{1}{2}\int P\,dx}$ is to be used (See art. 7.2).

9. If the above procedure fails, choose $z = f(x)$ such that

$$\left(\frac{dz}{dx}\right)^2 = \frac{\pm Q}{a^2}$$

(choosing the sign + or – before Q so that the RHS is positive).
Now compute

$$\frac{\dfrac{d^2 z}{dx^2} + P\dfrac{dz}{dx}}{\left(\dfrac{dz}{dx}\right)^2}.$$

If this is a constant (or, in particular, zero), the transformation

$$z = \int \sqrt{\frac{\pm Q}{a^2}}\, dx$$

yields a linear equation with constant coefficients (art. 7.3, case 7.3.2).

10. There are special cases where the left member of the eq. (7.5.1) is operationally factorable; the problem is then reduced to that of solving two linear equations of order one (art. 7.4).

Examples VII(D)

Solve:

1. $\frac{d^2y}{dx^2} - \frac{3}{x}\frac{dy}{dx} + \frac{3}{x^2}y = 2x - 1.$
2. $x\frac{d^2y}{dx^2} + (1 - x)\frac{dy}{dx} - 2(1 + x)y = e^{-x}(1 - 6x).$
3. $[(x + 3)D^2 - (2x + 7)D + 2]y = (x + 3)^2 e^x.$
4. $(x + 1)y_2 + (x - 1)y_1 - 2y = 0.$
5. $xy_2 + (x^2 + 1)\, y_1 + 2xy = 2x$, given that $y = 2$ and $y_1 = 0$, when $x = 0$.
6. $(x^2 - 1)\,\frac{d^2y}{dx^2} - (4x^2 - 3x - 5)\,\frac{dy}{dx} + (4x^2 - 6x - 5)\,y = e^{2x}$, given that $y = 1$ and $y_1 = 2$, when $x = 0$.

Answers

1. $y = c_1x^3 + c_2x + x^2(1 + x\ln x).$
2. $y = x\,e^{-x} + c_1\, e^{2x} \int \frac{e}{x}^{-2x}\, dx + c_2\, e^{2x}.$
3. $y = -x\,e^x - 4e^x + c_1(2x + 7) + c_2\, e^{2x}.$
4. $y = c_1\,(x^2 + 1) + c_2\, e^{-x}.$
5. $y = 1 + e^{-x^{1/2}}.$
6. $y = e^{2x}.$

Chapter 8

Simple Eigenvalue Problems

In this chapter we shall find the solution of ordinary differential equations containing a parameter and subject to certain boundary conditions. We shall start with very simple problems and then discuss briefly Sturm-Liouville systems. The presentation is intended to be rather more a collection of useful facts than a detailed introduction to the subject.

8.1 Eigenfunctions and eigenvalues

Let us try to find a *nontrivial* solution (i.e. not identically zero) of the equation

$$\frac{d^2y}{dx^2} + \lambda y = 0, \tag{8.1.1}$$

that satisfies the *boundary conditions*

$$y(0) = 0 \text{ and } y(\pi) = 0. \tag{8.1.2}$$

The parameter λ in (8.1.1) is free to assume any real value whatsoever and our problem is to obtain the λ's for which the problem can be solved.

We have previously considered *initial value problems* in which the solution of a second order differential equation is sought that satisfies two conditions at a single value of the independent variable.

For instance, if we are to solve $\frac{d^2x}{dt^2} + n^2x = 0$ given that when $t = 0$, $x = a$, $\frac{dx}{dt} = u$, the solution becomes $x = a\cos nt + \frac{u}{n}\sin nt$.

Also we have solved problems like $\frac{d^2y}{dx^2} + y = 0$, given $y = 2$ for $x = 0$, and $y = -2$, when $x = \frac{\pi}{2}$, the solution of which is $y = 2(\cos x - \sin x)$.

Here we will have entirely different situation for we wish to satisfy one condition at each of the two distinct values of x, for example, at $x = 0$, $y(0) = 0$ and at $x = \pi$, $y(\pi)$ also equals to 0 and that the solution is nontrivial.

Problems of this kind are called *boundary value problems* and in general, they are more difficult and far-reaching—in both theory and practice—than the initial value problems.

Let us put this boundary value problem in the form of an example.

Example 8.1.1. *Find the eigenvalues λ_n and the eigenfunctions $y_n(x)$ for the differential equation*

$$\frac{d^2y}{dx^2} + \lambda y = 0, \tag{8.1.3}$$

satisfying the boundary conditions

$$y(0) = 0 \text{ and } y(\pi) = 0. \tag{8.1.4}$$

Solution: The auxiliary equation is $m^2 + \lambda = 0$, or, $m^2 = -\lambda$.

Case 8.1.1. *When $\lambda = 0$, $m = 0, 0$ and hence the solution is $y = e^{0x}(c_1 + c_2x) = c_1 + c_2x$.*

But $y(0) = 0$ and $y(\pi) = 0$, whereby $0 = c_1$, $0 = c_1 + c_2\pi$ and hence $c_1 = 0 = c_2$. And there results a *trivial solution* $y = 0$.

Case 8.1.2. *For $\lambda < 0$, the solution is (here $m = \pm\sqrt{-\lambda}$), $y = c_1e^{\sqrt{-\lambda}x} + c_2e^{-\sqrt{-\lambda}x}$, where $(-\lambda)$ and $\sqrt{-\lambda}$ are positive. Now apply $y(0) = 0$, $y(\pi) = 0$.*

Then $c_1 + c_2 = 0$ and $c_1e^{\sqrt{-\lambda}\pi} + c_2e^{-\sqrt{-\lambda}\pi} = 0$.

The only solutions are $c_1 = 0 = c_2$. Again, these result to a *trivial solution* $y = 0$.

Case 8.1.3. *When $\lambda > 0$, the solution is (here $m = \pm i\sqrt{\lambda}$), $y = c_1 \cos\sqrt{\lambda}x + c_2 \sin\sqrt{\lambda}x$.*

But $y(0) = 0$, that is, the solution is to vanish when $x = 0$, thus $0 = c_1 \cos 0 + c_2 \sin 0 = c_1$, or, $c_1 = 0$.

Therefore, the solution becomes $y = c_2 \sin\sqrt{\lambda}x$.

Further $y(\pi) = 0$, that is, the solution is to vanish when $x = \pi$, then $0 = c_2 \sin\sqrt{\lambda}\pi$.

This will be true, of course, if $c_2 = 0$. But that leads to y becoming identically zero, which is a *trivial solution*. Since we are interested in a *nontrivial solution*, the only alternative is that $\sin\sqrt{\lambda}\pi = 0$. Thus, $\sin\sqrt{\lambda}\pi = 0$ implies $\sqrt{\lambda}\pi = n\pi$ or, $\sqrt{\lambda} = n$, that is, $\lambda = n^2$ for $n = 1, 2, 3, \ldots$.

In other words, λ must be equal to one of the numbers $1^2, 2^2, 3^2, \ldots, n^2, \ldots$ and no others.

These values of λ to be denoted by λ_n are called the *eigenvalues* of the problem and the corresponding nontrivial solutions $\sin x, \sin 2x, \sin 3x, \ldots, \sin nx, \ldots$ and in general, $A_n \sin nx$ are called *eigenfunctions*. Here the notation A_n signifies that the arbitrary constant A_n can be different for different values of n.

Conclusion. Collecting the results of all three cases, we conclude that the eigenvalues are $\lambda_n = n^2$ and the corresponding eigenfunctions are $y_n = A_n \sin nx$ for $n = 1, 2, 3, \ldots, n, \ldots$; and the problem is completely solved.

Observation

I. A value of λ for which the problem 8.1.1 has a nontrivial solution is called an *eigenvalue* and the nontrivial solution is called an *eigenfunction*. The set of eigenvalues of the problem is called the *spectrum* of the problem.

II. The eigenvalues form an increasing sequence of positive numbers that approaches ∞.

III. The nth function, $\sin nx$, vanishes at the end points of the interval $[0, \pi]$ and has exactly $(n - 1)$ zeros in the interior of this interval.

Let us now work out a few more examples.

Example 8.1.2. *Find the eigenvalues and eigenfunctions for the differential equation $\frac{d^2y}{dx^2} + \lambda y = 0$, satisfying the boundary conditions $y(0) = 0$ and $y(1) = 0$.*

Solution: The auxiliary equation is $m^2 + \lambda = 0$.

Case 8.1.4. *When $\lambda = 0$, the solution is $y = e^{0x}(c_1 + c_2x) = c_1 + c_2x$, since $m = 0, 0$.*

But $y(0) = 0$, $y(1) = 0$, whereby $0 = c_1$, $0 = c_1 + c_2$. Thus, $c_1 = 0 = c_2$, which results in the *trivial solution* $y = 0$.

Case 8.1.5. *For $\lambda < 0$, the solution is $y = c_1e^{\sqrt{-\lambda}x} + c_2e^{-\sqrt{-\lambda}x}$, where $(-\lambda)$ and $\sqrt{-\lambda}$ are positive. Now apply $y(0) = 0$ and $y(1) = 0$. $c_1 + c_2 = 0$ and $c_1e^{\sqrt{-\lambda}} + c_2e^{-\sqrt{-\lambda}} = 0$.*

The only solutions are $c_1 = 0 = c_2$. Again, a trivial solution $y = 0$.

Case 8.1.6. *When $\lambda > 0$, the solution is $y = c_1 \cos\sqrt{\lambda}x + c_2 \sin\sqrt{\lambda}x$.*

Apply the boundary conditions $y(0) = 0$ and $y(1) = 0$, then $c_1 = 0$ and $c_2 \sin\sqrt{\lambda} = 0$.

Note that $\sin\theta = 0$, if and only if $\theta = n\pi$, $n = 0, \pm1, \pm2, \ldots$.

Further, if θ be positive, then n must also be positive.

Observe that $c_2 \sin\sqrt{\lambda} = 0$ implies either $c_2 = 0$ or, $\sin\sqrt{\lambda} = 0$.

But as in cases I and II, if both $c_1 = 0$ and $c_2 = 0$, the solution becomes trivial. Hence, for a nontrivial solution $y = c_2 \sin\sqrt{\lambda}x$, we have $\sin\sqrt{\lambda} = 0$, whereby $\sqrt{\lambda} = n\pi$, n being a positive integer.

Thus, $\lambda_n = n^2\pi^2$. In other words, λ_n must be equal to one of the numbers $1^2\pi^2, 2^2\pi^2, 3^2\pi^2, 4^2\pi^2, \ldots$.

These values of λ are called the *eigenvalues* of the differential equation and the corresponding (nontrivial) solutions are $y_n = c_2 \sin\sqrt{\lambda}x = c_2 \sin n\pi x = A_n \sin n\pi x$, where n is a positive integer. That means the corresponding solution or in other words, the corresponding *eigenfunctions* are $A_1 \sin\pi x$, $A_2 \sin 2\pi x$, $A_3 \sin 3\pi x, \ldots$.

Example 8.1.3. *Find the eigenvalues and eigenfunctions of* $\frac{d^2y}{dx^2} + \lambda y = 0$, *where* $y(0) = 0$ *and* $\frac{dy}{dx} = 0$ *at* $x = \pi$.

Solution: Proceed as in examples 8.1.1 and 8.1.2 and see that only trivial solution $y = 0$ is found when $\lambda = 0$ or λ is negative.

So we consider the case when, $\lambda > 0$.

The solution is $y = c_1 \cos\sqrt{\lambda}x + c_2 \sin\sqrt{\lambda}x$.

Given $y(0) = 0$, that is, $y = 0$ at $x = 0$. Then $c_1 = 0$.

Then the solution reduces to $y = c_2 \sin\sqrt{\lambda}x$.

But $\frac{dy}{dx} = 0$ at $x = \pi$, then $\sqrt{\lambda}c_2 \cos\sqrt{\lambda}\pi = 0$.

Since for $\theta > 0$, $\cos\theta = 0$, if and only if θ is a positive odd multiple of $\frac{\pi}{2}$, that is, for $\theta = (2n-1)\frac{\pi}{2}$, when $n = 1, 2, 3, \ldots$.

Thus, to satisfy boundary condition we must have $c_1 = 0$ and either $c_2 = 0$ or, $\cos\sqrt{\lambda}\pi = 0$. The choice $c_2 = 0$ results in the trivial solution $y = 0$. But $\cos\sqrt{\lambda}\pi = 0$ is equivalent to $\sqrt{\lambda} = n - \frac{1}{2}$.

Thus, the eigenvalues are $\lambda_n = \left(n - \frac{1}{2}\right)^2$ for $n = 1, 2, 3, \ldots$, and the corresponding eigenfunctions are $y_n = A_n \sin\left(n - \frac{1}{2}\right)x$, when $n = 1, 2, 3, \ldots$.

Example 8.1.4. *Find the eigenvalues and eigenfunctions for* $\frac{d^2y}{dx^2} + \lambda y = 0$, $(\lambda > 0)$ *satisfying the boundary conditions* $y_1(0) = 0$ *and* $y_1(1) = 0$.

Solution: Proceed as before and see that, since $\lambda > 0$, the solution is $y = c_1 \cos\sqrt{\lambda}x + c_2 \sin\sqrt{\lambda}x$.

Since it is given that $y_1(0) = 0$ and $y_1(1) = 0$, we have from

$$y_1 = \sqrt{\lambda}\left\{-c_1 \sin\sqrt{\lambda}x + c_2 \cos\sqrt{\lambda}x\right\},$$

$$0 = c_2\sqrt{\lambda} \text{ and } 0 = \sqrt{\lambda}\left\{-c_1 \sin\sqrt{\lambda} + c_2 \cos\sqrt{\lambda}\right\}.$$

Now c_2 being equal to 0, $c_1 \sin\sqrt{\lambda} = 0$, or, $\sqrt{\lambda} = n\pi$, or, $\lambda = n^2\pi^2$.

Therefore, the eigenvalues are $\lambda_n = n^2\pi^2$, $n = 1, 2, 3, \ldots$ and the eigenfunctions will be (from $y = c_1 \cos\sqrt{\lambda}x$), $y_n = A_n \cos n\pi x$, where $n = 1, 2, 3, \ldots$.

Example 8.1.5. *Find the eigenvalues for the differential equation* $\frac{d^2y}{dx^2} + \lambda y = 0$, $(\lambda > 0)$ *satisfying the boundary conditions* $y(0) = 0$ *and* $y(\pi/2) = 0$.

Solution: As before, since $\lambda > 0$, the solution is $y = c_1 \cos\sqrt{\lambda}x + c_2 \sin\sqrt{\lambda}x$.

When $y = 0$ at $x = 0$, that is, $y(0) = 0$, $c_1 = 0$ and the solution becomes $y = c_2 \sin\sqrt{\lambda}x$.

Next $y(\pi/2) = 0$, whereby $0 = c_2 \sin\left(\sqrt{\lambda}\pi/2\right)$ and therefore, $\left(\sqrt{\lambda}\pi/2\right) = n\pi$, $\sqrt{\lambda} = 2n$.

Thus, the eigenvalues are $\lambda_n = 4n^2$ for $n = 1, 2, 3, \ldots$.

8.2 Regular Sturm-Liouville problems

A second order *Sturm-Liouville problem* is a homogeneous boundary value problem of the form:

$$\frac{d}{dx}\left\{p(x)\frac{dy}{dx}\right\} + \{q(x) + \lambda r(x)\}\, y = 0 \text{ or, } (py')' + (q + \lambda r)y = 0, \tag{8.2.1}$$

that satisfies the *boundary conditions*

$$\alpha_1 y(a) + \beta_1 y'(a) = 0,\ \alpha_2 y(b) + \beta_2 y'(b) = 0, \tag{8.2.2}$$

where $p(x)$, $q(x)$, $r(x)$ and $p'(x)$ are real-valued continuous functions on $[a, b]$ and both $p(x)$ and $r(x)$ are positive on $[a, b]$, the constant λ being an arbitrary parameter. The parameter λ appears in Sturm-Liouville problem only as indicated above. That is, the real-valued functions p, q and r in the differential equation are independent of λ and the real numbers α_1, α_2, β_1, β_2 are also independent of λ. Also α_1, α_2, β_1, β_2 are any constants such that α_1 and α_2 are not both zero and β_1 and β_2 are not both zero.

Let us now see whether the following two boundary value problems are Sturm-Liouville problems.

Example 8.2.1. *Determine whether the boundary value problem* $\frac{d^2y}{dx^2} + \lambda y = 0$; $y(0) = 0$, $y(\pi) = 0$ *of Ex. 8.1.1 is a Sturm-Liouville problem.*

Solution: Here $p(x) = 1$, $q(x) = 0$, $r(x) = 1$ and on the interval of interest $[0, \pi]$ both $p(x)$ and $r(x)$ are positive.

Hence, it is a Sturm-Liouville problem.

Example 8.2.2. *Determine whether the boundary value problem* $(xy')' + \left(x^2 + 1 + \lambda e^x\right) y = 0$; $y(1) + 2y'(1) = 0$, $y(2) - 3y'(2) = 0$ *is a Sturm-Liouville problem.*

Solution: Here $p(x) = x$, $q(x) = x^2 + 1$, $r(x) = e^x$ and on the interval of interest $[1, 2]$, both $p(x)$ and $r(x)$ are positive.

Hence, it is a Sturm-Liouville problem.

Example 8.2.3. *Determine whether the boundary value problem* $\frac{d^2y}{dx^2} + \lambda(1 + x)y = 0$; $y'(0) = 0$, $y(1) + y'(1) = 0$ *is a Sturm-Liouville problem.*

Solution: Here $p(x) = 1$, $q(x) = 0$, $r(x) = 1 + x$ and on the interval of interest $[0, 1]$, both $p(x)$ and $r(x)$ are positive.

Hence, it is a *Sturm-Liouville* problem.

Example 8.2.4. *Determine whether the boundary value problem* $\left(\frac{1}{x}y'\right)' + (x + \lambda)y = 0$; $y(0) + 3y'(0) = 0$, $y(2) = 0$ *is a Sturm-Liouville problem.*

Solution: Here $p(x) = \frac{1}{x}$, $q(x) = x$, $r(x) = 1$. Since $p(x)$ is not continuous on $[0, 2]$, in particular at $x = 0$, this is not a *Sturm-Liouville* problem.

8.3 Conversion to Sturm-Liouville form

A second order differential equation of the form:

$$P_0(x)\frac{d^2y}{dx^2} + P_1(x)\frac{dy}{dx} + P_2(x)y + \lambda r(x)y = 0 \tag{8.3.1}$$

can be changed to a second order Sturm-Liouville form where $P_0(x)$ and $r(x)$ be positive in the interval of interest.

Multiply (8.3.1) by

$$I(x) = e^{\int \frac{P_1(x)}{P_0(x)}dx}.$$

Then

$$I(x)P_0(x)y_2 + I(x)P_1(x)y_1 + I(x)P_2(x)y + \lambda I(x)r(x)y = 0$$

$$\text{or,}\quad P_0(x)\cdot \frac{d}{dx}\{I(x)\cdot y_1\} + I(x)P_2(x)y + \lambda I(x)r(x)y = 0.$$

Dividing by $P_0(x)$, we have

$$\frac{d}{dx}\{I(x)y_1\} + \left\{I_x\frac{P_2(x)}{P_0(x)} + \lambda I(x)\frac{r(x)}{P_0(x)}\right\}y = 0,$$

which is of Sturm-Liouville form. Note that since $I(x)$ is an exponential and $P_0(x)$ does not vanish, $I(x)$ is positive.

Example 8.3.1. *Convert* $\frac{d^2y}{dx^2} + 2x\frac{dy}{dx} + (x+\lambda)y = 0$ *into the form required for a Sturm-Liouville problem.*

Solution: Here $p_0(x) = 1$, $p_1(x) = 2x$, $p_2(x) = x$, $r(x) = 1$.

Now

$$I(x) = e^{\int \frac{p_1(x)}{p_0(x)}dx} = e^{\int 2x\,dx} = e^{x^2}.$$

Multiplying the given differential equation by $I(x)$,

$$e^{x^2}y'' + 2xe^{x^2}y' + xe^{x^2}y + \lambda e^{x^2}y = 0 \text{ or, } \left(e^{x^2}y'\right)' + xe^{x^2}y + \lambda e^{x^2}y = 0$$

with $p(x) = e^{x^2}$, $q(x) = xe^{x^2}$. $r(x) = e^{x^2}$, a form of Sturm-Liouville.

Example 8.3.2. *Find all the eigenvalues and eigenfunctions of the Sturm-Liouville problem*

$$4\left(e^{-x}y'\right)' + (1+\lambda)e^{-x}y = 0,\ y(0) = 0,\ y(1) = 0.$$

Solution: This is same as

$$\frac{d}{dx}\left\{4e^{-x}\frac{dy}{dx}\right\} + \{e^{-x} + \lambda e^{-x}\}y = 0.$$

Here $p(x) = 4e^{-x}$, $q(x) = e^{-x}$, $r(x) = e^{-x}$ and the interval is $[0, 1]$, where $p(x)$ and $r(x)$ are positive.

Hence, it is a Sturm-Liouville problem.

Simplifying the given equation, we have

$$4e^{-x}\frac{d^2y}{dx^2} - 4e^{-x}\frac{dy}{dx} + e^{-x}y + \lambda e^{-x}y = 0 \text{ or, } 4\frac{d^2y}{dx^2} - 4\frac{dy}{dx} + (1+\lambda)y = 0,$$

where $y(0) = 0$, $y(1) = 0$.

Case 8.3.1. *Let* $\lambda = 0$. *Then the equation becomes* $\left(4D^2 - 4D + 1\right)y = 0$.

Auxiliary equation is $4m^2 - 4m + 1 = 0 = (2m-1)^2$, thus $m = \frac{1}{2}, \frac{1}{2}$.

Solution is $y = e^{\frac{1}{2}x}(c_1 + c_2x)$.

Put $y(0) = 0$, then $c_1 = 0$ and solution becomes $y = e^{\frac{1}{2}x}c_2x$.

Put $y(1) = 0$, then $e^{\frac{1}{2}}\cdot(c_1 + c_2) = 0$, whereby $c_2 = 0$.

Thus, $y = 0$ becomes a trivial solution.

Case 8.3.2. *Let* λ *be negative. Make* $\lambda = -\mu^2$, *where* μ *is* $\neq 0$.

Then the given equation reduces to $\left\{4D^2 - 4D + \left(1-\mu^2\right)\right\}y = 0$.

The auxiliary equation is $4m^2 - 4m + \left(1-\mu^2\right) = 0$, hence $m = \frac{1}{2} \pm \frac{1}{2}\mu$.

Solution is $y = c_1e^{\left(\frac{1}{2}+\frac{1}{2}\mu\right)x} + c_2e^{\left(\frac{1}{2}-\frac{1}{2}\mu\right)x}$.

But $y(0) = 0$ and $y(1) = 0$. Then $c_1 = 0 = c_2$ and again, $y = 0$ becomes a trivial solution.

Case 8.3.3. *Let λ be positive. Choose $\lambda = \mu^2, \mu \neq 0$.*

Then the equation becomes $\{4D^2 - 4D + (1 + \mu^2)\} y = 0$.

Here auxiliary equation is $4m^2 - 4m + 1 + \mu^2 = 0$, hence $m = \frac{1}{2} \pm i\frac{1}{2}\mu$.

Solution is $y = e^{\frac{x}{2}} \{c_1 \cos \frac{1}{2}\mu x + c_2 \sin \frac{1}{2}\mu x\}$.

But $y(0) = 0$, then $c_1 = 0$. Then solution becomes $y = e^{\frac{x}{2}} \{c_2 \sin \frac{1}{2}\mu x\}$.

Next $y(1) = 0$, therefore, $e^{\frac{1}{2}} \{c_2 \sin \frac{1}{2}\mu\} = 0$.

Thus, for a nontrivial solution $\sin \frac{1}{2}\mu = 0 = \sin n\pi$.

Hence, $\mu = 2n\pi$ and $\lambda = \mu^2 = 4n^2\pi^2$.

The eigenvalues are, therefore, $\lambda_n = 4n^2\pi^2$, $n = 1, 2, 3, \ldots$ and the eigenfunctions are $y_n = A_n e^{\frac{x}{2}} \sin n\pi x$, $n = 1, 2, 3, \ldots$.

8.4 Reduction to Cauchy-Euler form

Example 8.4.1. *Find the eigenvalues and eigenfunctions of $\frac{d}{dx}\left(x\frac{dy}{dx}\right) + \frac{\lambda}{x}y = 0$, $(\lambda > 0)$ under boundary conditions $y(1) = 0$ and $y(e^\pi) = 0$.*

Solution: The equation is

$$x\frac{d^2y}{dx^2} + \frac{dy}{dx} + \frac{\lambda}{x}y = 0, \text{ or, } x^2\frac{d^2y}{dx^2} + x\frac{dy}{dx} + \lambda y = 0.$$

This being homogeneous linear equation, we put $z = \ln x$, or, $x = e^z$.

Then the transformed equation becomes, taking $D \equiv \frac{d}{dz}$, $\{(D(D-1) + D + \lambda\} y = 0$, or, $(D^2 + \lambda) y = 0$ under boundary conditions $y(1) = 0$, or, $y = 0$, when $x = 1$ and $y(e^\pi) = 0$, or, $y = 0$, when $x = e^\pi$.

Thus, $y = 0$, when $z = \ln 1 = 0$ and $y = 0$ when $z = \ln(e^\pi) = \pi$.

Hence, the problem becomes $\frac{d^2y}{dz^2} + \lambda y = 0$, $y(0) = 0$ and $y(\pi) = 0$.

This is worked out in Ex. 8.1.1. Thus, the eigenvalues are $\lambda_n = n^2$ and the corresponding eigenfunctions are $y_n = A_n \sin nz = A_n \sin(n \ln x)$, where in both cases $n = 1, 2, 3, \ldots$.

Example 8.4.2. *Find the eigenvalues and eigenfunctions of $\frac{d}{dx}\left(x\frac{dy}{dx}\right) + \frac{\lambda}{x}y = 0$; $y'(1) = 0 = y'(e^{2\pi})$, where $\lambda > 0$.*

Solution: As in example 8.3.3, it reduces to

$$x^2\frac{d^2y}{dx^2} + x\frac{dy}{dx} + \lambda y = 0; \; y'(1) = 0 = y'(e^{2\pi})$$

and again after putting $x = e^z$, or, $z = \ln x$, it becomes $\frac{d^2y}{dz^2} + \lambda y = 0$.

Now λ being > 0, the solution is $y = c_1 \cos\sqrt{\lambda}z + c_2 \sin\sqrt{\lambda}z = c_1 \cos\sqrt{\lambda}(\ln x) + c_2 \sin\sqrt{\lambda}(\ln x)$.

Now

$$y_1 = -\frac{c_1}{x}\sqrt{\lambda}\sin\left\{\sqrt{\lambda}(\ln x)\right\} + \frac{c_2}{x}\left\{\sqrt{\lambda}\cos\sqrt{\lambda}(\ln x)\right\}.$$

Put $y'(1) = 0$ and $y'(e^{2\pi}) = 0$.

Hence, $c_2 = 0$, $-\frac{c_1}{e^{2\pi}}\sqrt{\lambda}\sin\left(2\sqrt{\lambda}\pi\right) = 0$.

Hence, $\sin\left(2\sqrt{\lambda}\pi\right) = 0 = \sin n\pi$ and $\sqrt{\lambda} = \frac{n}{2}$, $\lambda = \frac{n^2}{4}$.

$\therefore$ eigenvalues are $\lambda_n = \frac{n^2}{4}$; $n = 1, 2, 3, \ldots$ and eigenfunctions are $y_n = A_n \cos\left(\frac{n}{2}\ln x\right)$; $n = 1, 2, 3, \ldots$.

8.5 Related problems

Let us study some more examples.

Example 8.5.1. *Find the eigenvalues and eigenfunctions of* $\frac{d^2y}{dx^2}+\lambda y = 0$ *satisfying the boundary conditions* $y_1(0) = 0$, $y_1(\pi) = 0$.

Solution: The auxiliary equation is $m^2 + \lambda = 0$.

Case 8.5.1. *When* $\lambda < 0$, *the solution is* $y = c_1 e^{\sqrt{-\lambda}x} + c_2 e^{-\sqrt{-\lambda}x}$, *where* $(-\lambda)$ *and* $\sqrt{-\lambda}$ *are positive with* $y_1(0) = 0$ *and* $y_1(\pi) = 0$.

Next, $y_1 = \sqrt{-\lambda}\left\{c_1 e^{\sqrt{-\lambda}x} - c_2 e^{-\sqrt{-\lambda}x}\right\}$.

When $y_1(0) = 0$, we have $c_1 - c_2 = 0$ and when $y_1(\pi) = 0$, $c_1 e^{\sqrt{-\lambda}\pi} - c_2 e^{-\sqrt{-\lambda}\pi} = 0$.

For a nontrivial solution for c_1 and c_2, we must have

$$\begin{vmatrix} 1 & -1 \\ e^{\sqrt{-\lambda}\pi} & -e^{-\sqrt{-\lambda}\pi} \end{vmatrix} = 0, \text{ or, } e^{\sqrt{-\lambda}\pi} - e^{-\sqrt{-\lambda}\pi} = 0.$$

That is $\cosh\sqrt{-\lambda}\pi = 0$, which has no solution, hence $c_1 = 0 = c_2$ and there results a trivial solution $y = 0$.

Case 8.5.2. *For* $\lambda = 0$, *the solution is* $y = c_1 + c_2 x$, *and* $y_1 = c_2$.

But $y_1(0) = 0$ and $y_1(\pi) = 0$; hence, $c_2 = 0$.

And the problem has solution $y = c_1 =$ a constant.

Case 8.5.3. *When* $\lambda > 0$, *the solution is* $y = c_1 \cos\sqrt{\lambda}x + c_2 \sin\sqrt{\lambda}x$ *and* $y_1 = \sqrt{\lambda}\left\{-c_1 \sin\sqrt{\lambda}x + c_2 \cos\sqrt{\lambda}x\right\}$.

When $y_1(0) = 0$, we have $c_2 = 0$ and the solution reduces to $y = c_1 \cos\sqrt{\lambda}x$, whereby $y_1 = -\sqrt{\lambda}c_1 \sin\sqrt{\lambda}x$.

But $y_1(\pi) = 0$. Hence, $c_1 \sin\sqrt{\lambda}\pi = 0$.

To obtain a nontrivial solution, we must have $c_1 \neq 0$, then $\sqrt{\lambda} = n$, or, $\lambda = n^2$.

Therefore, the eigenvalues are $\lambda_n = n^2$, $n = 1, 2, 3, \ldots$ and the eigenfunctions will be $y_n = A_n \cos nx$, $n = 1, 2, 3, \ldots$.

Example 8.5.2. *Find the eigenvalues and eigenfunctions for* $\frac{d^2y}{dx^2} + \lambda y = 0$, $(\lambda > 0)$, *where* $y_1(0) = 0$ *and* $y_1(c) = 0$.

Solution: As before, for $\lambda > 0$, we have the solution $y = c_1 \cos\sqrt{\lambda}x + c_2 \sin\sqrt{\lambda}x$.

Differentiating, we have $y_1 = \sqrt{\lambda}\left\{-c_1 \sin\sqrt{\lambda}x + c_2 \cos\sqrt{\lambda}x\right\}$. Since $y_1(0) = 0$ and $y_1(c) = 0$, we have the relation $c_1 \sin\sqrt{\lambda}c = 0$, whereby $\sqrt{\lambda}c = n\pi$, $n \in \mathbb{N}$.

Thus, the eigenvalues are $\lambda_n = \frac{n^2\pi^2}{c^2}$, $n = 1, 2, 3, \ldots$, and the corresponding eigenfunctions are $y_n = A_n \cos\frac{n\pi}{c}x$, $n = 1, 2, 3, \ldots$.

Example 8.5.3. *Find the eigenvalues and the corresponding eigenfunctions of the following boundary problem:* $\frac{d^2y}{dx^2} + \lambda y = 0 (\lambda > 0)$ *with* $y(0) + y'(0) = 0$ *and* $y(1) + y'(1) = 0$.

Solution: As before, for $\lambda > 0$ we have the solution $y = c_1 \cos\sqrt{\lambda}x + c_2 \sin\sqrt{\lambda}x$.

Differentiating, we have $y' = \sqrt{\lambda}\left\{-c_1 \sin\sqrt{\lambda}x + c_2 \cos\sqrt{\lambda}x\right\}$.

Now $y(0) + y'(0) = 0$ implies $c_1 + \sqrt{\lambda}c_2 = 0$.

Also, $y(1) + y'(1) = 0$ means $\left(c_1 \cos\sqrt{\lambda} + c_2 \sin\sqrt{\lambda}\right) + \sqrt{\lambda}\left(-c_1 \sin\sqrt{\lambda} + c_2 \cos\sqrt{\lambda}\right) = 0$.

Putting $c_1 = -\sqrt{\lambda}c_2$ and simplifying, $c_2(1 + \lambda)\sin\sqrt{\lambda} = 0$.

But $(1+\lambda) \neq 0$ and for a nontrivial solution we cannot accept $c_2 = 0$, since then c_1 also will be 0. Hence, $\sin\sqrt{\lambda} = 0$ so that $\sqrt{\lambda} = n\pi, \lambda = 1, 2, 3, \ldots$.

And the solution becomes $y = -\sqrt{\lambda}c_2 \cos\sqrt{\lambda}x + c_2 \sin\sqrt{\lambda}x = c_2\left(\sin\sqrt{\lambda}x - \sqrt{\lambda}\cos\sqrt{\lambda}x\right)$.

Thus, the eigenvalues are $\lambda_n = n^2\pi^2$, $n = 1, 2, 3, \ldots$ and the corresponding eigenfunctions are

$$y_n = A_n\{\sin n\pi x - n\pi \cos n\pi x\}, \quad n = 1, 2, 3, \ldots$$

Example 8.5.4. *Find the eigenvalues and eigenfunctions of $\frac{d^2y}{dx^2} + \lambda y = 0$ satisfying the boundary conditions $y(0) = 0$ and $y'(l) = 0$.*

Solution: The auxiliary equation is $m^2 + \lambda = 0$.

Case 8.5.4. *When $\lambda = 0$, the solution is $y = c_1 + c_2x$; $y' = c_2$.*

But $y(0) = 0, y'(l) = 0$, whereby $c_1 = 0 = c_2$, which results in the trivial solution $y = 0$.

Case 8.5.5. *When $\lambda < 0$, $y = c_1e^{\sqrt{-\lambda}x} + c_2e^{-\sqrt{-\lambda}x}$ and $y' = \sqrt{-\lambda}\left\{c_1e^{\sqrt{-\lambda}x} - c_2e^{-\sqrt{-\lambda}x}\right\}$.*

Putting, $y(0) = 0, y'(l) = 0$; $c_1 + c_2 = 0$, $c_1e^{\sqrt{-\lambda}l} - c_2e^{-\sqrt{-\lambda}l} = 0$.

Solving, $c_1 = 0 = c_2$ and a trivial solution comes in.

Case 8.5.6. *When $\lambda > 0$, the solution is $y = c_1 \cos\sqrt{\lambda}x + c_2 \sin\sqrt{\lambda}x$.*

When $y(0) = 0$, $c_1 = 0$, then the solution becomes $y = c_2 \sin\sqrt{\lambda}x$.

But $y'(l) = 0$, that is, $y'(l) = \sqrt{\lambda}c_2 \cos\sqrt{\lambda}l = 0$.

Since for $\theta > 0$, $\cos\theta = 0$, if and only if θ is a positive odd multiple of $\frac{\pi}{2}$, that is, for $\theta = (2n-1)\frac{\pi}{2}$, when $n = 1, 2, 3, \ldots$.

Thus, to satisfy boundary conditions we must have $c_2 = 0$, or, $\cos\sqrt{\lambda}l = 0$. The choice $c_2 = 0$ results in the trivial solution $y = 0$. But $\cos\sqrt{\lambda}l = 0$ is equivalent to $\sqrt{\lambda} = (2n-1)\frac{\pi}{2l}$, or, $\lambda = (2n-1)^2\frac{\pi^2}{4l^2}$.

Thus, the eigenvalues are $\lambda_n = (2n-1)^2 \cdot \frac{\pi^2}{4l^2}$ for $n = 1, 2, 3, \ldots$ and the corresponding eigenfunctions are $y_n = A_n \sin(2n-1)\frac{\pi x}{2l}$, $n = 1, 2, 3, \ldots$.

Example 8.5.5. *Find the eigenvalues and eigenfunctions of $\frac{d^2y}{dx^2} + \lambda y = 0$, $(\lambda > 0)$ satisfying the boundary conditions $y(0) = 0$ and $y(a) = 0$.*

Solution: As usual since $\lambda > 0$, the solution is $y = c_1 \cos\sqrt{\lambda}x + c_2 \sin\sqrt{\lambda}x$.

Next $y(0) = 0$ gives $c_1 = 0$ and the solution becomes $y = c_2 \sin\sqrt{\lambda}x$.

Again, $y(a) = 0$ gives $c_2 \sin\sqrt{\lambda}a = 0$, but $c_2 = 0$ leads to a trivial solution, hence $\sin\sqrt{\lambda}a = 0$.
$\therefore \sqrt{\lambda} = \frac{n\pi}{a}$, $n \in \mathbb{N}$.

Thus, the eigenvalues are $\frac{n^2\pi^2}{a^2}$, $n = 1, 2, 3, \ldots$ and the eigenfunctions will be (from $y = c_2 \sin\sqrt{\lambda}x$)

$$y_n = A_n \sin\frac{n\pi x}{a}; \quad n = 1, 2, 3, \ldots.$$

Example 8.5.6. *Find the eigenvalues and eigenfunctions of $\frac{d^2y}{dx^2} + \lambda y = 0$ $(\lambda > 0)$, for $0 < x < c$, satisfying the boundary condition $y'(0) = 0$, $hy(c) + y'(c) = 0$, where h is a positive constant.*

Solution: Since $\lambda > 0$, the solution is $y = c_1 \cos\sqrt{\lambda}x + c_2 \sin\sqrt{\lambda}x$.

$$y' = \sqrt{\lambda}\left\{c_2 \cos\sqrt{\lambda}x - c_1 \sin\sqrt{\lambda}x\right\}.$$

When $y'(0) = 0$, $c_2 = 0$; hence $y = c_1 \cos\sqrt{\lambda}x$ and $y' = -\sqrt{\lambda}c_1 \sin\sqrt{\lambda}x$.

When $hy(c) + y'(c) = 0$, then $c_1(h\cos\sqrt{\lambda}c - \sqrt{\lambda}\sin\sqrt{\lambda}c) = 0$.

For a nontrivial solution, c_1 must be non-zero. Hence, $h\cos\sqrt{\lambda}c = \sqrt{\lambda}\sin\sqrt{\lambda}c$.

Since $\lambda > 0$, we may write $\lambda = \alpha^2 (\alpha > 0)$, then the solution becomes $y = c_1 \cos\alpha x$ and $h\cos\alpha c = \alpha\sin\alpha c$, or, $\tan\alpha c = \frac{h}{\alpha}$.

That is, if we are to have an eigenvalue $\lambda = \alpha^2 (\alpha > 0)$, the number α must be a positive root of the equation $\tan\alpha c = \frac{h}{\alpha}$.

The eigenvalues are the numbers $\lambda_n = \alpha_n^2, n = 1, 2, 3, \ldots$, we identify them by simply writing $\lambda_n = \alpha_n^2$, where $\tan\alpha_n c = \frac{h}{\alpha_n} (\alpha_n > 0)$, $n = 1, 2, 3, \ldots$ and the corresponding eigenfunctions are $y_n = A_n \cos\alpha_n x$; $n = 1, 2, 3, \ldots$.

Example 8.5.7. *Find the eigenvalues and eigenfunctions of* $\frac{d^2y}{dx^2} + 2\frac{dy}{dx} + (1 + \lambda)y = 0$, $(\lambda > 0)$ *satisfying the boundary conditions* $y(0) = 0$ *and* $y'(a) = 0$.

Solution: Auxiliary equation $m^2 + 2m + 1 + \lambda = 0$, whereby $m = -1 \pm i\sqrt{\lambda}$.

Solution is $y = e^{-x}(c_1\cos\sqrt{\lambda}x + c_2\sin\sqrt{\lambda}x)$.

$$y' = -e^{-x}(c_1\cos\sqrt{\lambda}x + c_2\sin\sqrt{\lambda}x) + \sqrt{\lambda}e^{-x}(c_2\cos\sqrt{\lambda}x - c_1\sin\sqrt{\lambda}x).$$

Now $y(0) = 0$ gives $c_1 = 0$, hence $y = c_2 e^{-x}\sin\sqrt{\lambda}x$.

Now $y' = \sqrt{\lambda}c_2 e^{-x}\cos\sqrt{\lambda}x - c_2 e^{-x}\sin\sqrt{\lambda}x$.

$y'(a) = 0$ implies $c_2(\sqrt{\lambda}\cos\sqrt{\lambda}a - \sin\sqrt{\lambda}a) = 0$, since $e^{-a} \neq 0$.

For a nontrivial solution $c_2 \neq 0$, hence $\sin\sqrt{\lambda}a = \sqrt{\lambda}\cos\sqrt{\lambda}a$, or, $\tan\sqrt{\lambda}a = \sqrt{\lambda}$.

That is, if we are to take an eigenvalue $\lambda = \alpha^2 (\alpha > 0)$, the number α must be a positive root of the equation $\tan\alpha a = \alpha$.

The eigenvalues are then the numbers $\lambda_n = \alpha_n^2, n = 1, 2, 3, \ldots$, where $\tan\alpha_n a = \alpha_n$, $(\alpha_n > 0)$ and the corresponding eigenfunctions are $y_n = A_n e^{-x}\sin\alpha_n x$; $n = 1, 2, 3, \ldots$.

Examples VIII

1. Find the eigenvalues λ_n and the eigenfunctions $y_n(x)$ for the differential equation $\frac{d^2y}{dx^2} + \lambda y = 0$, $(\lambda > 0)$ satisfying the boundary conditions:
   (a) $y(0) = 0$ and $y(2\pi) = 0$.
   (b) $y_1(0) = 0$ and $y_1(1) = 0$, where $y_1 = \frac{dy}{dx}$.
   (c) $y(0) = 0$ and $y(l) = 0$ when $l > 0$.
   (d) $y(-l) = 0$, $y(l) = 0$, where $l > 0$.
   (e) $y(a) = 0$, $y(b) = 0$ when $a < b$.
2. Find the eigenvalues and eigenfunctions for $y'' + \lambda y = 0$ $(\lambda > 0)$ satisfying the following boundary conditions:
   (a) $y(0) = 0$ and $y'(l) = 0$.
   (b) $y'(0) = 0$ and $hy(c) + y'(c) = 0$.
3. Find all the eigenvalues and eigenfunctions of the Sturm-Liouville problems:
   (a) $4\left(e^{-x}y'\right)' + (1 + \lambda)e^{-x}y = 0$; $y(0) = 0$, $y(1) = 0$.
   (b) $\frac{d}{dx}\left(x\frac{dy}{dx}\right) + \frac{\lambda}{x}y = 0 (\lambda > 0)$; $y(1) = 0$, $y(e^{\pi}) = 0$.
   (c) $\frac{d}{dx}\left(x\frac{dy}{dx}\right) + \frac{\lambda}{x}y = 0 (\lambda > 0)$; $y'(1) = 0 = y'(e^{2\pi})$.
4. Determine the eigenvalues and the corresponding eigenfunctions of the boundary value problems:
   (a) $\frac{d^2y}{dx^2} + 2\frac{dy}{dx} + (1 + \lambda)y = 0$; $y(0) = 0$ and $y'(a) = 0$.
   (b) $\frac{d^2y}{dx^2} + \lambda y = 0$ with $y(0) + y'(0) = 0$ and $y(1) + y'(1) = 0$.
   (c) $\frac{d^2y}{dx^2} - (\lambda + h)y(x) = 0$, $0 < x < c$, where h, c are real constants.

5. Show that the eigenvalues obtained from $y''(x) + \lambda y(x) = 0$ with $y\left(\frac{\pi}{2}\right) = 0$, $y(0) = -3y'(0)$ may be given by $\lambda_n = 4n^2\pi^2$, $(n = 1, 2, 3, \ldots)$.
6. Find the eigenvalues and corresponding eigenfunctions for the problem $y''(x) + \lambda y(x) = 0$:
   (a) Given $\ln \pi \leq x \leq e$ with $y'(\ln \pi) = 0 = y'(e)$.
   (b) $y'\left(\sin^{-1}[e^{0 \cot \alpha}]\right) = y'\left\{-\cos^{-1}\left(e^{0 \cot \alpha}\right)\right\} = 0.$
7. Show that the non-zero eigenvalues of the eigenvalue problem $y''(x) + \lambda y(x) = 0$, $0 \leq x \leq 1$, with $y(0) = 0$, $y(1) = by'(1)$ are the solutions of the transcendental equation $\tan\sqrt{\lambda} = b\sqrt{\lambda}$, where b is real.
8. Consider the eigenvalue problem $y''(x) + \lambda y(x) = 0$ with $0 \leq x \leq 1$ and $y(1) = \alpha y(0)$, $y'(1) = -\alpha y'(0)$, where α is real. Prove that there are no complex eigenvalues, if $\alpha^2 \neq 1$, and all the eigenvalues are complex if $\alpha^2 = 1$.

Answers

1. (a) $\frac{1}{4}n^2$; $A_n \sin\frac{n\pi}{2}$.
   (b) WO Ex. 8.2.4.
   (c) $n^2\pi^2/l^2$; $A_n \sin(n\pi x/l)$.
   (d) $n^2\pi^2/4l^2$; $A_n \sin n\pi(x + l)/2l$.
   (e) $n^2\pi^2/(b - a)^2$; $A_n \sin n\pi(x - a)/(b - a)$.
6. (a) $\lambda_n = \frac{n^2\pi^2}{(\ln \pi - e)^2}$; $n = 1, 2, 3, \ldots, n$.
   (b) $\lambda_n = 4n^2$; $n = 1, 2, \ldots, n$.

Chapter 9

Simultaneous Linear Equations

We have so far considered only single differential equations with two variables—one independent and the other dependent. In the present chapter we proceed to consider system of equations with more than one dependent variable, all depending on only one independent variable. *For the separate and complete determination of each of these dependent variables, the number of equations in the system must be equal to the number of dependent variables. We shall be concerned with systems of ordinary linear differential equations with constant coefficients.*

9.1 Method of solution of simultaneous linear equations with constant coefficients

Method of Solution of Simultaneous Linear Equations with Constant Coefficients: Operator Method.

Let us start by introducing the general linear system of two first-order linear differential equations in two unknown functions $x(t)$ and $y(t)$. The most general form of this system is given below:

$$\left.\begin{aligned} a_1(t)\frac{dx}{dt} + a_2(t)\frac{dy}{dt} + a_3(t)x + a_4(t)y = f_1(t) \\ b_1(t)\frac{dx}{dt} + b_2(t)\frac{dy}{dt} + b_3(t)x + b_4(t)y = f_2(t) \end{aligned}\right\}, \tag{9.1.1}$$

where $a_i(t)$ and $b_i(t)$, $i = 1, 2, 3, 4$ and $f_j(t)$, $j = 1, 2$ are functions of t.

But in this chapter we will be concerned with this type of systems having constant coefficients. Let us consider a_i and b_i, $i = 1, 2, 3, 4$ as constants.

Then, a solution of the system (9.1.1) will be an ordered pair $(f(t), g(t))$ of real functions such that both of the equations of system (9.1.1) must be satisfied by $x = f(t)$, $y = g(t)$ on some closed interval $[a, b]$.

Now we shall illustrate the method by the following example:

Example 9.1.1. *Suppose, there are two dependent variables x and y, each depending on the independent variable t. We denote by D the operatoı* $\frac{d}{dt}$.

Consider the simultaneous equations:

$$(5D + 4)x - (2D + 1)y = e^{-t} \tag{9.1.2}$$

$$(D + 8)x - 3y = 5e^{-t}. \tag{9.1.3}$$

Solution: Eliminate y as in simultaneous linear equations of elementary algebra. We multiply eq. (9.1.2) by 3 and operate on eq. (9.1.3) by $(2D + 1)$ and then subtract the results. We then get

$$\{3(5D + 4) - (2D + 1)(D + 8)\}\ x = 3e^{-t} - (2D + 1)5e^{-t}$$

$$\text{or, } \left(-2D^2 - 2D + 4\right) x = 8e^{-t}$$

$$\text{or, } \left(D^2 + D - 2\right) x = -4e^{-t}.$$

Solving this in the usual way, we get $x = 2e^{-t} + Ae^{t} + Be^{-2t}$.

From (9.1.3)

$$(D + 8)\left\{2e^{-t} + Ae^{t} + Be^{-2t}\right\} - 3y = 5e^{-t}$$

$$\text{or, } \left\{-2e^{-t} + Ae^{t} - 2Be^{-2t}\right\} + \left\{16e^{-t} + 8Ae^{t} + 8Be^{-2t}\right\} - 3y = 5e^{-t}$$

$$\text{or, } y = 3e^{-t} + 3Ae^{t} + 2Be^{-2t}.$$

Normal form

Let us consider the special type of linear system (9.1.1) in the following form:

$$\left.\begin{aligned} \frac{dx}{dt} &= a_{11}(t)x + a_{12}(t)y + f_1(t) \\ \frac{dy}{dt} &= a_{21}(t)x + a_{22}(t)y + f_2(t) \end{aligned}\right\}. \tag{9.1.4}$$

The linear system (9.1.4) is called the *normal form* in the case of two linear differential equations in two unknown functions.

We can express (9.1.4) in matrix form as $\begin{pmatrix} \dot{x} \\ \dot{y} \end{pmatrix} = \begin{pmatrix} a_{11}(t) & a_{12}(t) \\ a_{21}(t) & a_{22}(t) \end{pmatrix} \begin{pmatrix} x \\ y \end{pmatrix} + \begin{pmatrix} f_1(t) \\ f_2(t) \end{pmatrix}$.

Or, in vector form as $\dot{\boldsymbol{x}}(t) = \boldsymbol{A}(t)\boldsymbol{x}(t) + \boldsymbol{f}(t)$, where

$$\boldsymbol{x}(t) = \begin{pmatrix} x(t) \\ y(t) \end{pmatrix}, \quad \boldsymbol{A}(t) = \begin{pmatrix} a_{11}(t) & a_{12}(t) \\ a_{21}(t) & a_{22}(t) \end{pmatrix}, \quad \boldsymbol{f} = \begin{pmatrix} f_1(t) \\ f_2(t) \end{pmatrix}.$$

For example, let us consider the following system:

$$\left.\begin{aligned} \frac{dx}{dt} &= 2x + 3y + 5t \\ \frac{dy}{dt} &= 4x + y - 2t^2 + 1 \end{aligned}\right\}.$$

Observe that the above system is in the normal form in the case of two linear differential equations with constant coefficients in two unknown functions.

The normal form in the general case of a linear system of n differential equations in n unknown functions, namely, $x_1, x_2, \ldots, x_n$, is given below:

$$\left.\begin{aligned} \frac{dx_1}{dt} &= a_{11}(t)x_1 + a_{12}(t)x_2 + \cdots + a_{1n}(t)x_n + f_1(t) \\ \frac{dx_2}{dt} &= a_{21}(t)x_1 + a_{22}(t)x_2 + \cdots + a_{2n}(t)x_n + f_2(t) \\ &\ \ \vdots \\ \frac{dx_n}{dt} &= a_{n1}(t)x_1 + a_{n2}(t)x_2 + \cdots + a_{nn}(t)x_n + f_n(t) \end{aligned}\right\}. \tag{9.1.5}$$

It is possible to express the system (9.1.5) as a single nth-order linear differential equation in one unknown function that is of the so-called normalized form:

$$\frac{d^n x}{dt^n} + a_1(t)\frac{d^{n-1}x}{dt^{n-1}} + a_2(t)\frac{d^{n-2}x}{dt^{n-2}} + \cdots + a_{n-1}(t)\frac{dx}{dt} + a_n(t)x = F(t).$$

Theorem 9.1.1. *Let the functions $a_{11}, a_{12}, a_{21}, a_{22}, f_1$ and f_2 in the system (9.1.4) be continuous on the closed interval $[a, b]$ and let t_0 be an arbitrary point in $[a, b]$. Then, there exists unique solution $x = f(t)$, $y = g(t)$ of the system (9.1.4), defined on $[a, b]$ such that $f(t_0) = c_1$ and $g(t_0) = c_2$, where c_1, c_2 are two constants.*

Important observation

If there be three dependent variables x, y, z given by the three equations:

$$\left.\begin{aligned} f_1(D)x + \phi_1(D)y + \psi_1(D)z = h_1(t) \\ f_2(D)x + \phi_2(D)y + \psi_2(D)z = h_2(t) \\ f_3(D)x + \phi_3(D)y + \psi_3(D)z = h_3(t) \end{aligned}\right\}$$

the number of independent arbitrary constants entering into the complete solution is the index of the highest power of D in the determinant

$$\Delta = \begin{vmatrix} f_1(D) & \phi_1(D) & \psi_1(D) \\ f_2(D) & \phi_2(D) & \psi_2(D) \\ f_3(D) & \phi_3(D) & \psi_3(D) \end{vmatrix}.$$

Similar result is also true when there are n dependent variables, given by n simultaneous linear differential equations having constant coefficients.

If $\Delta = 0$, the system is dependent; such system will not be considered here.

For the example given above the number of independent arbitrary constants is the highest degree of D in the determinant $\begin{vmatrix} 5D+4 & -(2D+1) \\ D+8 & -3 \end{vmatrix} = 2D^2 + 2D - 4$, i.e. there should be only two (degree of D being two) independent arbitrary constants in the complete solution for x and y. This agrees with our solution.

Another observation

In solving the systems of given equations we first eliminated y and then solved for x. Then substituting this value of x, we obtained y. After having obtained x, we could obtain y exactly in the same way (by eliminating x). This would give $\{-(2D+1)(D+8) + 3(5D+4)\}\, y = (D+8)e^{-t} - (5D+4)5e^{-t}$ or, $\{-2D^2 - 2D + 4\}y = 12e^{-t}$ and hence, $y = 3e^{-t} + Ee^{t} + Fe^{-2t}$ (E, F are arbitrary constants).

But E and F are not independent. In fact, they can be expressed in terms of A and B. To find these relations substitute in either of the original equations, say (9.1.2), the expressions for x and y. Thus,

$$(D+8)\left\{2e^{-t} + Ae^{t} + Be^{-2t}\right\} - 3\left\{3e^{-t} + Ee^{t} + Fe^{-2t}\right\} = 5e^{-t} \text{ or, } (9A - 3E)e^{t} + (6B - 3F)e^{-2t} = 0$$

whence $9A - 3E = 0$ and $6B = 3F$ or, $3A = E$ and $2B = F$, so that $y = 3e^{-t} + 3Ae^{t} + 2Be^{-2t}$.

This agrees with our previous solution.

Example 9.1.2. *Solve the system* $\left(D \equiv \frac{d}{dt}\right)$

$$(D^2 - 2)\,x - 3y = e^{2t} \tag{9.1.6}$$

$$(D^2 + 2)\,y + x = 0. \tag{9.1.7}$$

Find the particular solution if the initial conditions are $x = y = 1$, $Dx = Dy = 0$, when $t = 0$.

Solution: Operate on (9.1.6) with D^2 to obtain

$$D^4x - 2D^2x - 3D^2y = 4e^{2t}. \tag{9.1.8}$$

From (9.1.6), $D^2x = e^{2t} + 2x + 3y$ and from (9.1.7), $D^2y = -x - 2y$.
Making these replacements, (9.1.8) becomes $(D^4 - 1)\,x = 6e^{2t}$.
Then

$$x = c_1e^t + c_2e^{-t} + c_3\cos t + c_4\sin t + \tfrac{2}{5}e^{2t} \tag{9.1.9}$$

and then using (9.1.6),

$$y = \tfrac{1}{3}\left[\left(D^2 - 2\right)x - e^{2t}\right] = -\tfrac{1}{3}\left(c_1e^t + c_2e^{-t}\right) - \left(c_3\cos t + c_4\sin t\right) - \tfrac{1}{15}e^{2t}. \tag{9.1.10}$$

For the particular solution:
When $t = 0$,

$$\left.\begin{aligned} x &= c_1 + c_2 + c_3 + \tfrac{2}{5} = 1 \\ y &= -\tfrac{1}{3}(c_1 + c_2) - c_3 - \tfrac{1}{15} = 1 \\ Dx &= c_1 - c_2 + c_4 + \tfrac{4}{5} = 0 \\ Dy &= -\tfrac{1}{3}(c_1 - c_2) - c_4 - \tfrac{2}{15} = 0 \end{aligned}\right\}$$

Then $c_1 = 3/4$, $c_2 = 7/4$, $c_3 = -19/10$, $c_4 = 1/5$.
Putting these values in (9.1.9) and (9.1.10) the required particular solution is obtained.

Alternative method

Since the determinant of the coefficients $\begin{vmatrix} D^2 - 2 & -3 \\ 1 & D^2 + 2 \end{vmatrix} = \left(D^4 - 1\right) \neq 0$, we may proceed as in Cramer's Rule and obtain

$$x = \frac{\begin{vmatrix} e^{2t} & -3 \\ 0 & D^2 + 2 \end{vmatrix}}{\begin{vmatrix} D^2 - 2 & -3 \\ 1 & D^2 + 2 \end{vmatrix}} \quad \text{or,}\ \left(D^4 - 1\right)x = \left(D^2 + 2\right)e^{2t} = 6e^{2t},$$

and then proceed as before.

Homogeneous linear systems

In the system (9.1.4), let us put $f_1(t) = 0$ and $f_2(t) = 0$ for all $t \in [a, b]$. Then the system becomes

$$\left.\begin{aligned} \frac{dx}{dt} &= a_{11}(t)x + a_{12}(t)y \\ \frac{dy}{dt} &= a_{21}(t)x + a_{22}(t)y \end{aligned}\right\}. \tag{9.1.11}$$

The system (9.1.11) is so-called homogeneous linear system of two linear differential equations in two unknown functions. It can be reduced to the single 2nd-order homogeneous linear differential equation of the form:

$$\frac{d^2x}{dt^2} + a_1(t)\frac{dx}{dt} + a_2(t)x = 0.$$

Similarly, corresponding to the system (9.1.5) we have the homogeneous system of n linear differential equations in n unknown functions as

$$\left.\begin{aligned}\frac{dx_1}{dt} &= a_{11}(t)x_1 + a_{12}(t)x_2 + \cdots + a_{1n}(t)x_n\\ \frac{dx_2}{dt} &= a_{21}(t)x_1 + a_{22}(t)x_2 + \cdots + a_{2n}(t)x_n\\ &\vdots\\ \frac{dx_n}{dt} &= a_{n1}(t)x_1 + a_{n2}(t)x_2 + \cdots + a_{nn}(t)x_n\end{aligned}\right\} \tag{9.1.12}$$

Equivalent single nth-order homogeneous linear differential equation is of the form:

$$\frac{d^n x}{dt^n} + a_1(t)\frac{d^{n-1}x}{dt^{n-1}} + a_2(t)\frac{d^{n-2}x}{dt^{n-2}} + \cdots + a_{n-1}(t)\frac{dx}{dt} + a_n(t)x = 0.$$

Theorem 9.1.2. *Let $x = f_1(t)$, $y = g_1(t)$ and $x = f_2(t)$, $y = g_2(t)$ be two solutions of the homogeneous linear system (9.1.11). Then, $x = c_1f_1(t) + c_2f_2(t)$, $y = c_1g_1(t) + c_2g_2(t)$ is also a solution of the system (9.1.11), where c_1, c_2 are two arbitrary constants.*

The proof is left to the readers as an exercise.

Corollary 9.1.1. *Any linear combination of two solutions of the homogeneous linear system (9.1.11) is itself a solution of the system.*

Definition 9.1.1. *Let $x = f_1(t)$, $y = g_1(t)$ and $x = f_2(t)$, $y = g_2(t)$ be two solutions of the homogeneous linear system (9.1.11). Then, the Wronskian of these solutions is denoted by the symbol $W(t)$ and defined by* $W(t) = \begin{vmatrix} f_1(t) & f_2(t)\\ g_1(t) & g_2(t)\end{vmatrix}$, $t \in [a, b]$.

Note 9.1.1. *Let $W(t)$ be the Wronskian of two solutions of the homogeneous linear system (9.1.11) on the closed interval $[a, b]$. Then, either $W(t) = 0\ \forall\ t \in [a, b]$ or, $W(t) \neq 0\ \forall\ t \in [a, b]$.*

Definition 9.1.2. *Let $x = f_1(t)$, $y = g_1(t)$ and $x = f_2(t)$, $y = g_2(t)$ be two solutions of the homogeneous linear system (9.1.11). These two solutions are said to be linearly dependent on the closed interval $[a, b]$, if there exists two constants, namely, c_1 and c_2, not both zero, such that $c_1f_1(t)+c_2f_2(t) = 0$, $c_1g_1(t)+c_2g_2(t) = 0$ for all $t \in [a, b]$.*

The solutions $x = f_1(t)$, $y = g_1(t)$ and $x = f_2(t)$, $y = g_2(t)$ are said to be linearly independent on $[a, b]$, if there exists two constants, namely, c_1 and c_2 such that $c_1f_1(t) + c_2f_2(t) = 0$, $c_1g_1(t) + c_2g_2(t) = 0$ for all $t \in [a, b]$ implies that $c_1 = c_2 = 0$.

Theorem 9.1.3. *Let $x = f_1(t)$, $y = g_1(t)$ and $x = f_2(t)$, $y = g_2(t)$ be two solutions of the homogeneous linear system (9.1.11). These two solutions are linearly independent on the closed interval $[a, b]$, if and only if the Wronskian,* $W(t) = \begin{vmatrix} f_1(t) & f_2(t)\\ g_1(t) & g_2(t)\end{vmatrix} \neq 0\ \forall\ t \in [a, b]$.

Theorem 9.1.4. *Every solution of the homogeneous system (9.1.11) can be expressed as a linear combination of any two linearly independent solutions of (9.1.11).*

Definition 9.1.3. *Let $x = f_1(t)$, $y = g_1(t)$ and $x = f_2(t)$, $y = g_2(t)$ be two linearly independent solutions of the homogeneous linear system (9.1.11). Let c_1, c_2 be two arbitrary constants. Then, the solution $x = c_1f_1(t) + c_2f_2(t)$, $y = c_1g_1(t) + c_2g_2(t)$ is called a general solution of the system (9.1.11).*

Example 9.1.3. *Let us consider the following homogeneous linear system: $\frac{dx}{dt} = x + 3y$, $\frac{dy}{dt} = 4x + 5y$. Observe that $x = 3e^{-t}$, $y = -2e^{-t}$ and $x = e^{7t}$, $y = 2e^{7t}$ be two solutions of the above homogeneous linear system.*

Now the Wronskian, $W(t) = \begin{vmatrix} 3e^{-t} & e^{7t}\\ -2e^{-t} & 2e^{7t}\end{vmatrix} = 6e^{-t}\cdot e^{7t} + 2e^{-t}\cdot e^{7t} = 8e^{6t} \neq 0\ \forall\ t \in \mathbb{R}$. It confirms that the set of solutions are linearly independent on the set $\mathbb{R}$ of all real numbers.

Hence, the general solution of the given homogeneous linear system is $x = 3c_1e^{-t} + c_2e^{7t}$, $y = -2c_1e^{-t} + 2c_2e^{7t}$, where c_1, c_2 are arbitrary constants.

Examples IX(A)

Solve the following system of simultaneous equations (D stands for $\frac{d}{dt}$) :

1. $Dx - 7x + y = 0$
   $Dy - 2x - 5y = 0.$
2. $\frac{dy}{dx} + 2y - 3z = x$
   $\frac{dz}{dx} + 2z - 3y = e^{2x}.$
3. $Dy - z = 0$
   $(D-1)y - (D+1)z = 0.$
4. $(D-17)x + (2D-8)y = 0$
   $(13D - 53)x - 2y = 0.$
5. $4\frac{dx}{dt} + 9\frac{dy}{dt} + 44x + 49y = t$
   $3\frac{dx}{dt} + 7\frac{dy}{dt} + 34x + 38y = e^t.$
6. $\left(D^2 - 3\right)x - 4y = 0$
   $\left(D^2 + 1\right)y + x = 0.$
7. (a) $(D-2)x + 2(D+1)y = 3e^t$
   $(3D+2)x + (D+1)y = 4e^{2t}.$
   (b) $\frac{dx}{dt} + 4x + 3y = t$
   $\frac{dy}{dt} + 2x + 5y = e^t.$
   (c) $\frac{dx}{dt} + 5x + y = e^t$
   $\frac{dy}{dt} - x + 3y = e^{2t}.$
8. $\left(D^2 + 4\right)x + y = te^{3t}$
   $\left(D^2 + 1\right)y - 2x = \cos^2 t.$
9. $\frac{d^2x}{dt^2} + m^2y = 0;\ \frac{d^2y}{dt^2} - m^2x = 0.$
10. $2\frac{d^2y}{dx^2} - \frac{dz}{dx} - 4y = 2x;\ 2\frac{dy}{dx} + 4\frac{dz}{dx} - 3z = 0.$
11. $2\frac{d^2y}{dx^2} - \frac{dy}{dx} + 9y - \frac{d^2z}{dx^2} - \frac{dz}{dx} - 3z = 0$
    $2\frac{d^2y}{dx^2} + \frac{dy}{dx} + 7y - \frac{d^2z}{dx^2} + \frac{dz}{dx} - 5z = 0.$
12. (a) $\frac{dx}{dt} = 2y;\ \frac{dy}{dt} = 2z;\ \frac{dz}{dt} = 2x.$
    (b) $\frac{dx}{dt} = -2y;\ \frac{dy}{dt} = x.$
13. $t^2\frac{d^2x}{dt^2} + t\frac{dx}{dt} + 2y = 0;\ t^2\frac{d^2y}{dt^2} + t\frac{dy}{dt} - 2x = 0.$
14. $\left(D^2 - 4D + 4\right)x - y = 0$
    $\left(D^2 + 4D + 4\right)y - 25x = 16e^t.$
15. $4\frac{dx}{dt} + 9\frac{dy}{dt} + 2x + 31y = e^t$
    $3\frac{dx}{dt} + 7\frac{dy}{dt} + x + 24y = 3.$
16. $(D+4)x + 3y = t;\ (D+5)y + 2x = e^t.$
17. Prove that the particular solution of the system $(D+1)^2x + 2Dy + 3Dz = 1$; $Dx + z = 0$ $x - Dy - Dz = 0$, when $x = z = 1$, $y = 0$, if $t = 0$ is given by $x = \frac{1}{3} + \frac{2}{3}e^{-3t/2}$, $y = \frac{1}{3}t - \frac{13}{9}e^{-3t/2} + \frac{13}{9}$, $z = e^{-3t/2}$.
18. $(D+5)x + y = e^t;\ (D+3)y - x = e^{2t}.$

Answers

1. $x = e^{6t}(A\cos t + B\sin t),$
   $y = e^{6t}[(A-B)\cos t + (A+B)\sin t].$
2. $y = c_1e^{-5x} + c_2e^x + \frac{3}{7}e^{2x} - \frac{2}{5}x - \frac{13}{25},$
   $z = -c_1e^{-5x} + c_2e^x + \frac{4}{7}e^{2x} - \frac{3}{5}x - \frac{12}{25}.$
3. $y = A\cos(t-\alpha);\ z = -A\sin(t-\alpha).$
4. $x = Ae^{5t} + Be^{3t},$
   $y = 6Ae^{5t} - 7Be^{3t}.$
5. $x = c_1e^{-t} + c_2e^{-6t} + \frac{19}{3}t - \frac{56}{9} - \frac{29}{7}e^t,$
   $y = -c_1e^{-t} + 4c_2e^{-6t} - \frac{17}{3}t + \frac{55}{9} + \frac{24}{7}e^t.$
6. $x = (c_1 + c_2t)e^t + (c_3 + c_4)e^{-t},$
   $2y = (c_2 - c_1 - c_2t)e^t - (c_3 + c_4 + c_4t)e^{-t}.$
7. (a) $x = c_1e^{-\frac{6}{5}t} + \frac{1}{2}e^{2t} - \frac{3}{11}e^t,$
   $x + \frac{y}{8} = c_2e^{-t} + \frac{1}{2}e^{2t} - \frac{3}{16}e^t.$
   (b) $x = c_1e^{-2t} + c_2e^{-7t} - \frac{31}{196} + \frac{5}{14}t - \frac{1}{8}e^t.$
   $y = -\frac{2}{3}c_1e^{-2t} + c_2e^{-7t} - \frac{9}{98} - \frac{1}{7}t + \frac{5}{24}t.$
   (c) $x = (c_1 + c_2t)e^{-4t} + \frac{4}{25}e^t - \frac{1}{36}e^{2t},$
   $y = -\{c_1 + c_2(1+t)\}e^{-4t} + \frac{7}{36}e^{2t} + \frac{1}{25}e^t.$
8. $x = c_1\cos\sqrt{3}t + c_2\sin\sqrt{3}t + c_3\cos\sqrt{2}t + c_4\sin\sqrt{2}t - \frac{49}{1452}e^{3t} + \frac{5}{66}te^{3t} - \frac{1}{12} - \frac{1}{4}\cos 2t,$
   $y = -3c_1\cos\sqrt{3}t - c_2\sin\sqrt{3}t - 2c_3\cos\sqrt{2}t - 2c_4\sin\sqrt{2}t + \frac{1}{66}te^{3t} + \frac{1}{3} - \frac{33}{1452}e^{3t}.$
9. $x = e^{\frac{mt}{\sqrt{2}}}\left(c_1\cos\frac{mt}{\sqrt{2}} + c_2\sin\frac{mt}{\sqrt{2}}\right) + e^{-\frac{mt}{\sqrt{2}}}\left(c_3\cos\frac{mt}{\sqrt{2}} + c_4\sin\frac{mt}{\sqrt{2}}\right),$

$$y = e^{\frac{mt}{\sqrt{2}}}\left(c_1 \sin\frac{mt}{\sqrt{2}} - c_2 \sin\frac{mt}{\sqrt{2}}\right) + e^{-\frac{mt}{\sqrt{2}}}\left(c_4 \cos\frac{mt}{\sqrt{2}} - c_3 \sin\frac{mt}{\sqrt{2}}\right).$$

10. $y = (c_1 + c_2x)\, e^x + 3c_3e^{-\frac{3}{2}x} - \frac{1}{2}x,$

$z = 2(3c_2 - c_1 - c_2x)\, e^x - c_3e^{-\frac{3}{2}x} - \frac{1}{3}.$

11. $y = Ae^x + B\cos(2x - \alpha);$

$z = 2Ae^x - B\cos(2x - \alpha).$

12. (a) $x = Ae^{2t} + Be^{-t}\cos\left(\sqrt{3}t - \alpha\right);$

$y = Ae^{2t} + Be^{-t}\cos\left(\sqrt{3}t - \alpha + 2\pi/3\right);$

$z = Ae^{2t} + Be^{-t}\cos\left(\sqrt{3}t - \alpha + 4\pi/3\right).$

(b) $x^2 + 2y + 2c = 0.$

13. $x = At\cos(\ln t - \alpha) + Bt^{-t}\cos(\ln t - \beta);$

$y = At\sin(\ln t - \alpha) - Bt^{-t}\sin(\ln t - \beta).$

14. $x = Ae^{3t} + Be^{-3t} + E\cos t + F\sin t - e^t.$

$y = Ae^{3t} + 25Be^{-3t} + (3E - 3F)\cos t + (3F + 4E)\sin t - e^t.$

15. $x = (c_1 \sin t + c_2 \cos t)\, e^{-4t} + \frac{31e^t}{26} - \frac{93}{17};$

$y = [(c_1 - c_2)\sin t - (c_2 + c_1)\cos t]e^{-4t} - \frac{2}{13}e^t + \frac{6}{17}.$

16. $x = c_1e^{-2t} + c_2e^{-7t} - \frac{31}{196} + \frac{5}{14}t - \frac{1}{3}e^t;$

$y = -\frac{2}{3}c_1e^{-2t} + c_2e^{-7t} + \frac{9}{98} - \frac{1}{7}t + \frac{5}{24}e^t.$

18. $x = (At + B)e^{-4t} + \frac{4}{25}e^t - \frac{1}{36}e^{2t}$. Now obtain y from $(D + 5)x + y = e^t$.

9.2 Simultaneous equations of the form

$$\frac{dx}{P} = \frac{dy}{Q} = \frac{dz}{R}$$

(P, Q, R are functions of x, y, z).

I. The general type of a set of simultaneous equations of first order having three variables is

$$\left.\begin{aligned} P_1\,dx + Q_1\,dy + R_1\,dz = 0 \\ P_2\,dx + Q_2\,dy + R_2\,dz = 0 \end{aligned}\right\} \tag{9.2.1}$$

where the coefficients are functions of x, y, z.

Solving these equations simultaneously, we get

$$\frac{dx}{Q_1R_2 - Q_2R_1} = \frac{dy}{R_1P_2 - R_2P_1} = \frac{dz}{P_1Q_2 - P_2Q_1}$$

which is of the form:

$$\frac{dx}{P} = \frac{dy}{Q} = \frac{dz}{R},$$

where P, Q and R are functions of x, y, z.

The given equations are said to be completely solved when we get a solution of the form $u_1(x, y, z) = c_1$ and $u_2(x, y, z) = c_2$, where u_1 and u_2 are two independent integrals of the given equations.

u_1 and u_2 are said to be independent integrals if u_1/u_2 is not merely a constant, e.g. $u_1 = x^2 + y^2 + z^2$, $u_2 = x + y + z$ *are independent integrals,* whereas $u_1 = 2x + 2y + 2z$, $u_2 = 2(x + y + z)$ *are not independent.*

II. Rules for solving:

$$\frac{dx}{P} = \frac{dy}{Q} = \frac{dz}{R}. \tag{9.2.2}$$

Rule I

Of the three ratios of (9.2.2) we may easily obtain an equation of any two variables. Solving we get one of the relations in the general solution of (9.2.2). Similar method is there to obtain another relation with the help of the other two ratios of (9.2.2). The two relations together give the general solution of the eq. (9.2.2).

Example 9.2.1. $\frac{xdx}{y^2z} = \frac{dy}{xz} = \frac{dz}{y^2}$.

Solution: Taking first two ratios, we write

$$\frac{xdx}{y^2z} = \frac{dy}{xz} \Rightarrow \frac{xdx}{y^2} = \frac{dy}{x} \text{ (cancelling } z \text{ from the denominator)}$$
$$\Rightarrow x^2dx = y^2dy.$$

On integration, $x^3 - y^3 = c_1$ is a relation.

Now taking the first and the third ratios, we get

$$\frac{xdx}{y^2z} = \frac{dz}{y^2} \Rightarrow \frac{xdx}{z} = dz \Rightarrow xdx = zdz.$$

On integration, $x^2 - z^2 = c_2$ is another relation.

The two integrals $x^3 - y^3$ and $x^2 - z^2$ are independent; the required general solution is given by $x^3 - y^3 = c_1$ and $x^2 - z^2 = c_2$.

Rule II

Suppose, one relation is obtained by the method of Rule I. We may use this relation to express one variable in terms of the others. This may help to obtain an equation in two variables. The solution of this equation will give the second relation for the general solution.

Example 9.2.2. $\frac{dx}{xz(z^2+xy)} = \frac{dy}{-yz(z^2+xy)} = \frac{dz}{x^4}$.

Solution: From first two ratios,

$$\frac{dx}{x} = -\frac{dy}{y} \text{ [cancelling } z(z^2+xy)\text{] or, } xy = c_1. \quad (9.2.3)$$

Using (9.2.3) from first and third ratios, we get

$$\frac{dx}{xz(z^2+c_1)} = \frac{dz}{x^4} \text{ or, } x^3dx = (z^3 + c_1z)dz \text{ or, } \frac{x^4}{4} - \frac{z^4}{4} - \frac{c_1z^2}{2} = \text{constant}$$
$$\text{or, } x^4 - z^4 = 2c_1z^2 + c_2 \text{ or, } x^4 - z^4 - 2xyz^2 = c_2 \text{ [using (9.2.3), } xy = c_1.] \quad (9.2.4)$$

∴ the relations (9.2.3) and (9.2.4) together give the required general solution.

Rule III

Let P_1, Q_1, R_1 be some suitable functions of x, y, z. Then we can write

$$\frac{dx}{P} = \frac{dy}{Q} = \frac{dz}{R} = \frac{P_1dx + Q_1dy + R_1dz}{PP_1 + QQ_1 + RR_1}. \quad (9.2.5)$$

If $PP_1 + QQ_1 + RR_1 = 0$, then the numerator $P_1dx + Q_1dy + R_1dz = 0$, which can be integrated to give $u_1(x, y, z) = c_1$.

This method can be repeated to obtain a second relation.

The two relations then give the general solution of the given equation.

Example 9.2.3. *Solve:*

$$\frac{adx}{(b-c)yz} = \frac{bdy}{(c-a)zx} = \frac{cdz}{(a-b)xy}. \tag{9.2.6}$$

Solution: Multiply both numerator and denominator of the first ratio by x, the second ratio by y, the 3rd ratio by z. Then each ratio of (9.2.6) is equal to

$$\frac{axdx + bydy + czdz}{(b-c)xyz + (c-a)xyz + (a-b)xyz}$$

so that the denominator = 0 and hence, $axdx + bydy + czdz = 0$.

On integration,

$$ax^2 + by^2 + cz^2 = c_1. \tag{9.2.7}$$

Again, choosing ax, by, cz as multipliers in the first, second, third ratios of (9.2.6), we find

$$\text{each ratio} = \frac{a^2xdx + b^2ydy + c^2zdz}{a(b-c)xyz + b(c-a)yzx + c(a-b)zxy}$$

so that the denominator = 0 and hence, $a^2xdx + b^2ydy + c^2zdz = 0$.

On integration,

$$a^2x^2 + b^2y^2 + c^2z^2 = c_2. \tag{9.2.8}$$

The required general solution (also called complete solution) is given by the relations (9.2.7) and (9.2.8).

Rule IV

Let P_1, Q_1, R_1 be suitable functions of x, y, z such that

$$\frac{dx}{P} = \frac{dy}{Q} = \frac{dz}{z} = \frac{P_1dx + Q_1dy + R_1dz}{P_1P + Q_1Q + R_1R} \text{ and } P_1dx + Q_1dy + R_1dz$$

is exact differential of $P_1P + Q_1Q + R_1R$.

Then this ratio is combined with a suitable ratio $\frac{dx}{P}$ or $\frac{dy}{Q}$ or $\frac{dz}{R}$ to give an integral.

Another integral may be obtained by **Rule I** or by repeating the procedure stated above.

Example 9.2.4. *Solve:*

$$\frac{dx}{y^2(x-y)} = \frac{dy}{-x^2(x-y)} = \frac{dz}{z(x^2+y^2)}. \tag{9.2.9}$$

Solution: One integral can be easily obtained by using **Rule I.**

Taking the first two ratios (cancelling $x - y$),

$$\frac{dx}{y^2} = -\frac{dy}{x^2} \text{ or, } x^2dx + y^2dy = 0 \text{ i.e. } x^3 + y^3 = c_1. \tag{9.2.10}$$

Now choosing $1, -1, 0$ as multipliers, each ratio of (9.2.9)

$$= \frac{dx - dy}{y^2(x-y) + x^2(x-y)} = \frac{dx - dy}{(x-y)(x^2+y^2)}.$$

Now taking 3rd ratio of (9.2.9) with this ratio, we write

$$\frac{dz}{z(x^2+y^2)} = \frac{dx-dy}{(x-y)(x^2+y^2)} \text{ or, } \frac{dz}{z} = \frac{d(x-y)}{x-y} \text{ (cancelling } x^2+y^2)$$

$$\text{or, } \ln z - \ln(x-y) = \text{constant or, } \frac{z}{x-y} = c_2 \text{ or, } z = c_2(x-y). \quad (9.2.11)$$

The required solution is given by the relations (9.2.10) and (9.2.11).

III. Solved Examples (Miscellaneous Type)

Example 9.2.5. $\frac{dx}{yz} = \frac{dy}{zx} = \frac{dz}{xy}$ **(Rule I)**

Solution: Taking the first two ratios, we get

$$\frac{dx}{y} = \frac{dy}{x} \text{ (cancelling } z) \text{ or, } xdx - ydy = 0.$$

On integration,

$$x^2 - y^2 = c_1. \quad (9.2.12)$$

Taking the first and the third ratios, we get

$$\frac{dx}{z} = \frac{dz}{x} \text{ (cancelling } y) \text{ or, } xdx - zdz = 0.$$

On integration,

$$x^2 - z^2 = c_2. \quad (9.2.13)$$

$\therefore$ (9.2.12) and (9.2.13) together give the required solution.

Example 9.2.6. $\frac{dx}{xy} = \frac{dy}{y^2} = \frac{dz}{xyz-2x^2}$ **(Rule II)**

Solution: Taking the first two ratios, we get

$$\frac{dx}{x} = \frac{dy}{y} \text{ (cancelling } y) \text{ or, } \ln x - \ln y = \ln c_1 \text{ (say). i.e. } x = c_1 y. \quad (9.2.14)$$

$\therefore$ from second and third ratios we get

$$\frac{dy}{y^2} = \frac{dz}{c_1 y^2 z - 2c_1^2 y^2} \text{ or, } c_1 dy = \frac{dz}{z - 2c_1}.$$

On integration, $c_1 y = \ln(z - 2c_1) + c_2$.
Substituting the value of c_1,

$$x = \ln\left(z - 2\frac{x}{y}\right) + c_2, \quad x = \ln(yz - 2x) - \ln y + c_2. \quad (9.2.15)$$

The required solution is given by (9.2.14) and (9.2.15).

Example 9.2.7. $\frac{dx}{z(x+y)} = \frac{dy}{z(x-y)} = \frac{dz}{x^2+y^2}$ **(Rule III)**

Solution: Multiply the first ratio by x, the second by $-y$ and the third by $-z$, each ratio

$$= \frac{xdx - ydy - zdz}{xz(x+y) - yz(x-y) - z(x^2+y^2)} = \frac{xdx - ydy - zdz}{0}.$$

$\therefore$ $xdx - ydy - zdz = 0$; on integration,

$$x^2 - y^2 - z^2 = c_1. \tag{9.2.16}$$

Now choosing the multipliers as $y, x, -z$, we get each ratio

$$= \frac{ydx + xdy - zdz}{yz(x+y) + xz(x-y) - z(x^2+y^2)} = \frac{ydx + xdy - zdz}{0} \text{ or, } d(xy) - d\left(\frac{z^2}{2}\right) = 0 \text{ i.e. } xy - \frac{z^2}{2} = c_2. \tag{9.2.17}$$

(9.2.16) and (9.2.17) together give the required solution.

Example 9.2.8. $\frac{dx}{mz-ny} = \frac{dy}{nx-lz} = \frac{dz}{ly-mx}$. **(Rule III)**

Solution: Choosing l, m, n as multipliers of first, second and third ratios, we get each ratio

$$= \frac{ldx + mdy + ndz}{lmz - lny + mnx - mlz + nly - mnx} = \frac{ldx + mdy - ndz}{0}.$$

$\therefore$ $ldx + mdy + ndz = 0$.

Integrating,

$$lx + my + nz = c_1. \tag{9.2.18}$$

Again, choosing x, y, z as multipliers, each ratio

$$= \frac{xdx + ydy + zdz}{mxz - nxy + nxy - lyz + lyz - mxz} = \frac{xdx + ydy + zdz}{0}$$

Hence, $xdx + ydy + zdz = 0$.

$$\text{i.e. } x^2 + y^2 + z^2 = c_2. \tag{9.2.19}$$

$\therefore$ (9.2.18) and (9.2.19) together give the complete solution.

Example 9.2.9. $\frac{dx}{x^2-y^2-z^2} = \frac{dy}{2xy} = \frac{dz}{2xz}$ **(Rule IV)**

Solution: Taking the 2nd and the 3rd ratios, we get

$$\frac{dy}{y} = \frac{dz}{z} \text{ (cancelling } 2x) \text{ i.e. } \ln y - \ln z = \ln c_1 \text{ or, } \frac{y}{z} = c_1. \tag{9.2.20}$$

Choosing x, y, z as multipliers, each ratio

$$= \frac{xdx + ydy + zdz}{x(x^2 - y^2 - z^2) + 2xy^2 + 2xz^2} = \frac{xdx + ydy + zdz}{x^3 + xy^2 + xz^2}.$$

$$\therefore \frac{dz}{2xz} = \frac{xdx + ydy + zdz}{x(x^2 + y^2 + z^2)} \text{ i.e. } \frac{dz}{z} = \frac{2xdx + 2ydy + 2zdz}{x^2 + y^2 + z^2} = \frac{d(x^2 + y^2 + z^2)}{x^2 + y^2 + z^2}.$$

On integration,

$$\ln z - \ln(x^2 + y^2 + z^2) = \ln c_2 \text{ i.e. } \frac{z}{x^2 + y^2 + z^2} = c_2 \text{ or, } (x^2 + y^2 + z^2)c_2 = z. \tag{9.2.21}$$

$\therefore$ (9.2.20) and (9.2.21) give the required solution.

Example 9.2.10. $\frac{dx}{x^2-yz} = \frac{dy}{y^2-zx} = \frac{dz}{z^2-xy}$. **(Rule IV)**

Solution: Choosing x, y, z as multipliers,

$$\text{each ratio} = \frac{xdx + ydy + zdz}{x^3 + y^3 + z^3 - 3xyz} = \frac{xdx + ydy + zdz}{(x + y + z)(x^2 + y^2 + z^2 - xy - yz - zx)}.$$

Again, choosing 1, 1, 1 as multipliers, each ratio $= \frac{dx+dy+dz}{x^2+y^2+z^2-xy-yz-zx}$.

$$\therefore \frac{\frac{1}{2}d(x^2 + y^2 + z^2)}{(x + y + z)(x^2 + y^2 + z^2 - xy - yz - zx)} = \frac{d(x + y + z)}{x^2 + y^2 + z^2 - xy - yz - zx}$$

$$\text{or, } \tfrac{1}{2}d(x^2 + y^2 + z^2) = (x + y + z)d(x + y + z).$$

Integrating,

$$\tfrac{1}{2}(x^2 + y^2 + z^2) = \frac{(x + y + z)^2}{2} + \text{constant}$$

$$\text{or, } (x^2 + y^2 + z^2) - (x + y + z)^2 = \text{constant}$$

$$\text{or, } xy + yz + zx = c_1. \qquad (9.2.22)$$

Again, choosing $1, -1, 0$ and $0, 1, -1$ as multipliers, we get each ratio $= \frac{dx-dy}{x^2-yz-y^2+xz} = \frac{dy-dz}{y^2-zx-z^2+xy}$.

$$\text{i.e. } \frac{d(x - y)}{(x - y)(x + y + z)} = \frac{d(y - z)}{(y^2 - z^2) + x(y - z)} \text{ or, } \frac{d(x - y)}{(x - y)(x + y + z)} = \frac{d(y - z)}{(y - z)(y + z + x)} \text{ or, } \frac{d(x - y)}{x - y} = \frac{d(y - z)}{y - z}.$$

On integration, $\ln(x - y) - \ln(y - z) = \ln c_2$.

$$\text{i.e. } \frac{x - y}{y - z} = c_2. \qquad (9.2.23)$$

$\therefore$ (9.2.22) and (9.2.23) together give the required solution.

Example 9.2.11. $\frac{dx}{x+y-xy^2} = \frac{dy}{x^2y-x-y} = \frac{dz}{z(y^2-x^2)}$.

Solution: Each ratio $= \frac{xdx+ydy}{x(x+y-xy^2)+y(x^2y-x-y)} = \frac{xdx+ydy}{x^2+xy-x^2y^2+x^2y^2-xy-y^2} = \frac{xdx+ydy}{x^2-y^2}$.

Now take this ratio = the last ratio, i.e. $\frac{xdx+ydy}{x^2-y^2} = -\frac{dz}{z(x^2-y^2)}$, or, $\frac{1}{2}d(x^2 + y^2) + \frac{dz}{z} = 0$.

On integration,

$$x^2 + y^2 + 2\ln z = c_1. \qquad (9.2.24)$$

Again each ratio

$$= \frac{ydx + xdy}{y(x + y - xy^2) + x(x^2y - x - y)} = \frac{ydx + xdy}{xy + y^2 - xy^3 + x^3y - x^2 - xy}$$

$$= \frac{ydx + xdy}{(y^2 - x^2) - xy(y^2 - x^2)} = \frac{ydx + xdy}{(y^2 - x^2)(1 - xy)}.$$

Combining this ratio with the last ratio, we obtain

$$\frac{ydx + xdy}{(y^2 - x^2)(1 - xy)} = \frac{dz}{z(y^2 - x^2)} \text{ or, } \frac{d(xy)}{1 - xy} = \frac{dz}{z}.$$

On integration, $-\ln(1 - xy) - \ln z = \text{constant}$

$$\text{or, } z(1 - xy) = c_2. \qquad (9.2.25)$$

$\therefore$ (9.2.24) and (9.2.25) together give the required solution.

Example 9.2.12. $\frac{dx}{y^2+yz+z^2} = \frac{dy}{z^2+zx+x^2} = \frac{dz}{x^2+xy+y^2}$.

Solution: Multipliers 1, −1, 0 will give

$$\text{each ratio} = \frac{dx - dy}{y^2 + yz + z^2 - z^2 - xz - x^2} = \frac{dx - dy}{(y^2 - x^2) + z(y - x)} = \frac{dx - dy}{(y - x)(x + y + z)}.$$

Again, choosing the multipliers 0, 1, −1, each ratio

$$= \frac{dy - dz}{z^2 + xz + x^2 - x^2 - xy - y^2} = \frac{dy - dz}{(z^2 - y^2) + x(z - y)} = \frac{dy - dz}{(z - y)(z + y + x)}.$$

$$\therefore \frac{d(x - y)}{(y - x)(x + y + z)} = \frac{d(y - z)}{(z - y)(x + y + z)} \text{ or, } \frac{d(x - y)}{x - y} = \frac{d(y - z)}{y - z}.$$

On integration, $\ln(x - y) - \ln(y - z) = \ln c_1$,

$$\text{i.e. } \frac{x - y}{y - z} = c_1. \tag{9.2.26}$$

Again, choosing x, y, z as multipliers, each ratio

$$= \frac{xdx + ydy + zdz}{xy^2 + xyz + xz^2 + yz^2 + yzx + yx^2 + zx^2 + xyz + zy^2}$$

$$= \frac{xdx + ydy + zdz}{(xy^2 + xyz + y^2z) + (yx^2 + xyz + x^2z) + (xz^2 + xyz + z^2y)}$$

$$= \frac{xdx + ydy + zdz}{(x + y + z)(xy + yz + zx)} = \frac{1}{2}\frac{d(x^2 + y^2 + z^2)}{(x + y + z)(xy + yz + zx)}.$$

Again, choosing $y + z, z + x, x + y$ as multipliers, each ratio

$$= \frac{(y + z)dx + (z + x)dy + (x + y)dz}{(y + z)(y^2 + z^2 + yz) + (z + x)(z^2 + zx + x^2) + (x + y)(x^2 + xy + y^2)}$$

$$= \frac{(ydx + xdy) + (ydz + zdy) + (zdx + xdz)}{2(x^3 + xy^2 + xz^2 + y^3 + yx^2 + yz^2 + z^3 + zx^2 + zy^2)} = \frac{d(xy + yz + zx)}{2(x + y + z)(x^2 + y^2 + z^2)}.$$

$$\therefore \frac{1}{2}\frac{d(x^2 + y^2 + z^2)}{(x + y + z)(xy + yz + zx)} = \frac{1}{2}\frac{d(xy + yz + zx)}{(x + y + z)(x^2 + y^2 + z^2)}$$

$$\text{or, } (x^2 + y^2 + z^2)d(x^2 + y^2 + z^2) = (xy + yz + zx)d(xy + yz + zx)$$

$$\text{or, } \frac{(x^2 + y^2 + z^2)^2}{2} - \frac{(xy + yz + zx)^2}{2} = \text{constant.}$$

$$\text{i.e. } (x^2 + y^2 + z^2)^2 - (xy + yz + zx)^2 = c_2. \tag{9.2.27}$$

∴ (9.2.26) and (9.2.27) together give the required solution.

Example 9.2.13. $\frac{dx}{y^2} = \frac{dy}{x^2} = \frac{dz}{x^2y^2z^2}$.

Solution: Taking the first two ratios, we get

$$x^2dx = y^2dy \Rightarrow x^3 - y^3 = c_1. \tag{9.2.28}$$

Taking the first and third ratios, we get

$$x^2dx = \frac{dz}{z^2} \text{ (cancelling } y^2) \Rightarrow \frac{x^3}{3} + \frac{1}{z} = \text{constant or, } x^3 + \frac{3}{z} = c_2. \tag{9.2.29}$$

∴ (9.2.28) and (9.2.29) together give the required solution.

Example 9.2.14. $\frac{dx}{1} = \frac{dy}{-2} = \frac{dz}{3x^2\sin(y+2x)}$.

Solution: Taking the first two ratios, we easily get

$$x + \frac{1}{2}y = \text{constant or, } y + 2x = c_1. \tag{9.2.30}$$

Taking first and third ratios and using (9.2.30), we get

$$dx = \frac{dz}{3x^2 \sin c_1} \text{ or, } x^3 \sin c_1 = z \text{ i.e. } x^3\sin(y+2x) = 3. \tag{9.2.31}$$

(9.2.30) and (9.2.31) together give the required solution.

Example 9.2.15. $\frac{dx}{x(y^2-z^2)} = \frac{dy}{y(z^2-x^2)} = \frac{dz}{z(x^2-y^2)}$.

Solution: Choosing x, y, z as multipliers, each ratio

$$= \frac{xdx + ydy + zdz}{x^2(y^2-z^2) + y^2(z^2-x^2) + z^2(x^2-y^2)} = \frac{xdx + ydy + zdz}{0}$$

$$\text{or, } xdx + ydy + zdz = 0 \text{ i.e. } x^2 + y^2 + z^2 = c_1. \tag{9.2.32}$$

Again, choosing $\frac{1}{x}, \frac{1}{y}, \frac{1}{z}$ as multipliers each ratio

$$= \frac{\frac{dx}{x} + \frac{dy}{y} + \frac{dz}{z}}{(y^2-z^2) + (z^2-x^2) + (x^2-y^2)} = \frac{\frac{dx}{x} + \frac{dy}{y} + \frac{dz}{z}}{0}$$

so that

$$\frac{dx}{x} + \frac{dy}{y} + \frac{dz}{z} = 0 \text{ or, } \ln x + \ln y + \ln z = \ln c_2 \text{ i.e. } xyz = c_2. \tag{9.2.33}$$

The required solution is given by (9.2.32) and (9.2.33).

Example 9.2.16. $\frac{dx}{y+z} = \frac{dy}{z+x} = \frac{dz}{x+y}$.

Solution: Choosing multipliers $1, -1, 0$, we get

$$\text{each ratio} = \frac{dx - dy}{y - x} \tag{9.2.34}$$

and choosing multipliers $0, 1, -1$, we get each ratio $= \frac{dy-dz}{z-y}$.

$$\therefore \quad \frac{dx-dy}{y-x} = \frac{dy-dz}{z-y} \text{ i.e. } \frac{dx-dy}{-(x-y)} = \frac{dy-dz}{-(y-z)} \text{ or, } \frac{d(x-y)}{x-y} = \frac{d(y-z)}{y-z}. \tag{9.2.35}$$

On integration, $\ln(x-y) - \ln(y-z) = \ln c_1$,

$$\text{i.e. } \frac{x-y}{y-z} = c_1. \tag{9.2.36}$$

Again, choosing $1, 1, 1$ as multipliers,

$$\text{each ratio} = \frac{dx + dy + dz}{2(x+y+z)} = \frac{d(x+y+z)}{2(x+y+z)}.$$

Taking this ratio and the ratio (9.2.34), we write

$$\frac{d(x-y)}{-(x-y)} = \frac{d(x+y+z)}{2(x+y+z)} \text{ or, } \ln(x-y) + \frac{1}{2}\ln(x+y+z) = \ln c_2,$$

$$\text{i.e. } (x-y)(x+y+z)^{1/2} = c_2 \text{ or, } (x-y)^2(x+y+z) = \text{constant } c_3 \text{ (say)}. \tag{9.2.37}$$

$\therefore$ (9.2.35) and (9.2.36) together give the required solution.

Example 9.2.17. $\frac{dx}{y(x+y)+az} = \frac{dy}{x(x+y)-az} = \frac{dz}{z(x+y)}$.

Solution: First choose 1, 1, 0 as multipliers; then

$$\text{each ratio} = \frac{dx+dy}{(x+y)^2+az-az} = \frac{d(x+y)}{(x+y)^2}.$$

This ratio and the last ratio of the given equation give

$$\frac{d(x+y)}{(x+y)^2} = \frac{dz}{z(x+y)} \text{ or, } \frac{d(x+y)}{x+y} = \frac{dz}{z}$$

On integration, we get $\ln(x+y) - \ln z = \ln c_1$

$$\text{or, } (x+y)/z = c_1. \tag{9.2.38}$$

Again, choosing $x, -y, 0$ as multipliers, we get each ratio

$$= \frac{xdx - ydy}{xy(x+y)+azx-yx(x+y)+azy} = \frac{xdx-ydy}{az(x+y)}.$$

This ratio and the third ratio of the given equation give

$$\frac{xdx-ydy}{az(x+y)} = \frac{dz}{z(x+y)} \text{ or, } \frac{1}{2}d(x^2-y^2) = adz \text{ or, } x^2-y^2 = 2az + c_2. \tag{9.2.39}$$

(9.2.37) and (9.2.38) together give the required solution.

Example 9.2.18.

$$\frac{dx}{x^2+a^2} = \frac{dy}{xy-az} = \frac{dz}{xz+ay}. \tag{9.2.40}$$

Solution: Choosing $0, z, -y$ as multipliers, each ratio

$$= \frac{zdy-ydz}{xyz-az^2-xyz-ay^2} = \frac{zdy-ydz}{-a(y^2+z^2)}. \tag{9.2.41}$$

Choosing $0, y, z$ as multipliers, each ratio

$$= \frac{ydy+zdz}{xy^2-ayz+xz^2+ayz} = \frac{ydy+zdz}{x(y^2+z^2)}. \tag{9.2.42}$$

From (9.2.39) and (9.2.40), we obtain

$$\frac{dx}{x^2+a^2} = \frac{zdy-ydz}{-a(y^2+z^2)} \text{ i.e. } \frac{adx}{x^2+a^2} + \frac{zdy-ydz}{y^2+z^2} = 0 \text{ or, } \frac{adx}{x^2+a^2} = \frac{\frac{zdy-ydz}{z^2}}{1+\left(\frac{y}{z}\right)^2} = \frac{d\left(\frac{y}{z}\right)}{1+\left(\frac{y}{z}\right)^2}.$$

On integration, we get

$$a \cdot \frac{1}{a}\tan^{-1}\frac{x}{a} = \tan^{-1}\frac{y}{z} + c_1 \text{ or, } \tan^{-1}\frac{x}{a} - \tan^{-1}\frac{y}{z} = c_1. \tag{9.2.43}$$

Taking the first ratio of (9.2.39) and the ratio (9.2.41), we get

$$\frac{dx}{x^2+a^2} = \frac{\frac{1}{2}d(y^2+z^2)}{x(y^2+z^2)} \text{ or, } \frac{2xdx}{x^2+a^2} = \frac{d(y^2+z^2)}{y^2+z^2} \text{ or, } \ln(x^2+a^2) - \ln(y^2+z^2) = \ln c_2 \text{ or, } \frac{x^2+a^2}{y^2+z^2} = c_2. \tag{9.2.44}$$

Thus, (9.2.42) and (9.2.43) together give the required solution.

Example 9.2.19. $\frac{dx}{x(x+y)} = \frac{dy}{-y(x+y)} = \frac{dz}{-(x-y)(2x+2y+z)}$.

Solution: Taking the first two ratios, we get

$$\frac{dx}{x} = -\frac{dy}{y} \text{ (cancelling } x+y) \text{ or, } \ln x + \ln y = \ln c_1 \text{ i.e. } xy = c_1. \quad (9.2.45)$$

Choosing 1, 1, 0 as multipliers, each ratio

$$= \frac{dx + dy}{(x^2 + xy) + (-xy - y^2)} = \frac{dx + dy}{(x+y)(x-y)} = \frac{d(x+y)}{(x+y)(x-y)}.$$

Again, choosing 1, 1, 1 as multipliers, each ratio

$$= \frac{dx + dy + dz}{(x^2 + xy) + (-xy - y^2) - (x-y)(2x+2y+z)}$$

$$= \frac{dx + dy + dz}{(x-y)(x+y-2x-2y-z)} = \frac{dx+dy+dz}{-(x-y)(x+y+z)} = \frac{d(x+y+z)}{-(x-y)(x+y+z)}.$$

$$\therefore \frac{d(x+y)}{(x+y)(x-y)} = -\frac{d(x+y+z)}{(x-y)(x+y+z)} \text{ or, } \ln(x+y) + \ln(x+y+z) = \ln c_2 \text{ or, } (x+y)(x+y+z) = c_2. \quad (9.2.46)$$

Thus, (9.2.44) and (9.2.45) together give the required solution.

Example 9.2.20. $\frac{dx}{\cos(x+y)} = \frac{dy}{\sin(x+y)} = \frac{dz}{z}$.

Solution: From the first two ratios

$$\frac{dy}{dx} = \frac{\sin(x+y)}{\cos(x+y)}. \text{ Put } x+y = v; \text{ then } \frac{dy}{dx} = \frac{dv}{dz} - 1$$

$$\text{or, } \frac{dv}{dx} = 1 + \tan v \text{ or, } \frac{dv}{1+\tan v} = dx.$$

Integrating,

$$x = \int \frac{dv}{1+\tan v} = \int \frac{\cos v dv}{\sin v + \cos v} = \frac{1}{2} \int \frac{(\cos v - \sin v) + (\sin v + \cos v)}{\sin v + \cos v} dv$$

$$= \frac{1}{2} \ln(\cos v + \sin v) + \frac{1}{2} v + \text{constant}.$$

$$\therefore 2x = \ln\{\cos(x+y) + \sin(x+y)\} + x + y + \text{constant}$$

$$\text{or, } x - y = \ln\{\cos(x+y) + \sin(x+y)\} + \ln c_1$$

$$\text{or, } e^{x-y} = c_1\{\cos(x+y) + \sin(x+y)\}. \quad (9.2.47)$$

Also

$$\frac{dz}{z} = \frac{dx}{\cos(x+y)} = \frac{dy}{\sin(x+y)} = \frac{d(x+y)}{\cos(x+y) + \sin(x+y)}.$$

$$\therefore \ln z + c_2 = \int \frac{dt}{\cos t + \sin t} \quad (t = x+y) = \int \frac{\frac{1}{\sqrt{2}} dt}{\sin\frac{\pi}{4}\cos t + \cos\frac{\pi}{4}\sin t} = \frac{1}{\sqrt{2}} \int \frac{dt}{\sin\left(\frac{\pi}{4} + t\right)}$$

$$\text{or, } \ln z + c_2 = \frac{1}{\sqrt{2}} \int \operatorname{cosec}\left(\frac{\pi}{4} + t\right) dt = \frac{1}{\sqrt{2}} \ln\tan\left(\frac{\pi}{8} + \frac{t}{2}\right)$$

$$\text{or, } \sqrt{2}\ln z + \ln\cot\left(\frac{\pi}{8} + \frac{t}{2}\right) = \ln c_3 \quad \text{(say)}$$

$$\text{or, } z^{\sqrt{2}} \cot\left(\frac{\pi}{8} + \frac{x+y}{2}\right) = c_3. \quad (9.2.48)$$

The relations (9.2.46) and (9.2.47) constitute the complete solution.

Example 9.2.21. $\frac{dx}{y^3x-2x^4} = \frac{dy}{2y^4-x^3y} = \frac{dz}{9z(x^3-y^3)}$.

Solution: Choosing $\frac{1}{x}, \frac{1}{y}, \frac{1}{3z}$ as multipliers,

$$\text{each ratio} = \frac{\frac{1}{x}dx + \frac{1}{y}dy + \frac{1}{3z}dz}{y^3 - 2x^3 + 2y^3 - x^3 + 3(x^3 - y^3)} = \frac{\frac{dx}{x} + \frac{dy}{y} + \frac{1}{3}\frac{dz}{z}}{0}, \text{ i.e. } \frac{dx}{x} + \frac{dy}{y} + \frac{1}{3}\frac{dz}{z} = 0.$$

Integrating, $\ln x + \ln y + \frac{1}{3}\ln z = \ln c_1$ (say);

$$\text{i.e. } xyz^{1/3} = c_1. \tag{9.2.49}$$

Again, taking the first two ratios, $y(2y^3 - x^3)dx - x(y^3 - 2x^3)dy = 0$.
Dividing by x^3y^3, we get

$$\frac{2y}{x^3}dx - \frac{1}{y^2}dx - \frac{1}{x^2}dy + \frac{2x}{y^3}dy = 0 \text{ or, } \frac{2y}{x^3}dy - \frac{1}{x^2}dy + \frac{2x}{y^3}dy - \frac{1}{y^2}dx = 0$$

$$\text{or, } \frac{2ydx - xdy}{x^3} + \frac{2xdy - ydx}{y^3} = 0 \text{ or, } \frac{2xydx - x^2dy}{x^4} + \frac{2xydy - y^2dx}{y^4} = 0$$

$$\text{or, } -d\left(\frac{y}{x^2}\right) - d\left(\frac{x}{y^2}\right) = 0 \text{ or, } \frac{y}{x^2} + \frac{x}{y^2} = \text{constant } c_2 \text{ (say)}. \tag{9.2.50}$$

(9.2.48) and (9.2.49) together give the complete solution.

Introduction of a new variable t (A New Method)

Example 9.2.22. *Solve:* $\frac{dx}{x+y-z} = \frac{dy}{y+z-x} = \frac{dz}{z+x-y}$.

Solution: Introduce a new variable t such that

$$\frac{dt}{t} = \frac{dx}{x+y-z} = \frac{dy}{y+z-x} = \frac{dz}{z+x-y} = \frac{ldx + mdy + ndz}{l(x+y-z) + m(y+z-x) + n(z+x-y)}$$

$$= \frac{ldx + mdy + ndz}{x(l-m+n) + y(l+m-n) + z(-l+m+n)}.$$

Choose l, m, n such that

$$\left.\begin{aligned} l - m + n &= \lambda l \\ l + m - n &= \lambda m \\ -l + m + n &= \lambda n \end{aligned}\right\} \text{ or, } \left.\begin{aligned} l(1-\lambda) - m + n &= 0 \\ l + m(1-\lambda) - n &= 0 \\ -l + m + n(1-\lambda) &= 0 \end{aligned}\right\}. \tag{9.2.51}$$

Eliminating l, m, n from the set of equations (9.2.50),

$$\begin{vmatrix} 1-\lambda & -1 & 1 \\ 1 & 1-\lambda & -1 \\ -1 & 1 & 1-\lambda \end{vmatrix} = 0$$

or, $\lambda^3 - 3\lambda^2 + 6\lambda - 4 = 0$ (expanding the left side determinant)

or, $(\lambda - 1)(\lambda^2 - 2\lambda + 4) = 0$ or, $\lambda = 1, \lambda = 1 \pm i\sqrt{3}$,

i.e. $\lambda = 1, -2\omega, -2\omega^2$,

where ω, ω^2 are the imaginary cube roots of unity ($\omega^3 = 1$).

For $\lambda = 1$, the equations (9.2.50) give $l = m = n = 1$.

$$\therefore\ \frac{dt}{t} = \frac{d(x+y+z)}{x+y+z} \Rightarrow \ln t = \ln(x+y+z) + \text{constant}. \tag{9.2.52}$$

For $\lambda = -2\omega$,

$$\left.\begin{aligned} l(1+2\omega) - m + n &= 0 \\ l + m(1+2\omega) - n &= 0 \end{aligned}\right\}.$$

They give

$$\frac{l}{1-1-2\omega} = \frac{m}{1+1+2\omega} = \frac{n}{(1+2\omega)^2+1} \text{ or, } \frac{l}{-2\omega} = \frac{m}{2(1+\omega)} = \frac{n}{2(1+2\omega+2\omega^2)}$$

$$\text{or, } \frac{l}{-2\omega} = \frac{m}{-2\omega^2} = \frac{n}{-2} \text{ (using } 1+\omega+\omega^2 = 0\text{) or, } \frac{l}{\omega} = \frac{m}{\omega^2} = \frac{n}{1}.$$

$$\therefore\ \frac{dt}{t} = \frac{d(\omega x + \omega^2 y + z)}{-2\omega(\omega x + \omega^2 y + z)} = -\frac{1}{2\omega}\frac{d(x+\omega y+\omega^2 z)}{x+\omega y+\omega^2 z}$$

Integrating,

$$\ln t = -\frac{1}{2\omega}\ln(x+\omega y+\omega^2 z) + \text{constant} \tag{9.2.53}$$

From equations (9.2.50), (9.2.51) and (9.2.52), we obtain

$$\left.\begin{aligned} 2\ln(x+y+z) + \omega^2\ln(x+\omega^2 y+\omega z) &= c_1 \\ \text{and}\quad 2\ln(x+y+z) + \omega\ln(x+\omega^2 y+\omega z) &= c_2 \end{aligned}\right\},\ c_1, c_2 \text{ are constants.}$$

They give the required solution.

Try in a similar way:

Example 9.2.23. *Solve:* $\frac{dx}{-1} = \frac{dy}{3y+4z} = \frac{dz}{2y+5z}$.

The required Solution: $y - z = c_1 e^{-x}, y + 2z = c_2 e^{-7x}$.

Examples IX(B)

Solve the following simultaneous equations:

1. $\frac{dx}{x^2+2y^2} = \frac{dy}{-xy} = \frac{dz}{zx}$.
2. $\frac{dx}{z} = \frac{dy}{-z} = \frac{dz}{z^2+(x+y)^2}$.
3. $\frac{dx}{xz} = \frac{dy}{yz} = \frac{dz}{xy}$.
4. $\frac{dx}{y} = \frac{dy}{-x} = \frac{dz}{bx-ay}$.
5. $\frac{dx}{z-y} = \frac{dy}{x-z} = \frac{dz}{y-x}$.
6. $\frac{yzdx}{y-z} = \frac{zxdy}{z-x} = \frac{xydx}{x-y}$.
7. $\frac{dx}{x(y^2-z^2)} = \frac{dy}{-y(z^2+x^2)} = \frac{dz}{z(x^2+y^2)}$.
8. $\frac{dx}{y^2+z^2} = \frac{dy}{-xy} = \frac{dz}{-xz}$.
9. $\frac{dx}{4y-3z} = \frac{dy}{4x-2z} = \frac{dz}{2y-3x}$.
10. $\frac{dx}{y} = \frac{dy}{-x} = \frac{dz}{2x-3y}$.
11. $\frac{dx}{x(x^2+3y^2)} = \frac{dy}{-y(3x^2+y^2)} = \frac{dz}{2z(y^2-x^2)}$
12. $\frac{dx}{1+y} = \frac{dy}{1+x} = \frac{dz}{z}$
13. $\frac{dx}{xz} = \frac{dy}{yz} = \frac{dz}{xy}$
14. $\frac{dx}{x^2+y^2+yz} = \frac{dy}{x^2+y^2-xz} = \frac{dz}{z(x+y)}$
15. $\frac{dx}{x^2} = \frac{dy}{y^2} = \frac{dz}{z(x+y)}$.
16. $\frac{dx}{x} = \frac{dy}{z} = \frac{dz}{y}$
17. $\frac{-dx}{x(x+y)} = \frac{dy}{y(x+y)} = \frac{dz}{(x-y)(2x+2y+z)}$.
18. $\frac{dx}{y^2+yz+z^2} = \frac{dy}{z^2+zx+x^2} = \frac{dz}{x^2+xy+y^2}$.
19. $\frac{dx}{1} = \frac{dy}{2} = \frac{dz}{5z+\tan(y-2x)}$
20. $\frac{dx}{\cos(x+y)} = \frac{dy}{\sin(x+y)} = \frac{dz}{z+\frac{1}{z}}$.

Answers

1. $yz = c_1$ and $x^2y^2 + y^4 = c_2$.
2. $x + y = c_1$ and $z^2 + (x+y)^2 = c_2 e^{2x}$.
3. $x = c_1 y$ and $xy - z^2 = c_2$.
4. $ax + by + z = c_1$ and $x^2 + y^2 = c_2$.
5. $x + y + z = c_1$ and $x^2 + y^2 + z^2 = c_2$.
6. $xyz = c_1$ and $x + y + z = c_2$.
7. $x^2 + y^2 + z^2 = c_1, x = c_2 yz$.
8. $y = c_1 z; x^2 + y^2 + z^2 = c_2$.
9. $2x - 3y - 4z = c_1; x^2 - y^2 - z^2 = c_2$.
10. $x^2 + y^2 = c_1, 3x + 2y + z = c_2$.
11. $\frac{xy}{z} = c_1, z(x^2 + y^2) = c_2$.
12. $(1+x)^2 - (1+y)^2 = c_1; (2 + x + y)/z = c_2$.
13. $x = c_1 y; xy - z^2 = c_2$.
14. $z - x + y = c_1, x^2 + y^2 = c_2 z^2$.
15. $\frac{1}{y} - \frac{1}{x} = c_1$; multipliers $1, -1, 0$ and last ratio gives $\frac{x-y}{z} = c_2$.
16. $y^2 - z^2 = c_1; (y+z)/x = c_2$.
17. $xy = c_1; z(x+y) + x^2 + y^2 = c_2$.
18. $\frac{y-x}{y-z} = c_1; (x^2 + y^2 + z^2)^2 = (xy + yz + zx)^2 + c_2$.
19. $y - 2x = c_1, 5z + \tan(y - 2x) = c_2 e^{5x}$.
20. $e^{y-x}\{\cos(x+y) + \sin(x+y)\} = c_1$, $(z^2+1)^{1/\sqrt{2}} \cot\left(\frac{\pi}{8} + \frac{x+y}{2}\right) = c_2$.

9.3 Total differential equations

I. We now propose to discuss single equation having one independent variable and more than one dependent variables. The usual form is $Pdx + Qdy + Rdz$, where P, Q, R are functions of x, y, z.

II. Pfaffian differential form is an expression like

$$\sum_{i=1}^{n} F_i(x_1, x_2, \cdots, x_n)dx_i,$$

where F_i are functions of some or all of the n variables $x_1, x_2, x_3, \cdots, x_n$.

The equation

$$\sum_{i=1}^{n} F_i dx_i = 0$$

is called a *Pfaffian differential equation.*

In two variables x and y, Pfaffian differential equation is of the form $Mdx + Ndy = 0$ (M, N are functions of x, y).

A *Pfaffian differential equation* in two variables always possesses an integrating factor.

For three variables x, y, z, the Pfaffian differential equation is of the form:

$$\mathbf{Pdx + Qdy + Rdz = 0\ (P, Q, R \text{ are functions of } x, y, z).} \tag{9.3.1}$$

This is also called *a total differential equation in x, y, z*. Such an equation can be directly integrated if there exists a function $\phi(x, y, z)$ whose total differential $d\phi = Pdx + Qdy + Rdz$. If no such function $\phi(x, y, z)$ is available, then the eq. (9.3.1) may or may not be integrable. We now proceed to find the condition which P, Q, R must satisfy so that (9.3.1) may be integrable. This will be called the *condition of integrability* of the single differential eq. (9.3.1).

III. Condition of Integrability

Theorem 9.3.1. *A necessary and sufficient condition for integrability of the single differential equation*

$$Pdx + Qdy + Rdz = 0\ (P, Q, R \text{ are functions of } x, y, z) \tag{9.3.2}$$

is that

$$P\left(\frac{\partial Q}{\partial z} - \frac{\partial R}{\partial y}\right) + Q\left(\frac{\partial R}{\partial x} - \frac{\partial P}{\partial z}\right) + R\left(\frac{\partial P}{\partial y} - \frac{\partial Q}{\partial x}\right) = 0.$$

Proof.

The condition is necessary

Suppose, the eq. (9.3.2) has an integral

$$\phi(x, y, z) = c \tag{9.3.3}$$

so that the expression

$$d\phi = \frac{\partial \phi}{\partial x}dx + \frac{\partial \phi}{\partial y}dy + \frac{\partial \phi}{\partial z}dz = Pdx + Qdy + Rdz$$

or equal to it multiplied by a factor $\mu(x, y, z)$, say.

Thus,

$$\frac{\frac{\partial \phi}{\partial x}}{P} = \frac{\frac{\partial \phi}{\partial y}}{Q} = \frac{\frac{\partial \phi}{\partial z}}{R} = \mu(x, y, z), \text{ i.e. } \frac{\partial \phi}{\partial x} = \mu P, \ \frac{\partial \phi}{\partial y} = \mu Q, \ \frac{\partial \phi}{\partial z} = \mu R. \tag{9.3.4}$$

From the first two relations of (9.3.4), we get

$$\frac{\partial}{\partial y}(\mu P) = \frac{\partial^2 \phi}{\partial y \partial x} = \frac{\partial^2 \phi}{\partial x \partial y} = \frac{\partial}{\partial x}(\mu Q) \text{ or, } \mu\frac{\partial P}{\partial y} + P\frac{\partial \mu}{\partial y} = \mu\frac{\partial Q}{\partial x} + Q\frac{\partial \mu}{\partial x}$$

$$\text{or, } \mu\left(\frac{\partial P}{\partial y} - \frac{\partial Q}{\partial x}\right) = Q\frac{\partial \mu}{\partial x} - P\frac{\partial \mu}{\partial y}. \tag{9.3.5}$$

Similarly taking two other relations of (9.3.4), we easily obtain

$$\mu\left(\frac{\partial Q}{\partial z} - \frac{\partial R}{\partial y}\right) = R\frac{\partial \mu}{\partial y} - Q\frac{\partial \mu}{\partial z} \tag{9.3.6}$$

$$\text{and } \mu\left(\frac{\partial R}{\partial x} - \frac{\partial P}{\partial z}\right) = P\frac{\partial \mu}{\partial z} - R\frac{\partial \mu}{\partial x}. \tag{9.3.7}$$

Multiplying (9.3.5), (9.3.6), (9.3.7) by R, P and Q respectively and adding, we obtain

$$\mu\left[R\left(\frac{\partial P}{\partial y} - \frac{\partial Q}{\partial x}\right) + P\left(\frac{\partial Q}{\partial z} - \frac{\partial R}{\partial y}\right) + Q\left(\frac{\partial R}{\partial x} - \frac{\partial P}{\partial z}\right)\right] = 0$$

$$\text{or, } P\left(\frac{\partial Q}{\partial z} - \frac{\partial R}{\partial y}\right) + Q\left(\frac{\partial R}{\partial x} - \frac{\partial P}{\partial z}\right) + R\left(\frac{\partial P}{\partial y} - \frac{\partial Q}{\partial x}\right) = 0, \tag{9.3.8}$$

which is the condition to be satisfied by P, Q, R for eq. (9.3.2) to possess an integral of the form (9.3.3).

The condition is sufficient

Suppose that the coefficients P, Q, R in the equation

$$Pdx + Qdy + Rdz = 0 \tag{9.3.2}$$

satisfy the condition

$$P\left(\frac{\partial Q}{\partial z} - \frac{\partial R}{\partial y}\right) + Q\left(\frac{\partial R}{\partial x} - \frac{\partial P}{\partial z}\right) + R\left(\frac{\partial P}{\partial y} - \frac{\partial Q}{\partial x}\right) = 0. \tag{9.3.9}$$

We shall prove that an integral of (9.3.2) can be found when the relation (9.3.9) holds. This will prove that the condition is sufficient.

We first observe that if relation (9.3.9) holds for the coefficients of (9.3.2), then *a similar relation* also holds for the coefficients of the relation:

$$\mu P dx + \mu Q dy + \mu R dz = 0, \tag{9.3.10}$$

where μ is a function of (x, y, z).

This can be verified by direct substitution. See the working given below:

$$\frac{\partial}{\partial z}(\mu Q) - \frac{\partial}{\partial y}(\mu R) = \mu\left(\frac{\partial Q}{\partial z} - \frac{\partial R}{\partial y}\right) + Q\frac{\partial \mu}{\partial z} - R\frac{\partial \mu}{\partial y}. \tag{9.3.11}$$

Similarly,

$$\frac{\partial}{\partial x}(\mu R) - \frac{\partial}{\partial z}(\mu P) = \mu\left(\frac{\partial R}{\partial x} - \frac{\partial P}{\partial z}\right) + R\frac{\partial \mu}{\partial x} - P\frac{\partial \mu}{\partial z} \tag{9.3.12}$$

$$\text{and } \frac{\partial}{\partial y}(\mu P) - \frac{\partial}{\partial x}(\mu Q) = \mu\left(\frac{\partial P}{\partial y} - \frac{\partial Q}{\partial x}\right) + P\frac{\partial \mu}{\partial y} - Q\frac{\partial \mu}{\partial x}. \tag{9.3.13}$$

Multiply (9.3.11) by μP, (9.3.12) by μQ and (9.3.13) by μR and then add.

We thus obtain

$$\mu P\left\{\frac{\partial}{\partial z}(\mu Q) - \frac{\partial}{\partial y}(\mu R)\right\} + \mu Q\left\{\frac{\partial}{\partial x}(\mu R) - \frac{\partial}{\partial z}(\mu P)\right\} + \mu R\left\{\frac{\partial}{\partial y}(\mu P) - \frac{\partial}{\partial x}(\mu Q)\right\}$$

$$= \mu^2\left\{P\left(\frac{\partial Q}{\partial z} - \frac{\partial R}{\partial y}\right) + Q\left(\frac{\partial R}{\partial x} - \frac{\partial P}{\partial z}\right) + R\left(\frac{\partial P}{\partial y} - \frac{\partial Q}{\partial x}\right)\right\} \text{ (other terms cancel and give zero)}$$

Hence, if the relation

$$\sum P\left(\frac{\partial Q}{\partial z} - \frac{\partial R}{\partial y}\right) = 0 \text{ is true, then } \sum \mu P\left\{\frac{\partial}{\partial z}(\mu Q) - \frac{\partial}{\partial y}(\mu R)\right\} = 0 \text{ is also true.}$$

Consider the expression $Pdx + Qdy$, which is *either* an exact differential with respect to x and y or if it is not an exact differential, then an integrating factor μ can be found for it and in that case the equation $\mu Pdx + \mu Qdy + \mu Rdz = 0$ can then be considered and the observations mentioned in the last page should be used.

Therefore, there is no loss of generality regarding $Pdx + Qdy$ as an exact differential. For this the condition is

$$\frac{\partial P}{\partial y} = \frac{\partial Q}{\partial x}.$$

Let $V = \int Pdx + Qdy$, then it follows that

$$P = \frac{\partial V}{\partial x},\ Q = \frac{\partial V}{\partial y}\ \left(\because dV = \frac{\partial V}{\partial x}dx + \frac{\partial V}{\partial y}dy = Pdx + Qdy\right)$$

$$\text{and } \frac{\partial P}{\partial z} = \frac{\partial^2 V}{\partial z \partial x},\ \frac{\partial Q}{\partial z} = \frac{\partial^2 V}{\partial z \partial y}.$$

Hence, from the condition (9.3.9) it follows that

$$\frac{\partial V}{\partial x}\left(\frac{\partial^2 V}{\partial z \partial y} - \frac{\partial R}{\partial y}\right) + \frac{\partial V}{\partial y}\left(\frac{\partial R}{\partial x} - \frac{\partial^2 V}{\partial z \partial x}\right) = 0.$$

$$\left[\because R\left(\frac{\partial P}{\partial y} - \frac{\partial Q}{\partial x}\right) = R\left(\frac{\partial^2 V}{\partial y \partial x} - \frac{\partial^2 V}{\partial x \partial y}\right) = R\left(\frac{\partial^2 V}{\partial y \partial x} - \frac{\partial^2 V}{\partial y \partial x}\right) = 0\right]$$

This may be written as

$$\frac{\partial V}{\partial x}\frac{\partial}{\partial y}\left(\frac{\partial V}{\partial z}-R\right)-\frac{\partial V}{\partial y}\frac{\partial}{\partial x}\left(\frac{\partial V}{\partial z}-R\right)=0$$

or the determinant

$$\begin{vmatrix} \frac{\partial V}{\partial x} & \frac{\partial}{\partial x}\left(\frac{\partial V}{\partial z}-R\right) \\ \frac{\partial V}{\partial y} & \frac{\partial}{\partial y}\left(\frac{\partial V}{\partial z}-R\right) \end{vmatrix}=0,$$

which implies that a relation independent of x and y exists between V and $\frac{\partial V}{\partial z}-R$.

Therefore, $\frac{\partial V}{\partial z}-R$ can be expressed as a function of z and V alone.

Suppose that $\frac{\partial V}{\partial z}-R=\phi(z,V)$.

Since

$$Pdx+Qdy+Rdz=\frac{\partial V}{\partial x}dx+\frac{\partial V}{\partial y}dy+\frac{\partial V}{\partial z}dz+\left(R-\frac{\partial V}{\partial z}\right)dz.$$

$\therefore$ the eq. (9.3.2), namely $Pdx+Qdy+Rdz=0$, may be written as $dV-\phi(z,V)dz=0$.

This is an equation in two variables. Its integration will lead to an equation of the form $F(V,z)=0$.

Hence, the condition (9.3.9) is sufficient for (9.3.2) to have an integral.

Note 9.3.1. *Let* $\vec{A}=(P,Q,R)$.

Then

$$\operatorname{curl}\vec{A}=\nabla\times\vec{A}=\begin{vmatrix}\hat{\imath} & \hat{\jmath} & \hat{k}\\ \frac{\partial}{\partial x} & \frac{\partial}{\partial y} & \frac{\partial}{\partial z}\\ P & Q & R\end{vmatrix}=\hat{\imath}\left(\frac{\partial R}{\partial y}-\frac{\partial Q}{\partial z}\right)+\hat{\jmath}\left(\frac{\partial P}{\partial z}-\frac{\partial R}{\partial x}\right)+\hat{k}\left(\frac{\partial Q}{\partial x}-\frac{\partial P}{\partial y}\right).$$

Condition of Integrability (in vector form) is $\vec{A}\cdot\operatorname{curl}\vec{A}=0$.

$$\text{i.e. } P\left(\frac{\partial R}{\partial y}-\frac{\partial Q}{\partial z}\right)+Q\left(\frac{\partial P}{\partial z}-\frac{\partial R}{\partial x}\right)+R\left(\frac{\partial Q}{\partial x}-\frac{\partial P}{\partial y}\right)=0.$$

Note 9.3.2. *The total differential equation* $Pdx+Qdy+Rdz=0$ *is said to be* **exact,** *if*

$$\frac{\partial P}{\partial y}=\frac{\partial Q}{\partial x},\quad \frac{\partial Q}{\partial z}=\frac{\partial R}{\partial y}\ \text{and}\ \frac{\partial R}{\partial x}=\frac{\partial P}{\partial z}.$$

When these conditions are satisfied, the condition for integrability of $Pdx+Qdy+Rdz=0$ *is obviously satisfied.*

IV. Solved Examples

Example 9.3.1. *Solve:*

$$(2x^2+2xy+2xz^2+1)dx+dy+2zdz=0. \qquad (9.3.14)$$

Solution:

Step 1. Comparing the given equation with $Pdx+Qdy+Rdz=0$, we write $P=2x^2+2xy+2xz^2+1$, $Q=1$, $R=2z$.

Now

$$P\left(\frac{\partial Q}{\partial z}-\frac{\partial R}{\partial y}\right)+Q\left(\frac{\partial R}{\partial x}-\frac{\partial P}{\partial z}\right)+R\left(\frac{\partial P}{\partial y}-\frac{\partial Q}{\partial z}\right)$$
$$=(2x^2+2xy+2xz^2+1)(0-0)+1(0-4xz)+2z(2x-0)=0.$$

Thus, the condition of integrability is satisfied and hence, the given equation is integrable.

Step 2. Let x be treated as constant, so that $dx = 0$. Then (9.3.14) becomes $dy + 2zdz = 0$.

Integrating,

$$y+z^2=\text{constant}=f(x),\ \text{say,} \tag{9.3.15}$$

where the constant of integration has been taken as a function $f(x)$ of x, because we have treated x as a constant.

Differentiating (9.3.15), we get

$$dy+2zdz-f'(x)dx=0. \tag{9.3.16}$$

Comparing (9.3.16) with the given eq. (9.3.14), we have

$$f'(x)=-[2x^2+2xy+2xz^2+1]$$
$$\text{or,}\quad f'(x)=-(2x^2+1)-2x(y+z^2)$$
$$\text{or,}\quad f'(x)=-(2x^2+1)-2xf(x)$$
$$\text{or,}\quad \frac{d}{dx}f(x)+2xf(x)=-(2x^2+1).$$

$\left(\textit{This is a linear equation in } f(x) \textit{ with } \text{IF} = e^{\int 2xdx}=e^{x^2}\right)$

Hence, multiplying by the IF, we get

$$\frac{d}{dx}\left\{f(x)\cdot e^{x^2}\right\}=-\int(2x^2+1)e^{x^2}dx=-\int xd(e^{x^2})-\int e^{x^2}dx$$

On integration

$$f(x)\cdot e^{x^2}=-xe^{x^2}+K \text{ (where } K \text{ is an arbitrary constant)}$$
$$\text{i.e.}\quad (f(x)+x)e^{x^2}=K$$
$$\text{or,}\quad (y+z^2+x)e^{x^2}=K\ (\textit{Required primitive of the given equation}).$$

Example 9.3.2. *Solve:*

$$3x^2dx+3y^2dy-(x^3+y^3+e^{2z})dz=0. \tag{9.3.17}$$

Solution:

Step 1. Comparing the given eq. (9.3.17) with $Pdx + Qdy + Rdz = 0$, we write $P = 3x^2$, $Q = 3y^2$, $R = -x^3 - y^3 - e^{2z}$.

Now

$$P\left(\frac{\partial Q}{\partial z}-\frac{\partial R}{\partial y}\right)+Q\left(\frac{\partial R}{\partial x}-\frac{\partial P}{\partial z}\right)+R\left(\frac{\partial P}{\partial y}-\frac{\partial Q}{\partial z}\right)=0$$
$$3x^2(0+3y^2)+3y^2(-3x^2-0)+(-x^3-y^3-e^{2z})(0-0)=0$$

so that the given equation is integrable.

Step 2. Let z be treated as constant so that $dz = 0$.

Then (9.3.17) becomes $x^3 + y^3 = \text{constant} = f(z)$, say, or, $3x^2dx + 3y^2dy - f'(z)dz = 0$.

Comparing it with the given equation $3x^2dx + 3y^2dy - (x^3 + y^3 + e^{2z})dz = 0$, we obtain

$$f'(z) = x^3 + y^3 + e^{2z} = f(z) + e^{2z}$$

$$\frac{d}{dz}f(z) - f(z) = e^{2z} \text{ (Linear equation in } f(z) \text{ with IF} = e^{-z}).$$

Multiplying by the IF, we get

$$\frac{d}{dz}(f(z) \cdot e^{-z}) = e^{2z} \cdot e^{-z} = e^z.$$

Integrating,

$$f(z) \cdot e^{-z} = e^z + K \text{ (where } K \text{ is an arbitrary constant)}$$

$$(x^3 + y^3)e^{-z} = e^z + K$$

$$\text{or,} \quad x^3 + y = e^{2z} + Ke^z \text{ (*Required solution of the given equation*).}$$

Example 9.3.3. *Solve:* $(x^2 + y^2 + z^2)dx - 2xydy - 2xzdz = 0$.

Solution: It is easy to verify that

$$\sum P\left(\frac{\partial Q}{\partial z} - \frac{\partial R}{\partial y}\right) = 0,$$

where $P = x^2 + y^2 + z$, $Q = -2xy$, $R = -2xz$.

Let x be regarded as constant so that $dx = 0$. Then the given equation becomes

$$2xydy + 2xzdz = 0 \text{ or, } 2ydy + 2zdz = 0.$$

Integrating, $y^2 + z^2 = \text{constant} = f(x)$, say, where $f(x)$ is taken as constant of integration, since x is treated as constant or, $2ydy + 2zdz = f'(x)dx$.

Comparing with the given equation,

$$\frac{x^2 + y^2 + z^2}{x}dx - f'(x)dx = 0 \text{ or, } f'(x) - \left(x + \frac{y^2 + z^2}{x}\right) = 0 \text{ or, } f'(x) - \frac{f(x)}{x} = x \text{ or, } \frac{d}{dx}\{f(x)\} - \frac{1}{x}f(x) = x.$$

This is a linear equation in $f(x)$ with IF $e^{-\int \frac{1}{x}dx} = \frac{1}{x}$.

Multiplying by the IF, we obtain $\frac{d}{dx}\left\{\frac{1}{x}f(x)\right\} = 1$.

On integration, we get

$$\frac{1}{x}f(x) = x + K \quad (K \text{ is an arbitrary constant})$$

$$\text{or,} \quad \frac{1}{x}(y^2 + z^2) = x + K$$

$$\text{or,} \quad y^2 + z^2 - x^2 = Kx \quad \text{(*Required solution*).}$$

Example 9.3.4. *Solve:* $y^2z(x\cos x - \sin x)dx + x^2z(y\cos y - \sin y)dy + xy(y\sin x + x\sin y + xy\cos z)dz = 0$.

Solution: First verify that the condition of integrability is satisfied.

Let z be treated as constant so that $dz = 0$. Then the given equation becomes

$$y^2 z(x\cos x - \sin x)dx + x^2 z(y\cos y - \sin y)dy = 0$$

$$\text{or, } \frac{(x\cos x - \sin x)}{x^2}dx + \frac{y\cos y - \sin y}{y^2}dy = 0 \text{ (} z \text{ cancelled)}$$

$$\text{or, } d\left(\frac{\sin x}{x}\right) + d\left(\frac{\sin y}{y}\right) = 0.$$

Integrating,

$$\frac{\sin x}{x} + \frac{\sin y}{y} = \text{constant} = f(z), \text{ say,} \tag{9.3.18}$$

where $f(z)$ is taken as the constant of integration since z is treated as constant.

Differentiating (9.3.18),

$$\frac{x\cos x - \sin x}{x^2}dx + \frac{y\cos y - \sin y}{y^2}dy = f'(z)dz$$

$$\text{or, } y^2 z(x\cos x - \sin x)dx + x^2 z(y\cos y - \sin y)dy - x^2 y^2 z f'(z)dz = 0. \tag{9.3.19}$$

Comparing (9.3.19) with the given equation, we can write $x^2 y^2 z f'(z) = -xy(y\sin x + x\sin y + xy\cos z)$.

$$zf'(z) = -\frac{\sin x}{x} - \frac{\sin y}{y} - \cos z$$

$$\text{or, } -zf'(z) = f(z) + \cos z$$

$$\text{or, } \frac{d}{dz}f(z) + \frac{f(z)}{z} = -\frac{\cos z}{z} \text{ (a linear equation in } f(z) \text{ with IF} = z)$$

$$\text{or, } z\frac{df}{dz} + f(z) = -\cos z \text{ or, } \frac{d}{dz}(f(z)\cdot z) = -\cos z.$$

Integrating, $f(z)\cdot z = -\sin z + K$, where K is an arbitrary constant

$$z\left[\frac{\sin x}{x} + \frac{\sin y}{c}\right] = -\sin z + K \text{ (Required solution).}$$

Example 9.3.5. *Verify that the following equation is integrable and find its primitive:* $zydx + (x^2y - zx)dy + (x^2z - xy)dz = 0$.

Solution: Let $P = yz$, $Q = x^2y - zx$, $R = x^2z - xy$.

$$P\left(\frac{\partial Q}{\partial z} - \frac{\partial R}{\partial y}\right) + Q\left(\frac{\partial R}{\partial x} - \frac{\partial P}{\partial z}\right) + R\left(\frac{\partial P}{\partial y} - \frac{\partial Q}{\partial x}\right)$$

$$= yz(-x + x) + (x^2y - zx)(2xz - y - y) + (x^2z - xy)(z - 2xy + z)$$

$$= yz\cdot 0 + (2x^3yz - 2x^2z^2 - 2y^2x^2 + 2xyz) + (2x^2z^2 - 2xyz - 2x^3yz + 2x^2y^2) = 0.$$

Hence, the given equation is integrable. Let x be treated as constant so that $dx = 0$ and the given equation becomes

$$(x^2y - xz)dy + (x^2z - xy)dz = 0 \quad \text{or, } x^2(ydy + zdz) - x(zdy + ydz) = 0$$

$$\text{or, } \frac{x^2}{2}d(y^2 + z^2) - xd(yz) = 0. \tag{9.3.20}$$

Integrating, (9.3.20) and remembering that x is being regarded as constant, we get

$$\tfrac{1}{2}x^2(y^2+z^2)-xyz=\text{constant}=f(x),\text{ say.} \qquad (9.3.21)$$

Differentiating (9.3.21), we get

$$\tfrac{1}{2}2xdx(y^2+z^2)+x^2(ydy+zdz)-xydz-yzdx-xzdy=f'(x)dx$$
$$\text{or, }\{x(y^2+z^2)-yz-f'(x)\}dx+(x^2y-xz)dy+(x^2z-xy)dz=0. \qquad (9.3.22)$$

Comparing (9.3.22) with the given equation, we can write

$$x(y^2+z^2)-yz-f'(x)=yz\text{ or, }x(y^2+z^2)-2yz=f'(x)$$
$$\text{or, }x^2(y^2+z^2)-2xyz=xf'(x)\text{ or, }2f(x)=xf'(x)\text{ [see (9.3.21)]}$$
$$\text{or, }\frac{df}{f}=\frac{2}{x}dx\text{ or, }f(x)=Kx^2.\ (K\text{ is an arbitary constant}).$$

$\therefore$ the required solution is $\tfrac{1}{2}x^2(y^2+z^2)-xyz=Kx^2$ or, $x^2(y^2+z^2-2K)=2xyz$.

Examples IX(C)

Solve the following differential equations:

1. $xz^3dx-zdy+2ydz=0$.
2. $(z+z^3)\cos x dx-(z+z^3)dy+(1-z^2)(y-\sin x)dz=0$.
3. $(mz-ny)dx+(nx-lz)dy+(ly-mx)dx=0$.
4. $y^2dx-2x^2dy+(xy-zy^3)dx=0$.
5. $yz(1+4xz)dx-xz(1+2xz)dy-xydz=0$.
6. $(1+y)(2+2x+y+z)dx-x(1+z+x)dy+(1+y)(1+x+y)dz=0$.
7. $\cos(2x+2y+z)(dx+dy)+\cos(x+y)\cos(x+y+z)dz=0$.
8. $z(x^2-yz-z^2)dx+(x+z)xzdy+x(z^2-x^2-xy)dz=0$.

Answers

[K = an arbitrary constant]

1. $2y=x^2z^2+2Kz^2$.
2. $(\sin x-y)(z^2+1)=Kz$.
3. $nx-lz=K(mz-ny)$.
4. $\frac{2}{y}-\frac{1}{x}=Kx^2$.
5. $yz=Kx(1+2xz)$.
6. $(1+x+y)(1+z+x)=K(1+y)$.
7. $\sin(x+y+z)\cos(x+y)=K$.
8. $z^{-1}(x+y)+x^{-1}(y+z)=K$.

V. Solution by Inspection

When the condition of integrability is satisfied, it is often possible and convenient to obtain the solution by inspection.

We give below some illustrations: (K is taken to represent an arbitrary constant.)

Example 9.3.6. *Solve:* $(y+z)dx+(z+x)dy+(x+y)dz=0$.

Solution: $P=y+z$, $Q=z+x$ and $R=x+y$.

See that

$$\sum P\left(\frac{\partial Q}{\partial z}-\frac{\partial R}{\partial y}\right)=0,$$

i.e. condition of integrability is satisfied.

Rearranging the terms of the given equation,

$$(ydx + xdy) + (zdx + xdz) + (zdy + ydz) = 0 \text{ or, } d(xy) + d(zx) + d(zy) = 0.$$

On integration, $xy + yz + zx = K_1$ (*Required primitive*).

Example 9.3.7. *Solve:* $(y+z)(x^2y^2z^2-1)dx + x^2y^2z^2(xy+xz)\left(\frac{dy}{y}+\frac{dz}{z}\right) + x(dy+dz) = 0.$

Solution: First verify that the condition of integrability is satisfied.

Now divide the given equation by $x(y+z)$. We obtain

$$(x^2y^2z^2-1)\frac{dx}{x} + x^2y^2z^2\left(\frac{dy}{y}+\frac{dz}{z}\right) + \frac{d(y+z)}{y+z} = 0$$

$$\text{or, } (xy^2z^2dx + x^2yz^2dy + x^2y^2zdz) - \frac{dx}{x} + \frac{d(y+z)}{y+z} = 0$$

$$\text{or, } \tfrac{1}{2}d(x^2y^2z^2) - d(\ln x) + d\ln(y+z) = 0.$$

On integration, we get the required solution: $\frac{1}{2}x^2y^2z - \ln x + \ln(y+z) = K.$

Example 9.3.8. *Solve:* $yz\ln z dx - zx\ln z dy + xydz = 0.$

Solution: First verify that the condition of integrability is satisfied.

Now dividing throughout by $xyz\ln z$, we get

$$\frac{dx}{x} - \frac{dy}{y} + \frac{dz/z}{\ln z} = 0.$$

Integrating, $\ln x - \ln y + \ln(\ln z) = \ln K$ (say), i.e. $x\ln z = Ky$ (*Required solution*).

Example 9.3.9. $(x^2y - y^3 - y^2z)dx + (xy^2 - x^2z - x^3)dy + (xy^2 + x^2y)dz = 0.$

Solution: First verify that the given equation satisfies the condition of integrability.

$$\sum P\left(\frac{\partial Q}{\partial z} - \frac{\partial R}{\partial y}\right) = 0.$$

Now, divide the given equation by x^2y^2. The given equation becomes

$$\frac{dx}{y} - \frac{ydx}{x^2} - \frac{zdx}{x^2} + \frac{dy}{x} - \frac{zdy}{y^2} - \frac{xdy}{y^2} + \left(\frac{dz}{x}+\frac{dz}{y}\right) = 0$$

$$\text{or, } \left(\frac{dx}{y} - \frac{xdy}{y^2}\right) + \frac{xdy - ydx}{x^2} + \frac{xdz - zdx}{x^2} + \frac{ydz - zdy}{y^2} = 0$$

$$\text{or, } \frac{ydx - xdy}{y^2} + \frac{xdy - ydx}{x^2} + \frac{xdz - zdx}{x^2} + \frac{ydz - zdy}{y^2} = 0$$

$$\text{or, } d\left(\frac{x}{y}\right) + d\left(\frac{y}{x}\right) + d\left(\frac{z}{x}\right) + d\left(\frac{z}{y}\right) = 0.$$

On integration,

$$\frac{x}{y} + \frac{y}{x} + \frac{z}{x} + \frac{z}{y} = K \text{ or, } x^2 + y^2 + z(x+y) = Kxy \text{ (Required solution).}$$

Example 9.3.10. *Find* $f(y)$ *such that the total differential* $\frac{yz+z}{x}dx - zdy + f(y)dz = 0$ *is integrable. Hence, solve it.*

Solution: Multiplying throughout by x, we get

$$(yz + z)dx - xzdy + f(y)xdz = 0. \qquad (9.3.23)$$

Comparing (9.3.23) with $Pdx + Qdy + Rdz = 0$, we have

$$P = yz + z, \quad Q = -zx, \quad R = xf(y). \qquad (9.3.24)$$

Since (9.3.23) is supposed to be integrable, therefore,

$$\sum P\left(\frac{\partial Q}{\partial z} - \frac{\partial R}{\partial y}\right) = 0$$

or, $(yz + z)(-x - xf'(y)) + (-zx)[f(y) - (y + 1)] + xf(y)[z + z] = 0$
or, $-xz(y + 1) - xz(y + 1)f'(y) - xzf(y) + (y + 1)zx + 2xzf(y) = 0$
or, $xz(1 + y)f'(y) = xzf(y)$
or, $\dfrac{\frac{df(y)}{dy}}{f(y)} = \dfrac{xz}{xz(1 + y)}$, i.e. $\dfrac{df(y)}{f(y)} = \dfrac{dy}{1 + y}$
or, $\ln f(y) = \ln(1 + y) + \ln c$, i.e. $f(y) = c(1 + y)$.

Putting this value of $f(y)$ in (9.3.22) the equation becomes

$$\frac{z(y + 1)}{x}dx - zdy + c(1 + y)dz = 0 \text{ i.e. } \frac{dx}{x} - \frac{dy}{y + 1} + c\frac{dz}{z} = 0$$

or, $\ln x - \ln(y + 1) + c\ln z = \ln k$ i.e. $xz^c = k(y + 1)$ is the *required solution.*

Example 9.3.11. *Solve:* $\frac{yzdx}{x^2+y^2} - \frac{xzdy}{x^2+y^2} - \tan^{-1}\frac{y}{x}dz = 0.$

Solution: First verify that the given equation is integrable.
On rearranging the terms of the given equation, we write

$$z\left(\frac{ydx - xdy}{x^2 + y^2}\right) - \tan^{-1}\frac{y}{x}dz = 0 \text{ or, } -\frac{1}{\tan^{-1}\frac{y}{x}}\frac{\frac{xdy-ydx}{x^2}}{1 + \left(\frac{y}{x}\right)^2} - \frac{dz}{z} = 0$$

$$\text{or, } \frac{1}{\tan^{-1}\frac{y}{x}}\frac{d\left(\frac{y}{x}\right)}{1 + \left(\frac{y}{x}\right)^2} + \frac{dz}{z} = 0 \text{ or, } \frac{d\left(\tan^{-1}\frac{y}{x}\right)}{\tan^{-1}\frac{y}{x}} + \frac{dz}{z} = 0.$$

Integrating,

$$\ln\tan^{-1}\frac{y}{x} + \ln z = \ln K \text{ i.e. } \tan^{-1}\frac{y}{x} = \frac{K}{z} \text{ or, } \frac{y}{x} = \tan\frac{K}{z} \text{ i.e. } y = x\tan\frac{K}{z}.$$

Examples IX(D)

Solve the following differential equations:

1. $(x^2z - y^3)dx + 3xy^2dy + x^3dz = 0.$
2. $\frac{xdx+ydy+zdz}{\sqrt{x^2+y^2+z^2}} + \frac{zdx-xdz}{z^2+x^2} + 3ax^2dx + 2bydy + cdz = 0.$
3. $x(y^2 - a^2)dx + y(x^2 - z^2)dy - z(y^2 - a^2)dz = 0$

4. $(x^2 - y^2 - z^2 + 2xy + 2xz)dx + (y^2 - z^2 - x^2 + 2yz + 2xy)dy + (z^2 - x^2 - y^2 + 2xz + 2yz)dz = 0.$
5. $x^2dx^2 + y^2dy^2 - z^2dz^2 + 2xydxdy = 0.$
6. Find $f(z)$ such that $\frac{[y^2+z^2-x^2]}{2x}dx - ydy + f(z)dz = 0$ is integrable. Hence, solve the equation.
7. $(y^2 + yz + z^2)dx + (z^2 + zx + x^2)dy + (x^2 + xy + y^2)dz = 0.$
8. $(2xyz + y^2z + z^2y)dx + (x^2z + 2xyz + xz^2)dy + (x^2y + xy^2 + 2xyz)dz = 0.$

Answers

[K = an arbitrary constant]

1. $x^2z + y^3 = Kx.$
2. $\sqrt{x^2 + y^2 + z^2} + \tan^{-1}\frac{x}{z} + ax^3 + by^2 + cz = K.$
3. $(x^2 - z^2)(y^2 - a^2) = K.$
4. $x + y + z = K(x^2 + y^2 + z^2).$
5. $(x^2 + y^2 + z^2 - K_1)(x^2 + y^2 - z^2 - K_2) = 0$, where K_1 and K_2 are arbitrary constants.
6. $f(z) = -z$. Soln: $xK = x^2 + y^2 + z^2.$
7. $xy + yz + zx = K(x + y + z).$
8. $xyz(x + y + z) = K.$

VI. Homogeneous Equations

The equation $Pdx + Qdy + Rdz = 0$ is called a homogeneous equation, if P, Q, R are all homogeneous functions in x, y, z of same degree.

We state below two Working Rules of solving such homogeneous equations:

Rule 1. If $D = Px + Qy + Rz \neq 0$, then take $\frac{1}{D}$ as the Integrating factor (IF).

Rule 2. ($D = 0$ and the equation is homogeneous) Put $x = zu$, $y = zv$ so that $dx = zdu + udz$ and $dy = zdv + vdz$.

Substituting these values in the given equation, we can solve the differential eq. (See the actual working in the illustrated examples).

In all the following problems, the conditions of integrability are to be first verified.

Example 9.3.12. $(yz + z^2)dx - xzdy + xydz = 0.$

Solution: (Use of Rule 1) First, verify that the equation is homogeneous.

With usual notations, $P = yz + z^2, Q = -xz, R = xy.$

$D = Px + Qy + Rz = x(yz + z^2) - xyz + xyz = xz(y + z) \neq 0.$

$\therefore$ we take $\frac{1}{D} = \frac{1}{xz(y+z)}$ as the IF.

Multiplying the given equation by the IF, we have

$$\frac{(yz + z^2)dx - xzdy + xydz}{xz(y + z)} = 0. \qquad (9.3.25)$$

We observe that

$$d(D) = d[xz(y + z)] = z(y + z)dx + xzdy + 2xzdz + xydz = z(y + z)dx + xzdy + x(y + 2z)dz.$$

Then the **Numerator** of (9.3.25) becomes

$$\begin{aligned} z(y + z)dx - xzdy + xydz &= \{z(y + z)dx + xzdy + x(y + 2z)dx\} - 2xzdy - 2xzdx \\ &= d(D) - 2xzd(y + z). \end{aligned}$$

$\therefore$ eq. (9.3.25) becomes

$$\frac{d(D)-2xzd(y+z)}{D}=0 \text{ or, } \frac{d(D)}{D}-\frac{2xzd(y+z)}{D}=0 \text{ or, } \frac{d(D)}{D}-\frac{2xzd(y+z)}{xz(y+z)}=0 \text{ or, } \frac{d(D)}{D}-2\frac{d(y+z)}{y+z}=0.$$

Integrating,

$$\ln D - 2\ln(y+z) = \ln c \ (c = \text{constant})$$

$$\text{or, } \frac{D}{(y+z)^2} = c \text{ or, } \frac{xz(y+z)}{(y+z)^2} = c,$$

i.e. $\boldsymbol{xz = c(y+z)}$ *is the required solution.*

Alternative method (using Rule II)

Let $x = uz$ so that $dx = udz + zdu$, and

$y = vz$ so that $dy = vdz + zdv$.

Putting these values in the given equation, namely, $(yz + z^2)dx - xzdy + xydz = 0$, we get $(vz^2 + z^2)(udz + zdu) - uz^2(vdz + zdv) + uvz^2dz = 0$ or, $z^3(v+1)du - uz^3dv + (v+1)uz^2dz = 0$.

Dividing by $u(v+1)z^3$, we get $\frac{du}{u} - \frac{dv}{v+1} + \frac{dz}{z} = 0$.

Integrating,

$$\ln u - \ln(v+1) + \ln z = \ln c \quad (c = \text{constant})$$

$$\text{or, } \frac{uz}{v+1} = c \text{ or, } \frac{x}{y/z+1} = c$$

or, $\boldsymbol{xz = c(y+z)}$ *is the required solution.*

Example 9.3.13. *Solve:* $yz(y+z)dx + zx(x+z)dy + xy(x+y)dz = 0$.

Solution: First verify that the given equation satisfies the condition of integrability.

We use **Rule 2.** See that the equation is homogeneous. We put $x = uz$, $y = vz$ so that $dx = udz + zdu$ and $dy = vdz + zdv$.

Putting these values in the given equation, we get

$$[v(v+1)du + u(u+1)dv]z^4 + 2uv(u+v+1)z^3dz = 0$$

Dividing by $uv(u+v+1)z^4$, we get

$$\frac{(v+1)du}{u(u+v+1)} + \frac{(u+1)dv}{v(u+v+1)} + \frac{2dz}{z} = 0$$

$$\text{or, } \left(\frac{1}{u} - \frac{1}{u+v+1}\right)du + \left(\frac{1}{v} - \frac{1}{u+v+1}\right)dv + \frac{2dz}{z} = 0$$

$$\text{or, } \frac{du}{u} + \frac{dv}{v} + 2\frac{dz}{z} - \frac{1}{u+v+1}d(u+v+1) = 0.$$

On integration, $\ln u + \ln v + 2\ln z - \ln(u+v+1) = \ln c$, ($c$ = arbitrary constant)

$$\text{or, } \frac{uvz^2}{u+v+1} = c, \text{ i.e. } \frac{\frac{x}{z}\cdot\frac{y}{z}\cdot z^2}{\frac{x}{z}+\frac{y}{z}+1} = c, \text{ i.e. } xyz = c(x+y+z) \ (\textit{Required solution}).$$

Example 9.3.14. *Solve:* $(2xz - yz)dx + (2yz - xz)dy - (x^2 - xy + y^2)dz = 0$.

Solution: The equation is homogeneous. Verify that the equation is integrable.

We put $x = uz, y = vz$ so that $dx = udz + zdu, dy = vdz + zdv$. Putting these values in the given equation, we get

$$(2uz^2 - vz^2)(udz + zdu) + (2vz^2 - uz^2)(vdz + zdv) - (u^2z^2 - uvz^2 + v^2z^2)dz = 0$$

$$\text{or, } (2u - v)zdu + (2v - u)zdv + \{(2u - v)u + (2v - u)v - u^2 + uv - v^2)\}dz = 0$$

$$\text{or, } z\{(2u - v)du + (2v - u)dv\} + (u^2 + v^2 - uv)dz = 0.$$

$$\text{or, } \frac{d(u^2 + v^2 - uv)}{u^2 + v^2 - uv} + \frac{dz}{z} = 0.$$

On integration, we have

$$\ln(u^2 - uv + v^2) + \ln z = \ln c \ (c = \text{arbitary constant})$$

$$\text{or, } (u^2 - uv + v^2)z = c$$

$$\text{or, } \left(\frac{x^2}{z^2} - \frac{xy}{z^2} + \frac{y^2}{z^2}\right) z = c,$$

i.e. $\boldsymbol{x^2 - xy + y^2 = cz}$. *(Required solution).*

Example 9.3.15. *Solve:* $z(z - y)dx + z(z + x)dy + x(x + y)dz = 0.$

Solution: The equation is homogeneous. Verify that it is integrable.

Comparing the given equation with $Pdx + Qdy + Rdz = 0$, we get $P = z^2 - yz, Q = z^2 + xz, R = x^2 + xy$.

Here let us use Rule I.

$$D = Px + Qy + Rz = xz^2 - xyz + yz^2 + xyz + x^2z + xyz$$
$$= z(xz + yz + x^2 + xy) = z(x + y)(x + z) \neq 0.$$

$$\text{IF} = \frac{1}{D} = \frac{1}{z(x + y)(x + z)}$$

Multiplying the given equation by the IF, we get

$$\frac{(z - y)dx}{(x + y)(x + z)} + \frac{dy}{x + y} + \frac{xdz}{z(x + z)} = 0$$

$$\text{or, } \left(\frac{1}{x + y} - \frac{1}{x + z}\right)dx + \frac{dy}{x + y} + \left(\frac{1}{z} - \frac{1}{x + z}\right)dz = 0$$

$$\text{or, } \frac{d(x + y)}{x + y} - \frac{d(x + z)}{x + z} + \frac{1}{z}dz = 0$$

On integration,

$$\ln(x + y) - \ln(x + z) + \ln z = \ln c \ (c = \text{arbitrary constant})$$

or, $\dfrac{(x + y)z}{x + z} = c$ i.e. $\boldsymbol{z(x + y) = c(z + x)}$ ***is the required solution.***

Examples IX(E)

Solve the following homogeneous equations:

1. $z^2 dx + (z^2 - 2yz)dy + (2y^2 - yz - xz)dz = 0.$
2. $2(y + z)dx - (x + y)dy + (2y - x + z)dz = 0.$
3. $y(y + z)dx + z(x + z)dy + x(x + y)dy = 0.$
4. $(y^2 z - y^3 + x^2 y)dx - (x^2 z + x^3 - xy^2)dy + (x^2 y - xy^2)dz = 0.$
5. $(x^2 - y^2 - z^2 + 2xy + 2xz)dx + (y^2 - z^2 - x^2 + 2yz + 2xy)dy + (z^2 - x^2 - y^2 + 2xz + 2yz)dz = 0.$
6. $2(2y^2 + yz - z^2)dx + x(4y + z)dy + x(y - 2z)dz = 0.$

Answers

[c = an arbitrary constant]

1. $(x + y)z - y^2 = cz^2.$
2. $(x + z)^2 = c(y + z).$
3. $x(y + z) = c(x + y).$
4. $x^2 + y^2 - yz + zx = cxy.$
5. $x^2 + y^2 + z^2 = c(x + y + z).$
6. $x^2(2y^2 + yz - z^2) = c.$

VII. Use of Auxiliary Equations (A Special Method)

Let

$$Pdx + Qdy + Rdz = 0 \tag{9.3.26}$$

be the given equation.

The condition of integrability is

$$\sum P\left(\frac{\partial Q}{\partial z} - \frac{\partial R}{\partial y}\right) = 0, \tag{9.3.27}$$

$$\text{i.e. } P\left(\frac{\partial Q}{\partial z} - \frac{\partial R}{\partial y}\right) + Q\left(\frac{\partial R}{\partial x} - \frac{\partial P}{\partial z}\right) + R\left(\frac{\partial P}{\partial y} - \frac{\partial Q}{\partial x}\right) = 0.$$

Comparing (9.3.26) and (9.3.27), we obtain simultaneous equations, known as *auxiliary equations*:

$$\frac{dx}{\frac{\partial Q}{\partial z} - \frac{\partial R}{\partial y}} = \frac{dy}{\frac{\partial R}{\partial x} - \frac{\partial P}{\partial z}} = \frac{dz}{\frac{\partial P}{\partial y} - \frac{\partial Q}{\partial x}}. \tag{9.3.28}$$

We solve them by the method described in art. 9.2.

Let $u(x, y, z) = c_1$ and $v(x, y, z) = c_2$ be the two integrals so obtained.

With these u and v we formulate the equation

$$Au + Bv = 0. \tag{9.3.29}$$

Now compare (9.3.25) and (9.3.28) and obtain the values of A and B. Put these values of A and B in (9.3.28) and then integrate the resulting equation. Now substitute the values of u and v in the relation after integration. We thus obtain the required general solution.

Note 9.3.3. *This method cannot be applied if $Pdx + Qdy + Rdz = 0$ be an exact equation because then*

$$\frac{\partial Q}{\partial z} = \frac{\partial R}{\partial y}, \frac{\partial R}{\partial x} = \frac{\partial P}{\partial z} \text{ and } \frac{\partial P}{\partial y} = \frac{\partial Q}{\partial x}.$$

Note 9.3.4. *This method is applied when the* **method by inspection** *or the method of homogeneous* **equations** *is not convenient to apply. We give below some illustrations:*

Example 9.3.16. *Solve:* $(2xz - yz)dx + (2yz - zx)dy - (x^2 - xy + y^2)dz = 0$.

Solution: Comparing the given equation with $Pdx + Qdy + Rdz = 0$, we have $P = 2xz - yz$, $Q = 2yz - zx$ and $R = -(x^2 - xy + y^2)$. The auxiliary equations are

$$\frac{dx}{\frac{\partial Q}{\partial z} - \frac{\partial R}{\partial y}} = \frac{dy}{\frac{\partial R}{\partial x} - \frac{\partial P}{\partial z}} = \frac{dz}{\frac{\partial P}{\partial y} - \frac{\partial Q}{\partial x}}$$

$$\text{or, } \frac{dx}{(2y - x) - (x - 2y)} = \frac{dy}{(-2x + y) - (2x - y)} = \frac{dz}{(-z) - (-z)}$$

$$\text{or, } \frac{dx}{2(2y - x)} = \frac{dy}{2(y - 2x)} = \frac{dz}{0} \qquad (9.3.30)$$

Thus, $dz = 0$, or, $z = c_1 = u$ **(say).**

Taking the first and second ratios of (9.3.30), we have

$$ydx - 2xdx = 2ydy - xdy$$
$$\text{or, } d(xy) - d(x^2 + y^2) = 0$$
$$\text{or, } xy - x^2 - y^2 = c_2 = v \text{ (say).}$$

We now determine two functions A and B such that the given equation becomes identical with

$$Adu + Bdv = 0 \qquad (9.3.31)$$
$$\text{i.e. } Adz + Bd(xy - x^2 - y^2) = 0,$$
$$\text{or, } Adz + B\{xdy + ydx - 2xdx - 2ydy\} = 0$$
$$\text{or, } B(y - 2x)dx + B(x - 2y)dy + Adz = 0.$$

Comparing with the given equation $B(y - 2x) = 2xz - yz$, $B(x - 2y) = 2yz - zx$ and $A = -(x^2 - xy + y^2)$ giving, $A = v$, $B = -z = -u$, i.e. $vdu - udv = 0$, or, $(1/v)dv - (1/u)du = 0$.

Integrating, $\ln v - \ln u = \ln c$, or, $v = cu$, or, $xy - x^2 - y^2 = cz$ *(Required solution).*

Example 9.3.17. *Solve:* $(y^2 + yz + z^2)dx + (z^2 + zx + x^2)dy + (x^2 + xy + y^2)dz = 0$.

Solution: Here $P = y^2 + yz + z^2$, $Q = z^2 + zx + x^2$, $R = x^2 + xy + y^2$.

The auxiliary equations are

$$\frac{dx}{\frac{\partial Q}{\partial z} - \frac{\partial R}{\partial y}} = \frac{dy}{\frac{\partial R}{\partial x} - \frac{\partial P}{\partial z}} = \frac{dz}{\frac{\partial P}{\partial y} - \frac{\partial Q}{\partial x}}$$

$$\text{or, } \frac{dx}{(2z + x) - (x + 2y)} = \frac{dy}{(2x + y) - (y + 2z)} = \frac{dz}{(2y + z) - (z + 2x)}$$

$$\text{or, } \frac{dx}{z - y} = \frac{dy}{x - z} = \frac{dz}{y - x}.$$

Each ratio $= \frac{dx+dy+dz}{z-y+x-z+y-x} = \frac{d(x+y+z)}{0}$, i.e. $d(x + y + z) = 0$ or, $x + y + z = c_1 = u$ (say).

Again, using multipliers, $z + y, x + z, y + x$,

$$\text{each ratio} = \frac{(z + y)dx + (x + z)dy + (y + x)dz}{z^2 - y^2 + x^2 - z^2 + y^2 - x^2} = \frac{(xdy + ydx) + (ydz + zdy) + (zdx + xdz)}{0}$$

$$= \frac{d(xy + yz + zx)}{0}$$

so that $d(xy + yz + zx) = 0$, or, $xy + yz + zx = c_2 = v$ (say).

We now proceed to determine two functions A and B such that the given equation becomes identical with

$$Adu + Bdv = 0 \text{ or, } Ad(x + y + z) + Bd(xy + yz + zx) = 0$$
$$\text{or, } A(dx + dy + dz) + B\{xdy + ydx + ydz + zdy + xdz + zdx\} = 0$$
$$\text{or, } (A + By + Bz)dx + (A + Bx + Bz)dy + (A + By + Bx)dz = 0$$

and the given equation is $(y^2 + yz + z^2)dx + (z^2 + zx + x^2)dy + (x^2 + xy + y^2)dz = 0$.

Comparing, we write

$$A + By + Bz = y^2 + yz + z^2$$
$$A + Bx + Bz = z^2 + zx + x^2$$
$$A + By + Bx = x^2 + xy + y^2.$$

Subtracting the first two, $B(y - x) = y^2 - x^2 + z(y - x)$ and hence, $B = x + y + z = u$.

Also from the first relation,

$$A = y^2 + yz + z^2 - B(y + z) = y^2 + yz + z^2 - (x + y + z)(y + z) = -(xy + yz + zx) = -v,$$

i.e. $Adu + Bdv = 0$ becomes $-vdu + udv = 0$ or, $\frac{du}{u} = \frac{dv}{v}$, i.e. $\ln v = \ln u + \ln c$ (on integration), i.e. $v = cu$ or, $xy + yz + zx = c(x + y + z)$ is the required solution.

VIII. Solution of Pdx + Qdy + Rdz = 0, when it is exact and homogeneous of degree n ≠ −1.

We can easily verify that $xP + yQ + zR = c$ is the required solution in this case.

Assume that $xP + yQ + zR = c$. Differentiating, we obtain

$$\left(P + x\frac{\partial P}{\partial x} + y\frac{\partial Q}{\partial x} + z\frac{\partial R}{\partial x}\right)dx + \left(x\frac{\partial P}{\partial y} + Q + y\frac{\partial Q}{\partial y} + z\frac{\partial R}{\partial y}\right)dy + \left(x\frac{\partial P}{\partial z} + y\frac{\partial Q}{\partial z} + R + z\frac{\partial R}{\partial z}\right)dz = 0. \tag{9.3.32}$$

Since $Pdx + Qdy + Rdz = 0$ is exact, we have

$$\frac{\partial P}{\partial y} = \frac{\partial Q}{\partial x}, \quad \frac{\partial Q}{\partial z} = \frac{\partial R}{\partial y} \text{ and } \frac{\partial R}{\partial x} = \frac{\partial P}{\partial z}.$$

∴ (9.3.31) reduces to

$$\left(P + x\frac{\partial P}{\partial x} + y\frac{\partial P}{\partial y} + z\frac{\partial P}{\partial z}\right)dx + \left(Q + x\frac{\partial Q}{\partial x} + y\frac{\partial Q}{\partial y} + z\frac{\partial Q}{\partial z}\right)dy + \left(R + x\frac{\partial R}{\partial x} + y\frac{\partial R}{\partial y} + z\frac{\partial R}{\partial z}\right)dz = 0. \tag{9.3.33}$$

Since $Pdx + Qdy + Rdz = 0$ is homogeneous of degree, each of P, Q, R are homogeneous functions of degree n. We can use Euler's theorem. The eq. (9.3.33) reduces to

$$(n + 1)(Pdx + Qdy + Rdz) = 0 \text{ or, } Pdx + Qdy + Rdz = 0 \ (\because\ n \neq -1).$$

Example 9.3.18. *Solve:* $(x - 3y - z)dx + (2y - 3x)dy + (z - x)dz = 0$.

Solution: Given equation can be written as $Pdx + Qdy + Rdz = 0$, where $P = x - 3y - z$, $Q = 2y - 3x$, $R = z - x$.

Thus, the given equation is *homogeneous* of degree $n = 1$, $(\neq -1)$.

Moreover, see that

$$\frac{\partial P}{\partial y} = -3 = \frac{\partial Q}{\partial x}, \ \frac{\partial Q}{\partial z} = 0 = \frac{\partial R}{\partial y} \text{ and } \frac{\partial R}{\partial x} = -1 = \frac{\partial P}{\partial z},$$

which proves that the given equation is *exact.*

Hence, the solution of the given equation can at once be written as

$$xP + yQ + zR = c \text{ or, } x(x - 3y - z) + y(2y - 3x) + z(z - x) = c$$

$$\text{or, } x^2 + 2y^2 + z^2 - 6xy - 2xz = c \text{ (}c\text{ is an arbitrary constant).}$$

IX. Non-integrable Differential Single Equation Pdx + Qdy + Rdz = 0

If the condition of integrability is not satisfied by the equation

$$Pdx + Qdy + Rdz = 0, \tag{9.3.34}$$

then there is no single relation between x, y, z like $f(x, y, z) = c$ that will satisfy the eq. (9.3.34).

If, however, there exists an arbitrary relation

$$\phi(x, y, z) = c_1, \tag{9.3.35}$$

which on differentiation gives the differential equation

$$\frac{\partial \phi}{\partial x}dx + \frac{\partial \phi}{\partial y}dy + \frac{\partial \phi}{\partial z}dz = 0, \tag{9.3.36}$$

then two integral relations can be found which together satisfy (9.3.34) and (9.3.36). Of course (9.3.35) is one of these two relations.

Suppose that

$$F(x, y, z) = c_2 \tag{9.3.37}$$

is a relation which with (9.3.35) forms the complete solution of equations (9.3.34) and (9.3.36).

But we know that the locus of the complete solution of (9.3.34) and (9.3.36) consists of the curves of intersection of (9.3.35) and (9.3.37).

Hence, geometrically, this solution of (9.3.34) amounts to finding the curves satisfying (9.3.34) that lie on the surfaces represented by (9.3.35).

Example 9.3.19. *Find the most general solution of the equation*

$$ydx + (z - y)dy + xdz = 0, \tag{9.3.38}$$

which is consistent with the relation

$$2x - y - z = 1. \tag{9.3.39}$$

Solution: It can be verified that the condition of integrability is not satisfied by the coefficients of eq. (9.3.38).

Then from (9.3.39), we get $z = 2x - y - 1$ and $dz = 2dx - dy$.

Substituting these values of z and dz in (9.3.38), we obtain

$$ydx + (2x - y - 1 - y)dy + x(2dx - dy) = 0$$

$$\text{or, } (y + 2x)dx + (x - 2y - 1)dy = 0$$

$$\text{or, } ydx + xdy + 2xdx - 2ydy - dy = 0$$

$$\text{or, } d(xy) + d(x^2 - y^2) - dy = 0$$

$$\text{or, } xy + x^2 - y^2 - y = c \tag{9.3.40}$$

The equations (9.3.39) and (9.3.40) constitute the required general solution of (9.3.38).

Example 9.3.20. *Solve:*

$$xdx + ydy + c\left(1 - \frac{x^2}{a^2} - \frac{y^2}{b^2}\right)^{1/2} dz = 0 \tag{9.3.41}$$

consistent with the relation:

$$\frac{x^2}{a^2} + \frac{y^2}{b^2} + \frac{z^2}{c^2} = 1. \tag{9.3.42}$$

Solution: Using (9.3.42), eq. (9.3.41) can be written as

$$xdx + ydy + zdz = 0 \tag{9.3.43}$$

$$\text{whence } x^2 + y^2 + z^2 = c^2. \tag{9.3.44}$$

Thus, (9.3.42) with (9.3.44) gives a solution of (9.3.41).

Geometric interpretation is that the lines upon the ellipsoid (9.3.42) which satisfy (9.3.41), have been determined and they are the lines of intersection of the family of spheres (9.3.44) with the ellipsoid (9.3.43).

Examples IX(F)

1. Solve: (a) $(y^2 + yz)dx + (xz + z^2)dy + (y^2 - xy)dz = 0$ (b) $z(z - y)dx + (z + x)zdy + x(x - y)dz = 0$.
2. Solve: $ydx + xdy - (x + y + 2z) = 0$ consistent with $x + y + 2z = 0$.
3. Find the curves represented by the solution of $ydx + (z - y)dy + xdz = 0$, which lie on the plane $2x - y - z = 1$.
   [The required curves are given by the intersection of the plane $2x - y - z = 1$ and rectangular hyperbolic cylinders $xy + x^2 - y^2 - y = c$.]
4. Show that there is no single integral of $dz = 2ydx + xdy$. Prove that the curves of this equation that lie in the plane $z = x + y$ lie also on surfaces of the family $(x - 1)^2(2y - 1) = c$.
   [Use $dz = dx + dy$ and the given equation $dz = 2ydx + xdy$ to obtain $\frac{2dx}{x-1} + \frac{2dy}{2y-1} = 0$]
5. Find the orthogonal projection on the plane of $x - z$ of curves which lie on the paraboloid $3z = x^2 + y^2$ and satisfy the equation $2dz = (x + z)dx + ydy$.

Answers

1. (a) $(x + z)y = c(y + z)$
   (b) $z(x + y) = c(x + y)$.
2. $xy = c$.
3. *Hints.* Required curve are given by the intersection of the plane. $2x - y - z = 1$ and rectangular hyperbolic cylinders $xy + x^2 - y^2 - y = c$.
4. Use $dz = dx + dy$ and the given equation $dz = 2y\,dx + x\,dy$ to obtain $\frac{2dx}{x-1} + \frac{2dy}{2y-1} = 0$.
5. $3z = ce^{2x}$.
   [Verify that $(x + z)dx + ydy - 2dz = 0$ does not satisfy the condition of integrability.
   Now differentiating $x^2 + y^2 - 3z = 0$, we get $2xdx + 2ydy - 3dz = 0$.
   Obtain: $3(x + z)dx + 3ydy - 3(2xdx + 2ydy) = 0$, or, $-xdx - ydy + 3zdx = 0$
   or, $xdx + ydy = (x^2 + y^2)dx$ (using $x^2 + y^2 - 3z = 0$)
   or, $\frac{2xdx+2ydy}{x^2+y^2} = 2dx$.
   $\therefore$ on integration, obtain $x^2 + y^2 = ce^{2x}$, i.e. $3z = ce^{2x}$, which gives the required projection on the *xy*-plane.]

Chapter 10

Geometrical and Physical Applications

In the present chapter we shall consider some geometrical as well as physical problems leading to construction of differential equations and their solutions. In applications, it is necessary to remember that a differential equation expresses analytically certain property common to every member of the family of curves given by its general solution. Conversely, if a property whose analytic representation involving derivatives is given, the solution of the resulting differential equation represents a family of curves, all possessing the given property. Each curve of the family is called an integral curve and a particular integral curve may be obtained if some additional property peculiar to that particular curve be known.

10.1 Geometrical data: A recapitulation

We give below a list of some of the principal geometrical data of the Differential Calculus for reference:

Rectangular coordinates

Suppose that the equation of a curve, rectangular coordinates being chosen, is $y = f(x)$ or, $F(x, y) = 0$ and that $P(x, y)$ is any point on this curve (see Fig. 10.1). Take (X, Y) as current coordinates.

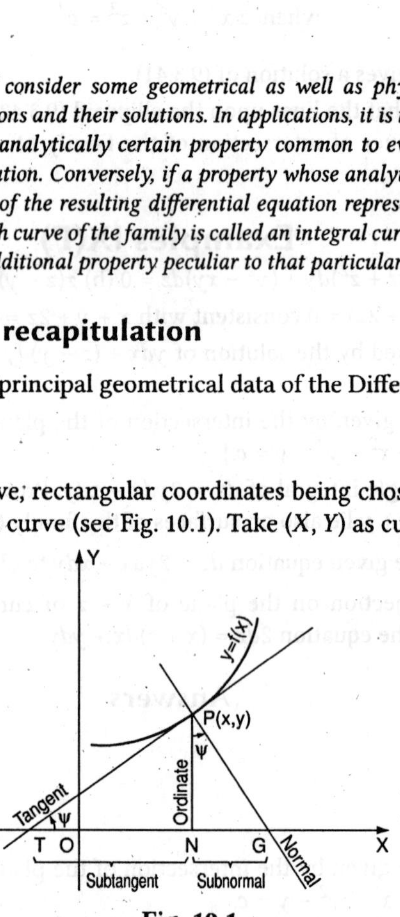

Fig. 10.1

1. $\frac{dy}{dx}$ = Slope of the tangent at $P = \tan\psi$.
2. $-\frac{dx}{dy}$ = Slope of the normal at P.
3. $Y - y = \frac{dy}{dx}(X - x)$ is the equation of the tangent PT.
4. $Y - y = -\frac{dx}{dy}(X - x)$ is the equation of the normal PG.
5. $x - y\frac{dx}{dy}$ and $y - x\frac{dy}{dx}$ are respectively the x- and y-intercepts of the tangent PT.
6. $x + y\frac{dy}{dx}$ and $y + x\frac{dx}{dy}$ are respectively the x- and y-intercepts of the normal PG.

7. Length of tangent PT = length of tangent between (x, y) and x-axis $= y\sqrt{1 + \left(\frac{dx}{dy}\right)^2}$.

Length of normal PG = length of normal between (x, y) and x-axis $= y\sqrt{1 + \left(\frac{dy}{dx}\right)^2}$.

8. Length of tangent between (x, y) and y-axis $= x\sqrt{1 + \left(\frac{dy}{dx}\right)^2}$.

Length of normal between (x, y) and y-axis $= x\sqrt{1 + \left(\frac{dx}{dy}\right)^2}$.

9. $y\frac{dx}{dy}$ = length of the subtangent TN.

$y\frac{dy}{dx}$ = length of the subnormal NG.

$\sqrt{x^2 + y^2}$ = radius vector OP.

10. Differential of the length of the arc $= ds = \sqrt{1 + \left(\frac{dx}{dy}\right)^2}\, dy$, or, $\sqrt{1 + \left(\frac{dy}{dx}\right)^2}\, dx$, differential of the area $= dA = y\, dx$, or, $x\, dy$.

$\sin\psi = \frac{dy}{ds}$ and $\cos\psi = \frac{dx}{ds}$.

11. Radius of curvature

$$\rho = \frac{\left\{1 + \left(\frac{dy}{dx}\right)^2\right\}^{3/2}}{\frac{d^2y}{dx^2}} = \frac{ds}{d\psi}.$$

Polar coordinates

Let the equation of the curve be given in polar coordinates (r, θ), say, $r = f(\theta)$, or, $F(r, \theta) = 0$.

12. $\tan\phi = r\frac{d\theta}{dr}$, where ϕ is the angle between the radius vector and the part of the tangent drawn towards the initial line OX.

13. $\psi = \theta + \phi$. $\left(\tan\psi = \frac{dy}{dx}\right)$.

14. $r\tan\phi$, or, $r^2\frac{d\theta}{dr}$ = length of the polar subtangent OT.

$r\cot\phi$, or, $\frac{dr}{d\theta}$ = length of the polar subnormal OG.

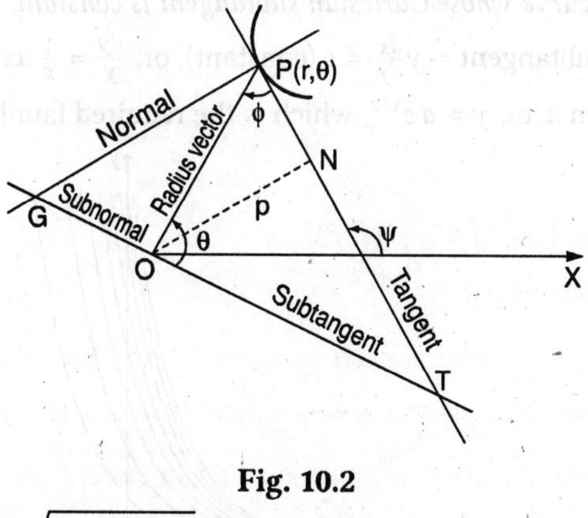

Fig. 10.2

15. $ds = \sqrt{1 + r^2\left(\frac{d\theta}{dr}\right)^2}\, dr$, or, $\sqrt{\left(\frac{dr}{d\theta}\right)^2 + r^2}\, d\theta$

$\sin\phi = r\frac{d\theta}{ds}$, $\cos\phi = \frac{dr}{ds}$.

16. $\frac{1}{p^2} = \frac{1}{r^2} + \frac{1}{r^4}\left(\frac{dr}{d\theta}\right)^2$ where p = length of the perpendicular from the pole O upon the tangent PT = $u^2 + \left(\frac{du}{d\theta}\right)^2$, if $u = \frac{1}{r}$.

17. Radius of curvature

$$\rho = r\frac{dr}{dp}, \text{ or, } \frac{\left\{r^2 + \left(\frac{dr}{d\theta}\right)^2\right\}^{3/2}}{r^2 + 2\left(\frac{dr}{d\theta}\right)^2 - r\frac{d^2r}{d\theta^2}}.$$

10.2 Illustrative examples

10.2.1 Applications of first order and first degree differential equations

Example 10.2.1. *Find the family of curves for which the angle between the radius vector and the tangent at (r, θ) is one-half of the vectorial angle.*

Solution: Refer to Fig. 10.2 and see that $\phi = \frac{1}{2}\theta$.

Therefore, $\tan\phi = r\frac{d\theta}{dr} = \tan\frac{1}{2}\theta$, or, $\frac{dr}{r} = \cot\frac{\theta}{2}\, d\theta$.

Integrating, $\ln r = 2\ln\sin\frac{1}{2}\theta + \ln c$ or, $r = c\sin^2(\theta/2) = a(1 - \cos\theta)$, if $a = c/2$.

Thus, the family of curves is the family of cardioides $r = a(1 - \cos\theta)$ (Fig. 10.3).

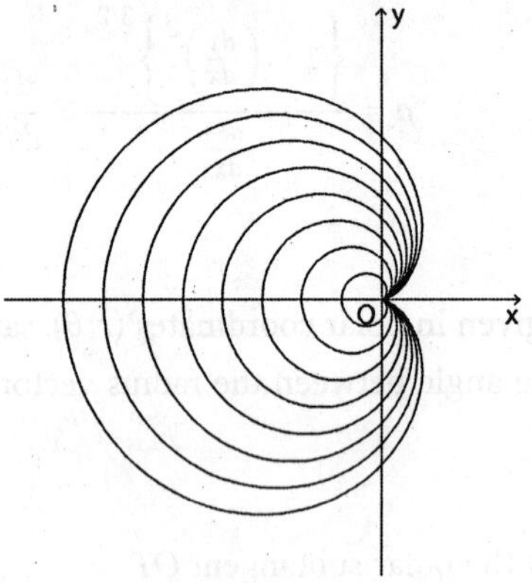

Fig. 10.3: $(a > 0)$.

Example 10.2.2. *Find the curve whose Cartesian subtangent is constant.*

Solution: From Fig. 10.1, subtangent = $y\frac{dx}{dy} = c$ (constant), or, $\frac{dy}{y} = \frac{1}{c}\, dx$.

Integrating, $\ln y = \frac{x}{c} + \ln a$, or, $y = a\, e^{x/c}$, which is the required family of curves (Fig. 10.4).

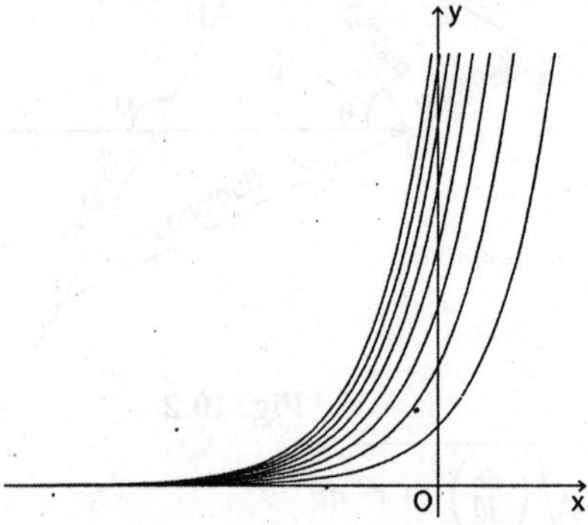

Fig. 10.4: $(a > 0)$.

Example 10.2.3. *The normal at any point P of a curve cuts OX at G and N in the foot of the ordinate of P. If NG varies as the square of the radius vector from O, find the family of curves, each member of which satisfies the property stated above.*

Solution: Referring to Fig. 10.1, we see that, radius vector = $OP = \sqrt{x^2+y^2}$, and subnormal = $NG = y\frac{dy}{dx}$.

Given: $NG \propto (OP)^2$.

$$\text{i.e.,}\quad y\frac{dy}{dx} = k\left(x^2+y^2\right),\ k = \text{ constant of variation}$$

$$\text{or,}\quad y\frac{dy}{dx} - ky^2 = kx^2 \text{ or, } \frac{dv}{dx} - 2kv = 2kx^2, \text{ if } v = y^2.$$

This is a linear equation in v with IF = e^{-2kx}.

Then

$$\frac{d}{dx}\left(v\,e^{-2kx}\right) = 2kx^2\,e^{-2kx}, \text{ or, } y^2\,e^{-2kx} = C - x^2\,e^{-2kx} - \frac{1}{k}x\,e^{-2kx} - \frac{1}{2k^2}\,e^{-2kx},$$

$$\text{or,}\quad y^2 = C\,e^{2kx} - x^2 - \frac{1}{k}x - \frac{1}{2k^2},$$

where C is the arbitrary constant of integration.

This gives the required family of curves.

Example 10.2.4. *Find the family of curves for which the length of the part of the tangent between the point of contact (x, y) and the y-axis is equal to the y-intercept of the tangent.*

Solution: Given: required length of tangent = y-intercept of the tangent.

Thus, from 8 and 5 of art. 10.1,

$$x\sqrt{1+\left(\frac{dy}{dx}\right)^2} = y - x\frac{dy}{dx}.$$

Squaring and simplifying,

$$x^2 = y^2 - 2xy\frac{dy}{dx}$$

which is a homogeneous equation. Putting $y = vx$,

$$\frac{dy}{dx} = \frac{y^2-x^2}{2xy}, \text{ or, } v + x\frac{dv}{dx} = \frac{v^2-1}{2v}, \text{ i.e., } \frac{dx}{x} + \frac{2v}{1+v^2}\,dv = 0.$$

Integrating, $\ln x + \ln\left(1+v^2\right) = \ln c$, or, $x^2+y^2 = cx$ is the required family of curves (Fig. 10.5).

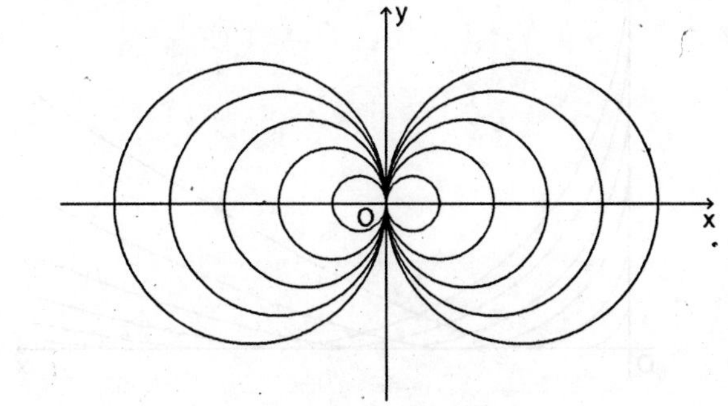

Fig. 10.5

Example 10.2.5. *Find the polar equation of the family of curves for which the sum of the radius vector and the polar subnormal varies as the kth power of the radius vector.*

Solution: Given:

$$(\text{radius vector} + \text{polar subnormal}) \propto (\text{radius vector})^k,$$

$$\text{or,} \quad r + \frac{dr}{d\theta} = \lambda r^k, \text{ where } \lambda \text{ is the constant of variation,}$$

$$\text{or,} \quad \frac{1}{r^k}\frac{dr}{d\theta} + \frac{1}{r^{k-1}} = \lambda,$$

$$\text{or,} \quad \frac{dv}{d\theta} - (k-1)v = -\lambda(k-1), \text{ putting } \frac{1}{r^{k-1}} = v.$$

This is a linear equation in v with IF $= e^{-(k-1)\theta}$. The required solution is

$$v\,e^{-(k-1)\theta} = C + \lambda\, e^{-(k-1)\theta}, \text{ or, } \frac{1}{r^{k-1}} = C\,e^{(k-1)\theta} + \lambda.$$

(C = arbitrary constant of integration).

10.2.2 Applications of first order and higher degree differential equations

Example 10.2.6. *Find the curve for which the sum of the intercepts of the tangent line on the coordinate axes is constant and equal to a.*

Solution: See Fig. 10.1 and look to 5 of art. 10.1, sum of intercepts = a.

$$\therefore \left\{x - y\frac{dx}{dy}\right\} + \left\{y - x\frac{dy}{dx}\right\} = a, \text{ i.e., } x - \frac{y}{p} + y - px = a, \text{ where } p = \frac{dy}{dx}.$$

Simplifying,

$$y = px + \frac{ap}{p-1},$$

which is in Clairaut's form, the primitive being the family of straight lines $y = cx + ac/(c-1)$. But, the required curve must be the *envelope* of the family of lines $y = cx + ac/(c-1)$, or, $xc^2 - (x + y - a)c + y = 0$, which is (by equating to zero, the discriminant) $(x + y - a)^2 = 4xy$.

On simplification, this may also be put in the form: $\sqrt{x} \pm \sqrt{y} = \sqrt{a}$ (Fig. 10.6).

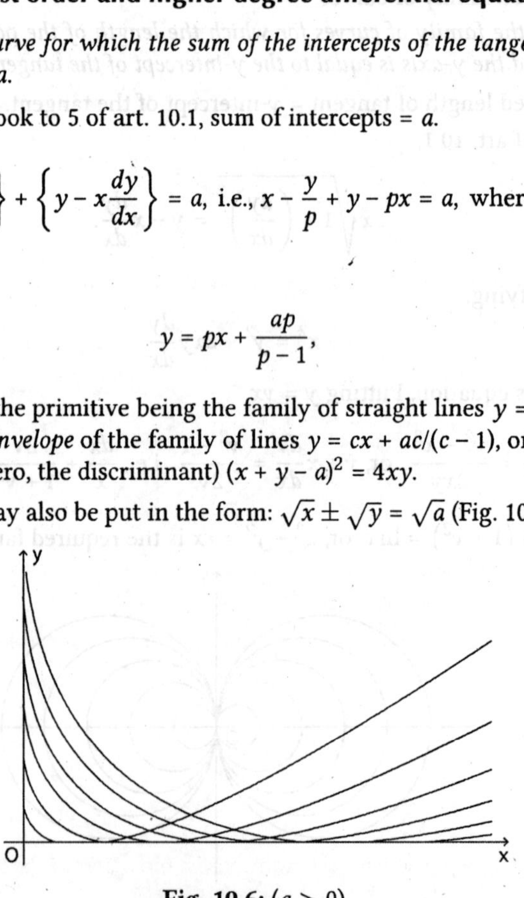

Fig. 10.6: ($a > 0$).

Observation

See that the required curve in Ex. 10.2.6 is an envelope (singular solution), since it satisfies the differential equation and cannot be obtained from the primitive by assigning a value to c.

Example 10.2.7. *Find the curve for which the portion of the tangent line intercepted by the coordinate axes is of constant length a.*

Solution: From the previous example (Ex. 10.2.6),

$$\left(x - \frac{y}{p}\right)^2 + (y - px)^2 = a^2.$$

After simplification,

$$y = px \pm \frac{ap}{\sqrt{1+p^2}}.$$

The primitive is the family of straight lines

$$y = cx \pm \frac{ac}{\sqrt{1+c^2}}.$$

For the singular solution, we differentiate with respect to c partially,

$$0 = x \pm \frac{a}{\left(1+c^2\right)^{3/2}}, \text{ or, } x = \mp \frac{a}{\left(1+c^2\right)^{3/2}}.$$

Thus,

$$y = cx \pm \frac{ac}{\sqrt{1+c^2}} = \pm \frac{ac^3}{\left(1+c^2\right)^{3/2}}.$$

And the envelope becomes $x^{2/3} + y^{2/3} = a^{2/3}$ (Fig. 10.7)

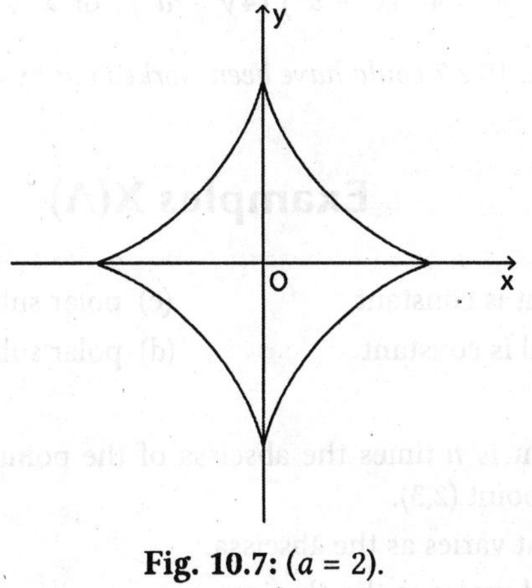

Fig. 10.7: (a = 2).

A different approach

Example 10.2.8. *Find the curve for which the product of the intercepts of the tangent line on the coordinate axes is equal to a.*

Solution: Let the equation of the tangent line be $y = px + f(p)$.

Then, the x-intercept $= -f(p)/p$ and the y-intercept $= f(p)$.

Hence, in this case,

$$f(p)\left\{-\frac{f(p)}{p}\right\} = a, \text{ or, } f(p) = \pm\sqrt{-ap}.$$

$\therefore$ the tangent line becomes $y = px \pm \sqrt{-ap}$.

This being in Clairaut's form, the primitive is $y = cx \pm \sqrt{-ac}$.

$$\text{i.e., } (y - cx)^2 = -ac, \text{ or, } x^2c^2 + (a - 2xy)c + y^2 = 0.$$

Hence, the curve (i.e., the singular solution) is $(a - 2xy)^2 = 4x^2y^2$, or, $4xy = a$, which is a rectangular hyperbola.

Example 10.2.9. *Find the curve for which the sum of the distances of the points $(h, 0)$ and $(-h, 0)$ from the tangent line is equal to a.*

Solution: Let the tangent line be $y = px + f(p)$, or, $px + f(p) - y = 0$.

Now, distance from $(h, 0)$+ distance from $(-h, 0) = a$,

$$\text{or, } \frac{ph + f(p)}{\sqrt{1 + p^2}} + \frac{-ph + f(p)}{\sqrt{1 + p^2}} = a, \text{ or, } \frac{2f(p)}{\sqrt{1 + p^2}} = a, \text{ i.e., } f(p) = \frac{1}{2}a\sqrt{1 + p^2}.$$

Thus, equation to tangent line is

$$y = px + \frac{a}{2}\sqrt{1 + p^2}.$$

This is in Clairaut's form, the primitive is

$$y = cx + \frac{a}{2}\sqrt{1 + c^2}, \text{ or, } \left(4x^2 - a^2\right)c^2 - 8xyc + 4y^2 - a^2 = 0.$$

The required curve, the envelope of the family of lines, has the equation:

$$64x^2y^2 = 4\left(4x^2 - a^2\right)\left(4y^2 - a^2\right), \text{ or, } x^2y^2 = a^2/4.$$

Note 10.2.1. *Ex. 10.2.6 and Ex. 10.2.7 could have been worked out by methods illustrated in Ex. 10.2.8 and Ex. 10.2.9 and vice versa.*

Examples X(A)

1. Find the curve whose
   (a) Cartesian subtangent is constant
   (b) Cartesian subnormal is constant
   (c) polar subtangent is constant
   (d) polar subnormal is constant.
2. Find the curve whose
   (a) Cartesian subtangent is n times the abscissa of the point of contact; also the curve which passes through the point (2,3).
   (b) Cartesian subtangent varies as the abscissa.
   (c) Cartesian subnormal varies as the abscissa.
   (d) Cartesian subtangent varies as the reciprocal of the square of the abscissa.
3. Find the curve
   (a) such that the angle between the tangent and the radius vector at any point is twice the vectorial angle.

(b) for which the tangent at any point makes a constant angle α with the radius vector.

(c) such that its length between any two points P and Q is proportional to the difference of the distances of Q and P from a fixed point O.

(d) Which satisfies the differential equation $(1 + y^2)dx - xydy = 0$ and passes through the point $(1, 0)$. Hence, find the foci.

4. Find the curve for which

(a) the length of the perpendicular from the pole to the tangent is constant.

(b) the length of the perpendicular from the pole to the tangent varies as the radius vector.

(c) the sum of the reciprocals of radius vector and the polar subtangent is a constant.

5. Determine the curve

(a) whose tangent cuts off from the coordinate axes intercepts whose sum is constant.

(b) Find the differential equation of all tangents to the parabola $y^2 = 4ax$. Solve the equation to find the Cartesian equation of that particular tangent which passes through the point $(0, 4)$.

(c) in which the perpendicular upon the tangent from the foot of the ordinate of the point of contact is constant and equals to a; determine the particular curve which will cut the axis of y at right angles.

(d) in which the perpendiculars from the origin upon the tangents of a curve are of constant length.

6. (a) Show that the curve in which the slope of the tangent at any point equals the ratio of the abscissa to the ordinate of the point is a rectangular hyperbola.

(b) The normal PG to a curve meets the x-axis at G. If OG = twice the abscissa of P, prove that the curve is a rectangular hyperbola. But if $OP = PG$, the curve is either a circle or a rectangular hyperbola.

(c) The tangent at any point P of a curve meets the x-axis in T. If T is on the positive side of the origin O and $OP = OT$, show that the family of curves having this property are parabolas whose common axis is the x-axis.

7. (a) The slope of the curve at any point is the reciprocal of twice the ordinate at that point. The curve also passes through the point $(4, 3)$. Show that its equation is $y^2 = x + 5$.

(b) A curve passes through $(2, 1)$ and is such that the square of the ordinate is twice the rectangle contained by the abscissa and the x-intercept of the normal. Find its equation.

8. (a) Find the curve in which the length of the portion of the normal intercepted between the curve and the x-axis varies as the square of the ordinate.

(b) Find the curve in which the portion of the x-axis between the origin and the tangent at any point is proportional to the ordinate of the point.

9. (a) The tangent at any point P of a curve meets the axis of y in A and parallel through P to the axis of y meet the axis of x at N; O is the origin. If the area of triangle AON is constant, show that the curve is a hyperbola.

(b) The normal at each point of a curve and the line from that point to the origin form an isosceles triangle with the base on the x-axis. Find the equation of the curve.

10. For a certain curve, $y = c\sec\psi$, where ψ is the inclination of the tangent at any point (x, y) to the positive direction of the x-axis. If $\psi = 0$ when $x = 0$, show that $x = c\ln(\sec\psi + \tan\psi)$.

11. Show that the curves for which the length of the perpendicular from the origin upon the tangent at any point is equal to the abscissa of the point of contact form a family of circles.

12. The tangent to a curve makes a triangle of a constant area with the coordinate axes. Prove that the curve is a rectangular hyperbola.

13. Determine the curve in which
    (a) the length of the subnormal is proportional to the square of the ordinate.
    (b) the part of the tangent intercepted by the axis is a constant k.
    (c) the length of the subnormal is proportional to the square of the abscissa.

Answers

1. (a) $y = c\,e^{x/\lambda}$
   (b) $y^2 = 4\lambda x + c$
   (c) $r = \lambda\theta + c$
   (d) $r(\theta + c) + \lambda = 0.$
2. (a) $y^n = cx;\ c = 3^n/2$
   (b) $x = cy^{\lambda}$
   (c) $y^2 = \lambda x^2 + c$
   (d) $x^3 + c = 3\lambda \ln y$
3. (a) $r^2 = c^2 \sin 2\theta$
   (b) $r = c\,e^{\theta \cot \alpha}$
   (c) $r = ce^{\lambda\theta}$
   (d) $x^2 - y^2 = 1; (\pm\sqrt{2}, 0).$
4. (a) $r = \lambda \sec(\theta + c)$
   (b) $r = c\,e^{\theta/\lambda}$
   (c) $\lambda r = 1 + cr\,e^{\theta}.$
5. (a) $(y - x)^2 - 2a(y + x) + a^2 = 0$
   (b) $x\left(\frac{dy}{dx}\right)^2 - y\frac{dy}{dx} + a = 0,\ y = \frac{1}{4}ax + 4$
   (c) [$y/a = \cosh(x/a + c)$; $c = 0$ in the particular case.]
   (d) $x^2 + y^2 = a^2.$
7. (b) $2y^2 + 4x^2 = 9x$
8. (a) $\lambda y = \cosh(\lambda x + \lambda c)$
   (b) $x = y \ln(c/y^{\lambda})$
9. (b) $x^2 - y^2 = c$
13. (a) $y = c\,e^{\lambda x}$
    (b) $x^{2/3} + y^{2/3} = k^{2/3}$
    (c) $3y^2 = 2\lambda(x^3 + c)$]

10.3 Trajectories

Definition 10.3.1. *Any curve which cuts every member of a given family of curves at a constant angle ω is called an ω-trajectory of the family. If $\omega = 90°$, the curve is called $90°$-trajectory or orthogonal trajectory of the family.*

To begin with, it will be easier to discuss orthogonal trajectories.

Orthogonal Trajectories

Definition 10.3.2. *Two families of curves with the property that every member of either family cuts each member of the other family at right angles are said to be orthogonal trajectories of each other.*

As a simple illustration we may consider a sheet of squared paper where the vertical family is orthogonal to the horizontal family of straight lines.

10.3.1 Rectangular Cartesian coordinates

Let

$$f(x, y, a) = 0 \tag{10.3.1}$$

represent one-parameter family of curves, a being the parameter.

Differentiating (10.3.1) with respect to x and eliminating a, we have the differential equation of the given family as

$$\phi\left(x, y, \frac{dy}{dx}\right) = 0, \text{ (say)}. \tag{10.3.2}$$

Considering the two curves, one from this family and the other from the family of orthogonal trajectories of this system, that pass through any point, the slope of the one curve is denoted by

$$m = \tan\psi = \frac{dy}{dx}.$$

If the slope of the other curve be denoted by m', the known condition of perpendicularity is $mm' = -1$. That is, $\frac{dy}{dx}$ for one curve is the *negative reciprocal* of $\frac{dy}{dx}$ for the other curve.

Thus, the differential equation of the system of orthogonal trajectories of (10.3.1) is given by

$$\phi\left(x, y, -\frac{dx}{dy}\right) = 0. \tag{10.3.3}$$

Integration then leads to the Cartesian form.

10.3.2 Polar coordinates

If, however, the curve be given by

$$f(r, \theta, a) = 0, \tag{10.3.4}$$

a being the arbitrary parameter, differentiating (10.3.4) with respect to θ and eliminating a, we have

$$\phi\left(r, \theta, \frac{dr}{d\theta}\right) = 0, \text{ (say)} \tag{10.3.5}$$

as the differential equation of the given family of curves.

Now, the angle ϕ between a curve and its radius vector is given by $\tan\phi = r\frac{d\theta}{dr}$.

If the same radius vector makes an angle ϕ' with the corresponding curve of the family of orthogonal trajectories, the relation between ϕ and ϕ' is $\phi' \sim \phi = \frac{1}{2}\pi$, or, $\tan\phi\tan\phi' = -1$. The corresponding procedure is, therefore, to replace [in (10.3.5)]

$$r\frac{d\theta}{dr} \text{ by } -\frac{1}{r}\frac{dr}{d\theta}; \text{ or, } \frac{d\theta}{dr} \text{ by } -\frac{1}{r^2}\frac{dr}{d\theta}.$$

Hence, the differential equation of the system of orthogonal trajectories is

$$\phi\left(r, \theta, -r^2\frac{d\theta}{dr}\right) = 0, \tag{10.3.6}$$

which on integration gives the polar form.

Working rule

Thus, for the determination of *orthogonal trajectories in rectangular Cartesian coordinates:*

1. Given $f(x, y, a) = 0$, where a is a variable parameter.
2. Differentiate $f(x, y, a) = 0$ with respect to x and eliminate a. We thus form a differential equation of the family of the form:
$$\phi\left(x, y, \frac{dy}{dx}\right) = 0.$$
3. Replace $\frac{dy}{dx}$ by $-\frac{dx}{dy}$. Then the differential equation of the family of orthogonal trajectories will be
$$\phi\left(x, y, -\frac{dx}{dy}\right) = 0.$$

4. The solution of 3 is the family of orthogonal trajectories.

For *Polar coordinates,*

1. Given $f(r, \theta, a) = 0$, a is a parameter.
2. Form a differential equation of the family of the form:

$$\phi\left(r, \theta, \frac{dr}{d\theta}\right) = 0, \text{ by eliminating } a.$$

3. Replace $\frac{dr}{d\theta}$ by $-r^2\frac{d\theta}{dr}$, whereby the differential equation of the family of orthogonal trajectories becomes

$$\phi\left(r, \theta, -r^2\frac{d\theta}{dr}\right) = 0.$$

4. Solve 3 which is the family of orthogonal trajectories.

10.3.3 ω-trajectories: Rectangular Cartesian coordinates

Let

$$f(x, y, a) = 0 \tag{10.3.7}$$

be the equation of the given system of curves, a being the variable parameter. We are interested in finding ω-trajectories of this system. Now refer to Fig. 10.8.

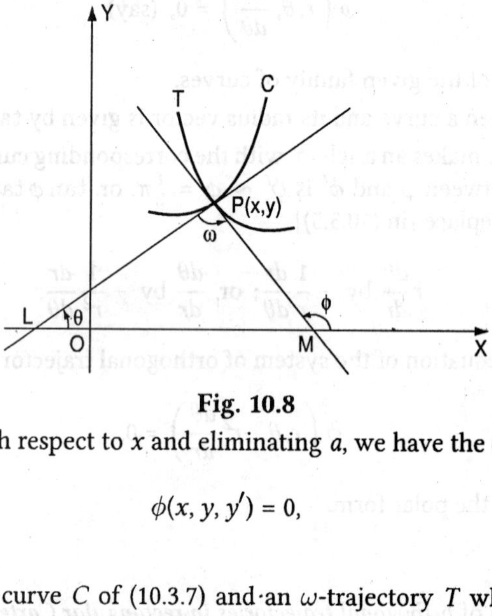

Fig. 10.8

Differentiating (10.3.7) with respect to x and eliminating a, we have the differential equation of the given family as

$$\phi(x, y, y') = 0, \tag{10.3.8}$$

where $y' = \frac{dy}{dx}$.

Now consider an integral curve C of (10.3.7) and an ω-trajectory T which intersect at $P(x, y)$ as shown in the figure. At each point of C for which (10.3.7) defines a value of y', we associate a triad of numbers (x, y, y'), the first two being the coordinates of the point and the third being the corresponding value of y' given by equation (10.3.7). Similarly, with each point of T for which there is a tangent line, we associate a triad of numbers (x, y, y'), the first two being the coordinates of the point and the third the slope of the tangent at (x, y) on T. Since $P(x, y)$ is on C as well as on T, to avoid confusion, let us write the triad associated with P on T as $(\bar{x}, \bar{y}, \bar{y}')$. Now from Fig. 10.8, we have $x = \bar{x}$, $y = \bar{y}$ at P while

$y' = \tan\theta$ and $\bar{y}' = \tan\phi$ are related by

$$y' = \tan\theta = \tan(\phi - \omega) = \frac{\tan\phi - \tan\omega}{1 + \tan\phi\tan\omega} = \frac{\bar{y}' - \tan\omega}{1 + \bar{y}'\tan\omega}.$$

Thus, for any point P on an ω-trajectory, the relation $\phi(x, y, y') = 0$ changes to

$$\phi\left(\bar{x}, \bar{y}, \frac{\bar{y}' - \tan\omega}{1 + \bar{y}'\tan\omega}\right) = 0$$

or, dropping dashes,

$$\phi\left(x, y, \frac{y' - \tan\omega}{1 + y'\tan\omega}\right) = 0$$

as the differential equation of the ω-trajectories.

Integration then leads to the Cartesian form.

Working rule

1. Given $f(x, y, a) = 0$, when a is a variable parameter.
2. Differentiate with respect to x and eliminate a, so that we form a differential equation of the given family by
$$\phi\left(x, y, \frac{dy}{dx}\right) = 0.$$
3. Replace $\frac{dy}{dx}$ by $\frac{(dy/dx)-\tan\omega}{1+(dy/dx)\tan\omega}$. Then the differential equation of the family of ω-trajectories will be
$$\phi\left(x, y, \frac{(dy/dx) - \tan\omega}{1 + (dy/dx)\tan\omega}\right) = 0.$$
4. Lastly solve 3.

10.3.4 Illustrative examples

Example 10.3.1. *Find the orthogonal trajectories of the series of hyperbolas $xy = a^2$.*

Solution: Differentiation gives $y + x\frac{dy}{dx} = 0$, as the differential equation of the given family. The differential equation of the orthogonal family would then be defined by $\left(\text{replace } \frac{dy}{dx} \text{ by } -\frac{dx}{dy}\right)$

$$y - x\frac{dx}{dy} = 0, \text{ or, } y\,dy - x\,dx = 0, \text{ or, } x\,dx - y\,dy = 0.$$

Integrating, $x^2 - y^2 = c^2$ becomes the required equation (Fig. 10.9).

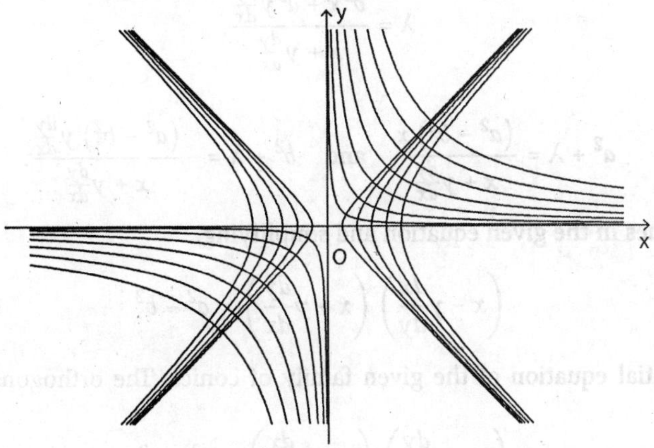

Fig. 10.9

Example 10.3.2. *Find the orthogonal trajectories of the family of straight lines $y = mx$.*

Solution: We have

$$\frac{dy}{dx} = m = \frac{y}{x}.$$

$\therefore$ orthogonal trajectories are given by $-\frac{dx}{dy} = \frac{y}{x}$.

Integrating, $x^2 + y^2 = c$ becomes the required equation (Fig. 10.10).

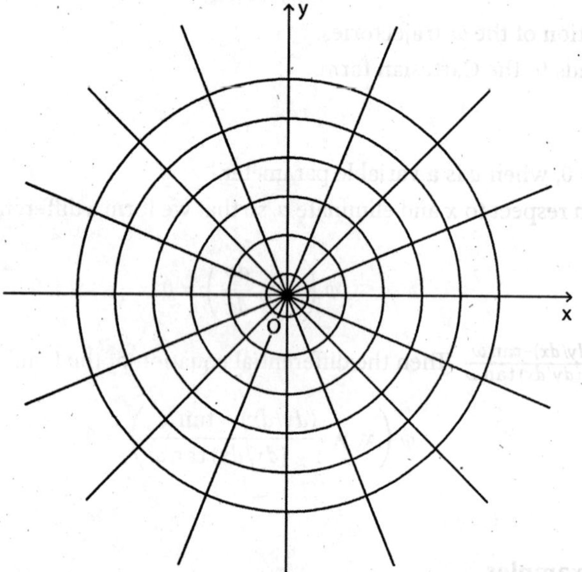

Fig. 10.10

Example 10.3.3. *Find the orthogonal trajectories of the family of confocal conics $\frac{x^2}{a^2+\lambda} + \frac{y^2}{b^2+\lambda} = 1$ (λ = parameter).*

Solution: Differentiating with respect to x, we have

$$\frac{x}{a^2 + \lambda} + \frac{y}{b^2 + \lambda}\frac{dy}{dx} = 0.$$

Then

$$\lambda = \frac{b^2 x + a^2 y\frac{dy}{dx}}{x + y\frac{dy}{dx}}.$$

Also

$$a^2 + \lambda = \frac{(a^2 - b^2)\,x}{x + y\frac{dy}{dx}} \quad \text{and} \quad b^2 + \lambda = -\frac{(a^2 - b^2)\,y\frac{dy}{dx}}{x + y\frac{dy}{dx}}.$$

Putting these values in the given equation and simplifying,

$$\left(x - y\frac{dx}{dy}\right)\left(x + y\frac{dy}{dx}\right) = a^2 - b^2 \qquad (10.3.9)$$

becomes the differential equation of the given family of conics. The orthogonal family will then be defined by

$$\left(x + y\frac{dy}{dx}\right)\left(x - y\frac{dx}{dy}\right) = a^2 - b^2, \qquad (10.3.10)$$

which is of the same form as (10.3.9). Hence, the required family becomes

$$\frac{x^2}{a^2+\mu}+\frac{y^2}{b^2+\mu}=1,$$

which defines a system of self-orthogonal confocal conics.

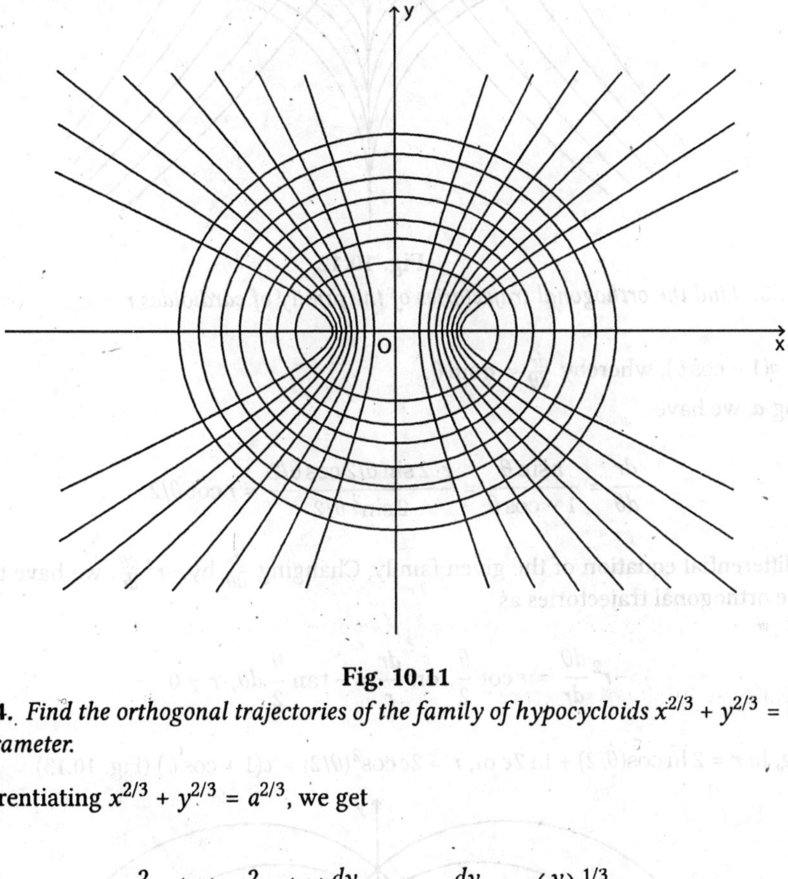

Fig. 10.11

Example 10.3.4. *Find the orthogonal trajectories of the family of hypocycloids $x^{2/3}+y^{2/3}=a^{2/3}$, where a is a variable parameter.*

Solution: Differentiating $x^{2/3}+y^{2/3}=a^{2/3}$, we get

$$\frac{2}{3}x^{-(1/3)}+\frac{2}{3}y^{-(1/3)}\frac{dy}{dx}=0, \text{ or, } \frac{dy}{dx}=-\left(\frac{y}{x}\right)^{1/3}. \tag{10.3.11}$$

Now $\frac{dy}{dx}$ for the given family is to be replaced by $-\frac{dx}{dy}$ for the trajectory. Making this change in (10.3.11), we get

$$-\frac{dx}{dy}=-\left(\frac{y}{x}\right)^{1/3}, \text{ or, } x^{1/3}\,dx-y^{1/3}\,dy=0,$$

$$\text{or,}\quad \frac{3}{4}x^{4/3}-\frac{3}{4}y^{4/3}=\frac{3}{4}C^{4/3} \text{ (say), or, } x^{4/3}-y^{4/3}=C^{4/3},$$

which gives the family of orthogonal trajectories (Fig. 10.12).

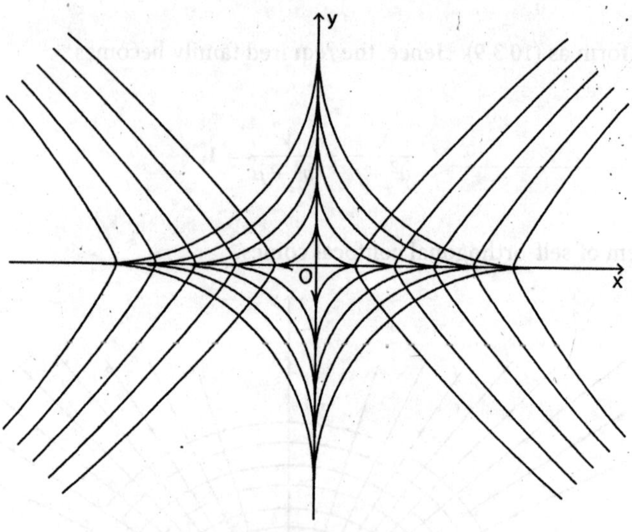

Fig. 10.12

Example 10.3.5. *Find the orthogonal trajectories of the family of cardioides $r = a(1 - \cos\theta)$ for different values of a.*

Solution: $r = a(1 - \cos\theta)$, whereby $\frac{dr}{d\theta} = a\sin\theta$.

Eliminating a, we have

$$\frac{dr}{d\theta} = \frac{r\sin\theta}{1-\cos\theta} = \frac{r\cdot 2\sin\theta/2\cos\theta/2}{2\sin^2\theta/2} = r\cot\theta/2.$$

which is the differential equation of the given family. Changing $\frac{dr}{d\theta}$ by $-r^2\frac{d\theta}{dr}$, we have the differential equation of the orthogonal trajectories as

$$-r^2\frac{d\theta}{dr} = r\cot\frac{\theta}{2}, \text{ or, } \frac{dr}{r} = -\tan\frac{\theta}{2}\,d\theta,\ r \neq 0.$$

Integrating, $\ln r = 2\ln\cos(\theta/2) + \ln 2c$ or, $r = 2c\cos^2(\theta/2) = c(1 + \cos\theta)$ (Fig. 10.13).

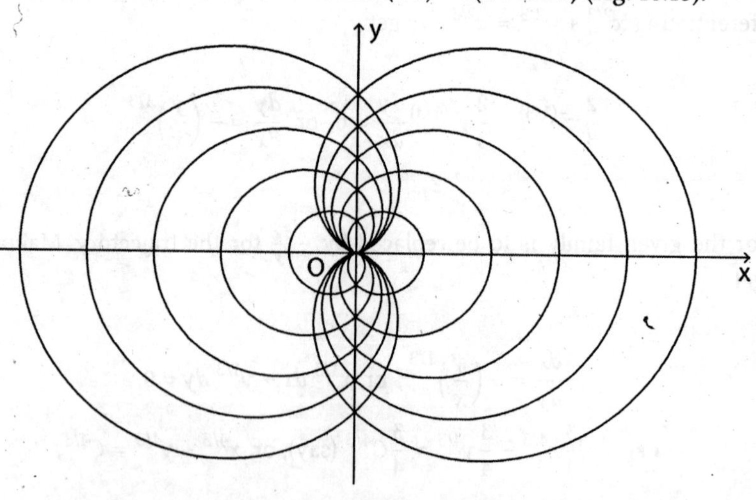

Fig. 10.13

Example 10.3.6. *Find the orthogonal trajectories of the circles defined by $r = a\cos\theta$, all of which pass through the origin and have their centres on the initial line, a being the variable diameter.*

Solution: By logarithmic differentiation,

$$\frac{dr}{rd\theta} = -\tan\theta$$

becomes the differential equation of the family. The orthogonal family will then be

$$r\frac{d\theta}{dr} = \tan\theta, \text{ or, } \frac{dr}{r} = \cot\theta\, d\theta.$$

Integrating, we have $\ln r = \ln\sin\theta + \ln c$, or, $r = c\sin\theta$. This is again a family of circles through the origin and touching the initial line (Fig. 10.14).

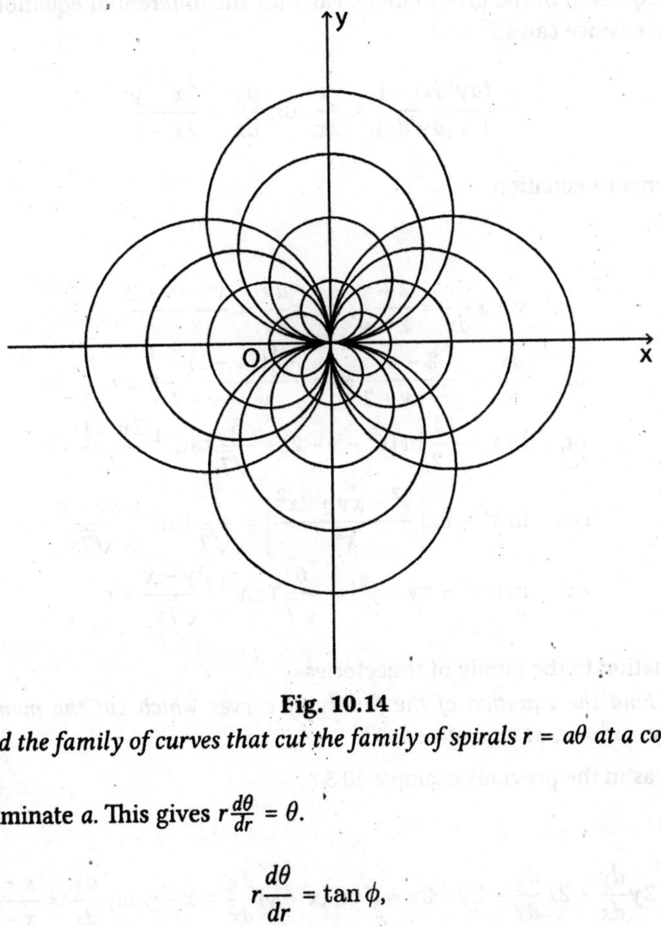

Fig. 10.14

Example 10.3.7. *Find the family of curves that cut the family of spirals $r = a\theta$ at a constant angle ω (a is a variable parameter).*

Solution: We first eliminate a. This gives $r\frac{d\theta}{dr} = \theta$.

Now

$$r\frac{d\theta}{dr} = \tan\phi,$$

where ϕ is the angle between the tangent and the radius vector. If ϕ' is the corresponding angle for the required family, then $\phi' = \phi \pm \omega$ and hence

$$\tan\phi' = \frac{\tan\phi \pm \tan\omega}{1 \mp \tan\phi\tan\omega} = \frac{\theta + k}{1 - k\theta},$$

putting the value of $\tan\phi$ and writing $k = \pm\tan\omega$.

Thus, for the second family,

$$r\frac{d\theta}{dr} = \frac{\theta + k}{1 - k\theta}.$$

Separating the variables and integrating, we get $r = c(\theta + k)^{k^2+1}\, e^{-k\theta}$, which gives the required family of curves.

Example 10.3.8. *Find the family of curves cutting the family of parabolas $y^2 = 4ax$ at 45°.*

Solution: Differentiating $y^2 = 4ax$, $2y\frac{dy}{dx} = 4a$.

Eliminating a, we have

$$\frac{dy}{dx} = \frac{4a}{2y} = \frac{y^2}{2xy} = \frac{y}{2x},\quad y \neq 0$$

as the differential equation of the given family. Now for the differential equation of the family of 45°-trajectories, we have (since $\tan 45° = 1$)

$$\frac{(dy/dx) - 1}{1 + (dy/dx)} = \frac{y}{2x},\ \text{or,}\ \frac{dy}{dx} = \frac{2x + y}{2x - y},$$

which is a homogeneous equation.

Let $y = vx$, then

$$v + x\frac{dv}{dx} = \frac{2 + v}{2 - v},\ \text{or,}\ x\frac{dv}{dx} = \frac{v^2 - v + 2}{2 - v},$$

$$\text{or,}\quad \frac{dx}{x} = \frac{2 - v}{v^2 - v + 2}dv = \frac{-\frac{1}{2}(2v - 1) + \frac{3}{2}}{v^2 - v + 2}dv,$$

$$\text{or,}\quad \ln x = -\frac{1}{2}\ln|v^2 - v + 2| + \frac{3}{\sqrt{7}}\tan^{-1}\frac{2v - 1}{\sqrt{7}} + c$$

$$\text{i.e.,}\quad \ln x^2 + \ln\left|\frac{y^2 - xy + 2x^2}{x^2}\right| = \frac{6}{\sqrt{7}}\tan^{-1}\frac{2y + x}{\sqrt{7}x} + c,$$

$$\text{or,}\quad \ln|2x^2 - xy + y^2| = \frac{6}{\sqrt{7}}\tan^{-1}\frac{2y - x}{\sqrt{7}x} + c,$$

as the required equation to the family of trajectories.

Example 10.3.9. *Find the equation of the family of curves which cut the members of the family of hyperbolas $y^2 + 2xy - x^2 = c$, at an angle of 45°.*

Solution: Proceed as in the previous example 10.3.8.

Differentiating,

$$2y\frac{dy}{dx} + 2x\frac{dy}{dx} + 2y - 2x = 0\ \text{or,}\ (x + y)\frac{dy}{dx} = x - y\ \text{or,}\ \frac{dy}{dx} = \frac{x - y}{x + y}.$$

45°-trajectories:

$$\frac{\frac{dy}{dx} - 1}{\frac{dy}{dx} + 1} = \frac{x - y}{x + y},\ \text{or,}\ x\,dx - y\,dx = 0.$$

Integrating, $x^2 - y^2 = c^2$ becomes the required equation, c being an arbitrary constant (Fig. 10.14).

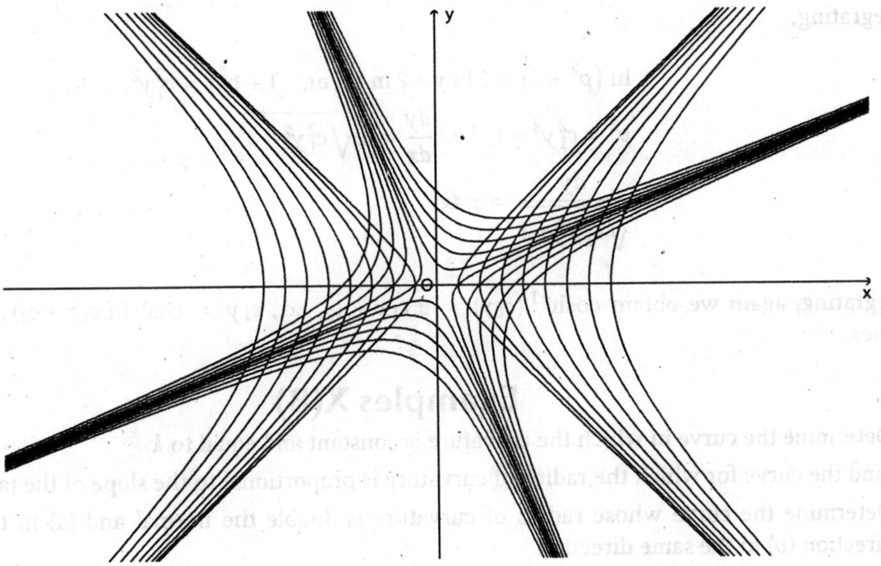

Fig. 10.14: $y^2 + 2xy - x^2 = a, x^2 - y^2 = c^2, a = \{-19, -13, -8, -4, -1, 1, 4, 8, 13, 19\}, c = \{1, 2, 3, 4, 5, 6, 7, 8\}$.

10.3.5 Further geometrical applications

We know that in rectangular coordinates, the radius of curvature ρ of a curve $y = f(x)$ at a point (x, y) on it is given by $\rho = \left(1 + y_1^2\right)^{3/2} / y_2$, where $y_1 = dy/dx$ and $y_2 = d^2y/dx^2$. Let the normal at (x, y) be drawn towards the x-axis. The normal and the radius of curvature have the same direction when y and d^2y/dx^2 have opposite signs, and have opposite directions when y and d^2y/dx^2 have the same signs.

Example 10.3.10. *Determine the curve whose radius of curvature at any point $P(x, y)$ is equal to the normal at P and they are* (a) *in the same direction,* (b) *in the opposite direction.*

Solution: (a) Here the normal and the radius of curvature are in the same direction; then y and d^2y/dx^2 have opposite signs. By normal at P we mean the length of the normal from the foot P up to the point where the normal cuts the x-axis.

$$\therefore \quad \frac{\left(1 + y_1^2\right)^{3/2}}{y_2} = -y\left(1 + y_1^2\right)^{1/2}, \text{ or, } yy_2 + y_1^2 + 1 = 0.$$

The equation is exact [since the LHS = $\frac{d}{dx}(yy_1) + 1$] the first integral gives

$$yy_1 + x = c_1, \quad \text{or,} \quad y\,dy + x\,dx = c_1\,dx, \text{ i.e., } \tfrac{1}{2}\left(x^2 + y^2\right) = c_1 x + c_2,$$

$$\text{or,} \quad y^2 + x^2 - 2c_1 x - 2c_2 = 0, \text{ or, } y^2 + (x - c_1)^2 = k,$$

where $k = c_1^2 - 2c_2$.

The primitive thus represents a family of circles with centres on the axis of x.

(b) In this case, obviously

$$\frac{\left(1 + y_1^2\right)^{3/2}}{y_2} = y\sqrt{1 + y_1^2},$$

$$\text{or,} \quad yy_2 - y_1^2 - 1 = 0,$$

$$\text{or,} \quad py\frac{dp}{dy} - p^2 - 1 = 0, \text{ where } y_1 = p \text{ and } y_2 = p\frac{dp}{dy},$$

$$\text{or,} \quad \frac{p\,dp}{p^2 + 1} = \frac{dy}{y}$$

Integrating,

$$\ln(p^2+1) = 2\ln y + 2\ln c_1, \text{ or, } (1+p^2) = c_1^2 y^2.$$

$$\therefore \quad p^2 = c_1^2 y^2 - 1, \text{ i.e., } \frac{dy}{dx} = \pm\sqrt{c_1^2 y^2 - 1}.$$

$$\therefore \quad \frac{dy}{\sqrt{c_1^2 y^2 - 1}} = \pm dx.$$

Integrating, again we obtain $\cosh^{-1}(c_1 y) = \pm c_1 x + c_2$, i.e., $c_1 y = \cosh(\pm c_1 x + c_2)$; a family of catenaries.

Examples X(B)

1. Determine the curve in which the curvature is constant and equal to k.
2. Find the curve for which the radius of curvature is proportional to the slope of the tangent.
3. Determine the curve whose radius of curvature is double the normal and (a) in the opposite direction (b) in the same direction.
4. Find the curve whose radius of curvature varies inversely as the abscissa.

Answers

1. Circle of radius $\frac{1}{k}$.
2. $y = \pm\sqrt{k^2-(x+c_1)^2} + k\ln\frac{\sqrt{k^2-(x+c_1)^2}-k}{x+c_1} + c_2.$
3. (a) $(x-a)^2 = 4c(y-c)$
   (b) the cycloid obtained by rolling any circle along the x-axis from any point.
4. $\{4k^2-(x^2-a^2)\}^{1/2}\,dy = \pm(x^2-a^2)\,dx.$

Examples X(C)

1. Find the orthogonal trajectories of the following families of curves (a and c being the variable parameters):

   (a) $x+2y=a$
   (b) $y=ax$
   (c) $xy=c^2$
   (d) $y^2=cx$
   (e) $y^2=4ax$
   (f) $y^2=4a(x+a)$
   (g) $x^2+2y^2=c$
   (h) $x^2+y^2-2ax=0$
   (i) $(x-1)^2+y^2+2ax=0$
   (j) $r=a\theta$
   (k) $r=a\cos\theta$
   (l) $r=a+\cos n\theta$.

2. Show that the orthogonal trajectories of (c, a are parameters)

   (a) the family of circles $x^2+y^2-k^2=2cy$ is $x^2+y^2+k^2=2bx$.
   (b) the series of astroids $x^{2/3}+y^{2/3}=c^{2/3}$ is $y^{4/3}-x^{4/3}=b^{4/3}$.
   (c) the Boyle's law curves PV = constant is P^2-V^2 = constant.
   (d) the family of curves $r^n\sin n\theta = a^n$ is $r^n\cos n\theta = c$.
   (e) the system of curves $r^2=a^2\cos 2\theta$ is $r^2=c^2\sin 2\theta$.
   (f) series of cardioides $r=a(1-\cos\theta)$ is $r=c(1+\cos\theta)$.
   (g) the system of curves $r=c\sin^2\theta$ is $r^2=a\cos\theta$.
   (h) the series of logarithmic spirals $r=a\theta$ is $r=e^{\frac{1}{2}(c^2-\theta^2)}$.
   (i) the system of curves $r^n=a^n\cos n\theta$ is $r^n=c^n\sin n\theta$.

(j) the system of curves $r^n = a^n \sin n\theta$ is $r^n = c^n \cos n\theta$.

(k) the system of curves $x^{4/3} + y^{4/3} = a^{4/3}$ is $x^{2/3} - y^{2/3} = c^{2/3}$.

3. Show that the family of semi-cubical parabolas $ay^2 = x^3$ gives a family of similar and similarly situated ellipses $3y^2 + 2x^2 = c^2$.
4. Prove that the equation of the system of orthogonal trajectories of a series of confocal and co-axial parabolas $r = 2a/(1 + \cos\theta)$ is $r = 2c/(1 - \cos\theta)$.
5. Show that the orthogonal trajectories of the system of co-axial circles $x^2 + y^2 + 2\lambda x + c = 0$ form another system of co-axial circles $x^2 + y^2 + 2\mu y - c = 0$ when λ and μ are parameters and c is a given constant.
6. Show that the orthogonal trajectories of

$$\frac{x^2}{a^2} + \frac{y^2}{a^2 + \lambda} = 1,$$

λ being arbitrary, is $x^2 + y^2 + c = 2a^2 \ln x$.

Find also when $a^2 = 1$.

7. (a) Obtain the differential equation of the system of confocal conics

$$\frac{x^2}{a^2 + \lambda} + \frac{y^2}{b^2 + \lambda} = 1 \text{ (λ is a variable parameter)}$$

and hence show that the system is self-orthogonal.

(b) Show that the system of co-axial parabolas $y^2 = 4a(x + a)$ is self-orthogonal.

8. Show that the family of confocal conics $x^2/A + y^2/(A - \lambda) = 1$, where A is an arbitrary constant, is self-orthogonal.
9. (a) Find the family of curves cutting the family of parabolas $y^2 = 4ax$ at constant angle 45°.

(b) Find the 45°-trajectories of the family of concentric circles $x^2 + y^2 = a^2$.

(c) Find the equation of the family of curves which cut the members of the family of hyperbolas $y^2 + 2xy - x^2 = c$, at an angle of 45°.

(d) Find the orthogonal trajectories of the family of curves $y^2 - x^2 + 2xy - 2ax = 0$ (a is a parameter).

(e) Find the curves that cut a system of concentric circles $x^2 + y^2 = a^2$ at a constant angle α.

(f) Find the equation of a set of curves, each member of which cuts every member of the family xy = constant at the angle $\pi/4$.

(g) Determine the family of curves which intersect the lemniscate $r^2 = a^2 \cos 2\theta$ at a constant angle α.

10. If $u + iv = f(x + iy)$, where u, v, x and y are all real, prove that the families u = constant and v = constant are orthogonal trajectories. Also prove that

$$\frac{\partial^2 u}{\partial x^2} + \frac{\partial^2 u}{\partial y^2} = 0, \quad \frac{\partial^2 v}{\partial x^2} + \frac{\partial^2 v}{\partial y^2} = 0.$$

Answers

1. (a) $y - 2x = c$

(b) $x^2 + y^2 = c$

(c) $x^2 - y^2 = a^2$

(d) $2x^2 + y^2 = a$

(e) $2x^2 + y^2 = c$

(f) Self-orthogonal

(g) $y = ax^2$

(h) $x^2 + y^2 = cy$

(i) $x^2 + (y - c)^2 = 1 + c^2$

(j) $x^2 = c\,e^{-\theta^2}$

(k) $r = c\sin\theta$

(l) $n^2 = r\{c + \ln|\operatorname{cosec} n\theta + \cot n\theta|\}$.

9. (a) $\ln(2x^2 - xy + y^2) = \frac{6}{\sqrt{7}}\tan^{-1}\frac{2y-x}{\sqrt{7}x} + c$

(b) $x^2 + y^2 = c\,e^{-2\tan^{-1}(y/x)}$

(c) $x^2 - y^2 = c^2$

(d) $x^2 + y^2 = c\,e^{-2\tan\alpha\tan^{-1}(y/x)}$

(e) $x^2 - 2xy - y^2 = c$

(f) $r^2 = b^2\cos(2\theta + \alpha)$ where $b^2 = -1/c^2\cos\alpha$.

Chapter 11

Vector Differential Equations

The present chapter is the extension of the Chapter 9 in which the system of ordinary linear differential equations has been introduced for solving several types of homogeneous and nonhomogeneous systems. Here, we are going to introduce another method for solving the homogeneous and nonhomogeneous systems of ordinary differential equations representing in matrix-vector form.

A vector differential equation contains more than one function arranged into the vector form along with the square matrix relating the functions to their derivatives. The vector differential equations play a very important role in the study of autonomous system that will be introduced in the next chapter. Here, we would like to present some basic theorems and definitions related to this chapter and also the most fundamental concepts about the method of solving the vector differential equations.

11.1 Homogeneous and nonhomogeneous systems

Let us consider a linear system of n first-order ordinary differential equations in n unknown functions $x_1(t), x_2(t), \ldots, x_n(t)$ in the following form:

$$\left.\begin{aligned} \dot{x}_1(t) &= a_{11}(t)x_1(t) + a_{12}(t)x_2(t) + \cdots + a_{1n}(t)x_n(t) + F_1(t) \\ \dot{x}_2(t) &= a_{21}(t)x_1(t) + a_{22}(t)x_2(t) + \cdots + a_{2n}(t)x_n(t) + F_2(t) \\ &\vdots \\ \dot{x}_n(t) &= a_{n1}(t)x_1(t) + a_{n2}(t)x_2(t) + \cdots + a_{nn}(t)x_n(t) + F_n(t) \end{aligned}\right\} \tag{11.1.1}$$

where the functions $a_{ij}(t)$ and $F_i(t)(i = 1, 2, \ldots, n; j = 1, 2, \ldots, n)$ are continuous for all $t \in [a, b]$, t being the only independent variable.

It can be expressed in more compact form given below:

$$\dot{x}_i(t) = \sum_{j=1}^{n} a_{ij}(t)x_j(t) + F_i(t), \quad i = 1, 2, \ldots, n,$$

$$\text{i.e. } \dot{\mathbf{x}}(\mathbf{t}) = \mathbf{A}(t)\mathbf{x}(t) + \mathbf{F}(t), \tag{11.1.2}$$

where $\mathbf{x}(t) = \begin{pmatrix} x_1(t) \\ x_2(t) \\ \vdots \\ x_n(t) \end{pmatrix}$, $\mathbf{F}(t) = \begin{pmatrix} F_1(t) \\ F_2(t) \\ \vdots \\ F_n(t) \end{pmatrix}$ and

$$\mathbf{A}(t) = \begin{pmatrix} a_{11}(t) & a_{12}(t) & \cdots & a_{1n}(t) \\ a_{21}(t) & a_{22}(t) & \cdots & a_{2n}(t) \\ & & \vdots & \\ a_{n1}(t) & a_{n2}(t) & \cdots & a_{nn}(t) \end{pmatrix}.$$

The system (11.1.1) is known as the *normal form* of a linear system of n first order linear ordinary differential equations. It is said to be homogeneous if for all $t \in [a, b]$, $F_i(t) = 0$, $i = 1, 2, \ldots, n$; otherwise it is called nonhomogeneous system.

Here the eq. (11.1.2) is known as the vector differential equation corresponding to the system (11.1.1). Conversely, the system (11.1.1) is called the scalar form of the vector differential eq. (11.1.2).

11.2 Wronskian of the vector functions

Let $\phi_1(t), \phi_2(t), \ldots, \phi_n(t)$ be n vector functions of $t \in [a, b]$ defined by

$$\phi_1(t) = \begin{pmatrix} \phi_{11}(t) \\ \phi_{21}(t) \\ \vdots \\ \phi_{n1}(t) \end{pmatrix}, \ \phi_2(t) = \begin{pmatrix} \phi_{12}(t) \\ \phi_{22}(t) \\ \vdots \\ \phi_{n2}(t) \end{pmatrix}, \ \ldots, \ \phi_n(t) = \begin{pmatrix} \phi_{1n}(t) \\ \phi_{2n}(t) \\ \vdots \\ \phi_{nn}(t) \end{pmatrix},$$

where $[a, b]$ is any closed interval.

Then the Wronskian of the vector functions $\phi_1(t), \phi_2(t), \ldots, \phi_n(t)$ is denoted by $W(\phi_1, \phi_2, \ldots, \phi_n)(t)$ and defined by

$$W(\phi_1, \phi_2, \ldots, \phi_n)(t) = \begin{vmatrix} \phi_{11}(t) & \phi_{12}(t) & \cdots & \phi_{1n}(t) \\ \phi_{21}(t) & \phi_{22}(t) & \cdots & \phi_{2n}(t) \\ \vdots & \vdots & & \vdots \\ \phi_{n1}(t) & \phi_{n2}(t) & \cdots & \phi_{nn}(t) \end{vmatrix}.$$

Theorem 11.2.1. *Let*

$$\phi_1(t) = \begin{pmatrix} \phi_{11}(t) \\ \phi_{21}(t) \\ \vdots \\ \phi_{n1}(t) \end{pmatrix}, \ \phi_2(t) = \begin{pmatrix} \phi_{12}(t) \\ \phi_{22}(t) \\ \vdots \\ \phi_{n2}(t) \end{pmatrix}, \ \ldots, \ \phi_n(t) = \begin{pmatrix} \phi_{1n}(t) \\ \phi_{2n}(t) \\ \vdots \\ \phi_{nn}(t) \end{pmatrix}$$

be n vector functions defined on $[a, b]$. If $\phi_1(t), \phi_2(t), \ldots, \phi_n(t)$ are linearly dependent on $[a, b]$, $W(\phi_1(t), \phi_2(t), \ldots, \phi_n(t)) = 0$ for all $t \in [a, b]$.

Proof. Let the vector functions $\phi_1(t), \phi_2(t), \ldots, \phi_n(t)$ be linearly dependent on $[a, b]$. Then there exist n numbers $c_1, c_2, \ldots, c_n$, not all zero such that

$$\sum_{j=1}^{n} c_j \phi_j(t) = 0,$$

i.e. $\sum_{j=1}^{n} c_j \phi_{ij}(t) = 0$ for all $t \in [a, b]$ and $i = 1, 2, \ldots, n$.

$\therefore$ for $t = t_0 \in [a, b]$,

$$\sum_{j=1}^{n} c_j \phi_{ij}(t_0) = 0 \ (i = 1, 2, \ldots, n). \tag{11.2.1}$$

Here (11.2.1) represents a homogeneous system of equations in $c_1, c_2, \ldots, c_n$.

Since, at least one of c_i's is non-zero, the system (11.2.1) has a nontrivial solution. It follows that the system (11.2.1) has infinitely many solutions. Thus the coefficient determinant must be zero,

$$\text{i.e.} \begin{vmatrix} \phi_{11}(t_0) & \phi_{12}(t_0) & \dots & \phi_{1n}(t_0) \\ \phi_{21}(t_0) & \phi_{22}(t_0) & \dots & \phi_{2n}(t_0) \\ \vdots & \vdots & \vdots & \vdots \\ \phi_{n1}(t_0) & \phi_{n2}(t_0) & \dots & \phi_{nn}(t_0) \end{vmatrix} = 0$$

i.e. $W(\phi_1, \phi_2, \dots, \phi_n)(t_0) = 0$. But t_0 arbitrary in $[a, b]$.

$\therefore\ W(\phi_1(t), \phi_2(t), \dots, \phi_n(t)) = 0$ for all $t \in [a, b]$.

Hence, the proof is complete.

11.3 Fundamental matrix

Let $\phi_1(t), \phi_2(t), \dots, \phi_n(t)$ be n linearly independent solutions of the linear homogeneous vector differential equation

$$\dot{\mathbf{x}}(t) = \mathbf{A}(t)\mathbf{x}(t). \tag{11.3.1}$$

Then the set $\{\phi_1(t), \phi_2(t), \dots, \phi_n(t)\}$ is called a *fundamental set of solutions* of the eq. (11.3.1), where each of the components of $\phi_1(t), \phi_2(t), \dots, \phi_n(t)$ is continuously differentiable on $[a, b]$.

An $n \times n$ matrix formed by a fundamental set of solutions of the system (11.3.1) is known as a *fundamental matrix.*

For example, let us consider the following linear system of first order differential equations:

$$\left.\begin{aligned} \dot{x}_1(t) &= x_1(t) + 3x_2(t) + t + 1 \\ \dot{x}_2(t) &= 3x_1(t) + x_2(t) + 3 \end{aligned}\right\}. \tag{11.3.2}$$

The vector differential equation corresponding to this system is

$$\dot{x}(t) = A(t)x(t) + F(t) \tag{11.3.3}$$

where $x(t) = \begin{pmatrix} x_1(t) \\ x_2(t) \end{pmatrix}$, $F(t) = \begin{pmatrix} t+1 \\ 3 \end{pmatrix}$ and $A(t) = \begin{pmatrix} 1 & 3 \\ 3 & 1 \end{pmatrix}$.

The linear system (11.3.2) is nonhomogeneous, because $F(t) \neq 0$ for all $t \in [a, b]$, whereas the vector differential equation $\dot{x}(t) = A(t)x(t)$ represents the linear homogeneous system corresponding to the system (11.3.2).

Now let us consider the vector functions $\phi_1(t)$ and $\phi_2(t)$ defined by $\phi_1(t) = \begin{pmatrix} e^{4t} \\ e^{4t} \end{pmatrix}$ and $\phi_2(t) = \begin{pmatrix} e^{-2t} \\ -e^{-2t} \end{pmatrix}$.

It can be easily verified that $\phi_1(t)$ and $\phi_2(t)$ are two linearly independent solutions of the homogeneous system corresponding to the system (11.3.3) for all $t \in [a, b]$.

Thus, $\phi_1(t)$ and $\phi_2(t)$ form a fundamental set of solutions of the vector equation $\dot{x}(t) = A(t)x(t)$. Consequently, a fundamental matrix of this system is $\begin{pmatrix} e^{4t} & e^{-2t} \\ e^{4t} & -e^{-2t} \end{pmatrix}$.

In general, it is very difficult task to construct a fundamental matrix corresponding to the system (11.3.1), if the elements of the matrix A are the functions of t.

But, in order to find a fundamental matrix corresponding to the homogeneous system represented by (11.3.1) with A as constant matrix, our task is easy. In this case, the members of the fundamental set of solutions can be assumed in the form $\phi(t) = ke^{\lambda t}$, where k and λ are two constants.

11.4 Conversion of linear differential equation to a normal linear system

Let us consider the following nth order linear differential equation

$$c_0(t)\frac{d^n}{dt^n}(x(t)) + c_1(t)\frac{d^{n-1}}{dt^{n-1}}(x(t)) + \cdots + c_{n-1}(t)\frac{d}{dt}(x(t)) + c_n(t)x(t) = G(t),$$

where $c_0(t) \neq 0$ for all $t \in [a, b]$.

Then the normalised nth order linear differential equation is

$$\frac{d^n}{dt^n}(x(t)) + \frac{c_1(t)}{c_0(t)}\frac{d^{n-1}}{dt^{n-1}}(x(t)) + \cdots + \frac{c_{n-1}(t)}{c_0(t)}\frac{d}{dt}(x(t)) + \frac{c_n(t)}{c_0(t)}x(t) = \frac{G(t)}{c_0(t)},$$

which is of the form:

$$\frac{d^n}{dt^n}(x(t)) + a_1(t)\frac{d^{n-1}}{dt^{n-1}}(x(t)) + \cdots + a_{n-1}(t)\frac{d}{dt}(x(t)) + a_n(t)x(t) = F(t), \tag{11.4.1}$$

where

$$a_i(t) = \frac{c_i(t)}{c_0(t)} (i = 1, 2, \ldots, n) \text{ and } F(t) = \frac{G(t)}{c_0(t)}, \quad t \in [a, b].$$

Let

$$x_1(t) = x(t), \quad x_2(t) = \frac{d}{dt}(x(t)), \quad x_3(t) = \frac{d^2}{dt^2}(x(t)), \ldots, x_n(t) = \frac{d^{n-1}}{dt^{n-1}}(x(t)).$$

Then

$$\left.\begin{aligned} \dot{x}_1(t) &= x_2(t) \\ \dot{x}_2(t) &= x_3(t) \\ \dot{x}_3(t) &= x_4(t) \\ &\vdots \\ \dot{x}_{n-1}(t) &= x_n(t) \\ \dot{x}_n(t) &= -a_n(t)x_1(t) - a_{n-1}(t)x_2(t) - \cdots - a_1(t)x_n(t) + F(t) \end{aligned}\right\} \tag{11.4.2}$$

This is the normal linear system of n equations in n unknown functions corresponding to the single nth order linear differential equation in one unknown function given by (11.4.1).

Here the system (11.4.2) is nonhomogeneous. It becomes homogeneous, if $F(t) = 0$ for all $t \in [a, b]$.

For example, let us consider the following linear differential equation of order three:

$$\frac{d^3x}{dt^3} + 3\frac{d^2x}{dt^2} - 4x = te^{-t}. \tag{11.4.3}$$

Let

$$x_1 = x, \; x_2 = \frac{dx}{dt}, \; x_3 = \frac{d^2x}{dt^2} \text{ and } x_4 = \frac{d^3x}{dt^3}.$$

Then $\dot{x}_1 = x_2$

$\dot{x}_2 = x_3$

$\dot{x}_3 = x_4$.

$$\text{i.e. } \left.\begin{aligned} \dot{x}_1(t) &= x_2(t) \\ \dot{x}_2(t) &= x_3(t) \\ \dot{x}_3(t) &= 4x_1(t) - 3x_3(t) + te^{-t} \end{aligned}\right\} \tag{11.4.4}$$

The system (11.4.4) is the normal linear system of three equations in three unknown functions corresponding to the linear eq. (11.4.3).

Note

1. Every linear differential equation of order n must be equivalent to a system of n first order linear differential equations. Though there exist multiple systems associated with every linear differential equation of order $n > 1$.
2. A system of n first order linear differential equations will be equivalent to a linear differential equation of order n if the coefficient matrix of the system is not diagonal matrix.

Theorem 11.4.1. *Let $A(t) = (a_{ij}(t))_{n\times n}$ and $F(t) = (F_i(t))_{n\times 1}$ be such that $a_{ij}(t)$ is continuously differentiable for $i = 1, 2, \ldots, n$; $j = 1, 2, \ldots, n$ and $F_i(t)$ is continuous for $i = 1, 2, \ldots, n$ on $[a, b]$.*

Let t_0 be any point in $[a, b]$ and $\mathbf{c} = \begin{pmatrix} c_1 \\ c_2 \\ \vdots \\ c_n \end{pmatrix}$.

Then there exists unique solution $\phi(t) = \begin{pmatrix} \phi_1(t) \\ \phi_2(t) \\ \vdots \\ \phi_n(t) \end{pmatrix}$ *of the vector differential equation $\dot{x}(t) = A(t)x(t) + F(t)$ such that $\phi(t_0) = \mathbf{c}$.*

Corollary 11.4.1. *Let $t_0 \in [a, b]$ be any point and let* $\phi(t) = \begin{pmatrix} \phi_1(t) \\ \phi_2(t) \\ \vdots \\ \phi_n(t) \end{pmatrix}$ *be a solution of the homogeneous vector differential equation $\dot{x}(t) = A(t)x(t)$ such that $\phi(t_0) = 0$, i.e. $\phi_i(t_0) = 0$ for $i = 1, 2, \ldots, n$. Then $\phi(t) = 0$ for all $t \in [a, b]$, i.e. $\phi_i(t) = 0$ for all $t \in [a, b]$.*

Theorem 11.4.2 (The Principle of Superposition). *Let the vector functions $\phi_i(t)$, $i = 1, 2, \ldots, k$, be k solutions of the homogeneous linear vector differential equation (11.3.1) and let c_i, $i = 1, 2, \ldots, k$ be, k numbers. Then the vector function*

$$\phi(t) = \sum_{i=1}^{k} c_i\phi_i(t)$$

is also a solution of the vector differential equation (11.3.1).

Proof. Since the vector functions $\phi_i(t)$, $i = 1, 2, \ldots, k$, are k solutions of (11.3.1),

$$\frac{d}{dt}(\phi_i(t)) = \mathbf{A}(t)\phi_i(t),\ i = 1, 2, \ldots, k \tag{11.4.5}$$

$$\therefore \frac{d}{dt}\left[\sum_{i=1}^{k} c_i\phi_i(t)\right] = \sum_{i=1}^{k}\left[\frac{d}{dt}(c_i\phi_i(t))\right] = \sum_{i=1}^{k} c_i\left[\frac{d}{dt}(\phi_i(t))\right] = \sum_{i=1}^{k} c_i\mathbf{A}(t)\phi_i(t), \text{ by (11.4.5)}$$

$$= \mathbf{A}(t)\sum_{i=1}^{k} c_i\phi_i(t).$$

$$\therefore \frac{d}{dt}(\phi(t)) = \mathbf{A}(t)\phi(t)$$

for all $t \in [a, b]$, $[a, b]$ being any closed interval.

It confirms that

$$\phi(t) = \sum_{i=1}^{k} c_i\phi_i(t)$$

is a solution of (11.3.1).

Hence, the proof is complete.

Example 11.4.1. *Let us consider the following vector differential equation:* $\dot{\mathbf{x}}(t) = \mathbf{A}(t)\mathbf{x}(t)$, *where* $\mathbf{A}(t) = \begin{pmatrix} 0 & 1 & 1 \\ 1 & 0 & 1 \\ 1 & 1 & 0 \end{pmatrix}$ *and* $\mathbf{x}(t) = \begin{pmatrix} x_1(t) \\ x_2(t) \\ x_3(t) \end{pmatrix}$.

Then it is quite easy to verify that the vector functions $\phi_1(t)$ and $\phi_2(t)$ defined by

$$\phi_1(t) = \begin{pmatrix} 0 \\ -\frac{1}{2}e^{-t} \\ \frac{1}{2}e^{-t} \end{pmatrix} \text{ and } \phi_2(t) = \begin{pmatrix} e^{-t} \\ -e^{-t} \\ 0 \end{pmatrix}$$

are the solutions of this vector differential equation for all $t \in [a, b]$, $[a, b]$ being any closed interval.

Clearly,

$$\phi_1(t) + \phi_2(t) = \begin{pmatrix} e^{-t} \\ -\frac{3}{2}e^{-t} \\ \frac{1}{2}e^{-t} \end{pmatrix}.$$

Now

$$\frac{d}{dt}(\phi_1(t) + \phi_2(t)) = \frac{d}{dt}\begin{pmatrix} e^{-t} \\ -\frac{3}{2}e^{-t} \\ \frac{1}{2}e^{-t} \end{pmatrix} = \begin{pmatrix} -e^{-t} \\ \frac{3}{2}e^{-t} \\ -\frac{1}{2}e^{-t} \end{pmatrix}$$

and

$$\mathbf{A}(t)(\phi_1(t) + \phi_2(t)) = \begin{pmatrix} 0 & 1 & 1 \\ 1 & 0 & 1 \\ 1 & 1 & 0 \end{pmatrix}\begin{pmatrix} e^{-t} \\ -\frac{3}{2}e^{-t} \\ \frac{1}{2}e^{-t} \end{pmatrix} = \begin{pmatrix} -e^{-t} \\ \frac{3}{2}e^{-t} \\ -\frac{1}{2}e^{-t} \end{pmatrix}.$$

$$\therefore \frac{d}{dt}(\phi_1(t) + \phi_2(t)) = \mathbf{A}(t)(\phi_1(t) + \phi_2(t))$$

holds for all $t \in [a, b]$.

Hence, $\phi_1(t) + \phi_2(t)$ is also a solution of the equation $\dot{\mathbf{x}}(t) = \mathbf{A}(t)\mathbf{x}(t)$.

Theorem 11.4.3. *Let* $\phi_1(t), \phi_2(t), \ldots, \phi_n(t)$ *be n vector functions defined by*

$$\phi_1(t) = \begin{pmatrix} \phi_{11}(t) \\ \phi_{21}(t) \\ \vdots \\ \phi_{n1}(t) \end{pmatrix}, \phi_2(t) = \begin{pmatrix} \phi_{12}(t) \\ \phi_{22}(t) \\ \vdots \\ \phi_{n2}(t) \end{pmatrix}, \ldots, \phi_n(t) = \begin{pmatrix} \phi_{1n}(t) \\ \phi_{2n}(t) \\ \vdots \\ \phi_{nn}(t) \end{pmatrix}$$

and let $\phi_i(t)$, $i = 1, 2, \ldots, n$ *be n solutions of the homogeneous linear vector differential equation* **(11.3.1).**

Then $\phi_1(t), \phi_2(t), \ldots, \phi_n(t)$ *are linearly dependent on* $[a, b]$, *if* $W(\phi_1, \phi_2, \ldots, \phi_n)(t_0) = 0$ *for some* $t_0 \in [a, b]$.

Proof. Let $c_1, c_2, \ldots, c_n$ be n unknown numbers such that

$$\sum_{j=1}^{n} c_j\phi_{ij}(t_0) = 0,\ i = 1, 2, \ldots, n. \tag{11.4.6}$$

Then the homogeneous system of equations in $c_1, c_2, \ldots, c_n$ given by (11.4.6) has trivial solution.

Let $W(\phi_1, \phi_2, \ldots, \phi_n)(t_0) = 0$. Then the coefficient determinant of the system (11.4.6) is $\det(\phi_{ij}(t_0))_{n\times n} = W(\phi_1, \phi_2, \ldots, \phi_n)(t_0) = 0$.

It confirms that the system (11.4.6) has a nontrivial solution.

So we can find n numbers $c_1, c_2, \ldots, c_n$, not all zero such that

$$\sum_{j=1}^{n} c_j\phi_j(t_0) = 0.$$

Since $\phi_1(t), \phi_2(t), \ldots, \phi_n(t)$ are n solutions of (11.3.1), for all $t \in [a, b]$,

$$\sum_{j=1}^{n} c_j\phi_j(t)$$

is also a solution of (11.3.1). But

$$\sum_{j=1}^{n} c_j\phi_j(t_0) = 0$$

for $t_0 \in [a, b]$. Therefore, by corollary 11.4.1

$$\sum_{j=1}^{n} c_j\phi_j(t) = 0$$

for all $t \in [a, b]$, where $c_1, c_2, \ldots, c_n$ are not all zero. It follows that $\phi_1(t), \phi_2(t), \ldots, \phi_n(t)$ are linearly dependent on $[a, b]$. Hence, the proof is complete.

Example 11.4.2. *Let us consider the vector differential equation $x(t) = A(t)x(t)$, where $A(t) = \begin{pmatrix} 0 & 1 & 1 \\ 1 & 0 & 1 \\ 1 & 1 & 0 \end{pmatrix}$ and $x(t) = \begin{pmatrix} x_1(t) \\ x_2(t) \\ x_3(t) \end{pmatrix}$.*

Then, it is not difficult to verify that the vector functions $\phi_1(t)$, $\phi_2(t)$ and $\phi_3(t)$ defined by

$$\phi_1(t) = \begin{pmatrix} e^{-t} \\ -\frac{1}{2}e^{-t} \\ -\frac{1}{2}e^{-t} \end{pmatrix},\ \phi_2(t) = \begin{pmatrix} 0 \\ -\frac{1}{2}e^{-t} \\ \frac{1}{2}e^{-t} \end{pmatrix} \text{ and } \phi_3(t) = \begin{pmatrix} e^{-t} \\ -e^{-t} \\ 0 \end{pmatrix}$$

are three solutions of this equation for $t \in [a, b]$, $[a, b]$ being any closed interval.

Now

$$W(\phi_1, \phi_2, \ldots, \phi_n)(t) = \begin{vmatrix} e^{-t} & 0 & e^{-t} \\ -\frac{1}{2}e^{-t} & -\frac{1}{2}e^{-t} & -e^{-t} \\ -\frac{1}{2}e^{-t} & \frac{1}{2}e^{-t} & 0 \end{vmatrix} = \frac{1}{4}e^{-3t} \begin{vmatrix} 2 & 0 & 1 \\ -1 & -1 & -1 \\ -1 & 1 & 0 \end{vmatrix} = 0 \text{ for all } t \in [a, b].$$

Again, $\phi_1(t) + \phi_2(t) + \phi_3(t) = 0$ for all $t \in [a, b]$.

It follows that $\phi_1(t), \phi_2(t), \phi_3(t)$ are linearly dependent for all $t \in [a, b]$.

Theorem 11.4.4. *Let the vector functions $\phi_1(t), \phi_2(t), \ldots, \phi_n(t)$ defined on $[a, b]$ be n solutions of the homogeneous vector differential equation* (11.3.1). *Then, either $W(\phi_1, \phi_2, \ldots, \phi_n)(t) = 0$, for all $t \in [a, b]$ or, $W(\phi_1, \phi_2, \ldots, \phi_n) \neq 0$ for all $t \in [a, b]$.*

Proof. Let $W(\phi_1, \phi_2, \ldots, \phi_n)(t_0) = 0$ for at least one $t_0 \in [a, b]$. Then, by the theorem 11.4.3, the solutions $\phi_1(t), \phi_2(t), \ldots, \phi_n(t)$ are linearly dependent on $[a, b]$. Consequently $W(\phi_1, \phi_2, \ldots, \phi_n)(t) = 0$, for all $t \in [a, b]$, by the theorem 11.4.3.

Hence, the proof is complete.

Theorem 11.4.5. *Let the vector functions $\phi_1(t), \phi_2(t), \ldots, \phi_n(t)$ defined on $[a, b]$ be n solutions of the homogeneous vector differential equation* (11.3.1). *Then $\phi_1(t), \phi_2(t), \ldots, \phi_n(t)$ form a linearly independent set of solutions of the equation* (11.3.1) *if and only if $W(\phi_1, \phi_2, \ldots, \phi_n)(t) \neq 0$ for all $t \in [a, b]$.*

The proof of the theorem is left to the reader.

Example 11.4.3. *Let us consider three solutions*

$$\phi_1(t) = \begin{pmatrix} e^{-t} \\ -\frac{1}{2}e^{-t} \\ -\frac{1}{2}e^{-t} \end{pmatrix}, \ \phi_2(t) = \begin{pmatrix} e^{2t} \\ e^{2t} \\ e^{2t} \end{pmatrix} \ \textit{and}\ \phi_3(t) = \begin{pmatrix} 0 \\ -\frac{1}{2}e^{-t} \\ \frac{1}{2}e^{-t} \end{pmatrix}$$

of the homogeneous equation given in example 11.4.2.

Now

$$W(\phi_1, \phi_2, \phi_3)(t) = \begin{vmatrix} e^{-t} & e^{2t} & 0 \\ -\frac{1}{2}e^{-t} & e^{2t} & -\frac{1}{2}e^{-t} \\ -\frac{1}{2}e^{-t} & e^{2t} & \frac{1}{2}e^{-t} \end{vmatrix} = 1 \neq 0 \cdot$$

for all $t \in [a, b]$, $[a, b]$ being arbitrary closed interval.

Hence, $\phi_1(t), \phi_2(t), \phi_3(t)$ are linearly independent solutions of the equation for all $t \in [a, b]$.

Theorem 11.4.6. *Every homogeneous linear first order vector differential equation has a fundamental set of solutions.*

Proof. Let the vector functions $\phi_1(t), \phi_2(t), \ldots, \phi_n(t)$ be n solutions of the linear homogeneous first order vector differential equation

$$\dot{\mathbf{x}}(t) = \mathbf{A}(t)\mathbf{x}(t), \tag{11.4.7}$$

where $t \in [a, b]$, $[a, b]$ being arbitrary closed interval.

Let us consider n constant vectors $\mathbf{e}_1, \mathbf{e}_2, \ldots, \mathbf{e}_n$ defined by

$$\mathbf{e}_1 = \begin{pmatrix} 1 \\ 0 \\ 0 \\ \vdots \\ 0 \end{pmatrix}, \ \mathbf{e}_2 = \begin{pmatrix} 0 \\ 1 \\ 0 \\ \vdots \\ 0 \end{pmatrix}, \ldots, \ \mathbf{e}_n = \begin{pmatrix} 0 \\ 0 \\ 0 \\ \vdots \\ 1 \end{pmatrix}.$$

Let $t_0 \in [a, b]$ be such that $\phi_i(t_0) = \mathbf{e}_i$, $i = 1, 2, \ldots, n$. Then $\phi_i(t_0) = \mathbf{e}_i$, $i = 1, 2, \ldots, n$ are n solutions of the eq. (11.4.7). These solutions are also unique, by the theorem 11.4.1.

Now

$$W(\phi_1,\phi_2,\ldots,\phi_n)(t_0) = \begin{vmatrix} 1 & 0 & 0 & \ldots & 0 \\ 0 & 1 & 0 & \ldots & 0 \\ \vdots & \vdots & \vdots & \ldots & \vdots \\ 0 & 0 & 0 & \ldots & 1 \end{vmatrix} = 1 \neq 0.$$

Therefore, by the theorem 11.4.4, $W(\phi_1,\phi_2,\ldots,\phi_n)(t) \neq 0$, for all $t \in [a, b]$.

It follows that $\phi_1(t), \phi_2(t), \ldots, \phi_n(t)$ are linearly independent on $[a, b]$.

Hence, $\phi_1, \phi_2, \ldots, \phi_n(t)$ form a fundamental solution set of the system given by (11.4.7).

Example 11.4.4. *The vector functions $\phi_1(t), \phi_2(t)$ and $\phi_3(t)$ defined by*

$$\phi_1(t) = \begin{pmatrix} e^{-t} \\ -\frac{1}{2}e^{-t} \\ -\frac{1}{2}e^{-t} \end{pmatrix}, \ \phi_2(t) = \begin{pmatrix} e^{2t} \\ e^{2t} \\ e^{2t} \end{pmatrix} \text{ and } \phi_3(t) = \begin{pmatrix} 0 \\ -\frac{1}{2}e^{-t} \\ \frac{1}{2}e^{-t} \end{pmatrix}$$

form a fundamental set of solutions of the homogeneous linear vector differential equation given in example 11.4.3, because $W(\phi_1,\phi_2,\ldots,\phi_n)\,(t) = 1 \neq 0$ for all $t \in [a, b]$.

Theorem 11.4.7. *Let the vector functions $\phi_1(t), \phi_2(t), \ldots, \phi_n(t)$ form a fundamental set of solutions of the homogeneous linear vector differential equation $\dot{x}(t) = A(t)x(t)$ on $[a, b]$ and let $\phi(t)$ be an arbitrary solution of this equation. Then $\phi(t)$ can be expressed as a linear combination of $\phi_1(t), \phi_2(t), \ldots, \phi_n(t)$ on $[a, b]$.*

Example 11.4.5. *The vector functions $\phi_1(t), \phi_2(t)$ and $\phi_3(t)$ defined in example 11.4.4 form a fundamental set of solutions of the homogeneous linear vector differential equation given in example 11.4.3.*

Thus, the general solution of this equation is given by

$$\begin{pmatrix} x_1(t) \\ x_2(t) \\ x_3(t) \end{pmatrix} = c_1 \begin{pmatrix} e^{-t} \\ -\frac{1}{2}e^{-t} \\ -\frac{1}{2}e^{-t} \end{pmatrix} + c_2 \begin{pmatrix} e^{2t} \\ e^{2t} \\ e^{2t} \end{pmatrix} + c_3 \begin{pmatrix} 0 \\ -\frac{1}{2}e^{-t} \\ \frac{1}{2}e^{-t} \end{pmatrix},$$

where c_1, c_2, c_3 are three arbitrary constants.

i.e. $x_1(t) = c_1 e^{-t} + c_2 e^{2t}$

$$x_2(t) = -\tfrac{1}{2}c_1 e^{-t} + c_2 e^{2t} - \tfrac{1}{2}c_3 e^{-t}$$

$$x_3(t) = -\tfrac{1}{2}c_1 e^{-t} + c_2 e^{2t} + \tfrac{1}{2}c_3 e^{-t}.$$

Theorem 11.4.8. *Let $\phi_0(t)$ be any solution of the nonhomogeneous linear vector differential equation*

$$\dot{x}(t) = A(t)x(t) + F(t) \tag{11.4.8}$$

and let $\phi_1(t), \phi_2(t), \ldots, \phi_n(t)$ form a fundamental set of solutions of the corresponding homogeneous linear vector differential equation

$$\dot{x}(t) = A(t)x(t). \tag{11.4.9}$$

Then the vector function

$$\psi(t) = \sum_{i=1}^{n} c_i\phi_i(t) + \phi_0(t), \ t \in [a, b]$$

is also a solution of the nonhomogeneous equation (11.4.8) for every chosen constants $c_1, c_2, \ldots, c_n$.

Proof. Since $\phi_1(t), \phi_2(t), \ldots, \phi_n(t)$ form a fundamental set of solutions of the homogeneous eq. (11.4.9), the eq. (11.4.9) must be satisfied by

$$\sum_{i=1}^{\infty} c_i\phi_i(t)$$

for every chosen constants $c_1, c_2, \ldots, c_n$.

$$\therefore \frac{d}{dt}\left[\sum_{i=1}^{n} c_i\phi_i(t)\right] = A(t)\left[\sum_{i=1}^{n} c_i\phi_i(t)\right] \tag{11.4.10}$$

Again, $\phi_0(t)$ is any solution of the nonhomogeneous linear eq. (11.4.8).

$$\therefore \frac{d}{dt}[\phi_0(t)] = \mathbf{A}(t)\phi_0(t) + \mathbf{F}(t). \tag{11.4.11}$$

Now

$$\frac{d}{dt}\left[\sum_{i=1}^{n} c_i\phi_i(t) + \phi_0(t)\right] = \frac{d}{dt}\left[\sum_{i=1}^{n} c_i\phi_i(t)\right] + \frac{d}{dt}[\phi_0(t)]$$

$$= \mathbf{A}(t)\left[\sum_{i=1}^{n} c_i\phi_i(t)\right] + \mathbf{A}(t)\phi_0(t) + \mathbf{F}(t), \text{ [by (11.4.10) and (11.4.11)]}$$

$$= \mathbf{A}(t)\left[\sum_{i=1}^{n} c_i\phi_i(t) + \phi_0(t)\right] + \mathbf{F}(t).$$

$\therefore \dot{\psi}(t) = \mathbf{A}(t)\psi(t) + \mathbf{F}(t)$, which shows that

$$\psi(t) = \sum_{i=1}^{n} c_i\phi_i(t) + \phi_0(t)$$

is a solution of the nonhomogeneous eq. (11.4.8).

Hence, the proof is complete.

Note 11.4.1. *For arbitrary constants* $c_1, c_2, \ldots, c_n$,

$$\sum_{i=1}^{n} c_i\phi_i(t) + \phi_0(t)$$

is the general solution of the equation (11.4.8).

Example 11.4.6. *Let us consider the nonhomogeneous linear vector differential equation*

$$\dot{x}(t) = A(t)x(t) + F(t), \tag{11.4.12}$$

where $A(t) = \begin{pmatrix} -2 & -4 \\ -1 & 1 \end{pmatrix}$, $x(t) = \begin{pmatrix} x_1(t) \\ x_2(t) \end{pmatrix}$ *and* $F(t) = \begin{pmatrix} 4t+1 \\ \frac{3}{2}t^2 \end{pmatrix}$.

It can be easily verified that the vector functions

$$\phi_1(t) = \begin{pmatrix} -e^{2t} \\ e^{2t} \end{pmatrix} \text{ and } \phi_2(t) = \begin{pmatrix} 4e^{-3t} \\ e^{-3t} \end{pmatrix}$$

are solutions of the homogeneous vector differential equation

$$\dot{\mathbf{x}}(t) = \mathbf{A}(t)\mathbf{x}(t). \tag{11.4.13}$$

Again

$$W(\phi_1, \phi_2)(t) = \begin{vmatrix} -e^{2t} & 4e^{-3t} \\ e^{2t} & e^{-3t} \end{vmatrix} = e^{2t} \cdot e^{-3t} \begin{vmatrix} -1 & 4 \\ 1 & 1 \end{vmatrix} = -5e^{-t} \neq 0 \text{ for all } t \in [a, b].$$

It confirms that $\phi_1(t)$ and $\phi_2(t)$ are linearly independent. Thus, $\{\phi_1(t), \phi_2(t)\}$ is a fundamental set of solutions of the eq. (11.4.13).

Observe that the vector function $\phi_0(t) = \begin{pmatrix} t^2 + t \\ -\frac{1}{2}t^2 \end{pmatrix}$ is a solution of the nonhomogeneous eq. (11.4.12).

Therefore, the general solution of the nonhomogeneous eq. (11.4.12) is given by

$$\begin{pmatrix} x_1(t) \\ x_2(t) \end{pmatrix} = c_1 \begin{pmatrix} -e^{2t} \\ e^{2t} \end{pmatrix} + c_2 \begin{pmatrix} 4e^{-3t} \\ e^{-3t} \end{pmatrix} + \begin{pmatrix} t^2 + t \\ -\frac{1}{2}t^2 \end{pmatrix}$$

i.e. $x_1(t) = -c_1 e^{2t} + 4c_2 e^{-3t} + t^2 + t$
$x_2(t) = c_1 e^{2t} + c_2 e^{-3t} - \frac{1}{2}t^2,$

where c_1 and c_2 are two arbitrary constants.

Theorem 11.4.9. *Let $\phi_0(t)$ be any solution of the nonhomogeneous linear vector differential equation* (11.4.8) *and let $\phi_1(t), \phi_2(t), \ldots, \phi_n(t)$ form a fundamental set of solutions of the corresponding homogeneous equation* (11.4.9). *Then every solution $\phi(t)$ of the nonhomogeneous equation* (11.4.8) *will be of the form*

$$\sum_{i=1}^{n} c_i\phi_i(t) + \phi_0(t),$$

where $c_1, c_2, \ldots, c_n$ are suitable constants.

Proof. Since $\phi_0(t)$ and $\phi(t)$ are solutions of the nonhomogeneous linear vector differential eq. (11.4.8),

$$\frac{d}{dt}(\phi_0(t)) = \mathbf{A}(t)\phi_0(t) + \mathbf{F}(t) \text{ and } \frac{d}{dt}(\phi(t)) = \mathbf{A}(t)\phi(t) + \mathbf{F}(t),$$

hold simultaneously.

$$\therefore \frac{d}{dt}(\phi(t)) - \frac{d}{dt}(\phi_0(t)) = \mathbf{A}(t)\phi(t) - \mathbf{A}(t)\phi_0(t) \text{ or, } \frac{d}{dt}(\phi(t) - \phi_0(t)) = \mathbf{A}(t)(\phi(t) - \phi_0(t))$$

which shows that $\phi(t) - \phi_0(t)$ is a solution of the homogeneous linear vector differential eq. (11.4.9).

Again, $\phi_1(t), \phi_2(t), \ldots, \phi_n(t)$ form a fundamental set of solutions of the homogeneous vector differential eq. (11.4.9).

Thus, for suitable choice of the constants $c_1, c_2, \ldots, c_n$, we must have

$$\phi(t) - \phi_0(t) = \sum_{i=1}^{n} c_i\phi_i(t), \text{ i.e. } \phi(t) = \sum_{i=1}^{n} c_i\phi_i(t) + \phi_0(t).$$

Hence, the proof is complete.

11.5 First order linear homogeneous systems with constant coefficients (Euler's method)

In this article, we would like to turn our attention to solve system of simultaneous first order linear homogeneous differential equations with constant coefficients. In order to study it, a fundamental concept of linear algebra is needed.

A first order linear homogeneous system of n equations and n unknown functions with constant coefficients is a system of differential equations of the following form:

$$\begin{aligned}\dot{x}_1(t) &= a_{11}x_1(t) + a_{12}x_2(t) + \cdots + a_{1n}x_n(t)\\ \dot{x}_2(t) &= a_{21}x_1(t) + a_{22}x_2(t) + \cdots + a_{2n}x_n(t)\\ &\vdots\\ \dot{x}_n(t) &= a_{n1}x_1(t) + a_{n2}x_2(t) + \cdots + a_{nn}x_n(t)\end{aligned}$$

i.e. $\dot{x}_i(t) = \sum_{j=1}^{n} a_{ij}x_j(t)$, $i = 1, 2, \ldots, n$, where a_{ij} $(i = 1, 2, \ldots, n; j = 1, 2, \ldots, n)$ are constants and $x_j(t)$ $(j = 1, 2, \ldots, n)$ are the desired functions of t.

In more compact form we have the following homogeneous vector differential equation:

$$\dot{\mathbf{x}}(t) = \mathbf{A}\mathbf{x}(t), \tag{11.5.1}$$

where $\mathbf{A} = \begin{pmatrix} a_{11} & a_{12} & \cdots & a_{1n}\\ a_{21} & a_{22} & \cdots & a_{2n}\\ \vdots & \vdots & \cdots & \vdots\\ a_{n1} & a_{n2} & \cdots & a_{nn}\end{pmatrix}$ and $\mathbf{x}(t) = \begin{pmatrix} x_1(t)\\ x_2(t)\\ \vdots\\ x_n(t)\end{pmatrix}$.

Here the nth order square matrix $\mathbf{A}$ is known as the constant coefficient matrix of the eq. (11.5.1).

Now in order to find nontrivial solutions of (11.5.1), let us put $x_i = k_ie^{\lambda t}$, where k_i and λ are constants for $i = 1, 2, \ldots, n$. Then the vector form of the solution of the vector eq. (11.5.1) is

$$\mathbf{x}(t) = \mathbf{k}e^{\lambda t}, \tag{11.5.2}$$

where $\mathbf{k} = \begin{pmatrix} k_1\\ k_2\\ \vdots\\ k_n\end{pmatrix}$.

$$\therefore\ \dot{\mathbf{x}}(t) = \lambda\mathbf{k}e^{\lambda t} \tag{11.5.3}$$

Using (11.5.2) and (11.5.3) in (11.5.1), we have $\lambda\mathbf{k}e^{\lambda t} = \mathbf{A}\mathbf{k}e^{\lambda t}$ or, $(\mathbf{A} - \lambda I_n)\mathbf{k} = 0$, I_n being the identity matrix of order n.

$$\text{i.e.}\quad \begin{pmatrix} a_{11}-\lambda & a_{12} & \cdots & a_{1n}\\ a_{21} & a_{22}-\lambda & \cdots & a_{2n}\\ \vdots & \vdots & \cdots & \vdots\\ a_{n1} & a_{n2} & \cdots & a_{nn}-\lambda\end{pmatrix}\begin{pmatrix} k_1\\ k_2\\ \vdots\\ k_n\end{pmatrix} = 0,$$

which gives the following homogeneous system of equations in $k_1, k_2, \ldots, k_n$:

$$\left.\begin{aligned}(a_{11}-\lambda)k_1 + a_{12}k_2 + \ldots + a_{1n}k_n &= 0\\ a_{21}k_1 + (a_{22}-\lambda)k_2 + \ldots + a_{2n}k_n &= 0\\ &\vdots\\ a_{n1}k_1 + a_{n2}k_2 + \ldots + (a_{nn}-\lambda)k_n &= 0\end{aligned}\right\}. \tag{11.5.4}$$

For the nontrivial solution of the above system we must have

$$\begin{vmatrix} a_{11}-\lambda & a_{12} & \cdots & a_{1n} \\ a_{21} & a_{22}-\lambda & \cdots & a_{2n} \\ \vdots & \vdots & \cdots & \vdots \\ a_{n1} & a_{n2} & \cdots & a_{nn}-\lambda \end{vmatrix} = 0, \text{ i.e. } \det(\mathbf{A}-\lambda I_n) = 0, \tag{11.5.5}$$

which is called the *characteristic equation* of the constant matrix **A**. It is a polynomial equation of degree n in λ having n roots $\lambda_1, \lambda_2, \ldots, \lambda_n$ (say). Then $\lambda_1, \lambda_2, \ldots, \lambda_n$ are the eigenvalues of **A**.

Now putting $\lambda = \lambda_i$ $(i = 1, 2, \ldots, n)$ in (11.5.4) we can find the corresponding nontrivial solutions in the form $k_1 = k_{1i}$, $k_2 = k_{2i}$, ..., $k_n = k_{ni}$, $i = 1, 2, \ldots, n$. Thus, the eigenvectors corresponding to eigenvalue $\lambda_i (i = 1, 2, \ldots, n)$ of **A** is given by

$$\mathbf{k}^{(i)} = \begin{pmatrix} k_{1i} \\ k_{2i} \\ \vdots \\ k_{ni} \end{pmatrix}, \quad i = 1, 2, \ldots, n.$$

Case 11.5.1. *All the eigenvalues are distinct.*

Let $\lambda_1, \lambda_2, \ldots, \lambda_n$ be n distinct eigenvalues of the matrix **A** and let $\mathbf{k}^{(i)}$ be an eigenvector corresponding to the eigenvalue λ_i, $i = 1, 2, \ldots, n$.

Sub-case 11.5.1(a)

Let $\lambda_1, \lambda_2, \ldots, \lambda_n$ be all real. Then the vector functions $\mathbf{x}^{(i)}(t)$ $(i = 1, 2, \ldots, n)$, all of which are distinct, defined by $\mathbf{x}^{(i)}(t) = \mathbf{k}^{(i)} e^{\lambda_i t}$ $(i = 1, 2, \ldots, n)$ must be the solutions of the eq. (11.5.1) on every closed interval $[a, b]$; because, for $i = 1, 2, \ldots, n$,

$$\dot{\mathbf{x}}^{(i)} = \lambda_i \mathbf{k}^{(i)} e^{\lambda_i t} = \mathbf{A}\mathbf{k}^{(i)} e^{\lambda_i t} \; (\because \lambda_i \mathbf{k}^{(i)} = \mathbf{A}\mathbf{k}^{(i)})$$
$$= \mathbf{A}\mathbf{x}^{(i)}(t), \text{ i.e. } \dot{\mathbf{x}}^{(i)} = \mathbf{A}\mathbf{x}^{(i)}(t),$$

which shows that the eq. (11.5.1) is satisfied by $\mathbf{x}^{(i)}(t)$.

Now

$$W\left(\mathbf{x}^{(1)}, \mathbf{x}^{(2)}, \ldots, \mathbf{x}^{(n)}\right)(t) = \begin{vmatrix} k_{11}e^{\lambda_1 t} & k_{12}e^{\lambda_2 t} & \cdots & k_{1n}e^{\lambda_n t} \\ k_{21}e^{\lambda_1 t} & k_{22}e^{\lambda_2 t} & \cdots & k_{2n}e^{\lambda_n t} \\ \vdots & \vdots & \cdots & \vdots \\ k_{n1}e^{\lambda_1 t} & k_{n2}e^{\lambda_2 t} & \cdots & k_{nn}e^{\lambda_n t} \end{vmatrix} = e^{(\lambda_1+\lambda_2+\cdots+\lambda_n)t} \begin{vmatrix} k_{11} & k_{12} & \cdots & k_{1n} \\ k_{21} & k_{22} & \cdots & k_{2n} \\ \vdots & \vdots & \cdots & \vdots \\ k_{n1} & k_{n2} & \cdots & k_{nn} \end{vmatrix}.$$

Clearly, $e^{(\lambda_1+\lambda_2+\cdots+\lambda_n)t} \neq 0$ for all $t \in [a, b]$.

Again, $d^{(i)} = \begin{pmatrix} k_{1i} \\ k_{2i} \\ \vdots \\ k_{ni} \end{pmatrix}$, $i = 1, 2, \ldots, n$ are linearly independent.

$$\therefore \begin{vmatrix} k_{11} & k_{12} & \cdots & k_{1n} \\ k_{21} & k_{22} & \cdots & k_{2n} \\ \vdots & \vdots & \cdots & \vdots \\ k_{n1} & k_{n2} & \cdots & k_{nn} \end{vmatrix} \neq 0.$$

$\therefore\ W\left(\mathbf{x}^{(1)}, \mathbf{x}^{(2)}, \ldots, \mathbf{x}^{(n)}\right)(t) \neq 0$ for all $t \in [a, b]$.

It follows that $\mathbf{x}^{(1)}, \mathbf{x}^{(2)}, \ldots, \mathbf{x}^{(n)}$ are linearly independent solutions of the vector differential eq. (11.5.1) on $[a, b]$. Hence, the general solution of the eq. (11.5.1) is given by

$$\mathbf{x}(t) = \sum_{i=1}^{n} c_i \mathbf{x}^{(i)},$$

where $c_1, c_2, \ldots, c_n$ are arbitrary constants.

Example 11.5.1. *Find the general solution of the vector differential equation* $\dot{x}(t) = \mathbf{A}x(t)$, *where* $A = \begin{pmatrix} 1 & 3 \\ 4 & 5 \end{pmatrix}$, $x(t) = \begin{pmatrix} x_1(t) \\ x_2(t) \end{pmatrix}$.

Solution: Let $\mathbf{x}(t) = \mathbf{k}e^{\lambda t}$ be a solution of the given vector differential equation, where $\mathbf{x}(t) = \begin{pmatrix} x_1(t) \\ x_2(t) \end{pmatrix}$ and $\mathbf{k} = \begin{pmatrix} k_1 \\ k_2 \end{pmatrix}$.

Then $x_1(t) = k_1 e^{\lambda t}$ and $x_2(t) = k_2 e^{\lambda t}$.

The characteristic equation of $\mathbf{A}$ is $\begin{vmatrix} 1-\lambda & 2 \\ 4 & 5-\lambda \end{vmatrix} = 0$ or, $\lambda^2 - 6\lambda - 7 = 0$.

$\therefore\ \lambda = -1, 7$.

Let $\mathbf{k}^{(1)} = \begin{pmatrix} k_1 \\ k_2 \end{pmatrix}$ be an eigenvector corresponding to the eigenvalue $\lambda_1 = -1$.

Then $2k_1 + 3k_2 = 0$
$4k_1 + 6k_2 = 0$.

The equivalent system is $2k_1 + 3k_2 = 0$ or, $\frac{k_1}{3} = \frac{k_2}{-2} = c$ (say), any non-zero number.
Choosing $c = 1$, we have $k_1 = 3, k_2 = -2$.

$$\therefore\ \mathbf{k}^{(1)} = \begin{pmatrix} 3 \\ -2 \end{pmatrix}.$$

Similarly, an eigenvector corresponding to the eigenvalue $\lambda_2 = 7$ is given by

$$\mathbf{k}^{(2)} = \begin{pmatrix} 1 \\ 2 \end{pmatrix}.$$

Thus, the fundamental set of solutions of the given equation is

$$\left\{\mathbf{k}^{(1)}e^{\lambda_1 t}, \mathbf{k}^{(2)}e^{\lambda_2 t}\right\}, \text{ i.e. } \left\{\begin{pmatrix} e^{-t} \\ -2e^{-t} \end{pmatrix}, \begin{pmatrix} 3e^{7t} \\ 2e^{7t} \end{pmatrix}\right\}.$$

Hence, the general solution of the equation is given by

$$\begin{pmatrix} x_1(t) \\ x_2(t) \end{pmatrix} = c_1 \begin{pmatrix} 3e^{-t} \\ -2e^{-t} \end{pmatrix} + c_2 \begin{pmatrix} e^{7t} \\ 2e^{7t} \end{pmatrix}$$

i.e. $x_1 = 3c_1 e^{-t} + c_2 e^{7t}$
$x_2 = -2c_1 e^{-t} + 2c_2 e^{7t}$,

where c_1, c_2 are two arbitrary constants.

Example 11.5.2. *Find the general solution of the vector differential equation $\dot{x}(t) = Ax(t)$, where*

$$A = \begin{pmatrix} 2 & -2 & 0 \\ -2 & 1 & -2 \\ 0 & -2 & 0 \end{pmatrix} \text{ and } x(t) = \begin{pmatrix} x_1(t) \\ x_2(t) \\ x_3(t) \end{pmatrix}.$$

Solution: Let $\mathbf{x}(t) = \mathbf{k}e^{\lambda t}$ be a solution of the given vector differential equation, where $\mathbf{k} = \begin{pmatrix} k_1 \\ k_2 \\ k_3 \end{pmatrix}$.

Then $x_1(t) = k_1 e^{\lambda t}, x_2(t) = k_2 e^{\lambda t}, x_3(t) = k_3 e^{\lambda t}$.

The characteristic equation of A is

$$\begin{vmatrix} 2-\lambda & -2 & 0 \\ -2 & 1-\lambda & -2 \\ 0 & -2 & 0-\lambda \end{vmatrix} = 0 \text{ or, } \lambda^3 - 3\lambda^2 - 6\lambda + 8 = 0,$$

$\therefore \lambda = 1, -2, 4.$

Let $\mathbf{k}^{(1)} = \begin{pmatrix} k_1 \\ k_2 \\ k_3 \end{pmatrix}$ be an eigenvector corresponding to the eigenvalue $\lambda_1 = 1$. Then

$$\begin{aligned} k_1 - 2k_2 &= 0 \\ -2k_1 - 2k_3 &= 0 \\ -2k_2 - k_3 &= 0 \end{aligned}$$

The equivalent system is

$$\begin{aligned} k_1 + k_3 &= 0 \\ 2k_2 + k_3 &= 0. \end{aligned}$$

Let $k_2 = c$, any non-zero number. Then $k_3 = -2c$ and $k_1 = 2c$.

$$\therefore \mathbf{k}^{(1)} = \begin{pmatrix} 2 \\ 1 \\ -2 \end{pmatrix}, \text{ choosing } c = 1.$$

Similarly, the eigenvectors corresponding to the eigenvalues $\lambda_2 = -2$ and $\lambda_3 = 4$ are given by

$$\mathbf{k}^{(2)} = \begin{pmatrix} 1 \\ 2 \\ 2 \end{pmatrix} \text{ and } \mathbf{k}^{(3)} = \begin{pmatrix} 2 \\ -2 \\ 1 \end{pmatrix} \text{ respectively.}$$

Thus, the fundamental set of solutions of the given equation is $\left\{\mathbf{k}^{(1)}e^{\lambda_1 t}, \mathbf{k}^{(2)}e^{\lambda_2 t}, \mathbf{k}^{(3)}e^{\lambda_3 t}\right\}$,

$$\text{i.e. } \left\{ \begin{pmatrix} 2e^t \\ e^t \\ -2e^t \end{pmatrix}, \begin{pmatrix} e^{2t} \\ 2e^{2t} \\ 2e^{2t} \end{pmatrix}, \begin{pmatrix} 2e^{4t} \\ -2e^{4t} \\ e^{4t} \end{pmatrix} \right\}.$$

Hence, the general solution of the given vector differential equation is

$$\begin{pmatrix} x_1(t) \\ x_2(t) \\ x_3(t) \end{pmatrix} = c_1 \begin{pmatrix} 2e^t \\ e^t \\ -2e^t \end{pmatrix} + c_2 \begin{pmatrix} e^{-2t} \\ 2e^{-2t} \\ 2e^{-2t} \end{pmatrix} + c_3 \begin{pmatrix} 2e^{4t} \\ -2e^{4t} \\ e^{4t} \end{pmatrix},$$

i.e. $x_1 = 2c_1e^t + c_2e^{-2t} + 2c_3e^{4t}$,

$x_2 = c_1e^t + 2c_2e^{-2t} - 2c_3e^{4t}$,

$x_3 = -2c_1e^t + 2c_2e^{-2t} + c_3e^{4t}$,

where c_1, c_2, c_3 are three arbitrary constants.

Sub-case 11.5.1(b)

Let $\lambda_1, \lambda_2, \ldots, \lambda_n$ be not all real. Here the square matrix **A** of order n is defined over the field $\mathbb{R}$, the set of all real numbers. Thus, the imaginary eigenvalues must occur in conjugate pairs. Let $2m$ be the number of imaginary eigenvalues of **A**. Then $2 \leqslant 2m \leqslant n$. Without loss of generality, let $\lambda_j = \alpha_j + (-1)^{j-1} i\beta_j$, where $i = \sqrt{-1}$ and $j = 1, 2, \ldots, 2m$ and let $\alpha_j = \alpha_{j+1}, \beta_j = \beta_{j+1}$ for $i, j = 1, 3, 5, \ldots, 2m - 1$.

Then the corresponding solutions are

$$\mathbf{x}^{(j)} = \mathbf{k}^{(j)} e^{(\alpha_j + (-1)^{j-1} i\beta_j)t}, \; j = 1, 2, \ldots, 2m$$
$$= \mathbf{k}^{(j)} e^{\alpha_j t} \left[\cos\left\{(-1)^{j-1}\beta_j t\right\} + i \sin\left\{(-1)^{j-1}\beta_j t\right\}\right], \; j = 1, 2, \ldots, 2m.$$

Remaining $(n - 2m)$ solutions (if $n \neq 2m$) are given by $\mathbf{x}^{(r)} = \mathbf{k}^{(r)} e^{\lambda_r t}$, $r = 2m+1, 2m+2, \ldots, n$. Since $\lambda_1, \lambda_2, \ldots, \lambda_n$ are distinct, their corresponding eigenvectors must be linearly independent. Thus, by the similar arguments in the sub-case 11.5.1(a),

$$W\left(\mathbf{x}^{(1)}, \mathbf{x}^{(2)}, \ldots, \mathbf{x}^{(n)}\right)(t) \neq 0, \text{ for all } t \in [a, b].$$

Consequently, $\mathbf{x}^{(1)}, \mathbf{x}^{(2)}, \ldots, \mathbf{x}^{(n)}$ forms a fundamental set of solutions of the eq. (11.5.1). Hence, the general solution of the equation is given by

$$\mathbf{x}(t) = \sum_{i=1}^{n} c_i \mathbf{x}^{(i)},$$

where $c_1, c_2, \ldots, c_n$ are arbitrary constants.

Example 11.5.3. *Find the general solution of the vector differential equation $\dot{x}(t) = \mathbf{A}x(t)$, where* $\mathbf{A} = \begin{pmatrix} 0 & -1 \\ 1 & 0 \end{pmatrix}$, $x(t) = \begin{pmatrix} x_1(t) \\ x_2(t) \end{pmatrix}$.

Solution: Let $\mathbf{x} = \mathbf{k}e^{\lambda t}$ be a solution of the given vector differential equation. Then $x_1(t) = k_1e^{\lambda t}$ and $x_2(t) = k_2e^{\lambda t}$, where $\mathbf{k} = \begin{pmatrix} k_1 \\ k_2 \end{pmatrix}$.

The characteristic equation of **A** is given by $\begin{vmatrix} -\lambda & -1 \\ 1 & -\lambda \end{vmatrix} = 0$, i.e. $\lambda^2 + 1 = 0$. $\therefore \lambda = \pm i$.

Let $\mathbf{k}^{(1)} = \begin{pmatrix} k_1 \\ k_2 \end{pmatrix}$ be an eigenvector corresponding to the eigenvalue $\lambda_1 = i$.

Then $-ik_1 - k_2 = 0$

$k_1 - ik_2 = 0$.

The equivalent system is $k_1 - ik_2 = 0$, i.e. $k_1 = ik_2$.

Let $k_2 = c$, c being any non-zero number. Then $k_1 = ic$. Choosing $c = 1$, we have $\mathbf{k}^{(1)} = \begin{pmatrix} i \\ 1 \end{pmatrix}$.

Similarly, an eigenvector corresponding to the eigenvalue, $\lambda_2 = -i$ is $\mathbf{k}^{(2)} = \begin{pmatrix} 1 \\ i \end{pmatrix}$.

Thus, a fundamental set of solutions of the given equation is

$$\left\{\mathbf{k}^{(1)}e^{\lambda_1 t}, \mathbf{k}^{(2)}e^{\lambda_2 t}\right\}, \text{ i.e. } \left\{\begin{pmatrix} ie^{it} \\ e^{it} \end{pmatrix}, \begin{pmatrix} e^{-it} \\ ie^{-it} \end{pmatrix}\right\}.$$

Hence, the general solution of the given vector differential equation is

$$\begin{pmatrix} x_1(t) \\ x_2(t) \end{pmatrix} = c_1 \begin{pmatrix} ie^{it} \\ e^{it} \end{pmatrix} + c_2 \begin{pmatrix} e^{-it} \\ ie^{-it} \end{pmatrix},$$

$$\begin{aligned}
\text{i.e. } x_1 &= c_1 ie^{it} + c_2 e^{-it} \\
x_2 &= c_1 e^{it} + c_2 ie^{-it}, \\
\text{i.e. } x_1 &= c_1 i(\cos t + i\sin t) + c_2(\cos t - i\sin t) \\
x_2 &= c_1(\cos t + i\sin t) + c_2 i(\cos t - i\sin t), \\
\text{i.e. } x_1 &= c_1(-\sin t + i\cos t) + c_2(\cos t - i\sin t) \\
x_2 &= c_1(\cos t + i\sin t) + c_2(\sin t + i\cos t).
\end{aligned}$$

Case 11.5.2. *The eigenvalues are not all distinct.*

Let the coefficient matrix $\mathbf{A}$ has a real eigenvalue of multiplicity m, where $1 < m \leqslant n$. Without loss of generality, let $\lambda_1 = \lambda_2 = \cdots = \lambda_m = \lambda_0$. If $m \neq n$, remaining eigenvalues are $\lambda_{m+1}, \lambda_{m+2}, \ldots, \lambda_n$.

Let there exist **r** linearly independent eigenvectors corresponding to the eigenvalue λ_0. Then $1 \leqslant \mathbf{r} \leqslant \mathbf{m}$ holds.

Sub-case 11.5.2(a): $1 \leqslant r < m$

Here the number of linearly independent eigenvectors corresponding to the eigenvalue λ_0 of multiplicity m is less than n. The solutions of (11.5.1) are of the form $\mathbf{k}^{(0)}e^{\lambda_0 t}$ corresponding to the eigenvalue λ_0. Thus the set of linearly independent solutions of the eq. (11.5.1) corresponding to the eigenvalue λ_0 fails to contain exactly m linearly independent solutions.

Consequently, we cannot have a fundamental set of solutions of (11.5.1) in the form $(\mathbf{k}^j)e^{\lambda_j t}$, where $k^{(j)}$ is the eigenvector corresponding to the eigenvalue λ_j. So we quest linearly independent solutions of the eq. (11.5.1) in another form.

In this book, we would like to consider only $m = 2$ and $r = 1$; because, otherwise the forms of m linearly independent solutions corresponding to λ_0 depend on $p = 1, 2, \ldots, m-1$. Then λ_0 is an eigenvalue of multiplicity 2. Two linearly independent solutions of (11.5.1) corresponding to the eigenvalue λ_0 can be considered in the form $\mathbf{k}^{(0)}e^{\lambda_0 t}$ and $\left(\mathbf{k}^{(0)}t + \mathbf{l}^{(0)}\right)e^{\lambda_0 t}$, where $\mathbf{k}^{(0)}$ is an eigenvector corresponding to λ_0 and $\mathbf{l}^{(0)}$ is a vector so that $(\mathbf{A} - \lambda_0 I_n)\mathbf{l}^{(0)} = \mathbf{k}^{(0)}$ holds.

Thus, $\mathbf{k}^{(0)}e^{\lambda_0 t}$, $\left(\mathbf{k}^{(0)}t + l^{(0)}\right)e^{\lambda_0 t}$, $\mathbf{k}^{(r)}e^{\lambda_r t}$, $r = 3, 4, \ldots, n$ form a fundamental set of solutions of (11.5.1).

Hence, the general solution of the eq. (11.5.1) is given by

$$\mathbf{x}(t) = c_1\mathbf{k}^{(0)}e^{\lambda_0 t} + c_2\left(\mathbf{k}^{(0)}t + \mathbf{l}^{(0)}\right)e^{\lambda_0 t} + \sum_{r=3}^{n} c_r\mathbf{k}^{(r)}e^{\lambda_r t},$$

where $c_1, c_2, \ldots, c_n$ are arbitrary constants.

Example 11.5.4. *Solve the equation* $\dot{x}(t) = Ax(t)$, *where* $A = \begin{pmatrix} 1 & 1 & 1 \\ -1 & -1 & -1 \\ 0 & 0 & 1 \end{pmatrix}$, $x(t) = \begin{pmatrix} x_1(t) \\ x_2(t) \\ x_3(t) \end{pmatrix}$.

Solution: Let $\mathbf{x} = \mathbf{k}e^{\lambda t}$ be a solution of the equation, where $\mathbf{k} = \begin{pmatrix} k_1 \\ k_2 \\ k_3 \end{pmatrix}$. Then $x_1(t) = k_1 e^{\lambda t}$, $x_2(t) = k_2 e^{\lambda t}$, $x_3(t) = k_3 e^{\lambda t}$.

The characteristic equation of $\mathbf{A}$ is given by $\begin{vmatrix} 1-\lambda & 1 & 1 \\ -1 & -1-\lambda & -1 \\ 0 & 0 & 1-\lambda \end{vmatrix} = 0.$

The roots of this equation are 0, 0, 1.

Let $\lambda_1 = \lambda_2 = 0 = \lambda_0$ and $\lambda_3 = 1$. Then an eigenvector corresponding to the eigenvalue 0 is $\mathbf{k}^{(0)} = \begin{pmatrix} -1 \\ 1 \\ 0 \end{pmatrix}$.

Let $\mathbf{l}^{(0)} = \begin{pmatrix} l_1 \\ l_2 \\ l_3 \end{pmatrix}$ be such that $(\mathbf{A} - \lambda_0 I)\mathbf{l}^{(0)} = \mathbf{k}^{(0)}$.

Then $\begin{pmatrix} 1 & 1 & 1 \\ -1 & -1 & -1 \\ 0 & 0 & 1 \end{pmatrix} \begin{pmatrix} l_1 \\ l_2 \\ l_3 \end{pmatrix} = \begin{pmatrix} -1 \\ 1 \\ 0 \end{pmatrix}$, which gives $\mathbf{l}^{(0)} = \begin{pmatrix} 0 \\ -1 \\ 0 \end{pmatrix}$.

Again, an eigenvector corresponding to the eigenvalue 1 is $\mathbf{k}^{(3)} = \begin{pmatrix} 1 \\ -1 \\ 1 \end{pmatrix}$.

$\therefore$ a fundamental set of solutions of the equation is given by

$$\left\{\mathbf{k}^{(0)}e^{\lambda_0 t}, \left(\mathbf{k}^{(0)}t + \mathbf{l}^{(0)}\right)e^{\lambda_0 t}, \mathbf{k}^{(3)}e^{\lambda_3 t}\right\}, \text{ i.e. } \left\{\begin{pmatrix} -1 \\ 1 \\ 0 \end{pmatrix}, \begin{pmatrix} -t \\ t-1 \\ 0 \end{pmatrix}, \begin{pmatrix} e^t \\ -e^t \\ e^t \end{pmatrix}\right\}.$$

Hence, required general solution is given by

$$\begin{pmatrix} x_1(t) \\ x_2(t) \\ x_3(t) \end{pmatrix} = c_1 \begin{pmatrix} -1 \\ 1 \\ 0 \end{pmatrix} + c_2 \begin{pmatrix} -t \\ t-1 \\ 0 \end{pmatrix} + c_3 \begin{pmatrix} e^t \\ -e^t \\ e^t \end{pmatrix},$$

i.e. $x_1 = -c_1 - c_2(t) + c_3 e^t$,
$x_2 = c_1 + c_2(t-1) - c_3 e^t$,
$x_3 = c_3 e^t$,
where c_1, c_2, c_3 are arbitrary constants.

Sub-case 11.5.2(b): $p = m$

In this sub-case, there are m linearly independent eigenvectors corresponding to the eigenvalue λ_0. Let these eigenvectors be $\mathbf{k}^{(j)}$, $j = 1, 2, \ldots, m$. Remaining $(n - m)$ (if $m \neq n$) eigenvectors are $\mathbf{k}^{(r)}$ corresponding the eigenvalue λ_r, $r = m+1, m+2, \ldots, n$. Thus, the linearly independent solutions of (11.5.1) are $\mathbf{k}^{(j)}e^{\lambda_0 t}, \mathbf{k}^{(r)}e^{\lambda_r t}$, where $j = 1, 2, \ldots, m$ and $r = m+1, m+2, \ldots, n$.

Hence, the general solution of the eq. (11.5.1) is given by

$$\mathbf{x}(t) = \sum_{j=1}^{m} c_j \mathbf{k}^{(j)} e^{\lambda_0 t} + \sum_{r=m+1}^{n} c_r \mathbf{k}^{(r)} e^{\lambda_r t},$$

where $c_1, c_2, \ldots, c_n$ are arbitrary constants.

Example 11.5.5. *Solve the equation $\dot{x}(t) = Ax(t)$, where* $\mathbf{A} = \begin{pmatrix} 3 & 2 & 1 \\ 2 & 3 & 1 \\ 0 & 0 & 1 \end{pmatrix}$ *and* $\mathbf{x}(t) = \begin{pmatrix} x_1(t) \\ x_2(t) \\ x_3(t) \end{pmatrix}$.

Solution: Let $\mathbf{x} = \mathbf{k}e^{\lambda t}$ be a solution of the equation $\dot{\mathbf{x}}(t) = Ax(t)$, where $\mathbf{k} = \begin{pmatrix} k_1 \\ k_2 \\ k_3 \end{pmatrix}$.

Then $x_1(t) = k_1 e^{\lambda t}, x_2(t) = k_2 e^{\lambda t}, x_3(t) = k_3 e^{\lambda t}$.

The characteristic equation of **A** is given by $\begin{vmatrix} 3-\lambda & 2 & 1 \\ 2 & 3-\lambda & 1 \\ 0 & 0 & 1-\lambda \end{vmatrix} = 0.$

The roots of this equation are 1, 1, 5. Two linearly independent eigenvectors corresponding to the eigenvalue $\lambda_0 = 1$ are

$$\mathbf{k}^{(1)} = \begin{pmatrix} -1 \\ 1 \\ 0 \end{pmatrix} \text{ and } \mathbf{k}^{(2)} = \begin{pmatrix} -1 \\ 0 \\ 2 \end{pmatrix}.$$

Again, $\mathbf{k}^{(3)} = \begin{pmatrix} 1 \\ 1 \\ 0 \end{pmatrix}$ is an eigenvector corresponding to the eigenvalue $\lambda_3 = 5$. Therefore, a fundamental set of solutions of the given equation is $\left\{ \begin{pmatrix} -e^t \\ e^t \\ 0 \end{pmatrix}, \begin{pmatrix} -e^t \\ 0 \\ 2e^t \end{pmatrix}, \begin{pmatrix} e^{5t} \\ e^{5t} \\ 0 \end{pmatrix} \right\}$.

Hence, required general solution of the equation is given by

$$\begin{pmatrix} x_1(t) \\ x_2(t) \\ x_3(t) \end{pmatrix} = c_1 \begin{pmatrix} -e^t \\ e^t \\ 0 \end{pmatrix} + c_2 \begin{pmatrix} -e^t \\ 0 \\ 2e^t \end{pmatrix} + c_3 \begin{pmatrix} e^{5t} \\ e^{5t} \\ 0 \end{pmatrix}$$

where c_1, c_2, c_3 are arbitrary constants.

11.6 First order Linear nonhomogeneous system with constant coefficients

In this article, we would like to quest the general solution of a first order linear nonhomogeneous system of n equations and n unknown functions with constant coefficients.

The system is given in the following form:

$$\begin{aligned} \dot{x}_1(t) &= a_{11}x_1(t) + a_{12}x_2(t) + \cdots + a_{1n}x_n(t) + F_1(t) \\ \dot{x}_2(t) &= a_{21}x_1(t) + a_{22}x_2(t) + \cdots + a_{2n}x_n(t) + F_2(t) \\ &\vdots \\ \dot{x}_n(t) &= a_{n1}x_1(t) + a_{n2}x_2(t) + \cdots + a_{nn}x_n(t) + F_n(t). \end{aligned}$$

The vector differential equation corresponding to this system is

$$\dot{\mathbf{x}}(t) = \mathbf{A}\mathbf{x}(t) + \mathbf{F}(t), \tag{11.6.1}$$

where $\mathbf{A} = \begin{pmatrix} a_{11} & a_{12} & \cdots & a_{1n} \\ a_{21} & a_{22} & \cdots & a_{2n} \\ & \vdots & & \\ a_{n1} & a_{n2} & \cdots & a_{nn} \end{pmatrix}$, $\mathbf{x}(t) = \begin{pmatrix} x_1(t) \\ x_2(t) \\ \vdots \\ x_n(t) \end{pmatrix}$ and $\mathbf{F}(t) = \begin{pmatrix} F_1(t) \\ F_2(t) \\ \vdots \\ F_n(t) \end{pmatrix}$.

191

Let $\phi_0(t)$ be any solution of the nonhomogeneous system represented by (11.6.1) and let $\phi_1(t), \phi_2(t), \ldots, \phi_n(t)$ form a fundamental set of solutions corresponding to the homogeneous system given by $\dot{\mathbf{x}}(t) = \mathbf{A}\mathbf{x}(t)$. Then, by the theorem 11.4.8,

$$\psi(t) = \sum_{i=1}^{n} c_i\phi_i(t) + \phi_0(t),\ t \in [a, b]$$

is the general solution of the nonhomogeneous system given by (11.6.1), where $c_1, c_2, \ldots, c_n$ are arbitrary constants.

Theorem 11.6.1. *Let $M(t)$ be a fundamental matrix of the homogeneous linear vector differential equation $\dot{x}(t) = \mathbf{A}x(t)$, $t \in [a, b], [a, b]$ being any closed interval. Then a solution $\phi_0(t)$ of the corresponding nonhomogeneous equation $\dot{x}(t) = \mathbf{A}x(t) + F(t)$ is given by*

$$\phi_0(t) = M(t)\int_{t_0}^{t} [M(u)]^{-1} F(u)\, du,\ t_0 \in [a, b].$$

Example 11.6.1. *Solve the nonhomogeneous equation $\dot{x}(t) = \mathbf{A}x(t) + F(t)$, where*

$$\mathbf{A} = \begin{pmatrix} 6 & -3 \\ 2 & 1 \end{pmatrix},\ x(t) = \begin{pmatrix} x_1(t) \\ x_2(t) \end{pmatrix},\ F(t) = \begin{pmatrix} e^{2t} \\ -e^{2t} \end{pmatrix}.$$

Solution: Let $\mathbf{x}(t) = \mathbf{k}e^{\lambda t}$ be a solution of the homogeneous equation $\dot{\mathbf{x}}(t) = \mathbf{A}\mathbf{x}(t)$ corresponding to the given nonhomogeneous equation, where $\mathbf{k} = \begin{pmatrix} k_1 \\ k_2 \end{pmatrix}$. Then $x_1(t) = k_1e^{\lambda t}$ and $x_2(t) = k_2e^{\lambda t}$.

The characteristic equation of $\mathbf{A}$ is given by $\begin{vmatrix} 6-\lambda & -3 \\ 2 & 1-\lambda \end{vmatrix} = 0.$

The solutions of this equation are 3 and 4.

$\therefore$ an eigenvector corresponding to the eigenvalue $\lambda_1 = 3$ is $\mathbf{k}^{(1)} = \begin{pmatrix} 1 \\ 1 \end{pmatrix}$ and an eigenvector corresponding to the eigenvalue $\lambda_2 = 4$ is $\mathbf{k}^{(2)} = \begin{pmatrix} 3 \\ 2 \end{pmatrix}$.

$\therefore$ a fundamental matrix of the homogeneous equation $\dot{\mathbf{x}}(t) = \mathbf{A}\mathbf{x}(t)$ is $\mathbf{M}(t) = \begin{pmatrix} e^{3t} & 3e^{4t} \\ e^{3t} & 2e^{4t} \end{pmatrix}$.

Now $\det \mathbf{M}(t) = 2e^{7t} - 3e^{7t} = -e^{7t} \neq 0$ for all $t \in [a, b], [a, b]$ being any closed interval.

$$\therefore\ [\mathbf{M}(t)]^{-1} = -\frac{1}{e^{7t}}\begin{pmatrix} 2e^{4t} & -3e^{4t} \\ -e^{3t} & e^{3t} \end{pmatrix} = \begin{pmatrix} -2e^{-3t} & 3e^{-3t} \\ e^{-4t} & -e^{-4t} \end{pmatrix}$$

and

$$[\mathbf{M}(t)]^{-1}\mathbf{F}(t) = \begin{pmatrix} -2e^{-3t} & 3e^{-3t} \\ e^{-4t} & -e^{-4t} \end{pmatrix}\begin{pmatrix} e^{2t} \\ -e^{2t} \end{pmatrix} = \begin{pmatrix} -5e^{-t} \\ 2e^{-2t} \end{pmatrix}.$$

$$\therefore\ \int_{t_0}^{t} [M(u)]^{-1}F(u)\, du = \int_{0}^{t} \begin{pmatrix} -5e^{-u} \\ 2e^{-2u} \end{pmatrix} du, \text{ choosing } t_0 = 0$$

$$= \begin{pmatrix} 5e^{-t} - 5 \\ -e^{-2t} + 1 \end{pmatrix}.$$

Thus, a solution $\phi_0(t)$ of the nonhomogeneous equation is given by

$$\phi_0(t) = \mathbf{M}(t)\int_{t_0}^{t}[\mathbf{M}(u)]^{-1}\mathbf{F}(u)\,du = \begin{pmatrix} e^{3t} & 3e^{4t} \\ e^{3t} & 2e^{4t}\end{pmatrix}\begin{pmatrix} 5e^{-t}-5 \\ -e^{-2t}+1\end{pmatrix} = \begin{pmatrix} 2e^{2t}-5e^{3t}+3e^{4t} \\ 3e^{2t}-5e^{3t}+2e^{4t}\end{pmatrix}$$

$$= \begin{pmatrix} 2e^{2t} \\ 3e^{2t}\end{pmatrix} - 5\begin{pmatrix} e^{3t} \\ e^{3t}\end{pmatrix} + \begin{pmatrix} 3e^{4t} \\ 2e^{4t}\end{pmatrix}.$$

Hence, required general solution of the nonhomogeneous equation is given by

$$\begin{pmatrix} x_1(t) \\ x_2(t)\end{pmatrix} = c_1\begin{pmatrix} e^{3t} \\ e^{3t}\end{pmatrix} + c_2\begin{pmatrix} 3e^{4t} \\ 2e^{4t}\end{pmatrix} + \begin{pmatrix} 2e^{2t} \\ 3e^{2t}\end{pmatrix},$$

i.e. $x_1 = c_1e^{3t} + 3c_2e^{4t} + 2e^{2t}$

$x_2 = c_1e^{3t} + 2c_2e^{4t} + 3e^{2t}$, where c_1, c_2 are arbitrary constants.

Examples

1. Examine whether the matrix $\mathbf{M}(t)$ is a fundamental matrix of the linear system $\dot{\mathbf{x}}(t) = \mathbf{A}\mathbf{x}(t)$:

(i) $\mathbf{M}(t) = \begin{pmatrix} e^{-5t} & e^{t} \\ -e^{-5t} & e^{t}\end{pmatrix}$; $\mathbf{A} = \begin{pmatrix} -2 & 3 \\ 3 & -2\end{pmatrix}$, $\mathbf{x}(t) = \begin{pmatrix} x_1(t) \\ x_2(t)\end{pmatrix}$

(ii) $\mathbf{M}(t) = \begin{pmatrix} 3e^{-2t} & e^{-7t} \\ -2e^{-2t} & e^{-7t}\end{pmatrix}$; $\mathbf{A} = \begin{pmatrix} 4 & 3 \\ 2 & 5\end{pmatrix}$, $\mathbf{x}(t) = \begin{pmatrix} x_1(t) \\ x_2(t)\end{pmatrix}$

(iii) $\mathbf{M}(t) = \begin{pmatrix} e^{2t} & 2e^{3t} \\ e^{2t} & e^{3t}\end{pmatrix}$; $\mathbf{A} = \begin{pmatrix} 4 & -2 \\ 1 & 1\end{pmatrix}$, $\mathbf{x}(t) = \begin{pmatrix} x_1(t) \\ x_2(t)\end{pmatrix}$

(iv) $\mathbf{M}(t) = \begin{pmatrix} \cos t & \sin t \\ -\sin t & \cos t\end{pmatrix}$; $\mathbf{A} = \begin{pmatrix} 0 & 1 \\ -1 & 0\end{pmatrix}$, $\mathbf{x}(t) = \begin{pmatrix} x_1(t) \\ x_2(t)\end{pmatrix}$

(v) $\mathbf{M}(t) = \begin{pmatrix} e^{-t} & e^{-3t} \\ e^{-t} & -e^{-3t}\end{pmatrix}$; $\mathbf{A} = \begin{pmatrix} 2 & 1 \\ 1 & 2\end{pmatrix}$, $\mathbf{x}(t) = \begin{pmatrix} x_1(t) \\ x_2(t)\end{pmatrix}$

(vi) $\mathbf{M}(t) = \begin{pmatrix} e^{-t} & e^{3t} \\ 2e^{-t} & -2e^{3t}\end{pmatrix}$; $\mathbf{A} = \begin{pmatrix} 1 & -1 \\ 1 & -4\end{pmatrix}$, $\mathbf{x}(t) = \begin{pmatrix} x_1(t) \\ x_2(t)\end{pmatrix}$

(vii) $\mathbf{M}(t) = \begin{pmatrix} e^{2t} & 0 & e^{t} \\ 2e^{2t} & e^{3t} & e^{t} \\ e^{2t} & e^{3t} & e^{t}\end{pmatrix}$; $\mathbf{A} = \begin{pmatrix} 1 & 1 & -1 \\ -2 & 3 & 0 \\ -2 & 1 & 2\end{pmatrix}$, $\mathbf{x}(t) = \begin{pmatrix} x_1(t) \\ x_2(t) \\ x_3(t)\end{pmatrix}$

(viii) $\mathbf{M}(t) = \begin{pmatrix} e^{t} & 0 & e^{3t} \\ e^{t} & e^{-t} & 2e^{3t} \\ 3e^{t} & e^{-t} & e^{3t}\end{pmatrix}$; $\mathbf{A} = \begin{pmatrix} 0 & 2 & 7 \\ 1 & -2 & 2 \\ 0 & 3 & 10\end{pmatrix}$, $\mathbf{x}(t) = \begin{pmatrix} x_1(t) \\ x_2(t) \\ x_3(t)\end{pmatrix}$.

2. Find a fundamental set of solutions of the linear homogeneous system $\dot{\mathbf{x}}(t) = \mathbf{A}\mathbf{x}(t)$ and hence solve it, where:

(i) $\mathbf{A} = \begin{pmatrix} 3 & 8 \\ -1 & -3\end{pmatrix}$, $\mathbf{x}(t) = \begin{pmatrix} x_1(t) \\ x_2(t)\end{pmatrix}$

(ii) $\mathbf{A} = \begin{pmatrix} 0 & -9 \\ 1 & 0\end{pmatrix}$, $\mathbf{x}(t) = \begin{pmatrix} x_1(t) \\ x_2(t)\end{pmatrix}$

(iii) $\mathbf{A} = \begin{pmatrix} 2 & 1 \\ -1 & 4\end{pmatrix}$, $\mathbf{x}(t) = \begin{pmatrix} x_1(t) \\ x_2(t)\end{pmatrix}$

(iv) $\mathbf{A} = \begin{pmatrix} -1 & 8 \\ 1 & 1\end{pmatrix}$, $\mathbf{x}(t) = \begin{pmatrix} x_1(t) \\ x_2(t)\end{pmatrix}$

(v) $\mathbf{A} = \begin{pmatrix} 1 & -1 \\ -1 & 1\end{pmatrix}$, $\mathbf{x}(t) = \begin{pmatrix} x_1(t) \\ x_2(t)\end{pmatrix}$

(vi) $\mathbf{A} = \begin{pmatrix} 2 & -1 & 1 \\ 1 & 2 & -1 \\ 1 & -1 & 2\end{pmatrix}$, $\mathbf{x}(t) = \begin{pmatrix} x_1(t) \\ x_2(t) \\ x_3(t)\end{pmatrix}$

(vii) $\mathbf{A} = \begin{pmatrix} 0 & -1 & 1 \\ 0 & 0 & 1 \\ -1 & 0 & 1\end{pmatrix}$, $\mathbf{x}(t) = \begin{pmatrix} x_1(t) \\ x_2(t) \\ x_3(t)\end{pmatrix}$

(viii) $\mathbf{A} = \begin{pmatrix} 0 & 1 & 1 \\ 1 & 0 & 1 \\ 1 & 1 & 0\end{pmatrix}$, $\mathbf{x}(t) = \begin{pmatrix} x_1(t) \\ x_2(t) \\ x_3(t)\end{pmatrix}$

3. Find a fundamental matrix of the homogeneous system $\dot{x}(t) = Ax(t)$ and hence solve, where

(i) $\mathbf{A} = \begin{pmatrix} 0 & 1 \\ 1 & 0 \end{pmatrix}$ (iv) $\mathbf{A} = \begin{pmatrix} 1 & -3 & 3 \\ 3 & -5 & 3 \\ 6 & -6 & 4 \end{pmatrix}$ (vi) $\mathbf{A} = \begin{pmatrix} 3 & 2 & 1 \\ 2 & 3 & 1 \\ 0 & 0 & 1 \end{pmatrix}$

(ii) $\mathbf{A} = \begin{pmatrix} -2 & -3 \\ -3 & -2 \end{pmatrix}$

(iii) $\mathbf{A} = \begin{pmatrix} 1 & -1 & -1 \\ 1 & 3 & 1 \\ -3 & 1 & -1 \end{pmatrix}$ (v) $\mathbf{A} = \begin{pmatrix} 4 & 2 & 2 \\ 2 & 4 & 2 \\ 2 & 2 & 4 \end{pmatrix}$ (vii) $\mathbf{A} = \begin{pmatrix} 2 & -2 & 0 \\ -2 & 1 & -2 \\ 0 & -2 & 0 \end{pmatrix}$.

4. Solve the nonhomogeneous equation $\dot{\mathbf{x}}(t) = \mathbf{A}\mathbf{x}(t) + \mathbf{F}(t)$, where

(i) $\mathbf{A} = \begin{pmatrix} 1 & -1 \\ -1 & 1 \end{pmatrix}$, $\mathbf{F}(t) = \begin{pmatrix} e^t \\ e^t \end{pmatrix}$ (iii) $\mathbf{A} = \begin{pmatrix} 0 & 1 \\ 1 & 0 \end{pmatrix}$, $\mathbf{F}(t) = \begin{pmatrix} t \\ -t \end{pmatrix}$

(ii) $\mathbf{A} = \begin{pmatrix} -2 & 3 \\ 3 & -2 \end{pmatrix}$, $\mathbf{F}(t) = \begin{pmatrix} t \\ e^{2t} \end{pmatrix}$ (iv) $\mathbf{A} = \begin{pmatrix} 6 & -3 \\ 2 & 1 \end{pmatrix}$, $\mathbf{F}(t) = \begin{pmatrix} e^{5t} \\ 4 \end{pmatrix}$.

Answers

1. (i) Yes (ii) No (iii) Yes (iv) Yes (v) No (vi) Yes (vii) Yes (viii) No.

2. (i) $x_1 = -4c_1e^t - 2c_2e^{-t}$
$x_2 = c_1e^t + c_2e^{-t}$

(ii) $x_1 = 3c_1 \cos 3t - 3c_2 \sin 3t$
$x_2 = c_1 \sin 3t + c_2 \cos 3t$

(iii) $x_1 = (c_1 + c_2t)e^{3t}$
$x_2 = (c_1 + c_2(1 + t))e^{3t}$

(iv) $x_1 = 2c_1e^{3t} - 4c_2e^{-3t}$
$x_2 = c_1e^{3t} + c_2e^{-3t}$

(v) $x_1 = c_1 + c_2e^{2t}$
$x_2 = c_1 - c_2e^{2t}$

(vi) $x_1 = c_1e^{2t} - c_2e^{3t}$
$x_2 = c_1e^{2t} - c_3e^t$
$x_3 = c_1e^{2t} - c_2e^{3t} - c_3e^t$

(vii) $x_1 = (c_1 - c_2)\cos t + (c_1 + c_2)\sin t$
$x_2 = c_1 \sin t - c_2 \cos t + c_3e^t$
$x_3 = c_1 \cos t + c_2 \sin t + c_3e^t$

(viii) $x_1 = c_1e^{-t} + c_2e^{2t}$
$x_2 = c_2e^{2t} + c_3e^{-t}$
$x_3 = -c_1e^{-t} + c_2e^{2t} - c_3e^{-t}$.

3. (i) $\begin{pmatrix} e^t & -e^{-t} \\ e^t & e^{-t} \end{pmatrix}$ (iv) $\begin{pmatrix} e^{-2t} & 0 & e^{4t} \\ e^{-2t} & e^{-2t} & e^{4t} \\ 0 & e^{-2t} & 2e^{4t} \end{pmatrix}$ (vi) $\begin{pmatrix} e^t & e^t & e^{5t} \\ -e^t & 0 & e^{5t} \\ 0 & -2e^t & 0 \end{pmatrix}$

(ii) $\begin{pmatrix} e^t & e^{-5t} \\ -e^t & e^{-5t} \end{pmatrix}$

(iii) $\begin{pmatrix} e^{2t} & e^{3t} & e^{-2t} \\ 0 & -e^{3t} & -e^{-2t} \\ -e^{2t} & -e^{3t} & 4e^{-2t} \end{pmatrix}$ (v) $\begin{pmatrix} e^{2t} & 0 & e^{8t} \\ 0 & e^{2t} & e^{8t} \\ -e^{2t} & -e^{2t} & e^{8t} \end{pmatrix}$ (vii) $\begin{pmatrix} -2e^t & e^{-2t} & 2e^{4t} \\ -e^t & 2e^{-2t} & -2e^{4t} \\ 2e^t & 2e^{-2t} & e^{4t} \end{pmatrix}$.

4. (i) $x_1 = c_1 + c_2e^{2t} + e^t$
$x_2 = c_1 - c_2e^{2t} + e^t$

(ii) $x_1 = -c_1e^{-5t} + c_2e^t + \frac{3}{7}e^{2t} + \frac{2}{5}t - \frac{13}{25}$
$x_2 = c_1e^{-5t} + c_2e^t + \frac{4}{7}e^{2t} - \frac{3}{5}t - \frac{12}{25}$

(iii) $x_1 = c_1e^t + c_2e^{-t} + t$
$x_2 = c_1e^t - c_2e^{-t} - t$

(iv) $x_1 = c_1e^{3t} + c_2e^{4t} + 2e^{5t} - 1$
$x_2 = c_1e^{3t} + c_2e^{4t} + e^{5t} - 2$.

Chapter 12

Autonomous System

An autonomous system is a system of ordinary differential equations not depending on the independent variable explicitly. It is actually closely related to the dynamical systems. Several physical problems can be expressed mathematically through ordinary differential equations which are nonlinear in nature. In a large number of cases, we can replace such nonlinear equations by suitable linear equations, which are able to give the approximate solutions to the original nonlinear equations.

Autonomous system has very important role in the study of mathematical modelling. Here we shall be mainly concerned with two-dimensional autonomous systems and the stability at their critical points. A brief introduction to the method of linearisation of a nonlinear plane autonomous system at a critical point is also given in the present chapter.

12.1 Autonomous system

Let us consider a system of ordinary differential equations in normal form as

$$\left.\begin{aligned} \frac{dx_1}{dt} &= F_1(x_1, x_2, x_3, \ldots, x_n, t) \\ \frac{dx_2}{dt} &= F_2(x_1, x_2, x_3, \ldots, x_n, t) \\ &\vdots \\ \frac{dx_n}{dt} &= F_n(x_1, x_2, x_3, \ldots, x_n, t) \end{aligned}\right\}, \tag{12.1.1}$$

where F_i, $i = 1, 2, 3, \ldots, n$ are defined on $I \times D$ and they are Lipschitz with respect to $(x_1, x_2, x_3, \ldots, x_n) \in D$, where $I \subseteq \mathbb{R}$ is an interval and $D \subseteq \mathbb{R}^n$ is a domain.

The system of equations (12.1.1) can be expressed in vector notation as

$$\frac{d\mathbf{x}}{dt} = \mathbf{F}(\mathbf{x}, t). \tag{12.1.2}$$

If t is dropped out from $\mathbf{F}(\mathbf{x}, t)$, the system (12.1.2) becomes

$$\frac{d\mathbf{x}}{dt} = \mathbf{F}(\mathbf{x}) \tag{12.1.3}$$

where $\mathbf{x} \in D \subseteq \mathbb{R}^n$. The eq. (12.1.3) represents a vector differential equation not depending on the independent variable t explicitly. Such a system is known as an n-dimensional *autonomous system.* It is said to be *time-invariant system,* if t represents time.

In the system of equations (12.1.2), we see that the independent variable t occurs explicitly. That is why, the system (12.1.1) or (12.1.2) is called an *nonautonomous system.*

12.2 Two-dimensional autonomous system

A normal system of two first order ordinary differential equations of the form:

$$\left.\begin{aligned} \dot{x} &= F_1(x, y) \\ \dot{y} &= F_2(x, y) \end{aligned}\right\} \tag{12.2.1}$$

where F_1 and F_2 have continuous first partial derivatives in the domain $D \subseteq \mathbb{R}^2$, except possibly at some points, is known as a two-dimensional autonomous system or *plane autonomous system*.

The set of all values of (x, y) corresponding to the plane autonomous system (12.2.1) is regarded as rectangular coordinates of points on the xy-plane that is called the *phase-plane*. As t increases, the point (x, y) traces out a smooth curve on the phase-plane. This curve is known as the *phase trajectory* or *phase path* of the autonomous system (12.2.1). The family of phase path or phase-plane orbit corresponding to the system (12.2.1) is called *phase portrait*. The existence and uniqueness theorem ensures that the trajectories do not cross each other. All the trajectories would fill the phase-plane if they were plotted.

Eliminating t from the equations in (12.2.1), we have

$$\frac{dy}{dx} = \frac{\dot{y}}{\dot{x}} = \frac{F_2(x, y)}{F_1(x, y)}, \text{ if } F_1(x, y) \neq 0 \text{ in } D.$$

It gives a one-parameter family of curves which are called the *integral curves*. Thus, we can say that the trajectory of the plane autonomous system (12.2.1) is the projection of an integral curve on the phase-plane.

Note 12.2.1. *If $F_1(x, y) = 0$ and $F_2(x, y) \neq 0$, the role of x and y will be interchanged.*

12.2.1 Critical point and ordinary point

A point $(x^*, y^*) \in D \subseteq \mathbb{R}^2$ is said to be a *critical point* or a *singular point* or *an equilibrium point* or a *reset point* of the autonomous system (12.2.1) if $F_1(x^*, y^*) = 0$ and $F_2(x^*, y^*) = 0$.

On the other hand, $(x', y') \in D$ is said to be an ordinary point corresponding to (12.2.1) if $F_1(x', y')$ and $F_2(x', y')$ are not simultaneously zero. The signs of $F_1(x', y')$ and $F_2(x', y')$ determine the direction of the phase trajectory at the point (x', y').

12.2.2 Isolated critical point

A critical point (x^*, y^*) of the plane autonomous system (12.2.1) is said to be,isolated if there exists a circle defined by $\|(x(t), y(t)) - (x^*, y^*)\| = r$ with centre (x^*, y^*) and radius r such that (x^*, y^*) is the only critical point within the circle.

12.2.3 Stability of a critical point

An isolated critical point (x^*, y^*) of the plane autonomous system (12.2.1) is said to be *stable*, if for any preassigned $\varepsilon > 0$, there exists a $\delta > 0$ such that every solution $(x(t), y(t))$ for which there is a t_0 such that $\|(x(t), y(t)) - (x^*, y^*)\| < \varepsilon$ whenever $\|(x(t_0), y(t_0)) - (x^*, y^*)\| < \delta$ holds for all $t \geqslant t_0$. Otherwise it is called *unstable critical point* of the system (12.2.1). (x^*, y^*) is said to be *asymptotically stable* if (x^*, y^*) is stable and if there exists a $\delta_0 > 0$ such that for some t_0, $\|(x(t_0), y(t_0)) - (x^*, y^*)\| < \delta_0$ implies that $\lim_{t \to \infty} (x(t), y(t)) = (x^*, y^*)$.

The point (x^*, y^*) is said to be *neutrally stable* if it is not asymptotically stable.

12.3 Linear two-dimensional autonomous system (linear plane autonomous system)

The plane autonomous system (12.2.1) is said to be linear if $F_1(x, y)$ and $F_2(x, y)$ can be expressed as $F_1(x, y) = ax + by$ and $F_2(x, y) = cx + dy$, where a, b, c, d are constants. Now we would like to investigate the critical points of the linear plane autonomous system given below:

$$\left.\begin{aligned} \dot{x} &= ax + by \\ \dot{y} &= cx + dy \end{aligned}\right\} \tag{12.3.1}$$

i.e. $\dot{\mathbf{x}} = \mathbf{A}\mathbf{x}$, where $\mathbf{x} = \begin{pmatrix} x \\ y \end{pmatrix}$ and $\mathbf{A} = \begin{pmatrix} a & b \\ c & d \end{pmatrix}$.

Let $\det \mathbf{A} \neq 0$. Then $\dot{x} = 0$ and $\dot{y} = 0$ give $x = 0$, $y = 0$. Therefore, $(0, 0)$ is the only critical point of the linear plane autonomous system (12.3.1).

Let $x = k_1 e^{\lambda t}$, $y = k_2 e^{\lambda t}$ be a solution of the system (12.3.1). Then $\begin{vmatrix} a-\lambda & b \\ c & d-\lambda \end{vmatrix} = 0$, the characteristic equation of $\mathbf{A}$.

$$\therefore \lambda^2 - (a + b)\lambda + (ad - bc) = 0 \text{ or, } \lambda^2 - T\lambda + D = 0, \tag{12.3.2}$$

where $T = a + d = \text{tr}(\mathbf{A})$, the trace of $\mathbf{A}$ and $D = ad - bc = \det \mathbf{A}$.

Therefore, λ_1 and λ_2, the eigenvalues of $\mathbf{A}$ are given by $\lambda = \frac{T \pm \sqrt{T^2 - 4D}}{2}$.

Now, let us consider the following cases:

12.3.1 Real and distinct eigenvalues

Let $T^2 - 4D > 0$, i.e. $T^2 > 4D$. Then λ_1 and λ_2 are real and distinct. Consequently, the general solution of the system is given by

$$\left.\begin{aligned} x &= c_1 k_{11} e^{\lambda_1 t} + c_2 k_{12} e^{\lambda_2 t} \\ y &= c_1 k_{21} e^{\lambda_1 t} + c_2 k_{22} e^{\lambda_2 t} \end{aligned}\right\}, \tag{12.3.3}$$

where c_1, c_2 are arbitrary constants and $k_{11}, k_{12}, k_{21}, k_{22}$ are fixed constants.

For $c_1 = 0$, the solution becomes

$$\left.\begin{aligned} x &= c_2 k_{12} e^{\lambda_2 t} \\ y &= c_2 k_{22} e^{\lambda_2 t} \end{aligned}\right\}. \tag{12.3.4}$$

Again, for $c_2 = 0$, the solution becomes

$$\left.\begin{aligned} x &= c_1 k_{11} e^{\lambda_1 t} \\ y &= c_1 k_{21} e^{\lambda_1 t} \end{aligned}\right\}. \tag{12.3.5}$$

Case 1 (stable node)

Let λ_1 and λ_2 be both negative. Then for $c_2 > 0$, (12.3.4) represents the phase-plane orbit consisting of half of the straight line given by $k_{22}x = k_{12}y$ and for $c_2 < 0$, it represents that of other half of the straight line. These two half-line orbits approach the critical point $(0, 0)$ as $t \to \infty$, because $\lambda_2 < 0$.

By similar arguments, the solution (12.3.5) represents another two phase-plane orbits consisting of half of the straight line given by $k_{21}x = k_{11}y$. These two orbits also approach $(0, 0)$ as $t \to \infty$ for $\lambda_1 < 0$.

Without loss of generality, let $\lambda_1 < \lambda_2 < 0$. Let $c_1 \neq 0$ and $c_2 \neq 0$. Then the orbits represented by (12.3.3) approach the critical point $(0, 0)$.

Clearly,

$$\frac{y}{x} = \frac{(c_1 k_{21}/c_2)e^{(\lambda_1-\lambda_2)t} + k_{22}}{(c_1 k_{11}/c_2)e^{(\lambda_1-\lambda_2)t} + k_{12}} \to \frac{k_{22}}{k_{12}} \text{ as } t \to \infty.$$

Consequently, all the phase-plane orbits represented by (12.3.3) enter the critical point (0, 0) with limiting slope $\frac{k_{12}}{k_{12}}$.

Thus, the critical point (0, 0) of the linear plane autonomous system (12.3.1) will be *stable node* if λ_1 and λ_2 are both negative. In other words, if the eigenvalues of **A** are real, distinct and negative, the critical point (0, 0) is stable node (Fig 12.1).

In this case, the flow along the phase-plane orbits will be toward the point (0, 0), that is actually asymptotically stable. The stable node (0, 0) is also known as *sink node.*

If $\lambda_1 = 0$ and $\lambda_2 < 0$ (or $\lambda_1 < 0$ and $\lambda_2 = 0$), the critical point (0, 0) is called *stable star* or *stable proper node* (Fig 12.2).

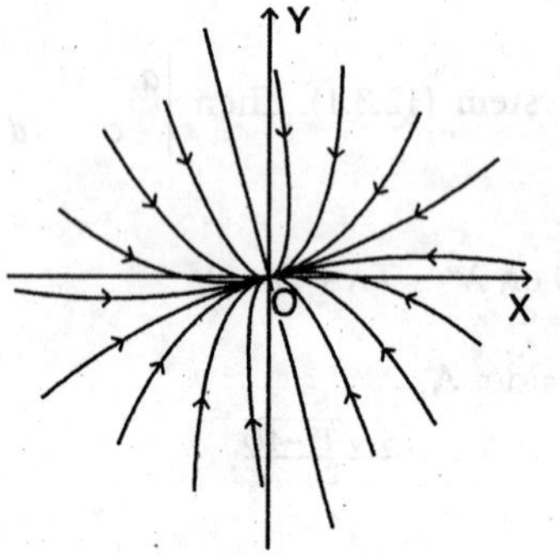

Fig 12.1 (Stable Node)

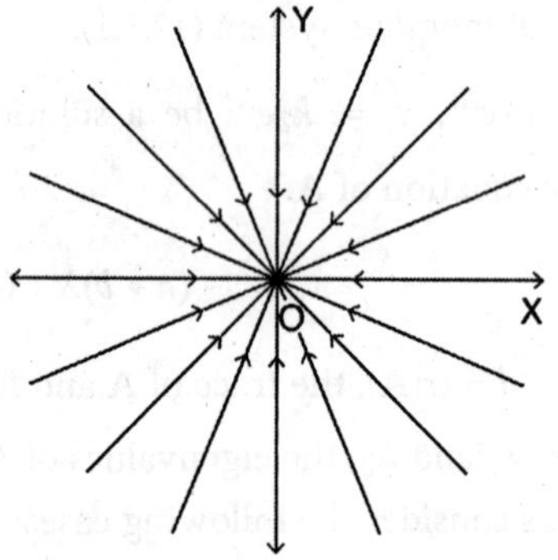

Fig 12.2 (Stable Star)

Example 12.3.1. *Let us consider the following linear plane autonomous system:*

$$\left.\begin{aligned} \dot{x} &= -2x + 6y \\ \dot{y} &= -4x - 12y \end{aligned}\right\}$$

i.e. $\dot{x} = Ax$, *where* $A = \begin{pmatrix} -2 & 6 \\ -4 & -12 \end{pmatrix}$ *and* $x = \begin{pmatrix} x \\ y \end{pmatrix}$.

Then det **A** = 48 ≠ 0. Thus, (0, 0) is the only critical point of the system.

Again, tr(A) = −2 − 12 = −14.

∴ the characteristic equation of A is $\lambda^2 + 14\lambda + 48 = 0$.

∴ $\lambda = -8, -6$, which are real, distinct and negative eigenvalues of A.

Thus, (0, 0) is the stable node of the system.

Again, an eigenvector corresponding to $\lambda_1 = -6$ is given by $\mathbf{k}^{(1)} = \begin{pmatrix} 3 \\ 2 \end{pmatrix}$ and that corresponding to $\lambda_2 = -8$ is given by $\mathbf{k}^{(2)} = \begin{pmatrix} -1 \\ 1 \end{pmatrix}$.

Case 2 (unstable node)

Let λ_1 and λ_2 be both positive. Without loss of generality, let us assume that $\lambda_1 > \lambda_2 > 0$. Then the general solution of the linear plane autonomous system (12.3.1) will be of the forms (12.3.3). Again, for $c_2 = 0$ and $c_1 = 0$ we have solutions of the form (12.3.4) and (12.3.5) respectively. Thus, we see that the phase-plane orbits of the autonomous system (12.3.1) approach and enter the critical point (0, 0) as $t \to \infty$.

Therefore, the critical point $(0,0)$ is said to be an *unstable node,* if λ_1 and λ_2 are both positive. In other words, if the eigenvalues of **A** are real, distinct and positive, the critical point $(0,0)$ is unstable node (Fig 12.3). Here the direction of the flow along every phase-plane orbit is away from the point $(0,0)$. The unstable node $(0,0)$ is also known as *source node.*

If $\lambda_1 = 0$ and $\lambda_2 > 0$ (or $\lambda_1 > 0$ and $\lambda_2 = 0$), the critical point $(0,0)$ is called *unstable star* or *unstable proper node* (Fig 12.4). On the other hand, it will be *stable star* or *stable proper node* if $\lambda_1 = 0$ and $\lambda_2 < 0$ (or $\lambda_1 < 0$ and $\lambda_2 = 0$).

Example 12.3.2. *Let us consider a linear plane autonomous system* $\dot{\boldsymbol{x}} = \boldsymbol{A}\boldsymbol{x}$, *where* $\boldsymbol{A} = \begin{pmatrix} 6 & -3 \\ 2 & 1 \end{pmatrix}$ *and* $\boldsymbol{x} = \begin{pmatrix} x \\ y \end{pmatrix}$.

Then $\det \mathbf{A} = 12 \neq 0$ and $\text{tr}(\mathbf{A}) = 6 + 1 = 7$. $\therefore$ $\dot{\mathbf{x}} = 0 \Rightarrow x = 0$ and $y = 0$ and $(0,0)$ is the only critical point of the system.

The characteristic equation of **A** is $\lambda^2 - 7\lambda^2 + 12 = 0$, which gives $\lambda = 3, 4$. Thus, the eigenvalues of **A** are real, distinct and positive. It ensures that the critical point $(0,0)$ is unstable node.

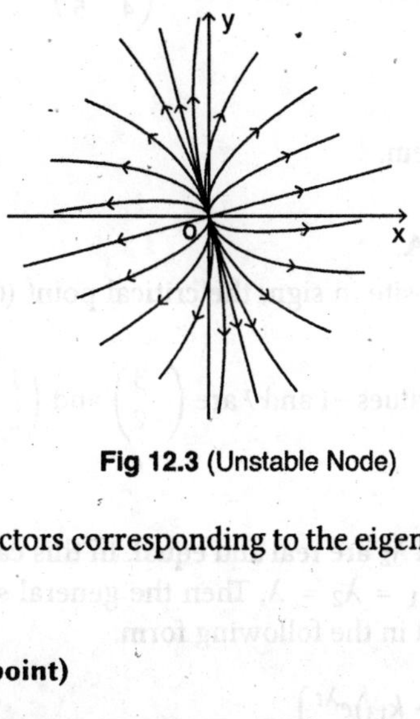

Fig 12.3 (Unstable Node)

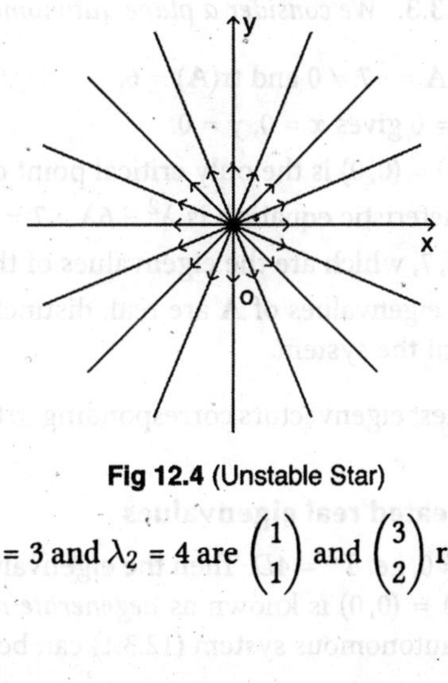

Fig 12.4 (Unstable Star)

The eigenvectors corresponding to the eigenvalues $\lambda_1 = 3$ and $\lambda_2 = 4$ are $\begin{pmatrix} 1 \\ 1 \end{pmatrix}$ and $\begin{pmatrix} 3 \\ 2 \end{pmatrix}$ respectively.

Case 3 (saddle point)

Let λ_1 and λ_2 be opposite in sign. Without loss of generality, let $\lambda_1 < 0$ and $\lambda_2 > 0$. Then the general solution of the linear plane autonomous system (12.3.1) is given by (12.3.3), that represents nonrectilinear phase-plane orbits for $c_1 \neq 0$ and $c_2 \neq 0$. None of these orbits is able to approach the critical point $(0,0)$ as $t \to \pm\infty$, because $\lambda_1 < 0$ and $\lambda_2 > 0$. However, they fail to pass through the point $(0,0)$ for $t \in [\alpha, \beta]$, $[\alpha, \beta]$ being any closed interval.

Again, the solutions (12.3.4) and (12.3.5) represents the rectilinear phase-plane orbits. Since $\lambda_1 < 0$, the half-line orbits (one for $c_1 > 0$ and another for $c_1 < 0$) represented by (12.3.5) approach and enter the critical point $(0,0)$ as $t \to \infty$. On the other hand, the half-line orbits (one for $c_2 > 0$ and other for $c_2 < 0$) represented by (12.3.4) approach and enter $(0,0)$ as $t \to -\infty$, because $\lambda_2 > 0$.

Thus, every nonrectilinear orbit becomes asymptotic to one of the half-line orbits represented by (12.3.5) as $t \to \infty$, because $\lambda_1 < 0$ and that represented by (12.3.4) as $t \to -\infty$, because $\lambda_2 > 0$.

Therefore, critical point $(0,0)$ is said to be a *saddle point* if λ_1 and λ_2 are opposite in sign (Fig 12.5). In other words, we can say that $(0,0)$ will be a *saddle point,* if the eigenvalues of A are real, unequal and opposite in sign.

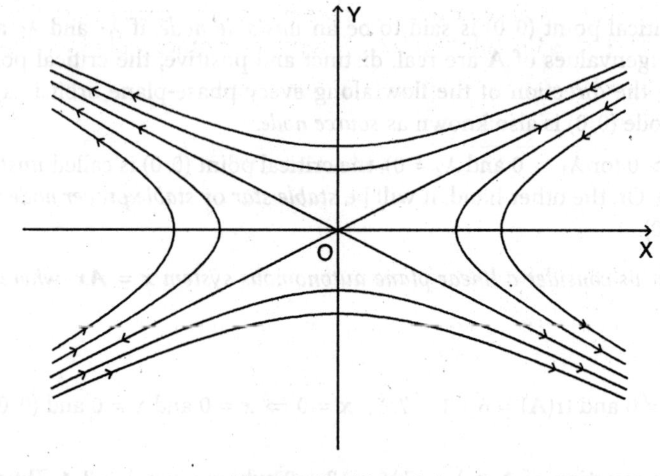

Fig 12.5 (Saddle Point)

Example 12.3.3. *We consider a plane autonomous system $\dot{\mathbf{x}} = \mathbf{A}\mathbf{x}$, where* $\mathbf{A} = \begin{pmatrix} 1 & 3 \\ 4 & 5 \end{pmatrix}$ *and* $\mathbf{x} = \begin{pmatrix} x \\ y \end{pmatrix}$.

Then $\det \mathbf{A} = -7 \neq 0$ and $\text{tr}(\mathbf{A}) = 6$.

Again, $\dot{\mathbf{x}} = 0$ gives $x = 0, y = 0$.

$\therefore$ $(x^*, y^*) = (0, 0)$ is the only critical point of the system.

The characteristic equation is $\lambda^2 - 6\lambda - 7 = 0$.

$\therefore$ $\lambda = -1, 7$, which are the eigenvalues of the matrix $\mathbf{A}$.

Since the eigenvalues of $\mathbf{A}$ are real, distinct and opposite in sign, the critical point $(0, 0)$ must be a saddle point of the system.

The simplest eigenvectors corresponding to the eigenvalues -1 and 7 are $\begin{pmatrix} 3 \\ -2 \end{pmatrix}$ and $\begin{pmatrix} 1 \\ 2 \end{pmatrix}$ respectively.

12.3.2 Repeated real eigenvalues

Let $T^2 - 4D = 0$, i.e. $T^2 = 4D$. Then the eigenvalues λ_1 and λ_2 are real and equal. In this case, the critical point $(x^*, y^*) = (0, 0)$ is known as *degenerate node*. Let $\lambda_1 = \lambda_2 = \lambda$. Then the general solution of the linear plane autonomous system (12.3.1) can be expressed in the following form:

$$\left.\begin{aligned} x &= c_1 k_{11} e^{\lambda t} + c_2(k_{12} t + k_{13}) e^{\lambda t} \\ y &= c_1 k_{21} e^{\lambda t} + c_2(k_{22} t + k_{23}) e^{\lambda t} \end{aligned}\right\}, \tag{12.3.6}$$

where c_1, c_2 are arbitrary constants and k_{ij}, $i = 1, 2$; $j = 1, 2, 3$, are fixed constants so that $k_{22}/k_{12} = k_{21}/k_{11}$.

Now, we would like to classify the degenerate nodes into two categories, namely (i) *Stable degenerate node* and (ii) *Unstable degenerate node.*

(i) Stable Degenerate Node (SDN): The degenerate node $(0, 0)$ is said to be stable, if $\lambda < 0$. Clearly, for $c_2 = 0$, the solution (12.3.6) becomes

$$\left.\begin{aligned} x &= c_1 k_{11} e^{\lambda t} \\ y &= c_1 k_{21} e^{\lambda t} \end{aligned}\right\} \tag{12.3.7}$$

$$\therefore\ y = \left(\frac{k_{21}}{k_{11}}\right) x. \tag{12.3.8}$$

Thus, for all $c_1 > 0$, (12.3.7) represents a phase-plane orbit consisting of half of the line given by (12.3.8). Similarly, for $c_1 < 0$, the solution (12.3.7) represents other half of the line.

Again, for $c_1 = 0$ and $c_2 \neq 0$, the solution (12.3.6) becomes

$$\left.\begin{aligned} x &= c_2(k_{12}t + k_{13})e^{\lambda t} \\ y &= c_2(k_{22}t + k_{23})e^{\lambda t} \end{aligned}\right\}, \tag{12.3.9}$$

which gives nonrectilinear phase-plane orbits.

Clearly,

$$\frac{y}{x} = \frac{k_{22}t + k_{23}}{k_{12}t + k_{13}}.$$

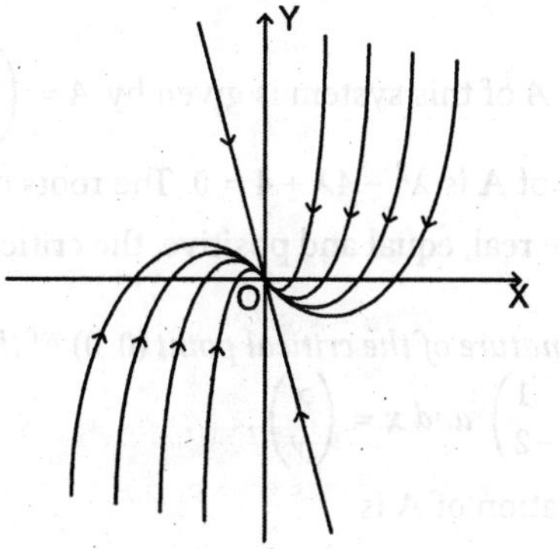

Fig 12.6 (Stable Degenerate Node)

$$\therefore \lim_{t\to\infty} \frac{y}{x} = \frac{k_{22}}{k_{12}}.$$

It confirms that all nonrectilinear phase-plane orbits enter the critical point (0, 0) with limiting slope $\frac{k_{22}}{k_{12}}$. This type of degenerate node is *asymptotically stable.*

Now for $c_1 \neq 0, c_2 \neq 0$, we have

$$\frac{y}{x} = \frac{c_1k_{21}e^{\lambda t} + c_2(k_{22}t + k_{23})e^{\lambda t}}{c_1k_{11}e^{\lambda t} + c_2(k_{12}t + k_{13})e^{\lambda t}} = \frac{(c_1k_{21} + c_2k_{23}) + c_2k_{22}t}{(c_1k_{11} + c_2k_{13}) + c_2k_{12}t}.$$

$$\therefore \lim_{t\to\infty} \frac{y}{x} = \frac{k_{22}}{k_{12}}$$

which confirms again that every nonrectilinear orbits enter the critical point (0, 0) with limiting slope $\frac{k_{22}}{k_{12}}$.

Note: SDN is also known as *coaxial node.*

(ii) **Unstable Degenerate Node (UDN):** The degenerate node (0, 0) is said to be unstable, if $\lambda > 0$. In this case, all the phase-plane orbits enter the critical point (0, 0) as $t \to -\infty$.

Note: For $\frac{k_{22}}{k_{12}} \neq \frac{k_{21}}{k_{11}}$, the critical point (0, 0) becomes star.

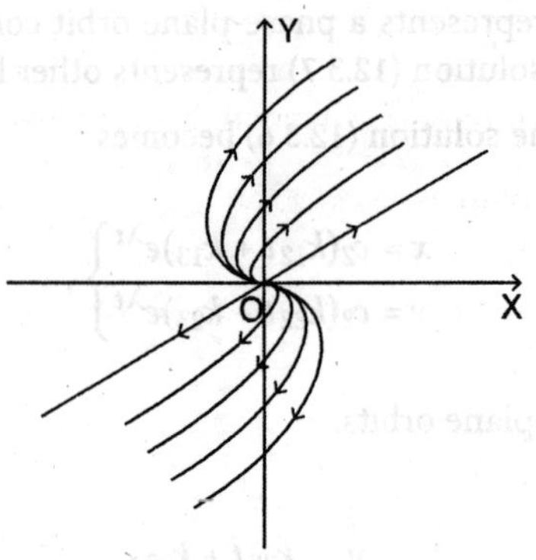

Fig 12.7 (Unstable Degenerate Node)

Example 12.3.4. *Determine the nature of the critical point* $(0,0)$ *of the following linear plane autonomous system:* $\dot{x} = 2x + 3y$, $\dot{y} = 2y$.

Solution: The coefficient matrix A of this system is given by $A = \begin{pmatrix} 2 & 3 \\ 0 & 2 \end{pmatrix}$.

$\therefore$ the characteristic equation of **A** is $\lambda^2 - 4\lambda + 4 = 0$. The roots of this equation are 2 and 2.

Since the eigenvalues of **A** are real, equal and positive, the critical point $(0,0)$ is unstable degenerate node.

Example 12.3.5. *Determine the nature of the critical point* $(0,0)$ *of the following linear plane autonomous system* $\dot{\mathbf{x}} = A\mathbf{x}$, *where* $\mathbf{A} = \begin{pmatrix} -4 & 1 \\ -1 & -2 \end{pmatrix}$ *and* $\mathbf{x} = \begin{pmatrix} x \\ y \end{pmatrix}$.

Solution: The characteristic equation of **A** is

$$\begin{vmatrix} -4-\lambda & 1 \\ -1 & -2-\lambda \end{vmatrix} = 0 \text{ or, } (\lambda+4)(\lambda+2)+1 = 0 \text{ or, } \lambda^2 + 6\lambda + 9 = 0.$$

$\therefore$ $\lambda = -3, -3$.

Therefore, the eigenvalues of **A** are real, equal and negative. Hence, the critical point $(0,0)$ is stable degenerate node.

12.3.3 Complex eigenvalues

Let $T^2 - 4D < 0$, i.e. $T^2 < 4D$. Then the roots λ_1 and λ_2 of the characteristic eq. (12.3.2) are imaginary. Let $\lambda_1 = \alpha + i\beta$, $\lambda_2 = \alpha - i\beta$, where α and β are real and $i = \sqrt{-1}$.

Then the general solution of the system is given by

$$x = e^{\alpha t}\left[c_1(k_{11}\cos\beta t - k_{12}\sin\beta t) + c_2(k_{12}\cos\beta t + k_{11}\sin\beta t)\right]$$
$$y = e^{\alpha t}\left[c_1(k_{21}\cos\beta t - k_{22}\sin\beta t) + c_2(k_{22}\cos\beta t + k_{21}\sin\beta t)\right],$$

where c_1, c_2 are arbitrary constants and $k_{ij}(i,j = 1,2)$ are fixed constants. We can express this solution in the form:

$$\left.\begin{aligned} x &= c' e^{\alpha t}\cos(\beta t + \varepsilon_1) \\ y &= c'' e^{\alpha t}\cos(\beta t + \varepsilon_2) \end{aligned}\right\}, \tag{12.3.10}$$

where $c', c'', \varepsilon_1, \varepsilon_2$ are arbitrary constants depending on a, b, c and d.

A special case

Let us consider the linear plane autonomous system in the following form $\dot{x} = \alpha x + \beta y$, $\dot{y} = -\beta x + \alpha y$.

Then the eigenvalues of the coefficient matrix $A = \begin{pmatrix} \alpha & \beta \\ -\beta & \alpha \end{pmatrix}$ are $\lambda_1 = \alpha + i\beta$, $\lambda_2 = \alpha - i\beta$.

$\therefore$ an eigenvector corresponding to the eigenvalue $\lambda_1 = \alpha + i\beta$ is $\begin{pmatrix} i \\ 1 \end{pmatrix}$ and that corresponding to the eigenvalue $\lambda_2 = \alpha - i\beta$ is $\begin{pmatrix} -i \\ 1 \end{pmatrix}$.

$\therefore$ the general solution of the system is given by

$$\begin{aligned}\begin{pmatrix} x \\ y \end{pmatrix} &= c_1 \begin{pmatrix} i \\ 1 \end{pmatrix} e^{(\alpha+i\beta)t} + c_2 \begin{pmatrix} -i \\ 1 \end{pmatrix} e^{(\alpha-i\beta)t}, \text{ where } c_1, c_2 \text{ are arbitrary constants} \\ &= c_1 \begin{pmatrix} i \\ 1 \end{pmatrix} e^{\alpha t}(\cos\beta t + i\sin\beta t) + c_2 \begin{pmatrix} -i \\ 1 \end{pmatrix} e^{\alpha t}(\cos\beta t - i\sin\beta t).\end{aligned}$$

$$\begin{aligned}\therefore\ x &= [i(c_1 - c_2)\cos\beta t - (c_1 + c_2)\sin\beta t]e^{\alpha t} \\ y &= [(c_1 + c_2)\cos\beta t + i(c_1 - c_2)\sin\beta t]e^{\alpha t},\end{aligned}$$

i.e.

$$\left.\begin{aligned} x &= (A\cos\beta t - B\sin\beta t)e^{\alpha t} \\ y &= (B\cos\beta t + A\sin\beta t)e^{\alpha t} \end{aligned}\right\}, \tag{12.3.11}$$

where $A = i(c_1 - c_2)$ and $B = c_1 + c_2$ are two arbitrary constants.

$$\therefore\ x^2 + y^2 = (A^2 + B^2)e^{2\alpha t}, \text{ i.e. } x^2 + y^2 = r^2 e^{2\alpha t}, \tag{12.3.12}$$

where $r = \sqrt{A^2 + B^2}$ is arbitrary constant.

Now, let us consider the following cases:

Case 1 (stable spiral or stable focus)

Let $\alpha < 0$. Then, by (12.3.10) and (12.3.11) we observe that $x \to 0$ and $y \to 0$ as $t \to +\infty$.

It confirms that all the phase-plane orbits approach the critical point $(0, 0)$ as $t \to +\infty$, but do not enter $(0, 0)$. In addition, equations (12.3.10) and (12.3.12) suggest that the family of curves on the phase-plane is one parameter family of spirals surrounding the critical point $(0, 0)$. In this case, the point is known as a *stable spiral* or *stable focus* (Fig 12.8). It is actually asymptotically stable.

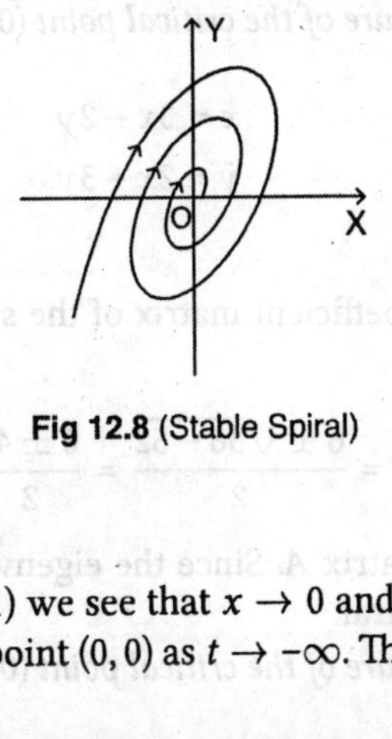

Fig 12.8 (Stable Spiral)

Case 2 (unstable spiral or unstable focus)

Let $\alpha > 0$. Then, by (12.3.10) and (12.3.11) we see that $x \to 0$ and $y \to 0$ as $t \to -\infty$. Therefore, all the phase-plane orbits approach the critical point $(0, 0)$ as $t \to -\infty$. The family of curves on the phase-plane is represented by the equations (12.3.10) and (12.3.12) that also gives one-parameter family of spiral for all $\alpha > 0$. Here the point $(0, 0)$ is called an *unstable spiral* or *unstable focus* (Fig 12.9).

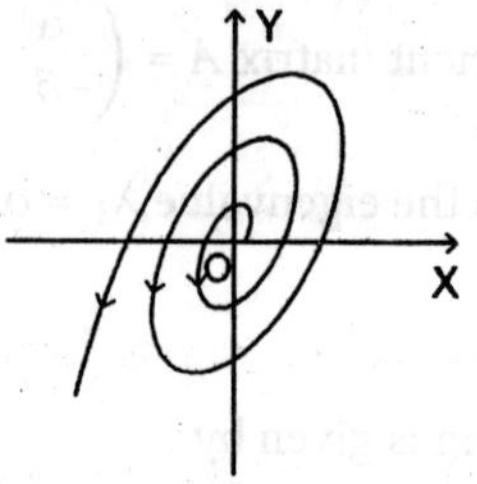

Fig 12.9 (Unstable Spiral)

Case 3 (centre)

Let $\alpha = 0$. Then the solution (12.3.10) must be of the form:

$$\left.\begin{aligned} x &= A\cos\beta t - B\sin\beta t \\ y &= B\cos\beta t + A\sin\beta t \end{aligned}\right\} \tag{12.3.13}$$

Clearly x and y are periodic functions of t and the phase-plane orbits fail to approach the critical point $(0, 0)$ as $t \to +\infty$ or $t \to -\infty$. Again, for $\alpha = 0$, the eq. (12.3.12) becomes $x^2 + y^2 = r^2$, which represents one-parameter family of circles surrounding the point $(0, 0)$. Whereas for $\alpha = 0$, (12.3.10) represents a family of ellipses, each member of which fails to approach the critical point $(0, 0)$. In (12.3.10), x and y are also periodic functions of t.

Here, the critical point $(0, 0)$ is called a *centre* (Fig 12.10).

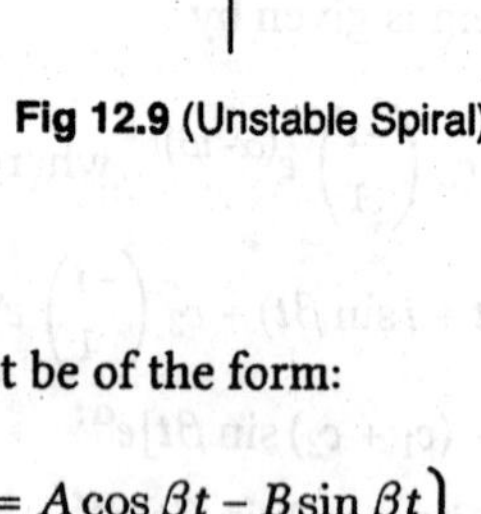

Fig 12.10 (Centre)

Example 12.3.6. *Determine the nature of the critical point* $(0, 0)$ *of the linear plane autonomous system*

$$\dot{x} = 3x - 2y$$
$$\dot{y} = 2x + 3y.$$

Solution: Let $A = \begin{pmatrix} 3 & -2 \\ 2 & 3 \end{pmatrix}$, the coefficient matrix of the system. Then the characteristic equation of A is $\lambda^2 - 6\lambda + 13 = 0$.

$$\therefore\ \lambda = \frac{6 \pm \sqrt{36 - 52}}{2} = \frac{6 \pm 4i}{2} = 3 \pm 2i,$$

which are the eigenvalues of the matrix A. Since the eigenvalues are complex with positive real part, the critical point $(0, 0)$ is unstable spiral.

Example 12.3.7. *Determine the nature of the critical point* $(0, 0)$ *of the system* $\dot{x} = x$, $\dot{y} = ky$ *for* $k \neq 0$.

Solution: The coefficient matrix of the system is $A = \begin{pmatrix} 1 & 0 \\ 0 & k \end{pmatrix}$.

$\therefore$ the characteristic equation of A is $\begin{vmatrix} 1-\lambda & 0 \\ 0 & k-\lambda \end{vmatrix} = 0$ or, $(1-\lambda)(k-\lambda) = 0$.

$\therefore$ $\lambda = 1, k$ are the eigenvalues of A.

Case 12.3.1. *Let $k > 0$. Then the eigenvalues of A are real, distinct and positive. Thus, the critical point $(0, 0)$ is an unstable node.*

Case 12.3.2. *Let $k = 1$. Then the eigenvalues of A are real, equal and positive. Consequently, $(0, 0)$ is an unstable degenerate node.*

Case 12.3.3. *Let $k < 0$. Then the eigenvalues of A are real, distinct and opposite in sign. It confirms that $(0, 0)$ is a saddle point.*

Example 12.3.8. *Determine the nature of the critical point $(0, 0)$ of the autonomous system $\dot{x} = -kx + y$, $\dot{y} = -x - ky$, where $k(\neq 0)$ is a parameter.*

Solution: The coefficient matrix of the system is given by $A = \begin{pmatrix} -k & 1 \\ -1 & -k \end{pmatrix}$.

$\therefore$ the characteristic equation of A is $\begin{vmatrix} -k-\lambda & 1 \\ -1 & -k-\lambda \end{vmatrix} = 0$, i.e. $(\lambda + k)^2 + 1 = 0$.

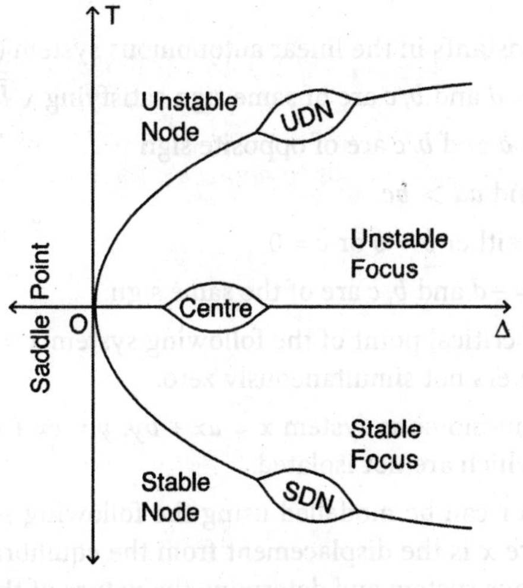

Fig 12.11 ($T^2 = 4\Delta$)

$\therefore$ $\lambda + k = \pm i$, i.e. $\lambda = -k \pm i$, which are the eigenvalues of A.

Since the eigenvalues are complex with real part $-k(\neq 0)$, the critical point $(0, 0)$ will be stable spiral, if $-k < 0$, i.e. $k > 0$ and it will be unstable spiral if $-k > 0$ i.e. $k < 0$.

Example 12.3.9. *Locate the critical point and find its nature for the system $\dot{x} = 5x - 6y + 2$, $\dot{y} = 4x - 5y + 3$. Also find the equation of the phase paths.*

Solution: $\dot{x} = 0$ and $\dot{y} = 0$ give $5x - 6y + 2 = 0$ and $4x - 5y + 3 = 0$. Solving, we obtain $x = 8$, $y = 7$. Therefore, $(8, 7)$ is the critical point of the system.

Let $X = x + 8$, $Y = y + z$. Then the system becomes $\dot{X} = 5X - 6Y$, $\dot{Y} = 4X - 5Y$.

Let $A = \begin{pmatrix} 5 & -6 \\ 4 & -5 \end{pmatrix}$. Then characteristic equation of A is given by $\begin{vmatrix} 5-\lambda & -6 \\ 4 & -5-\lambda \end{vmatrix} = 0$ or, $\lambda^2 = 1$, i.e. $\lambda = \pm 1$. Thus, the eigenvalues are real, distinct but opposite in sign. It confirms that the critical point $(X^*, Y^*) = (0, 0)$ is a saddle point. Hence, the critical point $(8, 7)$ of the given system is also saddle point.

Now,

$$\frac{dy}{dx} = \frac{\dot{y}}{\dot{x}} = \frac{4x - 5y + 3}{5x - 6y + 2}$$

gives $(5x - 6y + 2)dy = (4x - 5y + 3)dx$ or, $5d(xy) - 4xdx - 6ydy - 3dx + 2dy = 0$.

Integrating, we obtain $2x^2 - 5xy + 3y^2 - 2y + c = 0$, c being arbitrary constant.

Hence, required phase paths is given by $2x^2 - 5xy + 3y^2 + 3x - 2y + c = 0$.

Examples XII(A)

1. Determine the nature of the critical point (0, 0) of the following plane autonomous system:

(a) $\dot{x} = 2x - 3y$, $\dot{y} = x + 4y$
(b) $\dot{x} = -3x + 2y$, $\dot{y} = -x - 4y$
(c) $\dot{x} = -2y$, $\dot{y} = 2x$
(d) $\dot{x} = 3y$, $\dot{y} = 2y$
(e) $\dot{x} = 5x - 2y$, $\dot{y} = -2x + 5y$
(f) $\dot{x} = x + 3y$, $\dot{y} = 3x + y$
(g) $\dot{x} = -y$, $\dot{y} = 4x$
(h) $\dot{x} = -2x$, $\dot{y} = x - 3y$
(i) $\dot{x} = x - 2y$, $\dot{y} = y$
(j) $\dot{x} = -2x$, $\dot{y} = 5x - 2y$.

2. Determine the nature of the critical point (0, 0) of the system $\dot{x} = y$, $\dot{y} = kx$ for $k \neq 0$.

3. Determine the nature of the critical point (0, 0) of the autonomous system $\dot{x} = x - ky$, $\dot{y} = kx + y$, k being a parameter.

4. Let a, b, c, d be real constants in the linear autonomous system (12.3.1). Then show that (0, 0) is a
   (a) saddle point, if $a = d$ and b, c are of same sign satisfying $\sqrt{bc} > |a|$
   (b) spiral point, if $a = d$ and b, c are of opposite sign
   (c) centre, if $a = -d$ and $ad > bc$
   (d) node, if $a = d$ and either $b = 0$ or $c = 0$
   (e) saddle point, if $a = -d$ and b, c are of the same sign.

5. Find the nature of the critical point of the following system $\dot{x} = \alpha x + \beta y$, $\dot{x} = -\beta x + \alpha y$, where α and β are two parameters not simultaneously zero.

6. Show that the plane autonomous system $\dot{x} = ax + by$, $\dot{y} = cx + dy$ with $ad - bc = 0$ has infinitely many critical points which are not isolated.

7. A mechanical oscillator can be modelled using the following second order differential equation $\ddot{x} + k\dot{x} - 4x = 0$, where x is the displacement from the equilibrium at any time t. Express it as a linear plane autonomous system and determine the nature of the equilibrium point.

8. Define the phase path of a plane autonomous system. Explain the formation of a node of a linear autonomous system.

9. Explain the formation of (a) a saddle point, (b) a focus, (c) a centre and (d) a degenerate node of a linear autonomous system.

10. Explain the terms: (a) Phase plane, (b) Phase trajectory, (c) Phase portrait, (d) Critical point, (e) Isolated critical point, (f) Stable critical point, (g) Asymptotically stable critical point, (h) Neutrally stable critical point.

11. Locate the critical point and find its nature for the system given below. Also find the equation of the phase paths:

(a) $\dot{x} = x + y$, $\dot{y} = x - y + 1$
(b) $\dot{x} = 2x + 3y - 5$, $\dot{y} = -3x - 2y + 5$
(c) $\dot{x} = x + 1$, $\dot{y} = x - y$
(d) $\dot{x} = 4x - 5y$, $\dot{y} = 5x - 4y$
(e) $\dot{x} = x - 2y + 1$, $\dot{y} = 2x - y + 1$.

Answers

1. (a) Unstable spiral (b) Stable spiral (c) Centre (neutrally stable) (d) Unstable proper node (e) Unstable node (f) Saddle point (g) Centre (neutrally stable) (h) Stable node (i) Unstable degenerate node (j) Stable degenerate node.
2. Saddle point for $k < 0$, centre for $k > 0$.
3. Unstable spiral for $k \neq 0$, unstable degenerate node for $k = 0$.
5. Stable degenerate node, if $\alpha < 0, \beta = 0$; Unstable degenerate node, if $\alpha > 0, \beta = 0$; Centre, if $\alpha = 0, \beta \neq 0$; Stable spiral $\alpha < 0, \beta \neq 0$; Unstable spiral, if $\alpha > 0, \beta \neq 0$.
6. $\dot{x} = y, \dot{y} = -4x - ky$; equilibrium point: $(0, 0)$. It is unstable node for $k > 4$ and $k < -4$, unstable degenerate node for $k = 4$, stable degenerate node for $k = -4$, centre for $k = 0$, unstable spiral for $0 < k < 4$, stable spiral for $-4 < k < 0$.
7. $\dot{x} = y, \dot{y} = -4x - ky$; equilibrium point: $(0, 0)$; SDN, if $k = \pm 4$; Stable spiral if $-4 < k < 4$; stable node, if $k > 4$; unstable node, if $k < -4$.
11. (a) $\left(-\frac{1}{2}, \frac{1}{2}\right)$; saddle point; $x^2 - 2xy - y^2 + 2x + c = 0$.
 Hints: (a) $\frac{dy}{dx} = \frac{\dot{y}}{\dot{x}} = \frac{x-y+1}{x+y}$ or, $x\,dx - d(xy) - y\,dy + dx = 0$.
 (b) $(1, 1)$; centre or spiral; $3x^2 + 4xy + 3y^2 - 10x - 10y + c = 0$.
 (c) $(-1, -1)$; saddle point; $(x + 1)(x - 2y - 1) = 0$.
 (d) $(0, 0)$; centre; $5x^2 - 8xy + 5y^2 = c$.
 (e) $\left(-\frac{1}{3}, \frac{1}{3}\right)$; saddle point; $x^2 - xy + y^2 + x - y + c = 0$.

12.4 Linearisation of a nonlinear plane autonomous system at a critical point

In Article 12.3, we have already developed the theory only on plane autonomous system. But, in most of the mathematical models, we see that the autonomous systems are nonlinear. So, in order to linearise the nonlinear systems, we would like to develop the theory involving Jacobian matrix (or Variational matrix) at the critical points of the nonlinear systems.

Let us consider the nonlinear plane autonomous system

$$\left.\begin{aligned} \dot{x} &= P(x, y) \\ \dot{y} &= Q(x, y) \end{aligned}\right\}, \tag{12.4.1}$$

where $P(x, y)$ and $Q(x, y)$ have continuous first partial derivatives in the domain $D \subseteq \mathbb{R}^2$, except possibly at some points.

Let $P(x^*, y^*) = 0$, $Q(x^*, y^*) = 0$. Then (x^*, y^*) is a critical point of the system (12.4.1).

$\therefore$ by the Taylor's theorem for the functions of two variables, we have

$$P(x, y) = P(x^*, y^*) + P_x(x^*, y^*) \cdot (x - x^*) + P_y(x^*, y^*) \cdot (y - y^*) + R_1(x, y)$$
$$Q(x, y) = Q(x^*, y^*) + Q_x(x^*, y^*) \cdot (x - x^*) + Q_y(x^*, y^*) \cdot (y - y^*) + R_2(x, y),$$

where the functions $R_1(x, y)$, $R_2(x, y)$ are remainders having property

$$\lim_{(x,y)\to(0,0)} \frac{R_i(x, y)}{\sqrt{(x - x^*)^2 + (y - y^*)^2}} = 0 \text{ for } i = 1, 2.$$

$$\therefore\ P(x, y) = P_x(x^*, y^*) \cdot (x - x^*) + P_y(x^*, y^*) \cdot (y - y^*) + R_1(x, y)$$
$$Q(x, y) = Q_x(x^*, y^*) \cdot (x - x^*) + Q_y(x^*, y^*) \cdot (y - y^*) + R_2(x, y).$$

In vector notation, $\mathbf{F} = \mathbf{J}(\mathbf{x}^*)(\mathbf{x}-\mathbf{x}^*) + \mathbf{R}(\mathbf{x})$, where $\mathbf{F} = \begin{pmatrix} P \\ Q \end{pmatrix}$, $\mathbf{x} = \begin{pmatrix} x \\ y \end{pmatrix}$, $\mathbf{x}^* = \begin{pmatrix} x^* \\ y^* \end{pmatrix}$, $\mathbf{R} = \begin{pmatrix} R_1(x, y) \\ R_2(x, y) \end{pmatrix}$

and $\mathbf{J}(\mathbf{x}^*) = \begin{pmatrix} P_x(x^*, y^*) & P_y(x^*, y^*) \\ Q_x(x^*, y^*) & Q_y(x^*, y^*) \end{pmatrix}$

$\therefore$ the system (12.4.1) can be expressed in vector notation as

$$\dot{\mathbf{x}} = \mathbf{K}(\mathbf{x}^*) \cdot (\mathbf{x} - \mathbf{x}^*) + \mathbf{R}(\mathbf{x}). \tag{12.4.2}$$

In order to shift the origin (0, 0) to the critical point (x^*, y^*) let us use the transformation $\mathbf{u} = \mathbf{x} - \mathbf{x}^*$, where $\mathbf{u} = \begin{pmatrix} u_1 \\ u_2 \end{pmatrix}$ and let $|\mathbf{J}(\mathbf{x}^*)| \neq 0$.

Then $\dot{\mathbf{u}} = \dot{\mathbf{x}}$ and (12.4.2) becomes

$$\dot{\mathbf{u}} = \mathbf{J}(\mathbf{x}^*)\mathbf{u} + \mathbf{R}(\mathbf{u} + \mathbf{x}^*). \tag{12.4.3}$$

If $\mathbf{u}$ is very closed to $\mathbf{0}$, the behaviour of the solutions of the system (12.4.3) and that of the system $\dot{\mathbf{u}} = \mathbf{J}(\mathbf{x}^*)\mathbf{u}$ will be the same, provided 0 is not an eigenvalue of the Jacobian matrix $\mathbf{J}(\mathbf{x}^*)$. In this case, we can neglect the remainder $\mathbf{R}(\mathbf{u} + \mathbf{x}^*)$ containing the nonlinear terms of $\mathbf{u}$.

Now, neglecting the remainder term $\mathbf{R}(\mathbf{u} + \mathbf{x}^*)$ of the system (12.4.3) we have the linear plane autonomous system given by

$$\dot{\mathbf{u}} = \mathbf{J}(\mathbf{x}^*)\mathbf{u}. \tag{12.4.4}$$

The nature of the critical point $(\mathbf{u}_1^*, \mathbf{u}_2^*) = (0, 0)$ of the system (12.4.4) is the same as that of the system (12.4.1) except the following cases:

(i) Let λ_1 and λ_2 be real and equal. Then the point (0, 0) may be either a node or a spiral point of (12.4.4), though it is a node of the system (12.3.1).

(ii) If λ_1 and λ_2 be purely imaginary, the point (0, 0) is either a centre or spiral of (12.4.4), though it is a centre of the system (12.3.1).

Example 12.4.1. *Find the equilibrium points of the following nonlinear autonomous system $\dot{x} = x$, $\dot{y} = x^2 + y^2 - 4$. Also determine the nature of the equilibrium points.*

Solution: Let $f(x, y) = x$ and $g(x, y) = x^2 + y^2 - 4$.

Then $f(x, y) = 0$ gives $x = 0$ and $g(x, y) = 0$ gives $x^2 + y^2 - 4 = 0$.

$\therefore\ y^2 = 4$ ($\because\ x = 0$), i.e. $y = \pm 2$.

Hence, the equilibrium points of the system are (0, −2) and (0, 2).

Now, $f_x(x, y) = 1 \quad g_x(x, y) = 2x$

$f_y(x, y) = 0 \quad g_y(x, y) = 2y.$

$\therefore$ the Jacobian matrix $J(x, y)$ is given by $J(x, y) = \begin{pmatrix} 1 & 0 \\ 2x & 2y \end{pmatrix}$.

$$\therefore\ J(0, -2) = \begin{pmatrix} 1 & 0 \\ 0 & -4 \end{pmatrix}.$$

The characteristic equation of $J(0, -2)$ is $\begin{vmatrix} 1-\lambda & 0 \\ 0 & -4-\lambda \end{vmatrix} = 0$, i.e. $(\lambda - 1)(\lambda + 4) = 0$.

$\therefore\ \lambda = 1, -4$, which are the eigenvalues of $J(0, -2)$.

Since the eigenvalues are real, distinct and opposite in sign, the équilibrium point (critical point) (0, −2) is a saddle point. Consequently, it is unstable.

Again, $J(0,2) = \begin{pmatrix} 1 & 0 \\ 0 & 4 \end{pmatrix}$.

The characteristic equation of J(0, 2) is $\begin{vmatrix} 1-\lambda & 0 \\ 0 & 4-\lambda \end{vmatrix} = 0$, i.e. $(1-\lambda)(4-\lambda) = 0$.

$\therefore \lambda = 1, 4$, which are the eigenvalues of J(0, 2).

We see that the eigenvalues are real, distinct and positive.

Therefore, (0, 2) is an unstable node.

Examples XII(B)

1. Find the critical points of the nonlinear autonomous system $\dot{x} = y, \dot{y} = x - x^3 + y$. Also determine the nature of the critical points.
2. Determine the nature of the critical points of the following nonlinear system: $\dot{x} = -5x + xy, \dot{y} = (x-1)(x-7)y - y^2$.
3. Determine the nature of the equilibrium points of the mathematical model given by the system $\dot{x} = x, \dot{y} = k^2 - y^2$, where k is a parameter.
4. A capacitor-resistor electrical circuit can be modelled using the following nonlinear plane autonomous system $\dot{x} = y, \dot{y} = -x + x^3 - (k + x)y, k \neq 0$, where x represents the current in the circuit at time t. Classify the critical points.
5. Express the second order non liner differential equation $\ddot{x} + k\dot{x} + 4x^2 = 0$ as a nonlinear plane autonomous system and determine the nature of the critical point.

Answers

1. Critical points: (0, 0), (1, 0), (−1, 0). (0, 0) is a saddle point; (±1, 0) are unstable focus.
2. Critical points: (0, 0), (0, −7), (6, 5), (2, 5).
   Saddle points: (0, −7), (2, 5)
   Stable node: (0, 0)
   Stable spiral: (6, 5).
3. $(0, k)$: Unstable node, if $k < 0$, saddle point, if $k > 0$ and unstable star, if $k = 0$.
   $(0, -k)$: Unstable node, if $k > 0$, saddle point, if $k < 0$ and unstable star, if $k = 0$.
4. Critical points: (0, 0), (1, 0), (−1, 0).
   $(0,0)$: $\begin{cases} \text{stable node for } k > 2 \text{ and } k < -2 \\ \text{stable degenerate node for } k = 2 \text{ (or stable spiral)} \\ \text{unstable degenerate node for } k = -2 \text{ (or unstable spiral)} \\ \text{stable spiral for } 0 < k < 2 \\ \text{unstable spiral for } -2 < k < 0. \end{cases}$
   $(1, 0)$: saddle point for $k \neq 0$.
   $(-1, 0)$: saddle point for $k \neq 0$.
5. $\dot{x} = y, \dot{y} = -4x^2 - ky$; (0, 0) is stable proper node for $k > 0$ and unstable proper node for $k < 0$.

Chapter 13

Power Series Solutions of Ordinary Differential Equations

In the previous chapters, several methods have been introduced to find the solutions of linear differential equations. But some linear differential equations belong to a collection of ordinary differential equations of particular form that cannot be solved by traditional method or their solutions cannot be expressed in the form of finite linear combinations of well-known elementary functions. Generally, higher-order linear differential equations may not have solutions that can be put in such a simple manner. In that case, we may find a convergent series arranged according to the powers of the independent variable such that this convergent series can be the approximate value of the dependent variable.

The present chapter is devoted to the method to obtain the solution in the form of infinite series of the second order linear differential equations about the ordinary points and regular singular points. In order to find the solutions about the regular singular points, here we would like to use the method of Frobenius named after Ferdinand Georg Frobenius (26 October 1849–3 August 1917), a German mathematician.

13.1 Power series

A power series in x is an infinite series of the form: $\sum_{n=0}^{\infty} a_n(x-x_0)^n = a_0 + a_1(x-x_0) + a_2(x-x_0)^2 + \ldots$, where $a_0, a_1, a_2, \ldots$ are real numbers and x_0 is the centre of the series.

This series will reduce to the simplest form $\sum_{n=0}^{\infty} a_n y^n = a_0 + a_1 y + a_2 y^2 + \ldots$, a power series in y about the point x_0, if we put $x - x_0 = y$.

A power series $\sum_{n=0}^{\infty} a_n x^n$ is said to be *everywhere convergent*, if it converges for all $x \in \mathbb{R}$. The series is called *nowhere convergent*, if it is convergent only for $x = 0$ and *absolutely convergent* for $x \in (-R, R)$ if $\lim_{n\to\infty} \left|\frac{a_n}{a_{n+1}}\right|$ exists and equal to R. Here R is called the *radius of convergence* and $(-R, R)$ is called the *interval of convergence* of the power series. In this case, the power series represents a continuous function which is differentiable term by term on the interval of convergence $(-R, R)$.

For example, let us consider a power series $\sum_{n=0}^{\infty} a_n x^n$, where $a_0 = 0$ and $a_n = (-1)^{n-1}\frac{1}{n}, n \geqslant 1$.

Then $\lim_{n\to\infty} \left|\frac{a_n}{a_{n+1}}\right| = \lim_{n\to\infty} \frac{n+1}{n} = 1$.

Thus, $R = 1$ and the series is convergent (absolutely) for $|x| < 1$.

At $x = 1$, the series becomes $1 - \frac{1}{2} + \frac{1}{3} - \ldots$, that is convergent but at $x = -1$, the series takes the form $-1 - \frac{1}{2} - \frac{1}{3} - \ldots$, which is divergent. Hence, the radius of convergence is 1 and region of convergence is $(-1, 1]$.

13.2 Analytic function

A function $f(x)$ is said to be *analytic* at a point x_0 if the Taylor's series of $f(x)$ about x_0,

$$\sum_{n=0}^{\infty} \frac{f^{(n)}(x_0)}{n!}(x - x_0)^n$$

exists and converges to the function $f(x)$ for all $x \in (x_0 - \delta, x_0 + \delta)$ for some $\delta > 0$.

Observe that all polynomial functions are analytic everywhere but a rational function of the form: $\frac{P(x)}{Q(x)}$, where $P(x)$ and $Q(x)$ are polynomials in x, is analytic except at those points, where the denominator $Q(x)$ vanishes.

For example, let us consider a rational function $f(x)$ defined by $f(x) = \frac{x+2}{x^2+x-2}$.

Here $f(x)$ is analytic except at the points -2 and 1, because the denominator $x^2 + x - 2$ vanishes at -2 and 1.

A single-valued function $f(x)$ will be analytic in some domain $D \subseteq \mathbb{R}$, if it is defined and differentiable at each point in D.

13.3 Ordinary point and Singular point

Let us consider a second-order homogeneous linear differential equation of the form:

$$P_0(x)\frac{d^2y}{dx^2} + P_1(x)\frac{dy}{dx} + P_2(x)y = 0. \tag{13.3.1}$$

Let the eq. (13.3.1) have no solution that can be expressed as a finite linear combination of elementary function but it is possible to have a power series solution. The equivalent normalised form of (13.3.1) is

$$\frac{d^2y}{dx^2} + P(x)\frac{dy}{dx} + Q(x)y = 0, \tag{13.3.2}$$

where $P(x) = \frac{P_1(x)}{P_0(x)}$ and $Q(x) = \frac{P_2(x)}{P_0(x)}$.

A point x_0 is said to be an *ordinary point* of the linear differential eq. (13.3.2), if $P(x)$ and $Q(x)$ are both analytic at x_0.

For example, let us consider a differential equation

$$(x^2 + 1)\frac{d^2y}{dx^2} + (x + 2)\frac{dy}{dx} + (x^2 - 1)y = 0.$$

The normalised form of the equation is $\frac{d^2y}{dx^2} + \frac{x+2}{x^2+1}\frac{dy}{dx} + \frac{x^2-1}{x^2+1}y = 0$.

Here $P(x) = \frac{x+2}{x^2+1}$ and $Q(x) = \frac{x^2-1}{x^2+1}$.

Observe that $P(x)$ and $Q(x)$ are both analytic everywhere. So, all the points in $\mathbb{R}$ are ordinary points of the equation.

A point x_0 is said to be a *singular point* of the differential eq. (13.3.2), if either or both of the functions $P(x)$ and $Q(x)$ are not analytic at x_0. It is called a *regular singular point*, if $(x - x_0)P(x)$ and $(x - x_0)^2Q(x)$ are both analytic at x_0, otherwise it is called an *irregular singular point* of the eq. (13.3.2).

For example, let us consider the differential equation $x^2(x + 1)y'' + (x - 2)y' + 4y = 0$.

Its normalised form is $y'' + \frac{x-2}{x^2(x+1)}y' + \frac{4}{x^2(x+1)}y = 0$.

Comparing it with (13.3.2), we obtain $P(x) = \frac{x-2}{x^2(x+1)}$ and $Q(x) = \frac{4}{x^2(x+1)}$, which are not analytic at the points 0 and −1. Thus, these two points are singular points.

We see that $xP(x) = \frac{x-2}{x(x+1)}$ is not analytic at $x = 0$ though $x^2Q(x) = \frac{4}{x+1}$ is analytic at $x = 0$. If follows that $x = 0$ is an irregular singular point. Again, $(x+1)P(x) = \frac{x-2}{x^2}$ and $(x+1)^2Q(x) = \frac{4(x+1)}{x^2}$, both of which are analytic at $x = -1$. Hence, $x = -1$ is a regular singular point.

Theorem 13.3.1. *Let x_0 be an ordinary point of the differential equation (13.3.2). Then it has two nontrivial linearly independent power series solutions of the form: $\sum_{n=0}^{\infty} a_n(x-x_0)^n$, which are convergent in some open interval $(x_0 - R, x_0 + R)$ about the ordinary point x_0 (where $R > 0$).*

13.4 Power series solution near the ordinary points

Let x_0 be an ordinary point of the eq. (13.3.2). Then, by the theorem 13.3.1, the eq. (13.3.2) has two nontrivial linearly independent power series solutions of the form: $\sum_{n=0}^{\infty} a_n(x-x_0)^n$ each of which is convergent in some open interval $(x_0 - R, x_0 + R)$ about the point x_0, where $R(> 0)$ is the radius of convergence of the power series.

Let

$$y = \sum_{n=0}^{\infty} a_n(x-x_0)^n. \tag{13.4.1}$$

Then, differentiating (13.4.1) term by term on the interval $(x_0 - R, x_0 + R)$ twice in successions, we obtain

$$\frac{dy}{dx} = \sum_{n=1}^{\infty} na_n(x-x_0)^{n-1} \text{ and } \frac{d^2y}{dx^2} = \sum_{n=2}^{\infty} n(n-1)a_n(x-x_0)^{n-2}. \tag{13.4.2}$$

Using (13.4.1) and (13.4.2) in (13.3.2) and simplifying, we have

$$C_0 + C_1(x-x_0) + C_2(x-x_0)^2 + \cdots + C_n(x-x_0)^n + \ldots, \tag{13.4.3}$$

where each $C_i (i = 0, 1, 2, \ldots)$ is the functions of the coefficients $a_n (n = 0, 1, 2, \ldots)$ of (13.4.1).

Observe that, for all $x \in (x_0 - R, x_0 + R)$, (13.4.3) is valid. Thus, (13.4.3) is an identity. Consequently,

$$C_0 = C_1 = C_2 = \cdots = 0. \tag{13.4.4}$$

Solving (13.4.4), we obtain the coefficients of (13.4.1) in terms of a_0 and a_1. Using these coefficients in (13.4.1), we get the desired power series solution of the differential eq. (13.3.2), where a_0 and a_1 become two arbitrary constants.

13.5 Illustrative examples

Example 13.5.1. *Locate and classify the singular point of the linear differential equation*

$$(x^2 + x)\frac{d^2y}{dx^2} - 2(x+1)\frac{dy}{dx} + 2y = 0.$$

Solution: The normalised form of the given equation is

$$\frac{d^2y}{dx^2} - \frac{2}{x}\frac{dy}{dx} + \frac{2}{x(x+1)}y = 0. \tag{13.5.1}$$

Comparing (13.5.1) with (13.3.2), we have $P(x) = -\frac{2}{x}$ and $Q(x) = \frac{2}{x(x+1)}$.

Clearly, 0 and −1 are the singular points of the eq. (13.5.1), because $P(x)$ and $Q(x)$ are not analytic at $x = 0$ and $Q(x)$ is not analytic at $x = -1$.

Again, $(x-0)P(x) = -2$ and $(x-0)^2Q(x) = \frac{2x}{x+1}$ are analytic at $x = 0$. Thus, 0 is a regular singular point of the eq. (13.5.1). On the other hand, $(x+1)P(x) = -\frac{2(x+1)}{x}$ and $(x+1)^2Q(x) = \frac{2(x+1)}{x}$ are also analytic at $x = -1$. Thus, −1 is another regular singular point of the given differential equation.

Example 13.5.2. *Solve the differential equation* $\frac{d^2y}{dx^2} - x^3\frac{dy}{dx} - y = 0$ *in powers of* x.

Solution: Comparing the given equation with (13.3.2), we have $P(x) = -x^3$ and $Q(x) = -1$, which are both analytic at $x = 0$. Thus, $x = 0$ is an ordinary point of the differential equation. Let the solution of the equation be of the form:

$$y = \sum_{n=0}^{\infty} a_n x^n. \tag{13.5.2}$$

Differentiating (13.5.2) twice in succession with respect to x, we obtain

$$\frac{dy}{dx} = \sum_{n=1}^{\infty} n a_n x^{n-1} \tag{13.5.3}$$

$$\text{and } \frac{d^2y}{dx^2} = \sum_{n=2}^{\infty} n(n-1)a_n x^{n-2}. \tag{13.5.4}$$

Using (13.5.2), (13.5.3) and (13.5.4) in the given equation, we have

$$\sum_{n=2}^{\infty} n(n-1)a_n x^{n-2} - x^3\sum_{n=1}^{\infty} n a_n x^{n-1} - \sum_{n=0}^{\infty} a_n x^n = 0$$

$$\text{or, } \sum_{n=2}^{\infty} n(n-1)a_n x^{n-2} - \sum_{n=1}^{\infty} n a_n x^{n+2} - \sum_{n=0}^{\infty} a_n x^n = 0$$

$$\text{or, } \sum_{n=0}^{\infty} (n+2)(n+1)a_{n+2}x^n - \sum_{n=3}^{\infty} (n-2)a_{n-2}x^n - \sum_{n=0}^{\infty} a_n x^n = 0$$

$$\text{or, } \sum_{n=0}^{2} (n+1)(n+2)a_{n+2}x^n + \sum_{n=3}^{\infty} (n+1)(n+2)a_{n+2}x^n - \sum_{n=3}^{\infty} (n-2)a_{n-2}x^n - \sum_{n=0}^{2} a_n x^n - \sum_{n=3}^{\infty} a_n x^n = 0$$

$$\text{or, } 2a_2 + 6a_3x + 12a_4x^2 - a_0 - a_1x - a_2x^2 + \sum_{n=3}^{\infty}[(n+1)(n+2)a_{n+2} - (n-2)a_{n-2} - a_n]x^n = 0$$

$$\text{or, } -(a_0 - 2a_2) - (a_1 - 6a_3)x - (a_2 - 12a_4)x^2 + \sum_{n=3}^{\infty}[(n+1)(n+2)a_{n+2} - (n-2)a_{n-2} - a_n]x^n = 0. \tag{13.5.5}$$

Relation (13.5.5) is valid for all $x \in (-R, R)$, where $R(> 0)$ is the radius of convergence of the power series (13.5.2).

Thus, the coefficient of each term in the left side of (13.5.5) must be zero.

$$\therefore a_0 - 2a_2 = 0, \text{ i.e. } a_2 = \tfrac{1}{2}a_0 \tag{13.5.6}$$

$$a_1 - 6a_3 = 0, \text{ i.e. } a_3 = \tfrac{1}{6}a_1 \tag{13.5.7}$$

$$a_2 - 12a_4 = 0, \text{ i.e. } a_4 = \tfrac{1}{12}a_2,$$

$$\text{i.e. } a_4 = \tfrac{1}{24}a_0 \text{ [by (13.5.6)]} \tag{13.5.8}$$

and $(n+1)(n+2)a_{n+2}-(n-2)a_{n-2}-a_n=0$,

$$\text{i.e. } a_{n+2}=\frac{(n-2)a_{n-2}+a_n}{(n+1)(n+2)} \text{ for all } n \geqslant 3 \tag{13.5.9}$$

Now, putting $n = 3$ in (13.5.9), we obtain

$$a_5=\frac{a_1+a_3}{20}=\frac{8}{120}a_1 \text{ [by (13.5.10)]}. \tag{13.5.10}$$

Putting $n = 4$ in (13.5.9), we obtain

$$a_6=\frac{2a_2+a_4}{30}=\frac{5}{144}a_0 \text{ [by (13.5.6) and (13.5.8).]}$$

In the same way, it is possible to express each coefficient in terms of a_0 and a_1. Putting all the values of $a_2, a_3, a_4, \ldots$ in (13.5.3), we have $y = a_0\left(1+\frac{1}{2}x^2+\frac{1}{24}x^4+\frac{5}{144}x^6+\ldots\right)+a_1\left(x+\frac{1}{6}x^3+\frac{7}{120}x^5+\ldots\right)$, which is the desired solution with two arbitrary constants a_0 and a_1.

Example 13.5.3. *Obtain the series solution of the differential equation* $\frac{d^2y}{dx^2}+3x\frac{dy}{dx}+3y=0$ *about the point* $x = 0$.

Solution: Comparing the given equation with (13.3.2), we have $P(x) = 3x$ and $Q(x) = 3$, which are both analytic at $x = 0$.

It follows that the point $x = 0$ is an ordinary point of the given differential equation.

Let the power series solution of the given equation be

$$y=\sum_{n=0}^{\infty}a_nx^n. \tag{13.5.11}$$

Differentiating (13.5.11) twice in succession with respect to x, we obtain

$$\frac{dy}{dx}=\sum_{n=1}^{\infty}na_nx^{n-1} \tag{13.5.12}$$

$$\text{and } \frac{d^2y}{dx^2}=\sum_{n=2}^{\infty}n(n-1)a_nx^{n-2}. \tag{13.5.13}$$

Using the results (13.5.11), (13.5.12) and (13.5.13) in the given equation, we obtain

$$\sum_{n=2}^{\infty}n(n-1)a_nx^{n-2}+3x\sum_{n=1}^{\infty}na_nx^{n-1}+3\sum_{n=0}^{\infty}a_nx^n=0$$

$$\text{or, } \sum_{n=0}^{\infty}(n+2)(n+1)a_{n+2}x^n+\sum_{n=1}^{\infty}3na_nx^n+\sum_{n=0}^{\infty}3a_nx^n=0$$

$$\text{or, } 2a_2+\sum_{n=1}^{\infty}(n+1)(n+2)a_{n+2}x^n+\sum_{n=1}^{\infty}3na_nx^n+3a_0+\sum_{n=1}^{\infty}3a_nx^n=0$$

$$\text{or, } (3a_0+2a_2)+\sum_{n=1}^{\infty}[(n+1)(n+2)a_{n+2}+3(n+1)a_n]x^n=0. \tag{13.5.14}$$

The relation (13.5.14) is valid for all $x \in (-R, R)$, where $R(> 0)$ is the radius of convergence of the power series (13.5.11).

Thus, the coefficient of each term in the left side of (13.5.14) must be zero.

$$\therefore 3a_0 + 2a_2 = 0, \text{ i.e. } a_2 = -\tfrac{3}{2}a_0 \tag{13.5.15}$$

$$\text{and } a_{n+2} = -\frac{3}{n+2}a_n, \text{ for all } n \geqslant 1. \tag{13.5.16}$$

Now putting $n = 1, 2, 3, \ldots$ in the recurrence relation (13.5.16), we obtain

$$\begin{aligned} a_3 &= -a_1 \\ a_4 &= \tfrac{9}{8}a_0 \\ a_5 &= \tfrac{3}{5}a_1 \\ a_6 &= -\tfrac{9}{16}a_0 \\ &\vdots \end{aligned}$$

Putting all the values of $a_2, a_3, a_4, \ldots$ in (13.5.11), we obtain

$$y = a_0\left(1 - \tfrac{3}{2}x^2 + \tfrac{9}{8}x^4 - \tfrac{9}{16}x^6 + \ldots\right) + a_1\left(x - x^3 + \tfrac{3}{5}x^5 - \ldots\right)$$

which is the desired solution with two arbitrary constants a_0 and a_1.

Example 13.5.4. *Find the power series solution of the equation* $(x^2+1)\frac{d^2y}{dx^2} + x\frac{dy}{dx} - xy = 0$ *about* $x = 0$.

Solution: The normalised form of the given equation is

$$\frac{d^2y}{dx^2} + \frac{x}{x^2+1}\frac{dy}{dx} - \frac{x}{x^2+1}y = 0 \tag{13.5.17}$$

Comparing (13.5.17) with (13.3.2), we have $P(x) = \frac{x}{x^2+1}$, $Q(x) = -\frac{x}{x^2+1}$, which are analytic at $x = 0$. Thus, $x = 0$ is an ordinary point of the eq. (13.5.17).

Let a solution of the eq. (13.5.17) be of the form:

$$y = \sum_{n=0}^{\infty} a_n x^n \tag{13.5.18}$$

Then differentiating (13.5.18) twice in succession with respect to x, we obtain

$$\frac{dy}{dx} = \sum_{n=1}^{\infty} n a_n x^{n-1} \tag{13.5.19}$$

$$\text{and } \frac{d^2y}{dx^2} = \sum_{n=2}^{\infty} n(n-1)a_n x^{n-2} \tag{13.5.20}$$

Using (13.5.18), (13.5.19) and (13.5.20) in the given equation, we have

$$(x^2+1)\sum_{n=2}^{\infty} n(n-1)a_n x^{n-2} + x\sum_{n=1}^{\infty} n a_n x^{n-1} - x\sum_{n=0}^{\infty} a_n x^n = 0$$

$$\text{or, } \sum_{n=2}^{\infty} n(n-1)a_n x^n + \sum_{n=2}^{\infty} n(n-1)a_n x^{n-2} + \sum_{n=1}^{\infty} n a_n x^n - \sum_{n=0}^{\infty} a_n x^{n+1} = 0$$

$$\text{or, } \sum_{n=2}^{\infty} n(n-1)a_n x^n + \sum_{n=0}^{\infty} (n+2)(n+1)a_{n+2} x^n + \sum_{n=1}^{\infty} n a_n x^n - \sum_{n=1}^{\infty} a_{n-1} x^n = 0$$

$$\text{or, } 2a_2 - (a_0 - a_1 - 6a_3)x + \sum_{n=2}^{\infty} [n(n-1)a_n + (n+2)(n+1)a_{n+2} + na_n - a_{n-1}]x^n = 0. \qquad (13.5.21)$$

The relation (13.5.21) is valid for all $x \in (-R, R)$, where $R(> 0)$ is the radius of converges of the power series (13.5.21).

∴ the coefficients of each term in the left side of (13.5.21) must be zero. It follows that $2a_2 = 0$, i.e. $a_2 = 0$, $a_0 - a_1 - 6a_3 = 0$, i.e. $a_3 = \frac{1}{6}(a_0 - a_1)$ and $n(n-1)a_n + (n+2)(n+1)a_{n+2} + na_n - a_{n-1} = 0$ for all $n \geqslant 2$,

$$\text{i.e. } a_{n+2} = \frac{a_{n-1} - n^2 a_n}{(n+1)(n+2)} \text{ for all } n \geqslant 2. \qquad (13.5.22)$$

Putting $n = 2, 3, \ldots$ in (13.5.22), we obtain $a_4 = \frac{1}{12}a_1$, $a_5 = -\frac{3}{40}(a_0 - a_1)$.

Now, substituting all the values of $a_2, a_3, \ldots$ in (13.5.18), we have

$$y = a_0\left(1 + \tfrac{1}{6}x^3 - \tfrac{3}{40}x^5 + \ldots\right) + a_1\left(x - \tfrac{1}{6}x^3 + \tfrac{1}{12}x^4 + \tfrac{3}{40}x^5 - \ldots\right),$$

which is the required solution of the given differential equation with two arbitrary constants a_0 and a_1.

Example 13.5.5. *Find the series solution of the differential equation* $\frac{d^2y}{dx^2} - 4y = 0$ *satisfying* $y(0) = 1$ *and* $y'(0) = -1$.

Solution: Comparing the given equation with (13.3.2), we have $P(x) = 0$ and $Q(x) = -4$, which are analytic at $x = 0$.

Thus, $x = 0$ is an ordinary point of the equation.

Let $y = \sum_{n=0}^{\infty} a_n x^n$ be a solution of the given equation.

Clearly,

$$\frac{dy}{dx} = \sum_{n=1}^{\infty} n a_n x^{n-1} \text{ and } \frac{d^2y}{dx^2} = \sum_{n=2}^{\infty} n(n-1)a_n x^{n-2}.$$

Using the above results in the given equation, we obtain

$$\sum_{n=2}^{\infty} n(n-1)a_n x^{n-2} - 4\sum_{n=0}^{\infty} a_n x^n = 0$$

$$\text{or, } \sum_{n=0}^{\infty} (n+2)(n+1)a_{n+2} x^n - \sum_{n=0}^{\infty} 4a_n x^n = 0$$

$$\text{or, } \sum_{n=0}^{\infty} [(n+1)(n+2)a_{n+2} - 4a_n]x^n = 0. \qquad (13.5.23)$$

The relation (13.5.23) is valid for all $x \in (-R, R)$, where $R(> 0)$ is the radius of convergence of the power series $\sum\limits_{n=0}^{\infty} a_n x^n$.

Thus, the coefficient of each term in the left side of (13.5.23) must be zero. It follows that

$$(n+1)(n+2)a_{n+2} - 4a_n = 0 \text{ i.e. } a_{n+2} = \frac{4a_n}{(n+1)(n+2)} \text{ for all } n \geqslant 0. \quad (13.5.24)$$

Putting $n = 0, 1, 2, \ldots$ in (13.5.24), we obtain $a_2 = 2a_0,\ a_3 = \frac{2}{3}a_1,\ a_4 = \frac{2}{3}a_0,\ a_5 = \frac{2}{15}a_1, \ldots$

Substituting all the values of $a_2, a_3, a_4, \ldots$ in terms of a_0 and a_1 in $y = \sum\limits_{n=0}^{\infty} a_n x^n$, we have

$$y = a_0\left(1 + 2x^2 + \tfrac{2}{3}x^4 + \ldots\right) + a_1\left(x + \tfrac{2}{3}x^3 + \tfrac{2}{15}x^5 + \ldots\right),$$

which is the general solution of the given equation with two arbitrary constants a_0 and a_1.

Clearly, $\frac{dy}{dx} = a_0\left(4x + \frac{8}{3}x^3 + \ldots\right) + a_1\left(1 + 2x^2 + \frac{2}{3}x^4 + \ldots\right)$.

Given that $y(0) = 1$ and $y'(0) = -1$.

$\therefore\ a_0 = 1$ and $a_1 = -1$.

Hence, required power series solution is $y = 1 - x + 2x^2 - \frac{2}{3}x^3 + \frac{2}{3}x^4 - \ldots$

Example 13.5.6. *Solve the differential equation* $\frac{d^2y}{dx^2} + (x-1)^2\frac{dy}{dx} - 4(x-1)y = 0$ *in series about the ordinary point* $x = 1$.

Solution: Here $x = 1$ is an ordinary point of the given differential equation, because $P(x) = (x-1)^2$ and $Q(x) = -4(x-1)$ are both analytic at $x = 1$.

Let $y = \sum\limits_{n=0}^{\infty} a_n (x-1)^n$ be a series solution about $x = 1$.

Then

$$\frac{dy}{dx} = \sum_{n=1}^{\infty} n a_n (x-1)^{n-1} \text{ and } \frac{d^2y}{dx^2} = \sum_{n=2}^{a} n(n-1)a_n(x-1)^{n-2}$$

Using the above results in the given equation, we have

$$\sum_{n=2}^{\infty} n(n-1)a_n(x-1)^{n-2} + (x-1)^2\sum_{n=1}^{\infty} n a_n (x-1)^{n-1} - 4(x-1)\sum_{n=0}^{\infty} a_n(x-1)^n = 0$$

$$\text{or, } \sum_{n=2}^{\infty} n(n-1)a_n(x-1)^{n-2} + \sum_{n=1}^{\infty} n a_n(x-1)^{n+1} - \sum_{n=0}^{\infty} 4a_n(x-1)^{n+1} = 0$$

$$\text{or, } \sum_{n=0}^{\infty}(n+2)(n+1)a_{n+2}(x-1)^n + \sum_{n=2}^{\infty}(n-1)a_{n-1}(x-1)^n - \sum_{n=1}^{\infty} 4a_{n-1}(x-1)^n = 0$$

$$\text{or, } 2a_2 + (6a_3 - 4a_0)(x-1) + \sum_{n=2}^{\infty}[(n+1)(n+2)a_{n+2} + (n-5)a_{n-1}](x-1)^n = 0. \quad (13.5.25)$$

The relation (13.5.25) is valid for all $x \in (1 - R, 1 + R)$, where $R(> 0)$ is the radius of convergence of the power series $\sum\limits_{n=0}^{\infty} a_n(x-1)^n$.

It follows that the coefficient of each term of the left side of (13.5.25) must be zero.

$\therefore\ a_2 = 0.$

$6a_3 - 4a_0 = 0$, i.e. $a_3 = \frac{2}{3}a_0$ and $(n+1)(n+2)a_{n+2} + (n-5)a_{n-1} = 0$ i.e. $a_{n+2} = -\frac{n-5}{(n+1)(n+2)}a_{n-1}$ for all $n \geqslant 2$.

Putting $n = 2, 3, 4, \ldots$ in the above recurrence relation, we obtain $a_4 = \frac{1}{4}a_1$, $a_5 = 0$, $a_6 = \frac{1}{45}a_0$, $a_7 = 0$, $a_8 = 0$, $a_9 = -\frac{1}{1620}a_0, \ldots$.

Putting all the values of $a_2, a_3, a_4, \ldots$ in terms of a_0 and a_1 in $y = \sum_{n=0}^{\infty} a_n(x-1)^n$, we obtain

$$y = a_0\left\{1 + \tfrac{2}{3}(x-1)^3 + \tfrac{1}{45}(x-1)^6 - \tfrac{1}{1620}(x-1)^9 + \ldots\right\} + a_1\left\{(x-1) + \tfrac{1}{4}(x-1)^4\right\},$$

which is the desired power series solution about $x = 1$ with two arbitrary constants a_0 and a_1.

Examples XIII(A)

1. Show that $x = 0$ is a regular singular point of the equation $x^2\frac{d^2y}{dx^2} + x\frac{dy}{dx} - 4y = 0$.
2. Show that $x = 0$ is an ordinary point of the equation $(x^2+1)\frac{d^2y}{dx^2} + x\frac{dy}{dx} - xy = 0$.
3. Show that $x = 0$ and $x = -1$ are singular points of the equation $x^2(x+1)\frac{d^2y}{dx^2} + (x^2-1)\frac{dy}{dx} + 2y = 0$. Hence, classify the singular points.
4. Locate and classify the singular points (if exist) of the following differential equations:
   (a) $x(x-1)^3\frac{d^2y}{dx^2} - 2(x-1)2\frac{dy}{dx} + 5y = 0$
   (b) $x^2(x+1)^2\frac{d^2y}{dx^2} + 2(x^2-1)\frac{dy}{dx} - 3y = 0$
   (c) $(1-x^2)\frac{d^2y}{dx^2} + 3x\frac{dy}{dx} - k(k-1)y = 0$, k being an arbitrary constant.
   (d) $(x^3+2x)\frac{d^2y}{dx^2} + \frac{dy}{dx} + 3xy = 0$
   (e) $(x^5+x^4-6x^3)\frac{d^2y}{dx^2} - x^2\frac{dy}{dx} - 2(x-2)y = 0$.
5. Find the power series solutions of the following differential equations about $x = 0$:
   (a) $y'' + xy' + (x^2+2)y = 0$
   (b) $(1-x^2)y'' + 2xy' - y = 0$
   (c) $y'' + xy' + x^2y = 0$
   (d) $y'' - y = x$
   (e) $y'' - xy' - ky = 0$, k being any constant.
   (f) $(x^2+1)y'' + xy' - xy = 0$.
6. Find the power series solution of the following differential equation about $x = 1$:
$$y'' + (x-1)^2y' - 4(x-1)y = 0.$$
7. Find the power series solution of the differential equation $y'' + (x-1)y' + y = 0$ near 2.
8. Find the power series solutions of the following initial value problems:
   (a) $y'' - 4y = 0$; $y(0) = 1$, $y'(0) = -1$
   (b) $y'' + xy' + 2y = 0$; $y(0) = 1$, $y'(0) = 0$
   (c) $xy'' + y' + 2y = 0$; $y(1) = 1$, $y'(1) = 2$
   (d) $y'' - xy' - y = 0$; $y(0) = 1$, $y'(0) = 0$.
9. Find the series solution of the following differential equations:
   (a) $y'' + (x-3)y' + y = 0$ about $x = 2$ (b) $y'' - (1-x)y' + y = 0$ about $x = 2$.

Answers

4. (a) $x = 0$: regular singular point; $x = 1$: irregular singular point
   (b) $x = 0$: irregular singular point; $x = -1$: regular singular point
   (c) $x = \pm 1$: regular singular points
   (d) $x = 0$: regular singular point
   (e) $x = 0$: irregular singular point; $x = 2, 3$: regular singular points.
5. (a) $y = a_0\left(1 - x^2 + \frac{1}{4}x^4 + \ldots\right) + a_1\left(x - \frac{1}{2}x^3 + \frac{3}{40}x^5 + \ldots\right)$

(b) $y = a_0\left(1 + \frac{1}{2}x^2 - \frac{1}{24}x^4 - \ldots\right) + a_1\left(x - \frac{1}{6}x^3 - \frac{1}{120}x^5 - \ldots\right)$

(c) $y = a_0\left(1 - \frac{1}{12}x^4 + \frac{1}{90}x^6 - \ldots\right) + a_1\left(x - \frac{1}{6}x^3 + \frac{1}{40}x^5 - \ldots\right)$

(d) $y = a_0\left(1 + \frac{1}{2}x^2 + \frac{1}{24}x^4 + \frac{1}{720}x^6 + \ldots\right) + a_1\left(x + \frac{1}{6}x^3 + \frac{1}{120}x^5 + \ldots\right) + \left(\frac{1}{6}x^3 + \frac{1}{120}x^5 + \ldots\right)$

(e) $y = a_0\left[1 + \frac{k}{2!}x^2 + \frac{k(k+2)}{4!}x^4 + \ldots\right] + a_1\left[x + \frac{k+1}{3!}x^3 + \frac{(k+1)(k+3)}{5!}x^5 + \ldots\right]$

(f) $y = a_0\left(1 + \frac{1}{6}x^3 - \frac{3}{40}x^5 + \ldots\right) + a_1\left(x - \frac{1}{6}x^3 + \frac{1}{12}x^4 + \ldots\right)$

6. $y = a_0\left\{1 + \frac{2}{3}(x-1)^3 + \frac{1}{45}(x-1)^6 + \ldots\right\} + a_1\left\{(x-1) + \frac{1}{4}(x-1)^4\right\}$

7. $y = a_0\left[1 - \frac{1}{2}(x-2)^2 + \frac{1}{6}(x-2)^3 + \frac{1}{12}(x-2)^4 + \ldots\right] + a_1\left[(x-2) - \frac{1}{2}(x-2)^2 - \frac{1}{6}(x-2)^3 + \frac{1}{6}(x-2)^4 + \ldots\right]$

8. (a) $y = 1 - x + 2x^2 - \frac{2}{3}x^3 + \frac{2}{3}x^4 - \frac{2}{15}x^5 + \ldots$

(b) $y = 1 - x^2 + \frac{1}{3}x^4 - \frac{1}{15}x^6 + \ldots$

(c) $y = 1 + 2(x-1) - 2(x-1)^2 + \frac{2}{3}(x-1)^3 - \frac{1}{6}(x-1)^4 + \frac{1}{15}(x-1)^5 + \ldots$

(d) $y = \sum_{n=0}^{\infty} \frac{1}{2^n . n!}x^{2n}$

9. (a) $y = a_0\left[1 - \frac{1}{2}(x-2)^2 - \frac{1}{6}(x-2)^3 - \frac{1}{12}(x-2)^4 + \ldots\right] + a_1\left[(x-2) + \frac{1}{2}(x-2)^2 - \frac{1}{6}(x-2)^3 - \ldots\right]$

(b) $y = a_0\left[1 - \frac{1}{2}(x-2)^2 + \frac{1}{6}(x-2)^3 + \frac{1}{12}(x-2)^4 + \ldots\right] + a_1\left[(x-2) - \frac{1}{2}(x-2)^2 - \frac{1}{6}(x-2)^3 + \frac{1}{6}(x-2)^4 + \ldots\right]$

13.6 Series solution about regular singular point

There are several methods to find series solutions of second order linear differential equations about regular singular points. But, here we would like to discuss the method of Frobenius only. If x_0 be a regular singular point of the eq. (13.3.2), the solution may be found by the method of Frobenius assuming a solution of the form:

$$y = (x - x_0)^k \sum_{n=0}^{\infty} a_n (x - x_0)^n, \; x \in (x_0, x_0 + R)$$

for some $R > 0$. In order to find the solution for $x \in (x_0 - R, x_0)$, we replace $x - x_0$ by $-(x - x_0)$.

Theorem 13.6.1. *Let x_0 be a regular singular point of the differential equation (13.3.2). Then the differential equation has at least one nontrivial solution that can be expressed as*

$$y = |x - x_0|^k \sum_{n=0}^{\infty} a_n (x - x_0)^n,$$

k being a constant. This solution is valid for all $x \in N'_R(x_0)$, $N'_R(x_0)$ being deleted neighbourhood of x_0 with radius $R(> 0)$.

The method of Frobenius

Let x_0 be a regular singular point and let the extended power series (Frobenius series) about x_0 given by

$$y = (x - x_0)^k \sum_{n=0}^{\infty} a_n (x - x_0)^n, \; x \in (x_0, x_0 + R)$$

be a solution of the differential eq. (13.3.2), where $a_0 \neq 0$. Then the solution may be written in the form:

$$y = \sum_{n=0}^{\infty} a_n (x - x_0)^{k+n}. \tag{13.6.1}$$

Differentiating (13.6.1) twice in succession with respect to x, we obtain

$$\frac{dy}{dx} = \sum_{n=0}^{\infty}(k+n)a_n(x-x_0)^{k+n-1} \tag{13.6.2}$$

$$\text{and } \frac{d^2y}{dx^2} = \sum_{n=0}^{\infty}(k+n)(k+n-1)a_n(x-x_0)^{k+n-2}. \tag{13.6.3}$$

Using (13.6.1), (13.6.2) and (13.6.3) in (13.3.2), we obtain an identity which is valid in some right neighbourhood $(x_0, x_0 + R)$ of the point x_0. Equating to zero the coefficient of the term of least degree of x, we obtain a quadratic equation in k known as indicial equation. Again, equating to zero the coefficient of x^{k+n} or x^{k+n-1} in the identity, we get a recurrence relation in terms of a_n and a_{n-1} or a_n and a_{n-2}. If the recurrence relation involves a_n and a_{n-2}, we determine a_1. It may be obtained by equating to zero the coefficient of the term of the next higher power of x. Obtaining all the coefficients in terms of a_0 and k from the recurrence relation we can find at least one Frobenius series solution (by the theorem 13.6.1). The number of independent Frobenius series solution (nontrivial) depends on the nature of the roots of the indicial equation.

Let α and β be the roots of the indicial equation. Then we would like to follow the three cases given below:

Case 13.6.1. *Let $\alpha \neq \beta$ and $\alpha - \beta$ be not an integer. Then*

$$u = |x-x_0|^{\alpha}\sum_{n=0}^{\infty} a_n(\alpha)(x-x_0)^n$$

$$\text{and } v = |x-x_0|^{\beta}\sum_{n=0}^{\infty} a_n(\beta)(x-x_0)^n, \; a_0 \neq 0$$

will be two linearly independent Frobenius solutions of the equation (13.3.2). *Thus, the general solution of the equation must be $y = c_1 u + c_2 v$, i.e.*

$$y = c_1|x-x_0|^{\alpha}\sum_{n=0}^{\infty} a_n(\alpha)(x-x_0)^n + c_2|x-x_0|^{\beta}\sum_{n=0}^{\infty} a_n(\beta)(x-x_0)^n,$$

where c_1 and c_2 are two arbitrary constants and $x \in N_R'(x_0)$.

Example 13.6.1. *Find the general solution of the differential equation $2x^2\frac{d^2y}{dx^2} - x\frac{dy}{dx} + (1-x^2)y = 0$ around $x = 0$.*

Solution: The normalised form of the given equation is $\frac{d^2y}{dx^2} - \frac{1}{2x}\frac{dy}{dx} + \frac{1-x^2}{2x^2}y = 0$.

Comparing the above equation with (13.3.2), we have $P(x) = -\frac{1}{2x}$ and $Q(x) = \frac{1-x^2}{2x^2}$, which are not analytic at $x = 0$.

But $xP(x) = -\frac{1}{2}$ and $x^2Q(x) = \frac{1}{2}(1-x^2)$ are both analytic at $x = 0$. It confirms that $x = 0$ is a regular singular point of the given equation.

Let the extended power series given by

$$y = x^k\sum_{n=0}^{\infty} a_n x^n, \; a \neq 0 \tag{13.6.4}$$

be a Frobenius solution of the given equation, where $x \in (0, R)$ for some $R > 0$.

Then differentiating (13.6.4) twice in succession with respect to x, we obtain

$$\frac{dy}{dx} = \sum_{n=0}^{\infty}(k+n)a_n x^{k+n-1} \tag{13.6.5}$$

$$\text{and } \frac{d^2y}{dx^2} = \sum_{n=0}^{\infty}(k+n)(k+n-1)a_n x^{k+n-2}. \tag{13.6.6}$$

Using the results (13.6.4), (13.6.5) and (13.6.6) in the given equation, we obtain

$$2x^2\sum_{n=0}^{\infty}(k+n)(k+n-1)a_n x^{k+n-2} - x\sum_{n=0}^{\infty}(k+n)a_n x^{k+n-1} + (1-x^2)\sum_{n=0}^{\infty}a_n x^{k+n} = 0$$

$$\text{or, } \sum_{n=0}^{\infty}2(k+n)(k+n-1)a_n x^{k+n} - \sum_{n=0}^{\infty}(k+n)a_n x^{k+n} + \sum_{n=0}^{\infty}a_n x^{k+n} - \sum_{n=0}^{\infty}a_n x^{k+n+2} = 0$$

$$\text{or, } \sum_{n=0}^{\infty}[2(k+n)^2 - 3(k+n) + 1]a_n x^{k+n} - \sum_{n=0}^{\infty}a_n x^{k+n+2} = 0$$

$$\text{or, } \sum_{n=0}^{\infty}(2k+2n-1)(k+n-1)a_n x^{k+n} - \sum_{n=0}^{\infty}a_n x^{k+n+2} = 0, \tag{13.6.7}$$

which is valid in the open interval $(0, R)$ for some $R > 0$. The relation (13.6.7) is an identity. Equating to zero the coefficient of the lowest degree term of x in (13.6.7), we obtain the indicial equation as a_0 $(2k-1)(k-1) = 0$.

$\therefore\ k = \frac{1}{2}, 1\ (\because\ a_0 \neq 0)$.

We see that the roots of the indicial equation are unequal and their difference is not an integer.

Now equating to zero the coefficient of x^{k+n}, we have the recurrence relation given below:

$$(2k+2n-1)(k+n-1)a_n - a_{n-2} = 0 \text{ i.e. } a_n = \frac{a_{n-2}}{(2k+2n-1)(k+n-1)},\quad n \geqslant 2. \tag{13.6.8}$$

Again, equating to zero the coefficient of x^{k+1}, we obtain

$$(2k+1)ka_1 = 0 \text{ i.e. } a_1 = 0 \text{ for } k = \tfrac{1}{2}, 1 \tag{13.6.9}$$

By (13.6.8) and (13.6.9), we have $a_1 = a_3 = a_5 = \cdots = 0$.

Putting $n = 2, 4, 6, \ldots$ in (13.6.8), we have

$$a_2 = \frac{1}{(2k+3)(k+1)}a_0,\ a_4 = \frac{1}{(2k+7)(2k+3)(k+3)(k+1)}a_0, \ldots$$

Putting the values of $a_1, a_2, a_3, a_4, \ldots$ in (13.6.4), we obtain.

$$y = a_0x^k\left[1 + \frac{x^2}{(k+1)(2k+3)} + \frac{x^4}{(k+1)(k+3)(2k+3)(2k+7)} + \ldots\right] \tag{13.6.10}$$

Putting $k = \frac{1}{2}$ in (13.6.10), we have

$$y = a_0\sqrt{x}\left(1 + \frac{x^2}{2\cdot 3} + \frac{x^4}{2\cdot 3\cdot 4\cdot 7} + \ldots\right) = a_0 u, \text{ say,}$$

where $u = \sqrt{x}\left(1 + \frac{x^2}{2\cdot3} + \frac{x^4}{2\cdot3\cdot4\cdot7} + \ldots\right)$ is a Frobenius series solution of the given equation.

Again, putting $k = 1$ in (13.6.10), we have $y = a_0 x\left(1+\frac{x^2}{2\cdot 5}+\frac{x^4}{2\cdot 4\cdot 5\cdot 9}+\ldots\right) = a_0 v$, say, where $v = x\left(1+\frac{x^2}{2\cdot 5}+\frac{x^4}{2\cdot 4\cdot 5\cdot 9}+\ldots\right)$ is another Frobenius series solution of the given equation. Clearly, u and v are linearly independent. Hence, desired general solution of the given equation is $y = c_1 u + c_2 v$, where c_1 and c_2 are two arbitrary constants.

Case 13.6.2. *Let $\alpha \neq \beta$ and $\alpha - \beta$ be a positive integer. Then the equation* (13.3.2) *has a nontrivial solution (Frobenius series solution) of the form:* $u = |x - x_0|^{\alpha} \sum_{n=0}^{\infty} a_n(\alpha)(x - x_0)^n$, $x \in N_R'(x_0)$ *and* $a_0 \neq 0$.

For another linearly independent Frobenius series solution, the following two sub-cases may arise:

Sub-case 1

Let $v = |x - x_0|^{\beta} \sum_{n=0}^{\infty} a_n(\beta)(x - x_0)^n$, $x \in N_R'(x_0)$ be another linearly independent Frobenius series solution (remember that β is the smaller root of the indicial equation).

Then the general solution of the eq. (13.3.2) is given by $y = c_1 u + c_2 v$, where c_1 and c_2 are two arbitrary constants.

Sub-case 2

Let $k = \beta$ fails to give such linearly independent solution or a finite solution. If it gives an infinite solution, a factor $(k - \beta)$ must be present in the denominators of the coefficients starting after finite number of terms.

In this sub-case, two linearly independent solutions of (13.3.2) will be of the form:

$$u = |x - x_0|^{\alpha} \sum_{n=0}^{\infty} a_n(\alpha)(x - x_0)^n \text{ and } \left[\frac{\partial w}{\partial k}\right]_{k=\beta},$$

where $w = (k - \beta)|x - x_0|^k \sum_{n=0}^{\infty} a_n(k)(x - x_0)^n$, $a_0 \neq 0$.

Thus, the general solution of the equation is given by $y = c_1 u + c_2 \left[\frac{\partial w}{\partial k}\right]_{k=\beta}$, where c_1 and c_2 are arbitrary constants.

Example 13.6.2. *Solve the equation* $(x - x^2)\frac{d^2y}{dx^2} + 4\frac{dy}{dx} + 2y = 0$ *in powers of* x.

Solution: The normalised form of the given equation is $\frac{d^2y}{dx^2} + \frac{4}{x(1-x)}\frac{dy}{dx} + \frac{2}{x(1-x)}y = 0$.

Comparing this equation with (13.3.2), we have $P(x) = \frac{4}{x(1-x)}$ and $Q(x) = \frac{2}{x(1-x)}$.

Clearly, $P(x)$ and $Q(x)$ are not analytic at $x = 0$, whereas $xP(x) = \frac{4}{1-x}$ and $x^2Q(x)\frac{2x}{1-x}$ are both analytic at $x = 0$. It follows that $x = 0$ is a regular singular point of the equation. Let the extended power series given by

$$y = x^k \sum_{n=0}^{\infty} a_n x^n, \quad a_0 \neq 0 \tag{13.6.11}$$

be a Frobenius solution of the given equation, where $x \in (0, R)$ for some $R > 0$.

Differentiating (13.6.11) twice in succession with respect to x, we have

$$\frac{dy}{dx} = \sum_{n=0}^{\infty}(k+n)a_n x^{k+n-1} \tag{13.6.12}$$

$$\text{and } \frac{d^2y}{dx^2} = \sum_{n=0}^{\infty}(k+n)(k+n-1)a_n x^{k+n-2}. \tag{13.6.13}$$

Using (13.6.11), (13.6.12) and (13.6.13) in the given equation, we have

$$(x-x^2)\sum_{n=0}^{\infty}(k+n)(k+n-1)a_n x^{k+n-2}+4\sum_{n=0}^{\infty}(k+n)a_n x^{k+n-1}+2\sum_{n=0}^{\infty}a_n x^{k+n}=0$$

$$\text{or, }\sum_{n=0}^{\infty}(k+n)(k+n+3)a_n x^{k+n-1}-\sum_{n=0}^{\infty}(k+n+1)(k+n-2)a_n x^{k+n}=0. \tag{13.6.14}$$

The relation (13.6.14) is valid in the open interval $(0, R)$ for some $R > 0$. Since (13.6.14) is an identity, equating to zero the coefficient of the lowest degree term of x, we obtain $k(k+3)a_0 = 0$.

$\therefore\ k = -3, 0$ ($\because\ a_0 \neq 0$).

We see that the roots of the indicial equation are unequal but their difference is an integer.

Equating to zero the coefficient of x^{k+n-1}, we have

$$(k+n)(k+n+3)a_n = (k+n)(k+n-3)a_{n-1} \text{ i.e. } a_n = \frac{k+n-3}{k+n+3}a_{n-1},\ n \geqslant 1.$$

Putting $n = 1, 2, 3, \ldots$ in the above recurrence relation, we obtain

$$a_1 = \frac{k-2}{k+4}a_0,\ a_2 = \frac{(k-1)(k-2)}{(k+5)(k+4)},\ a_3 = \frac{k(k-1)(k-2)}{(k+6)(k+5)(k+4)}a_0, \ldots$$

Using the above results in (13.6.13), we obtain

$$y = a_0 x^k\left[1+\frac{k-2}{k+4}x+\frac{(k-1)(k-2)}{(k+5)(k+4)}x^2+\frac{k(k-1)(k-2)}{(k+6)(k+5)(k+4)}x^3+\ldots\right]. \tag{13.6.15}$$

Putting $k = 0$ in (13.6.15), we have $y = a_0\left(1-\frac{1}{2}x+\frac{1}{10}x^2\right) = a_0 u$, say, where $u = 1-\frac{1}{2}x+\frac{1}{10}x^2$ is a solution of the given equation.

Again, putting $k = -3$ in (13.6.15), we obtain $y = a_0 x^{-3}(1-5x+10x^2-10x^3+\ldots) = a_0 v$, say, where $v = x^{-3}(1-5x+10x^2-10x^3+\ldots)$ is another Frobenius solution of the equation. Here u and v are linearly independent. Hence, the desired general solution is given by $y = c_1 u + c_2 v$, where c_1 and c_2 are two arbitrary constants.

Example 13.6.3. *Find the series solution of the differential equation* $(x-x^2)\frac{d^2y}{dx^2}-3x\frac{dy}{dx}-y=0$ *in powers of* x.

Solution: The normalised form of the given differential equation is $\frac{d^2y}{dx^2}-\frac{3}{1-x}\frac{dy}{dx}-\frac{1}{x(1-x)}y=0$.

Comparing the above equation with (13.3.2), we obtain $P(x) = -\frac{3}{1-x}$ and $Q(x) = -\frac{1}{x(1-x)}$.

We see that $Q(x)$ is not analytic at $x = 0$. But $xP(x) = -\frac{3x}{1-x}$ and $x^2Q(x) = -\frac{x}{1-x}$ are analytic at $x = 0$. It follows that $x = 0$ is a regular singular point of the differential equation.

Let the extended power series given by

$$y = x^k\sum_{n=0}^{\infty}a_n x^n,\ a_0 \neq 0 \tag{13.6.16}$$

be a Frobenius series solution of the given equation, where $x \in (0, R)$ for some $R > 0$.

Differentiating (13.6.16) twice in succession with respect to x, we have

$$\frac{dy}{dx} = \sum_{n=0}^{\infty}(k+n)a_n x^{k+n-1} \tag{13.6.17}$$

$$\text{and } \frac{d^2y}{dx^2} = \sum_{n=0}^{\infty}(k+n)(k+n-1)a_n x^{k+n-2}. \tag{13.6.18}$$

Using (13.6.16), (13.6.17) and (13.6.18) in the given equation, we have

$$(x-x^2)\sum_{n=0}^{\infty}(k+n)(k+n-1)a_n x^{k+n-2} - 3x\sum_{n=0}^{\infty}(k+n)a_n x^{k+n-1} - \sum_{n=0}^{\infty} a_n x^{k+n} = 0$$

$$\text{or, } \sum_{n=0}^{\infty}(k+n)(k+n-1)a_n x^{k+n-1} - \sum_{n=0}^{\infty}(k+n+1)^2 a_n x^{k+n} = 0. \tag{13.6.19}$$

This is an identity which is valid in the open interval $(0, R)$ for some $R > 0$.
Now, equating to zero the coefficient of the lowest degree term of x in (13.6.19), we obtain $k(k-1)a_0 = 0$.
$\therefore\ k = 0, 1$ ($\because\ a_0 \neq 0$).
Thus, the roots of the indicial equation are distinct but differ by an integer.
Again, equating to zero the coefficient of x^{k+n-1} in the identity (13.6.19), we obtain

$$(k+n)(k+n-1)a_n - (k+n)^2 a_{n-1} = 0.$$

$$\therefore\ a_n = \frac{k+n}{k+n-1}a_{n-1},\ n \geqslant 1.$$

Now putting $n = 1, 2, 3, \ldots$ in the above recurrence relation, we have

$$a_1 = \frac{k+1}{k}a_0,\ a_2 = \frac{k+2}{k}a_0,\ a_3 = \frac{k+3}{k}a_0,\ \ldots$$

Using the above results in (13.6.16), we obtain

$$y = a_0 x^k\left(1 + \frac{k+1}{k}x + \frac{k+2}{k}x^2 + \frac{k+3}{k}x^3 + \ldots\right).$$

Here $k = 1$ is the larger root of the indicial equation and also we see that, for $k = 0$, there does not exist finite solution.

For $k = 1$, we obtain $y = a_0 x(1 + 2x + 3x^2 + 4x^3 + \ldots) = a_0 u$, say, where $u = x + 2x^2 + 3x^3 + 4x^4 + \ldots$ is a Frobenius solution of the given equation.

For another linearly independent solution, let us consider

$$w = a_0 k x^k\left(1 + \frac{k+1}{k}x + \frac{k+2}{k}x^2 + \frac{k+3}{k}x^3 + \ldots\right)$$

$$= a_0 x^k[k + (k+1)x + (k+2)x^2 + (k+3)x^3 + \ldots].$$

$$\therefore\ \frac{\partial w}{\partial k} = a_0 x^k \ln x[k + (k+1)x + (k+2)x^2 + (k+3)x^3 + \ldots] + a_0 x^k(1 + x + x^2 + x^3 + \ldots).$$

$$\therefore\ \left[\frac{\partial w}{\partial k}\right]_{k=0} = a_0[\ln x(x + 2x^2 + 3x^3 + \ldots) + (1 + x + x^3 + x^3 + \ldots)]$$

$$= a_0(u\ln x + 1 + x + x^2 + x^3 + \ldots) = a_0 v,\ \text{say},$$

where $v = u\ln x + 1 + x + x^2 + x^3 + \ldots$ is another solution of the equation. Here u and v are two linearly independent solutions of the given equation.

Hence, required general solution is given by $y = c_1 u + c_2 v$, where c_1 and c_2 are two arbitrary constants.

Example 13.6.4. *Find the series solution of the differential equation* $x^2\frac{d^2y}{dx^2} + x\frac{dy}{dx} + (x^2-1)y = 0$ *in powers of* x.

Solution: The normalised form of the given equation is $\frac{d^2y}{dx^2} + \frac{1}{x}\frac{dy}{dx} + \frac{x^2-1}{x^2}y = 0$.

Comparing this equation with (13.3.2), we have $P(x) = \frac{1}{x}$ and $Q(x) = \frac{x^2-1}{x^2}$, which are not analytic at $x = 0$.

But $xP(x) = 1$ and $x^2Q(x) = x^2 - 1$ are analytic at $x = 0$.

It confirms that $x = 0$ is a regular singular point of the given equation.

Let the extended power series given by

$$y = x^k \sum_{n=0}^{\infty} a_n x^n, \; a_0 \neq 0 \tag{13.6.20}$$

be a Frobenius solution of the given equation, where $x \in (0, R)$ for some $R > 0$.

Differentiating (13.6.20) twice in succession with respect to x, we obtain

$$\frac{dy}{dx} = \sum_{n=0}^{\infty}(k+n)a_n x^{k+n-1} \tag{13.6.21}$$

$$\text{and } \frac{d^2y}{dx^2} = \sum_{n=0}^{\infty}(k+n)(k+n-1)a_n x^{k+n-2}. \tag{13.6.22}$$

Using (13.6.20), (13.6.21) and (13.6.22) in the given equation, we have

$$x^2\sum_{n=0}^{\infty}(k+n)(k+n-1)a_n x^{k+n-2} + x\sum_{n=0}^{\infty}(k+n)a_n x^{k+n-1} + (x^2-1)\sum_{n=0}^{\infty}a_n x^{k+n} = 0$$

$$\text{or, } \sum_{n=0}^{\infty}(k+n)(k+n-1)a_n x^{k+n} + \sum_{n=0}^{\infty}(k+n)a_n x^{k+n} + \sum_{n=0}^{\infty}a_n x^{k+n+2} - \sum_{n=0}^{\infty}a_n x^{k+n} = 0$$

$$\text{or, } \sum_{n=0}^{\infty}(k+n-1)(k+n+1)a_n x^{k+n} + \sum_{n=0}^{\infty}a_n x^{k+n+2} = 0. \tag{13.6.23}$$

The relation (13.6.23) is valid for all $x \in (0, R)$, for some $R > 0$.

Since (13.6.23) represents an identity, equating to zero the coefficient of the lowest degree term of x, we obtain $a_0(k-1)(k+1) = 0$.

$\therefore k = \pm 1$ ($\because a_0 \neq 0$).

Thus, the roots of the indicial equation are distinct and their difference is an integer.

Equating to zero the coefficient of x^{k+n}, we have

$$(k+n+1)(k+n-1)a_n + a_{n-2} = 0 \text{ i.e. } a_n = -\frac{1}{(k+n+1)(k+n-1)}a_{n-2}, \; n \geqslant 2, \tag{13.6.24}$$

which is a recurrence relation between a_n and a_{n-2}.

Now equating to zero the coefficient of the next higher power of x, we obtain $k(k+2)a_1 = 0$.

$\therefore a_1 = 0$ ($\because k = \pm 1$).

Thus, by (13.6.24), $a_1 = a_3 = a_5 = \cdots = a_{2n+1} = \cdots = 0$.

Now, putting $n = 2, 4, 6, \ldots$ in (13.6.24), we have

$$a_2 = -\frac{1}{(k+1)(k+3)}a_0; \quad a_4 = \frac{1}{(k+1)(k+3)^2(k+5)}a_0; \quad a_6 = -\frac{1}{(k+1)(k+3)^2(k+5)^2(k+7)}a_0, \ldots$$

Using the above results in (13.6.20), we obtain

$$y = a_0x^k\left[1 - \frac{1}{(k+1)(k+3)}x^2 + \frac{1}{(k+1)(k+3)^2(k+5)}x^4 - \frac{1}{(k+1)(k+3)^2(k+5)^2(k+7)}x^6 + \ldots\right]. \quad (13.6.25)$$

Here $k = 1$ is the larger root of the indicial equation and all the coefficients except the 1st are undefined at $k = -1$.

$\therefore$ exactly one independent Frobenius series solution exists for $k = 1$. Putting $k = 1$ in (13.6.25), we get

$$y = a_0x\left[1 - \frac{1}{2\cdot 4}x^2 + \frac{1}{2\cdot 4^2\cdot 6}x^4 - \frac{1}{2\cdot 4^2\cdot 6^2\cdot 8}x^6 + \ldots\right] = a_0u, \text{ say.}$$

Here

$$u = x - \frac{1}{2\cdot 4}x^3 + \frac{1}{2\cdot 4^2\cdot 6}x^5 - \frac{1}{2\cdot 4^2\cdot 6^2\cdot 8}x^7 + \ldots$$

is a solution of the given equation.

For another linearly independent solution, let us consider

$$w = a_0(k+1)x^k\left[1 - \frac{1}{(k+1)(k+3)}x^2 + \frac{1}{(k+1)(k+3)^2(k+5)}x^4 - \frac{1}{(k+1)(k+3)^2(k+5)^2(k+7)}x^6 + \ldots\right]$$

$$= a_0x^k\left[(k+1) - \frac{1}{k+3}x^2 + \frac{1}{(k+3)^2(k+5)}x^4 - \frac{1}{(k+3)^2(k+5)^2(k+7)}x^6 + \ldots\right].$$

$$\therefore \frac{\partial w}{\partial k} = a_0x^k\ln x\left[(k+1) - \frac{1}{k+3}x^2 + \frac{1}{(k+3)^2(k+5)}x^4 - \frac{1}{(k+3)^2(k+5)^2(k+7)}x^6 + \ldots\right]$$

$$+ a_0x^k\left[1 + \frac{1}{(k+3)^2}x^2 - \left\{\frac{1}{(k+3)^2(k+5)^2} + \frac{2}{(k+3)^3(k+5)}\right\}x^4 + \ldots\right]$$

$$\therefore \left[\frac{\partial w}{\partial k}\right]_{k=-1} = a_0\left[x^{-1}\ln x\left(-\frac{1}{2}x^2 + \frac{1}{2^2\cdot 4}x^4 - \frac{1}{2^2\cdot 4^2\cdot 6}x^6 + \ldots\right) + x^{-1}\left(1 + \frac{1}{2^2}x^2 - \frac{5}{2^2\cdot 4^2}x^4 + \ldots\right)\right]$$

$$= a_0\left[-\frac{1}{2}\ln x\left\{x - \frac{1}{2\cdot 4}x^3 + \frac{1}{2\cdot 4^2\cdot 6}x^5 - \ldots\right\} + x^{-1}\left\{1 + \frac{1}{2^2}x^2 - \frac{5}{2^2\cdot 4^2}x^4 + \ldots\right\}\right]$$

$$= a_0\left[x^{-1}\left(1 + \frac{1}{2^2}x^2 - \frac{5}{2^2\cdot 4^2}x^4 + \ldots\right) - \frac{1}{2}u\ln x\right] = a_0v, \text{ say,}$$

where $v = x^{-1}\left(1 + \frac{1}{2^2}x^2 - \frac{5}{2^2.4^2}x^4 + \ldots\right) - \frac{1}{2}u\ln x$ is another solution. Clearly, u and v are linearly independent solutions.

Hence, required general solution is given by $y = c_1u + c_2v$, where c_1 and c_2 are two arbitrary constants.

Case 13.6.3. *Let $\alpha = \beta$. Then two linearly independent solutions of the equation* (13.3.2) *are*

$$v_{k=\alpha} = u = |x - x_0|^\alpha \sum_{n=0}^{\infty} a_n(\alpha)(x - x_0)^n \text{ and } \left[\frac{\partial v}{\partial k}\right]_{k=\alpha}.$$

$\therefore$ the general solution of the equation is given by $y = c_1u + c_2\left[\frac{\partial v}{\partial k}\right]_{k=\alpha}$, where c_1 and c_2 are arbitrary constants.

Example 13.6.5. *Find the general solution of the differential equation* $x(1-x)\frac{d^2y}{dx^2}+(1-5x)\frac{dy}{dx}-4y=0$ *near* 0.

Solution: The normalised form of the given equation is $\frac{d^2y}{dx^2}+\frac{1-5x}{x(1-x)}\frac{dy}{dx}-\frac{4}{x(1-x)}y=0$.

Comparing the above equation with (13.3.2), we have $P(x)=\frac{1-5x}{x(1-x)}$ and $Q(x)=-\frac{4}{x(1-x)}$, which are not analytic at $x=0$. But $xP(x)=\frac{1-5x}{1-x}$ and $x^2Q(x)=-\frac{4x}{1-x}$ are both analytic at $x=0$. It follows that $x=0$ is a regular singular point of the equation.

Let the extended power series given by

$$y=x^k\sum_{n=0}^{\infty}a_nx^n,\ a_0\neq 0 \tag{13.6.26}$$

be a Frobenius solution of the given equation, where $x\in(0,R)$ for some $R>0$.

Differentiating (13.6.26) twice in succession with respect to x, we have

$$\frac{dy}{dx}=\sum_{n=0}^{\infty}(k+n)a_nx^{k+n-1} \tag{13.6.27}$$

$$\text{and}\quad \frac{d^2y}{dx^2}=\sum_{n=0}^{\infty}(k+n)(k+n-1)a_nx^{k+n-2}. \tag{13.6.28}$$

Using (13.6.26), (13.6.27) and (13.6.28) in the given equation, we have

$$x(1-x)\sum_{n=0}^{\infty}(k+n)(k+n-1)a_nx^{k+n-2}+(1-5x)\sum_{n=0}^{\infty}a_n(k+n)x^{k+n-1}-4\sum_{n=0}^{\infty}a_nx^{k+n}=0$$

$$\text{or,}\ \sum_{n=0}^{\infty}(k+n)^2a_nx^{k+n-1}-\sum_{n=0}^{\infty}(k+n+2)^2a_nx^{k+n}=0. \tag{13.6.29}$$

Relation (13.6.29) is an identity in x that is valid in the open interval $(0,R)$ for some $R>0$.

Equating to zero the coefficient of the lowest degree term of x in (13.6.29), we obtain $a_0k^2=0$, i.e. $k=0,0$ ($\because\ a_0\neq 0$). Thus, the roots of the indicial equation are equal. Now, equating to zero the coefficient of x^{k+n-1}, we get

$$a_n(k+n)^2-a_{n-1}(k+n+1)^2=0 \text{ i.e. } a_n=\frac{(k+n+1)^2}{(k+n)^2}a_{n-1}$$

Putting $n=1,2,3,\ldots$, we obtain

$$a_1=\frac{(k+2)^2}{(k+1)^2}a_0,\ a_2=\frac{(k+3)^2}{(k+1)^2}a_0,\ a_3=\frac{(k+4)^2}{(k+1)^2}a_0,\ldots$$

Using the above results in (13.6.26), we have

$$y=a_0x^k\left[1+\left(\frac{k+2}{k+1}\right)^2x+\left(\frac{k+3}{k+1}\right)^2x^2+\left(\frac{k+4}{k+1}\right)^2x^3+\ldots\right]. \tag{13.6.30}$$

$\therefore$ for $k=0$, (13.6.20) becomes $y=a_0(1+2^2x+3^2x^2+4^2x^3+\ldots)=a_0u$, say, where $u=1+2^2x+3^2x^2+4^2x^3+\ldots$ is the Frobenius solution of the equation.

Differentiating (13.6.30) partially with respect to k, we have

$$\frac{\partial y}{\partial k} = a_0 x^k \ln k \left[1 + \left(\frac{k+2}{k+1}\right)^2 x + \left(\frac{k+3}{k+1}\right)^2 x^2 + \left(\frac{k+4}{k+1}\right)^2 x^3 + \ldots\right]$$
$$+ a_0 x^k \left[\frac{2(k+1)(k+2) - 2(k+2)^2}{(k+1)^3} x + \frac{2(k+1)(k+3) - 2(k+3)^2}{(k+1)^3} x^2 + \ldots\right]$$

$$\therefore \left[\frac{\partial y}{\partial k}\right]_{k=0} = a_0 \left[\ln x(1 + 2^2 x + 3^2 x^2 + 4^2 x^3 + \ldots) + (-4x - 2 \cdot 6x^2 - \ldots)\right]$$
$$= a_0[u \ln x - 2(1 \cdot 2x + 2 \cdot 3x^2 + \ldots)]$$
$$= a_0 v, \text{ say,}$$

where $v = u \ln x - 2(1 \cdot 2x + 2 \cdot 3x^2 + \ldots)$ is another linearly independent solution.

Hence, required general solution is given by $y = c_1 u + c_2 v$, where c_1 and c_2 are two arbitrary constants.

13.7 Series solution about a point at infinity

In order to find the series solution of the differential eq. (13.3.1) about a point at infinity, we give a transformation $x = \frac{1}{z}$ in (13.3.1). This transformation makes the equation in the following form:

$$\frac{d^2y}{dz^2} + P_1(z)\frac{dy}{dz} + Q_1(z)y = 0. \tag{13.7.1}$$

The eq. (13.3.1) will have a regular singular point at infinity if the eq. (13.7.1) gains a regular singular point at $z = 0$.

Example 13.7.1. *Show that the point at infinity is a regular singular point of the following differential equation:* $x^2 \frac{d^2y}{dx^2} + (3x - 1)\frac{dy}{dx} + y = 0$.

Solution: Let $z = \frac{1}{x}$.

Then
$$\frac{dy}{dx} = \frac{dy}{dz} \cdot \frac{dz}{dx} = -\frac{1}{x^2}\frac{dy}{dz} = -z^2\frac{dy}{dz}$$

and

$$\frac{d^2y}{dx^2} = \frac{d}{dx}\left(\frac{dy}{dx}\right) = \frac{d}{dz}\left(-z^2\frac{dy}{dz}\right)\frac{dz}{dx} = -\frac{1}{x^2}\left(-2z\frac{dy}{dz} - z^2\frac{d^2y}{dz^2}\right) = z^2\left(2z\frac{dy}{dz} + z^2\frac{d^2y}{dz^2}\right) = z^4\frac{d^2y}{dz^2} + 2z^3\frac{dy}{dz}.$$

Thus, by the transformation $z = \frac{1}{x}$, given equation becomes

$$\frac{1}{z^2}\left(z^4\frac{d^2y}{dz^2} + 2z^3\frac{dy}{dz}\right) + \left(\frac{3}{z} - 1\right)\left(-z^2\frac{dy}{dz}\right) + y = 0 \text{ or, } z^2\frac{d^2y}{dz^2} + (z^2 - z)\frac{dy}{dz} + y = 0. \tag{13.7.2}$$

The normalised form of this equation is $\frac{d^2y}{dz^2} + \frac{z-1}{z}\frac{dy}{dz} + \frac{1}{z^2}y = 0$.

Comparing it with $\frac{d^2y}{dz^2} + P(z)\frac{dy}{dz} + Q(z)y = 0$, we have $P(z) = \frac{z-1}{z}$ and $Q(z) = \frac{1}{z^2}$, which are not analytic at $z = 0$.

But $zP(z) = z - 1$ and $z^2Q(z) = 1$ are analytic at that point. Consequently, $z = 0$ is a regular singular point of the eq. (13.7.2).

Hence, the point at infinity is a regular singular point of the given equation.

13.8 Bessel's equation

The second order differential equation of the form:

$$x^2\frac{d^2y}{dx^2} + x\frac{dy}{dx} + (x^2 - \gamma^2)y = 0 \tag{13.8.1}$$

is known as the *Bessel's equation* of order γ, where γ is a nonnegative constant.

The normalised form of this equation is

$$\frac{d^2y}{dx^2} + \frac{1}{x}\frac{dy}{dx} + \frac{x^2-\gamma^2}{x^2}y = 0. \tag{13.8.2}$$

Comparing (13.8.2) with (13.3.2), we have $P(x) = \frac{1}{x}$ and $Q(x) = \frac{x^2-\gamma^2}{x^2}$, which are not analytic at $x = 0$.
But $xP(x) = 1$ and $x^2Q(x) = x^2 - \gamma^2$ are both analytic at $x = 0$.
Therefore, $x = 0$ is a regular singular point of the Bessel's equation.
Let the extended power series given by

$$y = x^k\sum_{n=0}^{\infty} a_0x^n,\ a_0 \neq 0 \tag{13.8.3}$$

be a Frobenius solution of the Bessel's eq. (13.8.1), where $x \in (0, R)$ for some $R > 0$.
Now, differentiating (13.8.1) twice in succession with respect to x, we obtain

$$\frac{dy}{dx} = \sum_{n=0}^{\infty}(k+n)a_nx^{k+n-1} \tag{13.8.4}$$

$$\text{and } \frac{d^2y}{dx^2} = \sum_{n=0}^{\infty}(k+n)(k+n-1)a_nx^{k+n-2}. \tag{13.8.5}$$

Using (13.8.3), (13.8.4) and (13.8.5) in the Bessel's equation, we obtain

$$x^2\sum_{n=0}^{\infty} a_n(k+n)(k+n-1)x^{k+n-2} + x\sum_{n=0}^{\infty} a_n(k+n)x^{k+n-1} + (x^2-\gamma^2)\sum_{n=0}^{\infty} a_nx^{k+n} = 0$$

$$\text{or, } \sum_{n=0}^{\infty} a_n\{(k+n)(k+n-1) + (k+n) - \gamma^2\}x^{k+n} + \sum_{n=0}^{\infty} a_nx^{k+n+2} = 0$$

$$\text{or, } \sum_{n=0}^{\infty} a_n(k+n+\gamma)(k+n-\gamma)x^{k+n} + \sum_{n=0}^{\infty} a_nx^{k+n+2} = 0. \tag{13.8.6}$$

Here, (13.8.6) is an identity in x that is valid in the open interval $(0, R)$ for some $R > 0$.
Equating to zero the coefficient of the lowest degree term of x (i.e. the coefficient of x^k), we obtain $a_0(k+\gamma)(k-\gamma) = 0$ i.e. $(k+\gamma)(k-\gamma) = 0$ ($\because a_0 \neq 0$).
$\therefore k = \pm\gamma$, which are the roots of the indicial equation.
Now, equating to zero the coefficient of x^{k+1}, we have $a_1(k+1+\gamma)(k+1-\gamma) = 0$.
$\therefore a_1 = 0$ ($\because k = \pm\gamma$).
Finally, equating to zero the coefficient of x^{k+n}, we have

$$a_n(k+n+\gamma)(k+n-\gamma) + a_{n-2} = 0 \text{ i.e. } a_n = \frac{1}{(\gamma+k+n)(\gamma-k-n)}a_{n-2}. \tag{13.8.7}$$

Putting $n = 3, 5, 7, \ldots$ in (13.8.7), we obtain $a_1 = a_3 = a_5 = a_7 = \cdots = 0$.
Again, putting $n = 2, 4, 6, \ldots$, we obtain

$$a_2 = \frac{1}{(\gamma + k + 2)(\gamma - k - 2)} a_0$$
$$a_4 = \frac{1}{(\gamma + k + 4)(\gamma - k - 4)} a_2 = \frac{1}{(\gamma + k + 4)(\gamma - k - 4)(\gamma + k + 2)(\gamma - k - 2)} a_0.$$

Using all the values of $a_1, a_2, a_3, a_4, a_5, \ldots$ in (13.8.3), we obtain

$$y = a_0 x^k \left[1 + \frac{x^2}{(\gamma + k + 2)(\gamma - k - 2)} + \frac{x^4}{(\gamma + k + 2)(\gamma - k - 2)(\gamma + k + 4)(\gamma - k - 4)} + \ldots \right].$$

Case 13.8.1. *Let γ be neither zero nor an integer.*

Then, replacing k by γ, we obtain

$$y = a_0 x^{\gamma} \left[1 - \frac{x^2}{4(1 + \gamma)} + \frac{x^4}{4 \cdot 8(1 + \gamma)(2 + \gamma)} - \ldots \right]$$

and replacing k by $-\gamma$, we obtain

$$y = a_0 x^{-\gamma} \left[1 - \frac{x^2}{4(1 - \gamma)} + \frac{x^4}{4 \cdot 8(1 - \gamma)(2 - \gamma)} - \ldots \right].$$

Choosing $a_0 = \dfrac{1}{2^{\gamma}\Gamma(\gamma + 1)}$, we have two linearly independent Frobenius series solutions $J_{\gamma}(x)$ and $J_{-\gamma}(x)$ defined by

$$J_{\gamma}(x) = \frac{1}{\Gamma(\gamma + 1)} \left(\frac{x}{2}\right)^{\gamma} \sum_{m=0}^{\infty} \frac{(-1)^m x^{2m}}{2^{2m} m!(\gamma + 1)(\gamma + 2) \ldots (\gamma + m)}$$

$$\text{i.e. } J_{\gamma}(x) = \sum_{m=0}^{\infty} \frac{(-1)^m \left(\frac{x}{2}\right)^{\gamma + 2m}}{\Gamma(\gamma + m + 1)\Gamma(m + 1)} \text{ and } J_{-\gamma}(x) = \sum_{m=0}^{\infty} \frac{(-1)^m \left(\frac{x}{2}\right)^{-\gamma + 2m}}{\Gamma(-\gamma + m + 1)\Gamma(m + 1)}.$$

Hence, the general solution of the Bessel eq. (13.8.1) is $y = AJ_{\gamma}(x) + BJ_{-\gamma}(x)$, where A, B are two arbitrary constants.

Case 13.8.2. *Let the difference of the roots γ and $-\gamma$ of the indicial equation is either zero or a positive integer.*

Then the second independent solution of the Bessel equation will be of the form: $v(x)J_{\gamma}(x)$, where

$$\frac{dv}{dx} = \frac{1}{[J_{\gamma}(x)]^2} \exp\left[-\int \frac{1}{x} dx \right] = \frac{1}{x[J_{\gamma}(x)]^2}, \text{ i.e. } v(x) = \int \frac{dx}{x[J_{\gamma}(x)]^2}.$$

Bessel function

The *Bessel function* of first kind and of order n is denoted by $J_n(x)$ and defined by

$$J_n(x) = \sum_{m=0}^{\infty} \frac{(-1)^m \left(\frac{x}{2}\right)^{n + 2m}}{\Gamma(n + m + 1)\Gamma(m + 1)}.$$

For positive integer n,

$$J_{-n}(x) = \sum_{m=0}^{\infty} \frac{(-1)^m \left(\frac{x}{2}\right)^{-n+2m}}{\Gamma(-n+m+1)\Gamma(m+1)} = \sum_{m=n}^{\infty} \frac{(-1)^m \left(\frac{x}{2}\right)^{-n+2m}}{\Gamma(-n+m+1)\Gamma(m+1)}$$

($\because$ the terms in $J_{-n}(x)$ are zero for $m \leqslant n-1$)

$$= \sum_{r=1}^{\infty} \frac{(-1)^{n+r} \left(\frac{x}{2}\right)^{n+2r}}{\Gamma(r+1)\Gamma(n+r+1)}, \text{ for } -n+m=r$$

$$= (-1)^n \sum_{r=0}^{\infty} \frac{(-1)^r \left(\frac{x}{2}\right)^{n+2r}}{\Gamma(n+r+1)\Gamma(r+1)} = (-1)^n J_n(x).$$

$\therefore J_{-n}(x) = (-1)^n J_n(x).$

13.9 Legendre's equation

The second order differential equation of the form:

$$(1-x^2)\frac{d^2y}{dx^2} - 2x\frac{dy}{dx} + n(n+1)y = 0 \qquad (13.9.1)$$

is known as the *Legendre's equation,* where n is a positive integer.

The normalised form of the eq. (13.9.1) is

$$\frac{d^2y}{dx^2} - \frac{2x}{1-x^2}\frac{dy}{dx} + \frac{n(n+1)}{1-x^2}y = 0.$$

Now, we would like to find the series solution of the equation near origin.

Comparing the above equation with (13.3.2), we obtain $P(x) = -\frac{2x}{1-x^2}$ and $Q(x) = \frac{n(n+1)}{1-x^2}$, both of which are analytic at $x = 0$. It follows that the origin is an ordinary point of the Legendre's eq. (13.9.1).

Let us consider a solution of the eq. (13.9.1) in the form

$$y = \sum_{r=0}^{\infty} a_r x^r, \; a_0 \neq 0. \qquad (13.9.2)$$

Then differentiating (13.9.2) twice in succession with respect to x, we obtain

$$\frac{dy}{dx} = \sum_{r=1}^{\infty} r a_r x^{r-1} \qquad (13.9.3)$$

$$\text{and } \frac{d^2y}{dx^2} = \sum_{r=2}^{\infty} r(r-1)a_r x^{r-2}. \qquad (13.9.4)$$

Using (13.9.2), (13.9.3) and (13.9.4) in the eq. (13.9.1), we have

$$(1-x^2)\sum_{r=2}^{\infty} r(r-1)a_r x^{r-2} - 2x\sum_{r=1}^{\infty} r a_r x^{r-1} + n(n+1)\sum_{r=0}^{\infty} a_r x^r = 0$$

$$\text{or, } \sum_{r=2}^{\infty} r(r-1)a_r x^{r-2} - \sum_{r=2}^{\infty} r(r-1)a_r x^r - 2\sum_{r=1}^{\infty} r a_r x^r + n(n+1)\sum_{r=0}^{\infty} a_r x^r = 0$$

$$\text{or, } \sum_{r=0}^{\infty}(r+2)(r+1)a_{r+2}x^r - \sum_{r=2}^{\infty} r(r-1)a_r x^r - \sum_{r=1}^{\infty} 2ra_r x^r + \sum_{r=0}^{\infty} n(n+1)a_r x^r = 0$$

$$\text{or, } [n(n+1)a_0 + 2a_2] + [n(n+1)a_1 - 2a_1 + 6a_3]x + \sum_{r=2}^{\infty}[(r+1)(r+2)a_{r+2} - r(r-1)a_r - 2ra_r + n(n+1)a_r]x^r = 0$$

$$\text{or, } [n(n+1)a_0 + 2a_2] + [(n-1)(n+2)a_1 + 6a_3]x + \sum_{r=2}^{\infty}[(r+1)(r+2)a_{r+2} + (n+r+1)(n-r)a_r]x^r = 0. \quad (13.9.5)$$

Here (13.9.5) represents an identity in x and it is valid in the open interval $(-R, R)$, where $R(> 0)$ is the radius of convergence of the power series (13.9.4).

Thus, the coefficients of each term in the left side of (13.9.5) must be zero.

$$\therefore n(n+1)a_0 + 2a_2 = 0, \text{ i.e. } a_2 = -\frac{n(n+1)}{2!}a_0,$$

$$(n-1)(n+2)a_1 + 6a_3 = 0, \text{ i.e. } a_3 = -\frac{(n-1)(n+2)}{3!}a_1$$

$$\text{and } (r+1)(r+2)a_{r+2} + (n+r+1)(n-r)a_r = 0, \text{ i.e. } a_{r+2} = -\frac{(n+r+1)(n-r)}{(r+1)(r+2)}a_r, \; r \geqslant 2. \quad (13.9.6)$$

Putting $r = 2, 3, 4, \ldots$ in (13.9.6), we obtain

$$a_4 = -\frac{(n-2)(n+3)}{3\cdot 4}a_2 = \frac{n(n-2)(n+1)(n+3)}{4!}a_0,$$

$$a_5 = -\frac{(n-3)(n+4)}{4\cdot 5}a_3 = \frac{(n-1)(n-3)(n+2)(n+4)}{5!}a_0,$$

and so on.

Now, substituting all the values of $a_2, a_3, a_4, \ldots$ in (13.9.4), we obtain

$$y = a_0\left[1 - \frac{n(n+1)}{2!}x^2 + \frac{n(n-2)(n+1)(n+3)}{4!}x^4 - \ldots\right] + a_1\left[x - \frac{(n-1)(n+2)}{3!}x^3 + \frac{(n-1)(n-3)(n+2)(n+4)}{5!}x^5 - \ldots\right],$$

which is the general solution of the Legendre's equation with two arbitrary constants a_0 and a_1.

Note. We can solve the Legendre's equation assuming the existence of the series solution in descending powers of x, that is $y = \sum_{r=0}^{\infty} a_r x^{k-r}$, $a_0 \neq 0$.

Then two solutions of the equation will be

$$v = a_0\left[x^{-n-1} + \frac{(n+1)(n+2)}{2\cdot(2n+3)}x^{-n-2} + \frac{(n+1)(n+2)(n+3)(n+4)}{2\cdot 4\cdot(2n+3)(2n+5)}x^{-n-5} + \ldots\right]$$

$$\text{and } u = a_0\cdot\left[x^n - \frac{n(n-1)}{2(2n-1)}x^{n-2} + \frac{n(n-1)(n-2)(n-3)}{2\cdot 4\cdot(2n-1)(2n-3)}x^{n-4} - \ldots\right].$$

For $n \in \mathbb{Z}^+$ and $a_0 = \frac{1\cdot 3\cdot 5\ldots(2n-1)}{n!}$, u becomes a terminating series which is called *Legendre's function* of the first kind denoted by $P_n(x)$. On the other hand, for $a_0 = \frac{n!}{1\cdot 3\cdot 5\ldots(2n+1)}$, v remains an infinite series which is called *Legendre's function* of the second kind denoted by $Q_n(x)$.

Here $P_n(x)$ and $Q_n(x)$ are two linearly independent solutions of the Legendre's equation. Thus, the general solution of the equation is given by $y = AP_n(x) + BQ_n(x)$, where c_1 and c_2 are two arbitrary constants.

Examples XIII(A)

1. Find the series solutions of the following differential equations near the point $x = 0$:

   (a) $\frac{d^2y}{dx^2} + x^2y = 0$

   (b) $x(x-1)\frac{d^2y}{dx^2} + 3x\frac{dy}{dx} + y = 0$

   (c) $(1-x^2)\frac{d^2y}{dx^2} - 4x\frac{dy}{dx} - 2y = 0$

   (d) $(x^2 - x)\frac{d^2y}{dx^2} + (3x+1)\frac{dy}{dx} + y = 0$

   (e) $2x^2\frac{d^2y}{dx^2} + x\frac{dy}{dx} - (x+1)y = 0$

2. Find the series solution of the differential equation $(x - x^2)y'' - (1 + 3x)y' + y = 0$ near the point $x = 0$.
3. Solve the equation $x\frac{d^2y}{dx^2} + \frac{dy}{dx} + xy = 0$ in powers of x.
4. Find two Frobenius solutions of the differential equation $x^2\frac{d^2y}{dx^2} + x(1+x)\frac{dy}{dx} + (x-9)y = 0$ near $x = 0$.
   Hence, find its general solution.
5. Find the series solution of the differential equation $(x^2 - 1)y'' + 2xy' - 6y = 0$ about the point at infinity.
6. Find two Frobenius solutions (if exist) of the equation $x(x^2+2)\frac{d^2y}{dx^2} - \frac{dy}{dx} - 6xy = 0$ near $x = 0$.

Answers

1. (a) $y = c_1\left(1 - \frac{1}{3\cdot4}x^4 + \frac{1}{3\cdot4\cdot7\cdot8}x^8 + \ldots\right) + c_2\left(x - \frac{1}{4\cdot5}x^5 + \ldots\right)$

   (b) $y = c_1(x + 2x^2 + 3x^3 + \ldots) + c_2[\ln x(x + 2x^2 + 3x^3 + \ldots) + (1 + x + x^2 + \ldots)]$

   (c) $y = c_1(1 + x^2 + x^4 + \ldots) + c_2(x + x^3 + x^5 + \ldots)$

   (d) $y = c_1(1\cdot 2x^2 + 2\cdot 3x^3 + 3\cdot 4x^4 + \ldots) + c_2[\ln x(1\cdot 2x^2 + 2\cdot 3x^3 + 3\cdot 4x^4 + \ldots) + (-1 + x + 3x^2 + 5x^3 + \ldots)]$

   (e) $y = c_1x\left(1 + \frac{x}{5} + \frac{x^2}{70} + \ldots\right) + c_2\frac{1}{\sqrt{x}}\left(1 - x - \frac{1}{2}x^2 + \ldots\right)$

2. $y = (c_1 + c_2\ln x)(1 - 1\cdot 2x^2 + 2\cdot 3x^3 + \ldots) + c_2(-1 + x + 5x^2 + 11x^3 + \ldots)$
3. $y = c_1u + c_2[u\ln x + v]$, where $u = 1 - \frac{x^2}{2^2} + \frac{x^4}{2^2.4^2} - \frac{x^6}{2^2.4^2.6^2} + \ldots$ and $v = \frac{1}{2^2}x^2 - \left(\frac{2}{2^3.4^2} + \frac{2}{2^2.4^3}\right)x^4 + \left(\frac{2}{2^3.4^2.6^2} + \frac{2}{2^2.4^2.6^2} + \frac{2}{2^2.4^2.6^3}\right)x^6 + \ldots$
4. $u = x^3\left(1 - \frac{4}{2\cdot7}x + \frac{4\cdot5}{2!(7\cdot8)}x^2 - \frac{4\cdot5\cdot6}{3!(7\cdot8\cdot9)}x^3 + \ldots\right)$,
   $v = x^{-3}\left(1 - \frac{2}{5}x + \frac{1}{20}x^2\right)$; $y = c_1u + c_2v$.
5. $y = c_1x^{-3}\left(1 + \frac{3\cdot4}{2\cdot7}x^{-2} + \frac{3\cdot5\cdot6}{2\cdot7\cdot9}x^{-4} + \ldots\right) + c_2x^{-2}\left(1 - \frac{1}{3}x^{-2}\right)$.
6. $u = 1 + 3x^2 + \frac{3}{5}x^4 + \ldots$, $v = x\sqrt{x}\left(1 + \frac{3}{8}x^2 - \frac{3}{128}x^4 + \ldots\right)$.

Chapter 14

Mathematical Modelling

From the very beginning of human civilization, people always tried to explore and understand our mortal world by observations. But it was not enough. In order to build up our scientific knowledge, we need to explore and to understand several changes in our surroundings by observations, collecting data, and finding mathematical rules among the data and finally we try to explore the truth behind it. In addition, we try to apply this knowledge to predict the future. The rules obtained from data play important role to make mathematical models. There are several types of mathematical rules. A large number of rules can be expressed by ordinary differential equations that describe relations between variables and their derivatives. In this chapter, our main aim is to introduce mathematical modelling in ordinary differential equations, how to do mathematical modelling, how to solve the corresponding differential equations, how to interpret the results and how to develop general theory for prediction about behaviour. Here, we would like to focus mainly on two important classes of mathematical models by using ordinary differential equations, where time and state are continuous. These two classes are 'population dynamics' and 'dynamics in classical mechanics'. In addition, models on radioactive decay, heat conduction, electric circuits, chemical process are also included in this chapter.

14.1 Mathematical modelling

The concept of dynamical systems has its origin in Newtonian mechanics, though many people think that the founder of dynamical systems is Henri Poincaré (29 April 1854–17 July 1912), a French mathematician. A dynamical system is a system in which a function describes the time dependence of a point in a geometrical space. Roughly speaking, a mathematical model is a description of a system using mathematical rules and language. On the other hand, mathematical modelling is a process by which a real problem in mortal world is described by a mathematical formulation. A mathematical model stating some rules that govern the changes of the state of system with time is called dynamic model. In order to predict the future value, one may follow the basic idea given below:

Future value = present value + change, i.e. change = future value – present value.

If the independent variable (time) varies continuously, the idea gives a differential or, a differential equation (first order) of the form:

$$dx = f(x, t)dt \text{ or, } \frac{dx}{dt} = f(x, t), \ x \text{ being the state variable.}$$

14.2 Models of population growth

Let x be a state variable corresponding to a dynamical system. Then the system can be represented by an ordinary differential equation $\frac{dx}{dt} = f(x, t)$, where t denotes time. Observe that it is nonautonomous system. For autonomous system, the differential equation becomes $\frac{dx}{dt} = f(x)$. Models of population growth are continuous-time models of population. So we need three biological and mathematical postulates given below:

(a) *The postulate of parenthood*: This principle states that every living organism comes from at least one parent of like kind to itself. It follows that living things can come only from other living things by means of reproduction.
(b) *The postulate of upper limit*: It is the maximum population size of a biological species occupying the finite space that contains food, water, habitant and other resources.
(c) *The postulate of continuity*: For mathematical reasons, we agree with the convention that the variation of a population size behaves as a continuous variable.

14.2.1 Population growth

The functional behaviour of growth rate of a population is generally measured in the laboratory. The dynamics of the growth of a given population can be described when the functional behavior of growth-rate is known to us. Let $x(t)$ denote the population size at any time t. Then by postulate (c), $x(t)$ is a smooth function and $x(t) \geq 0 \ \forall \ t \in [0, \infty)$. Therefore, the general model equation of a single-species population growth can be expressed as $\frac{dx}{dt} = f(x)$.

Expanding f in Taylor's series, we obtain $f(x) = a_0 + a_1 x + a_2 x^2 + \ldots$.

By postulate (a), we have $f(0) = 0$. Then $[\frac{dx}{dt}]_{x=0} = f(0) = 0 \Rightarrow a_0 = 0$.

$$\therefore \frac{dx}{dt} = a_1 x + a_2 x^2 + a_3 x^3 + \cdots = x(a_1 + a_2 x + a_3 x^2 + \ldots)$$

$$\text{or,} \ \frac{dx}{dt} = xg(x), \ \text{where} \ g(x) = a_1 + a_2 x + a_3 x^2 + \ldots$$

$$\text{or,} \ \frac{1}{x}\frac{dx}{dt} = g(x), \tag{14.2.1}$$

where $\frac{1}{x}\frac{dx}{dt}$ is called the growth rate per capita and the function $g(x) = a_1 + a_2 x + a_3 x^2 + \ldots$ is called the *intrinsic growth rate* of the population.

14.2.2 Malthus population model (unlimited growth)

A population model is a particular type of mathematical model, which is applied to the study of population dynamics. Malthus population model, named after Thomas Robert Malthus (14 February 1766–23 December 1834), an English economist, is the simplest model of population growth. It describes that the rate of change of population size is directly proportional to the population size. In order to establish it, let the organisms live in an unrestricted environment and let the development of all individuals do not depend on one another. Let us also consider that there is no competition between two individuals and let migration be not occurred in the system. Then, we can build a mathematical model based on the following hypothesis:

$$\frac{dx}{dt} = \text{births–deaths}.$$

Clearly, the rate of change of population size in a closed system can be compounded if the birth rate and death-rate are both known. Let b be the birth rate and d be the death rate per capita. Let the population change in a small time interval $(t, t + \Delta t)$. Then

$$x(t + \Delta t) - x(t) \approx (b - d)x(t)\Delta t \ \text{or,} \ \frac{x(t + \Delta t) - x(t)}{\Delta t} \approx (b - d)x(t).$$

$$\therefore \lim_{\Delta t \to 0} \frac{x(t + \Delta t) - x(t)}{\Delta t} = (b - d)x(t) \ \text{or,} \ \frac{dx}{dt} = (b - d)x(t).$$

Let $r = (b - d)$, the net growth rate. Then the equation becomes

$$\frac{dx}{dt} = rx. \tag{14.2.2}$$

It proves that the growth-rate is directly proportional to the population size.

The differential eq. (14.2.2) represents the Malthus model equation corresponding to the population growth. Observe that eq. (14.2.1) coincides with (14.2.2) if $g(x) = r$ holds.

Now by eq. (14.2.2), we have

$$\int \frac{dx}{x} = r \int dt + \ln k, \ k \text{ being an integrating constant}$$

$$\text{or, } \ln x = rt + \ln k \text{ or, } \ln\left(\frac{x}{k}\right) = rt \text{ or, } \frac{x}{k} = e^{rt} \text{ or, } x = ke^{rt}. \tag{14.2.3}$$

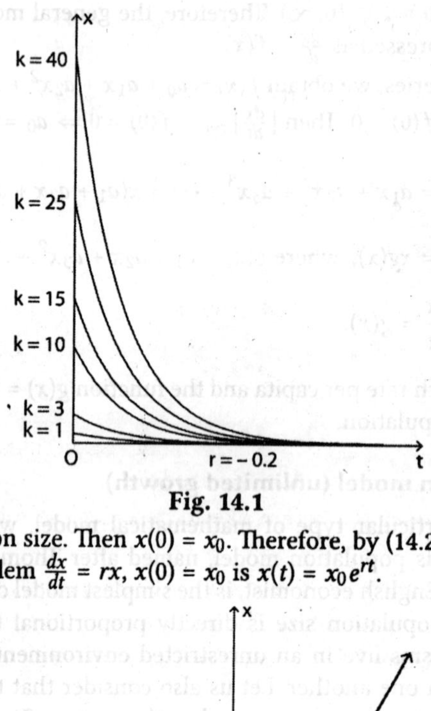

Fig. 14.1

Let x_0 be the initial population size. Then $x(0) = x_0$. Therefore, by (14.2.3), we have $k = x_0$. Thus, the solution of the initial-value problem $\frac{dx}{dt} = rx,\ x(0) = x_0$ is $x(t) = x_0 e^{rt}$.

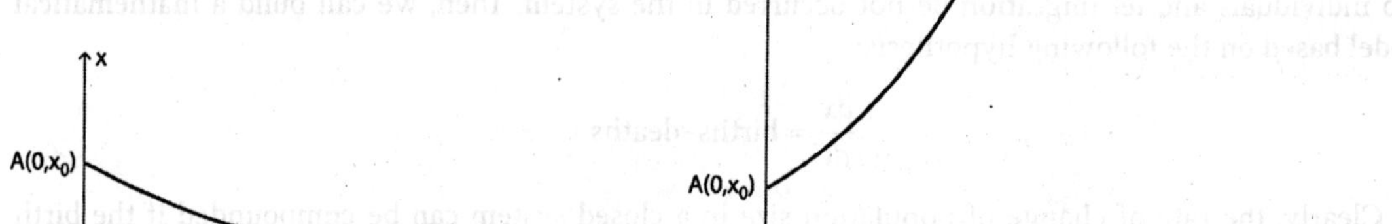

Fig. 14.2

It confirms that the growth of the population is exponential that cannot be realistic. Generally, growth rate tends to decrease as population size increases. For $b > d$ the model represents bacterial growth or the growth of any reproductive population, while for $b < d$ the model describes radioactive

decay. This model is the simplest population model, where exponential growth is considered for a very time.

If we take migration in account, we can build another mathematical population model based on the following hypothesis:

$$\frac{dx}{dt} = \text{births}-\text{deaths}+\text{migration}.$$

Example 14.2.1. *In a certain culture of bacteria, the rate of increase is proportional to the number present. If it is found that their number doubles in 4 hours, what should be their number at the end of 12 hours?*

Solution: Let the number of bacteria at the time t hours be x. According to Malthus population model $\frac{dx}{dt} = kx$, $k > 0$. Then

$$\int \frac{dx}{x} = k \int dt + c, \text{ } c \text{ being an integrating constant}$$

or, $\ln x = kt + c$.

Initially, when $t = 0$, let $x = A$. Then $c = \ln A$, i.e. $A = e^c$.

$\therefore$ $\ln x = kt + \ln A$, i.e. $x = Ae^{kt}$.

Next x becomes $2x$ when $t = 4$ and let it be y when $t = 12$.

$\therefore$ $2x = Ae^{4k}$ and $y = Ae^{12k}$.

$\therefore$ $2x = xe^{4k}$ or, $e^{4k} = 2$.

Then $\frac{y}{2x} = e^{8k} = 2^2 = 4$ or, $y = 8x$.

Hence, there will be an 8 times increase.

14.2.3 Logistic population model (limited growth)

Malthus population model is not realistic. In this purpose, we may consider the census of USA in the year 1800. We can find the growth rate r by fitting the data of the census. The data of the year 1800 yields that the growth rate $r = 0.03067$. It is observed that this value of r fits very well until 1820. But the data collected from each census after the year 1820, fails to fit and the discrepancy becomes larger and larger. This knowledge suggests us to improve the Malthus model.

Before going to Logistic model, we would like to consider the following two limitations in the previous model:

(a) The growth rate treated as a constant is valid only in short interval of time.

(b) The environmental limitation is not taken into account for making the model.

In Logistic population model, inter-species competition for limited food, habitant and other resources is considered.

Let $g(x) = a_1 + a_2x$.

Then eq. (14.2.1) becomes

$$\frac{1}{x}\frac{dx}{dt} = a_1 + a_2x, \text{ i.e. } \frac{1}{x}\frac{dx}{dt} = a_1 + a_2x = a_1\left(1 + \frac{a_2}{a_1}x\right). \tag{14.2.4}$$

Let $a_1 = r$ and $k = -\frac{a_1}{a_2}$. Then eq. (14.2.4) becomes

$$\frac{1}{x}\frac{dx}{dt} = r\left(1 - \frac{x}{k}\right), \text{ i.e. } \frac{dx}{dt} = rx\left(1 - \frac{x}{k}\right). \tag{14.2.5}$$

This equation is known as Logistic population model suggested by Pierre Francois Verhulst (28 October 1804–15 February 1849), a Belgian mathematician.

Observe that $\frac{dx}{dt} = 0 \Rightarrow x = 0, k$, a steady-state of the Logistic equation. Thus, eq. (14.2.5) has two equilibria, given by $x^* = 0$ and $x^* = k$ at which the growth rate for the population is zero.

Observe that $\frac{dx}{dt} > 0$ for $x < k$ and $\frac{dx}{dt} < 0$ for $x > k$. It follows that population will increase for $x < k$ and decrease for $x > k$. Here the quantity k is called the carrying capacity of the system.

Now by the eq. (14.2.5), we have

$$\frac{dx}{x(k-x)} = \frac{r}{k}dt.$$

$$\therefore -\int \frac{dx}{x(x-k)} = \frac{r}{k}\int dt - \frac{1}{k}\ln c, \text{ } c \text{ being an integrating constant}$$

$$\text{or, } -\int \frac{dx}{\left(x-\frac{k}{2}\right)^2 - \left(\frac{k}{2}\right)^2} = \frac{rt}{k} - \frac{1}{k}\ln c$$

$$\text{or, } -\frac{1}{k}\ln\left(\frac{x-k}{x}\right) = \frac{rt}{k} - \frac{1}{k}\ln c.$$

$$\therefore \ln\left(\frac{x-k}{x}\right) = -rt + \ln c \text{ or, } 1 - \frac{k}{x} = ce^{-rt} \text{ or, } \frac{k}{x} = 1 - ce^{-rt} \text{ or, } x = \frac{k}{1-ce^{-rt}}.$$

Initially, let $x(0) = x_0$. Then $x_0 = \frac{k}{1-c}$, i.e. $c = 1 - \frac{k}{x_0}$.

$$\therefore x(t) = \frac{kx_0}{x_0 + (k-x_0)e^{-rt}},$$

which is the solution of the Logistic equation. The curve corresponding to this function is called the *logistic curve* and the growth is referred to as *logistic growth*. We see that $x \to k$ as $t \to \infty$, i.e. population size approaches the carrying capacity as $t \to \infty$. Observe that the logistic model predicts rapid initial growth for $0 < x_0 < k$ and then a decrease in growth rate as time passes so that the population size approaches a limit. The value of r in the solution is called the intrinsic growth-rate.

Let $u = x - k$, where x is very closed to $x^* = k$.

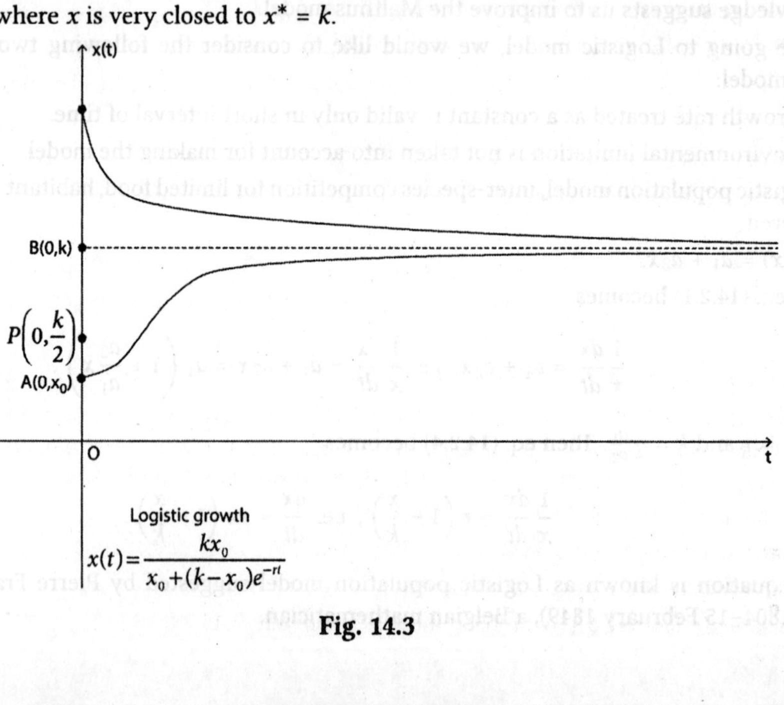

Fig. 14.3

Then eq. (14.2.5) becomes

$$\frac{du}{dt} = -ru - \frac{r}{k}u^2 \Rightarrow \frac{du}{dt} \approx -ru \text{ (since } u \text{ is very small).}$$

Observe that for $r > 0$, small perturbations about $x^* = k$ decay exponentially. Here the equilibrium $x^* = k$ is asymptotically stable and the solutions to the logistic equation are essentially a combination of exponential growth that is closed to zero and of exponential decay that is closed to k, the carrying capacity.

Note 14.2.1. *An equilibrium point x^* is said to be asymptotically stable if it is stable and if there exists $\delta > 0$ such that $\lim_{t\to\infty} |x(t) - x^*| = 0$ for all x_0 satisfying $|x_0 - x^*| \leqslant \delta$.*

14.2.4 A mathematical model of limited population growth with a threshold level

Let a given population of a species be able to keep increasing just if the initial population size is large enough; otherwise the species dies out. The level that allows increasing of the population size is called a threshold level.

In the population model with threshold level (T), the rate of increase of population is directly proportional to the initial population size (N) and the difference of the initial population size and the threshold level.

The model can be expressed as $\frac{dN}{dt} = kN(N - T)$.

It is an autonomous equation. Observe that $\frac{dN}{dt} > 0 \Rightarrow N > T$ and $\frac{dN}{dt} < 0 \Rightarrow N < T$. Thus, the population size decreases to zero if the initial population size $N_0 < T$ and it increases unboundedly if $N_0 > T$. For $N_0 = T$, the population size remains constant.

Example 14.2.2. *Let us consider a population of a species satisfying the following three properties:*

(i) *The population becomes extinct if the number of individuals falls below a minimum survival level m.*

(ii) *If the population size is above the minimum survival level, the growth is limited by the carrying capacity K.*

(iii) *If the population size is above the carrying capacity K, it will be declined due to the environment that cannot sustain such high population level.*

Find a mathematical model corresponding to the above situation.

Solution: Let $N(t)$ denote total number of individuals at any time t. In our model, there are two equilibrium points given by $N^* = m$ and $N^* = K$.

According to the given conditions, we have three conclusions given below:

(a) If $N_0 < m$ holds, N is to decrease towards zero. It confirms that $\frac{dN}{dt} < 0$, for $N_0 < m$.

(b) If $m < N_0 < K$ holds, N is to increase towards the carrying capacity K. It confirms that $\frac{dN}{dt} > 0$, for $m < N_0 < K$.

(c) If $N_0 > K$ holds, N is to decrease towards the carrying capacity K. It confirms that $\frac{dN}{dt} < 0$, for $N_0 > K$.

Therefore, the situation given above can be represented by the differential equation $\frac{dN}{dt} = r(K - N)(N - m)$, r being a positive constant. Here our mathematical model is $\frac{dN}{dt} = r(K - N)(N - m)$.

Example 14.2.3. *Let us consider an insect population of size N measured as biomass in kilograms. The population is increasing by 30 % per year. However, the population is also controlled by natural predator population that destroys 6 kg of insects per year.*

(a) *Find a mathematical model that describes the population size N at any time t.*

(b) *Sketch the graphs of the solutions and explain.*

(c) *Find the population size 4 years after if the initial biomass is 15 kg.*

Solution: (a) Here the population of insect is increasing by 30 % per year. It is destroyed by natural predation. Due to predation, the amount of destroyed biomass is 6 kg per year.

Then the rate of change of population = increasing rate – predation rate.

Thus, our mathematical model is $\frac{dN}{dt} = \frac{30N}{100} - 6$, i.e. $\frac{dN}{dt} = 0.3(N - 20)$.

Solution of the differential equation: $\frac{dN}{N-20} = 0.3dt$.

$$\therefore \int \frac{dN}{N-20} = 0.3\int dt + \ln c,\ \ln c \text{ being an integrating constant}$$

$$\text{or, } \ln(N - 20) = 0.3t + \ln c \text{ or, } N - 20 = ce^{0.3t} \text{ or, } N = 20 + ce^{0.3t}$$

Hence, the solution of the model is $N(t) = 20 + ce^{0.3t}$.

(b) Graphical representation:

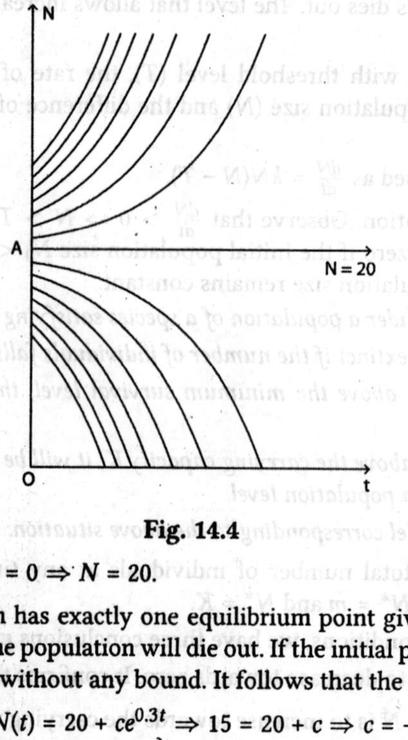

Fig. 14.4

Analysis: $\frac{dN}{dt} = 0 \Rightarrow 0.3(N - 20) = 0 \Rightarrow N = 20$.

Thus, the autonomous equation has exactly one equilibrium point given by $N^* = 20$. If the initial population size is less than 20 kg, the population will die out. If the initial population size is greater than 20 kg, the population will increase without any bound. It follows that the threshold level is 20 kg.

(c) Given that $N(0) = 15$. Then $N(t) = 20 + ce^{0.3t} \Rightarrow 15 = 20 + c \Rightarrow c = -5$.

$\therefore N(t) = 20 - 5e^{0.3t}$. Consequently, $N(4) = 20 - 5e^{0.3\times4} = 20 - 5.e^{1.2} \approx 3.4$.

Hence, the population size 4 years later is 3.4 kg.

14.2.5 Lotka-Volterra model

In this article, we would like to introduce a two-dimensional mathematical model corresponding to a prey-predator system. It is known as Lotka-Volterra model, named after an American mathematician Alfred James Lotka (2 March 1880–5 December 1949) and an Italian mathematician and physicist Vito Volterra (3 May 1860–11 October 1940). This model is represented by a system of two first-order nonlinear differential equations.

Let $x(t)$ and $y(t)$ be respectively the number of preys and predators at any time t in a system. Then Lotka-Volterra model equations are

$$\frac{dx}{dt} = \lambda x - bxy \tag{14.2.6}$$

$$\frac{dy}{dt} = -\mu y + cxy. \tag{14.2.7}$$

The first term on the right-hand side of the eq. (14.2.6) shows that the preys will grow exponentially in the absence of the predators and the second term that is known as the mass-action term, indicates the loss of preys due to predators. Loss of preys is directly proportional to both numbers of preys and predators. Again, the first term on the right-hand side of eq. (14.2.7) shows that the predator will decrease exponentially in the absence of the predators and the second term of this equation will increase the production of predators depending on the resulting number of preys in the system. The loss of preys is directly proportional to both numbers of preys and predators.

Now eliminating t from eqs (14.2.6) and (14.2.7), we obtain

$$\frac{dy}{dx} = \frac{y(cx-\mu)}{-x(by-\lambda)} \quad \text{or,} \quad \frac{by-\lambda}{y}dy + \frac{cx-\mu}{x}dx = 0.$$

$$\therefore \int\left(b - \frac{\lambda}{y}\right)dy + \int\left(c - \frac{\mu}{x}\right)dx = C, \ C \text{ being an integrating constant}$$

$$\text{or, } by + cx - \lambda \ln y - \mu \ln x = C.$$

Let $f(x, y) = by + cx - \lambda \ln y - \mu \ln x$. Then $f(x, y) = C$ represents the orbit of the prey-predator system.

Now $f_x(x, y) = c - \frac{\mu}{x}$ and $f_y(x, y) = b - \frac{\lambda}{y}$.

$$\therefore f_x(x, y) = 0 \Rightarrow c - \frac{\mu}{x} = 0 \Rightarrow x = \frac{\mu}{c} \text{ and } f_y(x, y) = 0 \Rightarrow b - \frac{\lambda}{y} = 0 \Rightarrow y = \frac{\lambda}{b}.$$

Thus, the equilibrium point of the system is $\left(\frac{\mu}{c}, \frac{\lambda}{b}\right)$ at which $f(x, y)$ will be minimum. The minimum value is given by $f(x^*, y^*) = f\left(\frac{\mu}{c}, \frac{\lambda}{b}\right) = \mu + \lambda - \mu \ln \frac{\mu}{c} - \lambda \ln \frac{\lambda}{b} = C_0$, say. Thus, the orbit of the system is given by $f(x, y) = C$ for $C \geq C_0$.

Let us consider the transformation $x = u + \frac{\mu}{c}$, $y = v + \frac{\lambda}{b}$, where (u, v) are the new coordinates.

Then $f(x, y) = C$ becomes

$$b\left(v + \frac{\lambda}{b}\right) + c\left(u + \frac{\mu}{c}\right) - \lambda \ln\left(v + \frac{\lambda}{b}\right) - \mu \ln\left(u + \frac{\mu}{c}\right) = C$$

$$\text{or, } b\left(v + \frac{\lambda}{b}\right) + c\left(u + \frac{\mu}{c}\right) - \lambda\left[\ln\frac{\lambda}{b} + \frac{bv}{\lambda} - \frac{b^2v^2}{\lambda^2} + \ldots\right] - \mu\left[\ln\frac{\mu}{c} + \frac{cu}{\mu} - \frac{c^2u^2}{\mu^2} + \ldots\right] = C$$

$$\text{or, } \frac{c^2u^2}{\mu^2} + \frac{b^2v^2}{\lambda^2} = C + \lambda \ln\frac{\lambda}{b} + \mu \ln\frac{\mu}{c} - \lambda - \mu$$

$$\text{or, } \frac{c^2u^2}{\mu^2} + \frac{b^2v^2}{\lambda^2} = C - C_0.$$

We see that the equation $\frac{c^2u^2}{\mu^2} + \frac{b^2v^2}{\lambda^2} = C - C_0$ represents a family of ellipses with centre at $\left(\frac{\mu}{c}, \frac{\lambda}{b}\right)$, when $C > C_0$ holds. Thus, the orbits are closed curves around the equilibrium point $\left(\frac{\mu}{c}, \frac{\lambda}{b}\right)$ for $C > C_0$. It confirms that the solution is periodic.

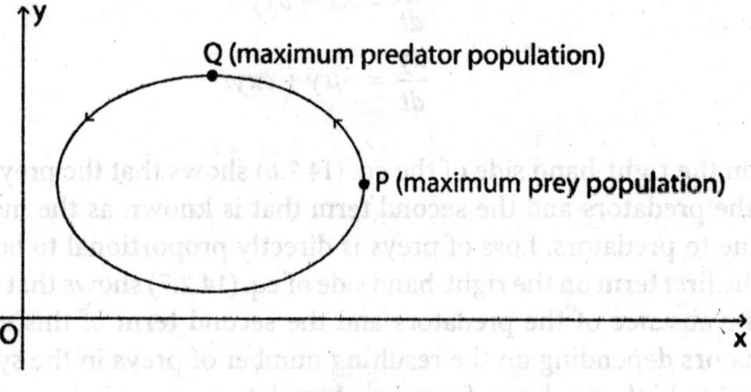

Fig. 14.5

Example 14.2.4. *Describe the orbit of the system* $\frac{dx}{dt} = x + y, \frac{dy}{dt} = x - y + 1.$

Solution: Let us consider y as a function of x.

Then $\frac{dx}{dt} = x + y, \frac{dy}{dt} = x - y + 1 \Rightarrow \frac{dy}{dx} = \frac{x-y+1}{x+y} \Rightarrow xdx - d(xy) - ydy + dx = 0.$

$$\therefore \int xdx - \int d(xy) - \int ydy + \int dx = \frac{c}{2},$$ c being an integrating constant

$$\text{or, } \frac{x^2}{2} - xy - \frac{y^2}{2} + x = \frac{c}{2} \text{ or, } x^2 - 2xy - y^2 + 2x = c,$$

which is the required orbit of the system.

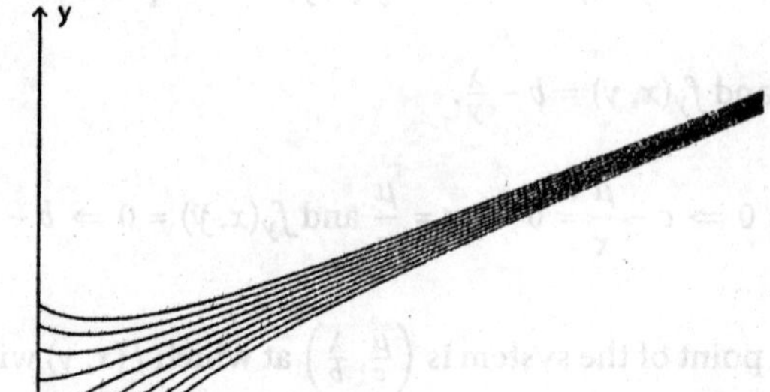

Fig. 14.6

14.2.6 Mutualism model

Mutualism population model is an important part of ecology. In our mortal world, we have a lot of examples of intersection of two species which are mutually beneficial. Mutualism in two species is a special type of interaction. In this case, species help one another to survive together in nature; each species extinct from the system without the assistance of other.

We can represent mutualism population model by the following pair of first-order nonlinear differential equations.

$$\frac{dx}{dt} = x(\lambda - ax + by), \quad \frac{dy}{dt} = y(\mu + cx - dy).$$

Here the mutualism of the interaction is modelled by the positive nature of the interaction terms, namely, cx and by.

Example 14.2.5. *Let us consider the following mutualism population model:* $\frac{dx}{dt} = x\left(-1 + \frac{y}{1+x}\right)$, $\frac{dy}{dt} = y\left(10 - y + \frac{x}{1+y}\right)$, $x \geq 0$, $y \geq 0$.

(a) *Sketch the curve* $\frac{y}{1+x} = 1$ *and* $10 - y + \frac{x}{1+y} = 0$ *in the xy-plane.*

(b) *Determine the equilibrium points.*

Solution:

$$\frac{dx}{dt} = 0 \Rightarrow x\left(-1 + \frac{y}{1+x}\right) = 0,$$

which gives $x = 0$, $y = x + 1$.

$$\frac{dy}{dt} = 0 \Rightarrow y\left(10 - y + \frac{x}{1+y}\right) = 0,$$

which gives $y = 0$, $x = (y+1)(y-10)$.

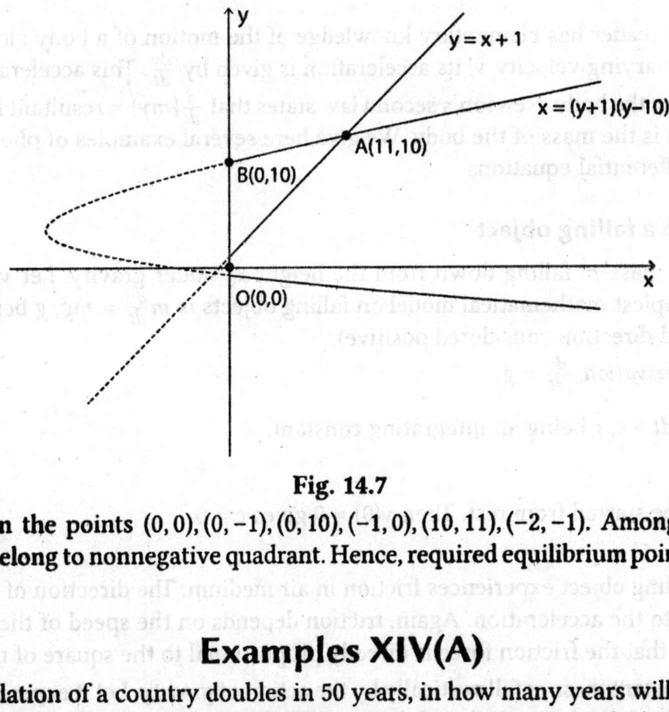

Fig. 14.7

Solving we obtain the points $(0,0), (0,-1), (0,10), (-1,0), (10,11), (-2,-1)$. Among these six points, $(0,0), (0,10), (10,11)$ belong to nonnegative quadrant. Hence, required equilibrium points are $(0,0), (0,10)$, $(10,11)$.

Examples XIV(A)

1. (a) If the population of a country doubles in 50 years, in how many years will it triple under the assumption that the rate of increase is proportional to the number of inhabitants?
   (b) Let a population have 10000 individuals on day zero and 22500 individuals at the end of 100 days. Find the population at the end of 150 days.
   (c) Let the growth rate of a given population be 0.21 per member per day. If the population size on day zero is 100, find the population size at the end of 100 days.
   (d) A population of bacteria develops with a constant relative growth rate of 0.7 per individual per day. If initial population size is 2, find the population size after six days.
2. Let us consider an insect population of size N measured as biomass in kilogram. The population is increasing by 50% per year. However, the population is also controlled by natural predator population that destroys 15 kg of insects per year.
   (a) Find a mathematical model that describes the population size N at any time t.

(b) Sketch the graphs of the solutions and explain.

(c) Find the population size 4 years later if the initial biomass is 28 kg.

3. Let a population satisfies a logistic model with $r = 0.4$. If the carrying capacity (k) is 100 and population has 5 members on day zero, find the population size at the end of 10th day.
4. The size of a population of Bengal fox in Bethuadahari forest is modelled by the differential equation $\frac{dN}{dt} = -kN\left(1 - \frac{N}{T}\right)$, where the threshold level $T = 100$ and $k = 0.8$ per year. Estimate the number of Bengal fox in the forest after a long period of time considering the initial size of the population as (a) 99, (b) 103.

Answers

1. (a) 79 years (b) 33750 (c) $100e^{21}$ (d) 133.
2. (a) $N(t) = 30 + ce^{0.5t}$ (c) 15.2 kg.
3. 74.
4. (a) 0 (b) infinity.

14.3 Dynamics in classical mechanics

We assume that the reader has elementary knowledge of the motion of a body along a straight line. If a body moves with varying velocity v, its acceleration is given by $\frac{dv}{dt}$. This acceleration is due to one or more forces acting on the body. Newton's second law states that $\frac{d}{dt}(mv)$ = resultant force in the direction of motion, where m is the mass of the body. We give here several examples of phenomena whose laws are expressed by differential equations.

14.3.1 Models on a falling object

(a) Let an object of mass m falling down from the height x_0 under gravity. Let $v(t)$ be its velocity at time t. Then the simplest mathematical model on falling objects is $m\frac{dv}{dt} = mg$, g being the gravitational constant (downward direction considered positive).

Mathematical derivation: $\frac{dv}{dt} = g$.

$\therefore \int dv = g \int dt + c$, c being an integrating constant.

$\therefore v(t) = gt + c$.

Let the motion be started from rest. Then $v(0) = 0$ gives $c = 0$.

$\therefore v(t) = gt$.

(b) Usually a falling object experiences friction in air medium. The direction of force due to friction should be opposite to the acceleration. Again, friction depends on the speed of the falling body. In our model, we consider that the friction force is directly proportional to the square of the velocity.

Let a body of constant mass m fall vertically by the action of gravity. Let it experience an atmospheric resistance kv^2, where its velocity is v and k is a constant. Then the downward force is mg and the atmospheric resistance upwards is kv^2. Thus, the mathematical model of the falling body in air medium is $\frac{d}{dt}(mv) = mg - kv^2$ (downward direction considered positive), i.e. $m\frac{dv}{dt} = mg - kv^2$.

Mathematical derivation: Separating the variables, we get

$$kdt = \frac{mdv}{mg/k} - v^2 \text{ or, } kdt = \frac{mdv}{a^2 - v^2},$$

where $a^2 = mg/k$.

$$k \int dt + c = m \int \frac{dv}{a^2 - v^2},$$

c being an integrating constant.

Let $a > v$. Then $c + kt = \frac{m}{a}\tanh^{-1}\left(\frac{v}{a}\right)$.

If we suppose the motion to have started from rest, i.e. $v(0) = 0$, c becomes zero and we have

$$kt = \frac{m}{a}\tanh^{-1}\left(\frac{v}{a}\right) \text{ or, } v = a\tanh\left(\frac{akt}{m}\right) \text{ or, } v = a\tanh bt,$$

where $b = ak/m = \sqrt{gk/m}$.

Note that $\tanh bt \approx 1$ as t increases and then $v \approx a = \sqrt{mg/k}$; that is called the terminal velocity.

The mass ceases to accelerate when the resistance just counterbalances the gravitational pull. This state gradually approaches as the velocity of descent increases. For this reason the arrival velocity of raindrops or hailstones is a little indication of the height from which they have fallen. Also, writing $v = \frac{ds}{dt} = a\tanh bt$, where s is the displacement at time t measured from the point of fall, we can easily obtain $s = \frac{a}{b}\ln(\cosh bt)$.

14.3.2 Model on resulting motion of a connected system

Let two particles A and B of masses m_1 and $m_2 (m_1 > m_2)$ respectively be connected by a light inextensible string passing over a light smooth fixed pulley C and let they be allowed to hang freely. Then the tension in the string is the same throughout its length. Let T be the tension of the string. Let the length of the strings from O, the highest point of the pulley to A and B be x_1, x_2 respectively at any instant. Then the accelerations of the particles A and B are respectively $\frac{d^2x_1}{dt^2}$ and $\frac{d^2x_2}{dt^2}$ downwards.

Thus, the equation of motion of the particle A is

$$m_1\frac{d^2x_1}{dt^2} = m_1g - T \tag{14.3.1}$$

and that of the particle B is

$$m_2\frac{d^2x_2}{dt^2} = m_2g - T. \tag{14.3.2}$$

Again, $x_1 + x_2 =$ constant (= the length of the string), which gives

$$\frac{d^2x_2}{dt^2} = -\frac{d^2x_1}{dt^2}. \tag{14.3.3}$$

Now by (14.3.1) and (14.3.2),

$$m_1\frac{d^2x_1}{dt^2} - m_2\frac{d^2x_2}{dt^2} = m_1g - T - m_2g + T = (m_1 - m_2)g$$

$$\text{or, } m_1\frac{d^2x_1}{dt^2} + m_2\frac{d^2x_1}{dt^2} = (m_1 - m_2)g, \text{ by (14.3.3)}$$

$$\text{or, } (m_1 + m_2)\frac{d^2x_1}{dt^2} = (m_1 - m_2)g$$

$$\text{i.e. } \frac{d^2x_1}{dt^2} = \frac{m_1 - m_2}{m_1 + m_2}g. \tag{14.3.4}$$

By (14.3.3) and (14.3.4)

$$-\frac{d^2x_2}{dt^2} = \frac{m_1 - m_2}{m_1 + m_2}g, \text{ i.e. } \frac{d^2x_2}{dt^2} = \frac{m_2 - m_1}{m_1 + m_2}g. \tag{14.3.5}$$

Here $m_1 > m_2$. Then $\frac{d^2x_1}{dt^2} = \frac{m_1-m_2}{m_1+m_2}g > 0$. It confirms that the particle A moves downwards. On the other hand, $\frac{d^2x_2}{dt^2} = \frac{m_2-m_1}{m_1+m_2}g < 0$. It follows that the particle B moves upwards with same acceleration as A.

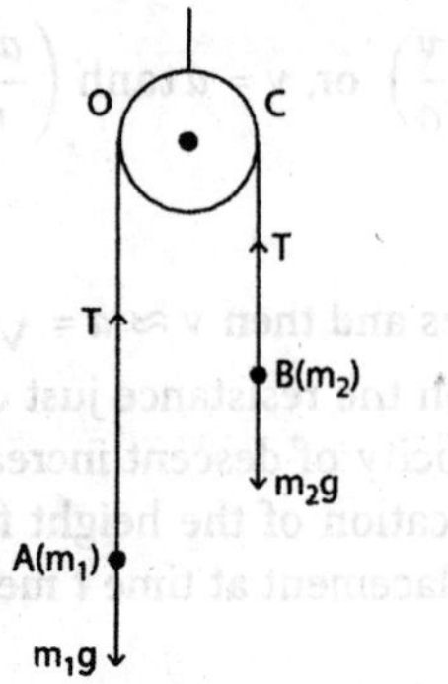

Fig. 14.8

14.3.3 Models on a particle moving along a straight line

A particle of constant mass m, free to move along the x-axis, is attracted towards the origin with a force proportional to its distance from the origin. Let it start from rest at $x = a$ and let it start with a velocity u at $x = a$ (moving away from the origin).

Let x denote the distance from the origin to the particle at time t. Then our model (equation of motion of the particle) is $m\frac{d^2x}{dt^2} = -kx$, k being a constant of proportionality.

Mathematical derivation

Let us write the model equation in the following form: $\frac{d^2x}{dt^2} = -n^2x$, where $k = mn^2$.

Multiplying both sides of the equation by $2\frac{dx}{dt}$, we obtain

$$2\frac{dx}{dt}\cdot\frac{d^2x}{dt^2} = -2n^2x\frac{dx}{dt} \text{ or, } \frac{d}{dt}\left\{\left(\frac{dx}{dt}\right)^2\right\} = -2n^2x\frac{dx}{dt}$$

Integrating, we get

$$\left(\frac{dx}{dt}\right)^2 = -n^2x^2 + c,$$

c being an integrating constant.

When $x = a$, $\frac{dx}{dt} = 0$. Thus, $c = n^2a^2$.

$$\therefore \left(\frac{dx}{dt}\right)^2 = n^2(a^2 - x^2) \text{ or, } \frac{dx}{\sqrt{a^2 - x^2}} = -ndt.$$

Here $\frac{dx}{dt}$ is negative since the particle is attracted towards the origin.

$$\therefore \int \frac{dx}{\sqrt{a^2 - x^2}} = -n\int dt + c_1,$$

c_1 being another integrating constant or, $\cos^{-1}\left(\frac{x}{a}\right) - nt + c_1 = 0$.

At $t = 0$, $x = a$. Then $c_1 = 0$.

$\therefore \cos^{-1}\left(\frac{x}{a}\right) = nt$, i.e. $x = a\cos nt$.

Observe that the motion is simple harmonic of amplitude a and period $\frac{2\pi}{n}$.

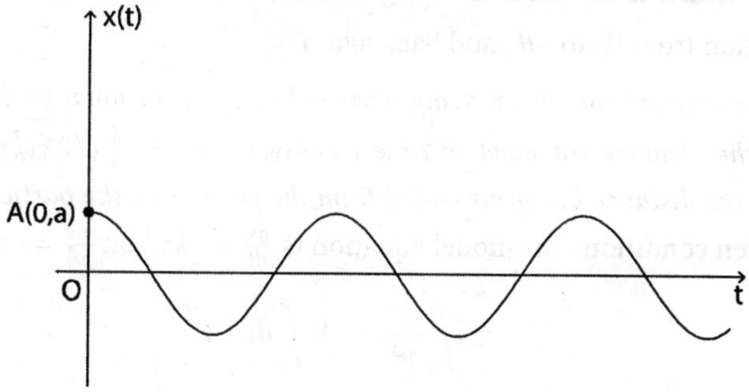

Fig. 14.9

Alternative method

Let us write the model equation in the following form $(D^2 + n^2)x = 0$, where $D \equiv \frac{d}{dt}$.

Let $x = e^{mt}$ be a trial solution of the equation. Then the auxiliary equation is $m^2 + n^2 = 0$.

$\therefore m = \pm ni$.

Thus, the general solution of the equation is $x = c_1 \cos nt + c_2 \sin nt$, where c_1, c_2 are two arbitrary constants.

Differentiating it once, we get $\frac{dx}{dt} = -nc_1 \sin nt + nc_2 \cos nt$.

Using the conditions $x = 0$, $t = 0$, $\frac{dx}{dt} = 0$, we obtain $c_1 = a$, $c_2 = 0$.

$\therefore x = a\cos nt$, which gives the displacement of the particle at time t.

14.3.4 Models on simple pendulum

For a simple pendulum of length l, we measure the position of the swinging arm by an angular coordinate θ. The angular component of acceleration is then $l\frac{d^2\theta}{dt^2}$ and the component of gravitational force in the same direction is $-mg\sin\theta$. Ignoring other forces, the equation of motion is $ml\frac{d^2\theta}{dt^2} = -mg\sin\theta$, i.e. $\frac{d^2\theta}{dt^2} = -\frac{g}{l}\sin\theta$.

This equation is nonlinear but it can be reduced to a linear equation.

Here we are considering small oscillations of the pendulum, i.e. motions where θ is small. Thus, we can make the approximation $\sin\theta \approx \theta$. Then the equation of motion becomes $\frac{d^2\theta}{dt^2} + \frac{g}{l}\theta = 0$, a linear first-order differential equation. Hence, our mathematical model corresponding to simple pendulum is $\frac{d^2\theta}{dt^2} + \frac{g}{l}\theta = 0$.

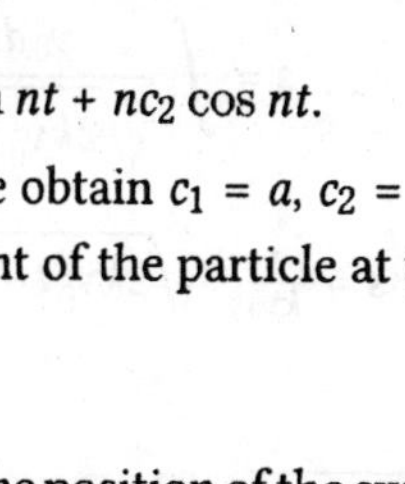

Fig. 14.10

Solving the model equation, we obtain $\theta = c_1 \cos\sqrt{\frac{gt}{l}} + c_2 \sin\sqrt{\frac{gt}{l}}$, where c_1, c_2 are two arbitrary constants.

Let $\theta = \theta_0$ and $\frac{d\theta}{dt} = 0$ when $t = 0$. Then $c_1 = \theta_0$, $c_2 = 0$ and hence $\theta = \theta_0 \cos\sqrt{\frac{gt}{l}}$, i.e. $t = \sqrt{\frac{l}{g}}\cos^{-1}\left(\frac{\theta}{\theta_0}\right)$, which is the time of swing from θ_0 to θ. If $\theta = -\theta_0$, $t = \pi\sqrt{\frac{l}{g}}$. Hence, the time of a complete oscillation from θ_0 to $-\theta_0$ and back again is $\frac{2\pi}{\sqrt{l/g}}$.

Example 14.3.1. *The acceleration of a moving particle being proportional to the cube of its velocity and negative, show that the distance travelled in time t is given by* $s = \left\{\sqrt{2kv_0^2 t + 1} - 1\right\}/kv_0$, *the initial velocity being* v_0 *and the distance being measured from the position of the particle at time* $t = 0$.

Solution: By the given condition, our model equation is $\frac{dv}{dt} = -kv^3$ or, $\frac{dv}{v^3} = -kdt$.

$$\therefore \int \frac{dv}{v^3} = -k\int dt + c,$$

c being an integrating constant,

or, $-\frac{1}{2v^2} = -kt + c$.

But $v = v_0$, when $t = 0$.

$\therefore\ c = -\frac{1}{2v_0^2}$.

$\therefore\ \frac{1}{2v^2} = kt + \frac{1}{2v_0^2}$.

On simplification, $v = \frac{v_0}{\sqrt{2kv_0^2 t+1}}$, i.e. $\frac{ds}{dt} = \frac{v_0}{\sqrt{2kv_0^2 t+1}}$.

$$\therefore\ s = \int \frac{v_0\,dt}{\sqrt{2kv_0^2 t + 1}} + c_1,$$

c_1 being an integrating constant, i.e. $s = \frac{1}{kv_0}\cdot\sqrt{2kv_0^2 t + 1} + c_1$.

Initially, $s = 0$. Then $c_1 = -\frac{1}{kv_0}$.

$$\therefore\ s = \frac{1}{kv_0}\sqrt{2kv_0^2 t + 1} - \frac{1}{kv_0}, \text{ i.e. } s = \left\{\sqrt{2kv_0^2 t + 1} - 1\right\}/kv_0.$$

Hence, the proof is complete.

14.3.5 Models on horizontal beams

Let E be the coefficient of elasticity of the material of a beam and let I be the moment of inertia of a cross section of the beam with respect to the neutral axis. Here our problem is that of determining the bending (or deflection) of a beam under given loadings. It is convenient to think of a beam as consisting of fibres running lengthwise. In a bent beam the fibres of the upper half are compressed and those of the lower half stretched, the two halves being separated by a neutral surface. The fibre which originally coincided with the horizontal axis of the beam lies (after the bending under given loadings) in the neutral surface along a curve, called the elastic curve. We consider a cross section of the beam at a distance x from one end. Let AB be its intersection with the neutral surface and $P(x, y)$ be its intersection with the elastic curve. It can be proved that the (bending) moment M with respect to AB of all external forces acting on either of the two segments into which the beam is separated by the cross section is given by $\frac{EI}{R} = M$, R being the radius of curvature of the elastic curve at P.

Since the slope $\frac{dy}{dx}$ of the elastic curve is small,

$$R = \frac{\left[1 + (dy/dx)^2\right]^{3/2}}{d^2y/dx^2} \approx \frac{1}{d^2y/dx^2}.$$

Hence, $EI\frac{d^2y}{dx^2} = M$.

Example 14.3.2. *A horizontal beam of length 2l is freely supported at both ends. Find the equation of its elastic curve and its maximum deflection when the load is ω per unit length.*

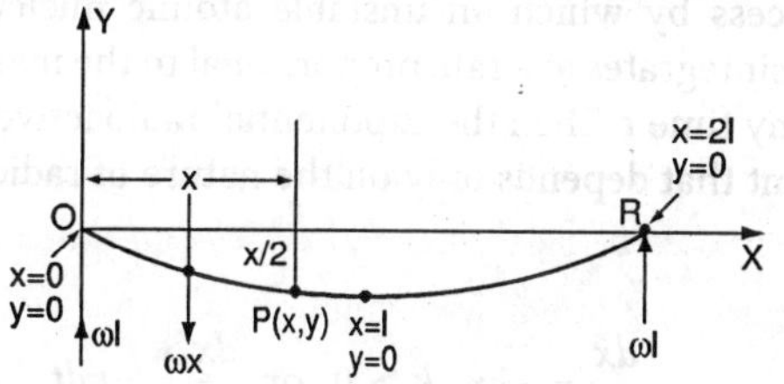

Fig. 14.11

Take the origin at the left end of the beam with the x-axis horizontal, y-axis vertical (see Fig. 14.11). Let $P(x, y)$ be a point on the elastic curve.

Consider the segment OP of the beam. There is an upward thrust ωl at O, x distance from P and the load ωx at the midpoint of OP, $\frac{1}{2}x$ from P. Then since $EI\frac{d^2y}{dx^2} = M$, we have

$$EI\frac{d^2y}{dx^2} = \omega lx - \omega x\left(\frac{x}{2}\right) = \omega lx - \frac{1}{2}\omega x^2.$$

Integrating twice, $Ely = \frac{1}{6}\omega lx^3 - \frac{1}{24}\omega x^4 + c_1x + c_2$, where c_1, $c_2$are two integrating constants.

At O, $x = 0$, $y = 0$, while at R, $x = 2l$, $y = 0$. Then $c_2 = 0$ and $c_1 = -\frac{1}{3}\omega l^3$.

The deflection of the beam at any distance x from O is given by $-y$. The maximum deflection occurs at the middle of the beam ($x = l$) and $-y_{\max} = -\frac{\omega}{24EI}(4l^4 - l^4 - 8l^4) = \frac{5\omega l^4}{24EI}$.

Examples XIV(B)

1. A mass m is projected vertically upward from the ground with initial velocity u. The atmospheric resistance is kv^2, when the velocity at time t is v and k is a constant. Prove that the time of ascent is $t = \frac{a}{g}\tan^{-1}\left(\frac{u}{a}\right)$, where $a = \sqrt{\frac{mg}{k}}$.
2. A body of mass m falls vertically from rest under gravity under an atmospheric resistance kv^2, when the velocity is v at time t. Assuming that the displacement x is measured from the point where it starts falling, prove that $\frac{2kx}{m} = \ln\left(\frac{a^2}{a^2-v^2}\right)\left[a = \sqrt{\frac{mg}{k}}\right.$.
3. A body falling from rest is subject to the force of gravity and air resistance proportional to the square of the velocity. If the resistance is equal to the weight of the body when the velocity is v at time t, show that the distance travelled by the body in t seconds is given by $\frac{b^2}{g}\ln\left(\cosh\left(\frac{gt}{b}\right)\right)$.
4. A moving body is opposed by a force proportional to the displacement and by a resistance proportional to the square of the velocity. Show that the velocity has the form given by $v^2 = ae^{-2bx} - \frac{cx}{b} + \frac{c}{2b^2}$.
5. A particle of mass m falls under gravity from rest in a medium which exerts a resistance mkv^2, where v is the velocity of the particle and k is a constant. Show that the velocity of the particle tends to a limit u. If the particle is projected vertically upwards in the same medium with velocity nu, show that it returns to the starting point with velocity $\frac{nu}{\sqrt{1+n^2}}$.

Answers

4. $v\frac{dv}{dx} + bv^2 + cx = 0$; make it linear in v^2.

14.4 Miscellaneous types of mathematical models

14.4.1 Models on radioactive decay

Radioactive decay is a process by which an unstable atomic nucleus loses energy by radiation. A substance such as radium disintegrates at a rate proportional to the mass remaining. Let x be the mass of a radioactive substance at any time t. Then the exponential radioactive decay model is $\frac{dx}{dt} = -kx, k > 0$, where k is the decay constant that depends only on the nature of radioactive substance.

Mathematical derivation:

$$\frac{dx}{dt} = -kx, \; k > 0, \text{ or, } \frac{dx}{x} = -kdt.$$

$$\therefore \int \frac{dx}{x} = -k \int dt + c_1,$$

c_1 being an integrating constant

or, $\ln x = -kt + c_1$.

Let $c_1 = \ln c$.

$$\therefore \ln x = -kt + \ln c, \text{ or, } \ln\left(\frac{x}{c}\right) = -kt \text{ or, } x = ce^{-kt},$$

which is the general solution of the exponential decay model.

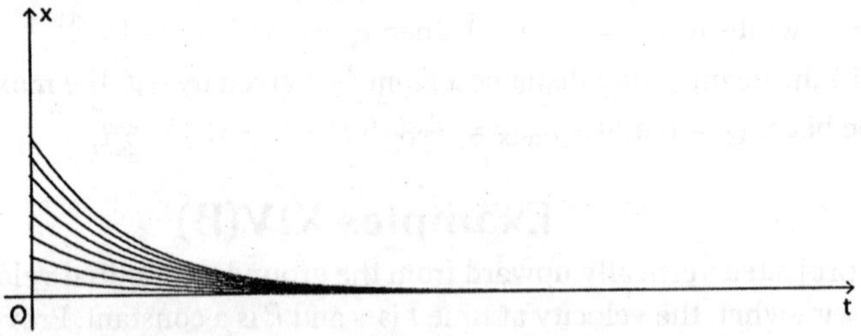

Fig. 14.12: $x = ce^{-0.2t}, c = \{1, 2, 3, 4, 5, 6, 7, 8\}$.

Let $x(0) = x_0$, initial amount of radioactive substance. Then $c = x_0$ and $x = x_0 e^{-kt}$. The amount that has been decayed in duration t is given by $x_t = x_0 - x = x_0(1 - e^{-kt})$.

Half-life: Half-life T_h is the time during which the total amount of active radioactive substance reduces to half of initial amount. Thus, at $t = T_h$, amount of active substance will be $\frac{x_0}{2}$.

Therefore, the decay equation $x = x_0 e^{-kt}$ gives $\frac{x_0}{2} = x_0 e^{-kT_h}$ or, $\frac{1}{2} = e^{-kT_h}$ or, $e^{kT_h} = 2$ or, $kT_h = \ln 2$, i.e. $T_h = \frac{\ln 2}{k} \approx \frac{0.7}{k}$.

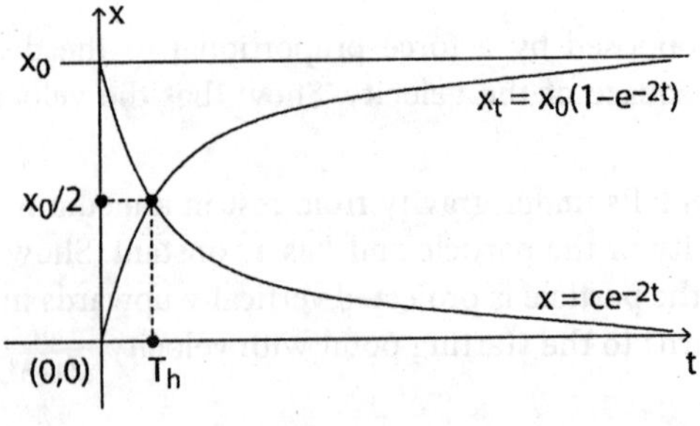

Fig. 14.13

If more of the substance is produced (e.g. in a nuclear reactor) at a rate $f(t)$, the total mass obeys the model equation $\frac{dx}{dt} + kx = f(t)$, a nonautonomous system.

Example 14.4.1. *Let k and 2k be respectively the decay constants of two radioactive substances. Initially they have equal number of active nuclei. Find the time to be the ratio e : 1 of active nuclei.*

Solution: Amount of active nuclei of given two radioactive substances are given by $x^{(1)} = x_0^{(1)} e^{-kt}$ and $x^{(2)} = x_0^{(2)} e^{-2kt}$.

$$\therefore \frac{x^{(1)}}{x^{(2)}} = \frac{x_0^{(1)} e^{-kt}}{x_0^{(2)} e^{-2kt}} \Rightarrow \frac{e}{1} = \frac{e^{-kt}}{e^{-2kt}} \left(\because x_0^{(1)} = x_0^{(2)}\right), \text{ i.e. } e^{kt} = e.$$

$\therefore kt = 1$, i.e. $t = k^{-1}$.

Hence, required time is k^{-1}.

Examples XIV(C)

1. The rate of decay of radium is proportional to the amount remaining. Prove that the amount at any time t is given by $x = x_0 e^{-kt}$, where $x(0) = x_0$.
2. Radium decomposes at a rate proportional to the amount present. If half the original amount disappears in 1600 years, find the percentage lost in 100 years.

Answers

2. 4.2 %.

14.4.2 Models on heat conduction

Newton's law of cooling states that the rate at which a substance cools in moving air is proportional to the difference between the temperature of the substance and that of the air. Let $x(t)$ be the temperature of the substance at time t and let α be the room temperature. Then the mathematical model on heat conduction is given by the following first-order linear differential equation $\frac{dx}{dt} = -k(x - \alpha)$, where k is conductivity coefficient that depends on object.

This mathematical model is called *Newton's law of cooling* or *heating*.

Observe that $\frac{dx}{dt} = 0 \Rightarrow x = \alpha$. Thus, the equilibrium point is given by $x^* = \alpha$.

Mathematical derivation:

$$\frac{dx}{x - \alpha} = -kdt.$$

$$\therefore \int \frac{dx}{x - \alpha} = -k \int dt + \ln c, \ \ln c \text{ being an integrating constant.}$$

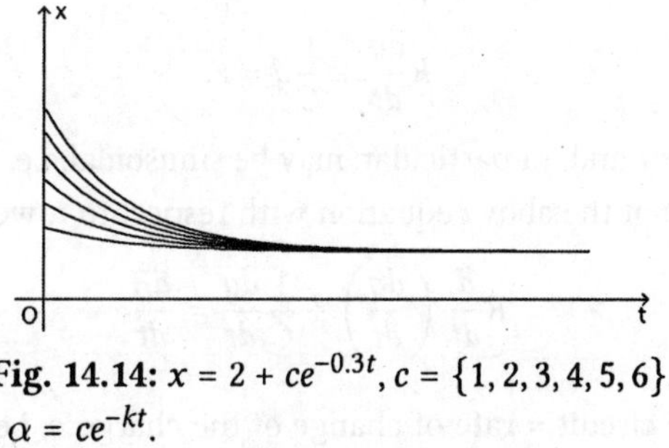

Fig. 14.14: $x = 2 + ce^{-0.3t}, c = \{1, 2, 3, 4, 5, 6\}$.

$\ln(x - \alpha) = -kt + \ln c$ or, $x - \alpha = ce^{-kt}$.

$\therefore x(t) = \alpha + ce^{-kt}$.

Let $x(0) = x_0$. Then $x_0 = \alpha + c$, i.e. $c = x_0 - \alpha$.

$\therefore\ x(t) = \alpha + (x_0 - \alpha)e^{-kt}$.

Observe that $x(t) = \alpha$, if $x(0) = \alpha$. Furthermore, $x(t) \to \alpha$ as $t \to \infty$ for any initial temperature x_0. Thus, the equilibrium is stable. It means that in any cooling or heating problem, $x(t)$ will eventually relax to the room temperature.

Examples XIV(D)

1. If the temperature of the air is 30 °C and the substance cools from 100 °C to 70 °C in 15 minutes, calculate the time on which the temperature will be 40 °C.
2. A thermometer which reads 75 °F indoors is taken outdoors. After five minutes it reads 65 °F; after another five minutes it reads 60 °F. Prove that the outdoor temperature is 55 °F.
3. *A* and *B* both order coffee and they receive their cups at the same time. *A* adds creams at once but does not drink his coffee until five minutes have elapsed. *B* waits five minutes, then adds cream and begins to drink. Who drinks the hotter coffee?
4. A dead body is found at 6.30 AM with temperature 18 °C. At 7.30 AM the body temperature is 16 °C. Suppose, the room temperature is 15 °C and the normal temperature of human body is about 37 °C. Estimate the time of death.
5. A dead body of temperature 82 °F is placed at time $t = 0$ in a place with room-temperature 55 °F. At the end of 10 minutes, the body has cooled down to a temperature of 60 °F.
   (a) What will be the temperature of the body at the end of 10 minutes?
   (b) When will the temperature of the body be 50 °F.

Answers

1. 52 minutes.
3. A.
4. 4.41 AM.

14.4.3 Models on electric circuits

One of the basic mathematical models in electrical circuits is

$$L\frac{di}{dt} + Ri = E(t), \tag{14.4.1}$$

where the circuit contains only an inductance *L*, a resistance *R* and an electromotive force (emf) *E*. It is actually a nonautonomous system.

If a circuit contains only a resistance *R*, and a conductance of capacitance *C* in series and an emf *R*, the charge *q* on the condenser varies with time *t* according to the linear equation given below:

$$R\frac{dq}{dt} + \frac{1}{C}q = E,$$

where *E* may depend on time *t* and, in particular, may be sinusoidal, i.e. $E = E_0 \sin \omega t$.

Differentiating both sides of the above equation with respect to *t*, we obtain

$$R\frac{d}{dt}\left(\frac{dq}{dt}\right) + \frac{1}{C}\frac{dq}{dt} = \frac{dE}{dt}$$

Since the current *i* in the circuit = rate of change of the charge *q*, i.e. $i = \frac{dq}{dt}$, the current obeys the model equation:

$$R\frac{di}{dt} + \frac{1}{C}i = \frac{dE}{dt}. \tag{14.4.2}$$

Example 14.4.2. *A series circuit consists of a battery of constant voltage E, a resistance R, and an inductance L. If the circuit, initially open, is closed at time t = 0, calculate the rise of the current.*

Solution: The current i obeys the eq. (14.4.1), i.e.

$$L\frac{di}{dt} + Ri = E\ (E = \text{constant})$$

and the variables are separable.

$$\frac{dt}{L} = \frac{di}{E - Ri}, \text{ whence } \frac{t}{L} = -\frac{1}{R}\ln(E - Ri) + \frac{1}{R}\ln C.$$

Initially, there is no current, so $t = 0 = i$ gives $C = E$.

$$\therefore\ \frac{t}{L} = \frac{1}{R}\ln\frac{E}{E - Ri}, \text{ or, } i = \frac{E}{R}\left\{1 - \exp\left(\frac{-Rt}{L}\right)\right\}.$$

As t becomes large i approaches a steady value E/R.

Simple electric circuits

We consider a circuit containing an inductance L (measured in henries), resistance R (ohms), capacitance C (farads), and a driving electromotive force E (volts) (see Fig. 14.15). The current i is measured in amperes, the charge q on the capacitor in coulombs and E, q and i are all considered to be functions of time t (seconds).

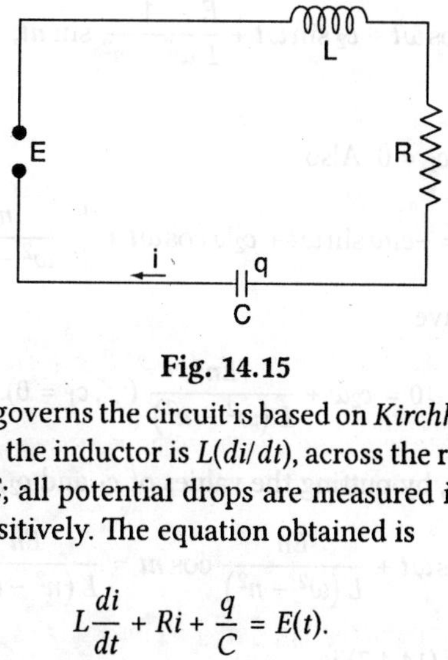

Fig. 14.15

The differential equation which governs the circuit is based on *Kirchhoff's Law: the total drop around a closed circuit is zero.* The drop across the inductor is $L(di/dt)$, across the resistor is Ri, across the capacitor is q/C, across the driving emf is $-E$; all potential drops are measured in volts, and with respect to the direction in which i is measured positively. The equation obtained is

$$L\frac{di}{dt} + Ri + \frac{q}{C} = E(t). \tag{14.4.3}$$

In addition, the current i is equal to the rate of change of q; $dq/dt = i$ and hence $di/dt = d^2q/dt^2$. Thus, the equation (14.4.3) becomes

$$L\frac{d^2q}{dt^2} + R\frac{dq}{dt} + \frac{q}{C} = E(t), \tag{14.4.4}$$

from which $q = q(t)$ can be found.

We can eliminate q by differentiating eq. (14.4.3) and using $i = dq/dt$:

$$L\frac{d^2 i}{dt^2} + R\frac{di}{dt} + \frac{i}{C} = \frac{dE}{dt}, \tag{14.4.5}$$

from which $i = i(t)$ can be found. The coefficients L, R, C are assumed to be positive constants, unless otherwise stated.

Suppose, in particular, $R = 0$ and $E(t) = E \sin nt$ (E, a constant). Then the differential equation becomes

$$L\frac{d^2 q}{dt^2} + \frac{q}{C} = E \sin nt. \tag{14.4.6}$$

Suppose, now the initial current and charge on the condenser are zero, to prove that the current

$$i = \frac{nE}{L\left(n^2 - \omega^2\right)}(\cos \omega t - \cos nt)$$

where $\omega^2 = \frac{1}{CL}, \omega \neq n$. What will happen if $n = \omega$?

Now the eq. (14.4.6) reduces to

$$\frac{d^2 q}{dt^2} + \frac{1}{CL}q = \frac{E}{L}\sin nt, \text{ or, } \frac{d^2 q}{dt^2} + \omega^2 q = \frac{E}{L}\sin nt. \tag{14.4.7}$$

The solution of this equation is

$$q = c_1 \cos \omega t + c_2 \sin \omega t + \frac{1}{D^2 + \omega^2}\frac{E}{L}\sin nt (D \equiv d/dt)$$

$$= c_1 \cos \omega t + c_2 \sin \omega t + \frac{E}{L}\frac{1}{\omega^2 - n^2}\sin nt,$$

where $n \neq \omega$.

Now, when $t = 0, q = 0$; then $c_1 = 0$. Also

$$i = \frac{dq}{dt} = -c_1\omega \sin \omega t + c_2\omega \cos \omega t + \frac{E}{L}\cdot\frac{n}{\omega^2 - n^2}\cos nt \tag{14.4.8}$$

and since $i = 0$ when $t = 0$, we have

$$0 = c_2\omega + \frac{En}{L\left(\omega^2 - n^2\right)} \quad (\because c_1 = 0).$$

$\therefore$ we finally obtain from (14.4.8), by putting the values of c_1 and c_2,

$$i = \frac{-En}{L\left(\omega^2 - n^2\right)}\cos \omega t + \frac{En}{L\left(\omega^2 - n^2\right)}\cos nt = \frac{En}{L\left(n^2 - \omega^2\right)}(\cos \omega t - \cos nt).$$

Now, if $n = \omega$, the solution of (14.4.7) is

$$q = c_1 \cos \omega t + c_2 \sin \omega t + \frac{1}{D^2 + \omega^2}\frac{E}{L}\sin \omega t \ (\because \ \omega = n)$$

$$= c_1 \cos \omega t + c_2 \sin \omega t + \frac{E}{L} \times \frac{-t\cos \omega t}{2\omega}$$

so that

$$i = -c_1\omega \sin \omega t + c_2\omega \cos \omega t - \frac{E}{2\omega L}(\cos \omega t - t\omega \sin \omega t). \tag{14.4.9}$$

Since $q = 0$, when $t = 0$, $c_1 = 0$.
Also $i = 0$, when $t = 0$, $c_2 = \frac{E}{2\omega^2 L}$.
$\therefore$ the eq. (14.4.9) becomes

$$i = \frac{E}{2\omega L}\cos\omega t - \frac{E}{2\omega L}(\cos\omega t - t\omega\sin\omega t).$$

Examples XIV(E)

1. If in equation (14.4.2), $E = E_0 \sin\omega t$, show that the current i at time t is given by

$$i = \frac{EC\omega}{1 + R^2C^2\omega^2}(\cos\omega t + RC\omega\sin\omega t) + C_1 e^{-t/RC}.$$

(assume that R, C, E_0, ω are constants; C_1 is the constant of integration).

2. In the equation $L\frac{di}{dt} + Ri = E$, find the current i at any time t, when
(a) $E = E_1\sin\omega t + E_2\sin(b\omega t + \theta)$, (b) $E = E_m\sin\omega t$.

Answers

2. (a) $i = ce^{-R_t/L} + \frac{E_1}{R^2+L^2\omega^2}(R\sin\omega t - L\omega\cos\omega t) + \frac{E_2}{R^2+L^2b^2\omega^2}\{R\sin(b\omega t + \theta) - Lb\omega\cos(b\omega t + \theta)\}$.
(b) $i = ce^{-(R/L)t} + \frac{E_m}{R^2+L^2\omega^2}\{R\sin\omega t - L\omega\cos\omega t\}$.

Examples XIV(F) (Miscellaneous)

1. Find the distance by a moving particle when its acceleration is directly proportional to its distance from a fixed point, the acceleration being directed
(a) towards the point from which the distance is measured;
(b) away from the point from which the distance is measured.
2. A particle moves in a straight line from rest at a distance a towards a centre of attraction, the attraction varying inversely as the cube of the distance. Find the whole time of motion.
3. A particle moves in a straight line under the action of an attraction varying inversely as 3/2th power of the distance. Show that the velocity acquired by falling from an infinite distance to a distance a from the centre is equal to the velocity which would be acquired in moving from rest at a distance a to a distance $a/4$.
4. Find the distance travelled by a particle when the acceleration is inversely proportional to the square of the distance from a fixed point.
5. Find the distance traversed by a body falling from rest, assuming that the resistance of the air is proportional to the square of the velocity.
6. An e.m.f. $E\sin pt$ is applied at $t = 0$ to a circuit containing a condenser C and inductance L in series. The current x satisfies the equation

$$L\frac{dx}{dt} + \frac{1}{C}\int x\,dt = E\sin pt (x = -dq/dt).$$

If $p^2 = 1/LC$ and initially the current x and the charge q are zero, show that the current in the circuit at time t is given by $\frac{E}{2L}t\sin pt$.
7. A simple pendulum, originally hanging in equilibrium, is disturbed by a force varying harmonically. The differential equation is $\frac{d^2x}{dt^2} + \omega_0^2 x = F_0\sin nt$.
Determine the motion, if the initial conditions are $t = 0$, $x = 0$, $dx/dt = 0$. What happens, if $\omega_0 = n$?

8. A simple pendulum of length l is oscillating through a small angle θ in a medium in which the resistance is proportional to the velocity. Show that the differential equation of its motion is

$$\frac{d^2\theta}{dt^2} + 2k\frac{d\theta}{dt} + \frac{g}{l}\theta = 0.$$

Discuss the motion and show that the period is $2\pi/\sqrt{\omega^2 - k^2}$, where $\omega^2 = g/l$.

9. The charge q of a condenser, capacity C, discharged in a circuit of resistance R and self-inductance L, satisfies the equation: $L\frac{d^2q}{dt^2} + R\frac{dq}{dt} + \frac{q}{C} = 0$.
Solve this equation, given that $q = q_0$ and $dq/dt = 0$ when $t = 0$, and that $CR^2 < 4L$.

10. The equation $\frac{d^2x}{dt^2} + n^2x = f \cos pt$ represents the motion of a point moving in a straight line with an acceleration n^2x towards a fixed centre in the straight line and with an additional acceleration $f \cos pt$. (*Harmonic motion disturbed by a periodic force varying as* $\cos pt$).
Solve this equation. Give physical interpretations of each term of the solution. What happens, if $p = n$?

11. (a) The differential equation $\frac{d^2x}{dt^2} + 2x\frac{dx}{dt} + \omega^2x = a\cos pt$ refers to a vibrating system damped by a force proportional to the velocity and acted upon by an external periodic force. Solve this equation and then state what each term in the solution represents. If $t \to \infty$, show that the solution takes the form:

$$x = \frac{a\cos(pt - \varepsilon)}{\sqrt{(\omega^2 - p^2)^2 + 4k^2p^2}},$$

where $\tan\varepsilon = \frac{2kp}{\omega^2 - p^2}$.

(b) In the present problem, prove that if p is a variable and k, ω and a are constants, then a/r is greatest, but when $p = \sqrt{(\omega^2 - 2k^2)} = \omega$, approx., if k is very small, and then $\varepsilon = \pi/2$ approx.
[This is the phenomenon of *Resonance* which has applications in Sound and Wireless Telegraphy.]

12. Show that the Particular Integral of $\frac{d^2y}{dt^2} + 2h\frac{dy}{dt} + (h^2 + p^2)\,y = k\,e^{-ht}\cos pt$ represents an oscillation with a variable amplitude $(k/2p)\,e^{-ht}$. Find the maximum value of this amplitude, and show that it is very large, if h is very small. What is the value of the amplitude, when $t \to \infty$?

13. Discuss the equation of forced vibrations $\frac{d^2x}{dt^2} + \mu x = F\sin pt$, and interpret the solution for values of $p >, =, < \mu$.

14. The free oscillations of a galvanometer needle, as affected by the viscosity of the surrounding air, which varies directly as the angular velocity of the needle, are determined by the equation

$$\frac{d^2\theta}{dt^2} + k\frac{d\theta}{dt} + \mu\theta = 0,$$

where k is the coefficient of viscosity and θ is the angular deflection of the needle at time t. Obtain θ in terms of t, considering the different possibilities.

15. A body executes damped forced vibrations given by the equation $\frac{d^2x}{dt^2} + 2k\frac{dx}{dt} + b^2x = e^{-kt}\sin\omega t$, obtain x in terms of t, for the two cases $\omega^2 \neq b^2 - k^2$ and $\omega^2 = b^2 - k^2$.

16. A particle of unit mass is attracted towards a fixed point by a force F. If (r, θ) are the polar co-ordinates of the particle at any instant and if $u = \frac{1}{r}$ the motion of the particle is given by

$$\frac{d^2u}{d\theta^2} + u = \frac{F}{h^2u^2},$$

where h is constant. Given $\theta = 0$, $du/d\theta = 0$, when $u = 1/a$, prove that the orbit is a rectangular hyperbola $r^2 \cos 2\theta = a^2$ if $F = -h^2/(a^4 u)$.

17. A particle of mass m is projected vertically upwards with an initial velocity v_0. The resisting force per unit mass at any time is k times the velocity. Formulate the differential equation of motion and show that the distance s covered by the particle at any time t is given by

$$s = \left(\frac{g}{k^2} + \frac{v_0}{u}\right)\left(1 - e^{-kt}\right) - \frac{g}{k}t.$$

18. Find the equation of the elastic curve for a cantilever beam of uniform cross section and length l, with a load P at the free end, the differential equation being $EI\frac{d^2y}{dx^2} = -Px$.
[Boundary conditions: $dy/dx = 0$ when $x = l$, and $y = 0$ when $x = 0$]

19. Find the elastic curve when the load is uniformly distributed over the beam described in the previous example, say ω per unit of length, the differential equation being $EI\frac{d^2y}{dx^2} = -\frac{\omega x^2}{2}$.
Also find the curve if further a horizontal tensile force Q is applied at the free end, the differential equation being $EI\frac{d^2y}{dx^2} = Qy - \frac{1}{2}\omega x^2$.

20. A horizontal tie rod of length $2l$ with concentrated load W at the centre, with ends freely hinged, satisfies the differential equation $EI\frac{d^2y}{dx^2} = Py - \frac{W}{2}x$.
With conditions $x = 0$, $y = 0$ and $x = l$, $dy/dx = 0$, prove that the deflection δ and the bending moment M at the centre $x = l$ are given by

$$\left.\begin{aligned} \delta &= \frac{W}{2Pn}(nl - \tanh nl) \\ M &= -\frac{W}{2n}\tanh nl. \end{aligned}\right\} \left(n^2 = \frac{P}{EI}\right).$$

21. If a uniform beam of length l is supported at its ends and bends under its own weight and end-thrusts P, the supports being at the same horizontal level, the equation of the bending moment is given by $EI\frac{d^2y}{dx^2} + Py = \frac{\omega}{2}\left(x^2 - lx\right)$.
Prove that the sag at the centre of the beam is $\frac{\omega}{Pn^2}\left(\sec\frac{nl}{2} - 1 - \frac{1}{8}n^2l^2\right)$ and the bending moment at the centre is $\frac{EI\omega}{P}\left(\sec\frac{nl}{2} - 1\right)$, where $n^2 = \frac{P}{EI}$.

Answers

1. (a) $s = a\sin(kt + b)$
(b) $s = c_1e^{kt} + c_2e^{-kt}$, where acceleration $= k^2$ (distance from the fixed point) or, $s = c_1\sin(kt + c_2)$, where acceleration $= -k^2$ (distance from the fixed point).

2. $\frac{a^2}{\sqrt{\mu}}$, if acceleration $= -\frac{\mu}{x^3}$.

4. $t = c_1 \pm \frac{\sqrt{c}}{\sqrt{2k}}\left\{\sqrt{s^2 - cs} + c\ln\left(\sqrt{s} + \sqrt{s - c}\right)\right\}$, according as the acceleration is $\frac{k^2}{s^2}$ or $-\frac{k^2}{s^2}$.

5. $s = (g/n^2)\ln(\cosh nt)$, if the resistance of the air is n^2/g times the square of the velocity.

7. $x = \frac{F_0}{\omega_0(\omega_0^2 - n^2)}[\omega_0 \sin\omega_0 t - n\sin\omega_0 t]$, when $\omega_0 \neq n$;
$x = \frac{F_0}{2n^2}[\sin nt - nt\cos nt]$, when $\omega_0 = n$.

9. $q = q_0\exp\left(-\frac{Rt}{2L}\right)\left(\cos\alpha t + \frac{R}{2\alpha L}\sin\alpha t\right)$, where $\alpha = \frac{\sqrt{C(4L - CR^2)}}{2CL}$.

10. Solution: $x = c_1 \cos(nt + c_2) + f \cdot \frac{1}{n^2 - p^2} \cos pt$, if $p \neq n$. The motion is compounded of the SHM, where periods are $2\pi/n$ and $2\pi/p$. The two terms occurring in the solution represent free and forced oscillations respectively. If $p = n$, $x = c_1 \cos(nt + c_2) + \frac{f}{2n} t \sin nt$, whence $\frac{dx}{dt} = -c_1 n \sin(nt + c_2) + \frac{f}{2n}(\sin nt + nt \cos nt)$. It shows that amplitude of the motion and also the velocity become very large when t is large.

14. For $k^2 = 4\mu, \theta = (A + Bt) \exp\left(-\frac{kt}{2}\right)$.

For $k^2 > 4\mu, \theta = A \exp\left[\left(\alpha - \frac{k}{2}\right) t\right] + B \exp\left[-\left(\alpha + \frac{k}{2}\right) t\right]$, where $\alpha = \frac{\sqrt{k^2 - 4\mu}}{2}$.

For $k^2 < 4\mu, \theta = \exp\left(-\frac{kt}{2}\right) [A \cos(\beta t) + B \sin(\beta t)]$, where $\beta = \frac{\sqrt{4\mu - k^2}}{2}$.

15. (a) $x = c_1 e^{-kt} \cos(nt + c_2) - \frac{t}{2\omega} e^{-kt} \cos \omega t$, where $n^2 = \omega^2 = b^2 - k^2$.

(b) $x = c_1 e^{-kt} \cos(nt + c_2) + \frac{1}{n^2 - \omega^2} e^{-kt} \sin \omega t$.

18. $6Ely = P(3l^2 x - x^2)$.

19. $24Ely = \omega(4l^3 x - x^4)$; $y = \frac{\omega EI}{Q^2}\left(1 - \cosh\sqrt{\frac{Q}{EI}}x\right) + \frac{\omega x^2}{2Q}$.

Part II

Partial Differential Equations

Chapter 1

Partial Differential Equations

Partial differential equations are mathematical equations (PDEs) *that involve at least two independent variables, an unknown function and partial derivatives of the unknown function with respect to the independent variables. Thus, ordinary differential equations are special class of partial differential equations corresponding to functions of a single variable. Partial differential equations are used to formulate mathematically several physical problems involving functions of several variables.*

The study of partial differential equations started way back in the 18th century in the work of Leonhard Euler (15 April 1707–18 September 1783), Joseph-Louis Lagrange (25 January 1736–10 April 1813), and Pierre-Simon Laplace (23 March 1749 –5 March 1827) as an important mathematical tool in mechanics. It is the foundational field of the modern scientific understanding of heat flow, sound, thermodynamics, fluid dynamics, elasticity, quantum mechanics, general theory of relativity, etc. Most of the physical phenomena can be described in general by partial differential equations. In part II of this Book, we would like to introduce several types of PDEs *with two independent variables, their usefulness, classification, solutions and applications. The main purpose of the first chapter is to define* PDE, *its order, degree, and linearity, construction of a* PDE *by elimination of arbitrary constants and that of a* PDE *by elimination of arbitrary functions.*

1.1 Definitions

Partial Differential Equations (PDE) are those equations which contain one or more partial derivatives and hence they must involve at least two independent variables.

The *order* of a PDE is the order of the derivative of the highest order in the equation.

The *degree* of a PDE is the greatest exponent of the highest order derivative involved in the equation.

[In what follows we shall (in almost all cases) take z as the dependent variable and x, y as independent variables. Further, the following symbols are universally used:

$$\frac{\partial z}{\partial x} = p,\ \frac{\partial z}{\partial y} = q,\ \frac{\partial^2 z}{\partial x^2} = r,\ \frac{\partial^2 z}{\partial x \partial y} = s,\ \frac{\partial^2 z}{\partial y^2} = t]$$

(i) $xp + yq = z$, i.e. $x\frac{\partial z}{\partial x} + y\frac{\partial z}{\partial y} = z$

is an example of a Partial Differential Equation of *first order* and of *first degree.*

(ii) $\frac{\partial^2 z}{\partial x^2} + 3\frac{\partial^2 z}{\partial x \partial y} + \frac{\partial^2 z}{\partial y^2} = 0$ or, $r + 3s + t = 0$

is an example of a Partial Differential Equation (PDE) of *second order* and of *first degree.*

First-order partial differential equations are classified into four categories, namely, linear, semi-linear, quasi-linear and nonlinear.

(a) **Liner equation:** A first-order partial differential equation $f(p, q, x, y, z) = 0$ is said to be *linear equation* if it is of the form $P(x, y)p + Q(x, y)q = R(x, y)z + S(x, y)$.

For example, let us consider the equation $(x + 2y)p - x(x - y)q = 2z(x + y) + xy$. It is of the form $P(x, y)p + Q(x, y)q = R(x, y)z + S(x, y)$, where $P(x, y) = (x + 2y)$, $Q(x, y) = x(x - y)$, $R(x, y) = 2z(x + y)$ and $S(x, y) = xy$. So it is an example of linear equation. Observe that every linear equation is linear in p, q and z.

(b) **Semi-linear equation:** A first-order partial differential equation $f(p, q, x, y, z) = 0$ is said to be *semi-linear equation* if it can be expressed in the form $P(x, y)p + Q(x, y)q = R(x, y, z)$.

For example, let us consider the equation $xp - yq = 2z^3(x - y)$. Observe that it is of the form $P(x, y)p + Q(x, y)q = R(x, y, z)$, where $P(x, y) = x$, $Q(x, y) = -y$ and $R(x, y) = 2z^3(x - y)$. So it is an example of semi-linear equation. A semi-linear equation is always linear in p and q and the coefficients of p and q are functions of x and y only.

(c) **Quasi-linear equation:** A first-order partial differential equation $f(p, q, x, y, z) = 0$ is said to be *Quasi-linear equation* if it is of the form: $P(x, y, z)p + Q(x, y, z)q = R(x, y, z)$.

For example, let us consider the equation $2xyzp - (x^2 + y^2 - z^2)q = z^2(x + y)$. Observe that it is of the form $P(x, y, z)p + Q(x, y, z)q = R(x, y, z)$, where $P(x, y) = 2xyz$, $Q(x, y) = -(x^2 + y^2 - z^2)$ and $R(x, y) = z^2(x + y)$. Therefore, the equation is a quasi-linear equation. A quasi-linear equation is linear in p and q and the coefficients of p and q are functions of x, y and z.

A quasi-linear equation is often called Lagrange's linear equation. Thus, we can say that the linear and semi-linear equations are two particular cases of Lagrange's linear equations.

(d) **Nonlinear equation:** A first-order partial differential equation$f(p, q, x, y, z) = 0$ is said to be nonlinear equation if it does not belong to the above three categories.

An example of nonlinear equation is$xp + yq = pq$.

A few more examples of PDE

(iii) $xp + yq = z \rightarrow$ PDE of *first order, linear.*

(iv) $r + t = 0 \rightarrow$ PDE of *second order, linear.*

(v) $r + 5s + t = 0 \rightarrow$ PDE of *second order, linear.*

(vi) $3px + qy = z^2 \rightarrow$ PDE of *first order, linear.*

We acquaint our readers with some important PDE used in Theoretical Physics:

(vii) **Laplace's Equation:** $\frac{\partial^2 w}{\partial x^2} + \frac{\partial^2 w}{\partial y^2} + \frac{\partial^2 w}{\partial z^2} = 0$, written as $\nabla^2 w = 0$, where ∇^2 (Laplacian) is an operator. $\frac{\partial^2}{\partial x^2} + \frac{\partial^2}{\partial y^2} + \frac{\partial^2}{\partial z^2}$ is an example of a PDE of *second order, linear.*

(viii) **Heat Equation:** $\nabla^2 w = \frac{1}{\sigma}\frac{\partial w}{\partial t}$ is an example of a PDE of *second order, linear.*

(ix) **Wave Equation:** $\nabla^2 w = \frac{1}{a^2}\frac{\partial^2 w}{\partial t^2}$ (second order, linear).

[In (viii), (ix), w is the dependent variable and x, y, z, t are four independent variables; t is the time variable, (x, y, z) are the rectangular Cartesian coordinates of a point in space.]

1.2 Construction of partial differential equations by the process of elimination of arbitrary constants

Let $\phi(x, y, z, a, b) = 0$ be a relation between three variables x, y, z and two arbitrary constants a, b. As usual, z is the dependent variable and x, y two independent variables. In order to eliminate a, b we require two other equations besides the given relation $\phi(x, y, z, a, b) = 0$.

Differentiating the given relation $\phi = 0$ with respect to x, we obtain

$$\phi_x + \phi_z \frac{\partial z}{\partial x} = 0. \tag{1.2.1}$$

Differentiating the given relation $\phi = 0$ with respect to y, we obtain

$$\phi_y + \phi_z \frac{\partial z}{\partial y} = 0. \tag{1.2.2}$$

Using the given relation $\phi = 0$, (1.2.1), (1.2.2) we can eliminate a, b and obtain $F(x, y, z, p, q) = 0$, where $p = \frac{\partial z}{\partial x}$, $q = \frac{\partial z}{\partial y}$.

This is a partial differential equation of order one.

Remember. If the number of arbitrary constants = the number of independent variables, then elimination of arbitrary constants will give rise to a PDE of first order.

Example 1.2.1. *Eliminate a, b from the relation* $z = ax^2 + by^2 + ab$.

Solution: Differentiating first with respect to x and then with respect to y, we get

$$\frac{\partial z}{\partial x} = 2ax; \ \frac{\partial z}{\partial y} = 2by, \text{ i.e. } a = \frac{p}{2x}; \ b = \frac{q}{2y}.$$

Substituting these values in the given relation, we obtain $z = \frac{p}{2x}x^2 + \frac{q}{2y}y^2 + \frac{pq}{4xy}$.

Simplifying, $4xyz = 2px^2y + 2qy^2x + pq$, which is the required PDE of first order.

Remember. If the number of arbitrary constants is less than the number of independent variables, then elimination of the arbitrary constants will give rise two distinct partial differential equations of first order, e.g.

Example 1.2.2. *Eliminate the arbitrary constant a from the given relation* $z = a(x + y)$.

Solution: As before

Differentiating w.r.t. x, we have $\frac{\partial z}{\partial x} = a$, i.e. $p = a$

Differentiating w.r.t. y, we have $\frac{\partial z}{\partial y} = a$, i.e. $q = a$.

We thus get two distinct PDE of order one as $z = p(x + y)$ and $z = q(x + y)$.

These two equations are the required eliminant.

Remember. If the number of arbitrary constants is more than the number of independent variables, then on elimination of the constants, a PDE (or equations) of order more than one can be obtained.

Example 1.2.3. *Eliminate the three arbitrary constants a, b, c from the relation* $z = ax + by + cxy$.

Solution: Differentiating the given relation:

$$\left.\begin{array}{l}\text{with respect to } x, \text{ we get } p = a + cy \\ \text{with respect to } y, \text{ we get } q = b + cx\end{array}\right\}$$

With these two equations we cannot eliminate all the three constants a, b, c.

So, we differentiate $p = a + cy$ with respect to x and obtain $\frac{\partial p}{\partial x} = \frac{\partial^2 z}{\partial x^2} = 0$, i.e. $r = 0$.

Similarly, we obtain $\frac{\partial q}{\partial y} = \frac{\partial^2 z}{\partial y^2} = 0$, i.e. $t = 0$ and $\frac{\partial p}{\partial y} = c$, i.e. $\frac{\partial^2 z}{\partial y \partial x} = s = c$.

Thus, we have $p = a + sy$ and $q = b + sx$, i.e. $a = p - sy$ and $b = q - sx$.

Finally, from the given relation $z = ax + by + cxy$, it follows $z = (p - sy)x + (q - sx)y + sxy$ or, $z = px + qy - sxy$.

∴ *required eliminant is* $r = 0$, $t = 0$ and $z = px + qy - sxy$.

Example 1.2.4. *Suppose, z depends on x, y, t. Eliminate three arbitrary constants a, b, c from* $z = a(x + y) + b(x - y) + abt + c$.

Solution: $\frac{\partial z}{\partial x} = a + b, \frac{\partial z}{\partial y} = a - b, \frac{\partial z}{\partial t} = ab.$

Since $(a + b)^2 = (a - b)^2 + 4ab$, hence, $\left(\frac{\partial z}{\partial x}\right)^2 = \left(\frac{\partial z}{\partial y}\right)^2 + 4\frac{\partial z}{\partial t}$.

The required eliminant is a partial differential equation of first order (but of degree two).

1.3 Construction of partial differential equations by the process of elimination of arbitrary functions

Suppose that u and v are two functions of x, y, z and suppose that there is a relation between u and v expressed

$$\left.\begin{array}{r}\text{either as } \phi(u, v) = 0 \\ \text{or as } u = f(v)\end{array}\right\} \phi \text{ and } f \text{ are arbitrary.} \tag{1.3.1}$$

We shall see that a partial differential equation is obtained on elimination of ϕ (or f). In fact, we shall obtain a linear PDE (i.e. first degree in p and q).

Differentiating (1.3.1) w.r.t. x and then w.r.t. y, we get

$$\frac{\partial \phi}{\partial u}\left[\frac{\partial u}{\partial x} + \frac{\partial u}{\partial z}\frac{\partial z}{\partial x}\right] + \frac{\partial \phi}{\partial v}\left[\frac{\partial v}{\partial x} + \frac{\partial v}{\partial z}\frac{\partial z}{\partial x}\right] \text{ or, } \frac{\partial \phi}{\partial u}\left(\frac{\partial u}{\partial x} + p\frac{\partial u}{\partial z}\right) + \frac{\partial \phi}{\partial v}\left(\frac{\partial v}{\partial x} + p\frac{\partial v}{\partial z}\right) = 0.$$

Similarly,

$$\frac{\partial \phi}{\partial u}\left(\frac{\partial u}{\partial y} + q\frac{\partial u}{\partial z}\right) + \frac{\partial \phi}{\partial v}\left(\frac{\partial v}{\partial y} + q\frac{\partial v}{\partial z}\right) = 0.$$

Eliminating $\frac{\partial \phi}{\partial u}$ and $\frac{\partial \phi}{\partial v}$ we easily get

$$\frac{\dfrac{\partial v}{\partial x} + p\dfrac{\partial v}{\partial z}}{\dfrac{\partial u}{\partial x} + p\dfrac{\partial u}{\partial z}} = \frac{\dfrac{\partial v}{\partial y} + q\dfrac{\partial v}{\partial z}}{\dfrac{\partial u}{\partial y} + q\dfrac{\partial u}{\partial z}}$$

$$\text{or, } \left(\frac{\partial u}{\partial x} + p\frac{\partial u}{\partial z}\right)\left(\frac{\partial v}{\partial y} + q\frac{\partial v}{\partial z}\right) - \left(\frac{\partial u}{\partial y} + q\frac{\partial u}{\partial z}\right)\left(\frac{\partial v}{\partial x} + p\frac{\partial v}{\partial z}\right) = 0$$

$$\text{or, } p\left[\frac{\partial u}{\partial z}\frac{\partial v}{\partial y} - \frac{\partial u}{\partial y}\frac{\partial v}{\partial z}\right] + q\left[\frac{\partial u}{\partial x}\frac{\partial v}{\partial z} - \frac{\partial v}{\partial x}\frac{\partial u}{\partial z}\right] = \frac{\partial u}{\partial y}\frac{\partial v}{\partial x} - \frac{\partial u}{\partial x}\frac{\partial v}{\partial y}$$

$$\text{or, } pP + qQ = R, \tag{1.3.2}$$

where $P = \frac{\partial u}{\partial z}\frac{\partial v}{\partial y} - \frac{\partial u}{\partial y}\frac{\partial v}{\partial z}$, etc.

Thus, from (1.3.1) involving an arbitrary function ϕ, we obtain a partial differential eq. (1.3.2). This PDE does not contain ϕ and it is linear in p and q.

Remember. When the relation between x, y, z contains two arbitrary functions, the partial differential equation derived therefrom will, in general, give a partial differential equation involving partial derivatives of order higher than the second.

Example 1.3.1. *Eliminate the arbitrary function ϕ from the relation: $z = \phi\left(\frac{y}{x}\right)$.*

Solution: Taking partial derivative first w.r.t. x and then w.r.t. y, we get

$$p = \frac{\partial z}{\partial x} = \phi'\left(\frac{y}{x}\right) \times \left(-\frac{y}{x^2}\right), q = \frac{\partial z}{\partial y} = \phi'\left(\frac{y}{x}\right) \times \left(\frac{1}{x}\right),$$

whence it follows: $\frac{p}{q} = \frac{-y}{x^2}/\frac{1}{x}$ or, $xp + yq = 0$ (first order PDE).

Example 1.3.2. *Eliminate the arbitrary functions f and ϕ from $y = f(x - at) + \phi(x + at)$.*

Solution: Here y depends on x and t. So we take partial derivatives of y w.r.t. x and t.

Thus,

$$\frac{\partial y}{\partial x} = f'(x - at) + \phi'(x + at) \qquad \frac{\partial y}{\partial t} = f'(x - at)\{-a\} + \phi'(x + at)\{a\}$$

$$\frac{\partial^2 y}{\partial x^2} = f''(x - at) + \phi''(x + at) \qquad \frac{\partial^2 y}{\partial t^2} = a^2 f''(x - at) + a^2\phi''(x + at)$$

whence it easily follows: $\frac{\partial^2 y}{\partial t^2} = a^2\frac{\partial^2 y}{\partial x^2}$ (a PDE of second order).

Remember. $\frac{\partial^2 y}{\partial t^2} = a^2\frac{\partial^2 y}{\partial x^2}$ is a very useful PDE of second order occurring in the study of a stretched string. It can be easily said that $y = f(x - at) + \phi(x + at)$ is the general solution of the second order PDE $\frac{\partial^2 y}{\partial t^2} = a^2\frac{\partial^2 y}{\partial x^2}$.

Example 1.3.3. *Find the partial differential equation arising from $\phi\left(\frac{z}{x^3}, \frac{y}{x}\right) = 0$, where ϕ is an arbitrary function of its arguments.*

Solution: Let us write the given relation in the form $\phi(u, v) = 0$, $\left[u = \frac{z}{x^3}, v = \frac{y}{x}\right]$.

Differentiating with respect to x and y, we write

$$\frac{\partial \phi}{\partial u}\cdot\frac{\partial u}{\partial x} + \frac{\partial \phi}{\partial v}\cdot\frac{\partial v}{\partial x} = 0 \text{ or, } \frac{\partial \phi}{\partial u}\left\{-\frac{3z}{x^4} + \frac{1}{x^3}\frac{\partial z}{\partial x}\right\} + \frac{\partial \phi}{\partial v}\left\{-\frac{y}{x^2}\right\} = 0. \quad (1)$$

Also

$$\frac{\partial \phi}{\partial u}\frac{\partial u}{\partial y} + \frac{\partial \phi}{\partial v}\cdot\frac{\partial v}{\partial y} = 0 \text{ or, } \frac{\partial \phi}{\partial u}\left\{\frac{1}{x^3}\frac{\partial z}{\partial y}\right\} + \frac{\partial \phi}{\partial v}\left(\frac{1}{x}\right) = 0. \quad (2)$$

Eliminating $\frac{\partial \phi}{\partial u}$ and $\frac{\partial \phi}{\partial v}$ from (1) and (2), we get

$$\begin{vmatrix} \frac{p}{x^3} - \frac{3z}{x^4} & -\frac{y}{x^2} \\ \frac{q}{x^3} & \frac{1}{x} \end{vmatrix} = 0 \quad \left[\text{writing } p = \frac{\partial z}{\partial x}, q = \frac{\partial z}{\partial y}\right]$$

$$\text{or, } \frac{p}{x^4} - \frac{3z}{x^5} + \frac{qy}{x^5} = 0 \text{ or, } px + qy = 3z. \text{ (Reqd. PDE)}$$

Note 1.3.1. *The given relation may not be written as $\phi\left(\frac{z}{x^3}, \frac{y}{x}\right) = 0$ but as $\frac{z}{x^3} = f\left(\frac{y}{x}\right)$, i.e. $z = x^3 f\left(\frac{y}{x}\right)$, where f is an arbitrary function of its argument.*

Let $v = \frac{y}{x}$. Then $z = x^3 f(v)$. Differentiation with respect to x and y yields

$$p = 3x^2 f(v) + x^3\frac{df}{dv}\cdot\frac{\partial v}{\partial x} = 3x^2 f(v) + x^3\frac{df}{dv}\cdot\left(-\frac{y}{x^2}\right) = 3x^2 f(v) - xyf'(v) \quad \left[\text{writing } \frac{df}{dv} = f'(v)\right]$$

$$\text{and } q = x^3 f'(v)\frac{\partial v}{\partial y} = x^3 f'(v)\cdot\frac{1}{x} = x^2 f'(v).$$

Eliminating $f'(v)$, we easily obtain $p = 3x^2 f(v) - xy\left(\frac{q}{x^2}\right)$ or, $xp + yq = 3x^3 f(v) = 3z$ [$\because z = x^3 f(v)$], i.e. the required PDE is as before $\boldsymbol{px + qy = 3z}$.

Example 1.3.4. *Eliminate the arbitrary function ϕ from $z = e^{my}\phi(x - y)$.*

Solution: Differentiating z w.r.t. x and y, we obtain

$$p = \frac{\partial z}{\partial x} = e^{my}\phi'(x - y)\cdot 1, \; q = \frac{\partial z}{\partial y} = me^{my}\phi(x - y) + e^{my}\phi'(x - y)\cdot\{-1\}.$$

Hence, $q = mz - p$, or, $p + q = mz$. (Reqd. eliminant).

1.4 Further solved problems

Example 1.4.1. *Eliminate arbitrary constants a and b from $z = (x^2 + a)(y^2 + b)$.*

Solution: Differentiating z w.r.t. x and y, we obtain $p = 2x(y^2 + b)$, $q = 2y(x^2 + a)$.

$\therefore\ pq = 4xy(x^2 + a)(y^2 + b) = 4xyz$. (Reqd. eliminant).

Example 1.4.2. *Obtain a partial differential equation by eliminating the arbitrary constants h and k from the relation $(x - h)^2 + (y - k)^2 + z^2 = c^2$ (c is a fixed constant).*

Solution: Differentiating w.r.t. x and y, we get

$$2(x - h) + 2pz = 0, \text{ i.e. } x - h = -pz$$

$$2(y - k) + 2qz = 0, \text{ i.e. } y - k = -qz.$$

Hence, from the given relation it follows $p^2z^2 + q^2z^2 + z^2 = c^2$, or, $(p^2 + q^2 + 1)z^2 = c^2$. (Reqd. eliminant).

The problem may be represented as:

Find the differential equation of all spheres of radius c, having centres on the xy-plane.

Example 1.4.3. *Show that the PDE obtained by eliminating b and c from $z = bx + \phi(b)y + c$, where ϕ is an arbitrary function of b, is free from x, y, z.*

Solution: Differentiating z w.r.t. x and y in the given relation, we get $p = b$ and $q = \phi(b)$.

The differential equation, obtained by eliminating b, is $q = \phi(p)$ or, may be expressed as $f(p, q) = 0$, where f is an arbitrary function of its arguments.

This eliminant does not contain x, y, z as was asked to obtain.

Example 1.4.4. *The equation of any cone with vertex at (a, b, c) is of the form $\phi\left(\frac{x-a}{z-c}, \frac{y-b}{z-c}\right) = 0$. Find the differential equation.*

Solution: Let $u = \frac{x-a}{z-c}$, $v = \frac{y-b}{z-c}$.

Then the given relation becomes

$$\phi(u, v) = 0. \tag{1}$$

Differentiating (1) w.r.t. x and y, we get

$$\frac{\partial \phi}{\partial u}\cdot\frac{\partial u}{\partial x} + \frac{\partial \phi}{\partial v}\cdot\frac{\partial v}{\partial x} = 0 \text{ or, } \frac{\partial \phi}{\partial u}\left\{\left(\frac{1}{z-c}\right) - \frac{x-a}{(z-c)^2}p\right\} + \frac{\partial \phi}{\partial v}\cdot\left\{-\frac{y-b}{(z-c)^2}p\right\} = 0, \left(p = \frac{\partial z}{\partial x}\right). \tag{2}$$

Also

$$\frac{\partial \phi}{\partial u}\cdot\frac{\partial u}{\partial y} + \frac{\partial \phi}{\partial v}\cdot\frac{\partial v}{\partial y} = 0, \text{ i.e. } \frac{\partial \phi}{\partial u}\left\{-\frac{x-a}{(z-c)^2}q\right\} + \frac{\partial \phi}{\partial v}\left\{\frac{1}{z-c} - \frac{y-b}{(z-c)^2}q\right\} = 0, \left(q = \frac{\partial z}{\partial y}\right). \tag{3}$$

Eliminating $\frac{\partial \phi}{\partial u}, \frac{\partial \phi}{\partial v}$ from (2) and (3), we obtain

$$\begin{vmatrix} \frac{1}{z-c} - \frac{x-a}{(z-c)^2}p & -\frac{y-b}{(z-c)^2}p + 0 \\ -q\frac{x-a}{(z-c)^2} + 0 & \frac{1}{z-c} - \frac{y-b}{(z-c)^2}q \end{vmatrix} = 0$$

$$\text{or, } \begin{vmatrix} \frac{1}{z-c} & \frac{-p(y-b)}{(z-c)^2} \\ \frac{-q(x-a)}{(z-c)^2} & \frac{1}{z-c} \end{vmatrix} + \begin{vmatrix} \frac{1}{z-c} & 0 \\ \frac{-q(x-a)}{(z-c)^2} & \frac{-q(y-b)}{(z-c)^2} \end{vmatrix}$$

$$\begin{vmatrix} \frac{-p(x-a)}{(z-c)^2} & \frac{-p(y-b)}{(z-c)^2} \\ 0 & \frac{1}{z-c} \end{vmatrix} + \begin{vmatrix} \frac{-p(x-a)}{(z-c)^2} & 0 \\ 0 & \frac{-q(y-b)}{(z-c)^2} \end{vmatrix} = 0$$

$$\text{or, } \frac{1}{(z-c)^2} - \frac{pq(x-a)(y-b)}{(z-c)^4} - \frac{q(y-b)}{(z-c)^3} - \frac{p(x-a)}{(z-c)^3} + \frac{pq(x-a)(y-b)}{(z-c)^4} = 0$$

$$\text{or, } \frac{1}{(z-c)^2} - \frac{1}{(z-c)^3}\{p(x-a) + q(y-b)\} = 0$$

or, $p(x-a) + q(y-b) = z - c$ is the required PDE.

Example 1.4.5. *Eliminating the arbitrary functions $f(x)$ and $g(y)$ from $z = yf(x) + xg(y)$ obtain the differential equation $xys = px + qy - z$, where $p = \frac{\partial z}{\partial x}$, $q = \frac{\partial z}{\partial y}$, and $s = \frac{\partial^2 z}{\partial x \partial y} = \frac{\partial^2 z}{\partial y \partial x}$.*

Solution: Differentiating $z = yf(x) + xg(y)$ w.r.t. x and y, we get

$$\left.\begin{aligned} p &= yf'(x) + g(y) \\ q &= f(x) + xg'(y) \end{aligned}\right\}. \quad (1)$$

In order to eliminate f, g, f', g' we find the second order partial derivatives

$$\left.\begin{aligned} r &= \frac{\partial^2 z}{\partial x^2} = \frac{\partial p}{\partial x} = yf''(x) \\ t &= \frac{\partial^2 z}{\partial y^2} = \frac{\partial q}{\partial y} = xg''(y) \\ \text{and} \quad s &= \frac{\partial^2 z}{\partial x \partial y} = \frac{\partial q}{\partial x} = f'(x) + g'(y) \end{aligned}\right\}. \quad (2)$$

From (1), we have $f'(x) = \frac{p - g(y)}{y}$ and $g'(y) = \frac{q - f(x)}{x}$.

$\therefore$ from (2) it follows:

$$s = f'(x) + g'(y) = \frac{p - g(y)}{y} + \frac{q - f(x)}{x}$$

or, $xys = px - xg(y) + qy - yf(x) = px + qy - [yf(x) + xg(y)] = px + qy - z$ (Proved).

Example 1.4.6. *Form a partial differential equation by eliminating the arbitrary functions f and ϕ from $z = f(x+iy) + \phi(x-iy)$, where $i^2 = -1$.*

Solution:

$$\frac{\partial z}{\partial x} = f'(x+iy) + \phi'(x-iy)$$

$$\frac{\partial^2 z}{\partial x^2} = f''(x+iy) + \phi''(x-iy)$$

$$\frac{\partial z}{\partial y} = if'(x+iy) - i\phi'(x-iy)$$

$$\frac{\partial^2 z}{\partial y^2} = i^2 f''(x+iy) + i^2 \phi''(x-iy) = -f''(x+iy) - \phi''(x-iy)$$

whence it follows $\frac{\partial^2 z}{\partial x^2} + \frac{\partial^2 z}{\partial y^2} = 0$, or, $\nabla^2 z = 0$ (Reqd. PDE).

Examples I

1. Eliminate the arbitrary constants a, b and obtain PDE: (a) $z = (x+a)(y+b)$ (b) $z = ax + a^2y^2 + b$ (c) $z = a(x+y) + b$.

2. Eliminating the arbitrary constants a, b, obtain from $z = (x-a)^2+(y-b)^2$ the PDE $\left(\frac{\partial z}{\partial x}\right)^2 + \left(\frac{\partial z}{\partial y}\right)^2 = 4z$.
3. Eliminate the arbitrary constants a and b from $z = ax + by + a^2 + b^2$ and obtain the PDE $z = px + qy + p^2 + q^2$.
4. Eliminate the constants k and A to obtain a second order partial differential equation $\frac{\partial^2 z}{\partial x^2} + \frac{\partial^2 z}{\partial t^2} = 0$ from the given relation $z = Ae^{kt} \sin kx$.
5. Find the differential equation of the set of all right circular cones whose axes coincide with z-axis.
   Remember. The general equation of the set of all right circular cones whose axes coincide with z-axis having semi-vertical angle α and vertex $(0, 0, c)$ is given by $x^2 + y^2 = (z - c)^2 \tan^2 \alpha$, c and α are arbitrary constants.
6. Eliminate a, b, c from $z = a(x + y) + b(x - y) + abt + c$.
7. Show that the differential equation of all cones which have their vertices at $(0, 0, 0)$ is $px + qy = z$. Verify that the surface $yz + zx + xy = 0$ satisfies the PDE.
   Remember. The equation of any cone with the vertex at the origin is $ax^2 + by^2 + cz^2 + 2fyz + 2gzx + 2hxy = 0$.
8. By eliminating three arbitrary constants a, b, c from the family of ellipsoids $\frac{x^2}{a^2} + \frac{y^2}{b^2} + \frac{z^2}{c^2} = 1$, obtain
$$zx\frac{\partial^2 z}{\partial x^2} + x\left(\frac{\partial z}{\partial x}\right)^2 - z\frac{\partial z}{\partial x} = 0,\ zy\frac{\partial^2 z}{\partial y^2} + y\left(\frac{\partial z}{\partial y}\right)^2 - z\frac{\partial z}{\partial y} = 0, \text{ or, } zs + pq = 0.$$
9. Eliminate the arbitrary constants a, b, c from $ax^2 + by^2 + cz^2 = 1$ and obtain $pq + zs = 0$.
10. Obtain a PDE of the family of spheres of radius 3 with centres on the plane $y = x$.
11. Form a PDE by eliminating the arbitrary function ϕ from $\phi(x + y + z, x^2 + y^2 - z^2) = 0$. What is the order of this PDE?
12. Eliminate the arbitrary function f and obtain a PDE:
    (a) $z = f(x^2 - y^2)$ (c) $z = y^2 + 2f\left(\frac{1}{x} + \ln y\right)$ (e) $z = f\left(\frac{xy}{z}\right)$.
    (b) $x + y + z = f(x^2 + y^2 + z^2)$ (d) $z = e^{ax+by} f(ax + by)$
13. Find the PDE of all surfaces of revolution, having z-axis as the axis of revolution.
    [The equation of any surface of revolution having z-axis as the axis of rotation may be taken as $z = \phi\left(\sqrt{x^2 + y^2}\right)$, where ϕ is an arbitrary function.]
14. Find a PDE by eliminating the arbitrary function F from the relation $F(x^2 + y^2 + z^2, z^2 - 2xy) = 0$.

Answers

1. (a) $z = pq$ (b) $q = 2yp^2$ (c) $q = p$.
5. $yp - xq = 0$.
6. $4\frac{\partial z}{\partial t} = p^2 - q^2$.
10. $(x - y)^2(p^2 + q^2 + 1) = 9(p - q)^2$.
11. $(y + z)p - (z + x)q = x - y$.
12. (a) $yp + xq = 0$ (b) $(y - z)p + (z - x)q = x - y$ (c) $x^2p + yq = 2y^2$ (d) $bp + aq = 2abz$ (e) $px - qy = 0$.
13. $yp = xq$.
14. $(p - q)z = y - x$.

Chapter 2

Lagrange's Method of Solving Partial Differential Equations

A partial differential equation of the form $Pp + Qq = R$, where P, Q and R are functions of x, y, z is sometimes called Lagrange's linear equation named after an Italian mathematician Joseph-Louis Lagrange. Lagrange made significant contribution to several mathematical and physical fields, especially to classical mechanics. In the present chapter our main aim is to introduce Lagrange method for solving Lagrange's linear equations that includes linear, semi-linear and quasi-linear equations. It is general method for solving first-order quasi-linear equations with two independent variables. Using this method we can find the general solution. In this chapter, we would like also to consider integral surfaces and orthogonal surfaces.

2.1 The general solution of a linear equation

We refer to Chapter 1, Art. 1.3: Given an arbitrary functional relation

$$\phi(u, v) = 0. \tag{2.1.1}$$

We can deduce a linear PDE of first order in the form:

$$Pp + Qq = R. \tag{2.1.2}$$

If (2.1.1) is deduced from (2.1.2), then we call (2.1.1), the general solution of (2.1.2). Since ϕ *is an arbitrary function,* the general solution (2.2.1) is more general than another solution of (2.1.2) that merely contains two arbitrary constants.

As for example, let $z = f(x^2 - y^2)$, where f is arbitrary. Then

$$p = \frac{\partial z}{\partial x} = f'(x^2 - y^2)2x,\ q = \frac{\partial z}{\partial y} = f'(x^2 - y^2)(-2y),$$

which leads to $py + qx = 0$, a linear PDE.

We shall call $z = f(x^2 - y^2)$ the general solution of $py + qx = 0$.

See that we could write

$$z = a(x^2 - y^2) + b(x^2 - y^2)^2 \quad (a, b \text{ are arbitrary constants})$$

$$\text{or, } z = a\sin(x^2 - y^2) + b \quad (a, b \text{ are arbitrary constants})$$

as the solution of $py + qx = 0$. But certainly $z = f(x^2 - y^2)$ is a more general solution; all these solutions with arbitrary constants are included in $z = f(x^2 - y^2)$.

2.2 An equation that is equivalent to $Pp + Qq = R$

A general type of a linear PDE in p and q is

$$Pp + Qq = R, \tag{2.2.1}$$

where P, Q, R are functions of x, y, z.

Suppose that $u(x, y, z) = c$ satisfies (2.2.1).

Differentiation w.r.t. x and y gives

$$\left.\begin{aligned} \frac{\partial u}{\partial x} + \frac{\partial u}{\partial z}p = 0 \\ \frac{\partial u}{\partial y} + \frac{\partial u}{\partial z}q = 0 \end{aligned}\right\} \quad p = -\frac{\dfrac{\partial u}{\partial x}}{\dfrac{\partial u}{\partial z}}; \quad q = -\frac{\dfrac{\partial u}{\partial y}}{\dfrac{\partial u}{\partial z}}.$$

Substituting these values of p and q in (2.2.1), we obtain

$$P\frac{\partial u}{\partial x} + Q\frac{\partial u}{\partial y} + R\frac{\partial u}{\partial z} = 0. \tag{2.2.2}$$

Therefore, if $u = c$ be an integral of (2.2.1), then $u = c$ also satisfies (2.2.2).

Conversely, if $u = c$ be an integral of (2.2.2), it is also an integral of (2.2.1).

Dividing (2.2.2) by $\frac{\partial u}{\partial z}$ and using the values of p and q, (2.2.2) reduces to $Pp + Qq = R$.

So we find eq. (2.2.2) is equivalent to eq. (2.2.1).

2.3 Lagrange's method of solving $Pp + Qq = R$

Lagrange's Rule: Statement: *The general solution of the linear partial differential equation*

$$Pp + Qq = R, \tag{2.3.1}$$

where P, Q, R are functions of x, y, z is given by

$$\phi(u, v) = 0, \tag{2.3.2}$$

where ϕ is an arbitrary function and

$$\left.\begin{aligned} u(x, y, z) = c_1 \\ v(x, y, z) = c_2 \end{aligned}\right\} \tag{2.3.3}$$

are two independent solutions of the Auxiliary Equations:

$$\frac{dx}{P} = \frac{dy}{Q} = \frac{dz}{R}. \tag{2.3.4}$$

This is known as Lagrange's solution of the linear equation.

Proof. Given $\phi(u, v) = 0$.

We consider z as dependent variable, and x and y as independent variables so that

$$\frac{\partial z}{\partial x} = p, \quad \frac{\partial z}{\partial y} = q, \quad \frac{\partial y}{\partial x} = 0, \quad \frac{\partial x}{\partial y} = 0.$$

Differentiating $\phi(u, v) = 0$ w.r.t. x, we get

$$\frac{\partial \phi}{\partial u}\left[\frac{\partial u}{\partial x}+\frac{\partial u}{\partial z}\cdot\frac{\partial z}{\partial x}\right]+\frac{\partial \phi}{\partial v}\left[\frac{\partial v}{\partial x}+\frac{\partial v}{\partial z}\cdot\frac{\partial z}{\partial x}\right]=0\,.$$

$$\frac{\dfrac{\partial \phi}{\partial u}}{\dfrac{\partial \phi}{\partial v}} = -\frac{\dfrac{\partial v}{\partial x}+\dfrac{\partial v}{\partial z}\dfrac{\partial z}{\partial x}}{\dfrac{\partial u}{\partial x}+\dfrac{\partial u}{\partial z}\dfrac{\partial z}{\partial x}} = -\frac{\dfrac{\partial v}{\partial x}+p\dfrac{\partial v}{\partial z}}{\dfrac{\partial u}{\partial x}+p\dfrac{\partial u}{\partial z}}\left(p=\frac{\partial z}{\partial x}\right).$$

Similarly, differentiating $\phi(u, v) = 0$ w.r.t. y, we get

$$\frac{\dfrac{\partial \phi}{\partial u}}{\dfrac{\partial \phi}{\partial v}} = -\frac{\dfrac{\partial v}{\partial y}+q\dfrac{\partial v}{\partial z}}{\dfrac{\partial u}{\partial y}+q\dfrac{\partial u}{\partial z}}\left(q=\frac{\partial z}{\partial y}\right).$$

Eliminating ϕ, we obtain

$$\frac{\dfrac{\partial v}{\partial x}+p\dfrac{\partial v}{\partial z}}{\dfrac{\partial u}{\partial x}+p\dfrac{\partial u}{\partial z}} = \frac{\dfrac{\partial v}{\partial y}+q\dfrac{\partial v}{\partial z}}{\dfrac{\partial u}{\partial y}+q\dfrac{\partial u}{\partial z}}$$

$$\left(\frac{\partial u}{\partial y}+q\frac{\partial u}{\partial z}\right)\left(\frac{\partial v}{\partial x}+p\frac{\partial v}{\partial z}\right)=\left(\frac{\partial u}{\partial x}+p\frac{\partial u}{\partial z}\right)\left(\frac{\partial v}{\partial y}+q\frac{\partial v}{\partial z}\right)$$

or, $Pp + Qq = R$,

where

$$\left.\begin{aligned} P &= \frac{\partial u}{\partial y}\cdot\frac{\partial v}{\partial z}-\frac{\partial u}{\partial z}\cdot\frac{\partial v}{\partial y},\\ Q &= \frac{\partial u}{\partial z}\cdot\frac{\partial v}{\partial x}-\frac{\partial u}{\partial x}\cdot\frac{\partial v}{\partial z},\\ R &= \frac{\partial u}{\partial x}\cdot\frac{\partial v}{\partial y}-\frac{\partial u}{\partial y}\cdot\frac{\partial v}{\partial x} \end{aligned}\right\}$$

Thus, $\phi(u, v) = 0$ is the general solution (integral) of $Pp + Qq = R$.

We now proceed to obtain u and v for substitution in $\phi(u, v) = 0$.

Consider $u(x, y, z) = c_1$ and $v(x, y, z) = c_2$, where c_1 and c_2 are arbitrary constants. Taking differentials, we get $du = 0$ and $dv = 0$

$$\frac{\partial u}{\partial x}dx+\frac{\partial u}{\partial y}dy+\frac{\partial u}{\partial z}dz=0 \text{ and } \frac{\partial v}{\partial x}dx+\frac{\partial v}{\partial y}dy+\frac{\partial v}{\partial z}dz=0.$$

Solving for dx, dy, dz,

$$\frac{dx}{\dfrac{\partial u}{\partial y}\dfrac{\partial v}{\partial z}-\dfrac{\partial u}{\partial z}\dfrac{\partial v}{\partial y}}=\frac{dy}{\dfrac{\partial u}{\partial z}\dfrac{\partial v}{\partial x}-\dfrac{\partial u}{\partial x}\dfrac{\partial v}{\partial z}}=\frac{dz}{\dfrac{\partial u}{\partial x}\dfrac{\partial v}{\partial y}-\dfrac{\partial u}{\partial y}\dfrac{\partial v}{\partial x}}.$$

Using the values of P, Q, R as given above, we may write $\frac{dx}{P}=\frac{dy}{Q}=\frac{dz}{R}$.

Thus, we find that $u = c_1$, $v = c_2$ form a solution of $\frac{dx}{P}=\frac{dy}{Q}=\frac{dz}{R}$. Hence, $u = c_1$, $v = c_2$ determine u and v for substitution in $\phi(u, v) = 0$.

This is what we wished to prove.

Note 2.3.1. *Equations* $\frac{dx}{P} = \frac{dy}{Q} = \frac{dz}{R}$ ***are known as Lagrange's Auxiliary (or Subsidiary) equations.***

Working Rule Solution of $Pp + Qq = R$ (Lagrange's Method)
1. Put the given linear equation in the form $Pp + Qq = R$.
2. Write down Lagrange's Auxiliary Equations $\frac{dx}{P} = \frac{dy}{Q} = \frac{dz}{R}$.
3. Taking two of the ratios at a time, and by using method of solving ordinary differential equations, obtain $u(x, y, z) = c_1$, $v(x, y, z) = c_2$ as two independent solutions of $\frac{dx}{P} = \frac{dy}{Q} = \frac{dz}{R}$.
4. The general solution of $Pp + Qq = R$ can be written as $\phi(u, v) = 0$ or in the form $u = f(v)$ or $v = F(u)$.

In Lagrange's Method of solving $Pp + Qq = R$, the most important part is to obtain $u = c_1$, $v = c_2$ from Lagrange's Auxiliary Equations $\frac{dx}{P} = \frac{dy}{Q} = \frac{dz}{R}$.

We ask our readers to remember four types of problems:

Solved Problems: (Type I – Type IV)

Type I

Example 2.3.1. *Solve:* $\frac{y^2z}{x}p + xzq = y^2$.

Solution: Lagrange's Auxiliary Equations are

$$\frac{dx}{\frac{y^2z}{x}} = \frac{dy}{xz} = \frac{dz}{y^2} \quad \left[\text{cf. } \frac{dx}{P} = \frac{dy}{Q} = \frac{dz}{R}\right]$$

From the first two ratios we see that the variable z cancels out and we are left with $\frac{dx}{y^2/x} = \frac{dy}{x}$ or, $x^2dx - y^2dy = 0$.

Integrating, we get $x^3 - y^3 = c_1$ [cf. $u = c_1$].

Taking the first and last ratios we see that y^2 cancels out and we obtain $\frac{dx}{z/x} = \frac{dz}{1}$ or, $x\,dx - z\,dz = 0$.

Integrating, we get $x^2 - z^2 = c_2$ [cf. $v = c_2$].

$\therefore$ the required general solution of the given equation is $\phi(u, v) = 0$, i.e. $\phi(x^3 - y^3, x^2 - z^2) = 0$, where ϕ is any arbitrary function of its arguments.

Example 2.3.2. *Solve:* $ap + aq = z$.

Solution: Lagrange's Auxiliary Equations are $\frac{dx}{a} = \frac{dy}{a} = \frac{dz}{z}$ $\left(\text{cf. } \frac{dx}{P} = \frac{dy}{Q} = \frac{dz}{R}\right)$.

First two ratios at once give $u = x - y = c_1$.

Again, second and third ratios give $v = y - a\ln z = c_2$.

$\therefore$ the required general solution is $\phi(u, v) = 0$, i.e. $\phi(x - y, y - a\ln z) = 0$, ϕ being an arbitrary function.

Example 2.3.3. *Solve:* $y^2p - xyq = x(z - 2y)$.

Solution: Lagrange's Auxiliary Equations are $\frac{dx}{y^2} = \frac{dy}{-xy} = \frac{dz}{x(z-2y)}$.

First two ratios give $\frac{dx}{y} = \frac{dy}{-x}$, i.e. $x\,dx + y\,dy = 0 \Rightarrow u \equiv x^2 + y^2 = c_1$.

Last two ratios give $\frac{dy}{-y} = \frac{dz}{z-2y}$, i.e. $\frac{dz}{dy} + \frac{1}{y}z = 2$ (ord. diff. eqn. linear form with IF $= e^{\int \frac{1}{y}dy} = e^{\ln y} = y$).

Multiplying by the IF y and integrating, we get $zy = y^2 + c_2$, i.e. $u = zy - y^2 = c_2$.

$\therefore$ the required general solution is $\phi(u, v) = 0$. $\phi(x^2 + y^2, zy - y^2) = 0$, ϕ being arbitrary.

A

Try Yourself (Examples for Practice)
Solve the following linear equations by using Lagrange's Auxiliary Equations (Lagrange's Method)

1. $2p + 3q = 1$
2. $p + q = \sin x$
3. $xzp + yzq = xy$
4. $p\tan x + q\tan y = \tan z$
5. $zp + x = 0$
6. $yzp + zxq = xy$
7. $x^2p + y^2q + z^2 = 0$
8. $x^2p + y^2q = z^2$
9. $xp + yq = z.$

Answers

1. $\phi(x - 2z, y - 3z) = 0$
2. $\phi(x - y, z + \cos x) = 0$
3. $\phi\left(z^2 - xy, \frac{y}{x}\right) = 0$
4. $\frac{\sin x}{\sin y} = \phi\left(\frac{\sin y}{\sin z}\right)$
5. $x^2 + z^2 = \phi(y)$
6. $\phi(x^2 - y^2, x^2 - z^2) = 0$
7. $\phi\left(\frac{1}{x} - \frac{1}{y}, \frac{1}{y} + \frac{1}{z}\right) = 0$
8. $\phi\left(\frac{1}{x} - \frac{1}{y}, \frac{1}{y} - \frac{1}{z} = 0\right)$
9. $\phi\left(\frac{x}{y}, \frac{x}{z}\right) = 0.$

Type II

Suppose that one integral of $\frac{dx}{P} = \frac{dy}{Q} = \frac{dz}{R}$ can be obtained easily by taking suitably chosen two ratios (as used in Type I problems) and suppose that another integral cannot be obtained by this same method.

Then one integral known to us is used to find another integral (see the solved examples given below).

In the second integral the constant of integration of the first integral should be removed.

Example 2.3.4. *Solve:* $p + 3q = 5z + \tan(y - 3x)$.

Solution: The Lagrange's Auxiliary Equations are $\frac{dx}{1} = \frac{dy}{3} = \frac{dz}{5z+\tan(y-3x)}$.

First see that one integral can be easily obtained from the first two ratios, namely $\frac{dx}{1} = \frac{dy}{3}$. Thus, $y - 3x = c_1$ (*first integral*). The second integral cannot be obtained in this manner. However, we take first ratio and the last ratio and obtain $\frac{dx}{1} = \frac{dz}{5z+\tan(y-3x)} = \frac{dz}{5z+\tan c_1}$ (using the known integral).

Writing $5dx = \frac{5dz}{5z+\tan c_1}$ and integrating, we obtain $5x - \ln(5z + \tan c_1) =$ arbitrary constant c_2 (say).

$\therefore$ the required general solution is $5x - \ln[5z + \tan(y - 3x)] = \phi(y - 3x)$, where ϕ is arbitrary.

Example 2.3.5. *Solve:* $xyp + y^2q = xyz - 2x^2$.

Solution: Lagrange's Auxiliary Equations are $\frac{dx}{xy} = \frac{dy}{y^2} = \frac{dz}{xyz-2x^2}$.

Taking the first two ratios (cancelling y), we get $\frac{dx}{x} = \frac{dy}{y}$ and hence, on integration $\frac{x}{y} = c_1$.

From second and third ratios

$$\frac{dy}{y^2} = \frac{dz}{xyz - 2x^2} = \frac{dz}{(c_1 y)yz - 2c_1^2 y^2} \quad \left(\because \frac{x}{y} = c_1 \text{ or, } x = c_1 y\right)$$

$$\Rightarrow \frac{dy}{y^2} = \frac{dz}{c_1 y^2 z - 2c_1^2 y^2} \Rightarrow c_1\,dy = \frac{dz}{(z - 2c_1)}$$

$$\Rightarrow c_1 y - \ln(z - 2c_1) = c_2 \Rightarrow x - \ln\left(z - \frac{2x}{y}\right) = c_2.$$

$\therefore$ the required general solution is $x - \ln\left(z - \frac{2x}{y}\right) = \phi\left(\frac{x}{y}\right)$, ϕ being an arbitrary function.

Example 2.3.6. *Solve:* $xzp + yzq = xy$.

Solution: The Lagrange subsidiary equations are $\frac{dx}{xz} = \frac{dy}{yz} = \frac{dz}{xy}$.

Taking the first two ratios (cancelling z), we get $\frac{dx}{x} = \frac{dy}{y}$ and this gives on integration $\ln x - \ln y = \ln c_1$ or, $\frac{x}{y} = c_1$.

The second and third ratios are $\frac{dy}{yz} = \frac{dz}{xy}$ or, $\frac{dy}{z} = \frac{dz}{x}$ or, $xdy = z\,dz$ or, $c_1 y\,dy = z\,dz$ (using the known integral $x = c_1 y$).

Hence, on integration we get $c_1\frac{y^2}{2} = \frac{z^2}{2}$ + constant or, $c_1 y^2 - z^2 = c_2$ or, $xy - z^2 = c_2$.

$\therefore$ the required general solution is $\phi(u, v) = 0$ or, $\phi\left(\frac{x}{y}, xy - z^2\right) = 0$, ϕ being arbitrary.

Example 2.3.7. *Solve:* $py + qx = xyz^2(x^2 - y^2)$.

Solution: Lagrange's subsidiary equations are $\frac{dx}{y} = \frac{dy}{x} = \frac{dz}{xyz^2(x^2-y^2)}$.

From first two ratios, we obtain $x^2 - y^2 = c_1$.

The last two ratios give

$$\frac{dy}{x} = \frac{dz}{xyz^2(x^2 - y^2)} \Rightarrow \frac{dy}{1} = \frac{dz}{yz^2 c_1} \quad (\because\ x^2 - y^2 = c_1)$$

$$\Rightarrow y\,dy = \frac{1}{c_1}\frac{dz}{z^2}.$$

On integration, $\frac{y^2}{2} = \frac{1}{c_1}\left(-\frac{1}{z}\right)$ + constant or, $c_1 y^2 = -\frac{2}{z}$+ constant or, $(x^2 - y^2)y^2 + \frac{2}{z} = c_2$.

Hence, the required general solution is $(x^2 - y^2)y^2 + \frac{2}{z} = \phi(x^2 - y^2)$, ϕ is arbitrary.

B

Try Yourself (Examples for Practice)
Solve by Lagrange's Method

1. $z(z^2 + xy)(px - qy) = x^4$
2. $p - q = \frac{z}{x+y}$
3. $p - 2q = 3x^2 \sin(y + 2x)$
4. $(x^2 - y^2 - z^2)p + 2xyq = 2xz$
5. $z(p - q) = z^2 + (x + y)^2$.

Answers

1. $\phi(xy, x^4 - z^4 - 2xyz^2) = 0$
2. $x - (x + y)\ln z = \phi(x + y)$
3. $x^3 \sin(y+2x) - z = \phi(y+2x)$
4. $\frac{x^2+y^2+z^2}{z} = \phi\left(\frac{y}{z}\right)$
5. $e^{2y}[z^2 + (x + y)^2] = \phi(x + y)$.

Type III

We may require to write (using a well-known Rule of Ratio and Proportion)

Lagrange's Subsidiary Equations are $\frac{dx}{P} = \frac{dy}{Q} = \frac{dz}{R} = \frac{P_1 dx + Q_1 dy + R_1 dz}{P_1 P + Q_1 Q + R_1 R}$.

Multipliers P_1, Q_1, R_1 should be chosen carefully. If $P_1 P + Q_1 Q + R_1 R = 0$, then write $P_1 dx + Q_1 dy + R_1 dz = 0$. Now on integration of $P_1 dx + Q_1 dy + R_1 dz = 0$, we may obtain an integral $u_1(x, y, z) = c_1$.

With the help of another set of suitably chosen multipliers, we may obtain another integral $u_2(x, y, z) = c_2$.

Sometimes only one integral is possible by use of suitable multipliers and the other integral may be obtained by methods of Type I and Type II.

Then the required general solution is written as $\phi(u_1, u_2) = 0$, where ϕ is an arbitrary function.

Example 2.3.8. *Solve:* $\frac{b-c}{a} yzp + \frac{c-a}{b} zxq = \frac{a-b}{c} xy$.

Solution: The Lagrange's subsidiary equations are

$$\frac{dx}{\left(\frac{b-c}{a}\right)yz} = \frac{dy}{\left(\frac{c-a}{b}\right)zx} = \frac{dz}{\left(\frac{a-b}{c}\right)xy} \text{ or, } \frac{a\,dx}{(b-c)yz} = \frac{b\,dy}{(c-a)zx} = \frac{c\,dz}{(a-b)xy}. \quad (1)$$

Multiply both numerator and denominator of first ratio by x, second ratio by y and third ratio by z. Then each ratio of (1)

$$= \frac{ax\,dx + by\,dy + cz\,dz}{xyz(b-c+c-a+a-b)} = \frac{ax\,dx + by\,dy + cz\,dz}{0}.$$

Hence, we may write $ax\,dx + by\,dy + cz\,dz = 0$.
On integration, we obtain

$$ax^2 + by^2 + cz^2 = \text{constant } c_1 \text{ (say)}. \quad (2)$$

Take another set of multipliers ax (first ratio), by (second ratio) and cz (third ratio). Then each ratio of (1) becomes

$$\frac{a^2x\,dx + b^2y\,dy + c^2z\,dz}{xyz[a(b-c) + b(c-a) + c(a-b)]} = \frac{a^2x\,dx + b^2y\,dy + c^2z\,dz}{0}.$$

Hence, $a^2x\,dx + b^2y\,dy + c^2z\,dz = 0$. Now integrating, we get

$$a^2x^2 + b^2y^2 + c^2z^2 = \text{constant } c_2 \text{ (say)}. \quad (3)$$

From (2) and (3), we write the general solution of the given equation as

$$\phi(ax^2 + by^2 + cz^2, a^2x^2 + b^2y^2 + c^2z^2) = 0,$$

where ϕ is an arbitrary function.

Example 2.3.9. *Solve:* $(mz - ny)p + (nx - lz)q = ly - mx$.

Solution: The Lagrange's subsidiary equations corresponding to the given equation are

$$\frac{dx}{mz - ny} = \frac{dy}{nx - lz} = \frac{dz}{ly - mx}. \quad (1)$$

Using multipliers x for the first ratio, y for the second and z for the third, we write each ratio of (1)

$$= \frac{x\,dx + y\,dy + z\,dz}{(mz-ny)x + (nx-lz)y + (ly-mx)z} = \frac{x\,dx + y\,dy + z\,dz}{0}.$$

Hence, we can write $x\,dx + y\,dy + z\,dz = 0$. On integration, we get

$$x^2 + y^2 + z^2 = \text{constant } c_1 \text{ (say)}. \quad (2)$$

Next we choose l, m, n as multipliers of first, second and third ratios of (1). Then each ratio of (1) becomes

$$= \frac{ldx + mdy + ndz}{l(mz-ny) + m(nx-lz) + n(ly-mx)} = \frac{ldx + mdy + ndz}{0}.$$

Hence,

$$ldx + mdy + ndz = 0 \Rightarrow lx + my + nz = \text{constant } c_2 \text{ (say)}. \quad (3)$$

The required general solution from (2) and (3) is

$$\phi(x^2 + y^2 + z^2, lx + my + nz) = 0,$$

ϕ being an arbitrary function.

Example 2.3.10. *Solve:* $x(y^2 - z^2)(p - y(z^2 + x^2)q = z(x^2 + y^2)$.

Solution: The Lagrange's Auxiliary Equations corresponding to the given equations are

$$\frac{dx}{x(y^2 - z^2)} = \frac{dy}{-y(z^2 + x^2)} = \frac{dz}{z(x^2 + y^2)}. \tag{1}$$

Using x, y, z as multipliers for first, second and third ratios of (1), we write

$$\text{each ratio of (1)} = \frac{x\,dx + y\,dy + z\,dz}{x^2(y^2 - z^2) - y^2(z^2 + x^2) + z^2(x^2 + y^2)} = \frac{x\,dx + y\,dy + z\,dz}{0}.$$

$$\therefore \quad x\,dx + y\,dy + z\,dz = 0, \text{ whence } x^2 + y^2 + z^2 = c_1. \tag{2}$$

Using $\frac{1}{x}, -\frac{1}{y}, -\frac{1}{z}$ as multipliers of ratios of (1), we again find each ratio of (1)

$$= \frac{\frac{1}{x}dx - \frac{1}{y}dy - \frac{1}{z}dz}{(y^2 - z^2) + (z^2 + x^2) - (x^2 + y^2)} = \frac{\frac{1}{x}dx - \frac{1}{y}dy - \frac{1}{z}dz}{0}.$$

$\therefore \quad \frac{1}{x}dx - \frac{1}{y}dy - \frac{1}{z}dz = 0$, which yields on integration,

$$\ln x - \ln y - \ln z = \ln c_2, \text{ or, } \frac{x}{yz} = c_2. \tag{3}$$

$\therefore$ the required general solution [from (2) and (3)] is $\phi\left(x^2 + y^2 + z^2, \frac{x}{yz}\right) = 0$, where ϕ is an arbitrary function.

Example 2.3.11. *Solve:* $x(y^2 + z)p - y(x^2 + z)q = z(x^2 - y^2)$.

Solution: The Lagrange's subsidiary equations corresponding to the given equation are

$$\frac{dx}{x(y^2 + z)} = \frac{dy}{-y(x^2 + z)} = \frac{dz}{z(x^2 - y^2)}. \tag{1}$$

Choosing $x, y, -1$ as multipliers of ratios of (1),

$$\text{each ratio of (1)} = \frac{x\,dx + y\,dy - dz}{x^2(y^2 + z) - y^2(x^2 + z) - z(x^2 - y^2)} = \frac{x\,dx + y\,dy - dz}{0}.$$

Hence, $x\,dx + y\,dy - dz = 0$ which, on integration, gives

$$x^2 + y^2 - 2z = c_1. \tag{2}$$

Again, choosing $\frac{1}{x}, \frac{1}{y}, \frac{1}{z}$ as multipliers of ratios of (1), we obtain each ratio of (1)

$$= \frac{dx/x + dy/y + dz/z}{y^2 + z - (x^2 + z) + x^2 - y^2} = \frac{dx/x + dy/y + dz/z}{0}.$$

Hence, $\frac{dx}{x} + \frac{dy}{y} + \frac{dz}{z} = 0$ or, $\ln xyz = \ln c_2$,

$$\text{i.e. } xyz = c_2. \tag{3}$$

$\therefore$ from (2) and (3) the required general solution is $\phi(x^2 + y^2 - 2z, xyz) = 0$, ϕ being an arbitrary function.

Example 2.3.12. *Solve:* $(y^2 + z^2)p - xyq = -zx$.

Solution: The Lagrange's auxiliary equations corresponding to the given equation are

$$\frac{dx}{y^2 + z^2} = \frac{dy}{-xy} = \frac{dz}{-zx}. \tag{1}$$

Using multipliers of the ratios of (1) as x, y, z respectively, we obtain

$$\text{each ratio of (1)} = \frac{x\,dx + y\,dy + z\,dz}{x(y^2 + z^2) - xy^2 - xz^2} = \frac{x\,dx + y\,dy + z\,dz}{0}.$$

$\therefore$ $x\,dx + y\,dy + z\,dz = 0$, which gives on integration

$$x^2 + y^2 + z^2 = c_1. \tag{2}$$

Again, from last two ratios of (1), we get $\frac{dy}{-xy} = \frac{dz}{-xz}$ or, $\frac{dy}{y} = \frac{dz}{z}$.
On integration this gives

$$\ln y - \ln z = \ln c_2 \text{ or, } \frac{y}{z} = c_2. \tag{3}$$

$\therefore$ the required general solution [from (2) and (3)] is $\phi\left(x^2 + y^2 + z^2, \frac{y}{z}\right) = 0$, where ϕ is an arbitrary function.

C

Try Yourself (Examples for Practice)
Solve by Lagrange's Method of Solution

1. $z(x + y)p + z(x - y)q = x^2 + y^2$
2. $x(y^2 - z^2)p + y(z^2 - x^2)q = z(x^2 - y^2)$
3. $x(y - z)p + y(z - x)q = z(x - y)$
4. $(y - zx)p + (x + yz)q = x^2 + y^2$
5. $(x + 2z)p + (4zx - y)q = 2x^2 + y$
6. $x^2(y - z)p + y^2(z - x)q = z^2(x - y)$
7. $(z^2 - 2yz - y^2)p + (xy + zx)q = xy - zx$
8. $z(xp - yq) = y^2 - x^4$
9. $(y^3x - 2x^4)p + (2y^4 - x^3y)q = 9z(x^3 - y^3)$
10. $(3x + y - z)p + (x + y - z)q = 2(z - y)$
11. $x(x^2 + 3y^2)p - y(3x^2 + y^2)q = 2z(y^2 - x^2)$.

[Here Lagrange's subsidiary equations are

$$\frac{dx}{x(x^2 + 3y^2)} = \frac{dy}{-y(3x^2 + y^2)} = \frac{dz}{2z(y^2 - x^2)}$$

Multipliers $\frac{1}{x}, \frac{1}{y}, -\frac{1}{z}$

$$\Rightarrow \frac{\frac{1}{x}dx + \frac{1}{y}dy - \frac{1}{z}dz}{x^2 + 3y^2 - 3x^2 - y^2 - 2(y^2 - x^2)} = \frac{\frac{1}{x}dx + \frac{1}{y}dy - \frac{1}{z}dz}{0}.$$

$\therefore$ $\frac{1}{x}dx + \frac{1}{y}dy + \frac{1}{z}dz = 0$ will give $\frac{xy}{z} = c_1$.

Now take the first two ratios $\frac{dx}{x(x^2+3y^2)} = \frac{dy}{-y(3x^2+y^2)}$

$$\text{or, } \frac{dy}{dx} = -\frac{y(3x^2 + y^2)}{x(x^2 + 3y^2)} \text{ (homogeneous equation)}$$

$$= -\frac{y}{x}\left[\frac{3 + \left(\frac{y}{x}\right)^2}{1 + 3\left(\frac{y}{x}\right)^2}\right].$$

Put $y = vx$. Then $v + x\frac{dv}{dx} = -v\frac{(3+v^2)}{1+3v^2}$.

or, $x\dfrac{dv}{dx} = -v\left[\dfrac{3 + v^2}{1 + 3v^2} + 1\right] = -\dfrac{4(1 + v^2)v}{1 + 3v^2}$ or, $4\dfrac{dx}{x} + \dfrac{1 + 3v^2}{v(1 + v^2)}dv = 0$ or, $4\dfrac{dx}{x} + \left(\dfrac{1}{v} + \dfrac{2v}{1 + v^2}\right) dv = 0$.

On integration, $4 \ln x + \ln v + \ln(1 + v^2) = \text{constant}$

$$\text{or, } x^4 v(1 + v^2) = \text{constant}$$

$$\text{or, } x^4 \frac{y}{x}\left(1 + \frac{y^2}{x^2}\right) = \text{constant}$$

$$\text{or, } xy(x^2 + y^2) = \text{constant}$$

$$\text{or, } c_1 z(x^2 + y^2) = \text{constant}$$

$$\text{or, } z(x^2 + y^2) = c_2 \text{ (say).}$$

$\therefore$ the required solution is $\phi\left(z(x^2 + y^2), \frac{xy}{z}\right) = 0$, ϕ being arbitrary.]

12. $(y - z)p + (z - x)q = x - y$.
13. $(y + zx)p - (x + yz)q + y^2 - x^2 = 0$. [Choose multipliers (i) x, y and $-z$ and (ii) y, x and 1.]
14. $y^2p + x^2q = x^2y^2z^2$.

Answers

1. $\phi(x^2 - y^2 - z^2, 2xy - z^2) = 0$
2. $\phi(x^2 + y^2 + z^2, xyz) = 0$
3. $\phi(x + y + z, xyz) = 0$
4. $\phi(x^2 - y^2 + z^2, xy - z) = 0$
5. $\phi(xy - z^2, x^2 - y - z) = 0$
6. $\phi\left(xyz, \frac{1}{x} + \frac{1}{y} + \frac{1}{z}\right) = 0$
7. $\phi(y^2 - z^2 - 2yz, x^2 + y^2 + z^2) = 0$
8. $\phi(x^2 + y^2 + z^2, xy) = 0$
9. $\phi\left(xyz^{1/3}, \frac{y}{x^2} + \frac{x}{y^2}\right) = 0$
10. $\phi\left(x - 3y - z, \frac{x-y+z}{\sqrt{x+y-z}}\right) = 0$
12. $\phi(x + y + z, xy + yz + zx) = 0$
13. $\phi(x^2 + y^2 - z^2, xy + z) = 0$
14. $\phi\left(x^3 - y^3, y^3 + \frac{3}{z}\right) = 0$.

Type IV

As in Type III, with one set of multipliers P_1, Q_1, R_1, we write

$$\frac{dx}{P} = \frac{dy}{Q} = \frac{dz}{R} = \frac{P_1\,dx + Q_1\,dy + R_1\,dz}{P_1P + Q_1Q + R_1R}.$$

Now if we see $P_1\,dx + Q_1\,dy + R_1\,dz$ is an exact differential of the denominator $P_1P + Q_1Q + R_1R$, then

$$\frac{P_1\,dx + Q_1\,dy + R_1\,dz}{P_1P + Q_1Q + R_1R}$$

may be combined with *a suitable ratio* of $\dfrac{dx}{P} = \dfrac{dy}{Q} = \dfrac{dz}{R}$ and this will give one integral.

Choose another set of multipliers, say P_2, Q_2, R_2, then

$$\text{each ratio} = \frac{P_2\,dx + Q_2\,dy + R_2\,dz}{P_2P + Q_2Q + R_2R}.$$

Suppose that the numerator is again an exact differential of $P_2P + Q_2Q + R_2R$. The two ratios

$$\frac{P_1\,dx + Q_1\,dy + R_1\,dz}{P_1P + Q_1Q + R_1R} \text{ and } \frac{P_2\,dx + Q_2\,dy + R_2\,dz}{P_2P + Q_2Q + R_2R}$$

are then combined to give a second integral.

Example 2.3.13. *Solve:* $(y+z)p + (z+x)q = (x+y)$.

Solution: The Lagrange's auxiliary equations corresponding to the given equation are

$$\frac{dx}{y+z} = \frac{dy}{z+x} = \frac{dz}{x+y}. \tag{1}$$

Choose multipliers $1, -1, 0$ for the first, second and third ratios of (1). Then, each ratio of (1)

$$= \frac{dx - dy}{y+z-z-x} = \frac{dx-dy}{y-x} = \frac{d(x-y)}{-(x-y)}.$$

[See that multipliers are so chosen that the numerator $dx - dy$ is an exact differential, namely $d(x-y)$]

Again, choosing multipliers as $1, 1, 1$, each ratio of (1)

$$= \frac{dx+dy+dz}{2(x+y+z)} = \frac{1}{2}\frac{d(x+y+z)}{x+y+z}.$$

Lastly choose $0, 1, -1$ as multipliers, then each ratio of (1)

$$= \frac{dy-dz}{z+x-(x+y)} = \frac{d(y-z)}{-(y-z)}.$$

So we may write

$$-\frac{d(x-y)}{(x-y)} = \frac{1}{2}\frac{d(x+y+z)}{x+y+z} = -\frac{d(y-z)}{y-z}. \tag{2}$$

First two ratios of (2) give on integration

$$-\ln(x-y) = \tfrac{1}{2}\ln(x+y+z) + \text{constant}$$

$$\text{i.e. } 2\ln(x-y) + \ln(x+y+z) = \text{constant}$$

$$\text{or, } (x-y)^2(x+y+z) = c_1.$$

Taking first and last ratios of (2), and integrating we easily obtain $\frac{x-y}{y-z} = c_2$.

$\therefore$ the required general solution is $\phi\left((x-y)^2(x+y+z), \frac{x-y}{y-z}\right) = 0$, where ϕ is an arbitrary function.

Example 2.3.14. *Solve:* $y^2(x-y)p + x^2(y-x)q = z(x^2+y^2)$.

Solution: Lagrange's auxiliary equations corresponding to the given equation are

$$\frac{dx}{y^2(x-y)} = \frac{dy}{x^2(y-x)} = \frac{dz}{z(x^2+y^2)}. \tag{1}$$

From the first two ratios we at once get $x^3 + y^3 = c_1$. Choosing $1, -1, 0$ as multipliers each ratio of (1)

$$= \frac{dx-dy}{y^2(x-y) - x^2(y-x)} = \frac{dx-dy}{(x-y)(x^2+y^2)}$$

Combining it with the third ratio of (1), we write

$$\frac{dx-dy}{(x-y)(x^2+y^2)} = \frac{dz}{z(x^2+y^2)} \text{ or, } \frac{d(x-y)}{x-y} - \frac{dz}{z} = 0.$$

On integration, we get $\ln(x-y) - \ln z = \ln c_2$ (say) or, $\frac{x-y}{z} = c_2$.

$\therefore$ the required general solution is $\phi\left(x^3+y^3, \frac{x-y}{z}\right) = 0$, where ϕ is any arbitrary function.

Example 2.3.15. *Solve:* $(x^2 - y^2 - z^2)p + 2xyq = 2xz$.

Solution: Lagrange's subsidiary equations corresponding to the given equation are

$$\frac{dx}{x^2 - y^2 - z^2} = \frac{dy}{2xy} = \frac{dz}{2xz}. \tag{1}$$

Taking the last two ratios we easily get on integration $y/z = c_1$.

Choosing x, y, z as multipliers of the ratios,

$$\text{each ratio of (1)} = \frac{x\,dx + y\,dy + z\,dz}{x(x^2 - y^2 - z^2) + 2xy^2 + 2xz^2} = \frac{x\,dx + y\,dy + z\,dz}{x(x^2 + y^2 + z^2)}$$

Combining this with third ratio of (1), we write

$$\frac{x\,dx + y\,dy + z\,dz}{x(x^2 + y^2 + z^2)} = \frac{dz}{2xz} \text{ or, } \frac{d(x^2 + y^2 + z^2)}{x^2 + y^2 + z^2} = \frac{dz}{z}.$$

On integration, we get $\ln(x^2 + y^2 + z^2) - \ln z = \ln c_2$ or, $\frac{x^2+y^2+z^2}{z} = c_2$.

Hence, the general solution is $\phi\left(\frac{y}{z}, \frac{x^2+y^2+z^2}{z}\right) = 0$, where ϕ is an arbitrary function.

Example 2.3.16. *Solve:* $\cos(x + y)p + \sin(x + y)q = z$.

Solution: Lagrange's auxiliary equations corresponding to the given equation are

$$\frac{dx}{\cos(x + y)} = \frac{dy}{\sin(x + y)} = \frac{dz}{z}. \tag{1}$$

Choosing $1, 1, 0$ as multipliers of the ratios of (1), each ratio of (1)

$$= \frac{d(x + y)}{\cos(x + y) + \sin(x + y)}. \tag{2}$$

Choosing $1, -1, 0$ as multipliers of the ratios of (1), each ratio of (1)

$$= \frac{d(x - y)}{\cos(x + y) - \sin(x + y)}. \tag{3}$$

From (1), (2) and (3), we write

$$\frac{dz}{z} = \frac{d(x + y)}{\cos(x + y) + \sin(x + y)} = \frac{d(x - y)}{\cos(x + y) - \sin(x + y)}. \tag{4}$$

Taking the first two ratios of (4),

$$\frac{dz}{z} = \frac{d(x + y)}{\cos(x + y) + \sin(x + y)}.$$

Putting $x + y = t$ on the right side, we write

$$\frac{dz}{z} = \frac{dt}{\cos t + \sin t} = \frac{dt}{\sqrt{2}\left\{\frac{1}{\sqrt{2}}\cos t + \frac{1}{\sqrt{2}}\sin t\right\}} = \frac{dt}{\sqrt{2}\sin\left(\frac{\pi}{4} + t\right)} \text{ or, } \sqrt{2}\frac{dz}{z} = \operatorname{cosec}\left(\frac{\pi}{4} + t\right) dt.$$

Hence, on integration,

$$\ln z^{\sqrt{2}} = \ln\left\{\tan\left(\frac{\pi}{8}+\frac{t}{2}\right)\right\} + \ln c_1$$

$\therefore\ z^{\sqrt{2}} = c_1 \tan\left(\frac{\pi}{8}+\frac{t}{2}\right)$, where $t = x + y$

$$\text{or, } z^{\sqrt{2}} \cot\left(\frac{\pi}{8}+\frac{t}{2}\right) = c_1. \tag{5}$$

Taking the last two ratios of (4), we get

$$d(x-y) = \frac{\cos(x+y)-\sin(x+y)}{\cos(x+y)+\sin(x+y)}d(x+y).$$

On the right side put $x + y = t$ and then integrate:

$$x - y = \int \frac{\cos t - \sin t}{\cos t + \sin t}dt + \text{constant}$$

$$\text{or, } x - y = \ln(\sin t + \cos t) + \ln c_2' = \ln c_2'(\sin t + \cos t)$$

$$\text{or, } e^{x-y} = c_2'[\sin(x+y)+\cos(x+y)]$$

$$\text{or, } [\sin(x+y)+\cos(x+y)]e^{y-x} = \frac{1}{c_2'} = c_2 \text{ (say)}.$$

Then the general solution can be written as $\phi\left(z^{\sqrt{2}}\cot\left(\frac{\pi}{8}+\frac{x+y}{2}\right), e^{y-x}\{\sin(x+y)+\cos(x+y)\}\right) = 0$, where ϕ is an arbitrary function.

Example 2.3.17. *Solve:* $px(x+y) - qy(x+y) + (x-y)(2x+2y+z) = 0.$

Solution: Given equation may be written as $x(x+y)p - y(x+y)q = -(x-y)((2x+2y+z)$.

Lagrange's subsidiary equations are

$$\frac{dx}{x(x+y)} = \frac{dy}{-y(x+y)} = \frac{dz}{-(x-y)(2x+2y+z)}. \tag{1}$$

The first two ratios of (1) at once give on integration $xy = c_1$.

Again, each ratio of (1) becomes

$$\frac{dx+dy}{x^2+y^2} = \frac{dx+dy+dz}{x(x+y)-y(x+y)-(x-y)(2x+2y+z)}$$

$$\text{or, } \frac{d(x+y)}{(x-y)(x+y)} = \frac{dx+dy+dz}{(x-y)[x+y-2x-2y-z]} = \frac{d(x+y+z)}{-(x+y+z)}.$$

Hence, on integration, we get $(x+y)(x+y+z) = c_2$.

$\therefore$ the required general solution is $\phi(xy, (x+y)(x+y+z)) = 0$, where ϕ is an arbitrary function.

D

Try Yourself (Examples for practice)
Solve the following linear partial differential equations

1. $(1 + y)p + (1 + x)q = z$
2. $xzp + yzq = xy$
3. $(x^2 - yz)p + (y^2 - zx)q = z^2 - xy$
4. $(x^2 - y^2 - yz)p + (x^2 - y^2 - zx)q = z(x - y)$
5. $xp + yq = z - a\sqrt{x^2 + y^2 + z^2}$
6. $(x^3 + 3xy^2)p + (y^3 + 3x^2y)q = 2z(x^2 + y^2)$
7. $p + q = x + y + z$
8. $(2x^2 + y^2 + z^2 - 2yz - zx - xy)p + (x^2 + 2y^2 + z^2 - yz - 2zx - xy)q = x^2 + y^2 + 2z^2 - yz - zx - 2xy$
9. $(x^2 + y^2 + yz)p + (x^2 + y^2 - xz)q = z(x + y)$
10. $\cos(x + y)p + \sin(x + y)q = z + \frac{1}{z}$

Answers

1. $\phi\left[(1 + x)^2 - (1 + y)^2, \frac{2+x+y}{z}\right] = 0$
2. $\phi(x/y, xy - z^2) = 0$
3. $\phi\left(xy + yz + zx, \frac{x-y}{y-z}\right) = 0$
4. $\phi\left(z - x + y, \frac{x^2+y^2}{z^2}\right) = 0$
5. $\phi\left[\frac{x^{a-1}}{z+\sqrt{x^2+y^2+z^2}}, \frac{x}{y}\right] = 0$
6. $\phi\left[(x - y)^{-2} - (x + y)^{-2}, \frac{xy}{z^2}\right] = 0$
7. $\phi[x - y, e^{-x}(2 + x + y + z)] = 0$
8. $\phi\left(\frac{x-y}{y-z}, \frac{y-z}{z^2}\right) = 0$
9. $\phi\left(z - x + y, \frac{x^2+y^2}{z^2}\right) = 0$
10. $\phi\left[(z^2 + 1)^{1/\sqrt{2}} \tan\left(\frac{3\pi}{8} - \frac{x+y}{z}\right), e^{y-x}\{\cos(x + y) + \sin(x + y)\}\right] = 0.$

A Few More Solved Problems (Miscellaneous Type)

Example 2.3.18. *Solve:* $(x + y - z)(p - q) + a(px - qy + x - y) = 0.$

Solution: Let $x + y = u$ and $x - y = v$.

$$p = \frac{\partial z}{\partial x} = \frac{\partial z}{\partial u}\cdot\frac{\partial u}{\partial x} + \frac{\partial z}{\partial v}\cdot\frac{\partial v}{\partial x} = \frac{\partial z}{\partial u}\cdot 1 + \frac{\partial z}{\partial v}\cdot 1, \text{ i.e. } p = \frac{\partial z}{\partial u} + \frac{\partial z}{\partial v}.$$

$$q = \frac{\partial z}{\partial y} = \frac{\partial z}{\partial u}\cdot\frac{\partial u}{\partial y} + \frac{\partial z}{\partial v}\cdot\frac{\partial v}{\partial y} = \frac{\partial z}{\partial u}\cdot 1 + \frac{\partial z}{\partial v}\cdot(-1), \text{ i.e. } q = \frac{\partial z}{\partial u} - \frac{\partial z}{\partial v}.$$

$$p - q = 2\frac{\partial z}{\partial v},\quad px - qy = x\left(\frac{\partial z}{\partial u} + \frac{\partial z}{\partial v}\right) - y\left(\frac{\partial z}{\partial u} - \frac{\partial z}{\partial v}\right) = (x - y)\frac{\partial z}{\partial u} + (x + y)\frac{\partial z}{\partial v}$$

$$\text{i.e. } px - q = yv\frac{\partial z}{\partial u} + u\frac{\partial z}{\partial v}.$$

The given equation thus reduces to

$$2(u - z)\frac{\partial z}{\partial v} + a\left(v\frac{\partial z}{\partial u} + u\frac{\partial z}{\partial v} + v\right) = 0 \text{ or, } av\frac{\partial z}{\partial u} + (2u - 2z + au)\frac{\partial z}{\partial v} = -av,$$

which is a linear equation in u, v, z (in place of x, y, z).

Lagrange's auxiliary equations are

$$\frac{du}{av} = \frac{dv}{2u - 2z + au} = \frac{dz}{-av}. \quad (1)$$

Considering the first and last ratios of (1), we easily obtain

$$u + z = c_1. \tag{2}$$

Considering the first and second ratios of (1),

$$\frac{du}{dv} = \frac{av}{2u - 2(c_1 - u) + au} = \frac{av}{(4+a)u - 2c_1} \quad \text{or, } (4+a)udu - 2c_1\,du = avdv.$$

Integrating, $(4+a)\frac{u^2}{2} - 2c_1 u = \frac{av^2}{2}$ + constant.

$$2u^2 + \frac{a}{2}u^2 - 2u(u+z) = \frac{av^2}{2} + \text{constant (using } c_1 = u + z)$$

$$\text{or,} av^2 + 4uz - au^2 = c_2. \tag{3}$$

From (2) and (3), we write the general solution as $\phi(u + z, av^2 + 4uz - au^2) = 0$ $(u = x + y, v = x - y)$, where ϕ is an arbitrary function.

Example 2.3.19. *Solve:* $(p - q)(e^{x+y} + xyz) + (yq - xp)(z + xye^{-x-y}) + (x - y)(1 - z^2) = 0.$

Solution: Put $e^{x+y} = u$ and $xy = v$. Proceed as in Ex. 2.3.18. Then the required solution is $\phi\left[\frac{u+v}{1-v}, \frac{u-v}{1+v}\right] = 0$, ϕ being an arbitrary function.

Example 2.3.20. *Find the surface whose tangent planes cut off an intercept of constant length k from the axis of z.*

Solution: The equation of the tangent plane at a point (x, y, z) to a surface $z = f(x, y)$ is

$$(X - x)p + (Y - y)q = Z - z, \tag{1}$$

where (X, Y, Z) are the current co-ordinates.

Since, by the given condition, (1) cuts an intercept k on the z-axis, it follows that (1) must pass through the point $(0, 0, k)$. Hence,

$$px + qy = z - k. \tag{2}$$

This is linear equation of the form $Pp + Qq = R$.

The Lagrange's auxiliary equations of (2) are

$$\frac{dx}{x} = \frac{dy}{y} = \frac{dz}{z - k}. \tag{3}$$

From the first two ratios of (3), we obtain $\ln x = \ln y + \ln c_1$ or, $\frac{x}{y} = c_1$.

Taking the first and the last ratios of (3), we write

$$\frac{dx}{x} = \frac{dz}{z-k}, \quad \text{or, } \ln x = \ln(z - k) + \ln c_2, \text{ i.e. } \frac{x}{z - k} = c_2.$$

Hence, the general solution of (2) is $\phi\left(\frac{x}{y}, \frac{x}{z-k}\right) = 0$, ϕ being arbitrary.

E

Try Yourself (Examples for Practice)
Solve the following differential equations

1. $(p+q)(x+y) = 1$
2. $(px - qy) = (z - xy)^2$
3. $xyp + y(2x - y)q = 2xz$
4. $x(y^m - z^m)p + y(z^m - x^m)q = z(x^m - y^m)$
5. $p - qy\ln y = z\ln y$
6. $x^2p + y^2q = x + y$
7. $z(p+q) = z^2 + (x-y)^2$
8. $(xz + y^2)p + (yz - 2x^2)q + 2xy + z^2 = 0$
9. $(x^2 + 2y^2)p - xyq = xz.$

Answers

1. $\phi(y - x, e^{-2x}y + x) = 0$
2. $\phi\left(xy, xe^{\frac{1}{z-xy}}\right) = 0$
3. $\phi(xy - x^2, z/xy) = 0$
4. $x^m + y^m + z^m = \phi(xyz)$
5. $\phi(yz, e^x \ln y) = 0$
6. $\phi\left[\frac{1}{y} - \frac{1}{x}, e^{-z}(x-y)\right] = 0$
7. $\ln\{z^2 + (x-y)^2\} - 2x = \phi(x-y)$
8. $\phi(yz + x^2, 2xz - y^2) = 0$
9. $\phi(x^2y^2 + y^4, yz) = 0.$

2.4 Integral surfaces through a given curve

Given: $Pp + Qq = R$, a linear PDE of first order. We obtain two independent solutions

$$u(x,y,z) = c_1 \text{ and } v(x,y,z) = c_2 \tag{2.4.1}$$

by using Lagrange's auxiliary equations.

Suppose, now we wish to obtain the integral surface which passes through a curve whose parametric equations are $x = x(t)$, $y = y(t)$, $z = z(t)$, where t is a parameter.

(2.4.1) is then written as $u[x(t), y(t), z(t)] = c_1$ and $v[x(t), y(t), z(t)] = c_2$.

We eliminate t and get a relation involving c_1 and c_2. Finally, putting $c_1 = u$ and $c_2 = v$, we get the required integral surface passing through the given curve.

Example 2.4.1. *Find the integral surface of the linear partial differential equation*

$$x(y^2 + z)p - y(x^2 + z)q = (x^2 - y^2)z,$$

which contains the straight line $x + y = 0$, $z = 1$.

Solution: Lagrange's auxiliary equations corresponding to the given PDE are

$$\frac{dx}{x(y^2 + z)} = \frac{dy}{-y(x^2 + z)} = \frac{dz}{(x^2 - y^2)z}. \tag{1}$$

Each ratio of (1)

$$= \frac{\frac{1}{x}dx + \frac{1}{y}dy + \frac{1}{z}dz}{y^2 + z^2 - (x^2 + z) + (x^2 - y^2)} = \frac{\frac{1}{x}dx + \frac{1}{y}dy + \frac{1}{z}dz}{0}.$$

$\therefore\ \frac{1}{x}dx + \frac{1}{y}dy + \frac{dz}{z} = 0$, i.e. $\ln x + \ln y + \ln z = \ln c_1$

$$\text{or, } xyz = c_1. \tag{2}$$

Each ratio of (1) (Multipliers $x, y, -1$)

$$= \frac{x\,dx + y\,dy - dz}{x^2(y^2 + z) - y^2(x^2 + z) - (x^2 - y^2)z} = \frac{x\,dx + y\,dy - dz}{0}$$

$$\text{or, } x\,dx + y\,dy - dz = 0 \text{ i.e. } x^2 + y^2 - 2z = c_2. \tag{3}$$

Now the straight line in the parametric forms: $x = t$, $y = -t$, $z = 1$.

$$\left.\begin{array}{l}\text{From (2) we then write } -t^2 = c_1\\ \text{From (3), we write } 2t^2 - 2 = c_2\end{array}\right\}.$$

$\therefore$ the required integral surface can be obtained as $-2c_1 - 2 = c_2$ or, $-2(xyz) - 2 = x^2 + y^2 - 2z$ or, $2xyz + x^2 + y^2 - 2z + 2 = 0$.

Example 2.4.2. *Find the equation of the integral surface of the linear differential equation*

$$2y(z-3)p + (2x-z)q = y(2x-3),$$

which passes through the circle $x^2 + y^2 = 2x$, $z = 0$.

Solution: Lagrange's auxiliary equations corresponding to the given equation are

$$\frac{dx}{2y(z-3)} = \frac{dy}{2x-z} = \frac{dz}{y(2x-3)}. \tag{1}$$

From first and the last ratios, we get by integration [i.e. by integrating $(2x-3)dx = 2(z-3)dz$]

$$x^2 - 3x - z^2 + 6z = c_1. \tag{2}$$

Again each ratio of (1)

$$= \frac{\frac{1}{2}dx + y\,dy - dz}{y(z-3) + y(2x-z) - y(2x-3)} \quad \begin{bmatrix}\text{Multipliers}\\ \text{are } \frac{1}{2}, y, -1\end{bmatrix}$$

$$= \frac{\frac{1}{2}dx + y\,dy - dz}{0}.$$

$$\therefore \frac{1}{2}dx + y\,dy - dz = 0 \Rightarrow \frac{x}{2} + \frac{y^2}{2} - z = \text{constant} \Rightarrow x + y^2 - 2z = c_2. \tag{3}$$

Now the given circle in the parametric form is $x = t$, $y = \sqrt{2t - t^2}$, $z = 0$.

From (2) and (3), we obtain $t^2 - 3t = c_1$ and $t + (2t - t^2) = c_2$.

Eliminating t, we get $c_1 + c_2 = 0$.

$\therefore$ $x^2 - 3x - z^2 + 6z + x + y^2 - 2z = 0$, i.e. $x^2 + y^2 - z^2 - 2x + 4z = 0$ [Reqd. Integral Surface].

We need not always write the equation of the given curve in parametric form. See the example given below:

Example 2.4.3. *Find the integral surface of the linear partial differential equation $(x-y)p+(y-x-z)q = z$ through the circle $x^2 + y^2 = 1$, $z = 1$.*

Solution: Given equation: $(x-y)p + (y-x-z)q = z$.

Lagrange's auxiliary equations are:

$$\frac{dx}{x-y} = \frac{dy}{y-x-z} = \frac{dz}{z}. \tag{1}$$

$$\text{Each ratio} = \frac{dx+dy+dz}{(x-y)+(y-x-z)+(z)} = \frac{dx+dy+dz}{0}.$$

$$\therefore \quad dx + dy + dz = 0 \text{ or, } x + y + z = c_1. \tag{2}$$

Taking the last two ratios of (1), we get $\frac{dy}{y-x-z} = \frac{dz}{z}$

$$\text{or,} \quad \frac{dy}{y-x-(c_1-x-y)} = \frac{dz}{z}, \text{ using (2)}$$

$$\text{or,} \quad \frac{dy}{2y-c_1} = \frac{dz}{z}.$$

On integration this gives $\frac{1}{2}\ln(2y-c_1) = \ln z + \text{constant}$

$$\text{or,} \quad \ln(2y-c_1) - 2\ln z = \ln c_2$$

$$\text{or,} \quad \frac{2y-c_1}{z^2} = c_2$$

$$\text{or,} \quad \frac{2y-(x+y+z)}{z^2} = c_2 \text{ [using (2)]}$$

$$\text{or,} \quad y-x-z = c_2 z^2. \tag{3}$$

Given curve: $x^2+y^2=1, z=1$
Using $z=1$, in (2) and (3),

$$x+y = c_1 - 1, \quad (y-x) = c_2+1.$$

$$\therefore \quad (x+y)^2 + (y-x)^2 = (c_1-1)^2 + (c_2+1)^2$$

$$\text{or,} \quad 2(x^2+y^2) = (c_1-1)^2 + (c_2+1)^2$$

$$\text{or,} \quad c_1^2 + c_2^2 - 2c_1 + 2c_2 = 0 \text{ (using } x^2+y^2=1 \text{ of the curve).}$$

From (2) and (3) it follows

$$(x+y+z)^2 + \frac{(y-x-z)^2}{z^4} - 2(x+y+z) + 2\frac{(y-x-z)}{z^2} = 0,$$

i.e. the required surface is $z^4(x+y+z)^2 + (y-x-z)^2 - 2z^4(x+y+z) + 2z^2(y-x-z) = 0$.

Example 2.4.4. *Find the integral surface of $x^2p + y^2q + z^2 = 0$, which passes through the hyperboloid $xy = x+y, z=1$.*

Solution: Given equation: $x^2p + y^2q + z^2 = 0$

$$\text{or,} \quad x^2p + y^2q = -z^2. \tag{1}$$

Given curve:

$$xy = x+y, \quad z=1. \tag{2}$$

Lagrange's auxiliary equations are

$$\frac{dx}{x^2} = \frac{dy}{y^2} = \frac{dz}{-z^2}. \tag{3}$$

Taking the first and third ratios,

$$x^{-2}dx + z^{-2}dz = 0.$$

Integrating, $-\frac{1}{x} - \frac{1}{z} = -c_1$ (say).

$$\text{or,} \quad \frac{1}{x} + \frac{1}{z} = c_1. \tag{4}$$

Taking the second and third ratios, $y^{-2}dy + z^{-2}dz = 0$

Integrating, $-\frac{1}{y} - \frac{1}{z} = -c_2$ (say)

$$\text{or,} \quad \frac{1}{y} + \frac{1}{z} = c^2. \tag{5}$$

Adding (4) and (5), $\frac{1}{x} + \frac{1}{y} + \frac{2}{z} = c_1 + c_2$

$$\text{or,} \quad \frac{x+y}{xy} + \frac{2}{z} = c_1 + c_2$$

$$\text{or,} \quad \frac{xy}{xy} + \frac{2}{1} = c_1 + c_2 \ \text{[using (2)]}$$

$$\text{or,} \quad c_1 + c_2 = 3. \tag{6}$$

Substituting the values of c_1 and c_2 from (4) and (5) in (6), we obtain

$$\frac{1}{x} + \frac{1}{z} + \frac{1}{y} + \frac{1}{z} = 3 \text{ or, } yz + 2xy + xz = 3xyz.$$

Surfaces orthogonal to a given system of surfaces

Let

$$f(x, y, z) = c \tag{2.4.2}$$

represent a system of surfaces with c as a parameter. We wish to obtain a system of surfaces which cut each surface of (2.4.2) at right angles.

Then the direction ratios of the normal at (x, y, z) to (2.4.2) which passes through the point are

$$\frac{\partial f}{\partial x}, \frac{\partial f}{\partial y} \text{ and } \frac{\partial f}{\partial z}.$$

Let the surface

$$z = \phi(x, y) \tag{2.4.3}$$

cut each surface of (2.4.2) at right angles. Then the normal at (x, y, z) to (2.4.3) has direction ratios

$$\frac{\partial z}{\partial x}, \frac{\partial z}{\partial y}, -1, \text{ i.e. } p, q, -1.$$

Since normals at (x, y, z) to (2.4.2) and (2.4.3) are at right angles, we have

$$p\frac{\partial f}{\partial x} + q\frac{\partial f}{\partial y} - \frac{\partial f}{\partial z} = 0, \tag{2.4.4}$$

which is of the form $Pp + Qq = R$.

Conversely, any solution of (2.4.4) is orthogonal to every surface of (2.4.2).

Example 2.4.5. *Find the surface which intersects the surface of the system $z(x+y) = c(3z+1)$ orthogonally and which passes through the circle $x^2 + y^2 = 1, z = 1$.*

Solution: The given system of surfaces is given by

$$f(x, y, z) = \frac{z(x+y)}{3z+1} = c. \tag{1}$$

$$\frac{\partial f}{\partial x} = \frac{z}{3z+1}, \quad \frac{\partial f}{\partial y} = \frac{z}{3z+1}, \quad \frac{\partial f}{\partial z} = (x+y)\left[\frac{1}{(3z+1)^2}\right].$$

The required orthogonal surface is solution of

$$p\frac{\partial f}{\partial x} + q\frac{\partial f}{\partial y} = \frac{\partial f}{\partial z} \text{ or } \frac{z}{3z+1}p + \frac{z}{3z+1}q = \frac{x+y}{(3z+1)^2} \text{ or } z(3z+1)p + z(3z+1)q = x + y. \quad (2)$$

Lagrange's auxiliary equations for (2) are

$$\frac{dx}{z(3z+1)} = \frac{dy}{3(3z+1)} = \frac{dz}{x+y}. \quad (3)$$

Taking the first two ratios of (3), we get $dx - dy = 0$ so that

$$x - y = c_1. \quad (4)$$

Choosing $x, y, -z(3z+1)$ as multipliers, each ratio of (3)

$$= \frac{x\,dx + y\,dy - z(3z+1)dz}{0} \text{ or, } x\,dx + y\,dy - 3z^2dz - z\,dz = 0.$$

Integrating, $\frac{x^2}{2} + \frac{y^2}{2} - 3 \times \frac{z^3}{3} - \frac{1}{2}z^2 = \frac{1}{2}c_2$ (say)

$$\text{or, } x^2 + y^2 - 2z^3 - z^2 = c_2. \quad (5)$$

Hence, any surface which is orthogonal to (1) has equation of the form:

$$x^2 + y^2 - 2z^3 - z^2 = \phi(x - y), \quad (6)$$

ϕ being any arbitrary function.

In order to get the required surface passing through the circle $x^2 + y^2 = 1$, $z = 1$, we must choose $\phi(x - y) = -2$. Thus, the required particular surface is $x^2 + y^2 - 2z^3 - z^2 = -2$.

Example 2.4.6. *What is the geometrical description of the solutions of $Pp + Qq = R$ and of the system of equations $dx/P = dy/Q = dz/R$ and establish the relationship between the two.*

Solution: Consider

$$Pp + Qq = R \quad (1)$$

$$\text{and } \frac{dx}{P} = \frac{dy}{Q} = \frac{dz}{R} \quad (2)$$

(P, Q, R are functions of x, y, z respectively.)

Let

$$z = \phi(x, y) \quad (3)$$

be the solution of (1). Then (3) represents a surface whose normal at (x, y, z) has direction ratios $\frac{\partial z}{\partial x}, \frac{\partial z}{\partial y}$ and -1, i.e. $p, q, -1$.

Also we know that the simultaneous equations given by (2) represent a family of curves such that the tangent at any point has direction ratios P, Q, R. Rewriting (1) as

$$Pp + Qq + R(-1) = 0, \quad (4)$$

showing that the normal to the surface (3) at any point is perpendicular to the member of family of curves (2) through that point. Hence, the member must touch the surface at that point. Since this holds for each point on (3), we conclude that the curves (2) lie completely on the surface (3), whose differential equation is (1).

Example 2.4.7. *Show that the surfaces represented by $Pp+Qq = R$ are orthogonal to the surface represented by $Pdx + Qdy + Rdz = 0$.*

Solution: We know that the curves whose equations are solutions of

$$\frac{dx}{P} = \frac{dy}{Q} = \frac{dz}{R} \tag{1}$$

are orthogonal to the system of the surfaces whose equation satisfies

$$Pdx + Qdy + Rdz = 0. \tag{2}$$

Again, by the previous example (**Ex. 2.4.6**) the curves of (1) lie completely on the surface represented by

$$Pp + Qq = R. \tag{3}$$

Hence, we conclude that surfaces represented by (2) and (3) are orthogonal.

Example 2.4.8. *Find the family of surfaces orthogonal to $\phi[z(x+y)^2, x^2 - y^2] = 0$.*

Solution: Given:

$$\phi[z(x+y)^2, x^2 - y^2] = 0. \tag{1}$$

Let

$$u = z(x+y)^2,\ v = x^2 - y^2. \tag{2}$$

Then (1) becomes

$$\phi(u, v) = 0. \tag{3}$$

Differentiating (3) w.r.t. x and y partially,

$$\frac{\partial \phi}{\partial u}\left(\frac{\partial u}{\partial x} + p\frac{\partial u}{\partial z}\right) + \frac{\partial \phi}{\partial v}\left(\frac{\partial v}{\partial x} + p\frac{\partial v}{\partial z}\right) = 0 \tag{4}$$

$$\frac{\partial \phi}{\partial u}\left(\frac{\partial u}{\partial y} + q\frac{\partial u}{\partial z}\right) + \frac{\partial \phi}{\partial v}\left(\frac{\partial v}{\partial y} + q\frac{\partial v}{\partial z}\right) = 0. \tag{5}$$

From (2)

$$\frac{\partial u}{\partial x} = 2z(x+y),\quad \frac{\partial u}{\partial y} = 2z(x+y),\quad \frac{\partial u}{\partial z} = (x+y)^2,$$
$$\frac{\partial v}{\partial x} = 2x,\quad \frac{\partial v}{\partial y} = -2y,\quad \frac{\partial v}{\partial z} = 0.$$

Using these values in (4) and (5), we get

$$\frac{\partial \phi}{\partial u}[2z(x+y) + p(x+y)^2] + \frac{\partial \phi}{\partial v}(2x + 0) = 0 \tag{6}$$

and $$\frac{\partial \phi}{\partial u}[2z(x+y) + q(x+y)^2] + \frac{\partial \phi}{\partial v}(-2y + 0) = 0, \tag{7}$$

whence

$$-\frac{\partial \phi}{\partial u}\Big/\frac{\partial \phi}{\partial v} = \frac{2x}{2z(x+y) + p(x+y)^2} = \frac{-2y}{2z(x+y) + q(x+y)^2}$$

or, $x(x+y)[(2z + q(x+y)] = -y(x+y)[2z + p(x+y)]$

or, $py(x+y) + qx(x+y) = -2z(x+y)$

or, $$py + qx = -2z, \tag{8}$$

which is the differential equation of the family of surfaces given by (1). So the differential equation of the family of surfaces orthogonal to (8) is given by [see Ex. 2.4.7]

$$ydx + xdy - 2z\,dz = 0 \text{ or, } d(xy) - d(z^2) = 0 \text{ or, } xy - z^2 = c \quad (c \text{ is a parameter}),$$

which is the required family of orthogonal surfaces.

Example 2.4.9. *Find the family of surfaces orthogonal to the family of surfaces given by the differential equation* $(y + z)p + (z + x)q = x + y$.

Solution: Let

$$P = y + z, \quad Q = z + x, \quad R = x + y. \tag{1}$$

Then the given differential equation becomes

$$Pp + Qq = R. \tag{2}$$

Now the differential equation of the family of surfaces orthogonal to the given family is given by

$$Pdx + Qdy + Rdz = 0$$

$$\text{or} \quad (y + z)dx + (z + x)dy + (x + y)dz = 0$$

$$\text{or} \quad (ydx + xdy) + (ydz + zdy) + (zdx + xdz) = 0.$$

Integrating $xy + yz + zx = c$ (c being a parameter) which is the required family of surfaces.

Examples II

1. Find the equation of the integral surface of the differential equation $(x^2 - yz)p + (y^2 - zx)q = z^2 - xy$, which passes through the line $x = 1$, $y = 0$.
   [Following the solved examples of art. 2.4 obtain $(x - y)(y - z) = c_1$ and $xy + yz + zx = c_2$. Using the equation of the line $x = 1$, $y = 0$, obtain $c_1 c_2 = -1$.
   $\therefore$ required surface $(x - y)(y - z)(xy + yz + zx) + 1 = 0$.]
2. Find the equation of the integral surface satisfying $4yzp + q + 2y = 0$ and passing through $y^2 + z^2 = 0, x + z = 2$.
3. Find the general integral of the partial differential equation $(2xy - 1)p + (z - 2x^2)q = 2(x - yz)$ and also the particular integral which passes through the lines $x = 1$, $y = 0$.
4. Find the equation of the integral surface of the PDE $2y(z - 3)p + (2x - z)q = y(2x - 3)$, which passes through the circle $z = 0$, $x^2 - 2x + y^2 = 0$.
5. Solve: $xp + yq = z$. Find a solution representing a surface meeting the parabola $y^2 = 4x$, $z = 1$.
6. Find the surface which is orthogonal to the one-parameter system $z = cxy(x^2 + y^2)$ and which passes through the hyperbola $x^2 - y^2 = a^2$, $z = 0$.

Answers

1. $(x - y)(y - z)(xy + yz + zx) + 1 = 0$
2. $y^2 + z^2 + x + z - 3 = 0$
3. $x^2 + y^2 + z - xz - y = 1$
4. $x^2 + y^2 - 2x = z^2 - 4z$
5. General solution $\phi\left(\frac{x}{z}, \frac{y}{z}\right) = 0$; surface $y^2 = 4xz$.
6. $(x^2 + y^2 + 4z^2)(x^2 - y^2)^2 = a^4(x^2 + y^2)$.

Chapter 3

Nonlinear Partial Differential Equations of Order One

Nonlinear partial differential equations describe several physical systems. These types of equations are not easy to study. Actually, no general methods exist that can be used for solving all such equations. We have several methods for solving nonlinear partial differential equations; Charpit's method is one of them. This method is named after a French mathematician Paul Charpit, one of the fascinating figures of French mathematics in the late 18th century. It is actually the general method for solving first-order nonlinear partial differential equations with two independent variables. Generally, this method is used to find the complete solution with two arbitrary constants.

3.1 Definitions of various classes of integrals

1. A solution or an integral of a differential equation is a relation between the variables involved, by means of which and the derivatives obtained therefrom, the given differential equation is satisfied.

Below are given definitions of various classes of integrals of a partial differential equation of order one.

2. Complete Integral: Singular Integral

Let the nonlinear partial differential equation of order one

$$f(x, y, z, p, q) = 0 \tag{3.1.1}$$

be derived from

$$\phi(x, y, z, a, b) = 0 \tag{3.1.2}$$

by eliminating the arbitrary constants a and b. Then (3.1.2) is called the complete integral or complete solution of (3.1.1).

This complete solution represents a two-parameter family of surfaces which may or may not have an envelope. To find the envelope (when exists) we eliminate a and b from

$$\left.\begin{array}{c} \phi(x, y, z, a, b) = 0 \\ \dfrac{\partial \phi}{\partial a} = 0; \quad \dfrac{\partial \phi}{\partial b} = 0 \end{array}\right\}.$$

If the eliminant

$$\lambda(x, y, z) = 0 \tag{3.1.3}$$

satisfies (3.1.1), it is called the singular solution of (3.1.1). If $\lambda(x, y, z) = \xi(x, y, z) + \eta(x, y, z)$, where $\xi(x, y, z) = 0$ satisfies (3.1.1), while $\eta(x, y, z) = 0$ does not, then $\eta(x, y, z) = 0$ is the singular solution of (3.1.1). As in case of ordinary differential equations, the singular solution may be obtained from the P.D.E. by eliminating p and q from

$$\left.\begin{array}{c} f(x, y, z, p, q) = 0 \\ \dfrac{\partial f}{\partial p} = 0, \quad \dfrac{\partial f}{\partial q} = 0 \end{array}\right\}.$$

Example 3.1.1. *$\phi = z - ax - by + (a^2 + b^2) = 0$ is a complete solution of $f = z - px - qy + (p^2 + q^2) = 0$ (can be easily checked).*

Eliminating a and b from

$$\left.\begin{array}{c} \phi = z - ax - by + a^2 + b^2 = 0 \\ \dfrac{\partial \phi}{\partial a} = -x + 2a = 0; \dfrac{\partial \phi}{\partial b} = -y + 2b = 0 \end{array}\right\}$$

we get $z = \frac{1}{2}x^2 + \frac{1}{2}y^2 - \frac{1}{4}(x^2 + y^2) = \frac{1}{4}(x^2 + y^2)$.

This satisfies the differential equation and is the singular solution.

See that the complete solution represents a two-parameter family of planes which envelopes the paraboloid $x^2 + y^2 = 4z$.

3. General solution: If, in the complete solution (3.1.2) $\phi\ (x, y, z, a, b) = 0$ one of the constants, say b, is replaced by a known function of the other, say $b = \theta(a)$, then $\phi(x, y, z, a, \theta(a)) = 0$ is a one-parameter family of the surfaces of (3.1.1).

If this family has an envelope, its equation may be found as usual by eliminating a from

$$\phi(x, y, z, a, \theta(a)) = 0 \text{ and } \frac{\partial}{\partial a}\phi(x, y, z, a, \phi(a)) = 0$$

and determining that part of the result which satisfies (3.1.1).

Example 3.1.2. *Let $b = \theta(a) = a$ in the complete solution of* Example (3.1.1), *given above.*

3.2 General method of solving the PDE of first order in two independent variables x and y: Charpit's method

Let

$$F(x, y, z, p, q) = 0 \tag{3.2.1}$$

be a given P.D.E. of first order in two independent variables x and y; z is a function of x and y; $p = \frac{\partial z}{\partial x}, q = \frac{\partial z}{\partial y}$. Since z depends on x and y, therefore,

$$dz = pdx + qdy. \tag{3.2.2}$$

Now, if another relation can be found among x, y, z, p, q such as

$$f(x, y, z, p, q) = 0, \tag{3.2.3}$$

then p and q can be eliminated:

The values of p and q deduced from (3.2.1) and (3.2.3) can be substituted in (3.2.2) and the elimination of p and q is then possible.

The integral of the ordinary differential equation thus formed involving x, y, z will satisfy the given eq. (3.2.1).

The problem thus reduces to find a relation of the form (3.2.3) which together with (3.2.1) will determine p and q that will render (3.2.2) integrable.

On differentiating (3.2.1) and (3.2.3) w.r.t. x and y, we shall obtain the following equations:

$$\left.\begin{array}{l}\text{(i)}\ \dfrac{\partial F}{\partial x}+\dfrac{\partial F}{\partial z}p+\dfrac{\partial F}{\partial p}\dfrac{\partial p}{\partial x}+\dfrac{\partial F}{\partial q}\dfrac{\partial q}{\partial x}=0\\ \text{(ii)}\ \dfrac{\partial F}{\partial y}+\dfrac{\partial F}{\partial z}q+\dfrac{\partial F}{\partial p}\dfrac{\partial p}{\partial y}+\dfrac{\partial F}{\partial q}\dfrac{\partial q}{\partial y}=0\end{array}\right\}\quad \left.\begin{array}{l}\text{(iii)}\ \dfrac{\partial f}{\partial x}+\dfrac{\partial f}{\partial z}p+\dfrac{\partial f}{\partial p}\dfrac{\partial p}{\partial x}+\dfrac{\partial f}{\partial q}\dfrac{\partial q}{\partial x}=0\\ \text{(iv)}\ \dfrac{\partial f}{\partial y}+\dfrac{\partial f}{\partial z}q+\dfrac{\partial f}{\partial p}\dfrac{\partial p}{\partial y}+\dfrac{\partial f}{\partial q}\dfrac{\partial q}{\partial y}=0\end{array}\right\}$$

The elimination of $\frac{\partial p}{\partial x}$ from (i) and (iii) gives

$$\left(\frac{\partial F}{\partial x}\frac{\partial f}{\partial p}-\frac{\partial F}{\partial p}\frac{\partial f}{\partial x}\right)+p\left(\frac{\partial F}{\partial y}\frac{\partial f}{\partial p}-\frac{\partial F}{\partial p}\frac{\partial f}{\partial z}\right)+\frac{\partial q}{\partial x}\left(\frac{\partial F}{\partial q}\frac{\partial f}{\partial p}-\frac{\partial F}{\partial p}\frac{\partial f}{\partial q}\right)=0;$$

and the elimination of $\frac{\partial q}{\partial y}$ from (ii) and (iv) gives

$$\left(\frac{\partial F}{\partial y}\frac{\partial f}{\partial q}-\frac{\partial F}{\partial q}\frac{\partial f}{\partial y}\right)+q\left(\frac{\partial F}{\partial z}\frac{\partial f}{\partial q}-\frac{\partial F}{\partial q}\frac{\partial f}{\partial z}\right)+\frac{\partial p}{\partial y}\left(\frac{\partial F}{\partial p}\frac{\partial f}{\partial q}-\frac{\partial F}{\partial q}\frac{\partial f}{\partial p}\right)=0.$$

On adding the LHS of these two equations we see that the last bracketed terms cancel each other, since

$$\frac{\partial q}{\partial x}=\frac{\partial^2 z}{\partial x\partial y}=\frac{\partial^2 z}{\partial y\partial x}=\frac{\partial p}{\partial y}.$$

Hence, adding and re-arranging, we get

$$\left(\frac{\partial F}{\partial x}+p\frac{\partial F}{\partial z}\right)\frac{\partial f}{\partial p}+\left(\frac{\partial F}{\partial y}+q\frac{\partial F}{\partial z}\right)\frac{\partial f}{\partial q}+\left(-p\frac{\partial F}{\partial p}-q\frac{\partial F}{\partial q}\right)\frac{\partial f}{\partial z}+\left(-\frac{\partial F}{\partial p}\right)\frac{\partial f}{\partial x}+\left(-\frac{\partial F}{\partial q}\right)\frac{\partial f}{\partial y}=0.\quad(3.2.4)$$

This is a linear equation of first order, which the auxiliary function f of eq. (3.2.3) must satisfy. Its integrals are integrals of the auxiliary equations:

$$\frac{dp}{\frac{\partial F}{\partial x}+p\frac{\partial F}{\partial z}}=\frac{dq}{\frac{\partial F}{\partial y}+q\frac{\partial F}{\partial z}}=\frac{dz}{-p\frac{\partial F}{\partial p}-q\frac{\partial F}{\partial q}}=\frac{dx}{-\frac{\partial F}{\partial p}}=\frac{dy}{-\frac{\partial F}{\partial q}}=\frac{df}{0}.\quad(3.2.5)$$

The equations (3.2.5) are known as Charpit's auxiliary equations. Any integrals of these equations involving p or q or both can be taken for the required second relation (3.2.3). Actually the simplest relation involving p or q or both is taken as relation (3.2.3). After obtaining the relation (3.2.3) p and q are obtained from (3.2.1) and (3.2.3) and these values are substituted in (3.2.2). On integrating it, we get the required complete solution of the given equation.

Example 3.2.1. *Find a complete integral of $px + qy = pq$.*

Solution: Here the given equation is

$$F(x, y, z, p, q) = px + qy - pq = 0.\quad(1)$$

$\therefore$ Charpit's auxiliary equations are

$$\frac{dp}{\frac{\partial F}{\partial x}+p\frac{\partial F}{\partial z}}=\frac{dq}{\frac{\partial F}{\partial y}+q\frac{\partial F}{\partial z}}=\frac{dz}{-p\frac{\partial F}{\partial p}-q\frac{\partial F}{\partial q}}=\frac{dx}{-\frac{\partial F}{\partial p}}=\frac{dy}{-\frac{\partial F}{\partial q}}$$

or,
$$\frac{dp}{p+p\cdot 0}=\frac{dq}{q+q\cdot 0}=\frac{dz}{-p(x-q)-q(y-p)}=\frac{dx}{-(x-q)}=\frac{dy}{-(y-p)}\quad(2)$$

Taking the first two ratios,

$$\frac{dp}{p} = \frac{dq}{q} \Rightarrow \ln p = \ln q + \ln a \Rightarrow p = aq \text{ (}a\text{ is an arbitrary constant.)} \quad (3)$$

Substituting this value of p in (1), we get

$$aqx + qy - aq^2 = 0 \text{ or, } aq = ax + y \ (q \neq 0). \quad (4)$$

From (3) and (4),

$$q = \frac{ax + y}{a}, \quad p = ax + y. \quad (5)$$

Putting these values of p and q in $dz = pdx + qdy$, we get

$$dz = (ax + y)dx + \frac{ax + y}{a}dy \text{ or, } adz = a(ax + y)dx + (ax + y)dy \text{ or, } adz = (ax + y)d(ax + y).$$

Integrating, $az = \frac{(ax+y)^2}{2} + b$, which is a complete integral, a, b being arbitrary constants.

Example 3.2.2. *Find a complete integral of* $q = 3p^2$.

Solution: Here the given equation is

$$f(x, y, z, p, q) = 3p^2 - q = 0. \quad (1)$$

$\therefore$ Charpit's auxiliary equations are

$$\frac{dp}{\frac{\partial F}{\partial x} + p\frac{\partial F}{\partial z}} = \frac{dq}{\frac{\partial F}{\partial y} + q\frac{\partial F}{\partial z}} = \frac{dz}{-p\frac{\partial F}{\partial p} - q\frac{\partial F}{\partial q}} = \frac{dx}{-\frac{\partial F}{\partial p}} = \frac{dy}{-\frac{\partial F}{\partial q}}$$

$$\text{or, } \frac{dp}{0 + p \cdot 0} = \frac{dq}{0 + q \cdot 0} = \frac{dz}{-6p^2 + q} = \frac{dx}{-6p} = \frac{dy}{1}. \quad (2)$$

Taking the first ratio of (2), $dp = 0$ so that

$$p = a. \quad (3)$$

Substituting this value of p in (1), we get

$$q = 3a^2. \quad (4)$$

Putting these values of p and q in $dz = pdx + qdy$, we obtain $dz = adx + 3a^2dy$ so that $z = ax + 3a^2y + b$ (a, b are arbitrary constants).

This is a complete integral.

Example 3.2.3. *Verify that a complete integral of* $z = pq$ *(by using Charpit's Method) is* $2\sqrt{z} = x\sqrt{a} + \left(\frac{1}{\sqrt{a}}\right) y + b$, *($a, b$ are arbitrary constants).*

Try yourself.

Example 3.2.4. *Find a complete integral of* $zpq - p - q = 0$.

Solution: Verify that Charpit's auxiliary equations are

$$\frac{dp}{p^2q} = \frac{dq}{pq^2} = \frac{dz}{-p(qz - 1) - q(pz - 1)} = \frac{dx}{-(qz - 1)} = \frac{dy}{-(pz - 1)}.$$

Now take first two ratios, obtain $p = \frac{1+a}{z}$ and $q = \frac{1+a}{az}$ and obtain a complete integral $z^2 = 2(1 + a)\left[x + \frac{1}{a}y\right] + b$, where a and b are two arbitrary constants.

Example 3.2.5. *Verify that a complete integral of* $p^2 - y^2 q = y^2 - x^2$ *is* $z = \frac{x}{2}\sqrt{a^2 - x^2} + \frac{a^2}{2}\sin^{-1}\frac{x}{a} - \frac{a^2}{y} - y + b$.

[Take the ratios, $\frac{dp}{2x} = \frac{dx}{-2p}$ and proceed as above.]

Example 3.2.6. *Find a complete integral of* $z^2(p^2 z^2 + q^2) = 1$.

Here $F(x, y, z, p, q) = p^2 z^4 + q^2 z^2 - 1 = 0$.

Charpit's auxiliary equations are

$$\frac{dp}{p(4p^2 z^3 + 2zq^2)} = \frac{dq}{q(4p^2 z^3 + 2zq^2)} = \frac{dz}{-2p^2 z^4 - 2q^2 z^2} = \frac{dx}{-2pz^4} = \frac{dy}{-2qz^2}.$$

Taking the first two ratios, $\frac{dp}{p} = \frac{dq}{q}$ so that $p = aq$.

Solving for p and q,

$$p = \frac{a}{z\sqrt{a^2 z^2 + 1}}, \quad q = \frac{1}{z\sqrt{a^2 z^2 + 1}}$$

so that $dz = pdx + qdy \;\Rightarrow\; adx + dy = z\sqrt{a^2 z^2 + 1}\,dz$.

Integrating,

$$ax + y = \int (a^2 z^2 + 1)^{1/2} \cdot z\, dz$$

or, $$ax + y + b = \frac{1}{3a^2}(a^2 z^2 + 1)^{3/2} \text{ (putting } a^2 z^2 + 1 = t^2)$$

or, $$9a^4(ax + y + b)^2 = (a^2 z^2 + 1)^3,$$

which is a complete integral, a and b being arbitrary constants.

Example 3.2.7. *Find a complete integral of* (i) $q = (z + px)^2$ (ii) $p = (z + qy)^2$.

(i) Here the given equation is $F(x, y, z, p, q) = (z + px)^2 - q = 0$.

Charpit's auxiliary equations are

$$\frac{dp}{2p(z + px) + 2p(z + px)} = \frac{dq}{2q(z + px)} = \frac{dz}{-2px(z + px) + q} = \frac{dx}{-2x(z + px)} = \frac{dy}{0}.$$

Taking the second and fourth ratios, $\frac{dq}{q} = -\frac{dx}{x}$.

Integrating, $\ln q = \ln a - \ln x$ so that $q = a/x$.

Hence, from the given equation $p = \frac{\sqrt{a}}{x\sqrt{x}} - \frac{z}{x}$.

$$\therefore\; dz = pdx + qdy = \left(\frac{\sqrt{a}}{x\sqrt{x}} - \frac{z}{x}\right)dx + \frac{a}{x}dy \text{ or, } xdz + zdx = \sqrt{a}\frac{dx}{\sqrt{x}} + ady \text{ or, } d(xz) = \sqrt{a}x^{-\frac{1}{2}}dx + ady.$$

Integrating, $xz = 2\sqrt{a}\sqrt{x} + ay + b$ (a, b being arbitrary constants).

(ii) Similar method: A complete integral: $yz = ax + \sqrt{ay} + b$.

Example 3.2.8. *Solve for a complete integral of* $yzp^2 - q = 0$.

Try yourself.

Solution: $z^2(a^2 - y^2) = (x + b)^2$ or, $z^2 = 2ax + a^2 y^2 + b$.

Example 3.2.9. *Find a complete integral, a singular solution and general solution of* $(p^2 + q^2)y = qz$.

Solution: Here the given equation is

$$F(x, y, z, p, q) = (p^2 + q^2)y - qz = 0. \tag{1}$$

$\therefore$ Charpit's auxiliary equations are

$$\frac{dp}{-pq} = \frac{dq}{p^2} = \frac{dz}{-2p^2y + qz - 2q^2y} = \frac{dx}{-2py} = \frac{dy}{-2q + z}. \tag{2}$$

Taking the first two ratios, we get $pdp + qdq = 0$ and hence (on integrating)

$$p^2 + q^2 = a^2. \tag{3}$$

(3) and (1) give $q = \frac{a^2y}{z}$ and $p = \frac{a}{z}\sqrt{z^2 - a^2y^2}$.

Putting these values of p and q in $dz = pdx + qdy$, we get

$$dz = \frac{a}{z}\sqrt{z^2 - a^2y^2}\,dx + \frac{a^2y}{z}dz \text{ or, } \frac{z\,dz - a^2y\,dy}{\sqrt{(z^2 - a^2y^2)}} = adx.$$

Integrating $(z^2 - a^2y^2)^{1/2} = ax + b$.

$$z^2 - a^2y^2 = (ax + b)^2 \tag{4}$$

which is a required complete integral.

To find singular integral we differentiate this complete integral partially w.r.t. a and b, and obtain

$$0 = 2ay^2 + 2(ax + b) \cdot x \tag{5}$$
$$0 = 2(ax + b). \tag{6}$$

Eliminating a and b between (4), (5) and (6), we get $z = 0$, which clearly satisfies (1) ($\because p = 0, q = 0$) and hence it is the singular integral.

General integral: Replacing b by some function of a, say $b = \phi(a)$ in (4), we get

$$z^2 - a^2y^2 = [ax + \phi(a)]^2. \tag{7}$$

Differentiating (7) partially w.r.t. a, we get

$$-2ay^2 = 2[ax + \phi(a)][x + \phi'(a)]. \tag{8}$$

The general integral is obtained by eliminating a from (7) and (8).

Example 3.2.10. *Find a complete and singular integrals of* $2xz - px^2 - 2qxy + pq = 0$.

Solution: Here the given equation is

$$F = 2xz - px^2 - 2qxy + pq = 0. \tag{1}$$

$\therefore$ Charpit's auxiliary equations are

$$\frac{dp}{2z - 2qy} = \frac{dq}{0} = \frac{dz}{px^2 + 2xyq - 2pq} = \frac{dx}{x^2 - q} = \frac{dy}{2xy - p}.$$

The second ratio gives $dq = 0$ or, $q = a$.

Putting $q = a$ in (1), we get $p = \frac{2x(z-ay)}{x^2-a}$.

Hence, from $dz = pdx + qdy$, we deduce

$$dz = \frac{2x(z-ay)}{x^2-a}dx + ady \text{ or, } \frac{dz - ady}{z-ay} = \frac{2x\,dx}{x^2-a}.$$

Hence, on integration,

$$\ln(z-ay) = \ln(x^2-a) + \ln b \text{ or, } z - ay = b(x^2-a) \text{ or, } z = ay + b(x^2-a) \tag{2}$$

is the required complete integral, a, b being arbitrary constants.

Differentiating the complete integral w.r.t. a and b we get $0 = y - b$ and $0 = x^2 - a$, i.e. $a = x^2$, $b = y$; substituting these values of a and b in (2) we get $z = x^2y$, which is the required singular integral.

Examples III

(Exercises based on Charpit's Method)

Find a complete integral of the following partial differential equations: (Using Charpit's auxiliary equations)

1. $q = px + p^2$
2. $pxy + pq + qy = yz$
3. $p^2 + q^2 - 2px - 2qy + 1 = 0$
4. $z = px + qy + p^2 + q^2$
5. $p^2 + q^2 - 2px - 2qy + 2xy = 0$
6. $p^2x + q^2y = z$
7. $2x(q^2z^2 + 1) = pz$
8. $2z + p^2 + qy + 2y^2 = 0$
9. $(p+q)(px+qy) = 1$
10. $2(z + px + qy) = yp^2$
11. $p - 3x^2 = q^2 - y$
12. $p(q^2+1) + (b-z)q = 0$.

Answers

1. $z = \frac{x^2}{4} \pm \frac{1}{2}\left[\frac{x}{2}\sqrt{x^2+4a} + 2a\ln\{x + \sqrt{x^2-4a}\}\right] + ay + b.$
2. $(z - ax)(y+a)^a = bc^y.$
3. $(a^2+1)z = \frac{1}{2}v^2 \pm \left\{\frac{1}{2}v\sqrt{v^2-(a^2+1)}\right\} - \frac{1}{2}(a^2+1)\ln\left\{v + \sqrt{v^2-(a^2+1)}\right\} + b$, where $v = ax + y$.
4. $z = ax + by + a^2 + b^2.$
5. $2z = x^2 + y^2 + ax + ay \pm \frac{1}{\sqrt{2}}\left\{(x-y)\sqrt{(x-y)^2 - \frac{a^2}{2}}\right\} - \frac{a^2}{2}\ln\left[(x-y) + \sqrt{(x+y)^2 - \frac{a^2}{2}}\right].$
6. $(1+a)^{1/2}\sqrt{z} = \sqrt{a}\sqrt{x} + \sqrt{y} + b.$
7. $z^2 + 2(a^2+1)x^2 + 2ay + b.$
8. $2y^2z + y^2(x-a)^2 + y^4 = b.$
9. $z(a+1)^{1/2} = 2(ax+b)^{1/2} + b.$
10. $yz - a(x/y) + (a^2/4y^2) = b.$
11. $z = x^2 + ax \pm \frac{2}{3}(y+a)^{3/2} + b.$
12. $2\sqrt{a(z-b)-1} = x + ay + b.$

Chapter 4

Classification of Partial Differential Equations of Second Order and Canonical Forms

In this chapter, our main aim is to study linear partial differential equations of second order that have been widely studied since the very beginning of the twentieth century. Generally, these types of equations are classified into three categories, namely, elliptic, parabolic and hyperbolic, which have a wide range of applications on several domains, especially on Physics. In order to examine whether an equation is elliptic or parabolic or hyperbolic, we need a standard form of the equation that is known as canonical form, a very simple and unique representation of the partial differential equation. Canonical form of an equation plays very important role in the study of partial differential equations. It gives the nature of the equation; however, we can find its solution without any complication by using the canonical form. An equation can be reduced to its canonical form by a suitable transformation of coordinates. Though, the transformation process corresponding to the three categories are different. The most relevant information on four topics, namely, (a) *second-order linear partial differential equations,* (b) *classification of linear partial differential equations of second order,* (c) *transformation of coordinates and* (d) *canonical forms, have been introduced here in consecutive four articles.*

4.1 Second-order linear partial differential equations

The general second-order linear partial differential equation in two independent variables x and y is of the form:

$$A(x,y)\frac{\partial^2 z}{\partial x^2} + B(x,y)\frac{\partial^2 z}{\partial x\partial y} + C(x,y)\frac{\partial^2 z}{\partial y^2} + D(x,y)\frac{\partial z}{\partial x} + E(x,y)\frac{\partial z}{\partial y} + F(x,y)z = G(x,y), \tag{4.1.1}$$

where A, B, C, D, E, F and G are the functions of x and y, defined in some domain $\Omega \subseteq \mathbb{R}^2$.

Here, A, B, C cannot vanish simultaneously in Ω, because (4.1.1) represents a second-order linear partial differential equation. In addition, $z(x, y)$ and all the coefficients of (4.1.1) are twice continuously differentiable in Ω.

Equation (4.1.1) can be expressed in more compact form as

$$Az_{xx} + Bz_{xy} + Cz_{yy} + F(x, y, z, z_x, z_y) = 0, \tag{4.1.2}$$

i.e. $Ar + Bs + Ct + F(x, y, z, p, q) = 0$.

4.2 Classification of linear partial differential equations of second order

The second-order linear (and quasilinear) partial differential equations can be classified (under some restrictions) into three categories namely, (i) Hyperbolic equations, (ii) Parabolic equations, (iii) Elliptic equations.

In analytical geometry, we have seen that the classification of the conic sections represented by the general equation of second degree

$$Ax^2 + Bxy + Cy^2 + Dx + Ey + F = 0 \tag{4.2.1}$$

depends on $D = B^2 - 4AC$ and $\Delta = ACF + \frac{1}{4}BDE - \frac{1}{4}AE^2 - \frac{1}{4}CD^2 - \frac{1}{4}FB^2$. The eq. (4.2.1) represents a parabola or a hyperbola or an ellipse according as $D = 0$ or $D > 0$ or $D < 0$, when $\Delta \neq 0$. So in order to classify eq. (4.1.2), we would like to use the terminology parabolic, hyperbolic and elliptic partial differential equation at a point $(x_0, y_0) \in \Omega$ depending on the sign of $D(x_0, y_0) = [B(x_0, y_0)]^2 - 4A(x_0, y_0)C(x_0, y_0)$, the discriminant of the eq. (4.1.2) at the point (x_0, y_0).

(i) *Hyperbolic Equation*: The partial differential eq. (4.1.2) is said to be hyperbolic at a point (x_0, y_0) if $D(x_0, y_0) > 0$. It will be hyperbolic everywhere in $\Omega \subseteq \mathbb{R}^2$ if $D(x, y) > 0$ for all $(x, y) \in \Omega$.

Example 4.2.1. *Show that the equation $z_{xx} - 2\sin x z_{xy} - \cos^2 x z_{yy} + \sin x z_y = 0$ is hyperbolic.*

Solution: Comparing the given equation with (4.1.2), we have $A = 1$, $B = -2\sin x$, $C = -\cos^2 x$.

$\therefore$ the discriminant of the equation is given by

$D(x, y) = B^2 - 4AC = 4\sin^2 x + 4\cos^2 x = 4 > 0$ for all $(x, y) \in \mathbb{R}^2$.

Hence, the given equation is hyperbolic in $\mathbb{R}^2$.

(ii) *Parabolic Equation*: The partial differential eq. (4.1.2) is said to be parabolic at a point (x_0, y_0) if $D(x_0, y_0) = 0$, i.e. $[B(x_0, y_0)]^2 - 4A(x_0, y_0)C(x_0, y_0) = 0$. If $D(x, y) = 0$ for all $(x, y) \in \Omega \subseteq \mathbb{R}^2$, the eq. (4.1.2) is called parabolic in Ω.

Example 4.2.2. *Show that the partial differential equation $x^2 z_{xx} - 2xyz_{xy} + y^2 z_{yy}^2 + xz_x + yz_y = 0$ is parabolic in $\mathbb{R}^2$.*

Solution: Comparing the given equation with (4.1.2), we have $A = x^2$, $B = -2xy$ and $C = y^2$.

Thus, the discriminant of the equation is given by

$D(x, y) = B^2 - 4AC = 4x^2y^2 - 4x^2y^2 = 0$ for all $(x, y) \in \mathbb{R}^2$.

It confirms that the given equation is parabolic in $\mathbb{R}^2$.

(iii) *Elliptic Equation*: The partial differential eq. (4.1.2) is said to be elliptic at a point (x_0, y_0), if $D(x_0, y_0) < 0$, i.e. $[B(x_0, y_0)]^2 - 4A(x_0, y_0)B(x_0, y_0) < 0$. It will be everywhere elliptic in $\Omega \subseteq \mathbb{R}^2$, if $D(x, y) < 0$ for all $(x, y) \in \Omega$.

Example 4.2.3. *Show that the following partial differential equation is elliptic in $\mathbb{R}^2$:*

$$(1 + x^2)z_{xx} + (1 + y^2)z_{yy} + 2yz_x + 2xz_y = 0.$$

Solution: Comparing the given equation with (4.1.2), we have $A = (1 + x^2)$, $B = 0$, $C = 1 + y^2$.

$\therefore$ the discriminant of the equation is $D(x, y) = B^2 - 4AC$, i.e. $D(x, y) = -4(1 + x^2)(1 + y^2) < 0$ for all $(x, y) \in \mathbb{R}^2$. It follows that the given equation is elliptic in $\mathbb{R}^2$.

Note 4.2.1. *It is remarkable that a particular partial differential equation may be different in nature in a domain $\Omega \subseteq \mathbb{R}^2$. For example, let us consider the Euler-Tricomi equation $z_{xx} + xz_{yy} = 0$.*

Comparing it with (4.1.2), we have $A = 1, B = 0$ and $C = x$.

$\therefore$ $D(x, y) = B^2 - 4AC = -4x$.

Thus, the Euler-Tricomi equation will be parabolic if $x = 0$, hyperbolic if $x < 0$ and elliptic if $x > 0$.

Note 4.2.2. *From the physical point of view, parabolic, hyperbolic and elliptic partial differential equations represent the time-dependent diffusion processes, wave propagation and equilibrium processes respectively.*

4.3 Transformation of coordinates

In order to explore the significance of $D(x, y)$ and to classify the partial differential eq. (4.1.2), we would like to reduce it to a normal or canonical form that is the simplest form of the eq. (4.1.2). Now, we introduce two new independent variables ξ and η by means of the transformation

$$\xi = \xi(x, y) \text{ and } \eta = \eta(x, y), \tag{4.3.1}$$

where ξ and η are both twice continuously differentiable and the Jacobian of transformation $J = \dfrac{\partial(\xi, \eta)}{\partial(x, y)} \neq 0$, i.e. $\begin{vmatrix} \xi_x & \xi_y \\ \eta_x & \eta_y \end{vmatrix} \neq 0$ in the region Ω.

It ensures the existence of a one-to-one transformation between the new and old variables.

Now,

$$p = \frac{\partial z}{\partial x} = \frac{\partial z}{\partial \xi}\frac{\partial \xi}{\partial x} + \frac{\partial z}{\partial \eta}\frac{\partial \eta}{\partial x} = z_\xi \xi_x + z_\eta \eta_x \tag{4.3.2}$$

$$q = \frac{\partial z}{\partial y} = \frac{\partial z}{\partial \xi}\frac{\partial \xi}{\partial y} + \frac{\partial z}{\partial \eta}\frac{\partial \eta}{\partial y} = z_\xi \xi_y + z_\eta \eta_y. \tag{4.3.3}$$

$$\therefore \quad \frac{\partial}{\partial x} \equiv \frac{\partial \xi}{\partial x}\frac{\partial}{\partial \xi} + \frac{\partial \eta}{\partial x}\frac{\partial}{\partial \eta} \text{ and } \frac{\partial}{\partial y} \equiv \frac{\partial \xi}{\partial y}\frac{\partial}{\partial \xi} + \frac{\partial \eta}{\partial y}\frac{\partial}{\partial \eta}.$$

$$r = \frac{\partial^2 z}{\partial x^2} = \frac{\partial}{\partial x}\left(\frac{\partial z}{\partial x}\right) = \left(\frac{\partial \xi}{\partial x}\frac{\partial}{\partial \xi} + \frac{\partial \eta}{\partial x}\frac{\partial}{\partial \eta}\right)\left(\frac{\partial z}{\partial \xi}\frac{\partial \xi}{\partial x} + \frac{\partial z}{\partial \eta}\frac{\partial \eta}{\partial x}\right)$$

$$= \frac{\partial \xi}{\partial x}\frac{\partial}{\partial \xi}\left(\frac{\partial z}{\partial \xi}\frac{\partial \xi}{\partial x} + \frac{\partial z}{\partial \eta}\frac{\partial \eta}{\partial x}\right) + \frac{\partial \eta}{\partial x}\frac{\partial}{\partial \eta}\left(\frac{\partial z}{\partial \xi}\frac{\partial \xi}{\partial x} + \frac{\partial z}{\partial \eta}\frac{\partial \eta}{\partial x}\right)$$

$$= \left(\frac{\partial \xi}{\partial x}\right)^2 \frac{\partial^2 z}{\partial \xi^2} + 2\frac{\partial \xi}{\partial x}\frac{\partial \eta}{\partial x}\frac{\partial^2 z}{\partial \xi \partial \eta} + \left(\frac{\partial \eta}{\partial x}\right)^2 \frac{\partial^2 z}{\partial \eta^2} + \frac{\partial z}{\partial \xi}\frac{\partial^2 \xi}{\partial x^2} + \frac{\partial z}{\partial \eta}\frac{\partial^2 \eta}{\partial x^2}.$$

$$\therefore \; r = z_{xx} = \xi_x^2 z_{\xi\xi} + 2\xi_x \eta_x z_{\xi\eta} + \eta_x^2 z_{\eta\eta} + z_\xi \xi_{xx} + z_\eta \eta_{xx}. \tag{4.3.4}$$

$$S = \frac{\partial^2 z}{\partial x \partial y} = \frac{\partial}{\partial x}\left(\frac{\partial z}{\partial y}\right) = \left(\frac{\partial \xi}{\partial x}\frac{\partial}{\partial \xi} + \frac{\partial \eta}{\partial x}\frac{\partial}{\partial \eta}\right)\left(\frac{\partial z}{\partial \xi}\frac{\partial \xi}{\partial y} + \frac{\partial z}{\partial \eta}\frac{\partial \eta}{\partial y}\right)$$

$$= \frac{\partial \xi}{\partial x}\frac{\partial}{\partial \xi}\left(\frac{\partial z}{\partial \xi}\frac{\partial \xi}{\partial y} + \frac{\partial z}{\partial \eta}\frac{\partial \eta}{\partial y}\right) + \frac{\partial \eta}{\partial x}\frac{\partial}{\partial \eta}\left(\frac{\partial z}{\partial \xi}\frac{\partial \xi}{\partial y} + \frac{\partial z}{\partial \eta}\frac{\partial \eta}{\partial y}\right)$$

$$= \frac{\partial \xi}{\partial x}\frac{\partial \xi}{\partial y}\frac{\partial^2 z}{\partial \xi^2} + \left(\frac{\partial \xi}{\partial x}\frac{\partial \eta}{\partial y} + \frac{\partial \xi}{\partial y}\frac{\partial \eta}{\partial x}\right)\frac{\partial^2 z}{\partial \xi \partial \eta} + \frac{\partial \eta}{\partial x}\frac{\partial \eta}{\partial y}\frac{\partial^2 z}{\partial \eta^2} + \frac{\partial z}{\partial \xi}\frac{\partial^2 \xi}{\partial x \partial y} + \frac{\partial z}{\partial \eta}\frac{\partial^2 \eta}{\partial x \partial y}.$$

$$\therefore \; S = z_{xy} = \xi_x \xi_y z_{\xi\xi} + (\xi_x \eta_y + \xi_y \eta_x)\, z_{\xi\eta} + \eta_x \eta_y z_{\eta\eta} + z_\xi \xi_{xy} + z_\eta \eta_{xy}. \tag{4.3.5}$$

and

$$t = \frac{\partial^2 z}{\partial y^2} = \frac{\partial}{\partial y}\left(\frac{\partial z}{\partial y}\right) = \left(\frac{\partial \xi}{\partial y}\frac{\partial}{\partial \xi} + \frac{\partial \eta}{\partial y}\frac{\partial}{\partial \eta}\right)\left(\frac{\partial z}{\partial \xi}\frac{\partial \xi}{\partial y} + \frac{\partial z}{\partial \eta}\frac{\partial \eta}{\partial y}\right)$$

$$= \frac{\partial \xi}{\partial y}\frac{\partial}{\partial \xi}\left(\frac{\partial z}{\partial \xi}\frac{\partial \xi}{\partial y} + \frac{\partial z}{\partial \eta}\frac{\partial \eta}{\partial y}\right) + \frac{\partial \eta}{\partial y}\frac{\partial}{\partial \eta}\left(\frac{\partial z}{\partial \xi}\frac{\partial \xi}{\partial y} + \frac{\partial z}{\partial \eta}\frac{\partial \eta}{\partial y}\right)$$

$$= \left(\frac{\partial \xi}{\partial y}\right)^2 \frac{\partial^2 z}{\partial \xi^2} + 2\frac{\partial \xi}{\partial y}\frac{\partial \eta}{\partial y}\frac{\partial^2 z}{\partial \xi \partial \eta} + \left(\frac{\partial \eta}{\partial y}\right)^2 \frac{\partial^2 z}{\partial \eta^2} + \frac{\partial z}{\partial \xi}\frac{\partial^2 \xi}{\partial y^2} + \frac{\partial z}{\partial \eta}\frac{\partial^2 \eta}{\partial y^2}.$$

$$\therefore \; t = z_{yy} = \xi_y^2 z_{\xi\xi} + 2\xi_y \eta_y z_{\xi\eta} + z_{\eta\eta}\eta_y^2 + z_\xi \xi_{yy} + z_\eta \eta_{yy}. \tag{4.3.6}$$

Using (4.3.2), (4.3.3), (4.3.4), (4.3.5) and (4.3.6) in (4.1.2), we obtain

$$az_{\xi\xi} + bz_{\xi\eta} + cz_{\eta\eta} = f(\xi, \eta, z, z_\xi, z_\eta), \tag{4.3.7}$$

where F becomes f and A, B, C become a, b, c respectively in new coordinates. Here, a, b, c and f are given by

$$\left.\begin{aligned} a &= A\xi_x^2 + B\xi_x\xi_y + C\xi_y^2 \\ b &= 2A\xi_x\eta_x + B(\xi_x\eta_y + \xi_y\eta_x) + 2C\xi_y\eta_y \\ c &= A\eta_x^2 + B\eta_x\eta_y + C\eta_y^2 \\ f &= -\left[z_\xi\left(A\xi_{xx} + B\xi_{xy} + C\xi_{yy}\right) + z_\eta\left(A\eta_{xx} + B\eta_{xy} + C\eta_{yy}\right) + F\right] \end{aligned}\right\} \tag{4.3.8}$$

The system (4.3.8) can be expressed in the following form:

$$\begin{pmatrix} a & b/2 \\ b/2 & c \end{pmatrix} = \begin{pmatrix} \xi_x & \xi_y \\ \eta_x & \eta_y \end{pmatrix} \begin{pmatrix} A & B/2 \\ B/2 & C \end{pmatrix} \begin{pmatrix} \xi_x & \xi_y \\ \eta_x & \eta_y \end{pmatrix}^T$$

$$\therefore \begin{vmatrix} a & b/2 \\ b/2 & c \end{vmatrix} = \begin{vmatrix} \xi_x & \xi_y \\ \eta_x & \eta_y \end{vmatrix} \begin{vmatrix} A & B/2 \\ B/2 & C \end{vmatrix} \begin{vmatrix} \xi_x & \xi_y \\ \eta_x & \eta_y \end{vmatrix}$$

$$\text{or, } ac - \frac{b^2}{4} = \left(AC - \frac{B^2}{4}\right)J^2$$

$$\text{or, } b^2 - 4ac = (B^2 - 4AC)J^2,$$

$$\text{i.e. } d = J^2 D, \tag{4.3.9}$$

where $d = b^2 - 4ac$ is called the discriminant of the eq. (4.3.7). The relation (4.3.9) ensures that the transformation $\xi = \xi(x, y)$, $\eta = \eta(x, y)$ is unable to change the type of the partial differential eq. (4.1.2), if the Jacobian of transformation $J \neq 0$ in Ω.

Now, we would like to find ξ and η so that the eq. (4.3.7) reduces to the simplex form. It will be quite easy if $B^2 - 4AC$, the discriminant of the quadratic equation

$$A\lambda^2 + B\lambda + C = 0 \tag{4.3.10}$$

is either everywhere zero or positive or negative in Ω. Here, (4.3.10) is known as characteristic polynomial equation in λ corresponding to the eq. (4.1.1).

4.4 Canonical forms

By a suitable transformation, the partial differential eq. (4.1.2) may reduce to a simple form like (4.3.7) in which one or possible two of the coefficients of the leading second-order term of the eq. (4.3.7) vanish in the region Ω. Such reduced equation is known as canonical form of the eq. (4.1.2). Generally, a canonical form specifies a unique representation of a partial differential equation. Now, we would like to introduce the canonical forms of the hyperbolic, parabolic and elliptic partial differential equations of second order.

4.4.1 Canonical forms of hyperbolic equations

Let the eq. (4.1.2) be hyperbolic. Then the discriminant $D = B^2 - 4AC > 0$ in the region Ω. Since the sign of the discriminant is invariant under the transformation $\xi = \xi(x, y), \eta = \eta(x, y)$, we must have $d = b^2 - 4ac > 0$.

Case 4.4.1. *Let $a = c = 0$. Then equation (4.3.7) becomes*

$$z_{\xi\eta} = \phi(\xi, \eta, z, z_\xi, z_\eta), \tag{4.4.1}$$

where $\phi = \dfrac{f}{b}$. The equation of the form (4.4.1) is known as the first canonical form of the hyperbolic equation.

Case 4.4.2. *Let $b = 0$ and $c = -a$. Then equation (4.3.7) becomes*

$$z_{\alpha\alpha} - z_{\beta\beta} = \phi(\alpha, \beta, z, z_\alpha, z_\beta) \tag{4.4.2}$$

where $\phi = \frac{f}{b}$. Here, the equation of the form (4.4.2) is called the second canonical form of the hyperbolic equation.

First canonical forms of the hyperbolic equations

In order to reduce the eq. (4.1.2) to the first canonical form, we have to choose the new variables ξ, η in such a way that the coefficients a and c vanish simultaneously in the region Ω of the eq. (4.3.7).

For $a = c = 0$, the system (4.3.8) gives

$$\left.\begin{aligned} A\xi_x^2 + B\xi_x\xi_y + C\xi_y^2 = 0 \\ A\eta_x^2 + B\eta_x\eta_y + C\eta_y^2 = 0 \end{aligned}\right\}. \tag{4.4.3}$$

This system can be expressed also in the following form:

$$A\left(\frac{\xi_x}{\xi_y}\right)^2 + B\left(\frac{\xi_x}{\xi_y}\right) + C = 0 \tag{4.4.4}$$

$$A\left(\frac{\eta_x}{\eta_y}\right)^2 + B\left(\frac{\eta_x}{\eta_y}\right) + C = 0. \tag{4.4.5}$$

Clearly, the eq. (4.4.4) is a quadratic in $\left(\frac{\xi_x}{\xi_y}\right)$. Let the roots of this equation be $\lambda_1(x, y)$ and $\lambda_2(x, y)$, which are given by

$$\lambda_1(x, y) = \frac{-B + \sqrt{B^2 - 4AC}}{2A} \text{ and } \lambda_2(x, y) = \frac{-B - \sqrt{B^2 - 4AC}}{2A}$$

provided $A(x, y) \neq 0$ in Ω. We see that $\lambda_1(x, y)$ and $\lambda_2(x, y)$ are also the roots of the eq. (4.4.5). Since the total number of possible distinct roots of the equations (4.4.4) and (4.4.5) is exactly two, we may consider λ_1 and λ_2 as the roots of the equations (4.4.4) and (4.4.5) respectively, where

$$\left.\begin{aligned} \lambda_1(x, y) = \frac{\xi_x}{\xi_y} = \frac{-B + \sqrt{B^2 - 4AC}}{2A} \\ \lambda_2(x, y) = \frac{\eta_x}{\eta_y} = \frac{-B - \sqrt{B^2 - 4AC}}{2A} \end{aligned}\right\}. \tag{4.4.6}$$

$\therefore$ λ_1 and λ_2 are the roots of the quadratic eq. (4.3.10). By (4.4.6), we obtain the following two equations:

$$\xi_x - \lambda_1(x, y)\xi_y = 0 \tag{4.4.7}$$

$$\eta_x - \lambda_2(x, y)\eta_y = 0. \tag{4.4.8}$$

The Lagrange's auxiliary equations of (4.4.7) are

$$\frac{dx}{1} = \frac{dy}{-\lambda_1} = \frac{d\xi}{0}.$$

Considering 3rd ratio, we have $d\xi = 0$, i.e. $\xi(x, y) =$ constant.
Again, considering 1st and 2nd ratios, we have

$$\frac{dy}{dx} = -\lambda_1(x, y), \text{ i.e. } \frac{dy}{dx} + \lambda_1(x, y) = 0. \tag{4.4.9}$$

By similar arguments, eq. (4.4.8) gives

$$\eta(x, y) = \text{constant and } \frac{dy}{dx} + \lambda_2(x, y) = 0. \tag{4.4.10}$$

Here, equations (4.4.9) and (4.4.10) are called the characteristic equations of (4.1.2). Thus, we see that the number of characteristic equations of a hyperbolic equation is two.

Now, solving the ordinary differential equations (4.4.9) and (4.4.10), we obtain $\xi(x, y) = c_1$ and $\eta(x, y) = c_2$ respectively, where c_1 and c_2 are two arbitrary constants.

The families of curves $\xi(x, y) = c_1$ and $\eta(x, y) = c_2$ are called the characteristic curves or characteristics of the hyperbolic partial differential equation. It follows that every second-order linear hyperbolic partial differential equation has two families of real characteristic curves on the xy-plane.

An observation

Let A, B, C be constants. Then

$$\lambda_1 = \frac{-B + \sqrt{B^2 - 4AC}}{2A} \text{ and } \lambda_2 = \frac{-B - \sqrt{B^2 - 4AC}}{2A}$$

are both constants.

Thus, solving (4.4.9) and (4.4.10), we obtain respectively

$$\left.\begin{aligned} y + \lambda_1 x = c_1 \\ y + \lambda_2 x = c_2 \end{aligned}\right\}, \tag{4.4.11}$$

where c_1 and c_2 are two arbitrary constants.

Equations in (4.4.11) represent two distinct families of parallel straight lines. Again, we have seen that the families of curves $\xi(x, y) = c_1$ and $\eta(x, y) = c_2$ are the characteristics.

$\therefore$ we can define ξ and η by

$$\xi = y + \frac{-B + \sqrt{B^2 - 4AC}}{2A} x = y + \lambda_1 x$$

$$\eta = y + \frac{-B - \sqrt{B^2 - 4AC}}{2A} x = y + \lambda_2 x.$$

$$\therefore\ \xi_x = \frac{-B + \sqrt{B^2 - 4AC}}{2A},\ \xi_y = 1,\ \eta_x = \frac{-B - \sqrt{B^2 - 4AC}}{2A},\ \eta_y = 1.$$

Thus, the first canonical form of the hyperbolic eq. (4.3.7) is $z_{\xi\eta} = \phi(\xi, \eta, z, z_\xi, z_\eta)$, where $\phi = f/b$ and

$$b = 2A\xi_x\eta_x + B(\xi_x\eta_y + \xi_y\eta_x) + 2C\xi_y\eta_y, \text{ [by (4.3.8)]}$$

$$= 2A\frac{B^2 - (B^2 - 4AC)}{4A^2} + B\left(-\frac{B}{A}\right) + 2C = 4C - \frac{B^2}{A} = -\frac{B^2 - 4AC}{A} = -\frac{D}{A}.$$

Second canonical forms of the hyperbolic equations

In order to reduce the eq. (4.1.2) to the second canonical form, we need to choose the new variables ξ, η in such a way that $b = 0$ and $c = -a$ hold. In this case, the hyperbolic partial differential equation reduces to the form $z_{\alpha\alpha} - z_{\beta\beta} = \psi(\alpha, \beta, z, z_\alpha, z_\beta)$ by the linear transformation $\alpha = \xi + \eta$, $\beta = \xi - \eta$. The reduced form $z_{\alpha\alpha} - z_{\beta\beta} = \psi(\alpha, \beta, z, z_\alpha, z_\beta)$ is known as the second canonical form of (4.1.2).

Working rule

(i) Confirm that $D = B^2 - 4AC > 0$ for hyperbolic equation.

(ii) Find the roots λ_1 and λ_2 of the quadratic equation $A\lambda^2 + B\lambda + C = 0$.

(iii) Solve the characteristic equation $\frac{dy}{dx} + \lambda_1 = 0$ and $\frac{dy}{dx} + \lambda_2 = 0$ to obtain characteristics $\xi(x, y) = c_1$ and $\eta(x, y) = c_2$.

(iv) Choose ξ, η as new variables.

(v) Find b and f using (4.3.8), here $a = c = 0$ that may be verified by (4.3.8).

(vi) Find that $\phi = \frac{f}{b}$.

(vii) Write the canonical form $z_{\xi\eta} = \phi$.

Example 4.4.1. *Show that the equation $z_{xx} - x^2 z_{yy} = 0$ $(x \neq 0)$ is hyperbolic. Obtain its characteristics and hence find its canonical form.*

Solution: Comparing the given equation with (4.1.2), we have $A = 1$, $B = 0$, $C = -x^2$ and $F = 0$.

$\therefore$ discriminant of the equation

$$z_{xx} - x^2 z_{yy} = 0 \tag{4.4.12}$$

is given by $D(x, y) = B^2 - 4AC = 4x^2 > 0$ $(\because\ x \neq 0)$.

Hence, the given equation is hyperbolic type.

The λ-quadratic equation corresponding to the eq. (4.4.12) is $A\lambda^2 + B\lambda + C = 0$, i.e. $\lambda^2 - x^2 = 0$.

$\therefore$ $\lambda = \pm x$, which are real and distinct.

$\therefore$ the characteristic equations are $\frac{dy}{dx} + x = 0$ and $\frac{dy}{dx} - x = 0$.

Solving, we have $y + \frac{1}{2}x^2 = c_1$ and $y - \frac{1}{2}x^2 = c_2$, where c_1 and c_2 are two arbitrary constants.

Hence, the characteristics of the given equation are $y + \dfrac{1}{2}x^2 = c_1$ and $y - \dfrac{1}{2}x^2 = c_2$ which represent two different families of curves in xy-plane.

$\therefore$ by the transformation, $\xi = y + \dfrac{1}{2}x^2$, $\eta = y - \dfrac{1}{2}x^2$ the given equation reduces to its canonical form.

Clearly, $\xi_x = x$, $\xi_y = 1$, $\xi_{xx} = 1$, $\xi_{xy} = 0$, $\xi_{yy} = 0$, $\eta_x = -x$, $\eta_y = 1$, $\eta_{xx} = -1$, $\eta_{xy} = 0$, $\eta_{yy} = 0$.

$\therefore$ $a = A\xi_x^2 + B\xi_x\xi_y + C\xi_y^2 = 0$, $c = A\eta_x^2 + B\eta_x\eta_y + C\eta_y^2 = 0$ and $b = 2A\xi_x\eta_x + B(\xi_x\eta_y + \xi_y\eta_x) + 2C\xi_y\eta_y = -4x^2$.

Again, $f = -\left[z_\xi(A\xi_{xx} + B\xi_{xy} + C\xi_{yy}) + z_\eta(A\eta_{xx} + B\eta_{xy} + C\eta_{yy}) + F\right] = -(z_\xi - z_\eta)$.

$$\therefore\ \phi = \frac{f}{b} = \frac{z_\xi - z_n}{4(\xi - \eta)}.$$

Hence, desired canonical form is

$$z_{\xi\eta} = \frac{z_\xi - z_\eta}{4(\xi - \eta)}.$$

Example 4.4.2. *Reduce the following equation to its canonical form and hence solve: $r - 2\sin x s - \cos^2 x t - \cos x q = 0$.*

Solution: Comparing the given equation with (4.1.2), we have $A = 1$, $B = -2\sin x$, $C = -\cos^2 x$ and $F = -\cos xq$.

$\therefore$ the discriminant of the equation is given by $D = B^2 - 4AC$, i.e. $D = 4\sin^2 x + 4\cos^2 x = 4 > 0$ for all $(x, y) \in \mathbb{R}^2$.

It follows that the equation is hyperbolic.

The λ-quadratic equation is $A\lambda^2 + B\lambda + C = 0$, i.e. $\lambda^2 - 2\sin x \cdot \lambda - \cos^2 x = 0$ or, $\lambda^2 - 2\sin x \cdot \lambda + \sin^2 x = 1$.

$\therefore\ \lambda = \sin x \pm 1$.

$\therefore$ the characteristic equations are $\frac{dy}{dx} + \sin x + 1 = 0$ and $\frac{dy}{dx} + \sin x - 1 = 0$.

Solving these two equations, we have $y - \cos x + x = c_1$ and $y - \cos x - x = c_2$, where c_1, c_2 are two arbitrary constants. These two families of curves are known as the characteristics of the given partial differential equation.

Now, let us choose the transformation $\xi = y - \cos x + x$, $\eta = y - \cos x - x$. Then $\xi_x = \sin x + 1$, $\xi_y = 1$, $\xi_{xx} = \cos x$, $\xi_{xy} = 0$, $\xi_{yy} = 0$, $\eta_x = \sin x - 1$, $\eta_y = 1$, $\eta_{xx} = \cos x$, $\eta_{xy} = 0$, $\eta_{yy} = 0$.

It is quite easy to verify that $a = 0$ and $c = 0$.

Again, $b = -4$ and $F = -\cos xq = -\cos x(z_\xi \xi_y + z_\eta \eta_y) = -\cos x(z_\xi + z_\eta)$.

$\therefore\ F = -\left[z_\xi(A\xi_{xx} + B\xi_{xy} + C\xi_{yy}) + z_\eta(A\eta_{xx} + B\eta_{xy} + C\eta_{yy}) + F\right]$
$= 0$.

Consequently, $\phi = f/b = 0$.

Hence, desired canonical form of the given equation is $z_{\xi\eta} = 0$. Now, integrating both sides of the equation $z_{\xi\eta} = 0$ w.r.t. ξ, we obtain $z_\eta = f_1(\eta)$, f_1 being an arbitrary function.

Integrating again w.r.t. η, we have

$$z = \int f_1(\eta)d\eta + \phi(\xi),$$

ϕ being another arbitrary function.

$\therefore\ z = \psi(\eta) + \phi(\xi)$, where

$$\psi(\eta) = \int f_1(\eta)\, d\eta, \text{ i.e. } z = \phi(y - \cos x + x) + \psi(y - \cos x - x),$$

which is the required general solution.

4.4.2 Canonical forms of the parabolic equations

Let the eq. (4.1.2) be parabolic. Then the discriminant is given by $D = B^2 - 4AC = 0$ in the region Ω.

$\therefore$ by (4.3.9), we have $d = b^2 - 4ac = 0$. In this case, a (or c) $= 0$ and $b = 0$. Thus, we need to choose the new variables ξ and η in such a way that a (or C) $= 0$ and b vanishes. Then the eq. (4.3.7) reduces to the form:

$$z_{\eta\eta} = \phi(\xi, \eta, z, z_\xi, z_\eta), \tag{4.4.13}$$

where $\phi = f/c$. Here, the eq. (4.4.23) is called the canonical form of the parabolic eq. (4.1.2).

Now, for $a = 0$, (4.3.8) gives

$$A\xi_x^2 + B\xi_x\xi_y + C\xi_y^2 = 0 \text{ or, } A\left(\frac{\xi_x}{\xi_y}\right)^2 + B\left(\frac{\xi_x}{\xi_y}\right) + C = 0. \tag{4.4.14}$$

The roots of this quadratic equation in $\left(\frac{\xi_x}{\xi_y}\right)$ are real and equal.

$$\therefore\ \lambda = \frac{\xi_x}{\xi_y} = \frac{-B \pm \sqrt{B^2 - 4AC}}{2A}, \text{ if } A(x, y) \neq 0 \text{ in } \Omega$$

$$= -\frac{B}{2A}, \ (\because\ B^2 - 4AC = 0),$$

which represents the root of the eq. (4.4.14).

$$\therefore\ \xi_x - \lambda\xi_y = 0 \tag{4.4.15}$$

The Lagrange's auxiliary equations of (4.4.15) are $\frac{dx}{1} = \frac{dy}{-\lambda} = \frac{d\xi}{0}$.
Considering 3rd ratio, we have $d\xi = 0$.
$\therefore\ \xi(x, y) = \text{constant}$.
Again, considering 1st and 2nd ratios,

$$\frac{dy}{dx} + \lambda(x, y) = 0. \tag{4.4.16}$$

Here, the new variable ξ is determined by the eq. (4.4.16) that is called the characteristic equation of (4.1.2). Equation (4.4.16) represents an ordinary differential equation corresponding to the family of curves (known as characteristics) in the xy-plane along which $\xi(x, y) = \text{constant}$.

In order to determine another variable η, we put $b = 0$ in (4.3.8) that gives

$$2A\xi_x\eta_x + B(\xi_x\eta_y + \xi_y\eta_x) + 2C\xi_y\eta_y = 0$$

$$\text{or, } 2A\left(\frac{\xi_x}{\xi_y}\right)\eta_x + B\left[\left(\frac{\xi_x}{\xi_y}\right)\eta_y + \eta_x\right] + 2C\eta_y = 0$$

$$\text{or, } 2A\left(-\frac{B}{2A}\right)\eta_x + B\left(-\frac{B}{2A}\cdot\eta_y + \eta_x\right) + 2C\eta_y = 0$$

$$\text{or, } (B^2 - 4AC)\eta_y = 0.$$

Since $B^2 - 4AC = 0$, η_y will be an arbitrary function of x and y. It confirms that the transformation variable η can be chosen arbitrarily satisfying $J = \frac{\partial(\xi,\eta)}{\partial(x,y)} \neq 0$ in Ω.

An observation

Let A, B, C be constants. Then $\lambda = -\frac{B}{2A}$ is constant.

$\therefore$ solving (4.4.26), we obtain $y + \lambda x = c_1$, c_1 being arbitrary constant, i.e. $y - \frac{B}{2A}x = c_1$.

Since $\xi(x, y) = \text{constant}$ represents the characteristic curves, we set $\xi = y - \frac{B}{2A}x$. Let us choose $\eta = x$.

$$\text{Then } J = \begin{vmatrix} \xi_x & \xi_y \\ \eta_x & \eta_y \end{vmatrix} = \begin{vmatrix} -\frac{B}{2A} & 1 \\ 1 & 0 \end{vmatrix} = -1 \neq 0.$$

By (4.3.8), we have

$$b = 2A\xi_x\eta_x + B(\xi_x\eta_y + \xi_y\eta_x) + 2C\xi_y\eta_y = 2A\left(-\frac{B}{2A}\right) + B = 0.$$

Thus, by $\xi = y - \frac{B}{2A}x$, $\eta = x$, eq. (4.3.7) reduces to the canonical form $z_{\eta\eta} = \phi(\xi, \eta, z, z_\xi, z_\eta)$, where $\phi = f/c$.

The canonical form depends on the choice of the new variable η. If we choose $c = 0$ instead of $a = 0$, the form will be $z_{\xi\xi} = \phi(\xi, \eta, z, z_\xi, z_\eta)$.

Working rule

(i) Confirm that $D = B^2 - 4AC = 0$ for parabolic equation.
(ii) Find the real and equal roots (λ_1) of the quadratic equation $A\lambda^2 + B\lambda + C = 0$.
(iii) Solve the characteristic equation $\frac{dy}{dx} + \lambda_1 = 0$ to obtain the characteristic $\xi(x, y) = c_1$.
(iv) Select ξ as a new variable.
(v) Choose another new variable η so that $J = \frac{\partial(\xi,\eta)}{\partial(x,y)} \neq 0$.
(vi) Find a, b, c and f using (4.3.8).
(vii) Find $\phi = f/c$, if $c \neq 0$, otherwise $\phi = f/a$.
(viii) Write the canonical equation $z_{\eta\eta} = \phi$ (or, $z_{\xi\xi} = \phi$).

Example 4.4.3. *Reduce the equation $z_{xx} - 4z_{xy} + 4z_{yy} + z = 0$ to its canonical form.*

Solution: Comparing the given equation with (4.1.2), we have $A = 1, B = -4, C = 4, F = Z$.
$\therefore$ the discriminant of the equation

$$z_{xx} - 4z_{xy} + 4z_{yy} = 0 \tag{4.4.17}$$

is given by $D = B^2 - 4AC = 0$. It follows that the eq. (4.4.15) is parabolic.
The λ-quadratic equation corresponding to the eq. (4.4.15) is $A\lambda^2 + B\lambda + C = 0$, i.e. $\lambda^2 - 4\lambda + 4 = 0$.
$\therefore$ $\lambda = 2, 2$, which are real and equal.
$\therefore$ the characteristic equation is $\frac{dy}{dx} + 2 = 0$, i.e. $dy + 2dx = 0$.
Integrating, we have $y + 2x = c_1$, c_1 being arbitrary constant.
Let us choose $\xi(x, y) = y + 2x$ and $\eta = y$.
Then $\xi_x = 2, \xi_y = 1, \xi_{xx} = 0, \xi_{xy} = 0, \xi_{yy} = 0, \eta_x = 0, \eta_y = 1, \eta_{xx} = 0, \eta_{xy} = 0, \eta_{yy} = 0$.

$$\therefore\ J = \frac{\partial(\xi, \eta)}{\partial(x, y)} = \begin{vmatrix} \xi_x & \xi_y \\ \eta_x & \eta_y \end{vmatrix} = 2 \neq 0.$$

Thus, the transformation $\xi = y + 2x, \eta = y$ is able to reduce the eq. (4.4.15) to its canonical form.
We see that

$$a = A\xi_x^2 + B\xi_x\xi_y + C\xi_y^2 = 0$$
$$b = 2A\xi_x\eta_x + B(\xi_x\eta_y + \xi_y\eta_x) + 2C\xi_y\eta_y = 0$$
$$c = A\eta_x^2 + B\eta_x\eta_y + C\eta_y^2 = 4$$
$$\text{and } f = -\left[z_\xi(A\xi_{xx} + B\xi_{xy} + C\xi_{yy}) + z_\eta(A\eta_{xx} + B\eta_{xy} + C\eta_{yy}) + F\right] = z.$$

$\therefore\ \phi = \frac{f}{c} = \frac{1}{4}z.$
Hence, desired canonical form is $z_{\eta\eta} = \frac{1}{4}z$.

4.4.3 Canonical forms of the elliptic equations

Let the partial differential eq. (4.1.2) be elliptic. Then the discriminant of this equation is $D = B^2 - 4AC < 0$ in the region Ω.

$\therefore$ by (4.3.9), $d = b^2 - 4ac = J^2D < 0$. It is possible only when $b = 0$ and $a = c$. Thus, in order to reduce the eq. (4.1.2) to its canonical form, we have to choose the new variables ξ, η in such a way that $b = 0$ and $a = c$, i.e. $a - c = 0$.

Now, from the system (4.3.8), we obtain the following equations:

$$\left.\begin{aligned} A(\xi_x^2 - \eta_x^2) + B(\xi_x\xi_y - \eta_x\eta_y) + C(\xi_y^2 - \eta_y^2) = 0 \\ \text{and } 2A\xi_n\eta_x + B(\xi_x\eta_y + \xi_y\eta_x) + 2C\xi_y\eta_y = 0 \end{aligned}\right\}.$$

Clearly, the above system of equations is represented by the equation

$$A(\xi_x + i\eta_x)^2 + B(\xi_x + i\eta_x)(\xi_y + i\eta_y) + C(\xi_y + i\eta_y)^2 = 0, \text{ where } i = \sqrt{-1}$$

$$\text{or, } A\left(\frac{\xi_x + i\eta_x}{\xi_y + i\eta_y}\right)^2 + B\left(\frac{\xi_x + i\eta_x}{\xi_y + i\eta_y}\right) + C = 0 \qquad (4.4.18)$$

$$\text{or, } A\lambda^2 + B\lambda + C = 0, \text{ where } \lambda = \frac{\xi_x + i\eta_x}{\xi_y + i\eta_y}.$$

$$\therefore \lambda = \frac{-B \pm \sqrt{B^2 - 4AC}}{2A}, \text{ if } A(x, y) \neq 0 \text{ in } \Omega$$

$$= \frac{-B \pm i\sqrt{4AC - B^2}}{2A} \quad (\because B^2 - 4AC < 0).$$

Let $\beta(x, y)$ be the complex conjugate of $\alpha(x, y)$ so that

$$\alpha(x, y) = \xi(x, y) + i\eta(x, y)$$

$$\beta(x, y) = \xi(x, y) - i\eta(x, y).$$

Then the roots of the eq. (4.4.18) are given by

$$\lambda_1 = \frac{\alpha_x}{\alpha_y} = \frac{-B + i\sqrt{4AC - B^2}}{2A} \text{ and } \lambda_2 = \frac{\beta_x}{\beta_y} = \frac{-B - i\sqrt{4AC - B^2}}{2A}.$$

$$\therefore \alpha_x - \lambda_1\alpha_y = 0 \qquad (4.4.19)$$

$$\text{and } \beta_x - \lambda_2\beta_y = 0. \qquad (4.4.20)$$

Now, the Lagrange's auxiliary equations of (4.4.19) are $\frac{dx}{1} = \frac{dy}{-\lambda_1} = \frac{d\alpha}{0}$.
Considering 3rd ratio, we have $d\alpha = 0$.

$\therefore$ α = constant.

Again, considering 1st and 2nd ratios, we have

$$\frac{dy}{dx} + \lambda_1 = 0. \qquad (4.4.21)$$

By similar arguments, β= constant and

$$\frac{dy}{dx} + \lambda_2 = 0. \qquad (4.4.22)$$

Equations (4.4.21) and (4.4.22) are known as the characteristic equations of (4.1.2). Clearly, the solutions of these two equations are necessarily complex-valued. Thus, in this case, there does not exist real characteristic of the eq. (4.1.2).

Solving the ordinary differential equations (4.4.21) and (4.4.22), we obtain $\alpha(x, y) = c_1$ and $\beta(x, y) = c_2$, where c_1 and c_2 are arbitrary complex constants. Here, Re α and Im β will give the new variables ξ and η.

$$\therefore \xi = \frac{1}{2}(\alpha + \beta) \text{ and } \eta = \frac{1}{2i}(\alpha - \beta)$$

by which the eq. (4.1.2) reduces to the canonical form $z_{\xi\xi} + z_{\eta\eta} = \phi(\xi, \eta, z, z_\xi, z_\eta)$, where $\phi = f/a$.

Working rule

(i) Confirm that $D = B^2 - 4AC < 0$ for elliptic equation.

(ii) Find the complex roots λ_1, λ_2 of the quadratic equation $A\lambda^2 + B\lambda + C = 0$.

(iii) Solve the equations $\frac{dy}{dx} + \lambda_1 = 0$ and $\frac{dy}{dx} + \lambda_2 = 0$ to find the solutions $\alpha(x, y) = c_1$ and $\beta(x, y) = c_2$ respectively.

(iv) Choose the new variables ξ and η such that $\xi = \frac{1}{2}(\alpha + \beta)$ and $\eta = \frac{1}{2i}(\alpha - \beta)$.

(v) Find $a(= c)$ and f by (4.3.8), where $b = 0$ may be verified by (4.3.8).

(vi) Find $\phi = f/a$ and write the canonical form $z_{\xi\xi} + z_{\eta\eta} = \phi$.

Example 4.4.4. *Reduce the equation $z_{xx} + x^2 z_{yy} = 0$ to its canonical equation for $x \neq 0$.*

Solution: Comparing the given equation with (4.1.2), we have $A = 1$, $B = 0$, $C = x^2$.

$\therefore$ the discriminant, $D = B^2 - 4AC = -4x^2 < 0$ for $x \neq 0$.

It follows that the given equation is elliptic.

Here, the λ-quadratic equation is $A\lambda^2 + B\lambda + C = 0$, i.e. $\lambda^2 + x^2 = 0$.

$\therefore \lambda = \pm ix$.

Thus, the characteristic equations are $\frac{dy}{dx} \pm ix = 0$.

Integrating, we obtain $y + \frac{1}{2}ix^2 = c_1$ and $y - \frac{1}{2}ix^2 = c_2$, where c_1 and c_2 are arbitrary constants.

Let $\alpha = y + \frac{1}{2}ix^2$ and $\beta = y - \frac{1}{2}ix^2$.

Then we choose $\xi = \frac{1}{2}(\alpha + \beta) = y$ and $\eta = \frac{1}{2i}(\alpha - \beta) = \frac{x^2}{2}$.

$\therefore \xi_x = 0, \xi_y = 1, \xi_{xx} = \xi_{xy} = \xi_{yy} = 0, \eta_x = x, \eta_y = 0, \eta_{xx} = 1, \eta_{xy} = \eta_{yy} = 0$.

Now,

$$a = A\xi_x^2 + B\xi_x\xi_y + C\xi_y^2 = x^2$$

$$b = 2A\xi_x\eta_x + B(\xi_x\eta_y + \xi_y\eta_x) + 2C\xi_y\eta_y = 0$$

$$c = A\eta_x^2 + B\eta_x\eta_y + C\eta_y^2 = x^2$$

$$\text{and } f = -[z_\xi(A\xi_{xx} + B\xi_{xy} + C\xi_{yy}) + z_\eta(A\eta_{xx} + B\eta_{xy} + C\eta_{yy}) + F] = -z_\eta.$$

$$\therefore \phi = \frac{f}{a} = -\frac{z_\eta}{x^2} = -\frac{1}{2\eta}z_\eta.$$

Hence, required canonical form is $z_{\xi\xi} + z_{\eta\eta} = -\frac{1}{2\eta}z_\eta$.

4.5 Further solved problems

Example 4.5.1. *Classify the following partial differential equations:*

(i) $z_{xx} - 2\cot x z_{xy} - z_{yy} = 0, x \neq n\pi, n \in \mathbb{Z}$

(ii) $(a - b)(z_{xx} - z_{yy}) + 4hz_{xy} = 0$, a, b, h are constants

(iii) $xz_{xx} - 4z_{xy} + z_{yy} + z_y = 0$.

Solution:

(i) Comparing the given equation with (4.1.2), we have $A = 1$, $B = -2\cot x$, $C = -1$.

$\therefore B^2 - 4AC = 4\cot^2 x + 4 = 4(\cos^2 x + 1) = 4\operatorname{cosec}^2 x > 0$.

Hence, the given equation is hyperbolic.

(ii) Here, $A = a - b$, $B = 4h$, $C = -(a - b)$.

$\therefore B^2 - 4AC = 16h^2 + 4(a - b)^2 > 0$, which shows that the given equation is hyperbolic.

(iii) Comparing the given equation with (4.1.2), we have $A = x$, $B = -4$, $C = 1$.

$\therefore\ B^2 - 4AC = 16 - 4x = 4(4 - x)$.

$$\therefore\ B^2 - 4AC \begin{cases} > 0, & \text{if } x < 4 \\ < 0, & \text{if } x > 4 \\ = 0, & \text{if } x = 4. \end{cases}$$

Hence, the given equation is hyperbolic in $\Omega_1 = \{(x, y) \in \mathbb{R}^2 : x < 4\}$, parabolic in $\Omega_2 = \{(x, y) \in \mathbb{R}^2 : x = 4\}$ and elliptic in $\Omega_2 = \{(x, y) \in \mathbb{R}^2 : x > 4\}$.

Example 4.5.2. *Find the nature of the partial differential equation $z_{xx} + (1 - k^2)z_{yy} = 0$. Reduce this equation to its canonical form and hence find the general solution of the equation for $k > 1$.*

Solution: Comparing the given equation with (4.1.2), we have $A = 1$, $B = 0$, $C = 1 - k^2$ and $F = 0$.

$\therefore$ the discriminant of the equation is given by $D = B^2 - 4AC = -4(1 - k^2)$.

Thus, the given equation will be hyperbolic if $D > 0$, i.e. $|k| > 1$, parabolic if $k = \pm 1$ and elliptic if $|k| < 1$.

The λ-quadratic equation is $A\lambda^2 + B\lambda + C = 0$, i.e. $\lambda^2 + (1 - k^2) = 0$ or, $\lambda^2 = k^2 - 1$.

$\therefore\ \lambda = \pm\sqrt{k^2 - 1}$, which is real and distinct for $|k| > 1$.

The characteristic equations are $\frac{dy}{dx} + \sqrt{k^2 - 1} = 0$ and $\frac{dy}{dx} - \sqrt{k^2 - 1} = 0$.

Solving, we obtain $y + \sqrt{k^2 - 1}x = c_1$ and $y - \sqrt{k^2 - 1}x = c_2$, where c_1 and c_2 are two arbitrary constants.

Let us choose new variables ξ and η such that $\xi = y + \sqrt{k^2 - 1} \cdot x$ and $\eta = y - \sqrt{k^2 - 1} \cdot x$.

Then $\xi_x = \sqrt{k^2 - 1}$, $\xi_y = 1$, $\xi_{xx} = \xi_{xy} = \xi_{yy} = 0$, $\eta_x = -\sqrt{k^2 - 1}$, $\eta_y = 1$, $\eta_{xx} = \eta_{xy} = \eta_{yy} = 0$.

$\therefore\ a = A\xi_x^2 + B\xi_x\xi_y + C\xi_y^2 = 0$, $c = A\eta_x^2 + \beta\eta_x\eta_y + C\eta_y^2 = 0$ and $b = -\frac{D}{A} = -4(k^2 - 1)$.

$\therefore\ f = -[z_\xi(A\xi_{xx} + B\xi_{xy} + C\xi_{yy}) + z_\eta(A\eta_{xx} + B\eta_{xy} + C\eta_{yy}) + F] = 0$.

$\therefore\ \phi = f/b = 0$.

Hence, the canonical form of the given equation is $z_{\xi\eta} = 0$.

Integrating w.r.t. ξ, we obtain $z_\eta = g(\eta)$, g being arbitrary function.

Integrating again w.r.t. η, we obtain

$$z = \int g(\eta)d\eta + \psi_2(\xi),\ \psi_2, \text{ being arbitrary function}$$

$$= \psi_1(\eta) + \psi_2(\xi), \text{ where } \psi_1(\eta) = \int g(\eta)d\eta$$

$$= \psi_1\left(y - \sqrt{k^2 - 1} \cdot x\right) + \psi_2\left(y + \sqrt{k^2 - 1} \cdot x\right).$$

Hence, required general solution is $z = \psi_1\left(y - \sqrt{k^2 - 1} \cdot x\right) + \psi_2\left(y + \sqrt{k^2 - 1} \cdot x\right)$, where ψ_1 and ψ_2 are two arbitrary functions.

Example 4.5.3. *Reduce the following equation to its canonical equation: $2z_{xx} - 4z_{xy} + 2z_{yy} + 3z = 0$.*

Solution: Comparing the given equation with (4.1.2), we have $A = 2$, $B = -4$, $C = 2$, $F = 3z$.

$\therefore$ the discriminant, $D = B^2 - 4AC = 16 - 4 \cdot 2 \cdot 2 = 0$.

It confirms that the equation is parabolic in $\mathbb{R}^2$.

The λ-quadratic equation is $A\lambda^2 + B\lambda + C = 0$, i.e. $2\lambda^2 - 4\lambda + 2 = 0$.

$\therefore\ \lambda = 1, 1$, which are real and equal.

Thus, the characteristic equation is $\frac{dy}{dx} + 1 = 0$ or, $dy + dx = 0$.
Integrating, we obtain $y + x = c_1$, c_1 being arbitrary constant.
Let us choose new variables ξ, η such that $\xi = y + x, \eta = y - x$.
Then $\xi_x = 1, \xi_y = 1, \xi_{xx} = \xi_{xy} = \xi_{yy} = 0, \eta_x = -1, \eta_y = 1, \eta_{xx} = \eta_{xy} = \eta_{yy} = 0$.

$$\therefore J = \frac{\partial(\xi, \eta)}{\partial(x, y)} = \begin{vmatrix} 1 & 1 \\ -1 & 1 \end{vmatrix} = 2 \neq 0.$$

Consequently,

$$b = 0$$
$$a = A\xi_x^2 + B\xi_x\xi_y + C\xi_y^2 = 0$$
$$c = A\eta_x^2 + B\eta_x\eta_y + C\eta_y^2 = 8$$
$$\text{and } f = -[z_\xi \cdot 0 + z_\eta \cdot 0 + 3z] = -3z.$$

$\therefore \phi = \frac{f}{c} = \frac{-3z}{8}$.
Hence, required canonical form is $z_{\eta\eta} = -\frac{3}{8}z$.

Alternative

Here, $A = 2, B = -4, C = 2$.
$\therefore$ the discriminant, $D = B^2 - 4AC = 0$, which shows that the given equation is parabolic in $\mathbb{R}^2$.
The λ-quadratic equation is $A\lambda^2 + B\lambda + C = 0$, i.e. $2\lambda^2 - 4\lambda + 2 = 0$.
$\therefore \lambda = 1, 1$, which are real and distinct.
The characteristic equation is $\frac{dy}{dx} + 1 = 0$.
The solution of this ordinary equation is $y + x = c_1$, c_1 being an arbitrary constant.
Let us choose the transformation $\xi = y + x, \eta = y - x$.
Then $\xi_x = 1, \xi_y = 1, \xi_{xx} = \xi_{xy} = \xi_{yy} = 0, \eta_x = -1, \eta_y = 1, \eta_{xx} = \eta_{xy} = \eta_{yy} = 0$.

$$\therefore J = \frac{\partial(\xi, \eta)}{\partial(x, y)} = 2 \neq 0.$$

$$\therefore z_x = z_\xi\xi_x + z_\eta\eta_x = z_\xi - z_\eta, \text{ by (4.3.2).}$$
$$z_y = z_\xi\xi_y + z_\eta\eta_y = z_\xi + z_\eta, \text{ by (4.3.3).}$$
$$z_{xx} = \xi_x^2 z_{\xi\xi} + 2\xi_x\eta_x z_{\xi\eta} + \eta_x^2 z_{\eta\eta} + z_\xi\xi_{xx} + z_\eta\eta_{xx}, \text{ by (4.3.4)}$$
$$= z_{\xi\xi} - 2z_{\xi\eta} + z_{\eta\eta}.$$
$$z_{xy} = \xi_x\xi_y z_{\xi\xi} + (\xi_x\eta_y + \xi_y\eta_x)z_{\xi\eta} + \eta_x\eta_y z_{\eta\eta} + z_\xi\eta_{xy} + z_\eta\eta_{xy}, \text{ by (4.3.5)}$$
$$= z_{\xi\xi} + (1 - 1)z_{\xi\eta} - z_{\eta\eta}$$
$$= z_{\xi\xi} - z_{\eta\eta}$$
$$\text{and } z_{yy} = \xi_y^2 z_{\xi\xi} + 2\xi_y\eta_y z_{\xi\eta} + z_{\eta\eta}\eta_y^2 + z_\xi\xi_{yy} + z_\eta\eta_{yy}, \text{ by (4.3.6)}$$
$$= z_{\xi\xi} + 2z_{\xi\eta} + z_{\eta\eta}.$$

Using above results in the given equation, we obtain

$$2(z_{\xi\xi} - 2z_{\xi\eta} + z_{\eta\eta}) - 4(z_{\xi\xi} - z_{\eta\eta}) + 2(z_{\xi\xi} + 2z_{\xi\eta} + z_{\eta\eta}) + 3z = 0. \text{ or, } 8z_{\eta\eta} + 3z = 0.$$

$\therefore z_{\eta\eta} = -\frac{3z}{8}$, which is the required canonical form.

Example 4.5.4. *Reduce the following equation to its canonical form and find its general solution:*

$$x^2 z_{xx} + 2xyz_{xy} + y^2 z_{yy} + xyz_x + y^2 z_y = 0,\ x \neq 0, y \neq 0.$$

Solution: Comparing the given equation with (4.1.2), we get $A = x^2$, $B = 2xy$, $C = y^2$, $F = xyz_x + y^2 z_y$.

$\therefore$ the discriminant of the equation is given by $D = B^2 - 4AC = 4x^2y^2 - 4x^2y^2 = 0$.

It follows that the given equation is parabolic.

The λ-quadratic equation is $A\lambda^2 + B\lambda + C = 0$, i.e. $x^2\lambda^2 + 2xy\lambda + y^2 = 0$ or, $(x\lambda + y)^2 = 0$.

$\therefore\ \lambda = -\frac{y}{x}, -\frac{y}{x}$, which are real and equal.

Thus, the characteristic equation is $\frac{dy}{dx} - \frac{y}{x} = 0$.

Solving this ordinary differential equation, we obtain $y/x = c_1$, c_1 being an arbitrary constant.

Let us choose the new variables ξ and η such that $\xi = y/x, \eta = y$.

$$\therefore\ \xi_x = -\frac{y}{x^2}, \xi_y = \frac{1}{x}, \xi_{xx} = \frac{2y}{x^3}, \xi_{xy} = -\frac{1}{x^2}, \xi_{yy} = 0, \eta_x = 0, \eta_y = 1, \eta_{xx} = \eta_{xy} = \eta_{yy} = 0.$$

$$\therefore\ J = \frac{\partial(\xi, \eta)}{\partial(x, y)} = -\frac{y}{x^2} \neq 0, \text{ for } x \neq 0, y \neq 0.$$

Consequently,

$$\begin{aligned}
b &= 0\\
a &= A\xi_x^2 + B\xi_x\xi_y + C\xi_y^2 = 0\\
c &= A\eta_x^2 + B\eta_x\eta_y + C\eta_y^2 = y^2\\
z_x &= z_\xi\xi_x + z_\eta\eta_x = -\frac{y}{x^2}z_\xi\\
z_y &= z_\xi\xi_y + z_\eta\eta_y = \frac{1}{x}z_\xi + z_\eta.
\end{aligned}$$

$$\therefore\ F = xy\left(-\frac{y}{x^2}z_\xi\right) + y^2\left(\frac{1}{x}z_\xi + z_\eta\right) = y^2 z_\eta = \eta^2 z_\eta.$$

$$\begin{aligned}
\therefore\ f &= -[z_\xi(A\xi_{xx} + B\xi_{xy} + C\xi_{yy}) + z_\eta(A\eta_{xx} + B\eta_{xy} + C\eta_{yy}) + F]\\
&= -\eta^2 z_\eta.
\end{aligned}$$

$\therefore\ \phi = f/c = -z_\eta$.

Hence, desired canonical form is $z_{\eta\eta} + z_\eta = 0$.

$$\therefore\ \frac{\partial}{\partial\eta}\left(\frac{\partial z}{\partial\eta}\right) + \frac{\partial z}{\partial\eta} = 0.$$

Integrating w.r.t. η, we have

$$\ln\left(\frac{\partial z}{\partial\eta}\right) + \eta = \ln\psi_1(\xi),\ \psi_1 \text{ being arbitrary function}$$

$$\text{or, } \frac{\partial z}{\partial\eta} = \psi_1(\xi)e^{-\eta}.$$

Integrating w.r.t. η, we have

$$\begin{aligned}
z &= -\psi_1(\xi)e^{-\eta} + \psi_2(\eta),\ \psi_2 \text{ being arbitrary function}\\
&= \psi_2\left(\frac{y}{x}\right) - \psi_1\left(\frac{y}{x}\right)e^{-y}.
\end{aligned}$$

Hence, required general solution is $z = \psi_2\left(\frac{y}{x}\right) - \psi_1\left(\frac{y}{x}\right)e^{-y}$.

Example 4.5.5. *Reduce the equation $z_{xx} - (1+x)^2 z_{yy} = 0$, $x \neq -1$ to its canonical form.*

Solution: Comparing the given equation with the eq. (4.1.2), we have $A = 1, B = 0, C = -(1+x^2), F = 0$.

$\therefore$ the discriminant, $D = B^2 - 4AC = 4(1+x)^2 > 0$.

It follows that the equation is hyperbolic.

The λ-quadratic equation is $A\lambda^2 + B\lambda + C = 0$, i.e. $\lambda^2 - (1+x)^2 = 0$.

$\therefore$ $\lambda = \pm(1+x)$, which are real and distinct.

Now, the characteristic equations are $\frac{dy}{dx} + 1 + x = 0$ and $\frac{dy}{dx} - 1 - x = 0$.

The solutions of these two ordinary differential equations are $y + x + \frac{1}{2}x^2 = c_1$ and $y - x - \frac{1}{2}x^2 = c_2$, where c_1 and c_2 are two arbitrary constants.

Now, let us choose the transformation $\xi = y + x + \frac{1}{2}x^2$ and $\eta = y - x - \frac{1}{2}x^2$.

Then $\xi_x = 1 + x, \xi_y = 1, \xi_{xx} = 1, \xi_{xy} = 0, \xi_{yy} = 0, \eta_x = -1 - x, \eta_y = 1, \eta_{xx} = -1, \eta_{xy} = 0, \eta_{yy} = 0$.

It is quite easy to verify that $a = 0, c = 0, b = -4(1+x)^2$.

$$\text{and } f = -\left[z_\xi(A\xi_{xx} + B\xi_{xy} + C\xi_{yy}) + z_\eta(A\eta_{xx} + B\eta_{xy} + C\eta_{yy}) + F\right] = -\left[z_\xi - z_\eta\right].$$

$$\therefore \phi = \frac{f}{b} = \frac{z_\xi - z_\eta}{4(1+x)^2} = \frac{z_\xi - z_\eta}{4(1+\xi-\eta)}.$$

Hence, desired canonical form is

$$z_{\xi\eta} = \frac{1}{4} \cdot \frac{z_\xi - z_\eta}{1+\xi-\eta}.$$

Examples IV

1. Find the nature of the following partial differential equations:
   (a) $z_{xx} - xz_{xy} + 4z_{yy} + 4z_y = 0$
   (b) $z_{xx} + yz_{xy} - 2z_{yy} = 0$
   (c) $az_{xx} + 2hz_{xy} + bz_{yy} = h(z_{xx} + z_{yy})$; a, b, h are real constant
   (d) $h(z_{xx} - z_{yy}) - (a-b)z_{xy} = 0$; a, b, h are real constant
   (e) $z_{xx} - 2\sec x z_{xy} + z_{yy} = 0$
   (f) $z_{xx} + 2z_{xy} + \operatorname{cosec}^2 y z_{yy} = 0$
   (g) $2(x^2+y^2)z_{xx} + 2(x+y)z_{xy} + z_{yy} = 0$.
2. Find the characteristics of the following equations:
   (a) $z_{xx} - 4z_{xy} + z_{yy} + 6z_x - 14z_y + 12z = 0$
   (b) $2z_{xx} - 3z_{xy} - 2z_{yy} + 7z_x - 9z_y = 0$
   (c) $4z_{xx} - 4z_{xy} + z_{yy} + 2z_x - 26z_y + 9z = 0$
   (d) $3z_{xx} + 10z_{xy} + 3z_{yy} - 14z_y = 0$
   (e) $9z_{xx} + 6\sin x z_{xy} - \cos^2 x z_{yy} - 2z_x = 0$
   (f) $\sin^2 x z_{xx} + 2\cos x z_{xy} - z_{yy} = 0$.
3. Show that the one-dimensional diffusion equation $k\frac{\partial^2 T}{\partial x^2} = \frac{\partial T}{\partial t}$, $-\infty < x < \infty$, $t > 0$ is parabolic.
4. Show that the one-dimensional wave equation $\frac{\partial^2 u}{\delta t^2} - c^2\frac{\partial^2 u}{\delta x^2} = 0$ is hyperbolic. Find the characteristics. Hence, solve it.
5. Show that the equation $z_{xx} + x^2 z_{yy} = 0$ is elliptic everywhere except on the straight line $x = 0$. Write the equation in equivalent canonical form.
6. Find the canonical form of the equation $z_{xx} - xz_{yy} = 0$ in the region, where it is hyperbolic.
7. Reduce the following equation to its canonical form: $x^2 z_{xx} - 2xyz_{xy} + y^2 z_{yy} + xz_x + yz_y = 0$, $x > 0$.
8. Find the canonical form of the equation $yz_{xx} + z_{yy} = 0$, $y > 0$.
9. Write the equation $x^2 z_{xx} - y^2 z_{yy} = 0$ ($x \neq 0, y \neq 0$) in equivalent canonical form.

10. Write the equation $yz_{xx} + (x + y)z_{xy} + xz_{yy} = 0$ in equivalent canonical form.
11. Write the equation $z_{xx} + 2z_{xy} + z_{yy} = 0$ in equivalent canonical form and hence solve it.
12. Find the canonical form of the equation $x^2 z_{xx} + 2xyz_{xy} + y^2 z_{yy} = 0$ and hence solve it.
13. Find the canonical form of the equation $y^2 z_{xx} + x^2 z_{yy} = 0$.
14. Write the following equation in equivalent canonical form: $(1 - c^2)z_{xx} + z_{yy} = 0,\ c > 1$.
15. Find the characteristics of the equation $u_{xx} - 2\sin x u_{xy} - \cos^2 x u_{yy} - \cos x u_y = 0$. Write the equation in equivalent canonical form and hence solve it.

Answers

1. (a) Hyperbolic in $\Omega = \{(x, y) \in \mathbb{R}^2 : |x| > 4\}$
   Parabolic in $\Omega = \{(x, y) \in \mathbb{R}^2 : x = \pm 4\}$
   Elliptic in $\Omega = \{(x, y) \in \mathbb{R}^2 : |x| < 4\}$
   (b) Hyperbolic
   (c) Hyperbolic, if $(a + b)h > ab$
   Parabolic, if $(a + b)h = ab$
   Elliptic, if $(a + b)h < ab$
   (d) Hyperbolic
   (e) Hyperbolic
   (f) Elliptic
   (g) Elliptic in $\Omega = \{(x, y) \in \mathbb{R}^2 : x \neq y\}$
   Parabolic in $\Omega = \{(x, y) \in \mathbb{R}^2 : x = y\}$.
2. (a) $y + (2 + \sqrt{3})x = c_1,\ y + (2 - \sqrt{3})x = c_2$
   (b) $y + 2x = c_1,\ y - \frac{1}{2}x = c_2$
   (c) $y + \frac{1}{2}x = c$
   (d) $y - 3x = c_1,\ y - \frac{1}{3}x = c_2$
   (e) $3y + \cos x + x = c_1,\ 3y + \cos x - x = c_2$
   (f) $y + \operatorname{cosec} x - \cot x = c_1$,
   $y + \operatorname{cosec} x + \cot x = c_2$.
4. $u_{\xi\eta} = 0;\ u = f(x - ct) + g(x + ct)$.
5. $z_{\xi\xi} + z_{\eta\eta} + \frac{1}{2\eta} z_\eta = 0$.
6. $z_{\xi\eta} = \frac{1}{6} \cdot \frac{z_\xi - z_\eta}{\xi - \eta}$ in the region $\Omega = \{(x, y) \in \mathbb{R}^2 : x > 0\}$.
7. $z_{\eta\eta} + \frac{1}{\eta} z_\eta = 0$.
8. $z_{\xi\xi} + z_{\eta\eta} + \frac{1}{3\xi} z_\xi = 0$.
9. $2\xi z_{\xi\eta} - z_\eta = 0$.
10. $\xi z_{\xi\eta} + z_\eta = 0$.
11. $z_{\eta\eta} = 0;\ z = (x + y)f(x - y) + g(x - y)$ (choosing $\eta = x + y$).
12. $z_{\eta\eta} = 0;\ z = yf\left(\frac{y}{x}\right) + g\left(\frac{y}{x}\right)$ (choosing $\eta = y$).
13. $z_{\xi\xi} + z_{\eta\eta} + \frac{1}{2}\left(\frac{1}{\xi} z_\xi + \frac{1}{\eta} z_\eta\right) = 0$.
14. $z_{\xi\eta} = 0$.
15. $u_{\xi\eta} = 0;\ u = f(\cos x + x - y) - g(\cos x - x - y)$.

Chapter 5

Initial and Boundary Value Problems on Wave Equation, Heat Equation and Laplace Equation

The most relevant information on the principle of superposition, wave equations, Fourier's method of separation of variables, Cauchy problem, heat equations and Laplace's equations have been introduced in the present chapter. Superposition principle and Fourier's method of separation of variables are very important for solving homogeneous linear partial differential equations. In our mortal world, we have several types of mechanical waves. All such waves can be described by second-order linear partial differential equations. A French mathematician Jean-Baptiste le Rond d'Alembert (16 November 1717–29 October 1783) was the first who introduced the one-dimensional wave equation on the way to study vibrating strings of musical instruments, though Leonhard Euler, Daniel Bernoulli and Joseph-Louis Lagrange also studied vibrating strings of musical instruments. On the other hand, the theory of heat equation was first introduced by a French mathematician Jean-Baptiste Joseph Fourier (21 March 1768–16 May 1830) for mathematical modelling on heat diffusions. In the last article Laplace equation that is named after a French mathematician Pierre-Simon Laplace (23 March 1749–5 March 1827), has been introduced. The solutions of Laplace equation are harmonic functions by nature. It is very important in several fields of mathematical physics. The general theory of solutions to this equation is often called potential theory.

5.1 The principle of superposition

Let u_1 and u_2 be two solutions of the linear partial differential equations $L[u] = f$ and $L[u] = g$ respectively. Then, for any constants c_1 and c_2, $c_1 f + c_2 g$ is a solution of the linear nonhomogeneous partial differential equation $L[u] = c_1 f + c_2 g$.

Note 5.1.1. *If $f = 0$ and $g = 0$, u_1 and u_2 will be two solutions of the homogeneous equation $L[u] = 0$. By the principle of superposition $c_1 u_1 + c_2 u_2$ must be another solution of the equation $L[u] = 0$.*

Note 5.1.2. *The principle of superposition is not valid for nonlinear partial differential equations. For example, let us consider a nonlinear partial differential equation*

$$\frac{\partial u}{\partial x}\frac{\partial u}{\partial y} - u\left(\frac{\partial u}{\partial x} + \frac{\partial u}{\partial y}\right) + u^2 = 0.$$

Observe that e^x and e^y are two solutions of this equation. But $c_1 e^x + c_2 e^y$ fails to be a solution of the equation unless $c_1 = c_2 = 0$.

5.2 One-dimensional wave equation

Let us consider one-dimensional plane progressive harmonic waves travelling in a medium with a phase velocity c. Then the equation representing a plane progressive harmonic wave is

$$u(x, t) = \alpha \sin(\omega t - kx), \tag{5.2.1}$$

where α is the amplitude, ω is the angular frequency and $k = \omega c$ is the wave number.

Now differentiating (5.1) partially twice with respect to t and x, we obtain respectively

$$\frac{\partial^2 u}{\partial t^2} = -\alpha\omega^2 \sin(\omega t - kx) \tag{5.2.2}$$

$$\text{and } \frac{\partial^2 u}{\partial x^2} = -\alpha k^2 \sin(\omega t - kx). \tag{5.2.3}$$

By (5.2.2) and (5.2.3), we obtain

$$\frac{\partial^2 u}{\partial t^2} - c^2 \frac{\partial^2 u}{\partial x^2} = 0, \tag{5.2.4}$$

where $c = \omega/k$ is the velocity of the wave and $-\infty < x < \infty, t > 0$.

The second-order linear partial differential eq. (5.2.4) is known as the *one-dimensional wave equation*. It was first introduced by Jean-Baptiste le Rond d'Alembert (16 November 1717–29 October 1783), a French mathematician and physicist.

Solution by canonical reduction

Comparing the eq. (5.2.4) with

$$A\frac{\partial^2 u}{\partial t^2} + B\frac{\partial^2 u}{\partial t\, \partial x} + C\frac{\partial^2 u}{\partial x^2} + F\left(x, t, u, \frac{\partial u}{\partial x}, \frac{\partial u}{\partial t}\right) = 0, \tag{5.2.5}$$

we have $A = -1$, $B = 0$, $C = -c^2$ and $F = 0$.

$\therefore$ the discriminant of the wave eq. (5.2.4) is given by $D = B^2 - 4AC = 4c^2 > 0$.

It confirms that the one-dimensional wave equation is hyperbolic type.

Now the λ-quadratic equation is $A\lambda^2 + B\lambda + C = 0$, i.e. $\lambda^2 - c^2 = 0$.

$\therefore \lambda = \pm c$.

$\therefore$ the characteristic equations are $\frac{dx}{dt} \pm c = 0$.

Integrating, we obtain $x + ct = c_1$ and $x - ct = c_2$, where c_1 and c_2 are two arbitrary constants. These two families of straight lines represent the characteristics of the wave eq. (5.2.4) on (x, t)-plane.

Now, let us choose the transformation $\xi = x - ct$ and $\eta = x + ct$. Then $\xi_t = -c$, $\xi_x = 1$, $\xi_{tt} = 0$, $\xi_{tx} = 0$, $\xi_{xx} = 0$, $\eta_t = c$, $\eta_x = 1$, $\eta_{tt} = 0$, $\eta_{tx} = 0$, $\eta_{xx} = 0$.

$\therefore$ $a' = 0$, $c' = 0$, $b' = -\frac{D}{A} = -4c^2$ and $f = 0$. Consequently, $\phi = f/b = 0$.

Thus, the wave equation reduces to the canonical form $w_{\xi\eta} = 0$.

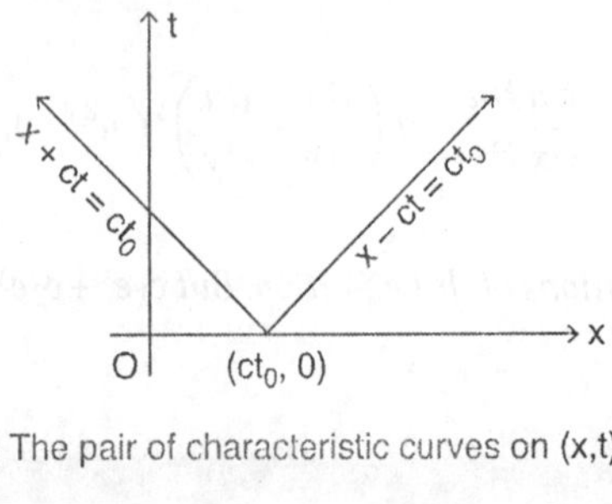

Fig 4.1: The pair of characteristic curves on (x,t)-plane.

Integrating both sides of the equation $w_{\xi\eta} = 0$ with respect to ξ, we obtain $w_\eta = f_1(\eta), f_1(\eta)$ being an arbitrary function.

Integrating again with respect to η, we have

$$w(\xi, \eta) = \int f_1(\eta)d\eta + g(\xi)$$

$g(\xi)$ being another arbitrary function, i.e. $w(\xi, \eta) = g(\xi) + h(\eta)$, where $h(\eta) = \int f_1(\eta)d\eta$.

Hence, the general solution of the wave equation is

$$u(x, t) = g(x - ct) + h(x + ct), \tag{5.2.6}$$

where g and h are two arbitrary functions.

We observe that, for a fixed $t_0 > 0$, the graph of the function $g(x - ct_0)$ has the same shape as the graph of $g(x)$, except that it is shifted to the right by a distance ct_0. Thus, the function $g(x - ct)$ represents a wave moving to the right with velocity c. This wave is known as a *forward wave.* On the other hand, the function $h(x + ct)$ represents a wave travelling to the left with the same speed. It is called a *backward wave.*

Thus, the solution (5.2.6) confirms that every solution of the one-dimensional wave equation is the sum of two such travelling waves.

Important observation

For any two real sectionally continuous functions g and h, (5.2.6) represents a sectionally continuous function $u(x, t)$, which is a superposition of a forward wave and a backward wave travelling in the opposite directions with the wave speed c. The families of straight lines represented by $x - ct$ = constant and $x + ct$ = constant are called the characteristics of the wave equation. Here the characteristics are straight lines in (x, t)-plane with slopes $\pm\frac{1}{c}$.

5.3 The Fourier's method of separation of variables

In the present article, we would like to introduce the method of separation of variables, a powerful basic method for solving partial differential equations. It is also used to obtain the solutions of a wide range of linear partial differential equations with initial and boundary conditions. Perhaps, it is the most well-known and oldest systematic technique for solving the partial differential equation.

Fourier's method for solving such equations is based on the technique of separation of variables. Here we are interested to use the method for solving heat equations, wave equations and Laplace's equations only.

Now, let us introduce the basic steps of this method for the case of two independent variables.

Step 1. First we have to search for solutions of the homogeneous partial differential equations that are called the product solutions. They are of the form $u(x, y) = X(x)Y(y)$ satisfying certain additional conditions, which are just homogeneous boundary conditions in many cases. In such cases, X and Y should be the solutions of two linear ordinary differential equations derived from the partial differential equation.

Step 2. We use the principle of superposition to generate out of the separated solutions, a more general solution of the partial differential equation in the form of an infinite series of the product solution $X(x)Y(y)$.

Step 3. In this step, we compute the coefficients of the infinite series.

If the given partial differential equation involves n independent variables $x_1, x_2, \ldots, x_n$ and one dependent variable u, we first assume that the equation possesses product solution of the form:

$$u(x_1, x_2, \ldots, x_n) = X_1(x_1)X_2(x_2)\ldots X_n(x_n),$$

X_i being a function of x_i only, $i = 1, 2, \ldots, n$. Substituting it into the given equation, we obtain n ordinary differential equations which give n solutions.

Solution of one-dimensional wave equation by separation of variables (Cartesian coordinates)

Vibrations of an infinite string

Let us consider the one-dimensional wave equation

$$\frac{\partial^2 u}{\partial t^2} - c^2\frac{\partial^2 u}{\partial x^2} = 0, \ -\infty < x < \infty, \ t > 0. \tag{5.3.1}$$

Let us assume that the wave eq. (5.3.1) has the solution in the form

$$u(x, t) = X(x)T(t). \tag{5.3.2}$$

Then

$$\frac{\partial^2 u}{\partial t^2} = X(x)T''(t), \ \frac{\partial^2 u}{\partial x^2} = X''(x)T(t)$$

Using the above results in (5.3.1), we obtain

$$X(x)T''(t) - c^2X''(x)T(t) = 0 \text{ or, } \frac{1}{X(x)}\frac{d^2X}{dx^2} = \frac{1}{c^2}\cdot\frac{1}{T(t)}\frac{d^2T}{dt^2} = \mu,$$

μ being a constant of separation.

$$\therefore \ \frac{d^2X}{dx^2} - \mu X = 0 \tag{5.3.3}$$

$$\text{and } \frac{d^2T}{dt^2} - c^2\mu T = 0. \tag{5.3.4}$$

Case 5.3.1. *Let $\mu = \lambda^2 > 0, \lambda$ being real. Then the solutions of the equations (5.3.3) and (5.3.4) are $X(x) = c_1e^{\lambda x} + c_2e^{-\lambda x}$ and $T(t) = c_3e^{c\lambda t} + c_4e^{-c\lambda t}$, where c_1, c_2, c_3, c_4 are arbitrary constants.*

Hence, the solution of the wave eq. (5.3.1) is

$$u(x, t) = \left(c_1e^{\lambda x} + c_2e^{-\lambda x}\right)\left(c_3e^{c\lambda t} + c_4e^{-c\lambda t}\right). \tag{5.3.5}$$

Case 5.3.2. *Let $\mu = 0$. Then the solutions of the equations (5.3.3) and (5.3.4) are $X(x) = c_1 + c_2x$ and $T(t) = c_3 + c_4t$, where c_1, c_2, c_3, c_4 are arbitrary constants. Thus, the solution of the wave equation (5.3.1) is*

$$u(x, t) = (c_1 + c_2x)(c_3 + c_4t). \tag{5.3.6}$$

Case 5.3.3. *Let $\mu = -\lambda^2 < 0$. Then the solutions of the equations (5.3.3) and (5.3.4) are $X(x) = c_1\cos\lambda x + c_2\sin\lambda x$ and $T(t) = c_3\cos c\lambda t + c_4\sin c\lambda t$, where c_1, c_2, c_3, c_4 are arbitrary constants.*

$\therefore$ the solution of the wave equation is

$$u(x, t) = (c_1\cos\lambda x + c_2\sin\lambda x)(c_3\cos c\lambda t + c_4\sin c\lambda t). \tag{5.3.7}$$

Example 5.3.1. *Solve the one-dimensional wave equation* $\frac{\partial^2 u}{\partial t^2} - c^2\frac{\partial^2 u}{\partial x^2} = 0, 0 < x < L, t > 0$ *subject to the boundary conditions* $u(0,t) = 0$, $u(L,t) = 0, t > 0$ *and the initial conditions* $u(x,0) = f(x)$, $u_t(x,0) = g(x)$.

Solution: Let the equation

$$\frac{\partial^2 u}{\partial t^2} - c^2\frac{\partial^2 u}{\partial x^2} = 0,\ 0 < x < L,\ t > 0 \tag{5.3.8}$$

has the solution of the form $u(x,t) = X(x)T(t)$. Then we have two ordinary differential equations given in (5.3.3) and (5.3.4). For $\mu = \lambda^2 > 0$, we have the solution given by (5.3.5). Now, using the boundary conditions $u(0,t) = 0$, $u(L,t) = 0$, $t > 0$ in (5.3.5), we obtain $c_1 + c_2 = 0$ and $c_1 e^{\lambda L} + c_2 e^{-\lambda L} = 0$.

We see that $\begin{vmatrix} 1 & 1 \\ e^{\lambda t} & e^{-\lambda L}\end{vmatrix} = 0 \Rightarrow e^{2\lambda L} = 1 \Rightarrow \lambda = 0 \Rightarrow \mu = 0$, which is impossible for the case $\mu = \lambda^2 > 0$.

$\therefore \begin{vmatrix} 1 & 1 \\ e^{\lambda L} & e^{-\lambda L}\end{vmatrix} \neq 0$. It confirms that $c_1 = c_2 = 0$.

Thus, for $\mu > 0$, we have trivial solution of the wave equation, that is not acceptable.

Next, we consider $\mu = \lambda^2 = 0$. In this case, the solution of the wave eq. (5.3.1) is given by (5.3.6). Using the given boundary conditions on (5.3.6), we obtain $c_3 = c_4 = 0$ ($\because L \neq 0$). It is not acceptable, because for $\mu = 0$, we have trivial solution $u(x,t) = 0$.

So, for the nontrivial solution, we are interested in the case $\mu = -\lambda^2 < 0$. In this case, the solution is given by (5.3.7).

Using the given boundary conditions on it, we obtain $c_1 = 0$ and $\sin \lambda L = 0$.

$\therefore \lambda L = n\pi$, $n = 1, 2, 3, \ldots$, i.e. $\lambda = \frac{n\pi}{L}$, which are known as the eigenvalues. The corresponding solutions are given by

$$\begin{aligned} u_n(x,t) &= c_2 \sin\frac{n\pi x}{L}\left(c_3\cos\frac{n\pi ct}{L} + c_4 \sin\frac{n\pi ct}{L}\right),\ n = 1,2,3,\ldots \\ &= \sin\frac{n\pi x}{L}\left(A_n\cos\frac{n\pi ct}{L} + B_n \sin\frac{n\pi ct}{L}\right), \end{aligned} \tag{5.3.9}$$

where $A_n = c_2c_3$ and $B_n = c_2c_4$.

Thus, by the principle of superposition, we have the solution

$$u(x,t) = \sum_{n=1}^{\infty} \sin\frac{n\pi x}{L}\left(A_n\cos\frac{n\pi ct}{L} + B_n \sin\frac{n\pi ct}{L}\right). \tag{5.3.10}$$

Now, using the initial condition $u(x,0) = f(x)$, we obtain

$$\sum_{n=1}^{\infty} A_n \sin\frac{n\pi x}{L} = f(x).$$

Again,

$$\begin{aligned} u_t(x,t) &= \sum_{n=1}^{\infty} \sin\frac{n\pi x}{L}\left(-A_n\frac{n\pi c}{L}\sin\frac{n\pi ct}{L} + B_n\frac{n\pi c}{L}\cos\frac{n\pi ct}{L}\right) \\ &= \sum_{n=1}^{\infty} \frac{n\pi c}{L}\sin\frac{n\pi x}{L}\left(B_n\cos\frac{n\pi ct}{L} - A_n \sin\frac{n\pi ct}{L}\right). \end{aligned}$$

Using second initial condition $u_t(x, 0) = g(x)$, we obtain

$$\sum_{n=0}^{\infty}\left(\frac{n\pi c}{L}B_n\right)\sin\frac{n\pi x}{L} = g(x), \text{ i.e. } \sum_{n=0}^{\infty} C_n \sin\frac{n\pi x}{L} = g(x), \text{ where } C_n = \frac{n\pi c}{L}B_n.$$

Clearly,

$$\sum_{n=0}^{\infty} A_n \sin\frac{n\pi x}{L} \text{ and } \sum_{n=0}^{\infty} C_n \sin\frac{n\pi x}{L} \text{ are half-range Fourier series.}$$

$$\therefore A_n = \frac{2}{L}\int_0^L f(x)\sin\frac{n\pi x}{L}\,dx \text{ and } B_n = \frac{2}{L}\int_0^L g(x)\sin\frac{n\pi x}{L}\,dx.$$

Hence, required solution of the wave eq. (5.3.8) is

$$u(x, t) = \sum_{n=1}^{\infty}\sin\frac{n\pi x}{L}\left[\left\{\frac{2}{L}\int_0^L f(x)\sin\frac{n\pi x}{L}\,dx\right\}\cos\frac{n\pi ct}{L} + \left\{\frac{2}{L}\int_0^L g(x)\sin\frac{n\pi x}{L}\,dx\right\}\sin\frac{n\pi ct}{L}\right].$$

Note 5.3.1. *$u_n(x, t)$ given by (5.3.9) are known as normal modes of vibration and $\omega_n = \frac{n\pi c}{L}$, $n = 1, 2, 3, \ldots$ are called normal frequencies.*

5.4 The Cauchy problem

For a partial differential equation of order m defined on $\mathbb{R}^n$, the Cauchy problem consists of finding the solution $u(x)$ that satisfies the following conditions:

$u(x) = f(x)$ for all $x \in S$.

$\frac{\partial^k u(x)}{\partial n^k} = g_k(x)$ for $k = 1, 2, 3, \ldots, m-1$ and $x \in S$, where $g_k(x)$ are given functions defined on the surface S and n is a normal vector to S. Here, the surface S is called the *Cauchy surface* and the conditions are called the *Cauchy conditions.*

The Cauchy problems have very important role in the study of initial and boundary value problems in partial differential equations. It is named after Augustin-Louis Cauchy (21 August 1789–23 May 1857), a great French mathematician and physicist.

Actually, a Cauchy problem asks for the solution of a partial differential equation satisfying certain conditions which are given on a hypersurface (known as Cauchy surface) in its domain.

5.4.1 Cauchy problem for one-dimensional homogeneous wave equation

The Cauchy problem for the one-dimensional homogeneous wave equation is given below:

$$\frac{\partial^2 u}{\partial t^2} - c^2\frac{\partial^2 u}{\partial x^2} = 0, \; -\infty < x < \infty, \; t > 0 \tag{5.4.1}$$

subject to the conditions

$$u(x, 0) = f(x), \; u_t(x, 0) = g(x). \tag{5.4.2}$$

It represents the vibrations of an infinite string.

The solution of the above Cauchy problem is a function $u(x, t)$ which is continuously twice differentiable for all $t > 0$ such that $u(x, t)$ and $\frac{\partial u}{\partial t}(x, t)$ are both continuous for $t \geqslant 0$ and by which (5.4.1) and (5.4.2) are satisfied.

The general solution of the one-dimensional wave equation is of the form:

$$u(x, t) = \phi(x - ct) + \psi(x + ct). \tag{5.4.3}$$

Now, we would like to find the functions ϕ and ψ satisfying the initial conditions given in (5.4.2). Putting $t = 0$ in (5.4.3), we obtain $u(x, 0) = \phi(x) + \psi(x) = f(x)$, by (5.4.2), i.e.

$$\phi(x) + \psi(x) = f(x). \tag{5.4.4}$$

Differentiating both sides of (5.4.3) partially with respect to t and putting $t = 0$, we obtain $u_t(x, 0) = -c\phi'(x) + c\psi'(x) = g(x)$, by (5.4.2),

$$\text{i.e. } c\phi'(x) - c\psi'(x) = -g(x). \tag{5.4.5}$$

Integrating over $[0, x]$, we obtain

$$c(\phi(x) - \psi(x)) = -\int_0^x g(r)\,dr + c'$$

where $c' = \phi(0) - \psi(0)$, i.e.

$$\phi(x) - \psi(x) = -\frac{1}{c}\int_0^x g(r)\,dr + c'', \tag{5.4.6}$$

where $c'' = \dfrac{c'}{c}$.

Solving (5.4.4) and (5.4.6), we have

$$\phi(x) = \frac{1}{2}f(x) - \frac{1}{2c}\int_0^x g(r)\,dr + \frac{c''}{2} \text{ and } \psi(x) = \frac{1}{2}f(x) - \frac{1}{2c}\int_0^x g(r)\,dr - \frac{c''}{2}.$$

$$\therefore\ \phi(x - ct) = \frac{1}{2}f(x - ct) - \frac{1}{2c}\int_0^{x-ct} g(r)\,dr + \frac{c''}{2} = \frac{1}{2}f(x - ct) + \frac{1}{2c}\int_{x-ct}^0 g(r)\,dr + \frac{c''}{2}$$

$$\text{and } \psi(x + ct) = \frac{1}{2}f(x + ct) + \frac{1}{2c}\int_0^{x+ct} g(r)\,dr + \frac{c''}{2}.$$

Using the above two results in (5.4.3), we obtain

$$u(x, t) = \frac{1}{2}[f(x - ct) + f(x + ct)] + \frac{1}{2c}\int_{x-ct}^{x+ct} g(r)\,dr, \tag{5.4.7}$$

which is called d'Alembert's formula named after Jean-Baptiste le Rond d'Alembert (16 November 1717–29 October 1783), a French mathematician.

Example 5.4.1. *Let $u(x, t)$ be the solution of the following Cauchy problem* $\frac{\partial^2 u}{\partial t^2} - 9\frac{\partial^2 u}{\partial x^2} = 0$, $-\infty < x < \infty$, $t > 0$ *subject to the initial conditions*

$$u(x, 0) = f(x) = \begin{cases} 1, & \text{if } |x| \leqslant 2 \\ 0, & \text{if } |x| > 2 \end{cases} \text{ and } u_t(x, 0) = g(x) = \begin{cases} 1, & \text{if } |x| \leqslant 2 \\ 0, & \text{if } |x| > 2. \end{cases}$$

(i) *Find* $u\left(0, \frac{1}{6}\right)$ *and* (ii) *Discuss the large time behaviour of the solution of the Cauchy problem.*

Solution: (i) Here the wave speed, $c = 3$.

$\therefore$ d'Alembert's formula (5.4.7) becomes

$$u(x, t) = \frac{1}{2}[f(x - 3t) + f(x + 3t)] + \frac{1}{6}\int_{x-3t}^{x+3t} g(r)\,dr. \tag{5.4.8}$$

$$\therefore u\left(0,\frac{1}{6}\right)=\frac{1}{2}\left[f\left(-\frac{1}{2}\right)+f\left(\frac{1}{2}\right)\right]+\frac{1}{6}\int_{-1/2}^{1/2}g(r)\,dr=\frac{1}{2}(1+1)+\frac{1}{6}\int_{-1/2}^{1/2}1\cdot dr$$

[using the given initial conditions]

$$=1+\frac{1}{6}=\frac{7}{6}.$$

(ii) For fixed $\xi\in\mathbb{R}$, $\lim_{t\to\infty}f(\xi-3t)=0=\lim_{t\to\infty}f(\xi+3t)$ and

$$\lim_{t\to\infty}\int_{\xi-3t}^{\xi+3t}g(r)\,dr=\int_{-\infty}^{-2}g(r)\,dr+\int_{-2}^{2}g(r)\,dr+\int_{2}^{\infty}g(r)\,dr=0+\int_{-2}^{2}1\cdot dr+0=4$$

$$\therefore\ \lim_{t\to\infty}u(\xi,t)=\frac{1}{6}\cdot 4=\frac{2}{3}\ \text{[by (5.4.8)]}$$

5.4.2 Cauchy problems for one-dimensional nonhomogeneous wave equation

Cauchy problem for one-dimensional nonhomogeneous wave equation is of the following form:

$$\frac{\partial^2 u}{\partial t^2}-c^2\frac{\partial^2 u}{\partial x^2}=F(x,t),\ \infty<x<\infty,\ t>0 \tag{5.4.9}$$

subject to the initial conditions,

$$u(x,0)=f(x),\ u_t(x,0)=g(x),\ -\infty<x<\infty. \tag{5.4.10}$$

This Cauchy problem models the vibration of a very long string under an external force F. The solution of this problem is given by the following d'Alembert's formula:

$$u(x,t)=\frac{1}{2}[f(x-ct)+f(x+ct)]+\frac{1}{2c}\int_{x-ct}^{x+ct}g(r)\,dr+\frac{1}{2c}\int_{\alpha=0}^{t}\int_{\beta=x-c(t-\alpha)}^{x+c(t-\infty)}F(\alpha,\beta)\,d\alpha\,d\beta. \tag{5.4.11}$$

Proof. Let us consider the homogeneous initial conditions given below

$$u(x,0)=u_t(x,0)=0,\ -\infty,<x<\infty. \tag{5.4.12}$$

Let $v(x,t;\alpha)$ satisfy the following equation for $t>\alpha$:

$$\frac{\partial^2 v}{\partial t^2}-c^2\frac{\partial^2 v}{\partial x^2}=0,\ -\infty<x<\infty,\ t>\alpha>0 \tag{5.4.13}$$

subject to the condition at $t=\alpha$,

$$v(x,\alpha;\alpha)=0,\ v_t(x,\alpha;\alpha)=F(x,\theta). \tag{5.4.14}$$

Then, by d'Alembert's formula (5.4.7), the solution of (5.4.12) is given by

$$v(x,t;\alpha)=\frac{1}{2c}\int_{x-c(t-\alpha)}^{x+c(t+\alpha)}F(r,\alpha)\,dr. \tag{5.4.15}$$

Let

$$u(x,t)=\int_0^t v(x,t;\alpha)\,d\alpha.$$

Then

$$u_t(x,t) = \int_0^t v_t(x,t;\alpha)\,d\alpha(x,t;t), \text{ by Leibniz's rule}$$
$$= \int_0^t v_t(x,t;\alpha)\,d\alpha \;(\because\; v(x,t;t) = 0)$$

and

$$u_{tt}(x,t) = \int_0^t v_{tt}(x,t;\alpha)\,d\alpha + v_t(x,t;t), \text{ by Leibniz's rule}$$
$$= F(x,t) + \int_0^t v_{tt}(x,t;\alpha)\,d\alpha.$$

Again,

$$u_{xx}(x,t) = \int_0^t v_{xx}(x,t;\alpha)\,d\alpha, \text{ by Leibniz's rule.}$$

$$\therefore\; \frac{\partial^2 u}{\partial t^2} - c^2\frac{\partial^2 u}{dx^2} = F(x,t) + \int_0^t \left(\frac{\partial^2 v}{\partial t^2} - c^2\frac{\partial^2 v}{\partial x^2}\right) d\alpha = F(x,t), \text{ by (5.4.13).}$$

Observe that $u(x,t)$ satisfies the conditions (5.4.12).
Now, let us replace the initial data (5.4.12) by

$$u(x,0) = f(x),\; u_t(x,0) = g(x),\; -\infty < x < \infty.$$

Then, by the principle of superposition, the desired solution of the nonhomogeneous eq. (5.4.9) is

$$u(x,t) = \frac{1}{2}[f(x-ct) + f(x+ct)] + \frac{1}{2c}\int_{x-ct}^{x+ct} g(r)\,dr + \frac{1}{2c}\int_{\alpha=0}^{t}\int_{\beta=x-c(t-\alpha)}^{x+c(t-\alpha)} F(\alpha,\beta)\,d\alpha\,d\beta.$$

Example 5.4.2. *Solve the following Cauchy problem* $\frac{\partial^2 u}{\partial t^2} - 9\frac{\partial^2 u}{\partial x^2} = e^x - e^{-x}$, $-\infty < x < \infty$, $t > 0$ *subject to the initial conditions:* (i) $u(x,0) = x$, $-\infty < x < \infty$ (ii) $u_t(x,0) = \sin x$, $-\infty < x < \infty$.

Solution: The d'Alembert's formula for nonhomogeneous wave equation

$$\frac{\partial^2 u}{\partial t^2} - c^2\frac{\partial^2 u}{\partial x^2} = F(x,t),\; -\infty < x < \infty,\; t > 0$$

with initial conditions $u(x,0) = f(x)$, $u_t(x,0) = g(x)$, $-\infty < x < \infty$ is of the form:

$$u(x,t) = \frac{1}{2}[f(x-ct) + f(x+ct)] + \frac{1}{2c}\int_{x-ct}^{x+ct} g(r)\,dr + \frac{1}{2c}\int_{\alpha=0}^{t}\int_{\beta=x-c(t-\alpha)}^{x+c(t-\alpha)} F(\alpha,\beta)d\alpha\,d\beta.$$

Here, the wave speed, $c = 3$, $f(x) = x$ and $g(x) = \sin x$.
Thus, using d'Alembert's formula, we obtain

$$u(x,t) = \frac{1}{2}(x - 3t + x + 3t) + \frac{1}{6}\int_{x-3t}^{x+3t} \sin r\,dr + \frac{1}{6}\int_{\alpha=0}^{t}\int_{\beta=x-3(t-\alpha)}^{x+3(t-\alpha)} \left(e^{\beta} - e^{-\beta}\right) d\alpha\,d\beta$$
$$= x + \frac{1}{3}\sin x\sin 3t - \frac{2}{9}(\sinh x - \sinh x\cosh 3t),$$

which is the desired solution.

5.4.3 Problem of a semi-infinite string with a fixed end

The motion of a semi-infinite string with a fixed end at $x = 0$ is given by the following equation

$$\frac{\partial^2 u}{\partial t^2} - c^2 \frac{\partial^2 u}{\partial x^2} = 0,\ 0 < x < \infty,\ t > 0 \tag{5.4.16}$$

subject to the initial conditions $u(x, 0) = f(x)$, $u_t(x, 0) = g(x)$, $x \geqslant 0$ and the boundary condition $u(0, t) = 0$ for $t \geqslant 0$.

The boundary condition $u(0, t) = 0$ follows that $u_t(0, t) = 0$ for $t \geqslant 0$.

Here $f(x - ct)$ does not exist for $t > x/c$. Thus, d'Alembert's formula (5.4.7) cannot be used for the problem (5.4.16) given above.

In order to solve the problem, let us consider the following modified problem for an infinite string: $u(x, 0) = F(x)$, $u_t(x, 0) = G(x)$, $-\infty < x < \infty$, where

$$F(x) = \begin{cases} f(x), & \text{if } x \geqslant 0 \\ -f(-x), & \text{if } x \leqslant 0 \end{cases} \text{ and } G(x) = \begin{cases} g(x), & \text{if } x \geqslant 0 \\ -g(-x), & \text{if } x \leqslant 0. \end{cases}$$

Then $F(x)$ and $G(x)$ are both odd functions of x satisfying the boundary conditions.

Thus, by the d'Alembert's formula (5.4.7), we obtain

$$u(x, t) = \tfrac{1}{2}[F(x - ct) + F(x + ct)] + \tfrac{1}{2c} \int_{x-ct}^{x+ct} G(r)\, dr.$$

$$\therefore\ u_t(x, t) = \tfrac{1}{2}[-cF'(x - ct) + cF'(x + ct)] + \tfrac{1}{2c}[G(x + ct) \cdot c + G(x - ct) \cdot (-c)],$$

by Leibniz's rule.

Thus, at $x = 0$,

$$u(0, t) = \tfrac{1}{2}[F(-ct) + F(ct)] + \tfrac{1}{2c} \int_{-ct}^{ct} G(r) dr$$

$$\text{and } u_t(0, t) = \tfrac{1}{2}[-cF'(-ct) + cF'(ct)] + \tfrac{1}{2}[G(ct) + G(-ct)].$$

Since F and G are odd functions

$$u(0, t) = \tfrac{1}{2}[-F(ct) + F(ct)] = 0$$

$$\text{and } u_t(0, t) = \tfrac{c}{2}[-f'(ct) + f'(ct)] + \tfrac{1}{2}[G(ct) - G(ct)] = 0.$$

Again, at $t = 0$ and $x > 0$, we have

$$u(x, 0) = \tfrac{1}{2}[F(x) + F(x)] = F(x) = f(x)$$

$$\text{and } u_t(x, 0) = \tfrac{1}{2}[cF'(x) - cF'(x)] + \tfrac{1}{2}[G(x) + G(x)] = G(x) = g(x).$$

It follows that the initial and boundary conditions are satisfied. Clearly, $g(x) = 0$ if the string is released from the rest. In this case, the solution of the problem becomes

$$u(x, t) = \begin{cases} \tfrac{1}{2}[f(x - ct) + f(x + ct)], & \text{if } x \geqslant ct \\ \tfrac{1}{2}[f(x + ct) - f(ct - x)], & \text{if } x \leqslant ct. \end{cases}$$

5.4.4 Problem of a string of finite length

The motion of a string of finite length l is given by the following equation

$$\frac{\partial^2 u}{\partial t^2} - c^2\frac{\partial^2 u}{\partial x^2} = 0,\ 0 < x < l,\ t > 0 \tag{5.4.17}$$

subject to the boundary conditions

$$\left.\begin{aligned} u(0,t) = u_t(0,t) = 0 \\ u(l,t) = u_t(l,t) = 0,\ t \geqslant 0 \end{aligned}\right\} \tag{5.4.18}$$

and the initial conditions

$$u(x,0) = f(x),\ u_t(x,0) = g(x),\ 0 \leqslant x \leqslant l. \tag{5.4.19}$$

Now, we would like to convert this problem into a problem of an infinite string:

Let us consider the following modified data to convert the problem to the vibration of an infinite string.

$$F(x) = \begin{cases} f(x), & \text{if } 0 \leqslant x \leqslant l \\ -f(-x), & \text{if } -l \leqslant x \leqslant 0, \end{cases}$$

$$F(x+2kl) = f(x),\ \text{if } |x| \leqslant l, k = \pm 1, \pm 2, \ldots.$$

We see that $F(x)$ is an odd function as well as periodic with period $2l$.

By similar way, we can define $G(x)$ as another periodic odd function with same period $2l$ such that $G(x+2kl) = g(x)$ for all $x \in [0,l]$.

Let the Fourier series expansion of the functions $F(x)$ and $G(x)$ be possible. Then

$$F(x) = \sum_{m=1}^{\infty} b_m(x)\sin\left(\frac{m\pi x}{l}\right) \text{ and } G(x) = \sum_{m=1}^{\infty} b'_m(x)\sin\left(\frac{m\pi x}{l}\right),$$

where

$$b_m = \frac{2}{l}\int_0^l f(r)\sin\left(\frac{m\pi r}{l}\right)dr \text{ and } b'_m = \frac{2}{l}\int_0^l g(r)\sin\left(\frac{m\pi r}{l}\right)dr.$$

$$\therefore \frac{1}{2}[F(x-ct)+F(x+ct)] = \sum_{m=1}^{\infty} b_m \sin\left(\frac{\mu\pi x}{l}\right)\cos\left(\frac{m\pi ct}{l}\right)$$

$$\text{and } \frac{1}{2c}\int_{x-ct}^{x+ct} G(r)dr = \frac{1}{\pi c}\sum_{m=1}^{\infty}\left(\frac{b'_m}{m}\right)\sin\left(\frac{m\pi x}{l}\right)\sin\left(\frac{m\pi ct}{l}\right).$$

Hence, the desired solution is given by

$$u(x,t) = \sum_{m=1}^{\infty} b_m \sin\left(\frac{m\pi x}{l}\right)\cos\left(\frac{m\pi ct}{l}\right) + \frac{1}{\pi c}\sum_{m=1}^{\infty}\left(\frac{b'_m}{m}\right)\sin\left(\frac{m\pi x}{l}\right)\sin\left(\frac{m\pi ct}{l}\right).$$

Example 5.4.3. *A tightly stretched homogeneous string of length l with its fixed ends at $x = 0$ and $x = l$ executes transverse vibrations. The motion is started with zero initial velocity by displacing the string into the form $f(x) = a\sin^2 \pi x$. Find the deflection $u(x,t)$ at any time t.*

Solution: By the d'Alembert's formula, the desired deflection is

$$\begin{aligned} u(x,t) &= \tfrac{1}{2}[f(x-ct)+f(x+ct)] = \tfrac{1}{2}a[\sin^2\pi(x-ct)+\sin^2\pi(x+ct)] \\ &= \tfrac{1}{4}a[1-\cos 2\pi(x-ct)+1-\cos 2\pi(x+ct)] = \tfrac{1}{4}a[2-\{\cos 2\pi(x-ct)+\cos 2\pi(x+ct)\}] \\ &= \tfrac{1}{4}a(2-2\cos 2\pi x\cos 2\pi ct) = \tfrac{1}{2}a(1-\cos 2\pi x\cos 2\pi ct). \end{aligned}$$

5.4.5 Uniqueness of the solution

The solution of the following Cauchy problem is unique: $\frac{\partial^2 u}{\partial t^2} - c^2 \frac{\partial^2 u}{\partial x^2} = F(x, t)$, $0 < x < l$; $t > 0$ subject to the initial conditions: $u(x, 0) = f(x)$, $0 \leqslant x \leqslant l$, $u_t(x, 0) = g(x)$, $0 \leqslant x \leqslant l$ and the boundary conditions $u(0, t) = u(l, t) = 0$, $t \geqslant 0$.

Proof. If possible, let u_1 and u_2 be two solutions of the given problem and let us put $v = u_1 - u_2$. Then it follows that

$$\frac{\partial^2 v}{\partial t^2} - c^2 \frac{\partial^2 v}{\partial x^2} = 0, \ 0 < x < l, \ t > 0 \tag{5.4.20}$$

subject to the conditions

$$\begin{aligned} v(x, 0) &= 0, 0 \leqslant x \leqslant l \\ v_t(x, 0) &= 0, 0 \leqslant x \leqslant l \\ v(0, t) &= 0, t \geqslant 0 \\ \text{and } v(l, t) &= 0, t \geqslant 0. \end{aligned}$$

Clearly,

$$\begin{aligned} v(0, t) = 0 &\Rightarrow v_t(0, t) = 0 \\ \text{and } v(l, t) = 0 &\Rightarrow v_t(l, t) = 0 \end{aligned} \tag{5.4.21}$$

for all $t \geqslant 0$.

Now, we would like to show that $v(x, t) = 0$.

Let us consider $E(t)$ defined by

$$E(t) = \frac{1}{2} \int_0^l (v_t^2 + c^2 v_x^2)\, dx. \tag{5.4.22}$$

Since $v(x, t)$ is twice differentiable, $E(t)$ must be a differentiable function of t.

Differentiating (5.4.22) with respect to t, we obtain

$$\begin{aligned} \frac{d}{dt}(E(t)) &= \frac{1}{2}\frac{d}{dt} \int_0^l (v_t^2 + c^2 v_x^2)\, dx = \frac{1}{2} \int_0^l (2v_t v_{tt} + 2c^2 v_x v_{xt}) dx = \int_0^l v_t v_{tt} dx + \left[c^2 v_x v_t\right]_0^l - \int_0^l c^2 v_t v_{xx} dx \\ &= \int_0^l v_t v_{tt} dx + c^2 v_x(l, t) v_t(l, t) - c^2 v_x(0, t) v_t(0, t) - \int_0^l c^2 v_t v_{xx} dt \\ &= \int_0^l v_t v_{tt} dx - \int_0^l c^2 v_t v_{xx} dx, \text{ by (5.4.21)} \\ &= \int_0^l v_t \left(v_{tt} - c^2 v_{xx}\right) dx = 0, \text{ by (5.4.20)}. \end{aligned}$$

It follows that $E(t)$ = constant.

We have the initial conditions $v(x, 0) = 0$ and $v_t(x, 0) = 0$, $0 \leqslant x \leqslant l$.

$\therefore$ $v_x(x, 0) = 0$.

Consequently, $E(0) = 0$, by (5.4.20).

$\therefore$ $E(t) \equiv 0$.

Thus, $v_x(x, t) \equiv 0$, $v_t(x, t) \equiv 0$ for all $t > 0$ and $0 < x < l$.

It follows that $v(x, t)$ = constant. Again, $v(x, 0) \equiv 0$.

$\therefore$ $v(x, t) \equiv 0$, i.e. $u_1 - u_2 \equiv 0$, i.e. $u_1 \equiv u_2$.

Hence, the proof is complete.

Cauchy-Kowalevski theorem

Cauchy-Kowalevski theorem states that every Cauchy problem has unique solution if the Cauchy data and the coefficients of the partial differential equation are real analytic functions.

Note 5.4.1. *A special case of this theorem was introduced by Cauchy in* 1842 *and the general result was proven by Sofia Vasilyevna Kowalevski* (1850–1891), *a Russian mathematician.*

5.4.6 Solution of the two-dimensional wave equation by the method of separation of variables (Cartesian coordinates)

The second-order linear homogeneous equation of the form:

$$\frac{\partial^2 u}{\partial x^2} + \frac{\partial^2 u}{\partial y^2} = \frac{1}{c^2}\frac{\partial^2 u}{\partial t^2} \tag{5.4.23}$$

is known as the two-dimensional wave equation.

Let us assume that the eq. (5.4.23) has the solution in the form:

$$u(x, y, t) = X(x)Y(y)T(t). \tag{5.4.24}$$

Substituting (5.4.24) in (5.4.23), we obtain

$$\frac{1}{X}\frac{d^2X}{dx^2} + \frac{1}{Y}\frac{d^2Y}{dy^2} = \frac{1}{c^2T}\frac{d^2T}{dt^2} = -\lambda^2, \text{ say.}$$

$-\lambda^2$ being a constant of separation for nontrivial solution.

Then

$$\frac{d^2T}{dt^2} + c^2\lambda^2T = 0 \text{ and } \frac{1}{X}\frac{d^2X}{dx^2} + \frac{1}{Y}\frac{d^2Y}{dy^2} = -\lambda^2 \text{ or, } \frac{1}{X}\frac{d^2X}{dx^2} + \lambda^2 = -\frac{1}{Y}\frac{d^2Y}{dy^2} = \mu^2, \text{ say,}$$

μ^2 being another constant of separation.

$$\therefore \quad \frac{d^2X}{dx^2} + \lambda_1^2X = 0, \quad \frac{d^2Y}{dy^2} + \mu^2Y = 0 \text{ and } \frac{d^2T}{dt^2} + c^2\lambda^2T = 0,$$

where $\lambda_1^2 = \lambda^2 - \mu^2$.

The solutions of the above three ordinary differential equations are

$X(x, y, t) = c_1\cos(\lambda_1x) + c_2\sin(\lambda_1x)$, $Y(x, y, t) = c_3\cos(\mu y) + c_4\sin(\mu y)$ and $T = c_5\cos(c\lambda t) + c_6\sin(c\lambda t)$ respectively, where $c_1, c_2, \ldots, c_6$ are arbitrary constants.

Hence, desired solution of the eq. (5.4.23) is given by

$u(x, y, t) = (c_1\cos(\lambda_1x) + c_2\sin(\lambda_1x))(c_3\cos(\mu y) + c_4\sin(\mu y))(c_5\cos(c\lambda t) + c_6\sin(c\lambda t))$.

Examples V(A)

1. Solve the one-dimensional wave equation $\frac{\partial^2u}{dt^2} - c^2\frac{\partial^2u}{\partial x^2} = 0$, $0 < x < \pi$, $t > 0$ subject to the boundary conditions $u(0, t) = u(\pi, t) = 0, t > 0$ and the initial conditions $u(x, 0) = \alpha$, $u_t(x, 0) = \beta$, where α, β are constants.
2. Let $u(x, t)$ be the solution of the following equation $\frac{\partial^2u}{\partial t^2} - 4\frac{\partial^2u}{\partial x^2} = 0$, $-\infty < x < \infty$, $t > 0$ subject to the initial conditions

$$u(x, 0) = f(x) = \begin{cases} 1, & \text{if } |x| \leqslant 1 \\ 0, & \text{if } |x| > 1 \end{cases} \text{ and } u_t(x, 0) = g(x) = \begin{cases} 1, & \text{if } |x| \leqslant 1 \\ 0, & \text{if } |x| > 1. \end{cases}$$

(i) Find $u\left(0, \frac{1}{4}\right)$. (ii) Discuss the large time behaviour of the solution of the equation.

3. Solve the following one-dimensional homogeneous wave equation $\frac{\partial^2 u}{\partial t^2} - 4\frac{\partial^2 u}{\partial x^2} = 0, -\infty < x < \infty$, $t > 0$ subject to the initial conditions $u(x, 0) = x$, $u_t(x, 0) = \cos x$.
4. Solve the following nonhomogeneous wave equation $\frac{\partial^2 u}{\partial t^2} - 4\frac{\partial^2 u}{\partial x^2} = e^{-x}, -\infty < x < \infty, t > 0$ subject to the initial conditions $u(x, 0) = x^2$ and $u_t(x, 0) = \cos x$.
5. Solve the following wave equation for a semi-infinite string with a fixed end $\frac{\partial^2 u}{\partial t^2} - 16\frac{\partial^2 u}{\partial x^2} = 0$, $0 < x < \infty$ $t > 0$ subject to the initial conditions $u(x, 0) = x$, $u_t(x, 0) = 0$, for $x \geqslant 0$ and the boundary conditions $u(0, t) = 0 = u_t(0)$ for $t \geqslant 0$.
6. Solve the following wave equation for a string of finite length $\frac{\partial^2 u}{\partial t^2} - 9\frac{\partial^2 u}{\partial x^2} = 0, 0 < x < 2, t > 0$ subject to the boundary conditions $u(0, t) = u_t(0, t) = 0$, $u(2, t) = u_t(2, t) = 0$, $t > 0$ and the initial conditions $u(x, 0) = x$ $u_t(x, 0) = 0$, $0 \leqslant x \leqslant 2$.
7. Show that the wave equation $\frac{\partial^2 u}{\partial t^2} - c^2\frac{\partial^2 u}{\partial x^2} = 0$ reduces to its canonical form $u_{\xi\eta} = 0$ under the transformation $\xi = x - ct, \eta = x + ct$. Using this information show that the general solution of the wave equation is $u(x, t) = f(x - ct) + g(x + ct)$, where f and g are two arbitrary functions.
8. Solve the following wave equation of a string of finite length using the method of separation of variables $\frac{\partial^2 u}{\partial t^2} - c^2\frac{\partial^2 u}{\partial x^2} = 0$, $0 < x < l, t > 0$ subject to the initial conditions $u(x, 0) = f(x)$, $u_t(x, 0) = g(x)$, $0 \leqslant x \leqslant l$ and the boundary conditions $u(0, t) = u(l, t) = 0$, $t \geqslant 0$.
9. Solve the following wave equation of a string of length 2 units using the method of separation of variables $\frac{\partial^2 u}{\sigma\partial t^2} - 9\frac{\partial^2 u}{\partial x^2}$, $0 < x < 2, t > 0$ subject to the conditions

$$u(x, 0) = x, \; 0 \leqslant x \leqslant 2$$
$$u_t(x, 0) = 0, \; 0 \leqslant x \leqslant 2$$
$$\text{and } u(0, t) = u(2, t) = 0, \; t \geqslant 0.$$

10. Find the solution of the Cauchy problem $\frac{\partial^2 u}{\partial t^2} - \frac{\partial^2 u}{\partial x^2} = 0, -\infty < x < \infty, t > 0$ subject to the initial conditions:

$$u(x,0) = f(x) = \begin{cases} x+1, & \text{if } -1 \leqslant x \leqslant 0 \\ 1 - x, & \text{if } 0 \leqslant x \leqslant 1 \\ 0, & \text{elsewhere} \end{cases} \quad \text{and } u_t(x,0) = g(x) = \begin{cases} 1, & \text{if } -1 \leqslant x \leqslant 1 \\ 0, & \text{elsewhere.} \end{cases}$$

Hence, find $u\left(1, \frac{1}{2}\right)$.

11. Solve the Cauchy problem $\frac{\partial^2 u}{\partial t^2} - \frac{\partial^2 u}{\partial x^2} = 0, -\infty < x < \infty, t > 0$ subject to the conditions $u(x, 0) = f(x) = 0, -\infty < x < \infty$ and $u_t(x, 0) = g(x) = \begin{cases} 0, & \text{if } x < 0 \\ 1, & \text{if } x \geqslant 0. \end{cases}$

Hence, find $u\left(1, \frac{1}{2}\right)$.

12. Show that the principle of superposition is not valid for nonlinear partial differential equation.

Answers

1. See example 4.7.1.
2. See example 4.81; $u\left(0, \frac{1}{4}\right) = \frac{5}{4}$.
3. $u(x, t) = x + \frac{1}{2}\cos x \sin 2t$; $\lim_{t\to\infty} u(\xi, \eta) = \frac{1}{2}$.
4. See example 4.82.
5. $u(x,t) = \begin{cases} x^2 + 16t^2, & \text{if } x \geqslant 4t \\ 8xt, & \text{if } x \leqslant 4t \end{cases}$

6. $u(x,t) = \sum_{m=1}^{\infty}(-1)^{m+1}\frac{4}{m\pi}\sin\frac{m\pi x}{2}\cos\frac{3m\pi t}{2}$.

8. $u(x,t) = \sum_{n=1}^{\infty}\left[a_n\cos\left(\frac{n\pi ct}{l}\right) + b_n\sin\left(\frac{n\pi ct}{l}\right)\right]\sin\frac{n\pi x}{l}$, where $a_n = \frac{2}{l}\int_0^l f(x)\sin\left(\frac{n\pi x}{l}\right)dx$ and $b_n = \frac{2}{n\pi c}\int_0^l g(x)\sin\left(\frac{n\pi x}{l}\right)dx$.

9. $u(x,t) = \frac{4}{\pi}\sum_{n=1}^{\infty}\left[(-1)^{n+1}\cdot\frac{1}{n}\sin\frac{n\pi x}{2}\cos\frac{3n\pi t}{2}\right]$.

10. Hint. By d'Alembert's formula, we have

$$u\left(1,\frac{1}{2}\right) = \frac{1}{2}\left[f\left(1-\frac{1}{2}\right) + f\left(1+\frac{1}{2}\right)\right] + \frac{1}{2}\int_{1-\frac{1}{2}}^{1+\frac{1}{2}} g(x)dx = \frac{1}{2}.$$

11. Hint. $u(x,t) = \frac{1}{2}\int_{x-t}^{x+t} g(r)dr = \frac{1}{2}\int_{x-t}^{0} g(r)dr + \frac{1}{2}\int_{0}^{x+t} g(r)dr = \frac{1}{2}\left[\max\{0,x+t\} - \max\{0,x-t\}\right]$.

5.5 One-dimensional heat equation (diffusion equation)

Let $u = u(x,t)$ be the thermal energy per unit volume in a heat-conducting medium. Then $u = \rho CT$, where ρ is the density, C is the specific heat and $T = T(x,t)$ is the temperature in degree.

In the absence of the sources, by the conservation law, we have

$$(\rho CT)_t + \phi_x = 0, \tag{5.5.1}$$

ϕ being the energy flux.

Again, by the Fourier's law of heat conduction, we have $\phi = -kT_x(x,t)$, k being the thermal conductivity of the region.

$\therefore\ \phi_x = -kT_{xx}(x,t)$.

Using it in (5.5.1), we obtain

$$T_t - kT_{xx} = 0, \text{ i.e. } k\frac{\partial^2 T}{\partial x^2} - \frac{\partial T}{\partial t} = 0, \tag{5.5.2}$$

where $k = \frac{K}{\rho C}$ is known as the diffusivity. The eq. (5.5.2) is called the *one-dimensional heat equation.*

In 1822, the heat equation was introduced by Jean-Baptiste Joseph Fourier (21 March 1768–16 May 1830), a French mathematician and physicist, in order to investigate a temperature distribution in materials.

5.5.1 The nature and characteristics

Comparing the heat eq. (5.5.2) with

$$A\frac{\partial^2 T}{\partial x^2} + B\frac{\partial^2 T}{\partial x\partial t} + C\frac{\partial^2 T}{\partial t^2} + F\left(x,t,T,\frac{\partial T}{\partial x},\frac{\partial T}{\partial t}\right) = 0,$$

we have $A = k, B = 0, C = 0$ and $F = -T_t$.

$\therefore$ the discriminant of the equation is given by $D = B^2 - 4AC = 0$.

Thus, the one-dimensional heat equation is parabolic type.

The λ-quadratic equations $A\lambda^2 + B\lambda + C = 0$, i.e. $k\lambda^2 = 0$.

$\therefore\ \lambda = 0, 0.$

$\therefore$ the characteristic equation is $\frac{dt}{dx} = 0$.

Integrating, we obtain $t = c_1$, c_1 being arbitrary constant.

Thus, $x = c_1$ represents the characteristics of the heat equation.

5.5.2 Solution of one-dimensional heat equation by the method of separation of variables (Cartesian coordinates)

Let us assume that the heat eq. (5.5.2) has the solution in the form $T(x, t) = X(x)\tau(t)$.

Then

$$\frac{\partial T}{\partial t} = X(x)\tau'(t) \text{ and } \frac{\partial^2 T}{\partial x^2} = X''(x)\tau(t)$$

Using the above results in (5.5.2), we obtain

$$kX''(x)\tau(t) = X(x)\tau'(t) \text{ or, } \frac{1}{X}\frac{d^2X}{dx^2} = \frac{1}{k\tau}\frac{d\tau}{dt} = \lambda, \text{ say,}$$

λ being a constant of separation.

$$\therefore\ \frac{d^2X}{dx^2} - \lambda X = 0 \tag{5.5.3}$$

$$\text{and } \frac{d\tau}{dt} - k\lambda\tau = 0. \tag{5.5.4}$$

Case 5.5.1. *Let $\lambda = \mu^2 > 0$. Then the solutions of the ordinary differential equations (5.5.3) and (5.5.4) are $X(x) = c_1 e^{\mu x} + c_2 e^{-\mu x}$ and $\tau(t) = c_3 e^{k\mu^2 t}$ respectively, where c_1, c_2, c_3 are arbitrary constants. Hence, desired solution of the heat equation (5.5.2) is given by $T(x, t) = (c_1 e^{\mu x} + c_2 e^{-\mu x})c_3 e^{k\mu^2 t} = (A' e^{\sqrt{\lambda}x} + B' e^{-\sqrt{\lambda}x})e^{k\lambda t}$, where $A' = c_1 c_3$ and $B' = c_2 c_3$.*

Case 5.5.2. *Let $\lambda = 0$. Then the solution of the equations (5.5.3) and (5.5.4) are $X(x) = c_1 x + c_2$ and $\tau(t) = c_3$ respectively, where c_1, c_2, c_3 are arbitrary constants. Hence, the solution of the heat equation is $T(x, t) = (c_1 x + c_2)c_3$, i.e. $T(x, t) = A'x + B'$, where $A' = c_1 c_3$ and $B' = c_2 c_3$.*

Case 5.5.3. *Let $\lambda = -\mu^2 < 0$. Then the solutions of the ordinary differential equation (5.5.3) and (5.5.4) are given by $X(x) = c_1 \cos \mu x + c_2 \sin \mu x$ and $\tau(t) = c_3 e^{-k\mu^2 t}$ respectively, where c_1, c_2, c_3 are arbitrary constants. Hence, the solution of the heat equation (5.5.2) is given by $T(x, t) = (c_1 \cos \mu x + c_2 \sin \mu x)c_3 e^{-k\mu^2 t} = (A' \cos \sqrt{-\lambda}x + B' \sin \sqrt{-\lambda}x)e^{k\lambda t}$, where $A' = c_1 c_3, B' = c_2 c_3$.*

5.5.3 Solution of heat equation for finite rod

We consider the problem of heat conduction in a finite rod of length l as

$$\frac{\partial T}{\partial t} - k\frac{\partial^2 T}{\partial x^2} = 0,\ 0 < x < l,\ t > 0 \tag{5.5.5}$$

subject to the boundary conditions

$$T(0, t) = T(l, t) = 0,\ t \geqslant 0 \tag{5.5.6}$$

and the initial condition

$$T(x, 0) = f(x),\ 0 \leqslant x \leqslant l. \tag{5.5.7}$$

Here we assume that (i) the rod is homogeneous and lies along the x-axis, (ii) the rod is sufficiently thin and the heat will be uniformly distributed over its cross section at a given time t, (iii) there will not be any loss of heat through the boundary of the rod.

Now, following the Fourier's method of separation of variable, let us put $T(x, t) = X(x)\tau(t)$.

Then eq. (5.5.5) gives $\frac{1}{X}\frac{d^2X}{dx^2} = \frac{1}{k\tau}\frac{d\tau}{dt} = \lambda$, λ being constant of separation.

For realistic situation from the physical point of view, λ must be negative. Thus, $X(x) = A' \cos\sqrt{-\lambda}x + B' \sin\sqrt{-\lambda}x$ and $\tau(t) = e^{k\lambda t}$.

$\therefore\ T(x, t) = (A' \cos\sqrt{-\lambda}x + B' \sin\sqrt{-\lambda}x)C' e^{k\lambda t}$.

Now, using the boundary condition (5.5.6), we obtain $X(0) = X(l) = 0$.

$X(0) = 0 \Rightarrow A' = 0$ and $X(l) = 0 \Rightarrow B' \sin\sqrt{-\lambda}l = 0$.

For nontrivial solutions, we have $\sin\sqrt{-\lambda}l = 0$.

$\therefore\ \sqrt{-\lambda}l = n\pi, n = 1, 2, 3, \ldots$, i.e. $-\lambda l^2 = n^2\pi^2$, i.e. $\lambda = -\frac{n^2\pi^2}{l^2}$.

It follows that $\lambda_n = -\frac{n^2\pi^2}{l^2}$, $n = 1, 2, 3, \ldots$.

$$\therefore\ T_n(x, t) = A_n \exp\left(-\frac{n^2\pi^2 kt}{l^2}\right) \sin\left(\frac{n\pi x}{l}\right),$$

where $A_n = B'C'$.

Thus, by the principle of superposition, we have

$$T(x, t) = \sum_{n=1}^{\infty} A_n \exp\left(-\frac{n^2\pi^2 kt}{l^2}\right) \sin\left(\frac{n\pi x}{l}\right).$$

Now, using the initial condition $T(x, 0) = f(x)$, we obtain

$$f(x) = \sum_{n=1}^{\infty} A_n \sin\left(\frac{n\pi x}{l}\right),\ 0 \leqslant x \leqslant l \text{ and } A_n = \frac{2}{l}\int_0^l f(x)\sin\left(\frac{n\pi x}{l}\right) dx.$$

Hence, the desired solution is given by

$$T(x, t) = \sum_{n=1}^{\infty}\left[\frac{2}{l}\int_0^l f(x)\sin\left(\frac{n\pi x}{l}\right) dx\right] \sin\left(\frac{n\pi x}{l}\right) \exp\left(-\frac{n^2\pi^2 kt}{l^2}\right)$$

Example 5.5.1. *Solve the following heat equation* $\frac{\partial T}{\partial t} - k\frac{\partial^2 T}{\partial x^2} = 0,\ 0 < x < \pi,\ t > 0$ *subject to the conditions* $T(0, t) = T(\pi, t) = 0,\ t \geqslant 0$ *and*

$$T(x, 0) = f(x) = \begin{cases} x, & \text{if } 0 \leqslant x \leqslant \frac{\pi}{2} \\ \pi - x, & \text{if } \frac{\pi}{2} \leqslant x \leqslant \pi \end{cases}$$

and $T(x, t) < \infty$ *as* $t \to \infty$.

Solution: Following the Fourier's method of separation of variables, let us put $T(x, t) = X(x)\tau(t)$.

Then the heat equation gives $\frac{1}{X}\frac{d^2X}{dx^2} = \frac{1}{k\tau}\frac{d\tau}{dt} = \lambda$, λ being the constant of separation.

$$\therefore\ \frac{d^2X}{\partial x^2} - \lambda X = 0 \text{ and } \frac{d\tau}{dt} - k\lambda\tau = 0.$$

For $\lambda > 0$, we have $T(x, t) = (A' e^{\sqrt{\lambda}t} + B' e^{-\sqrt{\lambda}t})e^{k\lambda t}$.

It cannot be accepted because the condition $T(x, t) < \infty$ as $t \to \infty$ is violated.

For $\lambda = 0$, we have $T(x, t) = A'x + B'$.

Using the boundary conditions $T(0, t) = T(\pi, t) = 0, t \geqslant 0$, we have $A' = B' = 0$ for which we get the trivial solution $T(x, t) = 0$.

Thus, for nontrivial solution, we consider $\lambda < 0$.

Then $T(x, t) = (A' \cos \sqrt{-\lambda}x + B' \sin \sqrt{-\lambda}x)e^{k\lambda t}$.

Using the boundary conditions $T(0, t) = T(\pi, t) = 0, t \geqslant 0$, we obtain $A' = 0$ and $\sin \sqrt{-\lambda}\pi = 0$, i.e. $\sqrt{-\lambda}\pi = n\pi, n = 1, 2, 3, \ldots$, i.e. $\sqrt{-\lambda} = n$, i.e. $\lambda = -n^2, n = 1, 2, 3, \ldots$.

$\therefore$ by the principle of superposition

$$T(x, t) = \sum_{n=1}^{\infty}(B_n \sin nx) \exp(-kn^2 t).$$

$$\therefore \; T(x, 0) = \sum_{n=1}^{\infty} B_n \sin nx,$$

which is a half-range Fourier sine series.

Thus, the Fourier coefficient B_n is given by

$$B_n = \frac{2}{\pi}\int_0^{\pi} f(x) \sin nx \, dx = \frac{2}{\pi}\left[\int_0^{\pi/2} x \sin nx \, dx + \int_{\pi/2}^{\pi} (\pi - x) \sin nx \, dx\right] = \frac{4}{n^2\pi} \sin\left(\frac{n\pi}{2}\right).$$

Hence, desired solution is given by

$$T(x, t) = \frac{4}{\pi}\sum_{n=1}^{\infty}\frac{1}{n^2} \sin nx \cdot \sin\left(\frac{n\pi}{2}\right) \exp(-kn^2 t).$$

Theorem 5.5.1 (Maximum-Minimum Principle). *Let the continuous function $T(x, t)$ be a solution of the heat equation $\frac{\partial T}{\partial t} - k\frac{\partial^2 T}{\partial x^2} = 0, 0 < x < l, 0 < t < t_0$, t_0 being a fixed time. Then $T(x, t)$ attains its maximum and minimum at $t = 0$ or at $x = 0$ and $x = l$ for some $t \in [0, t_0]$.*

Theorem 5.5.2. *Let $T(x, t)$ be a solution of the heat equation*

$$\frac{\partial T}{\partial t} - k\frac{\partial^2 T}{\partial x^2} = 0, \; 0 < x < l, \; 0 < t < t_0$$

subject to the initial condition $T(x, 0) = f(x)$ and the boundary condition $T(0, t) = g_1(t)$ and $T(l, t) = g_2(t)$, where $f(x)$, $g_1(t)$ and $g_2(t)$ are continuous functions on their domains of definitions. Then the solution of the equation is unique.

Proof. If possible, let $T_1(x, t)$ and $T_2(x, t)$ be two solutions of the heat equations

$$\frac{\partial T}{\partial t} - k\frac{\partial^2 T}{\partial x^2} = 0, \; 0 < x < l, 0 < t < t_0 \tag{5.5.8}$$

subject to the initial condition

$$T(x, 0) = f(x) \tag{5.5.9}$$

and the boundary conditions

$$T(0, t) = g_1(t) \text{ and } T(l, t) = g_2(t). \tag{5.5.10}$$

Let $u(x, t) = T_1(x, t) - T_2(x, t), 0 \leqslant x \leqslant l, 0 \leqslant t \leqslant t_0$.

Then $u(x, t)$ must be a solution of the eq. (5.8). Here $u(x, t)$ is a continuous function in its domain and $u(x, 0) = 0$ for $0 \leqslant x \leqslant l$ and $u(0, t) = u(l, t) = 0$ for $0 \leqslant t \leqslant t_0$.

Thus, $u(x, t)$ satisfies all the conditions of the maximum-minimum principle and so $u(x, t) \equiv 0$ for $0 \leqslant x \leqslant l$ and $0 \leqslant t \leqslant t$, i.e. $T_1(x, t) \equiv T_2(x, t)$. Hence, the proof is complete.

5.5.4 Solution of two-dimensional heat equation by the method of separation of variables (Cartesian coordinates)

The linear partial differential equation of the form:

$$\frac{\partial T}{\partial t} = k\left(\frac{\partial^2 T}{\partial x^2} + \frac{\partial^2 T}{\partial y^2}\right)$$

is known as the two-dimensional heat equation in Cartesian coordinates.

Let us assume that the solution of the equation is of the form $T(x, y, t) = X(x)Y(y)\tau(t)$.

Then the heat equation gives

$$X(x)Y(y)\tau'(t) = k(X''(x)Y(y)\tau(t) + X(x)Y''(y)\tau(t)) \text{ or, } \frac{1}{X}\frac{d^2X}{dx^2} + \frac{1}{Y}\frac{d^2Y}{dy^2} = \frac{1}{k\tau}\frac{d\tau}{dt} = \lambda, \text{ say,}$$

λ being a constant of separation.

For nontrivial realistic solution, we consider $\lambda = -\mu_1^2 < 0$.

$\therefore \frac{d\tau}{dt} + k\mu_1^2\tau = 0$, which gives $\tau(t) = c_1 e^{-k\mu_1^2 t}$, c_1 being arbitrary constant.

Again,

$$\frac{1}{X}\frac{d^2X}{dx^2} + \frac{1}{Y}\frac{d^2Y}{dy^2} = -\mu_1^2 \text{ or, } \frac{1}{X}\frac{d^2X}{dx^2} = -\left(\frac{1}{Y}\frac{d^2Y}{dy^2} + \mu_1^2\right) = -\mu_2^2, \text{ say.}$$

$$\therefore \frac{d^2X}{dx^2} + \mu_2^2 X = 0 \text{ and } \frac{d^2Y}{dy^2} + (\mu_1^2 - \mu_2^2)Y = 0, \text{ i.e. } \frac{d^2Y}{dy^2} + \mu_3^2 Y = 0,$$

where $\mu_3^2 = \mu_1^2 - \mu_2^2$.

The solutions of above two ordinary differential equations are $X(x) = c_2 \cos\mu_2 x + c_3 \sin\mu_2 x$ and $Y(y) = c_4\cos\mu_3 y + c_5 \sin\mu_3 y$, where c_2, c_3, c_4, c_5 are arbitrary constants.

Hence, desired solution is given by

$$T(x, y, t) = (A_1\cos\mu_2 x + A_2\sin\mu_2 x)(A_3\cos\mu_3 y + A_4\sin\mu_3 y)\exp(-k\mu_1^2 t)$$

where $A_1 = c_1c_2$, $A_2 = c_1c_3$, $A_3 = c_1c_4$ and $A_4 = c_1c_5$.

Examples V(B)

1. Solve the heat equation for a finite rod by the method of separation of variables: $\frac{\partial T}{\partial t} - 4\frac{\partial^2 T}{\partial x^2} = 0$, $0 < x < 2$, $t > 0$ subject to the boundary conditions $T(0, t) = T(2, t) = 0$, $t > 0$ and the initial condition $T(x, 0) = x, 0 < x < 2$.
2. Solve the heat equation: $\frac{\partial T}{\partial t} - 3\frac{\partial^2 T}{\partial x^2} = 0$, $0 < x < \pi$, $t > 0$ subject to the conditions
   (i) $T(0, t) = T(\pi, t) = 0$, $t \geqslant 0$
   (ii) $T(x, t) < \infty$ as $t \to \infty$
   (iii) $T(x, 0) = \begin{cases} x, & \text{if } 0 \leqslant x \leqslant \frac{\pi}{2} \\ \pi - x, & \text{if } \frac{\pi}{2} \leqslant x \leqslant \pi. \end{cases}$
3. Solve the one-dimensional heat equation $\frac{\partial T}{\partial t} - k\frac{\partial^2 T}{\partial x^2} = 0$, $0 < x < l$, $t > 0$ subject to the conditions: (i)] $T(x, 0) = x(l - x)$, $0 < x < l$; (ii) $T_x(0, t) = T_x(l, t) = 0$, $t \geqslant 0$; (iii) $T(x, t) < \infty$ as $t \to \infty$. Hence, evaluate $\lim_{t\to\infty} T(x, t)$.
4. Using the method of separation of variables find a formal solution $T(x, t)$ of the heat equation: $\frac{\partial T}{\partial t} - \frac{\partial^2 T}{\partial x^2} = 0$ subject to the boundary conditions: (i) $T(0, t) = 0, t > 0$; (ii) $T(l, t) = 0, t > 0$ and the initial condition (iii) $T(x, 0) = f(x)$, $0 \leqslant x \leqslant l (l > 0)$, where f is a specified function of $x \in [0, l]$.
5. Apply the method of separation of variables to find a formal solution $T(x, t)$ of the heat equation $\frac{\partial T}{\partial t} - \frac{\partial^2 T}{\partial x^2} = 0$ subject to the boundary conditions $T(0, t) = 0 = T(\pi, t)$, $t > 0$ and the initial condition $T(x, 0) = f(x)$, $0 \leqslant x \leqslant \pi$, where f is a specified function of $x \in [0, \pi]$.

Answers

1. $T(x,t) = \frac{4}{\pi}\sum_{n=1}^{\infty}(-1)^{n+1}\frac{1}{n}\sin\frac{n\pi x}{2}e^{-n^2\pi^2 t}$

2. $T(x,t) = \frac{4}{\pi}\sum_{n=1}^{\infty}\frac{1}{n^2}\sin nx\sin\frac{n\pi}{2}\exp(-3n^2 t)$

3. $T(x,t) = \frac{l^2}{6} - \frac{l^2}{\pi^2}\sum_{m=1}^{\infty}\frac{1}{m^2}\cos\frac{2m\pi x}{l}\exp\left(-\frac{4km^2\pi^2 t}{l}\right)$, $\lim_{t\to\infty} T(x,t) = \frac{l^2}{6}$.

4. $T(x,t) = \sum_{n=1}^{\infty}\left[\frac{2}{l}\int_0^l f(x)\sin\frac{n\pi x}{l}dx\right]\sin\frac{n\pi x}{l}\exp\left(-\frac{n^2\pi^2 t}{l^2}\right)$

5. $T(x,t) = \sum_{n=1}^{\infty}\left[\frac{2}{\pi}\int_0^{\pi} f(x)\sin nx\,dx\right]\sin nx\exp(-n^2 t)$

6. $T(x,t) = \sum_{n=1}^{\infty}(-1)^{n+1}\cdot\frac{2}{n}\cdot\sin nx\exp(-n^2 t)$.

5.6 Laplace's equation (in Cartesian coordinates)

The Laplace's equation (in three dimensions) is a second-order partial differential equation of the form:

$$\nabla^2\phi = 0, \tag{5.6.1}$$

where $\nabla^2 \equiv \frac{\partial^2}{\partial x^2}+\frac{\partial^2}{\partial y^2}+\frac{\partial^2}{\partial z^2}$ is called the Laplace operator named after Pierre Simon de Laplace (23 March 1749–5 March 1827), a French mathematician. Here ϕ is a scalar function which is twice-differentiable function of real variables x, y and z. Laplace's equation is a simplest example of the elliptic partial differential equation. The solutions of the Laplace's eq. (5.6.1) are called the *harmonic functions.*

The general theory of the solutions to the Laplace's equations is known as the potential theory that is the study of harmonic functions. The type of equation plays very important role in several fields of science, specially in Magnetostatics, Gravitation, Irrotational Motion of a perfect fluid, etc.

5.6.1 Boundary value problems

In the field of partial differential equations, a boundary value problem for Laplace's equation is a partial differential equation together with a set of additional constraints known as boundary conditions where the unknown potential function ϕ is to be determined. Here ϕ satisfies Laplace's equation in a bounded region R. However, it satisfies the constraints on the boundary S of the region R.

A boundary value problem has some conditions which are specified at the boundaries of the independent variables related to the partial differential equation whereas an initial value problem has the conditions which are specified only at the lower boundaries of the domain of independent variables.

For example, let us consider the Laplace's equation $\frac{\partial^2\phi}{\partial x^2}+\frac{\partial^2\phi}{\partial y^2} = 0$ defined on $[a,b]\times[c,d]$. Then the corresponding boundary value problem would specify the values of $\phi(x,y)$ at $x = a$ and $x = b$ (or at $y = c$ and $y = d$) whereas an initial value problem would specify a value of $\phi(x,y)$ and $\phi_x(x,y)$ (or $\phi_y(x,y)$) at $x = a$ (or $y = c$) only.

Now, we would like to introduce the following two types of boundary value problems.

Dirichlet problem (the first boundary value problem)

Let f be a continuous function defined on the boundary S of the region R. Then the problem of finding a harmonic function ϕ at any point within R satisfying the condition $\phi = f$ on S is called the *interior Dirichlet problem*. For exterior Dirichlet problem $\nabla^2\phi = 0$ holds outside R.

Example 5.6.1. *The following boundary value problem is an example of Dirichlet problem* $\frac{\partial^2\phi}{\partial x^2} + \frac{\partial^2\phi}{\partial y^2} = 0$, $0 < x < a, 0 < y < b$ *subject to the boundary conditions* $\phi(x, b) = \phi(a, y) = \phi(0, y) = 0, \phi(x, 0) = f(x)$.

Neumann problem (second boundary value problem)

Let f be a continuous function defined uniquely at any point of the boundary S of a finite region R. Then the problem of finding a harmonic function ϕ is said to be the interior Neumann problem if the normal derivative $\frac{\partial\phi}{\partial n} = f$ at every point of S. It is said to be exterior Neumann problem if $\nabla^2\phi = 0$ holds outside R.

Example 5.6.2. *The following boundary value problem is an example of Neumann problem* $\frac{\partial^2\phi}{\partial x^2} + \frac{\partial^2\phi}{\partial y^2}$, $0 < x < a, 0 < y < b$ *subject to the boundary conditions* $\phi_x(0, y) = \phi_x(a, y) = \phi_y(x, 0) = 0$ *and* $\phi_y(x, b) = f(x)$.

Example 5.6.3. *Show that the Laplace's equation* $\nabla^2\phi = 0$ *has no real characteristics on* (x, y)*-plane.*

Solution: Two-dimensional Laplace's equation is

$$\frac{\partial^2\phi}{\partial x^2} + \frac{\partial^2\phi}{\partial y^2} = 0. \tag{5.6.2}$$

Comparing it with

$$A\frac{\partial^2\phi}{\partial x^2} + B\frac{\partial^2\phi}{\partial x\partial y} + C\frac{\partial^2\phi}{\partial y} + F\left(x, y, \phi, \frac{\partial\phi}{\partial x}, \frac{\partial\phi}{\partial y}\right) = 0,$$

we have $A = 1, B = 0, C = 1, F = 0$.

$\therefore$ the discriminant of the Laplace's equation is given by $D = B^2 - 4AC = -4 < 0$.

It follows that the Laplace's eq. (5.6.2) is elliptic type.

The λ-quadratic equation is $A\lambda^2 + B\lambda + C = 0$, i.e. $\lambda^2 + 1 = 0$, i.e. $\lambda = \pm i$.

Thus, the characteristic equations are $\frac{dy}{dx} \pm i = 0$.

Integrating, we obtain $y + ix = c_1$ and $y - ix = c_2$, where c_1, c_2 are arbitrary constants.

Observe that the solutions of the characteristic equations are necessarily complex-valued and therefore, there is no real characteristic for the Laplace's equation.

5.6.2 Solution of two-dimensional Laplace's equation by separation of variables (Cartesian coordinates)

The two-dimensional Laplace's equation in Cartesian coordinates is given by

$$\nabla^2\phi = 0, \tag{5.6.3}$$

where $\nabla^2 \equiv \frac{\partial^2}{\partial x^2} + \frac{\partial^2}{\partial y^2}$.

Let us assume that the Laplace's eq. (5.6.3) has the solution in the form:

$$\phi(x, y) = X(x)Y(y). \tag{5.6.4}$$

Then $\frac{\partial^2\phi}{\partial x^2} = X''(x)Y(y)$ and $\frac{\partial^2\phi}{\partial y^2} = X(x)Y''(y)$.

Using these results in (5.6.3), we obtain $\frac{1}{X}\frac{d^2X}{dx^2} = -\frac{1}{Y}\frac{d^2Y}{dy^2} = \mu$, say, where μ is a constant of separation.

$$\therefore \frac{d^2X}{dx^2} - \mu X = 0 \text{ and } \frac{d^2Y}{dy^2} + \mu Y = 0. \tag{5.6.5}$$

Case 5.6.1. *Let $\mu = \lambda^2 > 0$, λ being real. The equations in (5.6.5) becomes $\frac{d^2X}{dx^2} - \lambda^2 X = 0$ and $\frac{d^2Y}{dy^2} + \lambda^2 Y = 0$. The solutions of these two equations are given by $X(x) = c_1 e^{\lambda x} + c_2 e^{-\lambda x}$ and $Y(y) = c_3 \cos \lambda y + c_4 \sin \lambda y$, where c_1, c_2, c_3, c_4 are arbitrary constants. Hence, the complete solution of the Laplace's equation (5.6.3) is*

$$\phi(x, y) = \left(c_1 e^{\lambda x} + c_2 e^{-\lambda x}\right)(c_3 \cos \lambda y + c_4 \sin \lambda y). \tag{5.6.6}$$

Case 5.6.2. *Let $\mu = 0$. Then the equations in (5.6.5) become $\frac{d^2X}{dx^2} = 0$ and $\frac{d^2Y}{dy^2} = 0$. The solutions of these two equations are $X(x) = c_1 x + c_2$ and $Y(y) = c_3 y + c_4$, where c_1, c_2, c_3, c_4 are arbitrary constants. Hence, the complete solution of the Laplace's equation (5.6.3) is*

$$\phi(x, y) = (c_1 x + c_2)(c_3 y + c_4). \tag{5.6.7}$$

Case 5.6.3. *Let $\mu = -\lambda^2 < 0$, λ being real. Then the solutions of the differential equations given in (5.6.5) are $X(x) = (c_1 \cos \lambda x + c_2 \sin \lambda x)$, $Y(y) = c_3 e^{\lambda y} + c_4 e^{-\lambda y}$, where c_1, c_2, c_3, c_4 are arbitrary constants. Hence, the complete solution of the Laplace's equation (5.6.3) is*

$$\phi(x, y) = (c_1 \cos \lambda x + c_2 \sin \lambda x)(c_3 e^{\lambda y} + c_4 e^{-\lambda y}). \tag{5.6.8}$$

Example 5.6.4. *Solve the following Laplace's equation at any interior point of the rectangle $0 < x < a$, $0 < y < b$, $\frac{\partial^2\phi}{\partial x^2} + \frac{\partial^2\phi}{\partial y^2} = 0$ subject to the boundary conditions $\phi(a, y) = \phi(x, b) = \phi(0, y) = 0$ and $\phi(x, 0) = f(x)$, where $f(x)$ can be expressed in Fourier sine series.*

Solution: The general solutions of the Laplace's equation $\frac{\partial^2\phi}{\partial x^2} + \frac{\partial^2\phi}{\partial y} = 0$ are given by (5.6.6) or (5.6.7) or (5.6.8).

Here, we are interested only for nontrivial solution.

Observe that the solutions (5.6.6) and (5.6.7) both reduce to the trivial solution for the given boundary conditions.

So, we consider the solution (5.6.8), i.e. $\phi(x, y) = (c_1 \cos \lambda x + c_2 \sin \lambda x)\,(c_3 e^{\lambda y} + c_4 e^{\lambda y})$.

Now, for the boundary condition $\phi(0, y) = 0$, we obtain $c_1\,(c_3 e^{-\lambda y} + c_4 e^{-\lambda y}) = 0$, which gives $c_1 = 0$.

Again, for another boundary condition $\phi(a, y) = 0$, we obtain $c_2 \sin \lambda a(c_3 e^{\lambda y} + c_4 e^{-\lambda y}) = 0$ ($\because c_1 = 0$).

It follows that $c_2 \sin \lambda a = 0$. Thus, for nontrivial solution, we have $\sin \lambda a = 0$.

$\therefore \lambda a = n\pi, n = 1, 2, 3, \ldots$, i.e. $\lambda = \frac{n\pi}{a}$.

$$\therefore \phi(x, y) = c_2 \sin \frac{n\pi x}{a}\left(c_3 e^{\frac{n\pi y}{a}} + c_4 e^{-\frac{n\pi y}{a}}\right) = \sin \frac{n\pi x}{a}\left(a_n e^{\frac{n\pi y}{a}} + b_n e^{-\frac{n\pi y}{a}}\right), \tag{5.6.9}$$

where $a_n = c_2 c_3$, $b_n = c_2 c_4$.

Hence, by the principle of superposition

$$\phi(x, y) = \sum_{n=1}^{\infty} \sin \frac{n\pi x}{a}\left(a_n e^{\frac{n\pi y}{a}} + b_n e^{-\frac{n\pi y}{a}}\right). \tag{5.6.10}$$

Now, using the boundary condition $\phi(x, b) = 0$ and (5.6.9), we obtain $a_n e^{\frac{n\pi b}{a}} + b_n e^{-\frac{n\pi b}{a}} = 0$.

$\therefore\ b_n = -a_n e^{\frac{2n\pi b}{a}}$.

Thus, (5.6.10) becomes

$$\phi(x,y) = \sum_{n=1}^{\infty} \sin\frac{n\pi x}{a}\left(a_n e^{\frac{n\pi y}{a}} - a_n e^{\frac{2n\pi b}{a}} \cdot e^{-\frac{n\pi y}{a}}\right) = \sum_{n=1}^{\infty} 2a_n e^{\frac{n\pi b}{a}} \sin\frac{n\pi x}{a}\left(\frac{e^{\frac{n\pi}{a}(y-b)} - e^{-\frac{n\pi}{a}(y-b)}}{2}\right),$$

i.e. $\phi(x,y) = \sum_{n=1}^{\infty} 2a_n e^{\frac{n\pi b}{a}} \sin\frac{n\pi x}{a} \sinh\frac{n\pi}{a}(y-b)$.

Now, by the nonhomogeneous boundary condition $\phi(x,0) = f(x)$, we obtain

$$f(x) = \sum_{n=1}^{\infty} 2a_n e^{\frac{n\pi b}{a}} \sin\frac{n\pi x}{a} \sinh\left(-\frac{n\pi b}{a}\right) = \sum_{n=1}^{\infty}\left[2a_n e^{\frac{n\pi b}{a}} \sinh\left(-\frac{n\pi b}{a}\right)\right] \sin\frac{n\pi x}{a}$$

$$= \sum_{n=1}^{\infty} A_n \sin\frac{n\pi x}{a}, \text{ where } A_n = 2a_n e^{\frac{n\pi b}{a}} \sinh\left(-\frac{n\pi b}{a}\right).$$

It is a half-range Fourier sine series.

$$\therefore\ A_n = \frac{2}{a}\int_0^a f(x)\sin\frac{n\pi x}{a}\,dx.$$

Consequently,

$$a_n = \frac{\displaystyle\int_0^a f(x)\sin\frac{n\pi x}{a}\,dx}{a e^{\frac{n\pi b}{a}} \sinh\left(-\frac{n\pi b}{a}\right)}.$$

Hence, desired solution is given by

$$\phi(x,y) = \sum_{n=1}^{\infty} \frac{2}{a\sinh\left(-\frac{n\pi b}{a}\right)} \sin\frac{n\pi x}{a} \sinh\frac{n\pi}{a}(y-b)\int_0^a f(x)\sin\frac{n\pi x}{a}\,dx.$$

Example 5.6.5. *Solve the following Laplace's equation at any interior point to the rectangle $0 \leqslant x \leqslant a$, $0 \leqslant y \leqslant b$, $\frac{\partial^2\phi}{\partial x^2} + \frac{\partial^2\phi}{\partial y^2} = 0$ subject to the boundary conditions $\phi_x(0,y) = \phi_x(a,y) = \phi_y(x,0) = 0$ and $\phi_y(x,b) = f(x)$.*

Solution: The general solutions of the Laplace's equation $\frac{\phi^2\phi}{\partial x^2} + \frac{\partial^2\phi}{\partial y^2} = 0$ are given by (5.6.6) or (5.6.7) or (5.6.8).

Here we would like to choose (5.6.8) for nontrivial suitable solution subject to the given boundary conditions.

By (5.6.8), we have $\phi(x,y) = (c_1\cos\lambda x + c_2\sin\lambda x)(c_3 e^{\lambda y} + c_4 e^{-\lambda y})$.

$\therefore\ \phi_x(x,y) = (-c_1\lambda\sin\lambda x + c_2\lambda\cos\lambda x)(c_3 e^{\lambda y} + c_4 e^{-\lambda y})$

and $\phi_y(x,y) = (c_1\cos\lambda x + c_2\sin\lambda x)(c_3\lambda e^{\lambda y} - c_4\lambda e^{-\lambda y})$.

$\therefore\ \phi_x(0,y) = c_2\lambda(c_3 e^{\lambda y} + c_4 e^{-\lambda y})$.

Since $\phi_x(0,y) = 0$, $c_2 = 0$.

Again, $\phi_x(a,y) = -c_1\lambda\sin\lambda a(c_3 e^{\lambda y} + c_4 e^{-\lambda y})$ ($\because\ c_2 = 0$).

Thus, $\phi_x(a,y) = 0 \Rightarrow c_1\sin\lambda a = 0$.

For nontrivial solution, $\sin \lambda a = 0$.

$\therefore\ \lambda a = n\pi,\ n = 0, 1, 2, 3, \ldots$, i.e. $\lambda = \dfrac{n\pi}{a}$.

$$\therefore\ \phi(x, y) = c_1 \cos \frac{n\pi x}{a}\left(c_3 e^{\frac{n\pi y}{a}} + c_4 e^{-\frac{n\pi y}{a}}\right) = \cos \frac{n\pi x}{a}\left(c_5 e^{\frac{n\pi y}{a}} + c_6 e^{-\frac{n\pi y}{a}}\right), \tag{5.6.11}$$

where $c_5 = c_1 c_3$ and $c_6 = c_1 c_4$.

Differentiating (5.6.11) partially with respect to y, we obtain

$$\phi_y(x, y) = \frac{n\pi}{a} \cos \frac{n\pi x}{a}\left(c_5 e^{\frac{n\pi y}{a}} - c_6 e^{-\frac{n\pi y}{a}}\right).$$

Now, using the boundary condition $\phi_y(x, 0) = 0$, we obtain $c_5 = c_6$.

Thus, (5.6.11) becomes

$$\phi(x, y) = 2c_5 \cos \frac{n\pi x}{a}\left(\frac{e^{\frac{n\pi y}{a}} + e^{-\frac{n\pi y}{a}}}{2}\right) = 2c_5 \cos \frac{n\pi x}{a} \cosh \frac{n\pi y}{a} = A_n \cos \frac{n\pi x}{a} \cosh \frac{n\pi y}{a},$$

where $2c_5 = A_n$.

Therefore, by the principle of superposition, we have

$$\phi(x, y) = \sum_{n=0}^{\infty} A_n \cos \frac{n\pi x}{a} \cosh \frac{n\pi y}{a}. \tag{5.6.12}$$

$$\therefore\ \phi_y(x, y) = \sum_{n=1}^{\infty} A_n \frac{n\pi}{a} \cos \frac{n\pi x}{a} \sinh \frac{n\pi y}{a}$$

Thus, the boundary condition $\phi_y(x, b) = f(x)$ gives

$$f(x) = \sum_{n=1}^{\infty}\left(A_n \frac{n\pi}{a}\right) \cos \frac{n\pi x}{a} \sinh \frac{n\pi b}{a}, \text{ i.e. } f(x) = \sum_{n=1}^{\infty} B_n \cos \frac{n\pi x}{a},$$

where $B_n = A_n \frac{n\pi}{a} \sinh \frac{n\pi b}{a}$.

Here, $\sum_{n=1}^{\infty} B_n \cos \frac{n\pi x}{a}$ is a half-range Fourier cosine series.

$$\therefore\ B_n = \frac{2}{a} \int_0^a f(x) \cos \frac{n\pi x}{a}\, dx.$$

$$\therefore\ A_n = \frac{2}{n\pi \sinh \frac{n\pi b}{a}} \int_0^a f(x) \cos \frac{n\pi x}{a}\, dx.$$

Hence, the desired solution is given by

$$\phi(x, y) = A_0 + \sum_{n=1}^{\infty}\left[\frac{2}{n\pi \sinh \frac{n\pi b}{a}} \int_0^a f(x) \cos \frac{n\pi x}{a}\, dx\right] \cos \frac{n\pi x}{a} \cosh \frac{n\pi y}{a},$$

where A_0 is a constant.

xample 5.6.6. *Find the solution of the Laplace's equation* $\frac{\partial^2 \phi}{\partial x^2} + \frac{\partial^2 \phi}{\partial y^2} = 0$ *subject to the following boundary onditions:* (i) $\phi(0, y) = \phi\left(\frac{\pi}{2}, y\right) = 0, 0 \leqslant y \leqslant \frac{\pi}{2}$; (ii) $\phi\left(x, \frac{\pi}{2}\right) = 0, 0 \leqslant x \leqslant \frac{\pi}{2}$; (iii) $\phi(x, 0) = f(x)$, $\leqslant x \leqslant \frac{\pi}{2}$, *where* $f(x)$ *is a specified function of* $x \in \left[0, \frac{\pi}{2}\right]$.

olution: The general solutions of the Laplace's equation $\frac{\partial^2 \phi}{\partial x^2} + \frac{\partial^2 \phi}{\partial y^2} = 0$ are given by (5.6.6) or (5.6.7) or .6.8).

In order to find the nontrivial solution, we consider only (5.6.8), i.e. $\phi(x, y) = (c_1 \cos \lambda x + c_2 \sin \lambda x)$ $_3 e^{\lambda y} + c_4 e^{-\lambda y})$.

For a boundary condition $\phi(0, y) = 0$, we obtain $c_1(c_3 e^{\lambda y} + c_4 e^{-\lambda y}) = 0$, which follows that $c_1 = 0$.

Again, for the boundary condition $\phi\left(\frac{\pi}{2}, y\right) = 0$, we obtain $c_2 \sin \lambda \frac{\pi}{2}\left(c_3 e^{\lambda y} + c_4 e^{-\lambda y}\right) = 0 \; (\because c_1 = 0)$, hich follows that $c_2 \sin \lambda \frac{\pi}{2} = 0$. Thus, for nontrivial solution, $\sin \lambda \frac{\pi}{2} = 0$.

$\therefore \lambda \frac{\pi}{2} = n\pi, n = 1, 2, 3, \ldots$, i.e. $\lambda = 2n$.

$$\therefore \phi(x, y) = c_2 \sin 2nx \left(c_3 e^{2ny} + c_4 e^{-2ny}\right) = \sin 2nx \left(a_n e^{2ny} + b_n e^{-2ny}\right), \tag{5.6.13}$$

here $a_n = c_2 c_3$ and $b_n = c_2 c_4$.

Thus, by the principle of superposition,

$$\phi(x, y) = \sum_{n=1}^{\infty} \sin 2nx \left(a_n e^{2ny} + b_n e^{-2ny}\right). \tag{5.6.14}$$

Now, the boundary condition $\phi\left(x, \frac{\pi}{2}\right) = 0$ gives $a_n e^{n\pi} + b_n e^{-n\pi} = 0$.

$\therefore b_n = -a_n e^{2n\pi}$.

$\therefore$ (5.6.14) becomes

$$\phi(x, y) = \sum_{n=1}^{\infty} \sin 2nx \left(a_n e^{2ny} - a_n e^{2n\pi} \cdot e^{-2ny}\right) = \sum_{n=1}^{\infty} 2n_n e^{n\pi} \sin 2nx \left(\frac{e^{2n\left(y-\frac{\pi}{2}\right)} - e^{-2n\left(y-\frac{\pi}{2}\right)}}{2}\right)$$

$$= \sum_{n=1}^{\infty} 2a_n e^{n\pi} \sin 2nx \sinh\left(2n\left(y - \frac{\pi}{2}\right)\right).$$

Again, the nonhomogeneous boundary condition $\phi(x, 0) = f(x)$ gives

$$f(x) = \sum_{n=1}^{\infty} 2a_n e^{n\pi} \sin 2nx \sinh(-n\pi) = \sum_{n=1}^{\infty} [2a_n e^{n\pi} \sinh(-n\pi)] \sin 2nx = \sum_{n=1}^{\infty} A_n \sin 2nx, \tag{5.6.15}$$

here $A_n = 2a_n e^{n\pi} \sinh(-n\pi)$.

Clearly (5.6.15) is a half-range Fourier sine series.

$$\therefore A_n = \frac{4}{\pi}\int_0^{\pi/2} f(x) \sin 2nx \, dx. \qquad \therefore a_n = \frac{2\displaystyle\int_0^{\pi/2} f(x) \sin(2nx) \, dx}{\pi e^{n\pi} \sinh(-n\pi)}.$$

Hence, the solution of the given problem is

$$\phi(x, y) = \sum_{n=1}^{\infty} \frac{4}{\pi \sinh(-n\pi)} \sin(2nx) \sinh 2n\left(y - \frac{\pi}{2}\right) \int_0^{\pi/2} f(x) \sin 2nx \, dx.$$

Examples V(C)

1. Find the solution of the following Laplace's equation at any interior point of the rectangle $0 \leqslant x \leqslant \pi, 0 \leqslant y \leqslant \pi$, $\frac{\partial^2\phi}{\partial x^2}+\frac{\partial^2\phi}{\partial y^2}=0$ subject to the boundary conditions $\phi(\pi, y)=\phi(x,\pi)=\phi(0,y)=0$ and $\phi(x,0)=f(x)$, a specified function of $x \in [0,\pi]$.
2. Find a formal solution of $\phi(x, y)$ of the Laplace's equation $\frac{\partial^2\phi}{\partial x^2}+\frac{\partial^2\phi}{\partial y^2}=0$ subject to the boundary conditions:
   (i) $\phi(0, y)=\phi\left(\frac{\pi}{3}, y\right)=0, 0 \leqslant y \leqslant \frac{\pi}{3}$
   (ii) $\phi\left(x, \frac{\pi}{3}\right)=0, 0 \leqslant x \leqslant \frac{\pi}{3}$
   (iii) $\phi(x,0)=f(x), 0 \leqslant x \leqslant \frac{\pi}{3}$, where f is a specified function of $x \in \left[0, \frac{\pi}{3}\right]$.
3. Find a formal solution $\phi(x, y)$ of the Laplace's equation $\frac{\partial^2\phi}{\partial x^2}+\frac{\partial^2\phi}{\partial y^2}=0$ subject to the boundary conditions:
   (i) $\phi(0, y)=\phi(\pi, y)=0, 0 \leqslant y \leqslant \frac{\pi}{2}$
   (ii) $\phi\left(x, \frac{\pi}{2}\right)=0, 0 \leqslant x \leqslant \pi$
   (iii) $\phi(x,0)=f(x), 0 \leqslant x \leqslant \pi$, where f is a specified function of $x \in [0,\pi]$.
4. Solve the following Laplace's equations by the method of separation of variables.
   (i) $\frac{\partial^2\phi}{\partial x^2}+\frac{\partial^2\phi}{\partial y^2}=0$ subject to the boundary conditions:
   $\phi(0, y)=\phi(\pi, y)=0, 0 \leqslant y \leqslant \frac{\pi}{2}, \phi\left(x, \frac{\pi}{2}\right)=0, 0 \leqslant x \leqslant \pi$ and $\phi(x,0)=x, 0 \leqslant x \leqslant \pi$
   (ii) $\frac{\partial^2\phi}{\partial x^2}+\frac{\partial^2\phi}{\partial y^2}=0$ subject to the boundary conditions:
   $\phi(\pi, y)=\phi(x,\pi)=\phi(0,y)=0, 0 \leqslant x \leqslant \pi, 0 \leqslant y \leqslant \pi$ and $\phi(x,0)=\pi-x, 0 \leqslant x \leqslant \pi$
   (iii) $\frac{\partial^2\phi}{\partial x^2}+\frac{\partial^2\phi}{\partial y^2}=0$ subject to the boundary conditions:
   $\phi(\pi, y)=\phi(x,\pi)=\phi(0,y)=0, 0 \leqslant x \leqslant \pi, 0 \leqslant y \leqslant \pi$ and $\phi(x,0)=\cos x, 0 \leqslant x \leqslant \pi$.
5. Solve the following Laplace's equation at any interior point to the rectangle formed by $0 \leqslant x \leqslant \pi, 0 \leqslant y \leqslant \pi$; $\frac{\partial^2\phi}{\partial x^2}+\frac{\partial^2\phi}{\partial y^2}=0$ subject to the boundary conditions: $\phi_x(0, y)=\phi_x(\pi, y)=\phi_y(x,0)$ and $\phi_y(x,\pi)=f(x)$, a function of $x \in [0,\pi]$.
6. Use the method of separation of variables to find a formal solution $\phi(x, y)$ of the Laplace's equation $\frac{\partial^2\phi}{\partial x^2}+\frac{\partial^2\phi}{\partial y^2}=0$ subject to the boundary conditions:
   $\phi_x(0, y)=0=\phi_x(\pi, y), 0 \leqslant y \leqslant \pi$
   $\phi_y(x,0)=0, 0 \leqslant x \leqslant \pi$
   $\phi_y(x,\pi)=x, 0 \leqslant x \leqslant \pi$.
7. Find the solution of the Laplace's equation $\frac{\partial^2\phi}{\partial x^2}+\frac{\partial^2\phi}{\partial y^2}=0$ subject to the boundary conditions: $\phi(x,0)=\phi(x,b)=0, 0 \leqslant x \leqslant a, \phi(0,y)=0$ and $\phi_x(a,y)=T\sin^3\frac{\pi y}{a}, 0 \leqslant y \leqslant b$.
8. A thin rectangular homogeneous thermally conducted plate occupies the rectangular region given by $0 \leqslant x \leqslant a, 0 \leqslant y \leqslant b$. The edge $y=0$ is held at the temperature $Tx(x-a)$, where T is a constant and the other edges are maintained at 0°. If the other faces are insulated and there is no heat source or sink inside the plate, find the steady-state temperature inside the plate.

Answers

1. $\phi(x, y)=\sum_{n=1}^{\infty}\frac{-2}{\pi\sin n\pi}\sin nx\sinh n(y-\pi)\int_0^{\pi}f(x)\sin nx\,dx.$

2. $\phi(x, y) = \sum_{n=1}^{\infty} \frac{-6}{\pi \sin n\pi} \sin(3nx) \sinh 3n\left(y - \frac{\pi}{3}\right) \int_0^{\pi/3} f(x) \sin 3nx \, dx.$

3. $\phi(x, y) = \sum_{n=1}^{\infty} \frac{-2}{\pi \sin \frac{n\pi}{2}} \sin nx \sinh n\left(y - \frac{\pi}{2}\right) \int_0^{\pi} f(x) \sin nx \, dx.$

4. (i) $\phi(x, y) = 2\sum_{n=1}^{\infty} \frac{(-1)^{n+1}}{n} \operatorname{cosec} \frac{n\pi}{2} \sin nx \sinh n\left(y - \frac{\pi}{2}\right).$

(ii) $\phi(x, y) = \sum_{n=1}^{\infty} \frac{2}{\pi n^2} \sin nx \sinh n(y - \pi).$

(iii) $\phi(x, y) = 2\sum_{n=1}^{\infty} \frac{(-1)^{n+1} - 1}{n + 1} \operatorname{cosec} n\pi \sin nx \sinh n(y - \pi).$

5. $\phi(x, y) = A_0 + \sum_{n=1}^{\infty} \left[\frac{2}{n\pi \sinh n\pi} \int_0^{\pi} f(x) \cos nx \, dx\right] \cos nx \cosh ny.$

6. $\phi(x, y) = A_0 + 2\sum_{n=1}^{\infty} \frac{(-1)^n - 1}{\pi n^3 \sinh n\pi} \cos nx \cosh ny.$

7. $\phi(x, y) = \frac{bT}{\pi} \sum_{n=1}^{\infty} \frac{1}{n} \operatorname{sech}\left(\frac{n\pi a}{b}\right) \sinh\left(\frac{n\pi x}{b}\right) \sin^3\left(\frac{\pi y}{a}\right).$

8. $\phi(x, y) = \frac{8Ta^2}{\pi^3} \sum_{n=1}^{\infty} \frac{1}{(2n + 1)^3} \cdot \operatorname{cosech}(2n + 1)\frac{\pi b}{a} \sin(2n + 1)\frac{\pi x}{a} \sinh(2n + 1)\frac{(y - b)\pi}{a}$

[Hint. The steady-state temperature ϕ satisfies the Laplace's equation $\frac{\partial^2 \phi}{\partial x^2} + \frac{\partial^2 \phi}{\partial y^2} = 0$ subject to the boundary conditions $\phi(0, y) = \phi(a, y) = \phi(x, b) = 0$ and $\phi(x, 0) = Tx(x - a)$.]

Part III

MCQ and Answers on Ordinary Differential Equations

Section 1

MCQ on chapter 1 and chapter 2

Select (a), (b), (c) or (d), whichever is correct:

1. Which one of the following differential equations is linear?
   (a) $\frac{dy}{dx} = 1 + x^2y^2$
   (b) $\left(\frac{dy}{dx}\right)^2 + \frac{dy}{dx} - 3y = 0$
   (c) $x^2\frac{d^2y}{dx^2} + x^3\frac{dy}{dx} + (\cos x)\sqrt{y} = 2x + 1$
   (d) $x^3\frac{d^2y}{dx^2} + \cos x\frac{dy}{dx} + (\sin x)y = 0.$
2. Let a curve be defined by the condition that the sum of x- and y-intercepts of its tangent at any point (x, y) is always equal to a. If we express the fact by a differential equation, which one of the following is NOT correct?
   (a) It is linear differential equation
   (b) It is nonlinear differential equation
   (c) Its order is one
   (d) Its degree is two.
3. Choose the INCORRECT answer:
   The differential equation corresponding to a family of confocal conics represented by the equation $\frac{x^2}{a^2+\lambda} + \frac{y^2}{b^2+\lambda} = 1$ (where a and b are fixed constants and λ is a parameter) is
   (a) a linear equation
   (b) nonlinear equation
   (c) of degree two
   (d) first-order linear.
4. Order of the differential equation corresponding to the family of curves represented by $ky = ae^{bx+c}$ (where a, b, c and k are arbitrary constants) is
   (a) 1
   (b) 2
   (c) 3
   (d) 4.
5. The differential equation of all circles each of which touches the axis of x at the origin is
   (a) $(x^2 + y^2)\frac{dy}{dx} = 2xy$
   (b) $(x^2 - y^2)\frac{dy}{dx} = 2y$
   (c) $(x^2 - y^2)\frac{dy}{dx} = 2xy$
   (d) $(x^2 + y^2)\frac{dy}{dx} = 2x.$
6. Which one of the following is NOT correct information about the differential equation of the family of curves given by $y = cx + c - c^3$ (c being a parameter)?
   (a) First order
   (b) 3rd degree
   (c) nonlinear
   (d) linear.
7. The order of the differential equation of the family of curves represented by $y = \frac{ax+b}{cx+d}$ (a, b, c, d are arbitrary constants) is

(a) 1 (c) 3
(b) 2 (d) 4.

8. The adiabatic law $pv^{\gamma} = c$, where γ is a fixed constant and c is an arbitrary constant, leads to a differential equation of
(a) first-order nonlinear (c) second-order linear
(b) first-order linear (d) second-order nonlinear.

9. The differential equation of all parabolas each of which has latus rectum $4a$ and whose axes are parallel to the x-axis is
(a) $\left(\frac{dy}{dx}\right)^3 + 2a\frac{d^2y}{dx^2} = 0$ (c) $\left(\frac{dy}{dx}\right)^3 - 2a\frac{d^2y}{dx^2} = 0$
(b) $\left(\frac{dy}{dx}\right)^2 + 4a\frac{d^2y}{dx^2} = 0$ (d) $\left(\frac{dy}{dx}\right)^2 - 4a\frac{d^2y}{dx^2} = 0$.

10. The differential equation of all hyperbolas which pass through the origin and have their asymptotes parallel to the coordinate axes is
(a) first-order linear (c) second-order linear
(b) first-order nonlinear (d) second-order nonlinear.

11. The equation of the integral curve through the point $(2, 3)$ corresponding to the differential equation $\frac{dy}{dx} = \frac{y}{x}$ is
(a) $3y = 2x$ (c) $y = 3x$
(b) $2y = 3x$ (d) none of these.

12. The differential equation $(x^2 - y^2)\frac{dy}{dx} = 2xy$ represents
(a) two-parameter family of curves (c) three-parameter family of curves
(b) one-parameter family of curves (d) one-parameter family of straight lines.

13. Which one of the following equations is NOT exact?
(a) $(\cos y + y\cos x)dx + (\sin x - x\sin y)dy = 0$ (c) $ydx + (x + e^y)dy = 0$
(b) $ydx + (x + \cos y)dy = 0$ (d) $x^2ydx - (x^3 + y^3)dy = 0$.

14. An integrating factor of the differential equation $(x^2y - 2xy^2)dx - (x^3 - 3x^2y)dy = 0$ is
(a) $\frac{1}{xy}$ (c) $\frac{1}{x^2y^2}$
(b) $\frac{1}{x^2+y^2}$ (d) $\frac{1}{x^2-y^2}$.

15. An integrating factor of the differential equation $x\cos x\frac{dy}{dx} + y(x\sin x + \cos x) = 1$ is
(a) $\frac{x}{\cos x}$ (c) $\frac{x}{y\cos x}$
(b) $\frac{1}{\cos x}$ (d) $\frac{x}{\sin x}$.

16. Which one of the following is Bernoulli's equation?
(a) $\frac{dy}{dx} - xy + e^{-x^2}y^3 = 0$ (c) $\frac{dy}{dx} - xy + e^{-x^2} = 0$
(b) $\frac{dy}{dx} - 2y^{-1} + e^{-x^2}y^3 = 0$ (d) $y\frac{dy}{dx} - xy + e^{-x^2}y^3 = 0$.

17. Let us consider the differential equation $(2xy + e^x)ydx - e^x dy = 0$. Then $\frac{e^x}{y} + x^2 = 1$ is
(a) the general solution (c) singular solution
(b) particular solution (d) not a solution of the equation.

18. The gradient of the tangent at any point to a curve is $-\frac{x+y}{x}$. If it passes through the point $(2, 1)$, the equation of the curve is

(a) $x^2 + 2xy = 8$
(b) $x^2 - 2xy = 8$
(c) $x^2 + 2xy - y^2 = 8$
(d) $x^2 + 2xy - x = 0$.

19. The general solution of the differential equation $x\cos\left(\frac{y}{x}\right)(ydx + xdy) = y\sin\left(\frac{y}{x}\right)(xdy - ydx)$ is
(a) $y = c\sec\left(\frac{y}{x}\right)$
(b) $y = c\cos\left(\frac{y}{x}\right)$
(c) $xy = c\cos\left(\frac{y}{x}\right)$
(d) $xy = c\sec\left(\frac{y}{x}\right)$.

20. The equation of the family of curves for which the Cartesian sub-tangent is inversely proportional to the square of the abscissa is
(a) $x^3 - k\ln y^3 = c$
(b) $x^3 + k\ln y^3 = c$
(c) $x^3 - k\ln y = c$
(d) $x^3 - \ln y^3 = c$.

21. If the angle between the tangent and the radius vector is half of the vectorial angle at any point to a curve, the equation of the curve is a
(a) circle
(b) spiral
(c) cardioid
(d) hyperbola.

22. The curves having constant curvature are
(a) circles
(b) straight lines
(c) ellipses
(d) circles and straight lines.

23. The general solution of the differential equation $\frac{dy}{dx} + y\frac{d\psi}{dx} = \psi(x)\frac{d\psi}{dx}$ is
(a) $y = ce^{-\psi(x)} + \psi(x)$
(b) $y = ce^{-\psi(x)} + \psi(x) - 1$
(c) $y = ce^{-\psi(x)} - \psi(x) - 1$
(d) $y = e^{-\psi(x)} + \psi(x) - c$.

24. The general solution of the differential equation $(2xy^4e^y + 2xy^3 + y)dx + (x^2y^4e^y - x^2y^2 - 3x)dy = 0$ is
(a) $x^2y^2(e^x + 1) + cy^3 + x = 0$
(b) $x^2y^3e^x + x^2y^2 + cy^3 + x = 0$
(c) $x^2y^3e^x + x^2y^2 + cy^3 - x = 0$
(d) $xe^x + \frac{x}{y^2} = c$.

25. If $(y - k)^2 = kx$ (k being an arbitrary constant) is the primitive of the differential equation $4x\left(\frac{dy}{dx}\right)^2 + 2x\frac{dy}{dx} - y = 0$, the number of integral curves passing through the point $(1, 0)$ is
(a) 1
(b) 2
(c) 0
(d) 3.

26. If $y = c(x - c)^2$ (c being an arbitrary constant) is the primitive of the differential equation $\left(\frac{dy}{dx}\right)^3 - 4xy\frac{dy}{dx} + 8y^2 = 0$, the number of integral curves passing through the point $(1, 4)$ is
(a) 1
(b) 0
(c) 3
(d) 2.

27. If $y^2 = 2cx + c^3$ (c being an arbitrary constant) is the primitive of the differential equation $8x\left(\frac{dy}{dx}\right)^3 = y\left[12\left(\frac{dy}{dx}\right)^2 - 9\right]$, the number of integral curves passing through the point $(1, 0)$ is
(a) 1
(b) 2
(c) 0
(d) 3.

28. An integral curve passing through the point $(1, 2)$ corresponding to the differential equation $x\frac{dy}{dx} = 2y$ also passes through the point

(a) (2, 1)
(b) (1, 1)
(c) (4, 1)
(d) (0, 0).

29. The differential equation $(4x + 3y + 1)dx + (3x + 2y + 1)dy = 0$ represents one-parameter family of
(a) parabolas
(b) circles
(c) hyperbolas
(d) ellipses.

30. The general solution of the differential equation $(2y^3 + x)\frac{dx}{dy} = y$ represents a family of
(a) circles
(b) ellipses
(c) hyperbolas
(d) parabolas.

31. The number of arbitrary constants in the general solution of the differential equation $F\left(x, y, \frac{dy}{dx}, \frac{d^2y}{dx^2}, \frac{d^4y}{dx^4}\right) = 0$ is
(a) 1
(b) 2
(c) 3
(d) 4.

32. The eccentricity of the curves represented by the differential equation $ydx - 2xdy = 0$ is given by
(a) $e = 0$
(b) $e = 1$
(c) $e > 1$
(d) $e < 1$.

33. The number of curves passing through the point (−1, 0) corresponding to the differential equation $(1 + x)\frac{dy}{dx} = 1 + e^{-y}$ is
(a) 0
(b) 1
(c) 2
(d) 3.

34. The value of k for which the differential equation $(3x^2 + ke^y)\frac{dx}{dy} + (2xe^y + 3y^2) = 0$ is exact, is
(a) 1
(b) 2
(c) 3
(d) 4.

35. The general solution of the differential equation $(x + 1)\frac{dy}{dx} - ny = e^x(x + 1)^{n+1}$ is
(a) $y = (c + e^x)(1 + x)^n$
(b) $y = (c + e^x)(1 + x)^{-n}$
(c) $y = (c + e^x)(1 - x)^n$
(d) $y = (c + e^{-x})(1 + x)^n$.

36. The value of k for which $(x + y)^k$ is an integrating factor of the differential equation $(4x^2 + 2xy + 6y)\frac{dx}{dy} + (2x^2 + 9y + 3x) = 0$ is
(a) −1
(b) 2
(c) 3
(d) 1.

37. $e^{\int f(x)dx}$ is an integrating factor of $Mdx + Ndy = 0$, if
(a) $\frac{1}{N}\left(\frac{\partial M}{\partial y} - \frac{\partial N}{\partial x}\right) = f(x)$
(b) $\frac{1}{N}\left(\frac{\partial M}{\partial x} - \frac{\partial N}{\partial y}\right) = f(x)$
(c) $\frac{1}{M}\left(\frac{\partial M}{\partial y} - \frac{\partial N}{\partial x}\right) = f(x)$
(d) $\frac{1}{M}\left(\frac{\partial M}{\partial x} - \frac{\partial N}{\partial y}\right) = f(x)$.

38. $\frac{e^x}{y} + x^2 = 1$ is a particular solution of the differential equation
(a) $(2xy + e^x)ydx - e^x dy = 0$
(b) $(2xy + e^x)ydx + e^x dy = 0$
(c) $(2xy - e^x)ydx - e^x dy = 0$
(d) $(2xy + e^x)dx - e^x dy = 0$.

39. The value of k for which x^k is an integrating factor of the differential equation $(y^2 - x)\frac{dx}{dy} - 2xy = 0$ is

(a) −1 (c) 3
(b) −2 (d) 1.

40. The value of k for which the differential equation $\left(\frac{1}{x} + x^k y^4\right) dx = y^3 x^{k+1} dy$ may be exact is
(a) 2 (c) −5
(b) 5 (d) 3.

41. The integrating factor of the differential equation $(x^2 + y^2 + 2x)\, dx + 2y dy = 0$ is
(a) 1 (c) e^x
(b) e^{-x} (d) e^{2x}.

42. The family of straight lines passing through the origin is represented by the equation
(a) $xdy - ydx = 0$ (c) $xdx - ydy = 0$
(b) $xdy + ydx = 0$ (d) $xdx + ydy = 0$.

43. The value of k for which the equation $(2xe^y + 3y^2)dy + (3x^2 + ke^y)dx = 0$ will be exact is
(a) 3 (c) 2
(b) 1 (d) 4.

44. The differential equation of the family of circles passing through the points of intersection of the circle $x^2 + y^2 = 1$ and the straight line $x - y = 0$ is
(a) $(x^2 - 2xy - y^2 + 1)dx + (x^2 + 2xy - y^2 - 1)dy = 0$
(b) $(x^2 + 2xy - y^2 + 1)dx + (x^2 + 2xy - y^2 - 1)dy = 0$
(c) $(x^2 - 2xy - y^2 + 1)dx + (x^2 + 2xy + y^2 - 1)dy = 0$
(d) $(x^2 - 2xy - y^2 + 1)dx - (x^2 + 2xy - y^2 - 1)dy = 0$.

45. Which one of the following is NOT correct? An equation of the form $\frac{dy}{dx} + P(x)y = Q(x)y^3$, where $P(x)$ and $Q(x)$ are functions of x is
(a) nonlinear equation (c) reducible to linear equation
(b) linear equation (d) Bernoulli's equation.

Answers

1. (d)	7. (c)	13. (d)	19. (d)	25. (b)	31. (d)	37. (a)	43. (c)
2. (a)	8. (b)	14. (c)	20. (a)	26. (c)	32. (b)	38. (a)	44. (a)
3. (b)	9. (a)	15. (a)	21. (c)	27. (a)	33. (a)	39. (b)	45. (b).
4. (b)	10. (d)	16. (a)	22. (d)	28. (d)	34. (b)	40. (c)	
5. (c)	11. (b)	17. (b)	23. (b)	29. (c)	35. (a)	41. (c)	
6. (d)	12. (b)	18. (a)	24. (b)	30. (d)	36. (d)	42. (a)	

Section 2

MCQ on chapter 3 and chapter 4

Select (a), (b), (c) or (d), whichever is correct:

1. Solution of the differential equation $x^2p^2 + 3xyp + 2y^2 = 0$ is
   (a) $(xy - c_1)(x^2y - c_2) = 0$
   (b) $(xy - c)(x^2y - c) = 0$
   (c) $(xy + c_1)(x^2y - c_2) = 0$
   (d) $(xy^2 - c)(x^2y - c) = 0$.
2. A particular solution of the differential equation $p^2 - 9p + 18 = 0$ is
   (a) $(y - 6x)(y - 3x) = 0$
   (b) $(y - 6x)(y + 3x) = 0$
   (c) $(y + 6x)(y - 3x) = 0$
   (d) $(y + 6x)(y + 3x) = 0$.
3. The number of curves passing through the point $(0, 0)$ corresponding to the differential equation $p^2 - 2p\cosh x + 1 = 0$ is
   (a) 0
   (b) 1
   (c) 2
   (d) 3.
4. If the differential equation $4xp^2 - (3x - 1)^2 = 0$ has p-discriminant relation as $x(3x - 1)^2 = 0$ and c-discriminant relation as $x(x - 1)^2 = 0$, the singular solution is
   (a) $x = 1$
   (b) $x = 2$
   (c) $x = \frac{1}{3}$
   (d) $x = 0$.
5. Which one of the following is correct?
   (a) Singular solution represents a particular solution of the differential equation
   (b) Singular solution represents the envelope of the family of curves represented by the general solution of a differential equation
   (c) p-discriminant relation always represents singular solution
   (d) c-discriminant relation always represents singular solution.
6. The differential equation $8ap^3 = 27y$ has singular solution
   (a) $y = c$
   (b) $y^2 = \frac{1}{a}(x - c)^2$
   (c) $y = 0$
   (d) $y = \frac{1}{a}(x - c)^2$.
7. The singular solution of the equation $p^2 + xp - y = 0$ is
   (a) $y^2 = 4x$
   (b) $x^2 = -4y$
   (c) $x^2 + y^2 = xy$
   (d) $xy = 4$.
8. The singular solution of the differential equation $(xp - y)^2 = p^2 - 1$ is

(a) a circle
(b) an ellipse
(c) a parabola
(d) a rectangular hyperbola.

9. The equation $4x^3 + 27y^2 = 0$ represents
(a) the general solution
(b) a particular solution
(c) the singular solution
(d) a particular solution at the point (0, 0) to the differential equation $y = px + p^3$.

10. Let $F(x, y, \frac{dy}{dx}) = 0$ represent a first-order non-linear differential equation. Three statements corresponding to the differential equation are given below:
(A) Singular solution of the equation is an envelope of its general solution.
(B) The general solution of the equation must contain exactly one arbitrary constant.
(C) Singular solution of the equation can be obtained by substituting particular value of the arbitrary constant in the general solution.
Select the correct answer using the given codes:
(a) A, C
(b) B, C
(c) A, B, C
(d) A, B.

11. The envelope of the family of curves represented by the general solution of the differential equation $y - px = a(1 + p^2)^{1/2}$ is
(a) a circle
(b) an ellipse
(c) a parabola
(d) a hyperbola.

12. The singular solution of the equation $y = 2px + p^2$
(a) is $x^2 + y = 0$
(b) is $x^2 - y = 0$
(c) is $x - y^2 = 0$
(d) does not exist.

13. Which one of the following is NOT correct corresponding to the differential equation $4xp^2 - (3x - 1)^2 = 0$?
(a) $x = 0$ is the singular solution
(b) $3x - 1 = 0$ is a tac-locus
(c) $x = 1$ is a nodal locus
(d) $3x - 1 = 0$ is an envelope locus.

14. The primitive of the differential equation $9yp^2 + 4 = 0$ is a
(a) family of parabolas
(b) family of semi-cubical parabolas
(c) family of circles
(d) family of rectangular hyperbolas.

15. The singular solution of the differential equation of the one-parameter family of curves $c^2 + 2cy - x^2 + 1 = 0$ (c being a parameter) is
(a) $x = 0$
(b) $x^2 + y^2 = 1$
(c) $x^2y^2 = x^2(1 - x^2)$
(d) none of these.

16. The general solution of the equation $p^2y + 2px = y$ is
(a) $y^2 = 4c^2 - 2cx$
(b) $y^2 = c^2 - cx$
(c) $y^2 = c^2 - 2cx$
(d) $y = c^2 - 2cx^2$.

17. The general solution of the equation $p^3 - 4xyp + 8y^2 = 0$ represents a family of
(a) parabolas
(b) ellipses
(c) circles
(d) semi-cubical parabolas.

18. The general solution of the equation $y = x^4p^2 - px$ represents a family of
(a) parabolas
(b) ellipses
(c) rectangular hyperbolas
(d) semi-cubical parabolas.

19. The general solution of the equation $y = y^2p^3 + 2px$ represents a family of
(a) parabolas whose vertices lie on x-axis
(b) parabolas whose vertices lie on y-axis
(c) rectangular hyperbolas whose foci lie on x-axis
(d) semi-cubical parabolas passing through origin.

20. The singular solution of the differential equation $y = px + 2\sqrt{1 + p^2}$ represents a conic section with eccentricity
(a) 0
(b) 1
(c) > 1
(d) < 1.

21. The singular solution of the differential equation $y = px + \sqrt{4 + 9p^2}$ represents a conic section with eccentricity
(a) $\frac{2}{3}$
(b) 1
(c) $\frac{\sqrt{5}}{2}$
(d) $\frac{\sqrt{5}}{3}$.

22. The singular solution of the differential equation $y = px + 9\sqrt{p^2 - 1}$ represents a conic section
(a) with constant curvature
(b) with eccentricity $\sqrt{2}$
(c) having no asymptotes
(d) having foci on y-axis.

23. The singular solution of the differential equation $y = px + \sqrt{p^2 + 1}$ represents a conic section
(a) with radius of curvature 1
(b) with eccentricity $\sqrt{2}$
(c) having no asymptotes
(d) having foci on y-axis.

24. The singular solution of the differential equation $y = px + 4\sqrt{p^2 - 1}$ represents a conic section
(a) with constant curvature
(b) with eccentricity 1
(c) having two asymptotes
(d) having foci on y-axis.

25. Which singular solution of the following differential equations has two perpendicular asymptotes?
(a) $y = px + \sqrt{16p^2 - 9}$
(b) $y = px + \sqrt{16p^2 + 9}$
(c) $y = px + 5\sqrt{p^2 + 1}$
(d) $y = px + 5\sqrt{p^2 - 1}$.

26. The general solution of the differential equation $\sin\left(x\frac{dy}{dx}\right)\cos y = \cos\left(x\frac{dy}{dx}\right)\sin y + \frac{dy}{dx}$ is a family of curves
(a) with constant curvature
(b) with finite radius of curvature
(c) having no members passing through the origin
(d) none of these.

27. Number of solutions of the form x^m to the equation $(x^3D^3 + 2x^2D^2 - 10xD - 8)y = 0$ is
(a) 0
(b) 1
(c) 2
(d) 3.

28. The set $\{e^{-x}, e^x, e^{2x}\}$ is a fundamental set of solutions of the equation
(a) $(D^3 - 2D^2 - D + 2)y = 0$
(b) $(D^3 + 2D^2 + D + 2)y = 0$
(c) $(D^3 - D + 2)y = 0$
(d) $(D^3 - 2D^2)y = 0$.

29. Number of solutions of the form x^m (m being an integer) to the equation $(x^3D^3 - x^2(x + 3)D^2 + 2x(x + 3)D - 2(x + 3))y = 0$ is

(a) 0 (c) 2

(b) 1 (d) 3.

Answers

1. (b)	5. (b)	9. (c)	13. (d)	17. (a)	21. (d)	25. (d)	29. (c).
2. (a)	6. (c)	10. (d)	14. (b)	18. (c)	22. (b)	26. (a)	
3. (c)	7. (b)	11. (a)	15. (b)	19. (a)	23. (a)	27. (d)	
4. (d)	8. (b)	12. (d)	16. (c)	20. (a)	24. (c)	28. (a)	

Section 3

MCQ on chapter 5 and chapter 6

Select (a), (b), (c) or (d), whichever is correct:

1. The general solution of the differential equation $(D^2 + 4)^2 y = 0$ is
   (a) $C_1 \cos 2x + C_2 \sin 2x$
   (b) $(C_1 + C_2 x) \cos 2x + (C_3 + C_4 x) \sin 2x$
   (c) $(C_1 + C_2) \cos 2x + (C_3 + C_4) \sin 2x$
   (d) $e^{2x}(C_1 \cos 2x + C_2 \sin 2x)$.
2. The number of linearly independent solutions of the form e^{mx} (m being a real constant) of the equation $(D^4 - D^3 - 3D^2 + 5D - 2)y = 0$ is
   (a) 1
   (b) 2
   (c) 3
   (d) 4.
3. An integrating factor of the differential equation $(x + 1)\frac{dy}{dx} - ny = e^x (x + 1)^{n+1}$ is
   (a) $x + 1$
   (b) $(x + 1)^{-1}$
   (c) e^x
   (d) $(x + 1)^{-n}$.
4. The principle of superposition is valid for
   (a) linear differential equations
   (b) nonlinear differential equations
   (c) linear and nonlinear differential equations
   (d) none of (a), (b) and (c).
5. Number of linearly independent solutions of the form e^{mx} corresponding to the differential equation $(D^4 - n^4)y = 0$ is
   (a) 1
   (b) 2
   (c) 3
   (d) 4.
6. A basic solution set of the equation $(D^3 - 6D^2 + 11D - 6)y = 0$ is
   (a) e^x, e^{2x}, e^{3x}
   (b) e^{-x}, e^{2x}, e^x
   (c) e^x, e^{-2x}, e^{3x}
   (d) e^x, e^{2x}, e^{-3x}.
7. A particular solution of the equation $(2D^2 - 3D + 1)y = 0$ is
   (a) e^x
   (b) $e^{x/2}$
   (c) $e^x + e^{x/2}$
   (d) $c_1 e^x + c_2 e^{x/2}$,

   c_1, c_2 being arbitrary constants.
8. Number of linearly independent solutions of the form x^r corresponding to the equation $(x^3 D^3 - 6xD + 12)y = 0$ is

(a) 1 (c) 3
(b) 2 (d) 4.

9. Particular integral of the differential equation $D^2y = \cos x$ is
(a) $\cos x$ (c) $\sin x$
(b) $-\cos x$ (d) $-\sin x$.

10. Particular integral of the differential equation $(D^2 - 5D + 6)y = (1 + e^{-x})^{-2}$ is
(a) $1 + e^{-x}\ln(1 + e^x)$ (c) $-1 + e^{-x}\ln(1 + e^x)$
(b) $-1 + e^x\ln(1 + e^x)$ (d) $-1 + e^x\ln(1 + e^{-x})$.

11. Particular integral of the differential equation $(D^2 + 4)y = \sin 2x$ is
(a) $\frac{x}{2}\cos 2x$ (c) $-\frac{x}{2}\cos 2x$
(b) $-\frac{x}{4}\cos 2x$ (d) $\frac{x}{4}\cos 2x$.

12. Which of the following equations is NOT exact?
(a) $(2x^2 + 3x)y'' + (6x + 3)y' + 2y = (x + 1)e^x$ (c) $\sin xy'' - \cos xy' + 2y\sin x = 0$
(b) $x^2yy'' + (xy' - y)^2 - 3y^2 = 0$ (d) $(1 + x + x^2)y''' + (3 + 6x)y'' + 6y' = 0$.

13. An integrating factor of the equation $x^3y''' + 5x^2y'' + (2x - x^3)y' - (2 + x^2)y = 40x^3 - 4x^5$ is
(a) $\frac{1}{x^2}$ (c) $\frac{1}{x^4}$
(b) $\frac{1}{x^3}$ (d) $\frac{1}{x}$.

14. The general solution of the equation $y'' = x^2 \sin x$ is
(a) $y = c_1 + c_2x + 6\sin x - 4x\cos x - x^2\sin x$ (c) $y = c_1 + c_2x - 6\sin x - 4x\sin x - x^2\cos x$
(b) $y = c_1 + c_2x + 6\cos x + 4x\sin x + x^2\cos x$ (d) $y = c_1 + c_2x - 6\cos x - 4x\cos x - x^2\sin x$.

15. The general solution of the equation $(D^4 - a^2D^2)y = 0$ is
(a) $y = c_1 + c_2 + c_3e^{ax} + c_4e^{-ax}$ (c) $y = (c_1 + c_2x)e^{-ax}$
(b) $y = c_1 + c_2x + (c_3 + c_4x)e^{ax}$ (d) $y = c_1 + c_2x + c_3e^{ax} + c_4e^{-ax}$.

16. The general solution of the equation $\frac{d^2y}{dx^2} = \sqrt{1 + \left(\frac{dy}{dx}\right)^2}$ is
(a) $y = \cos(a + x) + b$ (c) $y = \cos(a - x) + b$
(b) $y = \cosh(a + x) + b$ (d) $y = \cos(a + x) - b$.

17. Two linearly independent solutions of the equation $(x + a)^2\frac{d^2y}{dx^2} - 4(x + a)\frac{dy}{dx} + 6y = 0$ are
(a) $(x + a)^2, (x + a)^3$ (c) $(x + a), (x + a)^3$
(b) $(x + a)^2, (x + a)$ (d) $(x + a), x$.

18. Which of the following functions are linearly independent?
(a) $\{1, x, x^2\}$ (c) $\{e^x, e^{2x}, e^{3x}\}$
(b) $\{e^x, xe^x, x^2e^x\}$ (d) the functions in (a), (b) and (c).

19. The functions e^{ax}, e^{bx}, e^{cx} will be linearly independent, if
(a) $a = b = c$ (c) $a \neq b = c$
(b) $a = b \neq c$ (d) $a \neq b \neq c$.

20. If $y = x$ be a solution of the equation $x^2\frac{d^2y}{dx^2} - x(x + 2)\frac{dy}{dx} + (x + 2)y = 0$, the complete primitive of the equation $x^2\frac{d^2y}{dx^2} - x(x + 2)\frac{dy}{dx} + (x + 2)y = x^3$ is

(a) $y = (c_1e^x + c_2)x - x^2$
(b) $y = (c_1e^x + c_2)x + x^2$
(c) $y = (c_1x + c_2)e^x - x^2$
(d) $y = (c_1e^{-x} + c_2)x - x^2$.

21. A particular solution of the equation $x^2\frac{d^2y}{dx^2} - (x+2)\frac{dy}{dx} + 2y = 0$ is
(a) x
(b) x^{-1}
(c) e^x
(d) e^{-x}.

22. Complementary function of the equation $x^2\frac{d^2y}{dx^2} + x\frac{dy}{dx} - y = x^2e^{2x}$ is
(a) $c_1x + c_2$
(b) $c_1x + c_2x^{-1}$
(c) $c_1x + c_2\ln x$
(d) $c_1x + c_2e^x$.

23. A particular solution of the equation $(x+2)\frac{d^2y}{dx^2} - (2x+5)\frac{dy}{dx} + 2y = 0$ is
(a) x^2
(b) x^{-2}
(c) e^{2x}
(d) e^{-2x}.

24. The set $\{e^x, (x-1)e^x, e^x(2-e^x)\}$ is a fundamental set of solutions of the equation
(a) $(D^3 - 4D^2 + 5D - 2)y = 0$
(b) $(D^3 + 4D^2 + 5D + 2)y = 0$
(c) $(D^3 - 5D^2 + 4D - 2)y = 0$
(d) $(D^3 - 4D^2 + 6D - 3)y = 0$.

25. An independent set of solutions of the equation $(x^3D^3 - 3x^2D^2 + 6xD - 6)y = 0$ on $(0, \infty)$ is
(a) $\{x^2, x, 1\}$
(b) $\{x^3, x^2, x\}$
(c) $\{x^2, 2x, 1\}$
(d) $\{x^3, x, 1\}$.

26. The set $\{\cos x^3, \sin x^3\}$ is a basic set of solutions on $(0, \infty)$ for the differential equation
(a) $(xD^2 + D + 9x^3)y = 0$
(b) $(xD^2 - 2D + 9x)y = 0$
(c) $(xD^2 - 2D + x^3)y = 0$
(d) $(xD^2 - 2D + 9x^3)y = 0$.

27. The adjoint equation of the equation $(x^2D^2 + 7xD + 8)y = 0$ is
(a) $(x^2D^2 + 7xD + 8)y = 0$
(b) $(x^2D^2 - 3xD + 3)y = 0$
(c) $(x^2D^2 + 7xD - 3)y = 0$
(d) $(x^2D^2 + 3xD + 8)y = 0$.

28. Which one of the following equations is self-adjoint?
(a) $(x^2D^2 + 7xD + 8)y = 0$
(b) $(x^2D^2 + 3xD + 2)y = 0$
(c) $(x^3D^2 + 3x^2D + 1)y = 0$
(d) $(x^2D^2 + 4xD + 1)y = 0$.

Answers

1. (b)	6. (a)	11. (b)	16. (b)	21. (c)	26. (d)
2. (b)	7. (a),(b),(c)	12. (b)	17. (a)	22. (b)	27. (b)
3. (d)	8. (c)	13. (a)	18. (d)	23. (c)	28. (c).
4. (a)	9. (b)	14. (a)	19. (d)	24. (a)	
5. (b)	10. (c)	15. (d)	20. (a)	25. (b)	

Section 4

MCQ on chapter 7 and chapter 8

Select (a), (b), (c) or (d), whichever is correct:

1. The eigenvalues for the equation $\frac{d^2y}{dx^2} + \lambda y = 0$ satisfying the boundary conditions $y(0) = 0$, $y(\pi) = 0$ are given by
   (a) $\lambda_n = n^2, n = 1, 2, 3, \ldots$
   (b) $\lambda_n = n, n = 1, 2, 3, \ldots$
   (c) $\lambda_n = n^2\pi^2, n = 1, 2, 3, \ldots$
   (d) $\lambda_n = 4n^2, n = 1, 2, 3, \ldots$
2. The eigenvalues for the equation $\frac{d^2y}{dx^2} + \lambda y = 0$ satisfying the boundary conditions $y(0) = 0$, $y'(\pi) = 0$ are given by
   (a) $\lambda_n = n^2, n = 1, 2, 3, \ldots$
   (b) $\lambda_n = n, n = 1, 2, 3, \ldots$
   (c) $\lambda_n = n^2\pi^2, n = 1, 2, 3, \ldots$
   (d) $\lambda_n = \left(n - \frac{1}{2}\right)^2, n = 1, 2, 3, \ldots$
3. The eigenfunctions for the equation $\frac{d^2y}{dx^2} + \lambda y = 0$ satisfying the boundary conditions $y'(0) = 0$, $y'(1) = 0$ are given by
   (a) $y_n = A_n \cos n\pi x, n = 1, 2, 3, \ldots$
   (b) $y_n = A_n \sin n\pi x$
   (c) $y_n = A_n \cos nx$
   (d) $y_n = A_n \sin nx$.
4. Which one of the following boundary value problems is NOT a Sturm-Liouville problem?
   (a) $\frac{d^2y}{dx^2} + \lambda(1 + x)y = 0, y'(0) = 0, y(1) + y'(1) = 0$
   (b) $\frac{d^2y}{dx^2} + \lambda y = 0, y(0) = 0, y(\pi) = 0$
   (c) $\frac{d}{dx}\left(\frac{1}{x}\frac{dy}{dx}\right) + (\lambda + x)y = 0, y(0) + 3y'(0) = 0, y(1) + y(2) = 0$
   (d) $\frac{d}{dx}\left(x\frac{dy}{dx}\right) + (\lambda e^x + x^2 + 1)y = 0, y(1) + 2y'(1) = 0, y(2) - 3y'(2) = 0.$
5. The eigenvalues for the equation $\frac{d^2y}{dx^2} + \lambda y = 0, \lambda > 0$ satisfying the boundary conditions $y(0) = 0$, $y(a) = 0$ are given by
   (a) $\lambda_n = a^2n^2, n = 1, 2, 3, \ldots$
   (b) $\lambda_n = an, n = 1, 2, 3, \ldots$
   (c) $\lambda_n = \frac{n^2\pi^2}{a^2}, n = 1, 2, 3, \ldots$
   (d) $\lambda_n = \left(n - \frac{1}{a}\right)^2, n = 1, 2, 3, \ldots$
6. Which one of the following equations is NOT integrable?
   (a) $3x^2dx + 3y^2dy - (x^3 + y^3 + e^{2z})dz = 0$
   (b) $(x^2 + y^2 + z^2)dx - 2xydy - 2xzdz = 0$
   (c) $zydx + (x^2y - zx)dy + (x^2z - xy)dz = 0$
   (d) $(yz + z^2)dx - zxdy + xydz = 0.$
7. Solution of the differential equation $yz \ln zdx - zx \ln zdy + xydz = 0$ is

(a) $zx = cy$ (c) $z \ln x = cy$

(b) $x \ln z = cy$ (d) $y \ln z = cx$.

8. Solution of the differential equation $\frac{yzdx}{x^2+y^2} - \frac{zxdy}{x^2+y^2} - \tan^{-1} \frac{y}{x} dz = 0$ is

(a) $y = x \tan \frac{c}{z}$ (c) $z = x \tan \frac{c}{y}$

(b) $y = z \tan \frac{c}{x}$ (d) $z = y \tan \frac{c}{x}$.

9. The boundary value problem $\frac{d^2y}{dx^2} + \lambda y = 0$, $y(0) = 0$, $y(\pi) = 0$ has nontrivial solution for

(a) $\lambda \in (-\infty, 0)$ (c) $\lambda \in \{n^2 : n \in \mathbb{N}$

(b) $\lambda \in \{\sqrt{n} : n \in \mathbb{N}\}$ (d) $\lambda \in \mathbb{R}$.

10. The boundary value problem $\frac{d^2y}{dx^2} + \lambda y = 0$, $y(0) = 0$, $y(1) = 0$ has nontrivial solution for

(a) $\lambda \in (-\infty, 0)$ (c) $\lambda \in \{n^2\pi^2 : n \in \mathbb{N}\}$

(b) $\lambda \in \{\sqrt{n\pi} : n \in \mathbb{N}\}$ (d) $\lambda \in \mathbb{R}$.

11. The boundary value problem $\frac{d^2y}{dx^2} + \lambda y = 0$, $y(0) = 0$, $y\left(\frac{\pi}{2}\right) = 0$ has nontrivial solution for

(a) $\lambda \in (-\infty, 0)$ (c) $\lambda \in \{n^2\pi^2 : n \in \mathbb{N}\}$

(b) $\lambda \in \{4n^2 : n \in \mathbb{N}\}$ (d) $\lambda \in \mathbb{R}$.

12. The boundary value problem $\frac{d}{dx}\left(x\frac{dy}{dx}\right) + \frac{\lambda}{x}y = 0$, $y'(1) = 0$, $y'(e^{2\pi}) = 0$ has nontrivial solution for

(a) $\lambda \in (-\infty, 0)$ (c) $\lambda \in \{n^2 : n \in \mathbb{N}$

(b) $\lambda \in \left\{\frac{n^2}{4} : n \in \mathbb{N}\right\}$ (d) $\lambda \in \mathbb{R}$.

13. The boundary value problem $\frac{d^2y}{dx^2} + \lambda y = 0$, $y(0) = 0$, $y(a) = 0$ has nontrivial solution for

(a) $\lambda \in (-\infty, 0)$ (c) $\lambda \in \{n^2\pi^2a^2 : n \in \mathbb{N}\}$

(b) $\lambda \in \left\{\frac{n^2\pi^2}{a^2} : n \in \mathbb{N}\right\}$ (d) $\lambda \in \mathbb{R}$.

Answers

1. (a) 3. (a) 5. (c) 7. (b) 9. (c) 11. (b) 13. (b).
2. (d) 4. (c) 6. (d) 8. (a) 10. (c) 12. (b)

Section 5

MCQ on chapter 9 and chapter 10

Select (a), (b), (c) or (d), whichever is correct:

1. The family of curves for which the angle between the radius vector and the tangent at (r, θ) is half of the vectorial angle is

   (a) $r = a(1 + \cos\theta)$
   (b) $r = a(1 - \cos\theta)$
   (c) $r = a(1 + \sin\theta)$
   (d) $r = a(1 - \sin\theta)$.

2. The family of curves for which the Cartesian subtangent is constant at any point (x, y) is of the form

   (a) $y = ae^{x/c}$
   (b) $y = ae^{-x/c}$
   (c) $y = ae^{cx}$
   (d) $y = e^{ax}$.

3. The length of the part of the tangent between the point of contact (x, y) and the y-axis is equal to the y-intercept of the tangent corresponding to the family of

   (a) cardioids
   (b) parabolas
   (c) circles
   (d) spirals.

4. The curve for which the portion of the tangent line intercepted by the coordinate axes is of constant length is

   (a) a circle
   (b) a spiral
   (c) an asteroid
   (d) a cardioid.

5. The curve for which the product of the intercepts of the tangent line on the coordinate axes is a is

   (a) a parabola
   (b) a circle
   (c) an ellipse
   (d) a rectangular hyperbola.

6. The curve for which the tangent at any point makes a constant angle with the radius vector is

   (a) a spiral
   (b) an equiangular spiral
   (c) a cardioid
   (d) an asteroid.

7. The curve for which the perpendiculars from the origin upon the tangents of a curve are of constant length is

   (a) a circle
   (b) a parabola
   (c) a hyperbola
   (d) an ellipse.

8. The family of orthogonal trajector of a family of hyperbolas is
   (a) a family of circles
   (b) a family of straight lines
   (c) a family of parabolas
   (d) a family of hyperbolas.
9. The family of orthogonal trajector of a family of concentric circles is
   (a) a family of circles
   (b) a family of straight lines
   (c) a family of parabolas
   (d) a family of hyperbolas.
10. The family of orthogonal trajector of the family of curves represented by the equation $\frac{x^2}{a^2+\lambda}+\frac{y^2}{b^2+\lambda}=1$, λ being a parameter, is a family of
   (a) concentric circles
   (b) confocal conics
   (c) concurrent straight lines
   (d) coaxial parabolas.
11. The family of orthogonal trajector of the family of curves represented by $x^{2/3}+y^{2/3}=a^{2/3}$, a being a parameter, is given by
   (a) $x^{4/3}+y^{4/3}=c^{4/3}$
   (b) $x^{2/3}-y^{2/3}=c^{2/3}$
   (c) $x^{3/4}+y^{3/4}=c^{3/4}$
   (d) $x^{4/3}-y^{4/3}=c^{4/3}$.
12. The family of orthogonal trajector of a family of cardioids is a family of
   (a) cardioids
   (b) cycloids
   (c) circles
   (d) asteroids.
13. Which one of the following equations represents a self-orthogonal family of curves considering c as a parameter?
   (a) $x^{4/3}+y^{4/3}=c^{4/3}$
   (b) $y^2=4c(x+c)$
   (c) $y^2=4c(x^2-c)$
   (d) $x^{2/3}+y^{2/3}=c^{2/3}$.
14. The value of k for which the parabolas represented by $y=ax^2+k$ are the orthogonal trajectories of the family of curves $x^2+2y^2-y=b$ (a, b being parameters) is
   (a) 4
   (b) 2
   (c) 1/2
   (d) 1/4.

Answers

1. (b)	3. (c)	5. (d)	7. (a)	9. (b)	11. (d)	13. (b)
2. (a)	4. (c)	6. (b)	8. (d)	10. (b)	12. (a)	14. (d).

Section 6

MCQ on chapter 11 and chapter 12

Select (a), (b), (c) or (d), whichever is correct:

1. The fundamental set of solutions of the equation $\dot{\vec{x}}(t) = A\vec{x}(t)$, where $A = \begin{pmatrix} 1 & 3 \\ 4 & 5 \end{pmatrix}$, $\vec{x}(t) = \begin{pmatrix} x_1(t) \\ x_2(t) \end{pmatrix}$ is

(a) $\left\{\begin{pmatrix} 3e^{-t} \\ -2e^{-t} \end{pmatrix}, \begin{pmatrix} 3e^{7t} \\ -2e^{7t} \end{pmatrix}\right\}$

(b) $\left\{\begin{pmatrix} 3e^{-t} \\ -2e^{-t} \end{pmatrix}, \begin{pmatrix} e^{7t} \\ 2e^{7t} \end{pmatrix}\right\}$

(c) $\left\{\begin{pmatrix} e^{-t} \\ -2e^{-t} \end{pmatrix}, \begin{pmatrix} e^{7t} \\ 2e^{7t} \end{pmatrix}\right\}$

(d) $\left\{\begin{pmatrix} e^{-t} \\ -2e^{-t} \end{pmatrix}, \begin{pmatrix} e^{7t} \\ -2e^{7t} \end{pmatrix}\right\}$.

2. The fundamental set of solutions of the equation $\dot{\vec{x}}(t) = A\vec{x}(t)$, where

$$A = \begin{pmatrix} 0 & -1 \\ 1 & 0 \end{pmatrix}, \vec{x}(t) = \begin{pmatrix} x_1(t) \\ x_2(t) \end{pmatrix} \text{ is}$$

(a) $\left\{\begin{pmatrix} ie^{it} \\ e^{it} \end{pmatrix}, \begin{pmatrix} e^{-it} \\ ie^{-it} \end{pmatrix}\right\}$

(b) $\left\{\begin{pmatrix} 2ie^{it} \\ e^{it} \end{pmatrix}, \begin{pmatrix} e^{-it} \\ 2ie^{-it} \end{pmatrix}\right\}$

(c) $\left\{\begin{pmatrix} ie^{it} \\ -e^{it} \end{pmatrix}, \begin{pmatrix} -e^{-it} \\ ie^{-it} \end{pmatrix}\right\}$

(d) $\left\{\begin{pmatrix} 2ie^{it} \\ -e^{it} \end{pmatrix}, \begin{pmatrix} -e^{-it} \\ 2ie^{-it} \end{pmatrix}\right\}$.

3. Which one is a solution of the vector differential equation $\dot{\vec{x}}(t) = A\vec{x}(t)$, where

$$A = \begin{pmatrix} 3 & 2 & 1 \\ 2 & 3 & 1 \\ 0 & 0 & 1 \end{pmatrix}, \vec{x}(t) = \begin{pmatrix} x_1(t) \\ x_2(t) \\ x_3(t) \end{pmatrix}?$$

(a) $\begin{Bmatrix} -e^{t} \\ e^{t} \\ 0 \end{Bmatrix}$

(b) $\begin{Bmatrix} -e^{t} \\ 0 \\ 2e^{t} \end{Bmatrix}$

(c) $\begin{Bmatrix} e^{5t} \\ e^{5t} \\ 0 \end{Bmatrix}$

(d) All the options are correct.

4. Which one is a solution of the non-homogeneous vector differential equation $\dot{\vec{x}}(t) = A\vec{x}(t)$, where

$$A = \begin{pmatrix} 6 & -3 \\ 2 & 1 \end{pmatrix}, \vec{x}(t) = \begin{pmatrix} x_1(t) \\ x_2(t) \end{pmatrix}, \vec{F}(t) = \begin{pmatrix} e^{t} \\ e^{t} \end{pmatrix}?$$

(a) $\begin{Bmatrix} e^{-t} \\ e^{-t} \end{Bmatrix}$

(b) $\begin{Bmatrix} e^{t} \\ e^{t} \end{Bmatrix}$

(c) $\begin{Bmatrix} e^{-t} \\ 2e^{-t} \end{Bmatrix}$

(d) $\begin{Bmatrix} 2e^{-t} \\ e^{-t} \end{Bmatrix}$.

5. Let us consider the vector differential equation $\dot{\vec{x}}(t) = A\vec{x}(t)$, where $A = \begin{pmatrix} 0 & 1 \\ -1 & 0 \end{pmatrix}$, $\vec{x}(t) = \begin{pmatrix} x_1(t) \\ x_2(t) \end{pmatrix}$.

Then the fundamental matrix of the system is given by

(a) $M(t) = \begin{pmatrix} \cos t & \sin t \\ \sin t & \cos t \end{pmatrix}$ (c) $M(t) = \begin{pmatrix} \cos t & \sin t \\ -\sin t & \cos t \end{pmatrix}$

(b) $M(t) = \begin{pmatrix} -\cos t & \sin t \\ \sin t & -\cos t \end{pmatrix}$ (d) $M(t) = \begin{pmatrix} \cos t & 2\sin t \\ -2\sin t & \cos t \end{pmatrix}$.

6. Any linear combination of n solutions $\phi_1(t), \phi_2(t), \phi_3(t), \ldots \phi_n(t)$ of a homogeneous linear system $\dot{\vec{x}}(t) = A\vec{x}(t)$ is also a solution of the system

(a) because $w(\phi_1, \phi_2, \phi_3, \ldots, \phi_n) = 0$ (c) by the principle of superposition

(b) because $w(\phi_1, \phi_2, \phi_3, \ldots, \phi_n) \neq 0$ (d) none of these.

7. Let f be a real function satisfying the following conditions (i) f is continuous in its domain D; (ii) f satisfies Lipschitz condition in D and let (x_0, y_0) be an interior point of D and a, b be such that the rectangle $R : |x - x_0| \leq a, |y - y_0| \leq b$, lies in D : let $M = \max |f(x)|$, for $(x, y) \in \mathbb{R}$ and let $h = \min\left(a, \frac{b}{M}\right)$. Then the initial value problem $\frac{dy}{dx} = f(x, y)$, $y(x_0) = y_0$ for $|x - x_0| \leq h$ has

(a) unique solution of the equation

(b) no solution of the equation

(c) infinite number of solutions of the equation

(d) a set of linearly independent solutions of the equation.

8. Which of the following solution-set is linearly dependent?

(a) $\left\{\begin{pmatrix} 2e^{3t} \\ -4e^{3t} \\ -2e^{3t} \end{pmatrix}, \begin{pmatrix} e^{3t} \\ -2e^{3t} \\ -e^{3t} \end{pmatrix}, \begin{pmatrix} -3e^{3t} \\ 6e^{3t} \\ 3e^{3t} \end{pmatrix}\right\}$ (c) $\left\{\begin{pmatrix} -e^{t} \\ e^{t} \\ 0 \end{pmatrix}, \begin{pmatrix} -e^{t} \\ 0 \\ 2e^{t} \end{pmatrix}, \begin{pmatrix} e^{5t} \\ e^{5t} \\ 0 \end{pmatrix}\right\}$

(b) $\left\{\begin{pmatrix} -1 \\ 1 \\ 0 \end{pmatrix}, \begin{pmatrix} 1-t \\ 1+t \\ 0 \end{pmatrix}, \begin{pmatrix} e^{t} \\ -e^{t} \\ e^{t} \end{pmatrix}\right\}$ (d) $\left\{\begin{pmatrix} e^{t} \\ e^{t} \\ 3e^{t} \end{pmatrix}, \begin{pmatrix} 0 \\ e^{-t} \\ e^{-t} \end{pmatrix}, \begin{pmatrix} e^{3t} \\ 2e^{3t} \\ e^{3t} \end{pmatrix}\right\}$.

9. Which one is NOT correct? The set of solutions

$$\left\{\begin{pmatrix} e^{2t} \\ 2e^{2t} \\ e^{2t} \end{pmatrix}, \begin{pmatrix} 0 \\ e^{3t} \\ e^{3t} \end{pmatrix}, \begin{pmatrix} e^{t} \\ e^{t} \\ e^{t} \end{pmatrix}\right\}$$

corresponding to a homogeneous system is a

(a) basic set of solutions (c) linearly independent set of solutions

(b) fundamental set of solutions (d) linearly dependent set of solutions.

10. The critical point $(0, 0)$ corresponding to the system $\dot{x} = ax + by$, $\dot{y} = cx + dy$ will be a node, if $a = d$ and

(a) $b = 0$ (c) $b = 0$ and $c = 0$

(b) $c = 0$ (d) either $b = 0$ or, $c = 0$.

11. The critical point $(0, 0)$ corresponding to the system $\dot{x} = ax + by$, $\dot{y} = cx + dy$ will be a saddle point, if $a = -d$ and

(a) b and c are of same sign
(b) b and c are of opposite sign
(c) $b = c = 0$
(d) either $b = 0$ or $c = 0$.

12. The critical point $(0, 0)$ corresponding to the system $\dot{x} = ax + by$, $\dot{y} = cx + dy$ will be a spiral point, if $a = d$ and
(a) b and c are of same sign
(b) b and c are of opposite sign
(c) $b = c = 0$
(d) either $b = 0$ or $c = 0$.

13. The critical point $(0, 0)$ corresponding to the system $\dot{x} = ax + by$, $\dot{y} = cx + dy$ will be a center, if $a = -d$ and
(a) $ad = bc$
(b) $ad < bc$
(c) $ad > bc$
(d) either $b = 0$ or $c = 0$.

14. The critical point $(0, 0)$ corresponding to the system $\dot{x} = ax + by$, $\dot{y} = cx + dy$ will be a node, if $a = d$ and b, c are of same sign and
(a) $\sqrt{bc} > |a|$
(b) $\sqrt{bc} < |a|$
(c) $\sqrt{bc} = |a|$
(d) none of these.

15. The critical point $(0, 0)$ corresponding to the system $\dot{x} = ax + by$, $\dot{y} = cx + dy$ will be a saddle point, if $a = d$ and b, c are of same sign and
(a) $\sqrt{bc} > |a|$
(b) $\sqrt{bc} < |a|$
(c) $\sqrt{bc} = |a|$
(d) none of these.

16. Which one of the following is true? Saddle point is
(a) always stable
(b) always unstable
(c) asymptotically stable
(d) neither stable nor unstable.

17. Which one of the following is true? A node
(a) may be stable or unstable
(b) is always stable
(c) is always unstable
(d) is always asymptotically stable.

18. Which one of the following is true? A spiral point
(a) may be stable or unstable
(b) is always stable
(c) is always unstable
(d) is always asymptotically stable.

19. Which one of the following is true? A star
(a) is always stable
(b) is always unstable
(c) may be stable or unstable
(d) is always asymptotically stable.

20. Which one of the following is true? A degenerate node
(a) is always asymptotically stable
(b) is always stable
(c) is always unstable
(d) may be stable or unstable.

21. The critical point $(0, 0)$ corresponding to the system $\dot{x} = 2x - 3y$, $\dot{y} = x + 4y$ is
(a) stable spiral
(b) unstable spiral
(c) stable node
(d) saddle point.

22. The critical point $(0, 0)$ corresponding to the system $\dot{x} = -2x$, $\dot{y} = 5x - 4y$ is
(a) stable spiral
(b) unstable spiral
(c) stable degenerate node
(d) saddle point.

23. The critical point (0, 0) corresponding to the system $\dot{x} = x + 3y,\ \dot{y} = 3x + y$ is
    (a) stable spiral
    (b) unstable spiral
    (c) stable degenerate node
    (d) saddle point.
24. The critical point (0, 0) corresponding to the system $\dot{x} = y,\ \dot{y} = kx (k \neq 0)$ is
    (a) stable spiral for $k > 0$
    (b) unstable spiral for $k < 0$
    (c) stable degenerate node for $k > 0$
    (d) saddle point for $k < 0$.
25. The critical point (0, 0) corresponding to the system $\dot{x} = x - ky,\ \dot{y} = kx + y$, k being a parameter, is
    (a) unstable spiral for $k \neq 0$
    (b) unstable spiral for $k = 0$
    (c) unstable degenerate node for $k = 0$
    (d) options (a) and (c).
26. The critical point (0, 0) corresponding to the system $\dot{x} = \alpha x - \beta y,\ \dot{y} = -\beta x + \alpha y$, α, β being parameters, is
    (a) stable degenerate node for $\alpha < 0, \beta = 0$
    (b) unstable degenerate node for $\alpha > 0, \beta = 0$
    (c) unstable spiral for $\alpha > 0, \beta \neq 0$
    (d) options (a), (b) and (c).
27. Equilibrium points of the non-linear system $\dot{x} = x,\ \dot{y} = x^2 + y^2 - 4$ are
    (a) (0, 2)
    (b) (0, −2)
    (c) $(0, \pm 2)$
    (d) none of these.
28. Critical points of the non-linear system $\dot{x} = y,\ \dot{y} = x - x^3 + y$ are
    (a) (0, 0)
    (b) $(\pm 1, 0)$
    (c) $(0, \pm 2)$
    (d) options (a) and (b).

Answers

1. (b)	5. (c)	9. (d)	13. (c)	17. (a)	21. (b)	25. (d)
2. (a)	6. (c)	10. (d)	14. (b)	18. (a)	22. (c)	26. (d)
3. (d)	7. (a)	11. (a)	15. (a)	19. (c)	23. (d)	27. (c)
4. (b)	8. (a)	12. (b)	16. (b)	20. (d)	24. (d)	28. (d).

Section 7

MCQ on chapter 13

Select (a), (b), (c) or (d), whichever is correct:

1. Which one of the following is correct for the linear differential equation $(x^2+x)\frac{d^2y}{dx^2}-2(x+1)\frac{dy}{dx}+2y = 0$?
   (a) 0 is an ordinary point
   (b) -1 is a regular singular point
   (c) -1 is an irregular singular point
   (d) 0 is an irregular singular point.
2. Which one of the following is correct for the linear differential equation $x(x-1)^3\frac{d^2y}{dx^2}-2(x-1)^2\frac{dy}{dx}+5y = 0$?
   (a) 1 is an ordinary point
   (b) 1 is a regular singular point
   (c) 1 is an irregular singular point
   (d) 0 and 1 are two irregular singular points.
3. Which one of the following is correct for the linear differential equation $x^3(x^2+x-6)\frac{d^2y}{dx^2}-x^2\frac{dy}{dx}-2(x-2)y = 0$?
   (a) 3 is an ordinary point
   (b) 0 is a regular singular point
   (c) 2 is an irregular singular point
   (d) 2 and 3 are two regular singular points.
4. Let x_0 be a regular singular point of the differential equation $\frac{d^2y}{dx^2}+P(x)\frac{dy}{dx}+Q(x)y = 0$. Then the equation has
   (a) infinite number of solutions in some deleted neighbourhood of x_0
   (b) at least one solution in some deleted neighbourhood of x_0
   (c) no solution in some deleted neighbourhood of x_0
   (d) a trivial solution in some deleted neighbourhood of x_0.
5. The method of Frobenius is useful to find the series solution of a second-order differential equation near
   (a) an ordinary point
   (b) a regular singular point
   (c) an irregular singular point
   (d) a singular point.
6. The roots of the indicial equation with respect to the regular singular point $x = 0$ corresponding to the equation $2x^2\frac{d^2y}{dx^2}-x\frac{dy}{dx}+(1-x^2)y = 0$ are
   (a) $1, 2$
   (b) $2, \frac{1}{2}$
   (c) $1, \frac{1}{2}$
   (d) $0, 2$.
7. If the difference between the roots of the indicial equation corresponding to a second-order differential equation is not an integer, there exist(s)
   (a) exactly two linearly independent Frobenius solutions of the equation
   (b) exactly one Frobenius solution of the equation

(c) no Frobenius solution of the equation

(d) infinitely many Frobenius solutions of the equation.

8. If the difference between the roots of the indicial equation corresponding to a second-order differential equation is a non-zero integer, there exist(s)

(a) exactly two linearly independent Frobenius solutions of the equation

(b) exactly one Frobenius solution of the equation

(c) no Frobenius solutions of the equation

(d) either one or two Frobenius solutions of the equation.

9. If the difference between the roots of the indicial equation corresponding to a second-order differential equation is zero, there exist(s)

(a) exactly two linearly independent Frobenius solutions of the equation

(b) exactly one Frobenius solution of the equation

(c) no Frobenius solution of the equation

(d) infinitely many Frobenius solutions of the equation.

10. The equation $x^2\frac{d^2y}{dx^2} + 4x\frac{dy}{dx} + 2y = 0$ has a regular singular point at

(a) infinity

(b) $x = 0$

(c) $x = 2$

(d) none of these.

Answers

1. (b)	3. (d)	5. (b)	7. (a)	9. (b)
2. (c)	4. (b)	6. (c)	8. (d)	10. (a).

Section 8

MCQ and Answers on PDE

Select (a), (b), (c) or (d), whichever is correct:

1. The partial differential equation corresponding to the family of surfaces given by the equation $z = \phi\left(\frac{y}{x}\right)$, ϕ being arbitrary function, is
   (a) $xp - yq = 0$
   (b) $yp + xq = 0$
   (c) $yp - xq = 0$
   (d) $xp + yq = 0$.
2. The partial differential equation corresponding to the two-parameter family of surfaces given by the equation $z = ax^2 + by^2 + ab$ is
   (a) $4xyz = 2px^2y + 2qxy^2 + pq$
   (b) $xyz = px^2y + qxy^2 + pq$
   (c) $4xyz = 2px^2y - 2qxy^2 + pq$
   (d) $4xyz = 2px^2y + 2qxy^2 - pq$.
3. $y = f(x - ct) + g(x + ct)$, f and g being two arbitrary functions, is the general solution of
   (a) heat equation
   (b) wave equation
   (c) Laplace equation
   (d) none of these.
4. The partial differential equation arising from $z = e^{my}\phi(x - y)$, ϕ being an arbitrary function, is
   (a) $p + q = mz$
   (b) $p - q = mz$
   (c) $p + q = mx$
   (d) $p + q = my$.
5. $y = f(x - iy) + g(x + iy)$, f and g being two arbitrary functions is the general solution of
   (a) heat equation
   (b) wave equation
   (c) Laplace equation
   (d) none of these.
6. The differential equation of the family of all right circular cones, whose axes coincide with z-axis, is
   (a) $yq + xp = 0$
   (b) $yp - xq = 0$
   (c) $yp + xq = 0$
   (d) $yp - xq = z$.
7. The differential equation of all cones having their vertices at the origin is
   (a) $px + qy = z$
   (b) $px - qy = z$
   (c) $qx + py = z$
   (d) $qx - py = z$.
8. The partial differential equation of the family of all spheres of radius 3 units and centres on the plane $y = x$ is
   (a) $(x - y)^2(p^2 - q^2 + 1) = 9(p - q)^2$
   (b) $(x + y)^2(p^2 - q^2 + 1) = 9(p + q)^2$
   (c) $(x - y)^2(p^2 + q^2 + 1) = 9(p - q)^2$
   (d) $(x + y)^2(p^2 + q^2 + 1) = 9(p + q)^2$.
9. The partial differential equation of all surfaces of revolution having z-axis as the axis of revolution is

(a) $yp - xq = 0$ (c) $xp + yq = 0$

(b) $yp + xq = 0$ (d) $xp - yq = 0$.

10. The general solution of the equation $a(p + q) = z$ is

(a) $\phi(x + y, y - a \ln z) = 0$ (c) $\phi(x + y, y + a \ln z) = 0$

(b) $\phi(x - y, y + a \ln z) = 0$ (d) $\phi(x - y, y - a \ln z) = 0$.

11. The general solution of the equation $zp + x = 0$ is

(a) $\phi(x^2 + z^2, y) = 0$ (c) $y = \phi(x^2 + z^2)$

(b) $x^2 + z^2 = \phi(y)$ (d) options (a), (b) and (c).

12. The general solution of the equation $(y^2 + z^2)p - xyq + zx = 0$ is

(a) $\phi\left(x^2 + y^2 + z^2, \frac{y}{z}\right) = 0$ (c) $y = z\phi(x^2 + y^2 + z^2)$

(b) $x^2 + y^2 + z^2 = \phi\left(\frac{y}{z}\right)$ (d) options (a), (b) and (c).

13. The surface whose tangent plane cuts off an intercept of constant length k from the axis of z is given by

(a) $\phi\left(\frac{x}{y}, \frac{x}{z-k}\right) = 0$ (c) $x = (z - k)\phi\left(\frac{x}{y}\right)$

(b) $x = y\phi\left(\frac{x}{z-k}\right)$ (d) options (a), (b) and (c).

14. The family of surfaces orthogonal to the family of surfaces given by the differential equation $(y + z)p + (z + x)q = x + y$ is

(a) $xy + yz + zx = c$ (c) $xy + yz - zx = c$

(b) $xy - yz + zx = c$ (d) $xy - zx - yz = c$.

15. The singular solution of the differential equation $2zx - px^2 - 2qxy + pq = 0$ is

(a) $z = xy^2$ (c) $z = x^2 y$

(b) $z = x^2 + y^2$ (d) $z = xy$.

16. Which one of the following equations is NOT hyperbolic type?

(a) $z_{xx} + yz_{xy} - 2z_{yy} = 0$ (c) $z_{xx} - 2 \sec x \cdot z_{xy} + z_{yy} = 0$

(b) $h(z_{xx} - z_{yy}) - (a - b)z_{xy} = 0$ (d) $z_{xx} + 2z_{xy} + \operatorname{cosec}^2 y \cdot z_{yy} = 0$.

17. The equation $z_{xx} - xz_{xy} + 4z_{yy} + 4z_y = 0$ is elliptic in the region

(a) $\Omega = \{(x, y) \in \mathbb{R}^2 : |x| > 4\}$ (c) $\Omega = \{(x, y) \in \mathbb{R}^2 : |x| \geq 4\}$

(b) $\Omega = \{(x, y) \in \mathbb{R}^2 : |x| < 4\}$ (d) $\Omega = \{(x, y) \in \mathbb{R}^2 : |x| = 4\}$.

18. The equation $az_{xx} + 2hz_{xy} + bz_{yy} = h(z_{xx} + z_{yy})$ is hyperbolic in the region

(a) $(a + b)h = ab$ (c) $(a + b)h < ab$

(b) $(a + b)h > ab$ (d) $(a + b)h \geq ab$.

19. The equation $2(x^2 + y^2)z_{xx} + 2(x + y)z_{xy} + z_{yy} = 0$ is parabolic in the region

(a) $\Omega = \{(x, y) \in \mathbb{R}^2 : x > y\}$ (c) $\Omega = \{(x, y) \in \mathbb{R}^2 : x \neq y\}$

(b) $\Omega = \{(x, y) \in \mathbb{R}^2 : x < y\}$ (d) $\Omega = \{(x, y) \in \mathbb{R}^2 : x = y\}$.

20. The characteristics of the equation $z_{xx} - 4z_{xy} + z_{yy} + 6z_x - 14z_y + 12z = 0$ are

(a) $y + (2 + \sqrt{3})x = c_1, y + (2 - \sqrt{3})x = c_2$ (c) $y \pm 3x = c$

(b) $y = 2x + c_1, y = 3x + c_2$ (d) none of these.

21. The characteristics of the equation $4z_{xx} - 4z_{xy} + z_{yy} + 2z_x - 26z_y + 9z = 0$ are given by

(a) $y - \frac{1}{2}x = c$
(b) $y = x + c$
(c) $x + y = c$
(d) $y + \frac{1}{2}x = c$.

22. The canonical form of the equation $\frac{\partial^2 u}{\partial t^2} - c^2 \frac{\partial^2 u}{\partial x^2} = 0$ is

(a) $u_{\xi\xi} = 0$
(b) $u_{\eta\eta} = 0$
(c) $u_{\xi\eta} = 0$
(d) $u_{\xi\eta} + \frac{1}{c^2} u_\eta = 0$.

23. The solution of the equation $z_{xx} + 2z_{xy} + z_{yy} = 0$ is

(a) $z = (x + y)f(x - y) + g(x - y)$
(b) $z = (x + y)f(x - y) + g(x + y)$
(c) $z = (x - y)f(x - y) + g(x - y)$
(d) $z = f(x - y) + g(x + y)$.

24. The solution of the equation $x^2 z_{xx} + 2xyz_{xy} + y^2 z_{yy} = 0$ is

(a) $z = f\left(\frac{y}{x}\right) + xg\left(\frac{y}{x}\right)$
(b) $z = yf\left(\frac{y}{x}\right) + g\left(\frac{x}{y}\right)$
(c) $z = yf\left(\frac{y}{x}\right) + g\left(\frac{y}{x}\right)$
(d) $z = yf\left(\frac{y}{x}\right) + xg\left(\frac{y}{x}\right)$.

25. The principle of superposition is NOT valid for the equation

(a) $z_x z_y - z(z_x + z_y)z^2 = 0$
(b) $z_{xx} - x^2 z_{yy} = 0$
(c) $z_{xx} + z_y = 0$
(d) $z_{xx} + 2z_{xy} + z_{yy} = z$.

26. Let $u(x, t)$ be the solution of the problem $u_{tt} - 9u_{xx} = 0, -\infty < x < \infty, t > 0$ subject to the conditions $u(x, 0) = f(x) = \begin{cases} 1, |x| \leq 0 \\ 0, |x| > 2, \end{cases}$ $\quad u_t(x, 0) = g(x) = \begin{cases} 1, |x| \leq 0 \\ 0, |x| > 2. \end{cases}$

Then the value of $u\left(0, \frac{1}{9}\right)$ is

(a) $\frac{7}{6}$
(b) $\frac{29}{27}$
(c) $\frac{2}{27}$
(d) none of these.

27. Which one is NOT correct? The partial differential equation $xu_{xx} + yu_{yy} = 0$ is

(a) elliptic for $x > 0, y < 0$
(b) hyperbolic for $x > 0, y < 0$
(c) elliptic for $x < 0, y > 0$
(d) hyperbolic for $x > 0, y > 0$.

28. Let $u(x, t)$ be the solution of the problem $u_t - u_{xx} = 0, x \in (0, 1), t > 0$ subject to the conditions $u(x, 0) = f(x) = \sin(\pi x), x \in [0, 1]; u(0, t) = u(1, t) = 0, t > 0$.

Then for $x \in (0, 1)$, $u\left(x, \frac{1}{\pi^2}\right)$ is

(a) $e\sin(\pi x)$
(b) $e^{-1}\sin(\pi x)$
(c) $\sin(\pi^{-1}x)$
(d) $\sin(\pi x)$.

29. Let $D = \{(x, y) \in \mathbb{R}^2 : x^2 + y^2 \leq 1\}$. Then the partial differential equation $(x^2 - 1)z_{xx} + 2yz_{xy} - z_{yy} = 0$ is

(a) parabolic for all $(x, y) \in D$
(b) hyperbolic for all $(x, y) \in D^c$
(c) parabolic for all $(x, y) \in D^c$
(d) hyperbolic for all $(x, y) \in D$.

30. Let $u(x, t)$ be the solution of the problem $u_{tt} - u_{xx} = 0$ subject to the conditions $u(x, 0) = x^3$; $u_t(x, 0) = \sin x$. Then for $u(\pi, \pi)$ is

(a) 0
(b) 4
(c) $4\pi^3$
(d) π^3.

31. The general solution of the equation $4u_{xx} - u_{yy} = 0$ is of the form:
    (a) $u = f(x + 2y) + g(x - 2y)$
    (b) $u = f(x) + g(y)$
    (c) $u = f(2x + y) + g(2x - y)$
    (d) $u = f(x + 4y) + g(x - 4y)$

    where f and g are twice differentiable functions.

Answers

1. (a)	5. (c)	9. (a)	13. (d)	17. (a)	21. (d)	25. (a)	29. (a)
2. (d)	6. (b)	10. (d)	14. (b)	18. (b)	22. (c)	26. (b)	30. (c)
3. (b)	7. (a)	11. (d)	15. (c)	19. (d)	23. (a)	27. (b)	31. (a).
4. (a)	8. (c)	12. (d)	16. (d)	20. (a)	24. (c)	28. (b)	

Revision: O.D.E. Test Exercises

[For each set of questions (based on different Examination Papers); time allowed = half-an-hour]

[I]

1. Show that the differential equation of all circles touching the axis of x at the origin is $\frac{dy}{dx} = \frac{2xy}{x^2-y^2}$.
2. Solve: (a) $x\frac{dy}{dx} - y = \frac{x}{a}\sqrt{x^2+y^2}$ (b) $x\frac{dy}{dx} + y = y^2 \ln x$.
3. Examine for singular solution of the equation: $y = px + p - p^2$, where $p = \frac{dy}{dx}$.
4. Solve: (a) $\frac{d^2y}{dx^2} - 2\frac{dy}{dx} + y = x^2 e^{2x}$ (b) $\frac{d^2y}{dx^2} - 4\frac{dy}{dx} + 4y - x = 0$ (c) $\frac{d^2y}{dx^2} - 4\frac{dy}{dx} + 3y = 2e^{3x}$.
5. Find the orthogonal trajectories of the family of cardioides $r = a(1-\cos\theta)$, where a is a parameter.

[II]

1. Solve: $\frac{d^2x}{dt^2} + 4\frac{dx}{dt} + 4x = 3\sin 2t$, if $x = 0$, $\frac{dx}{dt} = 0$, when $t = 0$.
2. Solve: $x^2p^2 - 2xyp + y^2 = x^2y^2 + x^2$, where $p = dy/dx$.
3. Solve: $\frac{dy}{dx} = \frac{1}{xy+x^2y^2}$.
4. Find the equation of the curve in which the angle between the radius vector and the tangent is n times the vectorial angle.
5. Solve: (a) $\frac{d^2y}{dx^2} + 3\frac{dy}{dx} + 2y = \cos 2x$ (b) $\frac{d^2y}{dx^2} = \frac{1}{\sqrt{ay}}$ (c) $x^3\frac{d^3y}{dx^3} + 2x^2\frac{d^2y}{dx^2} - x\frac{dy}{dx} = x^2 \ln x$.

[III]

1. Solve: $\frac{dy}{dx} = \frac{1+y^2}{1+x^2}$.
2. Solve: $x\cos(y/x) + y\sin(y/x) = \{y\sin(y/x) - x\cos(y/x)\}x\frac{dy}{dx}$.
3. If $y = y_1(x)$, $y = y_2(x)$ are two solutions of the equation $\frac{d^2y}{dx^2} + P(x)\frac{dy}{dx} + Q(x)y = 0$, where $P(x)$ and $Q(x)$ are continuous functions of x, prove that $y_1\frac{dy_2}{dx} - y_2\frac{dy_1}{dx} = Ce^{-\int P dx}$, where C is a constant $\neq 0$.
4. Solve: $\frac{d^2y}{dx^2} + a^2y = \sec ax$.
5. Solve: (a) $\frac{d^2y}{dx^2} + \frac{dy}{dx} + y = \sin 2x$ (b) $(x-y)^2\frac{dy}{dx} = a^2$ (c) $y(xy + 2x^2y^2)dx + x(xy - x^2y^2)dy = 0$.

[IV]

1. Obtain the primitive and the singular solutions, if any, of the equation $\left(\frac{dy}{dx}\right)^4 = 4y\left(x\frac{dy}{dx} - 2y\right)$.

2. Solve: $y(xy+2x^2y^2)dx + x(xy - x^2y^2)dy = 0$.
3. Solve the simultaneous equations:

$$\left.\begin{aligned}\frac{d^2x}{dt^2} - a\frac{dy}{dt} + \mu^2 x = 0\\ \frac{d^2y}{dt^2} + a\frac{dx}{dt} + \mu^2 y = 0\end{aligned}\right\}.$$

4. Show that $3\frac{d^2y}{dx^2} + 4x\frac{dy}{dx} - 8y = 0$ has an integral which is a polynomial in x. Deduce the general solution.
5. Show that $x\sin x$ and $x\cos x$ are solutions of the equation $\frac{d^4y}{dx^4} + 2\frac{d^2y}{dx^2} + y = 0$. What is the general solution?

[V]

1. Solve: $\frac{dy}{dx} = \frac{1}{x^3y^3+xy}$.
2. Solve: $(ax+by+c)\frac{dy}{dx} = Ax + By + C$.
3. Solve: $\frac{d^2z}{dx^2} - 4\frac{dz}{dx} + 3z = e^x\cos 2x + \cos 3x$.
4. Solve: $x^2\frac{d^2y}{dx^2} + x\frac{dy}{dx} - y = x^3$.
5. The deflection y of a uniform beam of length l supported at its ends in a horizontal position is given by the equation $d^4y/dx^4 = \omega$, where ω is proportional to the weight per unit length of the beam and is constant. Find y, assuming that $y = d^2y/dx^2 = 0$ at both ends of the beam, the axis of x being horizontal.

[VI]

1. Solve the simultaneous equations:

$$\left.\begin{aligned}\frac{d^2x}{dt^2} - a\frac{dy}{dt} + k^2 x = 0\\ \frac{d^2y}{dt^2} + a\frac{dx}{dt} + k^2 y = 0\end{aligned}\right\}.$$

2. Solve: $2x\frac{d^2y}{dx^2} + 8\frac{dy}{dx} = 4x^2$. (Given: $y = 10/9$, when $x = 1$; $y = 8/9$, when $x = 2$.)
3. Solve: $x\cos x\frac{dy}{dx} + y(x\sin x + \cos x) = 2\cos^2\frac{1}{2}x$.
4. Reduce the equation $axyp^2 + (x^2 - ay^2 - b)p - xy = 0$ ($p = dy/dx$) to Clairaut's form and hence solve it.
5. Find the orthogonal trajectories of the family of conics $y^2 - x^2 + 2xy - 2ax = 0$ (a is a parameter).

[VII]

1. Solve: $(x - y - 2)dx - (2x - 2y - 3)dy = 0$.
2. Find for what values of α the differential equation $\frac{d^2y}{dt^2} + \alpha^2 y = 0$ has solutions which are not identically zero but satisfy the conditions $y'(0) = 0$, $y(0) = 0$.
3. If a differentiable function $f(x)$ satisfies $f(x+y) = f(x) + f(y)$, then prove that either $f(x) = 0$ or $f(x) = e^{ax}$ (a is an arbitrary constant).
4. Solve: $\dot{x} - y = e^t$, $\dot{y} + x = 1$ ($\dot{x} = dx/dt$, $\dot{y} = dy/dt$).
5. Find the general as well as singular solution of the equation $y = px + \frac{ap}{\sqrt{1+p^2}}\left(p = \frac{dy}{dx}\right)$.

[VIII]

1. The distance x described by a person falling by means of a parachute satisfies the differential equation $\left(\frac{dx}{dt}\right)^2 = k^2\left(1 - e^{-\frac{2gx}{k^2}}\right)$, where k and g are constants and $x = 0$, when $t = 0$. Show that $x = \frac{k^2}{g}\cosh\left(\frac{gt}{k}\right)$.
2. Find the integral of the equation $\frac{d^2x}{dt^2} + 2n\cos\alpha\frac{dx}{dt} + n^2x = a\cos nt$. (Given: when $t = 0$, $x = 0$, $dx/dt = 0$.)
3. Solve: $\frac{d^2y}{dx^2} + n^2y = x\cos ax$.
4. Solve: $y^2 + x^2\frac{dy}{dx} = xy\frac{dy}{dx}$.
5. Find the curve which is such that the portion of the axis of x cut off between the origin and the tangent at any point is proportional to the ordinate of that point.

[IX]

1. Solve: $L\frac{d^2Q}{dt^2} + R\frac{dQ}{dt} + \frac{Q}{C} = 0$ $(CR^2 < 4L)$. [Given: $Q = Q_0 \cdot dQ/dt = 0$, when $t = 0$; C, R, L are all constants.]
2. Solve: $x^2\frac{d^2y}{dx^2} + y = 3x^2$.
3. If $dy/dx + 2y\tan x = \sin x$ and $y = 0$, when $x = \pi/3$, show that the maximum value of y is 1/8.
4. Solve the system of simultaneous equations: $\frac{dx}{dt} = -\mu x$, $\frac{dy}{dt} = x$ (μ is a positive constant).
5. Integrate: $a\frac{d^2y}{dx^2} = \sqrt{1 + \left(\frac{dy}{dx}\right)^2}$ (a is a constant).

[X]

1. Find a family of curves for which the length of the part of the tangent between the point of contact (x, y) and the point where it meets the y-axis is equal to the length of the intercept made by the tangent on the y-axis.
2. Obtain the differential equation of all circles which have their centre at (a, b). What is the geometrical interpretation of the differential equation?
3. Solve: $1 + y^2 + \left(x - e^{-\tan^{-1}y}\right)\frac{dy}{dx} = 0$.
4. Solve: $e^{3x}(p - 1) + p^3e^{2y} = 0$ $(p = dy/dx)$.
5. Solve: $\frac{d^4y}{dx^4} - y = e^x\cos x$.

[XI]

1. Given that $f(x+y) = \frac{f(x)+f(y)}{1-f(x)f(y)}$, where x and y are independent variables and $f(t)$ is a differentiable function and $f(0) = 0$, show that $f(t) = \tan\alpha t$ (α = a constant).
2. Prove that the general solution of the simultaneous equations

$$\left.\begin{aligned} m\frac{d^2x}{dt^2} &= Ve - He\frac{dy}{dt} \\ m\frac{d^2y}{dt^2} &= He\frac{dx}{dt}\end{aligned}\right\} \; m, V, H, e \text{ are given constants}$$

is $x = A + B\cos(\omega t - \alpha)$, $y = (V/H)t + C + B\sin(\omega t - \alpha)$, where $\omega = He/m$ and A, B, C, α are arbitrary constants.

Given that $\frac{dx}{dt} = \frac{dy}{dt} = 0$; $x = y = 0$, when $t = 0$, show that the solutions reduce to $x = \frac{V}{\omega H}(1 - \cos\omega t)$, $y = \frac{V}{\omega H}(\omega t - \sin\omega t)$.

3. Determine the curve whose radius of curvature is proportional to the length of the normal.
4. Solve: $(x - y)^3 dx + 2xydy = 0$.
5. Solve: $p^3 - y^2(a - p) = 0$ $(p = dy/dx)$.

[XII]

1. The motion of a particle of mass m moving along a straight line is governed by the equation $m\frac{d^2x}{dt^2} + r\frac{dx}{dt} + kx = 0$, where r and k are constants. Find the condition under which the motion will be of oscillatory nature. Find x as a function of t in this case.
2. Show that the general solution of $\frac{dy}{dx} + Py = Q$, when P and Q are functions of x, is given by $y = K\{u(x) - v(x)\} + v(x)$, K being an arbitrary constant, where $y = u(x)$ and $y = v(x)$ are particular solutions of the above equation.
3. Solve: $x^4\frac{d^4y}{dx^4} + 2x^3\frac{d^3y}{dx^3} + x^2\frac{d^2y}{dx^2} - x\frac{dy}{dx} + y = \ln x$.
4. Reduce the equation $x^2p^2 + yp(2x+y) + y^2 = 0$ $(p = dy/dx)$ to Clairaut's form by the transformation $\xi = y, \eta = xy$. Hence, solve the equation and find the singular solution, if there be any.
5. Show how to find the orthogonal trajectories of a one-parameter family of curves $f(x, y, c) = 0$. Use your method to find the orthogonal trajectories of the family of parabolas $y = cx^2$ (c is the parameter) and of the family of coaxial circles $x^2 + y^2 + 2gx + c = 0$ (g is the only parameter, c is a fixed constant).

[XIII]

1. If y_1 and y_2 are two linearly independent integrals of $\frac{d^2y}{dx^2} + P(x)\frac{dy}{dx} + Q(x)y = 0$, where $P(x)$, $Q(x)$ are continuous functions of x, show that $y_1\frac{dy_2}{dx} - y_2\frac{dy_1}{dx} = Ce^{-\int Pdx}$ (C is a constant $\neq 0$).
2. Obtain the general and singular solutions of the equation $\frac{dy}{dx} = 3(y - x/2)^{2/3} + \frac{1}{2}$.
3. Solve: $a\left(x\frac{dy}{dx} + 2y\right) = xy\frac{dy}{dx}$.
4. Solve: $\frac{d^2s}{dt^2} + 4s = 2\cos 2t$ (given $s = 0$, $ds/dt = 2$ where $t = 0$).
5. Solve: $3\frac{d^2y}{dx^2} = y^{-5/3}$.

[XIV]

1. Solve: $\frac{dy}{dx} = \frac{2xy}{x^2-y^2}$.
2. Find the form to which the differential equation $\frac{d^2y}{dx^2} + \frac{1}{x}\frac{dy}{dx} + y = 0$ is reduced by the substitution $xy = u$. Hence, solve it.
3. Given: $(1 + x)\frac{dy}{dx} = \alpha y$ (α is a constant). Show that $\frac{d}{dx}\left\{\frac{y}{(1+x)\alpha}\right\} = 0$. Deduce that $(1 + x)\frac{dy}{dx} = \alpha y$ and $y(0) = 1$ determine y uniquely to be equal to $(1 + x)^{\alpha}$.
4. A plane curve is given by the differential equation $\frac{d^2y}{dx^2} + p^2y = 0$ (p = a constant $\neq 0$). It is given that the curve cuts the x-axis at $x = 0$ and the tangent to the curve at $x = l$ is parallel to the x-axis. Find the possible values for p.
5. Solve:

$$\left.\begin{aligned}\frac{dx}{dt} - 7x + y = 0\\ \frac{dy}{dt} - 2x - 5y = 0\end{aligned}\right\}$$

[XV]

1. Solve: $xp^2 - 2yp - x = 0$ $(p = dy/dx)$.
2. The normal *PG* to a curve meets the axis of *x* in *G*. If the distance of *G* from the origin is twice the abscissa of *P*, prove that the curve is a rectangular hyperbola.
3. Solve: $\frac{d^2y}{dx^2} + 4y = \sin 2x$.
4. Solve the system of simultaneous equations:

$$\left.\begin{array}{l}\dot{x} + 5x + y = e^t \\ \dot{y} + 3y - x = e^{2t}\end{array}\right\} \left(\dot{x} = \frac{dx}{dt}, \dot{y} = \frac{dy}{dt}\right).$$

5. If the differentiable function $f(x)$ satisfies the equation $f(x) + f(y) = f\left(x\sqrt{1-y^2} + y\sqrt{1-x^2}\right)$, then prove that $f(x) = c\sin^{-1} x$ for $|x| < 1$ (c is a constant).

[XVI]

1. The tangent to a curve makes a triangle of constant area with the coordinate axes. Prove that the curve is a rectangular hyperbola.
2. In the case of a stretched string which has one end fixed and a particle of mass *m* attached to the other end, the equation of motion is $m\frac{d^2s}{dt^2} = -\frac{mg}{e}(s - l)$, where *l* is the natural length of the string and *e* its elongation due to a weight *mg*. Find *s* and *v*, determining the constants so that $s = s_0$ at time $t = 0$ and $v = 0$, when $t = 0$.
3. Find the equation of the curve for any point of which the second derivative of the ordinate is inversely proportional to 3/2th power of the product of the sum and difference of the abscissa and a constant length *a*. Determine the curve so that it will cut the *y*-axis at right angle and the *x*-axis at a distance *a* from the origin.
4. A particle of mass *m* falls under gravity from rest in a medium which exerts a resistance mkv^2, where *v* is the speed of the particle and *k* is a constant. Show that the speed of the particle tends to a limit *u*.

   If the particle is projected vertically upwards in the same medium with speed *nu*, show that it returns to the starting point with speed $nu/\sqrt{1+n^2}$.
5. Solve: $(x + 2y^3)\frac{dy}{dx} = y$.

[XVII]

1. Solve: $x\frac{dy}{dx} - y = x\sqrt{x^2+y^2}$ and $y^2 + x^2\frac{dy}{dx} = xy\frac{dy}{dx}$.
2. Solve: $x^2\frac{d^2y}{dx^2} + 4x\frac{dy}{dx} + 2y = e^x$.
3. Solve: $y = 2px + y^2p^3$ $(p = dy/dx)$.
4. Solve: $\frac{d^2y}{dx^2} - 2\frac{dy}{dx} + 4y = e^x \cdot \cos x$.
5. Find the foci of the curve which satisfies the differential equation $(1 + y^2)dx - xydy = 0$ and passes through the point $(1, 0)$.

[XVIII]

1. Solve: (a) $2x^2\frac{dy}{dx} = xy + y^2$ (b) $2dy + ydx = y^3(x - 1)dx$.
2. Solve: $\frac{d^2y}{dx^2} + \frac{2}{x}\frac{dy}{dx} = 10$.

3. By using the transformation $z = \sin x$, find y, where $\frac{d^2y}{dx^2} + \frac{dy}{dx}\tan x + y\cos^2 x = 0$ and for $x = 0$, $y = 1$, $\frac{dy}{dx} = 0$.
4. Solve: $\frac{d^2y}{dx^2} - 5\frac{dy}{dx} + 6y = x^2e^{3x}$.
5. Solve: $x + yp^2 = p(1 + xy)$ $(p = dy/dx)$.

[XIX]

1. Find the family of curves cutting the family of parabolas $y^2 = 4ax$ at 45°.
2. Solve: $(D-1)^2(D^2+1)^2y = e^x + x(D \equiv d/dx)$.
3. Solve: $\left(\frac{dy}{dx}\right)^2 + 2x\frac{dy}{dx} = 3x^2$.
4. Solve: $x^2\frac{d^2y}{dx^2} - x\frac{dy}{dx} + y = 0$.
5. Find the complete primitive and singular solutions, if any, of the differential equation: $xp^2 - 2yp + 4x = 0$ $(p = dy/dx)$.

[XX]

1. Solve: $(D^5 - D)y = e^x + \sin x - x(D \equiv d/dx)$.
2. Solve: $2dy + ydx = y^3(x-1)dx$.
3. Solve: $x\frac{dy}{dx} + y = xy^2$.
4. Solve: $\frac{dy}{dx} = \frac{x+2y-3}{2x+y-3}$.
5. Solve: $(D^2 - 2D + 2)y = e^x \sin 2x(D \equiv d/dx)$.

[XXI]

1. Solve: $y = p^2x + p(p = dy/dx)$.
2. Solve: $D^2(D^2 + D + 1)y = x^2(D \equiv d/dx)$.
3. Use the substitution $x = \sinh z$ to solve the equation $(1+x^2)\frac{d^2y}{dx^2} + x\frac{dy}{dx} = 4y$. (See that under the substitution the equation becomes $d^2y/dz^2 = 4y$).
4. Solve: $x^3\frac{dy}{dx} = y^3 + y^2\sqrt{y^2 - x^2}$.
5. Solve: $(D^3 + 4D^2 + 4D)y = 8e^{-2x}(D \equiv d/dx)$.

[XXII]

1. Solve: (a) $\frac{dy}{dx} + \frac{y}{x}(\ln y) = \frac{y}{x^2}(\ln y)^2$ (b) $x^2y - x^3\frac{dy}{dx} = y^4 \cos x$.
2. Obtain the differential equation satisfied by the family of curves given by $c^2 + 2cy - x^2 + 1 = 0$, where c is the parameter of the family. Find the singular solution of this differential equation.
3. Solve: $y = \ln(p^3 + p)$ $(p = dy/dx)$.
4. Explain the method of finding the general solution and singular solution of a differential equation of the form $y = px + f(p)$, where $p = dy/dx$ and obtain an interpretation of the singular solution.
5. Solve: $\frac{d^3y}{dx^3} - y = x\sin x$.

[XXIII]

[*Time extended to* 90 *minutes*]

1. Solve by the method of undetermined coefficients:
   (a) $(D^2 + 6D + 9)y = 24e^{-3x}$.
   (b) $(D^2 - 2D + 1)y = xe^x$.
   (c) $(D^2 + 1)y = 2\cos x$.
   (d) $(D^2 - 4D + 1)y = x^2 - 2x + 2$.
2. Solve by the method of variation of parameters:
   (a) $(D^2 - 6D + 9)y = e^{3x}/x^2$.
   (b) $(D^2 - D - 2)y = 4x^2$.
   (c) $(D^2 + 4)y = \sin 2x$.
   (d) $\frac{d^2y}{dx^2} + \frac{1}{x}\frac{dy}{dx} - \frac{1}{x^2}y = \ln x \ (x > 0)$,

   if $y = x$ and $y = 1/x$ are two linearly independent solutions of the corresponding homogeneous equations.
3. Find the eigenvalues and eigenfunctions of:
   (a) $\frac{d^2y}{dx^2} + \lambda y = 0 (\lambda > 0)$; $y(0) = 0$, $y(\pi) = 0$.
   (b) $\frac{d}{dx}\left(x\frac{dy}{dx}\right) + \frac{\lambda}{x}y = 0 (\lambda > 0)$; $y'(1) = 0 = y'(e^{2\pi})$.

Index

Abel's theorem 120
Absolutely convergent 316
Analytic function 317
Asymptotically stable 302, 307
Autonomous system 301

Backward wave 423
Basic existence and uniqueness 25
Bernoulli's equation 63
Bessel's equation 335
Bessel's function 335
Boundary-value problem 24, 440
Bounded function 22

Canonical form 408
Cauchy conditions 426
Cauchy-Kowalevski theorem 433
Cauchy problem 24, 426
Cauchy surface 426
Cauchy-Lipschitz condition 22
c-discriminant relation 102
Centre 310
Charpit's method 398
Clairaut's equation 92
Clairaut's form 92
Coaxial node 307
Complementary function 131
Complete integral 10, 26, 397
Complete primitive 10, 26, 111
Condition of integrability 34
Continuity 22
Convex set 22
Critical point 302
Cusp 102
Cuspidal locus 103

d'Alembert's formula 427
Degree 4, 367
Dirichlet problem 441
Double point 102
Dynamics in classical mechanics 350

Eigenfunctions 212
Eigenvalues 212
Elliptic equation 405
Envelope locus 103
Equations of the second order 193
Equilibrium point 302
Everywhere convergent 316
Exact equation 33, 178
Extended form of Clairaut's equation 94
Extraneous loci 102

Forward wave 423
Fourier's method 423
Fundamental matrix 281
Fundamental set of solutions 119, 281

General solution 10, 26, 92, 111, 398

Half-life 356
Heat equation 368
Homogeneous equations 47, 250
Homogeneous linear equations 170
Homogeneous system 279
Hyperbolic equation 405

Ill-posed problem 24
Initial-value problem 23
Integral curve 18, 302
Integral surface 390
Integrating factors 37
Interval of convergence 316
Irregular singular point 317
Isolated critical point 302

Lagrange's method 376
Laplace equation 368, 440
Legendre's equation 337
Legendre's function 338
Linear combination 115
Linear differential equation 4
Linear differential operator 115
Linear equations 57, 367
Linearly dependent 115
Linearly independent 115
Lipschitz condition 22
Lipschitz constant 22
Logistic curve 344
Logistic growth 344
Logistic population model 343
Lotka-Volterra 346

Malthus population model 341
Mathematical modelling 340
Method of Frobenius 325
Method of undetermined coefficients 146
Method of variation of parameter 57, 158
Models of population growth 340
Models on electric circuits 358
Models on falling object 350
Models on heat conduction 357
Models on horizontal beams 354
Models on radioactive decay 356
Models on simple pendulum 353
Multiple point 102
Mutualism model 348

Neumann problem 441
Neutrally stable 302
Newton's law of cooling or heating 357
Nodal locus102
Nodal point 102
Node 102
Non-autonomous system 301
Non-homogeneous system 279
Non-linear differential equations 4
Non-linear equation 368
Non-linear exact equations 181
Normal form 203, 280

One-dimensional heat equation 435
One-dimensional wave equation 422
Order 4, 367
Ordinary differential equations 3
Ordinary point 302, 317
Orthogonal surface 393
Orthogonal trajectories 266

Parabolic equation 405
Partial differential equations 4, 367
Particular integral 133
Particular solution 10, 26
p-discriminant relation 102
Pfaffian differential 240
Pfaffian differential equation 240
Phase path 302
Phase plane 302
Phase portrait 302
Phase trajectory 302
Picard's approximation 29
Picard's theorem 26
Plane autonomous system 302
Population growth 341
Postulate of continuity 341
Postulate of parenthood 341
Postulate of upper limit 341
Power series 316
Principle of superposition 126, 127, 421

Quasi-linear equation 368

Radius of convergence 316
Reduced equation 115
Reducible 11
Reduction of order 121
Reduction to Cauchy-Euler form 217
Regular singular point 317
Regular Sturm-Liouville priblem 215
Rest point 302

Saddle point 305
Semi-linear equation 368
Singular integral 397
Singular point 18, 102, 302, 317
Singular solution 10, 92, 102, 104
Sink node 304
Source node 305
Stability of a critical point 302
Stable critical point 302
Stable degenerate node 306
Stable focus 309
Stable node 303
Stable proper node 304
Stable spiral 309
Stable star 304

Tac-locus 103
Time-invariant system 301
Total differential equation 240
Trajectories 266
Triple point 102
Two dimensional autonomous system 302

Uniqueness of solution 432
Unstable critical point 302
Unstable degenerate node 306, 307
Unstable focus 309
Unstable node 304
Unstable proper node 305
Unstable spiral 309
Unstable star 305

Wave equation 368
Well-posed problem 24
Wronskian 117, 280
ω-trajectories 268